Calculus, Concepts & Computers

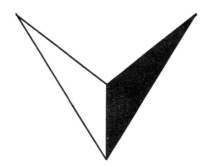

Second Edition

Ed Dubinsky
Purdue University

Keith E. Schwingendorf
Purdue University

David M. Mathews
Longwood College

McGraw-Hill, Inc.
College Custom Series

New York St. Louis San Francisco Auckland Bogotá
Caracas Lisbon London Madrid Mexico Milan Montreal
New Delhi Paris San Juan Singapore Sydney Tokyo Toronto

McGraw-Hill's College Custom Series consists of products that are produced from camera-ready copy. Peer review, class testing, and accuracy are primarily the responsibility of the author(s).

CALCULUS, **CONCEPTS** & COMPUTERS

1 2 3 4 5 6 7 8 9 0 SEM SEM 9 0 9 8 7 6 5 4

ISBN 0–07–041033-x

Editor: J.D. Ice
Text Preparation: M&N Toscano
Cover Design: Pat Koch
Cover Photo: ©Telegraph Colour Library/FPG International
Printer/Binder: Quebecor Printing Semline, Inc.

CONTENTS

CALCULUS, CONCEPTS AND COMPUTERS

1 COMPUTERS AND APPROXIMATE CALCULUS

2 FUNCTIONS

3 LIMITS OF FUNCTIONS

4 DERIVATIVES

▮ 5 INTEGRALS

6 SEQUENCES AND SERIES

▎ 7 CURVES, VECTORS, SURFACES AND CALCULUS

■ COMMENTS FOR THE STUDENT

Working to Learn Mathematics   This book is very different from other mathematical textbooks you have used in the past. Using any textbook represents an implicit "contract" between the writer and the reader, between the instructor and the student. In our "contract" we ask more from you and in turn we promise you will get more.

We ask you to work hard. And we really mean *work*. You will not be a passive recipient of mathematical knowledge. Rather, you'll be actively involved in *doing* things (mainly on the computer) and in discussing them with your fellow students. In fact, you and your colleagues will be, collectively, *constructing your communal and personal knowledge.* So when we say "work", we mean more than just putting in hours and doing the assignments. We are asking you to THINK — not only about the mathematics, but about what is going on in your mind as you try to learn Calculus. We are, in fact, asking you to take into your own hands the responsibility for your own learning and that of your colleagues.

And in return, we promise you that this book will make it possible for you to learn the mathematics meaningfully. If you use it properly and if your course is consistent with the principles on which this book is written, then you will begin to own your mathematical ideas, you will become an active learner, and you will learn a great deal of mathematics. You will see that abstract mathematical concepts start to make sense to you. You will be able to understand these concepts and even succeed in proving things about them. Less often will you find yourself stuck, staring at the symbols as if they were just so many ink stains on the paper. Indeed, we really can promise you that the symbols of mathematics will have meaning for you because, in working with the computer and reflecting together with your colleagues on that work, you will have constructed the meanings that the symbols represent. You will begin to control mathematics, rather than be controlled by it.

And perhaps we can promise you one more thing. If you are successful in this course, you will begin to see some of the beauty of mathematics. You will become initiated to a way of thinking that goes beyond utility and has for centuries fascinated some of the most powerful minds that the human species has produced. You will be touched by these minds and stand on some very tall shoulders. We think you will be very pleased with what you see.

Constructing Mathematical Ideas   In using this book to learn Calculus, much of what you will be doing is constructing mathematical ideas on a computer. You will write small pieces of code, or "programs" that get the computer to perform various mathematical operations. In getting the computer to work the mathematics, you will more or less automatically learn how the mathematics works! Anytime you construct something on a computer then, whether you know it or not, you are constructing something in your mind.

This is a fairly new approach to learning mathematics, but it has proved effective in a number of courses over a period of several years. It works a lot better if you get into the spirit of things. In order to learn mathematics you have to somehow figure out most ideas for yourself. This means that you will often find yourself in situations where you are asked to work a problem without ever having been given an explanation of how to do it. We do this intentionally. The idea is to get you to think for yourself, to make the mathematical ideas your own by constructing them yourself, not by listening to someone talk about them.

A lot of people in this situation have a natural question. "If I am asked to solve a problem, or do something on the computer, and I am not told how to do it, what should I do?" The answer is very simple. Make some guesses (better yet, conjectures) and try them out on the computer. Ask yourself if it worked — or what part of your guess worked and what part did not. Try to explain why. Then refine your guess and try again. And again. Keep repeating this cycle until you understand what is going on. The most important thing for you to remember is not to think of these explorations in terms of success and failure. Whenever the computer result is different from what you expected, think of this as an opportunity for you to improve your understanding. Remember: instead of just being stuck, not knowing what to do next, you now have an opportunity to experiment, to make conjectures and try them out, and to gradually refine your conjectures until you are satisfied with your understanding of the topic at hand.

Another piece of advice is to *talk* about your work. Talk about your problems, your solutions, your guesses, your successes and your failures. Talk to your fellow students, your teachers, to anyone. It is not enough in mathematics to do something right. You have to know what you are doing and why it was right. Talking about what you are doing is a good way of trying to understand what you did. When we teach this course, our students always work in teams, whether they are doing computer activities or paper and pencil homework assignments. And they always talk about what they are doing.

Executing Mathematical Expressions on the Computer and in Your Mind
Solving mathematical problems and writing definitions and proofs is like writing programs in a mathematical programming language and executing them in your head. Most people find this very hard to do. In this book we offer you a way to have the computer help in executing these mathematical expressions, so that you can always test your conjectures and compare your expectations with the results on the computer screen. Our experience shows that, as students become better at carrying out mathematical activities on the computer, their ability to "run" them in their mind improves as well.

Much of your computer work will be with a mathematical programming language, like **ISETL** or the programming facility in **MapleV**. The nice thing about mathematical programming languages is that the way they work is very close to the way mathematics works. When you have written some **ISETL**, or **MapleV**, code and you are running it, try to think about what the computer is doing and how it manipulates the objects you have given it. When you do this, you are actually figuring out how some piece of mathematics works. Another nice thing about this language is that learning to program in it is very similar to learning the mathematics involved. There's very little programming "overhead".

If you are not already familar with the computer system that you will be using, then you should expect to spend some hours practicing. If you are really new to this sort of thing, then, at first, it will seem very strange to you and things might go very slowly. Don't be discouraged. Everybody starts slowly in working with computers, but things get better very quickly.

Learning With This Book   The book consists of ten chapters and each chapter is divided into sections (usually three or four). The structure of individual sections in this book reflects our beliefs (supported by contemporary theory and research) on how people learn best. We believe (together with Dewey, Montessori, Papert, and Piaget) that people learn best by *doing* and by *thinking* about what they do. The abstract and the formal should be firmly grounded in *experience*.

Thus each section starts with a substantial list of *activities*, to be done in teams working on a computer. These are intended to create the experiential base for the next learning stages. The activities in each section are followed by discussions, introducing the "official" subject matter. Some of these discussions read like a standard mathematics book with definitions, examples, theorems and proofs. But there are important differences. Explanations in the text are often only partial, raising more questions than they answer. Sometimes, an issue is left hanging and only you, the reader can supply the missing link. Sometimes the discussion even uses ideas which will only be discussed officially some pages later. Once again, this style represents our realization that mathematical ideas can not be *given* to you. You must make them yourself. All we can do is try to create situations in which you are likely to construct appropriate mathematical ideas.

One reason why this kind of discussion works is because in reading the text you are not being introduced to totally unfamiliar material. Rather, it is just a more general and formal summary of what you have previously experienced and talked about in doing the activities. It summarizes and generalizes *your own experiences*. It is very important to remember that the activities are meant for *doing*, not just reading, and for thinking about *before* reading the discussions in the text (however, you will not be penalized if you decide to peek ahead before you do an activity). It is also very important to remember that the main benefit you get from the activities is due to the time and effort you have spent on them. It doesn't really matter so much whether you have actually found all the right answers. We repeat that the main role for the activities is to create an experiential base, an intuitive familiarity with the mathematical ideas. The right answers will come after you have read the text, or after you have discussed matters in class and with your colleagues. In a few cases, however, there are some "right answers" that represent very deep ideas and you may not get them for a long time. Another part of learning mathematics is to learn to live with ideas that you only understand partially, or not at all. Through activities and discussion, you will come to understand more and more aspects of a topic. Eventually, it will all fit together and you will begin to understand the subject as a whole.

After the discussion in a section come the exercises. These are fairly standard because they come after you have had every opportunity to construct the mathematics in the section. The purpose of the exercises is to help you solidify your knowledge, to challenge your thinking, and to give you a chance to relate to some mathematical ideas that were not included in the section. You will find no cases in which the text lays out a mechanical procedure for solving a certain class of problems and then asks you to apply this procedure mindlessly to solve numerous problems of the same class. We omit this kind of material because, as must be clear to you by now, we believe that sort of interaction is of little help in real learning of mathematics.

Last Words of Wisdom   We have had our say; now it's *your* turn to roll up your sleeves and start working. We hope that as you struggle through the activities, the text and the exercises, and especially as you struggle with the frustration which inevitably must accompany any meaningful learning of such deep material, you will not lose the Grand View of what you are doing — namely, learning successfully and meaningfully one of the most beautiful albeit difficult pieces of mathematics. Especially we hope that you will succeed in maintaining an attitude of play, exploration and wonderment, which is what the spirit of true mathematics is all about.

Ed Dubinsky
Keith E. Schwingendorf
David M. Mathews

October, 1994.

◼ COMMENTS FOR THE INSTRUCTOR

Teaching a Course With This Book   This book is intended to support a constructivist (in the epistemological, not mathematical sense) approach to teaching.   That is, it can be used in an undergraduate Calculus course to help create an environment in which students construct, for themselves, mathematical concepts appropriate to understanding and solving problems in this area.   Of course, the pedagogical ideas on which the book is based do not appear explicitly in the text, but rather are implicit in the structure and content.

The ideas in our text are not presented in a completed, polished form, adhering to a strict logical sequence, but roughly and circularly, with the student responsible for trying to straighten things out.   The student is given considerable help in making mathematical constructions to use in making sense out of the material.   This help comes from a combination of computer activities, leading questions and a conversational style of writing.   It should be noted that although it is assumed that each learning cycle begins with activities, the students are not expected to *discover* all the mathematics for themselves.   In fact, since the main purpose of the activities is to establish an *experiential base* for subsequent learning, anyone who spends considerable time and effort working on them will reap the benefits whether they have discovered the "right" answers or not.   Definitions, theorems, proofs and summaries are presented only after the student has had a chance to work with the ideas related to the concepts which are to be formalized.

It is also important to point out that the book is *not* primarily intended as a reference.   Our main concern has been writing a book that will best facilitate a student's first introduction to Calculus.   We see nothing wrong with a student using a more traditional text as a reference.

In teaching Calculus courses based on this book, we are finding that our approach appears to be extremely effective for most students, bringing them much more into an understanding of the ideas in this subject than one would think possible from the usual experience with this course.   For the superior students, the exercise set is strong enough to challenge them and whet their appetite for more mathematics.   For all levels, we find that the students who go through our course develop a more positive attitude towards mathematical abstraction and mathematics in general.

Finally, before describing the structure of the book, we should point out that the text is only part of the course.   We have available an instructor's resource package to aid instructors in using our approach.   This package includes: documentation for **ISETL** and introductions to **MapleV** and **Derive**; complete sets of assignments, class lesson plans, and sample exams; answer keys; information on dividing students into teams.   The **ISETL** software and a user's manual can be obtained from West Publishing Sales Representative.

A revised teaching package and an Instructor's Resource Manual will be available in early 1995.   These publications will include much of the above material and also papers which describe constructivism and its implications for teaching.

The ACE Cycle   The text is divided into sections, each intended to be covered by an average class in about a week.   Each section consists of a set of activities, class discussion material, and a set of exercises.

Activities. These are tasks that present problems which require students to write computer code in a mathematical programming language which represent mathematical constructs that can be used to solve the problems. Often, an activity will require use of mathematics not yet covered in the text. The student is expected to try to discover the mathematics or sometimes just make guesses, possibly reading ahead in the text for clues or explanations.

Class Discussion Material. These portions contain some explanations, some completed mathematics and many questions, all taking place under the assumption that the student has already spent considerable time and effort on the activities related to the same topics. Our experience indicates that with this background, students can relate much more meaningfully to formal definitions and theorems. Each unanswered question in the text is either answered later in the book or repeated as an explicit problem in the exercises. Our way of using this discussion material in a course is to have the students in a class working together in teams to solve paper and pencil problems, mainly suggested by the open questions in the text. This largely replaces lectures which occur only as summations after the students have had a chance, through the activities and discussions, to understand the material

Exercises. These are relatively traditional and are used to reinforce and solidify the ideas that the students have constructed up to this point. They occasionally introduce preliminary versions of topics that will be considered later.

Covering the Course Material

Though the teaching method supported by this book is novel, the selection of material is fairly standard. The first seven chapters contains all of the material usually covered in a standard calculus course on functions of a single variable together with an introduction to vectors, vector-valued functions and curvilinear motion. Chapters 8 through 10 covers multivariable calculus together with an introduction to vector calculus. One feature of our pedagogical approach is a "holistic spray" of the ideas of calculus, in a spiraling manner, through the activities and text discussion. This is done to provide students with ample opportunity to do the necessary mental constructions needed for a deeper understanding of the basic concepts. This text is intended to be used in a cooperative (small group) learning environment to foster reflective thinking. We feel that with a proper use of this text, students will have the opportunity to gain a deep understanding of calculus concepts.

The first chapter provides the reader with the opportunity to practice with a symbolic computer system and a mathematical programming language, both of which will be used throughout the book. In addition, Chapter 1 introduces the basic problems and ideas of both differential and integral calculus, or we should say "approximate calculus" and points toward the notion of limit. Chapter 2 refines the concept of function and attempts to deepen student understanding of this concept necessary for the study of calculus and other areas of mathematics. From our recent research, and that of others, on how students learn calculus, we believe that the time spent on functions and introducing the basic problems and ideas of calculus is well worth the effort. Chapter 3 covers the concept of the limit of a function at a point together with the notions of limits at infinity and infinite limits. In Chapter 4 we cover the basic concepts and related problem solving situations in differential calculus. In Chapter 5 we cover the concepts of integral calculus together with elementary differential equations. Chapter 6 deals with sequences and series of numbers together with power series and Taylor series representations of functions. Chapter 7 covers polar coordinates, an introduction to vectors, parametric equations, and curvilinear motion together with vector-valued functions. Chapter 8 extends the ideas of differentiation in Chapter 4 to functions of more than one variable. Chapter 9 extends the ideas of integration in Chapter 5 to functions of more than one variable. Finally, Chapter 10 provides a brief introduction to the ideas of vector calculus, culminating with the famous theorems of Green, Stokes and Gauss.

Our experience and that of the many faculty who have implemented this text using our pedagogical methods, indicates that this text provides the opportunity for students to gain a much deeper understanding of the concepts of Calculus than may be possible with a standard text for a course taught in a more traditional manner.

Ed Dubinsky
Keith E. Schwingendorf
David M. Mathews

October, 1994

Acknowledgements

We would like to express our deepest gratitude to our graduate assistants for their support in the preparation of the preliminary version of this book: Daniel H. Breidenbach, Julie Hoover, Devilyna Nichols and Draga Vidakovic. We would also like to say thank you to our many undergraduate assistants for their enthusiastic support of this project. A special thanks goes to Branislav Vidakovic for the many hours he so generously gave us in his preliminary draft of all the illustrations and to Draga Vidakovic for her assistance and most helpful suggestions for the preparation this text. Without the hardwork and cooperation of the above individuals, the completion of this text would not have been possible.

We would like to acknowledge the Department of Mathematics, the School of Education, the School of Science Administration and the Science Counseling Office of Purdue University, West Lafayette, Indiana; the Mathematics/Physics Section and Administration of Purdue University North Central, Westville, Indiana; the Mathematics/Computer Science Department and the Administration of Longwood College, Farmville, Virginia; and the National Science Foundation for their support over the past six years.

We would especially like to thank our Editor J. D. Ice and the fine, professional staff of McGraw-Hill for their enthusiastic support of this very innovative and revolutionary text. In addition, we are most grateful to the staff of M & N Toscano, Somerville, MA for the excellent work in the final preparation of this manuscript and related technical illustrations for publication. The suggestions of numerous reviewers have been very valuable.

A special thanks goes to the many faculty from across the United States and around the world who have used this text. Their support, commitment, hard work and helpful suggestions have inspired many fruitful changes in both this text and the design of the courses taught using this text.

We would like to thank the many hundreds of students who have learned with us during the past six years of this project. For their dedication, patience, and hard work we are most thankful. We dedicate this book to our students of the past and the future students, at Purdue and around the world, who, using this text, will have a chance to *learn* calculus as no other students have learned it.

Finally, we would like to thank those closest to us who have been most supportive and patient in all they have had to deal with as we prepared this text. Our heartfelt thanks goes to Jennie Dautermann, Lisa Schwingendorf and Janet Mathews for their understanding and loving support throughout this venture.

Ed Dubinsky
Keith Schwingendorf
David Mathews

October, 1994

COMPUTERS AND
APPROXIMATE CALCULUS

1.1 PRELIMINARIES

This is a textbook for a Calculus course that uses computers. It uses computers for more than just calculating answers and working problems. In using this text you will do computer activities which help you *learn Calculus*. Keep that in mind as you use this book. You will learn how to solve many math problems by using your head to think and your machine to compute. You will also learn to think in mathematical ways. That means dealing with a problem situation first by understanding what is going on and then by using the simplest method you can think of to handle the situation. Sometimes you will use a method that you used before. Sometimes you will have to modify an old one or make one up that is entirely new to you.

One approach of this book is to use computers to stimulate your thinking about strategies and methods for solving problems and to help you develop skills to implement them. Another is to confront you with the kind of questions that will force you to think about problems and how to solve them.

1.1.1 What You Need to Know About Computers

You don't have to know anything about computers in order to use this text. We start at the very beginning. The goal of this first chapter is to give you a chance to get to know the computer components you will be using, and to introduce you to the basic ideas and the problems, methods and mathematical concepts that you will be dealing with in your study of Calculus.

This chapter takes you through the use of the computer components necessary for using this text very slowly and carefully. This will give you a chance to really get to know both components and to develop an initial facility in using each of them.

1.1.2 Philosophy and Goals

Throughout this course, you will be faced with many mathematical problem situations. Some you will be able to handle with what you already know or are learning just now. Others you may not be able to solve until you develop more ideas and more tools. And we're not always going to tell you which is which! Part of learning is to decide which are the situations you can handle and which you need more knowledge to handle. It can be frustrating, but this is the way it goes with growing and learning. You have to get into the spirit of things and look upon learning Calculus as a great adventure. You have to guess and you have to try new things if you want to learn. Above all, remember this: *The time to conclude that you are presently unable to solve a problem is after you have given it a royal try — not before!*

1.1.3   Why Two Components?

The first thing for you to notice is that you will be using a computer in two different ways — a mathematical programming language (MPL) and a symbolic computer system (SCS). The reason we use these two components is that most mathematical ideas can be interpreted in (at least) two ways: first as a collection of mathematical objects and processes that you think about and, second, as a symbol system that you use to work with and communicate the ideas. A mathematical programming language tends to interpret mathematics in the first way and a symbolic computer system does it the second way. Knowing how to use these two different components will help you use the right tool for the right job, and each has different and important uses in the learning process.

Of course, there are software packages, such as **MapleV** or **Mathematica**, which are capable of both interpretations. On the other hand, there are some SCS packages, such as **Derive**, which have no mathematical programming language. For such systems, we think that the use of a MPL, like **ISETL** (an acronym which stands for **I**nteractive **SET** **L**anguage), is essential for learning mathematics. Conversely, there are software systems, like **ISETL**, which have no symbolic computation capabilities. We believe it is essential that you can use each component for what it does best. This will help you learn both interpretations, and either think about them separately, or together, depending on your choice of method to understand and solve a particular problem.

1.1.4   Programming Distractions

Anytime you work with computer systems you will have to pay attention to a number of details. Pressing the right keys, getting the syntax correct, dealing with errors and crashes are all part of life with computers. If the number of these difficulties is too large, they can create distractions that will get in the way of your learning. We have tried to minimize those difficulties. It may not feel that way at first, but after a very short time, things will become more and more familiar and you will become quite adept with using the hardware and software. Frustrations with the computer will never disappear completely. Sometimes what goes wrong will not really have to do with the computer, but with your understanding of the mathematics involved. This is a signal for you to think harder and get smarter — in mathematics. Other times (and we hope it is rare after the first few weeks) minor details will bug you. This is a fact of life with computers that is acceptable only if the value of using computers is high enough to compensate for it. You decide!

1.2   COMPUTER COMPONENTS IN THIS TEXT

We recommend that this book be used in a course that is connected with a computer laboratory, or at least that students have full access to the computer components we will be discussing. In this section you will have your first introduction to these computer components.

1.2.1   Your Computer

The first thing you need to know about is the computer you will be using. This text can be used on Macintosh computers, any PC compatible under MS DOS, and some mainframe computers and workstations (e.g., the Sun). In order to get going you need to have certain information about operating your computer. Specific details will be provided to you by handouts from your instructor. Figure 1.1 lists some of the questions that this information will answer.

- How do you turn the thing on?

- How do you turn the thing off?

- How do you enter information? Keyboard? Mouse? Disks?

- How do you move around the screen? Menus?

- How do you make files and how are they organized?

- How do you save files or discard them?

- How do you print from files? From windows or screens?

- How do you edit data in a file or window? Add, delete or change text?

- How do you make back-ups?

Figure 1.1. What you need to know about your computer.

1.2.2   Your Symbolic Computer System (SCS)

There are a number of Symbolic Computer systems that can be and have been used with this book. We can mention **MapleV**, **Mathematica**, **Derive**, and **MathCad**. Our experience has been that, for the purposes of this course, there is not much difference between them. The choice depends largely on which computer system you are using since not all of these systems are available on all computers. Most of the examples that we give in this book will be in **MapleV** which runs on Macintosh's, PC's (compatible with MS DOS), mainframes and some workstations.

As with your computer, there are certain things you need to know about using your SCS. Again, specific details will be provided to you by handouts from your instructor. Figure 1.2 lists some of the questions that this information will answer. As the course progresses, you will learn much more about this system.

1.2.3   Activities

Use your SCS to perform the following operations. If you use the SCS in **MapleV** to perform these operations then your screen will look something like Terminal Screen 1 in Figure 1.3. The prompts, > , at the beginning of each line in Figure 1.3 are not to be typed in. They are part of the **MapleV** system. The semicolon, ; , at the end of a line should be typed in. After typing the semicolon you should press the "Enter" key in **MapleV** (not the return key). Note that in **Derive**, and perhaps other systems a special command must be given if you wish to use multiple letters rather than single letters for variable names. Also the command structure for your system may be substantially different. Your instructor will provide you with a handout of Terminal Screen 1 for your particular SCS.

- How do you start a session?

- How do you end a session?

- How do you enter information to the system?

- How do you make graphs?

- How do you change things, correct errors, add or delete material?

- How do you save the work that you do in a session?

- How do you print from files or windows?

- How do you deal with difficulties that may arise?

Figure 1.2. *What you need to know about your SCS.*

We should also note, and we will remind you on occasion, that you are expected to read ahead in this text. In particular, as suggested in the directions for the (computer) activities above, we expect you to read ahead, especially if you get stuck with some of the (computer) activities. The discussion sections which follow the activities will provide you with questions, figures and other information intended to stimulate your thinking and help you to complete the activities.

1. Enter the following quantity and display its value.

$$1 + \frac{1}{4} + \frac{1}{16} + \frac{1}{256}$$

2. Assign the value of the following quantity to the variable d.

$$\frac{3^{50} + 4^{20}}{2^{80}}$$

   Display the value of the variable d.

3. Assign the value of the following quantity to the variable t.

$$\tan\left(\frac{3\pi}{10}\right)$$

   Display the value of the variable t.

4. Display the following expression.

$$(x+1)^7$$

   Expand this expression.

5. Display the following expression.

$$(y - x)\left(y^4 + y^3 x + y^2 x^2 + yx^3 + x^4\right)$$

    (a) Expand the above expression.

    (b) Factor the resulting expression in part (a).

6. Factor the integers 27 and 9615319.

7. Factor the following expression

$$x^{12} - 1$$

and assign it to the variable z. Then expand the value of the variable z.

```
> 1 + 1/4 + 1/16 + 1/256;
```

$$\frac{337}{256}$$

```
> d := (3^50 + 4^20)/2^80;
```

$$d := \frac{717897987692952100398025}{1208925819614629174706176}$$

```
> t := tan(3*Pi/10);
```

$$t := \frac{1}{5}\sqrt{5}\sqrt{5 + 2\sqrt{5}}$$

```
> (x + 1)^7;
```

$$(x+1)^7$$

```
> expand(");
```

$$x^7 + 7x^6 + 21x^5 + 35x^4 + 35x^3 + 21x^2 + 7x + 1$$

```
> (y - x) * (y^4 + y^3*x + y^2*x^2 + y*x^3 + x^4);
```

$$(y - x)\left(y^4 + y^3 x + y^2 x^2 + yx^3 + x^4\right)$$

```
> expand(");
```

$$y^5 - x^5$$

```
> factor(");
```

$$-(x - y)\left(y^4 + y^3 x + y^2 x^2 + yx^3 + x^4\right)$$

```
> ifactor(27); ifactor(9615319);
```

$$(3)^3$$

$$(7)^3 (17)^2 (97)$$

```
> z := factor(x^12 - 1);
```

$$z := (x - 1)(x + 1)\left(x^2 + x + 1\right)\left(x^2 - x + 1\right)\left(x^2 + 1\right)\left(x^4 - x^2 + 1\right)$$

```
> expand(z);
```

$$x^{12} - 1$$

Figure 1.3. Terminal Screen 1.

1.2.4 Discussion

The main things you will be doing with (symbolic) expressions in this course is entering them into your SCS, manipulating them to put them in more convenient forms, and drawing graphs that represent the values of an expression, or a function represented by an expression.

Entering Expressions   Most of the expressions you will give your symbolic computer system will be in standard mathematical form and the commands you use will be straightforward versions of standard mathematical operations. You should not have much difficulty figuring out how to enter these expressions and commands. The (computer) activities and exercises in this chapter, together with the handouts from your instructor plus the illustrated terminal screens should be all the information you need in order to use your SCS in the (computer) activities in this text.

If you do need more help, or if you would just like to learn more about your SCS, then you should get a copy of the users' manual for your SCS and study it. If the manual is not available in your lab, then you should be able to purchase at your campus bookstore. Also, virtually all SCS's have some on-line help facility which will answer many of your questions. Your instructor will show you how to access the help facility of your SCS.

Manipulating Expressions   Algebraic expressions can be manipulated using commands with names like **expand**, **factor**, and **simplify**. These are very similar to what a human would do in performing such operations. But they may not be identical to what you might do. Why not? What sorts of instructions do you think the people who made your symbolic computer system gave it for these commands? Can you think of some things that a human can learn to do easily but are very difficult to get a computer to do?

The goal in manipulating an expression is to get it into the simplest, most convenient form for the way in which you plan to use the expression. This could mean different things in different situations. Based on the experience you had manipulating expressions in the activities, how would you define "simplest, most convenient form"?

One consequence of all this has to do with the number of correct answers to a mathematics problem. An expression which is the answer to a problem can be in several different forms. Sometimes it can be difficult to recognize immediately that two expressions are the same. Therefore, a mathematics problem might have "more than one correct" (or acceptable) answer.

Graphing Expressions   In the exercises for this section you will make some graphs. Graphs are visual representations of mathematical expressions. Do you recall how to make a graph of an expression? The way that a computer makes graphs is first to evaluate the expression to be graphed at a large number of points. These points are plotted and then some method may or may not be used to connect the dots. Isn't that what you would do too? Can you think of any problems that might arise in connecting the dots? We should note that you may wish to use a graphics calculator to make graphs for some activities and exercises in this text.

Some instructors will arrange the system so that the only thing that actually appears on the screen are the dots and no connections are made. It is then up to you to connect the dots yourself after the screen has been printed. If there is more than one graph on the same picture (as is the case in some of the exercises) then you may have to do a little thinking to be sure of which dot goes with which graph.

Here are a few issues in graphing that you should keep in mind.

- Which portion of the graph do you wish to display?

- Could the picture look very different using one region rather than another?

- In your computer graphing system (or graphics calculator), how do you specify the region of the graph that will be displayed?

- What does this have to do with the ideas of domain and range of a function?

1.2.5 Exercises

1. Perform the following operations on your SCS. If your SCS is **MapleV**, then your screen will look something like Terminal Screen 2 in Figure 1.4. If not, your instructor will provide you with a handout similar to Terminal Screen 2 for your SCS.

   (a) Compute 172!

   (b) Display the following expression.

   $$\left(\frac{xy}{2} - \frac{y^2}{3} \right)(x - y)^2$$

   (c) Expand the expression in part (b).

   (d) Multiply the expression in part (c) by

   $$\frac{3x + y}{x - y}$$

   (e) Simplify the expression in part (d).

   (f) Divide the expression in part (e) by

   $$x^3 - x^2 y - xy + y^2$$

   (g) Simplify the resulting expression in part (f) and assign it to the variable z.

2. Terminal Screen 3 in Figure 1.5 shows some expressions and operations entered in **MapleV** on a Macintosh. Use your SCS to enter the same expressions and perform the same operations.

3. Use your SCS to do the following graphing operations. Your instructor will provide you handouts on how to make graphs in your particular SCS. If you are using **MapleV**, Terminal Screen 3 in Figure 1.5 shows how you can substitute expressions in **MapleV**.

   (a) Let

   $$f = x^{\frac{1}{2}} + x^{-\frac{1}{2}}$$
   $$g = x^{\frac{1}{2}} - x^{-\frac{1}{2}}$$

   (b) Graph f.

   (c) Graph f with the domain equal to the interval [2, 30]

   (d) Graph f with the domain equal to the interval [2, 30] and the range equal to the interval [−2, 30]

   (e) Graph f and g on the same coordinate system.

   (f) Graph f and g on the same coordinate system with the domain equal to the interval [0.01, 0.1].

   (g) Draw the graph of the expression obtained by substituting $3x - 1$ for x in f. Let the domain be the interval [1, 50]

   (h) Let $f1$ be the expression obtained by substituting $|x|$ for x in f and let $g1$ be the expression obtained by substituting $|x|$ for x in g.

   (i) Graph $f1$ and $g1$ on the same coordinate system.

4. Your SCS or your MPL (or both) may include a graphing system which is capable of producing a graph from a list of data points (pairs of coordinates). There will be a number of these in computer files provided for you by your instructor throughout this text and you will have opportunities to investigate them and determine some of their properties. We will call these *data functions* and refer to them by **data 1**, **data 2**, ... Your instructor will provide you with information on how to access these data functions on your system, and investigate and make graphs of them.

Make a graph of **data 1** using the graphing system in your SCS or MPL as directed by your instructor.

1.2.6 Your Mathematical Programming Language (MPL)

As mentioned previously, this text has been used successfully with the MPL **ISETL** and also the MPL which is part of **MapleV**. By the time you use this text, other MPL's may also be capable of doing the kinds of things this text supports to help you learn mathematical ideas and concepts. Examples of programming code given in this book will be from **ISETL** and **MapleV**. Your instructor will provide specific details to you by handouts for the small amount of programming required in this text.

Once again, as with your SCS, there are certain things you need to know about using your MPL. If your MPL is part of your SCS, then you may already know all that you need to know. If not, then as usual, specific details about your MPL will be provided by handouts from your instructor. Figure 1.6 lists some of the questions that this information will answer. As with your SCS you will, of course, learn much more about your MPL as the course progresses.

- How do you start a session?

- How do you end a session?

- How do you enter information to the system?

- How do you make graphs?

- How do you change things, correct errors, add or delete material?

- How do you save the work that you do in a session?

- How do you print from files or windows?

- How do you deal with difficulties that may arise?

Figure 1.6. What you need to know about your MPL.

> 172!;

2134551080774388656290725701457338867300561\5
9330291227886899710221263354938130981514\
7533402367238647191519730342873065730833\
0105569480225198097362954157931066140145\
5397074590303866009781148657954570396550\
7036184372108858758667410445754789899781\
9191200697052233479864975360000000000000\
00000000000000000000000000000

> (x*y/2 - (y**2)/3) * (x - y) ** 2;

$$\left(\frac{1}{2}xy - \frac{1}{3}y^2\right)(x-y)^2$$

> expand(");

$$\frac{1}{2}x^3y - \frac{4}{3}x^2y^2 + \frac{7}{6}xy^3 - \frac{1}{3}y^4$$

> "*(3 * x + y)/(x - y);

$$\frac{\left(\frac{1}{2}x^3y - \frac{4}{3}x^2y^2 + \frac{7}{6}xy^3 - \frac{1}{3}y^4\right)(3x+y)}{x-y}$$

> simplify(");

$$\frac{1}{6}y\left(3x^2 - 5xy + 2y^2\right)(3x+y)$$

> z := "/(x ** 3 - x ** 2 * y - x * y + y ** 2);

$$z := \frac{1}{6}\frac{y\left(3x^2 - 5xy + 2y^2\right)(3x+y)}{x^3 - x^2y - xy + y^2}$$

Figure 1.4. Terminal Screen 2.

1.2.7 Activities

1. Use your MPL to make your screen look similar to Terminal Screen 4. If you are not using **ISETL** as your MPL, then your instructor will provide you with a handout of Terminal Screen 4 in Figure 1.7 for use with your MPL.

2. Your instructor may provide you with an activity to help you get familiar with the editing feature of your MPL. If so, complete that activity.

> f := 2 * x ** 2 - x - 3;

$$f := 2x^2 - x - 3$$

> subs(x = 3, f); subs(x = x - 2, f);

$$12$$
$$2(x-2)^2 - x - 1$$

> simplify(");

$$2x^2 - 9x + 7$$

> subs(x = 1, sin(x) ** 2 - cos(x) ** 2);

$$\sin(1)^2 - \cos(1)^2$$

> evalf(");

$$.4161468365$$

> expand((sin(x) ** 2 + 1) ** 2);

$$\sin(x)^4 + 2\sin(x)^2 + 1$$

> subs(sin(x) ** 2 = 1 - cos(x) ** 2, *);

$$\sin(x)^4 + 3 - 2\cos(x)^2$$

> solve(x^2 - 5 * x + 3 = 0, x);

$$\frac{5}{2} + \frac{1}{2}\sqrt{13}, \frac{5}{2} - \frac{1}{2}\sqrt{13}$$

> evalf(");

$$4.302775638, .697224362$$

> 3.21 * 10^(-5);

$$.00003210000000$$

Figure 1.5. Terminal Screen 3.

3. Use your MPL to make your screen look similar to Terminal Screen 5 in Figure 1.8. If you are not using **ISETL** as your MPL, then your instructor will provide you with a handout of Terminal Screen 5 for use with your MPL.

4. Use your MPL and type in the code shown in Terminal Screen 6 in Figure 1.9. Before running the code, write down what you think will be your MPL's response. If you are not using **ISETL** as your MPL, then your instructor will provide you with a handout of Terminal Screen 6 for use with your MPL.

5. After experimenting with code like you used in Terminal Screens 4, 5, and 6, respectively in Figures 1.7, 1.8, and 1.9, write brief explanations of what each of the following predefined functions does: **abs**, **min**, **max**, **sqrt**, ******, **ceil** and **floor**.

```
>        7 + 18;
25;
>        13*(-233.8);
-3039.400000;
>        5=2.0+3;
true;
>        4>=2+3;
false;
>        17
>>        +23.7-46
>>        *2
>>        ;
-51.300000;
>        x := -23/37;
>        x;
-0.621622;
>        27/36
0.750000;
>        p := [3, -2]; q := [1, 4.5]; r := [0.5, -2, -3];
>        p; q; r;
[3,    -2]
[1,    4.500000];
[0.500000,    -2,    -3];
>        p(1); p(2); q(2); r(3); r(4);
3;
-2;
4.500000;
-3;
OM;
>        p(1)*q(1) + p(2)*q(2);
-6.000000;
>        length := 0;
>        for i in [1..3] do
>>            length := length + r(i)**2;
>>        end;
>        length := sqrt(length);
>        length;
3.640055;
```

Figure 1.7. Terminal Screen 4.

```
>       x := 10;
>       x;
10;
>       x := x+20; x;
30;
>       x := x-4; x;
26;
>       MaxEquals := (max(2,3)=2);
>       maxequals;
OM;
>       MaxEquals;
false;
>       MaxEquals or (0.009758 < 0.013432);
true;
>       4**5; 7.0**3; 7**3.0; 2**0.5; 17.5/3.8; sqrt(2); sqrt(3**2+4**2);
1024;
343.000000;
343.000000;
1.414214;
4.605263;
1.414214;
5.000000;
>       sqrt(7) = 7**(0.2); sqrt(7); 7**(0.5);
false;
2.645751;
2.645751;
>       (2/=3) and ((5.2/3.1)>0.9); (3<=3) impl (not(3=2+1));
true;
false;
>       float(23356); fix(373*10**(-2)); floor(19.455334);
23356.000000;
3;
19;
>       abs(4.6); abs(-3+4.7-1.9); abs(0);
4.600000;
0.200000;
0;
>       x := 3;
>       abs(x-2) < 1.001;
true;
>       sgn(2.7); sgn(-4); sgn(0);
1.000000;
-1;
0;
```

Figure 1.8. Terminal Screen 5.

```
>        min(sgn(-0.5),-0.97); abs(sgn(4.2)-3);

>        4**3**2; (4**3)**2; 4**(3**2);

>        2=3; (4+5)/=-123; (2**4**2)<1000;

>        abs(min(-10,12)) - max(-10,12);

>        max(sgn(34)*(124/(-11))-2**5+min(20,abs(-20)),-31);

>        x := 1; y := 0; max(y,min(x,y)); min(min(x,y),y);

>        sgn(48.678)=ceil(48.678)-floor(48.678);

>        abs(-67.0*10**(-4))+0.998;

>        min(sqrt(0.16),0.16);

>        a := 0; b := 1; c := 2;

>        abs(2*a) < 2; abs(2*b-3) < 2; abs(2*c-3) < 2;
```

Figure 1.9. Terminal Screen 6.

1.2.8 Discussion

Programming in Your MPL   In your MPL, you will work with a variety of mathematical constructs. These include numbers, intervals, functions and sequences. Because you need to learn about these somewhat complicated things in order to learn mathematics, it is very useful to do a small amount of programming. We have found that trying to construct a mathematical idea on a computer (that is, programming it) is a very good way to learn the idea or underlying concept.

So you will learn to do a little programming in this course. At first, this might seem a little awesome. But don't worry. We will introduce it to you very gradually. You will learn the programming as you need it for Calculus. You might find it slow going at first, but after two or three weeks, it shouldn't bother you very much at all. Very shortly, it should become an effective, streamlined way of thinking mathematically. This has been our experience with the many students who have learned mathematics in this way.

Evaluation   In the activities you encountered expressions like that below.

$$p(2);$$

This is called *evaluation of p at 2*. In Activity 1, $p = [3, -2]$ is a *tuple*, or *sequence*, with two *components* and $p(2)$ is the value of the second component of p (which, here, is -2). We speak here of *evaluation of* **p** *at* **2**.

Evaluation is a very general operation and will be used in many situations, especially for functions. In the exercises for this section there are a number of functions which are predefined in your MPL. You will be evaluating them all the time throughout this course.

Notice that a MPL calculates with numbers and not with symbols. Thus, all of the variables in a MPL expression must have values. You can usually give a value to a variable using the assignment (:=) operation. If you try to return the value of an expression whose variables don't all have values, you will get **OM** if you are using **ISETL**, which stands for undefined — or you might get an error message or no value in other MPL's . For example, if an expression such as **p(2)** has no value in **MapleV**, then **MapleV** returns nothing (unless you or your instructor makes other arrangements in the system).

Notice also that the value of a variable or an expression is not always a number. It can be a "boolean" value, i.e., true or false, or lots of other things. What are some things you can imagine as being the value of an expression?

1.3 USING YOUR SCS

1.3.1 Overview

The main purpose of this section is for you to get familiar with the kinds of things you can do with your SCS. You need to get up to speed and you will do this by getting some experience, by practicing. You will be working with problems that you will deal with throughout Calculus. In this way you can kill two birds with one stone: you will learn how to use your SCS and you will get an introduction to some of the problems of Calculus.

The most important thing you have to learn how to do in Calculus is to translate a problem situation into a mathematical expression, or a function represented by an expression. Then you will learn how to simplify expressions, combine several expressions into one, and transform an expression by substitution.

You'll be spending a lot of time with functions in this course. In this introductory section you will use the idea of evaluating an expression to make the connection with functions. Then you will be able to study properties of functions. We hope that one result of doing this work is that you will further develop your conception of the idea of function.

One important object connected with a function is its graph. You will learn about graphs, expressions, and functions which are really all different ways of looking at the same thing. In particular you will study properties of functions from all three of these points of view.

The last topic in this section is the difference quotient which is a preview of the derivative concept in Calculus. The derivative is one of the most important calculus concepts connected with functions. Don't worry if you don't completely understand about the difference quotient after you complete this section. It is here at this point to get you to start thinking about it.

The activities in this book are not like the math problems you might be used to with their fixed methods of solution and unique, correct, numerical answers. They are designed instead to help you develop a mental context for thinking through situations mathematically. It is like the painter who must "see" and "feel" a picture before he or she can set brush to canvas. You need to "see" a problem mathematically before you can begin to solve it. Then the solution will make sense, as will the mathematics involved. This is especially true if you figure a problem out yourself.

As you go through the Activities and Exercises, remember that you are quite likely to come up against things you can't do. Sometimes you will be able to learn how to do them by thinking hard and getting some ideas. Other times you will get explanations from other students or from the instructor. Occasionally, an explanation will be given in the text — so as we have advised previously, READ AHEAD. Finally you will come across some things that will just have to wait until later in the course before you can do them. The important thing is that you try to do as many things as you can. This is how you learn and how you grow intellectually. It doesn't really matter how many right answers you get. *What really matters is how much hard thinking you do.*

1.3.2 Activities

1. In each of the following situations, choose letters for the variables and enter the expression which is described into your SCS. Use your SCS to simplify each expression. Then answer each question (approximately).

   (a) Suppose you have a square piece of tin which is 18 inches on each side and you wish to cut squares out of each corner and then fold up the sides to make an open box. Find an expression, or function, which represents the volume of the box in terms of length of the sides of the squares you cut out.

      i. What are all possible values for the length of the squares cut out?

      ii. What value of the length of a side cut out results in the box with the largest volume?

      iii. What is the volume of the largest box?

   (b) The distance traveled by a falling object is approximately given by $16t^2$ feet t seconds after it is dropped from the top of a cliff. Find an expression, or function, which represents the average rate of change of the distance the object falls with respect to time (i.e., the change in distance over the change in time) of the object when it falls from t seconds to $t + 0.1$ seconds. Then simplify your expression for the average rate of change of the distance with respect to time. What are its units?

      i. Suppose you replaced the time interval 0.1 by 0.01 and then simplified the expression for the difference quotient. What is the resulting expression?

      ii. What do you think might be the meaning of replacing the length of your time interval by a negative quantity such as -0.1?

      iii. Suppose you kept replacing your time interval by smaller and smaller positive values: What do you think is the "ultimate expression" for the difference quotient? Can you think of a physical quantity the "ultimate expression" for the difference quotient might represent? What are its units?

   (c) A very large chain of variety stores sells skateboards. Because of variations in local conditions, the selling price is different at different stores and the marketing manager must work with average prices, average number of sales and so on.

      The manager observes that if the average price is $45 per board, then, on the average, skaters have been buying 75 boards per month. Experience has shown that at any store, if the price is raised from $45 then, for each $2 increase in price, 3 fewer boards will be sold each month. Find an expression, or function, which represents the average number of boards sold in terms of the selling price.

      i. What price will result in an average number of zero skateboards sold?

      ii. What selling price will generate the largest amount of money in sales, i.e., what selling price results in the greatest revenue?

      iii. What is the largest amount of money in sales, i.e., what is the greatest revenue?

(d) Suppose you want to fence in a rectangular region having a fixed area of 400 square feet. Use your SCS to find an expression, or function, which represents the length dimension of the region in terms of the amount of fencing. Think about what happens to the length of the region if you increase the amount of fencing. Will this length increase or will it decrease?

2. Enter the expression described in each of the following situations into your SCS. Simplify each by hand and then use your SCS to simplify each expression. Compare the results and explain any differences between what you obtained and what the computer obtained. Use your SCS to graph each expression.

   (a) The expression which is the sum of the cube of two less than a number and the square of three minus twice that number.

   (b) The expression which is the difference of the square root of one more than a number and the reciprocal of the square root of one more than that number.

   (c) The product of the tangent of an angle and the sum of the tangent of the angle and the cotangent of the angle.

   (d) The ratio of the sine of an angle and the cosecant of the angle plus the ratio of the cosine of the angle and the secant of the angle.

3. Use your SCS to sketch the function f given by the expression $x - 2x^2$. Draw rough hand sketches of graphs representing each of the following expressions, without using further calculations. Your sketches should be clearly labeled, including the scale and variables represented by each axis.

   (a) $f(x-2)$ (b) $f(x)+1$ (c) $f(2x)$

   (d) $3f(x)$ (e) $f(x/2)$ (f) $-f(x)$

   (g) $f(-x)$ (h) $f(x-2)+3$ (i) $f(x+1)-1$

   (j) $1-f(x+2)$

4. The following formula gives a reasonable estimate of the Heat Capacity C of a certain element at very low Fahrenheit temperatures F.

$$C = 1.715 \times 10^{-1} F^3 + 2.364 \times 10^2 F^2 + 1.086 \times 10^5 F + 1.662 \times 10^7$$

Use your SCS to graph this expression in the interval $-400 \le F \le -200$.

Now use your SCS to make the substitution

$$F = 1.8 T^{\frac{1}{3}} - 459$$

and simplify the resulting expression which gives the Heat Capacity in terms of the cube root of the absolute temperature. Use your SCS to graph the resulting expression for C in terms of T on the interval $3.0 \times 10^4 \le T \le 3.0 \times 10^6$.

5. The freezing point T of a certain chemical dissolved in a fluid depends on the density of the solution. Two chemicals A and B have their freezing points given by the following expressions in terms of the mole fraction m of the chemical in the solution.

$$T_A = 5m^4 - 146m^2 - 17$$

$$T_B = -23.6m^3 + 129m$$

If the two solutions are mixed, then the expression for the freezing point in terms of the mole fraction m of chemical A relative to B is given by the sum of these two expressions.

Use your SCS to make a graph of the two individual freezing point expressions and the combined expression all on the same coordinate system. Label each graph appropriately.

6. It is possible to use the graphics system of your SCS to get estimates of the zeros of a function by trying to determine where its graph crosses the horizontal axis. The first step would be to sketch the graph over a large interval for the independent variable to see where the zeros are and then to use smaller and smaller intervals to get more accurate estimates. Do this for the function g given by the expression $x^2 - 3x + 1$. Can you find the exact values of the zeros of g in this situation?

1.3.3 Discussion

The Source of Expressions — Situations If you want to apply any kind of mathematics, it is important first to express the problem mathematically. That's where expressions come from. Before you can apply the ideas and techniques of calculus, it is usually necessary to express a problem in terms of a function. In Activity 1 you had several examples of situations which you had put in mathematical form as expressions.

Unfortunately, there is no complete general recipe for doing this. Each problem is different and you just have to develop the general ability to express situations as functions. This is not always easy and that is one reason students have so much trouble with word problems.

There are, however, some general considerations that can be useful, such as the following list.

1. Determine the quantity that is free to vary. This is the *independent variable*. Give it a name and describe it.

2. Try to see how the situation describes what happens to the independent variable. Usually the result will be the value of some quantity. This is called the *dependent* variable (or variables). You should identify it and give it a name as well.

3. A good way to start getting a feeling for the situation is to try some numbers. Give a value to the independent variable and figure out what happens to it, that is, what is the corresponding value of the dependent variable(s). Now do this again for a different value of the independent variable.

4. It can be very useful to figure out what happens to several values of the independent variable. Organize this information in a table. Your table should have separate columns for the dependent and independent variables.

5. Now think about your results and try to express what happens to the independent variable in as mathematical a form as you can. That is, try to write an expression, or some sort of recipe for calculation that describes what is done to a value of the independent variable in the situation to obtain the value of the dependent variable.

An Example

Let's see how this works with an example from our activities. Look at the skateboard problem, Activity 1 (c). What is the quantity that is free to vary?

The quantity that is free to vary is the price. Other things, such as the number of boards sold are also varying, but not so freely. They depend on the price. Let's call the price p. What happens to p is that it determines (or we might say is transformed into) the (average) number of skateboards sold. Let's call the number sold N.

If $p = 45$ then the number sold is 75. If we increase p by 2, then the number sold is decreased by 3. Thus if $p = 47$ then the number sold is 72. You might see the formula right away, or you might want to do this for some more values and organize them in a table like Table 1.1.

p	N
45	75
47	72
49	69
51	66
53	63

Table 1.1. Skateboard Sales.

At this point, you should be able to realize that what happens is that the number sold begins with 75 when p is 45 and then is decreased by 3 every time p is increased by 2. That is, the number sold is decreased by $3\left(\frac{p-45}{2}\right) = 1.5(p-45)$. So the expression for the number sold is $75 - 1.5(p-45)$ and simplifying we have,

$$N = 142.5 - 1.5p.$$

Sometimes there are two or more (quantities free to vary) and appropriate adjustments must be made. This is the case with the box problem in Activity 1 (a).

In the next section, you will see that one way of getting good at expressing a situation as a function is to write a computer program that implements the situation. Many people have found that this is very helpful. Once you get used to writing **computer functions**, you will find that thinking about a situation as a **computer function** is more than half way to seeing how to express it mathematically as a function.

Once you have some expressions, you can use them to make other expressions. You can do this by simplifying an expression, or combining two or more expressions.

Simplifying Expressions — Identities   If you have a good SCS you can often take a very complicated expression, give a single command, and it will be transformed into a simple expression that has exactly the same meaning as the original. How do you think the computer does that? What do you think is going on inside its head (oops! excuse us, we mean inside its central processor or internal engine)?

One possibility is that the computer uses some pre-programmed strategies for applying various rules of computation. These can be very elementary like rules for removing parentheses, gathering like terms, canceling, and so on. You probably remember these from previous math courses, or at least you are able to use them, whether you are aware of it or not.

You might also remember (from other math courses) some more complicated rules that apply to trigonometric expressions. Trigonometric expressions are those that involve sin, cos, tan, cot, sec and csc. Rules for computation with trigonometric expressions are called trigonometric identities. Figure 1.10 provides a brief list of some of the most important ones. See the Appendix A on Trigonometry at the end of this text for a brief review of trigonometry which includes the identities below and others. The independent variable or *argument* in the expression is θ to remind us that the situation described by the expression involves an angle.

Can you use these rules to simplify some of the expressions in Activity 2?

$$\sin^2(\theta) + \cos^2(\theta) = 1$$
$$\tan(\theta) = \frac{\sin(\theta)}{\cos(\theta)}$$
$$\cot(\theta) = \frac{1}{\tan(\theta)}$$
$$\sec(\theta) = \frac{1}{\cos(\theta)}$$
$$\csc(\theta) = \frac{1}{\sin(\theta)}$$
$$1 + \tan^2(\theta) = \sec^2(\theta)$$
$$1 + 1 + \cot^2(\theta) = \csc^2(\theta)$$
$$\sin(2\theta) = 2\sin(\theta)\cos(\theta)$$

$$\cos(2\theta) = \cos^2(\theta) - \sin^2(\theta)$$
$$= 2\cos^2(\theta) - 1$$
$$= 1 - 2\sin^2(\theta)$$
$$\sin^2(\theta) = \frac{1 - \cos(2\theta)}{2}$$
$$\cos^2(\theta) = \frac{1 + \cos(2\theta)}{2}$$
$$\sin(-\theta) = -\sin(\theta)$$
$$\cos(-\theta) = \cos(\theta)$$
$$\tan(-\theta) = -\tan(\theta)$$

Figure 1.10. Some trigonometric Identities.

Can you think of other ways the computer might simplify expressions using methods other than following preset rules?

The Value of an Expression — Functions   In the beginning of this section when we considered how to represent a situation as an expression we thought about choosing and then naming an independent variable, giving it values and working out what happened to it in the situation. When you think about an expression like this, you are thinking about it as a function.

Although the independent variable in an expression is free to vary, the details of the expression or the situation itself can place restrictions on the values it can take. This is what we were thinking about in box problem in Activity 1 (a). Which of the other activities have independent variables whose values are restricted?

The set of values which the independent variable <u>can</u> take on is called the *domain* of the expression. If you want to give a name to the values of the expression, this is called the *dependent variable*. The set of values which the dependent variable can take on is called the *range* of the expression. For which of the situations in Activity 1 can you figure out the domains and the ranges of the variables? That is, given a variable, can you describe precisely which values it can take on and which, if any, it cannot?

You might also want to give a name to the function represented by the expression. If you call it f, then the notation $f(3)$, for example, refers to the value of the expression, or the function represented by the expression, when the independent variable has the value 3. Something like $f(x^2 + 3)$ stands for what you get if you *substitute* $x^2 + 3$ for the independent variable in the expression.

Graphs of Expressions — Functions Again   An expression is an invitation to calculate. For example, Activity 2(b) leads to the following sequence of symbols

$$\sqrt{x+1} - \frac{1}{\sqrt{x+1}}$$

and this expression should suggest the action of picking a numerical value for the argument x, substituting it in the expression and computing the value. It also refers to repeating this action, many times. Indeed, the idea is to repeat it for all numerical values of the argument for which you can actually get an answer. If there is a number for which the expression doesn't give you an answer (how could this happen?) we say that the expression is not defined for that number.

As we said before, when you think about an expression like this, you are beginning to think of it as a *function*. Calculus is really about functions and you will work with them often in this course. We can use the idea of a table once again here. One column lists possible values you could choose for the argument (recall that argument is another name for independent variable). The other column lists, alongside each value of the argument, the corresponding value of the expression, that is the value you get by substituting what is in the first column for the argument and calculating. You can imagine such a table in your mind, but you can't really draw a picture of it — because it is infinite. Table 1.2 is a picture of a finite portion of this table.

To save space, we call the expression g so that the label for the second column is $g(x)$.

You can "plot" these pairs in the table as points on a coordinate system. Your SCS has a specific syntax for making graphs. We recommend that you use the feature (if it is available) that just plots a large number of individual points without connecting them. Using such a feature the graph will look like that in Figure 1.11.

You can connect these points yourself to get a smooth curve as in Figure 1.12.

x	$g(x)$
−0.9	−2.846
−0.7	−1.278
−0.2	−0.224
0	0
0.4	0.338
1	0.707
1.5	0.949
2	1.155
3	1.5
6	2.268

Table 1.2. The function g.

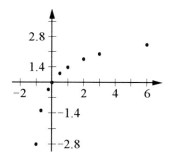

Figure 1.11. Points for g.

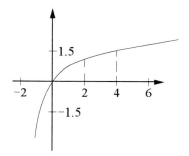

Figure 1.12. A graph for g.

This is what your graphing system does to produce all the pictures you made in the activities.

Wait a minute! What, exactly are you doing when you connect the points? What relationship do the line segments (or whatever) that you (or the computer) use for the connection have to do with the expression you are working with? Are they both the same thing? How do you know? How can you be sure? What assumption(s) are you making when you connect these dots?

In any case, we now have an expression, and this smooth curve, and the idea of a function we were just talking about. How are they all interrelated? A function is about picking a value for the argument, plugging it in and chugging to an answer. Or you can say it in fancier language. A function is a process that takes quantities of one kind and transforms them to quantities of another (or the same) kind. How would you interpret this way of thinking about a function when it is given by an expression? By a table? By a graph?

Seeing the Behavior of a Function from its Graph — Properties   If you think about an expression as representing a function, then this function will have a number of properties that are of particular interest. You already worked with some properties in the Activities, such as zeros, maximum

and minimum values. Usually you can figure out what a property is from its name together with an example or two. We will discuss a few here and give you a complete list in the exercises for you to consider.

Once you know what a property means, you can decide how it goes with a particular function. You can do this analytically, from the expression representing the function, or geometrically, from the graph of the function.

You can tell where a function has a zero by solving an equation or by seeing where it crosses the x-axis. How can you tell where a function is increasing? Decreasing? What properties of the graph will tell you that a function is even? By *even* we mean that for all x in the domain of the function the values of the function satisfy the following relation (where f is the name of the function).

$$f(-x) = f(x)$$

We also speak of a function being *odd* which means, if its name is g, then for all x in the domain of g the values of g satisfy the following relation

$$g(-x) = -g(x).$$

How would this property be manifested in the graph?

Many properties of a function are manifested in a graph of that function. How would this be for properties like maximum, minimum, or regions of concavity? (You may not know what some of these words mean, but try to guess.) Can you think of any other properties and their manifestations in graphs?

Substituting in Expressions

Suppose a function is represented by an expression of a variable, i.e., for each value of the variable, there is only one value of the expression. Then, if you replace the variable in the expression by another expression, many things could change in the underlying function represented by the original expression. For example, the graph may look very different and various properties of the function might change.

For example, in Activity 3 you saw how replacing x by $x - 2$ in an expression will shift the graph two units to the right. Can you explain exactly why this happens? If you know a fact like this and you have the graph of f then you can see without using any computers what the graph of the expression $f(x-2)$ will look like.

Do you see how to do this with the other examples in Activity 3?

Activity 4 is about substitution but the situation is different because the substitution is not so simple. The point here is that, beginning with a fairly complicated expression for C, it was possible to find a substitution that made the graph very simple (a straight line segment). This can be useful when you are trying to fit a curve to experimental data. Incidentally, do you see any meaning in the substitution?

Combining Expressions or Functions

Combining two or more expressions can also change graphs and properties. In Activity 5 all you had to do was add the two expressions and then make the graph to obtain the graph of the sum. Suppose that you only have the graphs of two expressions (or equivalently, the two functions represented by the expressions) to work with and you have to figure out by hand the graph of the difference and quotient of the two given expressions (or functions).

If you go back to thinking about how the graph of an expression (or function) comes from evaluating the expression (or function) then it is not hard to see what has to be done. You take a point and look at the graphs of the two expressions to find the values of the expressions (you might want to use a straight edge.) Then, if you are to subtract, you would simply subtract the two values. What would you do if you were going to divide the two expressions? Are there any problems you might encounter? If so, what do you do? After doing this for lots of points, you can make a table and then graph the resulting pairs of points.

What happens to properties? For example, what is the meaning of looking for the point at which the resulting graph crosses the horizontal axis? What about some of the other properties you are working with in this section?

A Special Case — the Difference Quotient   There is one way of constructing a new expression from a given one that is very important in Calculus. It is called the *difference quotient*. Consider the expression $16t^2$ which came from the situation in the Activity 1 (b).

Let's call the function represented by the expression $16t^2$, by the name f. Then we have, $f(t) = 16t^2$. Suppose that we make a new expression by substituting $3+h$ for t. If we call the resulting expression $f(3+h)$, subtract the number $f(3)$ from it, and divide by h, we are forming the difference quotient,

$$\frac{f(3+h) - f(3)}{h}$$

or simply the change $f(3+h) - f(3)$ in the values of f over the change $3+h-3 = h$ in the values of t. In this example, do you see that the difference quotient works out to be

$$\frac{16(3+h)^2 - 144}{h}.$$

Try simplifying the latter expression yourself. What happens "ultimately" to the value of the difference quotient if you let h gets closer and closer to zero?

How would you represent the difference quotient on the graph of f? What do you think is the meaning of letting h get closer and closer to 0?

Go back and read through the situation in Activity 1 (b). Can you give a physical interpretation to the difference quotient in that situation?

1.3.4 Exercises

1. Write the following expressions in the form x^n.

   (a) $x^3 x^2$ (b) $x^{-2} x^5$

   (c) $x^3 x^{1/3}$ (d) $x^{2/3} x^{1/2}$

   (e) $(\sqrt{x})^3$ (f) $x^3 \sqrt{x}$

   (g) $1/x^3$ (h) $(x^{-3} x^5)/(x^2)^3$

   (i) $x^{-1}/(x^{-2} x^{1/3})^3$

2. In each of the following situations, write an expression which represents the situation and then simplify it by hand. Then use your SCS to simplify them. Compare the results and explain any differences between what you obtained and what the computer obtained.

   (a) The expression which is the sum of 36 divided by a number and the reciprocal of the number.

   (b) The expression which is the difference of the cube root of one less than the square of a number all squared and the reciprocal of the cube root of one less than the square of the number.

   (c) The product of one minus the cosine of an angle and one plus the cosine of the angle.

(d) The ratio of the sine of an angle plus the cosine of the angle to the cosine of the angle.

(e) The sum of the sine of an angle and the product of the cosine and cotangent of the angle.

(f) The reciprocal of the sum of the tangent of an angle and the cotangent of the angle.

3. For each of the situations listed below, find an expression, or a function, which represents the desired quantity in each situation. Make a sketch of the function using your SCS. Then, if possible, answer the indicated questions (approximately).

Here are the situations and questions.

(a) Suppose you want to fence in a rectangular region using a 1000 feet of fencing. Find the area of the region as a function of the length of the region.

    i. Think about what happens to the area of the region if you increase the length of the region. Will the area increase or will it decrease?

    ii. Think about what happens to the area of the region if you decrease the length of the region. Will the area increase or will it decrease?

    iii. Can you find the largest area which you can enclose with 1000 feet of fencing?

    iv. Can you find the dimensions of the region which results in the largest area?

(b) A cattle rancher wants to enclose a rectangular region along a straight river bank so that the region is separated into two equal rectangular regions by a fence down the middle of the region that is perpendicular to the straight river bank. The rancher intends to graze horses in one region and cattle in the other. The rancher has 720 feet of fencing available. Find the area of the region as a function of the length of the fence perpendicular to the river.

    i. Can you find the area of the largest amount of grazing land that can be fenced as described?

    ii. Can you find the dimensions of the region which give the greatest amount of grazing land enclosed by the fence?

(c) An appliance manufacturer produces washing machines and determines that to sell x machines its unit selling price must be $s = S(x) = 300 - 0.3x$ dollars per machine. It is also determined that the manufacturer's total cost of producing x washing machines is given by $0.7x^2 + 5600$ dollars. Find the total profit as a function of the number of washing machines sold.

    i. Can you find the manufacturer's maximum total profit?

    ii. Can you find how many washing machines produced gives the greatest total profit?

    iii. Can you find the price of a washing machine so that the profit is the largest?

    iv. Can you find the manufacturer's *break even point(s)*, i.e., the number of washing machines the manufacturer must sell so that the profit is zero?

(d) The postal service restricts the size of packages which can be accepted for delivery. The restriction is that the length of a package plus the girth (perimeter around a cross section perpendicular to the length) cannot exceed 108 inches. Find the volume of a box package, with square cross sections perpendicular to the length, as a function of the length of a side of the square cross sections.

    i. Can you find the dimensions of the largest box package?

    ii. Can you find the volume of largest box package?

(e) A manufacturer of tin cans is to make a can holding 80 in^3. Find the surface area as a function of the radius of the base of the can.

    i. Can you find the least amount of tin needed to make the can?

    ii. Can you find the radius and height of the can requiring the least tin?

(f) Consider the situation in the previous Exercise where the top and bottom of the tin can cost $0.02 per in^2 and the sides cost $0.03 per in^2. Find the cost to make such a can as a function of the radius of the base of the can.

    i. Can you find the least cost of such a can?

    ii. Can you find the radius and height of the most economical can?

(g) A triangle has two sides of length 5 and 8. Express the area of the triangle in terms of the angle between the two sides.

    i. Can you find the angle between the two sides so that the area is the largest?

    ii. Can you find the perimeter of the triangle when its area is the largest?

(h) For the triangle in the previous Exercise, express the perimeter of the triangle as a function of the angle between the two sides.

    i. Can you find the angle between the sides so that the perimeter is the largest? smallest?

    ii. Can you find the triangle so that the perimeter is the largest? smallest?

(i) A charter airline runs a promotion on its late night flight to Reno. The fare per person is $100 with a reduced fare of $2 per person for every person over the 75th person. Each $2 reduction is for every person on the plane. Find the revenue as a function of the number of passengers who buy tickets when the maximum capacity of the plane is 120 passengers.

    i. Can you find the number of tickets sold so that the revenue is the greatest?

    ii. Can you find the largest revenue the airline can expect?

(j) The net charge q that has moved to the right in a conductor at a certain point x, where t is in seconds, is given by four times the time plus sixteen for $-4 < t \leq -2$, twice the square of the time for $-2 < t \leq 1$, eight less than five times the square root of the quantity time plus three for $1 < t \leq 6$, seven for $t > 6$ and zero otherwise.

    i. Can you give an explanation for why the time variable t can be negative?

ii. Can you find the largest net charge in the conductor at x?

iii. At what time t does the largest value of q occur?

(k) You wish to construct an open box with the length of the base three times the width. The cost of material for the base is $3 per square meter and the cost of material for sides is $2 per square meter. Find an expression for the cost of material for the box.

i. Is there anything unusual about this problem?

(l) A woman is in a boat, 3000 meters from her house on the shore on a line perpendicular to the (straight) shoreline. A lighthouse is 2000 meters down shore from her position. The woman intends to get to the lighthouse by rowing towards the shore at an angle so as to arrive at a point between her house and the lighthouse. She will walk the rest of the way. She can row at 2 kilometers per hour and walk at 4 kilometers per hour. Find an expression for the total time it takes the woman to get to the lighthouse in terms of the angle at which she rows.

i. Can you determine what are the possible values for this angle?

ii. What about the minimum time it takes her to make the trip?

iii. What is the maximum time it takes her to make the trip?

4. For each of the expressions listed below, do the following.

(a) Find all values of x, if any, where the expression is (i) zero and (ii) undefined.

(b) Simplify the expression by hand. Use your SCS to check your answer.

(c) Find all values of x, if any, where the simplified expression is (i) zero and (ii) undefined. Compare your answers in this part with your answers in part (a). Explain any differences.

(d) Find all values of x, if any, where the original expression and the simplified expression are both (i) zero and (ii) undefined.

Here are the expressions.

i. $\dfrac{2x^3}{1-x^2}+x$ ii. $\dfrac{\frac{1}{x^2}-\frac{1}{9}}{x-3}$

iii. $x\left(\dfrac{1}{2}\right)\left(4-x^2\right)^{-\frac{1}{2}}(2x)+\sqrt{4-x^2}$

iv. $\dfrac{\left(\sqrt{1+x^2}\right)(2x)-x^2\left(\frac{1}{2}\right)\left(1+x^2\right)^{-\frac{1}{2}}(2x)}{1+x^2}$

v. $\dfrac{1}{\tan(x)\sin(x)+\cos(x)}$

vi. $\dfrac{\sin(2x)-\cos(2x)}{\cos(2x)}+1$

5. Use your SCS, or a graphics calculator, to sketch each function represented by the given expression on the indicated interval(s).

(a) $\dfrac{23.675}{14.26x-37.98}+18.63,\quad -5\le x\le 5$ and $-50\le y\le 50$

(b) $x^3-13.7x^2+40.87,\ -5\le x\le 15$

(c) $\dfrac{x^2-3.7x}{x+1.23},\ -5\le x\le 5$

(d) $\sec(x),\qquad -15\le x\le -15,\qquad$ and $-40\le y\le 40$

6. Use your SCS, or a graphics calculator, to sketch each of the following pairs of expressions both on the same coordinate axes. What interesting observation can you make? Are you surprised? Why or why not?

    (a) $\sin^2(x) + \cos^2(x)$ and $\sec^2(x) - \tan^2(x)$

    (b) $\sin(-x)$ and $-\sin(x)$

    (c) $\cos(2x)$ and $\cos(-2x)$

    (d) $1 - \sin^2(x)$ and $\cos^2(x)$

    (e) $1 + \tan^2(2x)$ and $\sec^2(2x)$

    (f) $2\sin(x)\cos(x)$ and $\sin(2x)$

    (g) $\cos(2x)$ and $\cos^2(x) - \sin^2(x)$

    (h) $2\cos^2(x)$ and $1 + \cos(2x)$

    (i) $1 - 2\sin^2(0.5x)$ and $\cos(x)$

7. Make a sketch of the graphs of the functions represented by the given expressions on the same coordinate axes. What interesting observations can you make about the expressions and their graphs?

    (a) x^3 (b) $x^3 + 1$

    (c) $x^3 - 2$ (d) $2x^3 + 1$

    (e) $0.5x^3 - 2$ (f) $(x-1)^3 - 1$

    (g) $(x+2)^3 - 1$ (h) $2(x-1)^3 + 1$

    (i) $2(x+1)^3 - 1$

8. After the holiday season, a store has 57 strings of tree lights and 43 boxes of tree ornaments left over. There are two ways to package these items for quick sale and the store manager must decide how many of each way to use. One way makes a package of 5 strings of lights and 2 boxes of ornaments. It sells for $6.50. The other way has 4 strings of lights and 3 boxes of ornaments. It sells for $5.00. Assume that no matter what she does, all packages will be sold and only the items that didn't fit in a package will be left.

    Find an expression, in terms of the number of packages of the first kind, for the minimum number of packages of the second kind that she would have to make in using up as many items as possible.

    (a) Can you determine the possible values for the number of packages of the first kind in this expression?

    (b) Can you figure out what is the best thing for the manager to do?

9. For each of the following pairs of expressions, use your SCS or a graphics calculator to sketch both of their graphs on the same coordinate axes. Then print your graph or reproduce it by hand on a sheet of paper. Use the indicated intervals where given. Use the indicated letter for the domain variable and anything you like for the range variable. Connect the dots and label the two graphs to distinguish them.

    (a) The expressions

    $$\frac{3u-5}{1+u} + 1, \quad \frac{u^2 - 2u + 7}{u} - 1$$

    for u from -2 to 2 and the dependent variable from -50 to 50.

    (b) $\left|\sqrt{x} - x\right|$, $\left|\sqrt{x-1} - x + 1\right|$ for x from -1 to 5.

    (c) $\sin(F)$, $1.2\sin(F)$.

    (d) $\sin(T)$, $\sin(1.2T)$.

10. Write a brief essay to explain each of the following.

    (a) What happens to the graph of a function represented by an expression in x if a number c is added (or subtracted) from the expression representing $f(x)$?

(b) What happens to the graph of a function represented by an expression in x if the expression representing $f(x)$ is multiplied by a number a?

(c) What happens to the graph of a function represented by an expression in x if x is replaced by $x+b$?

(d) What happens to the graph of a function represented by an expression in x if the expression is multiplied by a and then c is added to the resulting expression? What if x is also replaced by $x+b$?

11. For each of the following functions find an expression for the difference quotient

$$\frac{f(x+h)-f(x)}{h}$$

at the indicated value of x with the specified value of h. Simplify your answer. Represent the difference quotient at the indicated value of x on a graph of the function by drawing the secant line between $(x, f(x))$ and $(x+h, f(x+h))$. Indicate what you think is the "geometric" meaning of the numerator, the denominator and the whole difference quotient by labeling on your sketch appropriately.

(a) $f(x) = 0.5x^2$, $x = -1$, $h = 0.3$

(b) $f(x) = 1/x$, $x = 1$, $h = 0.2$

(c) $g(x) = \sqrt{x}$, $x = 1$, $h = 0.5$

(d) $h(x) = 4 - x^2$, $x = -2$, $h = 0.1$

12. Make a sketch using your SCS or a graphics calculator for each of the following functions on the indicated interval. Then print your graph or reproduce it by hand on a sheet of paper. Draw a small segment of an approximate tangent line by hand at each of the indicated points, estimate the slope of each tangent line and then make a table of values with the x values and slope values $s(x)$ for the "slope" function. Finally,

make a sketch of the slope function using your table.

(a) $f(x) = 0.5x^2$, $-2 \leq x \leq 2$, $x = -2$, -1.5, -1, -0.5, 0, 0.5, 1, 1.5, 2

(b) $g(x) = 4/x$, $1 \leq x \leq 4$, $x = 1$, 1.5, 2, 2.5, 3, 3.5, 4

(c) $h(x) = 1 - \frac{1}{3}x^3$, $-1 \leq x \leq 3$, $x = -1$, -0.5, 0, 0.5, 1, 1.5, 2, 2.5, 3

13. For the functions f and g in represented by the sketches below, make a sketch (by hand) of each of the following functions.

(a) $f + g$ (b) $f - g$

(c) $f \cdot g$ (d) f/g

(e) $1/f$ (f) $1/g$

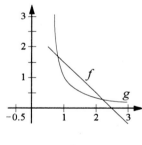

i.

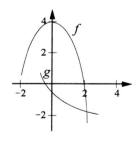

ii.

▌ 1.4 USING YOUR MPL

1.4.1 Overview

The main purpose of this section is to get you familiar with the kinds of things you will do with your MPL. The strength of a SCS lies in its ability to manipulate mathematical symbols. A MPL, however, is more concerned with the values that these symbols can have. In Calculus, those values are usually numbers. When you think about plugging in numbers to evaluate expressions, then you are really getting into functions and thinking of functions as processes. The idea of a function as a process is one of the strong points of a good MPL. In this book, we will give examples of **computer functions** in both the MPL, **ISETL** and the MPL which is part of **MapleV**. Your instructor will provide you with any necessary handouts if you use another MPL.

So this section will be concerned with numbers, functions and graphs. You will begin to get familiar with using a MPL to deal with these mathematical objects and the underlying processes. Many of the problems will be similar to the ones you worked with using your SCS. There will also be some new problems. Again we have put in some things that you are still not able to handle but must wait for more mathematical tools and greater understandings. We hope you are getting used to this!

Because a MPL is more powerful, there are a few things you must learn about its syntax. By *syntax* we mean the rules for making expressions. Once you have that taken care of you will be able to express situations in your MPL as **computer functions**. You will deal with evaluating functions, study changes of signs and zeros, and finally put it all together with a new way of making graphs, which hopefully will deepen your understanding of the concept of function. Sometimes you may find it convenient to make graphs of functions on a graphics calculator. That will be a decision we shall leave for your and/or your instructor to make when appropriate.

1.4.2 Activities

1. Use your MPL to make your screen look similar to Terminal Screen 7 in Figure 1.13. Before running the code for each **computer function** and the indicated function values in Figure 1.13, write down what you think will be your MPL's response. If you are not using **ISETL** as your MPL, then your instructor will provide you with a handout of Terminal Screen 7 for use with your MPL.

2. Use your MPL to make your screen look similar to Terminal Screen 8 in Figure 1.14. Before running the code, write down what you think will be your MPL's response. If you are not using **ISETL** as your MPL, then your instructor will provide you with a handout of Terminal Screen 8 for use with your MPL. Print a copy of what is printed on your screen as a result of performing the indicated operations.

3. The Terminal Screens that you worked with in the previous two activities use **computer functions** (those shown in Terminal Screen 7 are referred to in **ISETL** as a **func**s) and **smaps** (an **smap** is a set of ordered pairs) to represent functions. Give a verbal description of the process of each type of function representation, i.e., describe how each function value (or output) is evaluated for a particular domain (or input) value.

```
>       g := func(x);
>>          return 2*x - x**3;
>>          end;
>       g(3); g(-2); g(0);

>       p := |x->2*x - x**3|;
>       p(3); p(-2); p(0);

>       f := func(x);
>>          if x > 0 then return x**2-1;
>>          elseif x = 0 then return 1;
>>          else return x + 1;
>>          end;
>>          end;
>       f(2); f(5); f(0); f(-1); f(-3);

>       F := func(t);
>>          return 3-4*t < 2*t + 5;
>>          end;
>       F(-2); F(-0.5); F(0); F(0.3); F(1);

>       G := func(y,z);
>>          return abs(2*y + 1) < abs(3-2*z);
>>          end;
>       G(-10,8); G(4,-1); G(-0.5,0); G(0.1,1);

>       h := func(x);
>>          if x /= -1 then return x*(x-2)/(x+1);
>>          else return;
>>          end;
>>          end;
>       for i in [1..10] do
>>          print h(-3+i/2);
>>          end;
```

Figure 1.13. Terminal Screen 7.

```
>        f := func(y);
>>              if 0 <= y and y < 3 then return sin(y);
>>                   elseif y <= 5 then return y-1;
>>                   elseif y < 8 then return 1-2*sqrt(y);
>>                   else return y**2-3
>>              end;
>>         end;
>        f(-2); f(1); f(4); f(10); f(15);

>        sk := {[45,75], [47,72], [49,69], [51,66], [53,63]};
>        sk(47); sk(48); sk(49);

>        h := {[45,112], [47,18.6], [49,-70], [51,35], [53,-68.734]};
>        h(47); h(49);

>        g := {[x,sk(x)+h(x)] : x in domain(sk)};
>        g(47); g(49);
>        g;
```

Figure 1.14. Terminal Screen 8.

4. The distance traveled by a falling object is approximately given by $16t^2$ feet t seconds after it is dropped from the top of a cliff. Write a **computer function** that computes the average rate of change of the distance that the object has fallen with respect to time (i.e., the change in distance divided by the corresponding change in time), or the average velocity, when it falls from t seconds to $t+h$ seconds. What is the average velocity when the time changes from $t = 3$ seconds to $t = 3.1$ seconds? What are the units of the average velocity?

   (a) Suppose you replaced the time changes from $t = 3$ seconds to $t = 3.01$ seconds. What is the average velocity then?

   (b) What would be the meaning of replacing your time interval by a negative quantity such as -0.1, i.e., say $t = 3$ seconds to $t = 2.9$ seconds?

   (c) Suppose you kept replacing your time increment, h, by smaller and smaller positive values. What would be the "ultimate value" of the average velocity in this situation? What are its units?

   (d) Find and simplify by hand an expression for the average velocity when the object falls from t to $t+h$ seconds. What are its units? What happens the expression for the average velocity when h is made closer and closer to zero? What is the "ultimate expression" for the average velocity? What do you think the "ultimate expression" represents?

5. A radio transmitter sends out signals at certain voltage strengths. This voltage is measured in *volts*. The amount of voltage will usually change as time goes on. Consider a particular signal which varies as follows. It is considered that nothing happens before time begins. In the first second, the voltage is equal to twice the time. In the next second it is 3 volts minus the time. From then until the fourth second, the voltage is sinusoidal, that is, it is given by the expression $\sin(\pi(t-2)/2)$, after which it disappears completely.

Write a **computer function** which represents the function which gives voltage as a function of time.

(a) Where is the voltage increasing? decreasing?

(b) Is the voltage increasing or decreasing at the following times:

$$t = 0,\ 0.5,\ 1,\ 2,\ 2.5,\ 3,\ 3.5,\ 4\ ?$$

6. Write a **computer function** that will accept two numbers a and b, and which returns **true** or **false** depending on whether or not

$$|a+b| < |a| + |b|.$$

(If you are using **ISETL**, look at Figure 1.13, page 30 for suggestions. In particular, notice that you don't have to use **if...then...** clauses and return **true** or **false** explicitly. It is enough to return the boolean value of an expression).

Apply your **computer function** in at least two cases, one of which gives **true** and the other, **false**. Explain exactly when you get **true** or **false**.

7. For each of the following write a **computer function**, using basic trigonometric identities (e.g., $\sin(x) = 1/\csc(x)$, etc.) and the Pythagorean identities (e.g., $\sin^2(x) + \cos^2(x) = 1$, etc.), that will return the value of the indicated trigonometric function of a real number x which satisfies the given condition. Run your **computer function** for the given data.

(a) $\cos(x)$ given $\sin(x)$. Run it for $\sin(x) = 3/5, 0.628, 0.001$, assuming that $\cos(x) > 0$.

(b) $\sin(x)$ given $\tan(x)$. Run it for $\tan(x) = 0.12, 1, 17$, assuming that $\sin(x) > 0$.

(c) $\tan(x)$ given $\cos(x)$. Run it for $\cos(x) = 1/3, 0, 1.2, 3$, assuming that $\tan(x) > 0$.

8. In Figure 1.15, plots are shown of absolute viscosity v as a function of temperature T for SAE 10W-30 oil and for Glycerin. Suppose you made a mixture that was thirty parts SAE 10W-30 oil to one part Glycerin. Make a reasonable guess about how you might calculate the viscosity of the mixture?

Make a set of ordered pairs to represent the viscosity of SAE 10W-30 oil by reading off 10 points and the corresponding viscosity from the graph. Do the same for Glycerin, using the same domain points. Now write a computer expression that will construct set of ordered pairs representing the function that gives the viscosity of the mixture. (If you are using **ISETL**, you might want to take a look at Figure 1.14 for suggestions.)

9. The compressibility z of a fixed number of gas molecules can be expressed as a function Z of the volume v which the gas occupies, i.e., $z = Z(v)$. The usual practice is to obtain experimental data giving the compressibility at various volumes and then to try to find an expression that gives the same information as the data. Then the expression can be used to estimate the compressibility at any volumes.

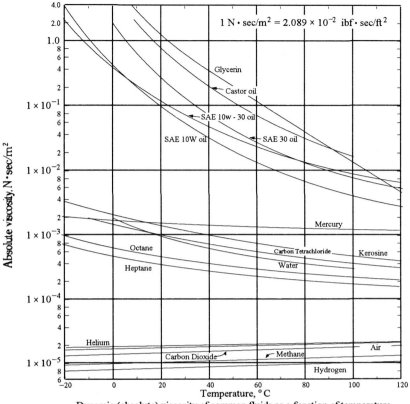

Dynamic (absolute) viscosity of common fluids as a function of temperature.

Figure 1.15. Graphs of viscosity versus temperature.

It can turn out that different expressions work in different regions of values of the volume. For example, the compressibility of one gas was found to be very close to the following expression

$$\frac{v^3}{3 \times 10^3} - 0.048v^2 + 1.73v - 7.724$$

if the volume was less than 90 and the following expression

$$6.432 - \frac{4v^{\frac{3}{2}}}{5\sqrt{v^3 - 7 \times 10^5}} - \frac{27}{v - 200}$$

otherwise.

(a) Write a **computer function Z** that will represent the above compressibility function Z, where $z = Z(v)$.

(b) Now imagine the volume changing as a function **V** of time **t** in such a way that the volume is given by the square of the time up to $t = 10$ and by the cube of the time thereafter. What would do expressions like the following represent?

$$\mathbf{Z(V(17))};\qquad\qquad\mathbf{Z(V(63))};$$

Write a **computer function** that will represent such a "composition" of functions, i.e., a **computer function** which accepts a value of **t** and returns the value of **Z(V(t))**, thus evaluating the compressibility for a given time at a value $V(t)$

(c) Use your computer to make a graph each function, Z, V and the composition Z of V on the interval $[0, 250]$. Then graph all three functions on the same coordinate axes on the interval $[0, 250]$.

1.4.3 Discussion

Numbers   Functions in calculus work on real numbers. In previous sections you used the computer to work with integers, fractions, decimals and radicals. These, so called algebraic numbers, together with the transcendental numbers (such as e and π) make up the *real numbers*. How would you explain the difference between integers, fractions, decimals, radicals and transcendentals?

Inequalities between numbers are very important in Calculus, especially when absolute values are involved. Take a look at the inequality in Activity 6. Do you understand why this relation is always false when a and b have the same sign (or both are zero) and true only if their signs are different? Why does replacing $<$ with $\leq$ change things so that the statement is always true? You should be able to read inequalities and think about what they mean.

Absolute values have to do with distance. For example, the statement $|1-3x| \leq 2$ asserts that the distance between 1 and 3 times the value of x is less than or equal to 2. What does the statement $|2x+5| > 3$ assert?

Computer Functions   A **computer function**, or procedure, describes an algorithm for transforming a number (or numbers) which it can accept into a number (or other expression) which it returns.

Here is a question to see how you think about functions right now. Look at the following equation.

$$x = y^2 + 1$$

Is there any function in this situation?

The Syntax of a Computer Function — Process   The description of an algorithm or process can get quite complicated if the function is complicated. Fortunately, the syntax of your MPL is usually powerful enough to handle most complications you will encounter.

Below is the **ISETL** code for the **computer function f** given in Figure 1.13.

```
f := func(x);
    if x > 0 then return x**2 - 1;
    elseif x = 0 then return 1;
    else return x + 1;
    end;
end;
```

The same **computer function** in **MapleV** can be written as follows (actually only the last semicolon is required in the code for a **MapleV** computer function):

$$\textbf{f} := \textbf{proc(x);}$$

$$\textbf{if x > 0 then x**2 - 1;}$$

$$\textbf{elif x = 0 then 1;}$$

$$\textbf{else x + 1;}$$

$$\textbf{fi;}$$

$$\textbf{end;}$$

You can read the process off easily from either computer code above as follows. The process of this function is set in motion by executing a command of the form **f(a)** where **a** is some number. The first thing the function does is set **x** equal to the value **a**. This is the value that the function has been given to work on. For example, if you enter **f(2)** you are giving x the value **2**. If you enter $\textbf{f}(-3)$ you are giving x the value -3. And so on.

For a number **x**, the process decides if it is greater than zero, equal to zero, or none of the above. In the first case (greater than zero), it would return the value obtained by squaring the value of **x** and subtracting **1**. In the second case (equal to zero), it would return **1**. Finally, in the third case (less than zero), it would return one more than the value of **x**.

Notice the syntax of such a **computer function**. It always begins with a key word, like **func** or **proc**, followed by a list in parentheses of the name(s) of the variable(s) it will accept, followed by a semicolon. It always ends with the key word **end** followed by a semicolon. In between is the code that describes the process of the function.

Notice the syntax of the **if** clause. It always begins with the key word **if** followed by an expression whose value is **true** or **false**. Next comes the key word **then**, followed by the word **return** in **ISETL**, while in **MapleV** the word **return** is omitted, followed by an expression to evaluate followed by a semicolon (the semicolon may be omitted in **MapleV**). This is what is returned if the condition is true. If the condition is not true, then it goes on. The **elseif**, or **elif**, clause is optional and there can be as many as you like. The **else** clause is also optional, but there can only be one and it must come last.

Finally notice the syntax of the **for** loop after the **func h** illustrated in **ISETL** in Figure 1.13. A **for** loop is for repeated actions. In **ISETL** it always begins with the key word **for**, followed by an iteration specifier, followed by the key word **do**. The loop terminates with the key word **end**. The iteration specifier names a variable and gives its possible values. In between **do** and **end** is a description of the action of the loop. This description can have as many statements as you like. The action is repeated once for each value of the variable in the iteration specifier. A similar syntax for a **for** loop exists in **MapleV** and other MPL's.

The Syntax of Smaps — Process

The ISETL construct in the latter three examples shown in Figure 1.14 is called an **smap** or a *set of ordered pairs*. It is another way to represent a function in **ISETL** when you think about the function as a table. The first **smap** in Figure 1.14 corresponds to the data concerning skateboard sales in Table 1.1, page 18. An **smap** can also be used in **MapleV** — see your instructor for details on how to use them.

In a table, you list the data in two columns. The first represents the independent variable and the second represents the dependent variable. In an **smap**, the idea is exactly the same, except instead of columns, you use a set of ordered pairs (i.e., sets of **tuples** with exactly two components). The first quantity in the pair corresponds to something that appears in the first column of a table and the second quantity in that same pair corresponds to what appears across from it in the second column. Thus, one of the pairs in the **smap sk** in Figure 1.14 is **[49, 69]**. Look at the table on page 18 and see that 49 appears in the first column and 69 appears right across from it in the second column.

Once an **smap**, or set of ordered pairs, has been defined and given a name then it is treated just like a **computer function** as far as operations like evaluation and graphing are concerned.

Notice that the last example in Figure 1.14 illustrates that an **smap** can be formed, not only by listing the pairs, but by an expression, called a *set former*, which constructs a set according to some rules.

Sources of Computer Functions and Smaps — Situations

Computer functions and **smaps** come from mathematical situations, just like the expressions that you worked with in Section 1.3. Notice that the **if** clause we use in a MPL allows us to construct a function whose value is given by one expression for some of its values and other expressions for other values. We refer to such a function as a *function defined in parts*. Some people call it a *function with split domain*.

As with expressions, the big job with situations is to translate them into mathematical language — in the case of a MPL, using **computer functions** or **smaps**. The advice that we can give you is another list of general considerations, very similar to the one in Section 1.3.3.

1. Determine the quantity that is free to vary. This is the *independent variable* and you should give it a name.

2. Try to see how the situation describes what happens to the independent variable. Identify and name the *dependent* variable(s).

3. Start getting a feeling for the situation by trying some numbers. Give a value to the independent variable and figure out what happens, that is, what is the corresponding value of the dependent variable(s).

4. It can be very useful to figure out what happens to several values of the independent variable. Organize this information in a table. Figure out what are the possible values of these variables.

5. Now think about your results and try to list in a step-by-step manner what happens to the independent variable. Organize this list in categories. Think in terms of expressions, conditions and cases (**if** clauses), repeated actions (**for** loops) and values returned (remember in **ISETL** you use the key word **return**).

6. Try to express each category of action in terms of a **computer function** structure such as an **if** clause, **for** loop, or individual code for expressions.

7. Organize everything in the syntax of a **computer function** or **smap**. Assign this expression to a variable. This variable is the name of your dependent variable.

An Example

Let's see how this works with the radio signal in Activity 5.

In this case there is one variable, time. Let's call it *t*. Since nothing happens before time begins we consider that *t* must be greater than or equal to zero. Obviously there are cases so we need an **if** clause. If we call the voltage *v*, then *v* is given by a function *V* of *t*, i.e., $v = V(t)$. Then in the first second we have

$v = V(t) = 2t$. As t goes from 1 to 2 we have $v = V(t) = 3 - t$. For the next 2 seconds, we have the sine function $v = V(t) \sin(\pi(t-2)/2)$. After $t = 4$ the value of v is 0, i.e., $V(t) = 0$ for $t \geq 4$.

We will illustrate how the translation can be done as an **ISETL computer function** whose syntax is given as follows:

> **V** : = **func(t);**
>> **if t < 0 then return;**
>> **elseif t <= 1 then return 2*t;**
>> **elseif t <= 2 then return 3 - t;**
>> **elseif t <= 4 then return sin(Pi*(t - 2)/2);**
>> **else return 0;**
>> **end;**
> **end;**

Notice that if $t < 0$, then **ISETL** will return nothing. In this case, if you try to evaluate the function V for a negative value of t, then **ISETL** will return **OM** which stands for "no value" or "undefined".

The **MapleV** code is similar and is given as follows:

> **V** : = **proc(t);**
>> **if t < 0 then;**
>> **elif t <= 1 then 2*t;**
>> **elif t <= 2 then 3 - t;**
>> **elif t <= 4 then sin(Pi*(t - 2)/2);**
>> **else 0;**
>> **fi;**
> **end;**

For the **MapleV** code above, if you try to evaluate the function V for a negative value of t, then **MapleV** will return nothing, i.e., no value.

Evaluating Computer Functions and Smaps Since a **computer function** or an **smap** represents transformations of quantities that are performed by a function, it is natural to use **computer functions** and **smaps** to evaluate a function — that is, to express what is the result of the function's process applied to a particular value. This is what you were doing in Figures 1.13 and 1.14.

Recall that back in Section 1.3 we suggested that you begin thinking about an expression to represent a situation by making a table of values. What is the connection between that and what results from the last three lines of code in Figure 1.13? Can you imagine how you might use a table to try to figure out a point in the domain of a function for which the function's value is some given number, say 0?

If you have a good strong hold on the idea of a process connected with a function (whether it be represented by an expression, a **computer function**, an **smap** or a graph), you can determine lots of useful information. This is what you were practicing in Activity 7. Here you must *reverse* the process. For example, in Activity 7 (a), what you must think about is something like the following. You are given

$\sin(x)$ and want to find $\cos(x)$ provided $\cos(x) > 0$. Since $\sin^2(x) + \cos^2(x) = 1$, you know that $\cos(x) = \pm\sqrt{1 - \sin^2(x)}$. So for Activity 7 (a) we can use a **computer function** like the following one in **ISETL**:

```
c : = func(s);
    if s < -1 or s > 1 then return;
    else return sqrt (1 - s **2);
    end;
end;
```

Graphs of Computer Functions and Smaps

In Section 1.3 you made graphs of functions represented by expressions (or functions) in your SCS. You can also make graphs of functions represented by a **computer func** or an **smap** in a MPL. We have a number of powerful tools for doing this and getting lots of information about a function. You will learn about them in succeeding chapters. For now, you only need to work on the simplest versions. Suppose that you are in **ISETL** and **f** is the name of a **func** or an **smap**. Then you can enter the following line of code to get a graph where a and b are numbers in the domain of **f** with $a \le b$.

plot(f, a, b);

You will see a graph of (a portion of) the function **f** which will include all those points between a and b which the **func** can evaluate or which are in the domain of the **smap**. (What is the domain of an **smap**?)

We should note that the syntax for graphing a **computer function f** with independent variable x in **MapleV** on an interval $a \le x \le b$ is similar to that in **ISETL** and is given by a statement similar to the following (depending on the version of **MapleV** you are using):

plot(f(x), x = a . . b);

In **ISETL** the graph consists of dots only and they are not connected. This option of graphs consisting of dots only can also be used in **MapleV** and other graphics systems. We believe it is best for you to do the connecting of the dots or not, depending on what you want to study in a particular situation.

The connection between a function and its graph is very important in mathematics and you will be spending a lot of your time in calculus studying it. We have only one question about this connection right now, but it is so important for your understanding that we will keep coming back to it.

> We have been talking about the process of a function. If all you have to work with for a particular function is its graph, what is it on the graph that represents the process?

Combinations of Functions — Arithmetic

Regardless of how you represent a function (i.e., as a computer function, graph, table, set of ordered pairs, etc.) you can form arithmetic combinations of functions. If f and g are two functions and a and b are numbers, then you can form the combination (addition and scalar multiplication) $af + bg$ which is a function whose process is given by the formula

$$(af + bg)(x) = af(x) + bg(x)$$

In other words, if you have a value x in the domain of both f and g then the value $(af + bg)(x)$ is obtained by multiplying a times $f(x)$, b times $g(x)$ and adding the two.

This idea provides one approach to Activity 8 where f was the function for the viscosity of SAE 10W-30 oil and g was the function for the viscosity of glycerin. These two functions can each be represented by an **smap** if you simply read some values off of their graphs. Then an **smap** to represent the combination might be constructed using a set former similar to the construction of the **smap g** in Figure 1.14, page 31.

In the exercises you will see how all this looks in the graphs. Of course, if you were willing to do some hand calculations, you could get the graph of $af + bg$ directly from the graphs of f and g by reading off points and forming the calculations by hand.

Do you see how all of this would work if you were multiplying two functions, or dividing them? Also, do you see that everything would work in exactly the same way if one or both of the functions were represented by a **func** instead of an **smap**?

Combinations of Functions — Composition

Composition or "a function of a function" is another way to combine two functions f and g. This time, you begin with a value x in the domain of g and evaluate $g(x)$. Then, if this result is a value in the domain of f you can apply it to get $f(g(x))$. This new function is called the *composition of f and g*. We write it as $f \circ g$ and $f \circ g(x) = f(g(x))$.

In Activity 9 you saw how to represent a composition as a **computer function**. The composition works exactly the same way if one or both of the functions is represented by an **smap**.

Can you figure out how to get points on the graph of $f \circ g$ directly from the graphs of f and g?

1.4.4 Exercises

1. Write, in your own words and as simply as possible, a description of the process represented by each of the following **computer functions**.

   (a) The **Maple proc** given by

   g : = proc(y);

      if y > = -5 and y < -3 then -(y**2);

      elif y >= -3 and y < 1 then y**3 - 1;

      elif y >= 1 and y < 2 then y + 4;

      else;

      fi;

   end;

   (b) The **ISETL func** given by

   h : = func(n);

      if is_integer(n) and n > 0 then

      if n = 1 then return 2;

      else return 0.5*(h(n - 1) + 2/h(n - 1));

      end;

   end;

   end;

   What happens to the values of this function as n gets larger and larger?

2. Suppose you want to fence in a rectangular region having a fixed area of 400 square feet.

   (a) Write a **computer function** that accepts the value of the amount of fencing and returns the length dimension of the region.

   (b) Think about what happens to the length of the region if you increase the amount of fencing. Will this length increase or will it decrease?

3. After the holiday season, a store has 57 strings of tree lights and 43 boxes of tree ornaments leftover. There are two ways to package these items for quick sale and the store manager must decide how many of

each way to use. One way makes a package of 5 strings of lights and 2 boxes of ornaments. It sells for $6.50. The other way has 3 strings of lights and 4 boxes of ornaments. It sells for $5.00. Assume that no matter what she does, all packages will be sold and only the items that didn't fit in a package will be left.

Write a **computer function** that will accept the number of packages of the first kind and will compute the number of packages of the second kind that she would have to make in order to use up as many items as possible.

(a) Can you determine the possible values for the number of packages of the first kind in this expression?

(b) Can you figure out what is the best thing for the manager to do?

4. Write a **computer function** which represents each of the following functions. Note that a mathematical statement like: if $-3 \le x < 1$ is written in computer code as if $-3 <= x$ and $x < 1$.

(a) $g(y) = 3|y| - \sqrt{y} + y^3$

(b) $F(u) = \dfrac{2}{u^2} + u^{1/3} + \text{sgn}(u)$

(c) $s(x) = \begin{cases} \dfrac{2}{x} & \text{if } x \le -1 \\ |\sin(x)| & \text{if } -1 < x \end{cases}$

(d) $h(x) = \begin{cases} -(x+4)^2 - 2 & \text{if } x < -3 \\ (x+2)^2 - 4 & \text{if } |x+1| < 2 \\ x+4 & \text{if } |x-1.5| < 0.5 \\ -2x+5 & \text{if } x \ge 2 \end{cases}$

5. Take a look at the three lines following the **computer function h** at the bottom of Figure 1.13 on page 30. This is a **for** loop which evaluates the function represented by **h** beginning at -3 and proceeding in steps of $\frac{1}{2}$ until it reaches 4.5. Figure out how you can adjust lines like this to home in on an approximation to a zero of a function (or, indeed any particular value) once you

can locate and bracket the point you are looking for between two numbers. Use such an approach to estimate the solution to the question about zero sales in the Skateboard problem in Activity 1 (c) part i on p. 15.

6. Print the graph you constructed with your **computer function** in the tree ornaments problem (Exercise 3). Shade in the region which excludes the points representing possible choices for the number of packages of the two kinds. Write a **computer function** that accepts two integers, checks if they represent a possible choice for the number of packages of the two kinds and then returns the amount of money the store would get if all of these packages were sold.

Use your **computer function** and your shaded graph to try to figure out what choice would bring in the most amount of money.

7. Write out a verbal description of the process of the function represented by the following **computer function**. You might look at some of its values or graph it to help give you an idea of what it looks like.

s := func(t);

if abs(t) <= 1 then return sin ((Pi/2)*t);

else return 2*s(t/2);

end;

end;

8. In each of the following situations there is a function. In some cases one or both of the variables are mentioned. In each situation, write a **computer function** which represents the function.

(a) You have 2000 feet of fencing to enclose a rectangular field, separated by fencing down the middle so that there are two equal areas. A dimension of the field and its area.

(b) You want to fence in a rectangular garden having an area of 1000 square feet. A dimension of the garden and the total amount of fencing.

(c) You have a rectangular piece of cardboard 30 in by 40 in and you wish to cut squares out of each corner and then fold up the sides to make an open box. The volume of the box.

(d) An isosceles triangle has a perimeter of 60 meters. The area of the triangle.

(e) The distance traveled by a falling object is approximately given by $4.9t^2$ meters t seconds after it is dropped from the top of a cliff. The average rate (distance over time) of the object when it falls from t seconds to $t+0.01$ seconds.

(f) The side of a cube has length x cm. The average change in the volume is the ratio of the change in the volume and the change in the length of a side. The average change when each side is decreased by 0.2 cm.

(g) An ideal gas in a container satisfies Boyle's Law

$$PV = C$$

where P is the pressure, V is the volume and C is a constant which depends on how much gas is in the container. The pressure P.

(h) A point on the line $y = x+1$ and its distance to the point $(2, -3)$.

(i) Two lights with intensities of 600 watts and 1000 watts are 300 feet apart. The illumination from each light is directly proportional to the intensity of the light and inversely proportional to the square of the distance from the light.

(j) You wish to build an open rectangular box having volume 60 m^3 so that the length of the base is three times the width. The cost of the base is \$3 per square meter and the cost for each side is \$2 per square meter. The total cost of the box.

(k) You drive your car 30 miles to a train station and then ride a train 70 miles to another town. The train averages 15 miles per hour faster than you drive. The total time it takes you to make the trip.

(l) An island is 1 mile from a straight shoreline and there is a village 10 miles down the shoreline. A shipping company can ship goods at a rate of 6 mph on water and truck goods at a rate of 40 mph on shore. The company is about to build a pier. The position of the pier on the shore and the total shipping time between the village and the island.

(m) The postal rate for first class letters in 1990 weighing no more than 1 ounce was 25 cents plus 20 cents more for each additional ounce or fraction of an ounce up to 2 pounds.

(n) The federal income tax rate in 1986 for a single taxpayer with an income of

over \$16,190 but not over \$19,640 was \$2,160 plus 23% of the excess over \$16,190,
over \$19,640 but not over \$25,360 was \$2,954 plus 26% of the excess over \$19,640,
over \$25,360 but not over \$31,080 was \$4,441 plus 30% of the excess over \$25,360.

9. Use your computer or a graphics calculator to make sketches of the function which represents the situation in each of the following exercises.

   (a) Exercise 1 (a).   (b) Exercise 1 (b).

   (c) Exercise 4 (c).   (d) Exercise 4 (d).

   (e) Exercise 8 (a).   (f) Exercise 8 (b).

   (g) Exercise 8 (c).   (h) Exercise 8 (d).

   (i) Exercise 8 (e).   (j) Exercise 8 (f).

   (k) Exercise 8 (g).   (l) Exercise 8 (h).

   (m) Exercise 8 (i).   (n) Exercise 8 (j).

   (o) Exercise 8 (k).   (p) Exercise 8 (l).

   (q) Exercise 8 (m).   (r) Exercise 8 (n).

10. On a circle with radius r units, the *radian measure* of a central angle θ subtended by an arc of length s units along the circle is the ratio of the arc length s and the length of the radius r. Make a sketch of this situation. For this situation write and test the following **computer functions**.

    (a) A **computer function** which accepts s and r, and returns the radian measure of θ. What about the units for s and r?

    (b) A **computer function** which accepts θ and r, and returns the length s. What about the units for θ? What are the units of s?

    (c) A **computer function** which accepts θ and s, and returns the length r. What about the units for θ? What are the units for r?

    (d) Two **computer functions** one which converts radian to degree measure and one which converts degree to radian measure using the relationship π radians = 180 degrees. You can use Pi for π in **ISETL** and **MapleV**.

11. Write a **computer function** which computes the value of the indicated trigonometric function.

    (a) $\sec(x)$ given $\sin(x)$. Run it for $\sin(x) = 0.4$, -0.53, 4, assuming $\sec(x) > 0$.

    (b) $\tan(x)$ given $\sec(x)$. Run it for $\sec(x) = 2.3$, -4, 0.63, assuming $\tan(x) < 0$.

    (c) $\sin(x)$ given $\cos(x)$. Run it for $\cos(x) = -0.23$, 0.8, 1.7, assuming $\sin(x) < 0$.

    (d) $\sin(x)$ given $\sec(x)$. Run it for $\sec(x) = 2$, -0.53, -4.21.

    (e) $\sec(x)$ given $\tan(x)$. Run it for $\tan(x) = 23.7$, -4, 0.63.

12. Use your **computer functions** from the activities and/or your mind to solve the following problems, assuming in each case that x is to be an angle in a triangle.

    (a) Find $\tan(x)$ given that $\cos(x) = -1/2$.

    (b) Find $\sin(x)$ given that $\tan(x) = -1$.

    (c) Find $\cos(x)$ given that $\sin(x) = -0.6$.

13. The distance traveled by a falling object is given by $16t^2$ feet t seconds after it is dropped from the top of a cliff. Express this function as a **computer function** and make a sketch of it. Pick 10 values in the domain of this function and for each of them, estimate the slope of the curve at the point corresponding to the value.

    (a) Use an **smap** to represent the function that has as its domain these 10 values and gives the slope of the curve.

    (b) Work out a formula for this function and represent it as a **computer function**.

(c) Sketch graphs of the two representations in parts (a) and (b). How do they compare?

(d) How does all this compare to what you did in Activity 4?

14. Use your computer or a graphics calculator to graph the three **smap**s you constructed in Activity 8. How close did the first two come to the corresponding graphs in Figure 1.15?

15. For the equation $y = 3(x + 2)^2 - 5$

   (a) Write a **computer function** that will accept a value for x and return a value y for which the equation is satisfied.

   (b) Write a **computer function** that will accept a value for y and return a value x for which the equation is satisfied. Are there any ambiguities

you must consider in this case? If so, discuss them briefly. Run your **computer functions** for several test values or use a calculator if you wish.

16. For each of the equations and given points, write a **computer function** that will accept any ordered pair $[x, y]$ check that the numbers satisfy the equation and, if so, will return the slope of the line segment connecting $[x, y]$ and the point listed in the problem. Check your computer functions for several test values by hand or use a computer if you wish.

   (a) $y = x^2$, $[1, 1]$

   (b) $y = x^3$, $[0, 0]$

   (c) $y = x^3$, $[-1, -1]$

   (d) $y = 1/x$, $[2, 1/2]$

   (e) $y = 1/x^2$, $[-2, 1/4]$

■ 1.5 REGIONS IN ONE AND TWO DIMENSIONS

1.5.1 Overview

The main purpose of studying this section is to help you to develop your mathematical intuition about points, lines and regions. We do this in two ways.

You will study different ways of representing these mathematical objects. You can specify each of them analytically in terms of coordinates, equations and inequalities. You can also show them in a picture or graph on a coordinate system. Neither is better than the other. Mathematical understanding implies a working knowledge of both analytic and visual representations of mathematical objects. Then you will look at some properties of these objects and various operations that can be performed with them and on them.

Informative pictorial representations of mathematical objects are usually obtained by using a coordinate system. You will study several in this text — rectangular, polar, cylindrical and spherical — in some detail. However, we will only consider rectangular coordinate systems in one and two dimensions in this section. The other coordinate systems will be studied in Chapters 7-10. Making transformations between coordinate systems and changing the picture can help solve certain problems you will work with later in this text and in many applications.

Once you have a coordinate line or rectangular coordinate system you can represent points. Intervals and other regions of numbers can be represented on a coordinate line. Mathematical expressions and equations can then be used to represent lines and curves in two dimensions. Regions in two dimensions are generally specified by inequalities.

What is important here is for you to think about the connection between the mathematical formulas and the pictures they represent, that is, between the analytical and graphical descriptions of mathematical objects.

Once you develop a pretty strong intuitive feel or conceptual sense of points, lines and regions: What do you think can be done with them? It has been said that to know something is to act on it, to transform it. In this section you will also get to know these mathematical objects better by looking at some of their properties and learning to perform certain operations with them.

In previous mathematics courses you learned how to find the length of a line segment, its midpoint and its slope. One of the interesting things you can do with calculus is to solve certain "size" problems such as finding the length of a curve, the area of flat and curved regions and the volume of solid regions. In this section you will think about such problems intuitively. This will give you a basis for learning about more powerful tools later.

Finally, at the end of this section we introduce two new and quite different concepts — windows and partitions — that are not directly part of the rest of the section. In this book we like to take a topic and spread it out over several sections. The purpose is to ease you gradually into learning some things that can be quite difficult. This way, you learn them first and only later realize what a difficult thing you have accomplished!

As usual, this section begins with a set of activities. In many cases, terminology will be used in the activities that you may or may not have learned about in other math courses. Everything is explained in the discussion which follows the activities. If there is a term that you don't understand, look for it in the Discussion section — once again we advise you to READ AHEAD through the text before and as you work on the activities. If you see something you don't understand, look it up in the index, or thumb through the discussion following the activities to find an explanation.

1.5.2 Activities

1. If you know two distinct points which do not lie on a vertical line, then you can figure out an equation of the form $y = mx + b$ which represents that line. You can then use a computer function to show a portion of graph of the line. Use a **computer function** to draw the graph of a portion of the line passing through the points P and R in Figure 1.16.

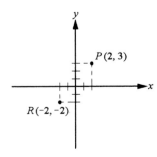

Figure 1.16. An xy-rectangular coordinate system.

2. A line can also be represented by an equation of the form $Ax + By + C = 0$. You can use the tool **graphR** in **ISETL** to produce a graph of the line or you can use an "implicit function" graphics tool, like **implicitplot** in **MapleV**, to plot an equation in two variables. Your instructor will provide you with the necessary syntax and directions to complete this activity.

The graph of the relationship between Centigrade temperature, $°C$, and Fahrenheit temperature, $°F$, is known to be a straight line (or linear). It is known that $0°C$ corresponds to $32°F$, and $100°C$ corresponds to $212°F$. Use your computer to draw a portion of the graph of the (linear) relationship between Centigrade and Fahrenheit temperatures. Your first step is to find the equation of this line in the form $Ax + By + C = 0$.

3. Figure 1.17 shows portions of the graphs of the three functions represented by the expressions below using only a set of closely spaced dots to indicate points on the graph of each function.

$$2x - \frac{1}{x^2}, \quad \left|1 - x^2\right|, \quad \text{and} \quad x^2 + \frac{1}{x}$$

Use the expressions to decide how to connect the dots to form smooth curves representing portions of the graph of each function.

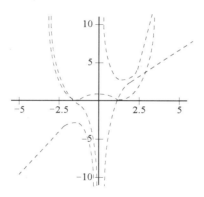

Figure 1.17. Graphs for Activity 3.

4. The graph of the following expression specifies a "region" in an xy-coordinate system if, for example, you restrict the values of x, say to an interval like $[0, 2]$ and consider the set of points that are below the graph and above the x-axis.

$$y = x^{\frac{1}{10}}$$

(a) Make a sketch of the "region" between the graph of $y = x^{1/10}$ and the x-axis with $0 \le x \le 2$. Show the region by shading it in on your graph. If your MPL is **ISETL**, then the **ISETL** tool **graphs** can be used to show this region. Your instructor will provide documentation.

(b) How can you be sure that a definite region is determined? For example, what if the curve falls below the horizontal axis?

5. The graphs of the following two expressions on the same rectangular coordinate system

$$y = \sin(x), \quad y = x^{10}$$

intersect in exactly two points so that there is a well-defined region "between" them, that is, their graphs enclose a definite region.

(a) Make a sketch of the "region" between the graphs of the curves $y = \sin(x)$ and $y = x^{10}$. Show the region by shading it in on your graph. If your MPL is **ISETL**, then the **ISETL** tool **graphs** can be used to show this region. Your instructor will provide documentation.

(b) How can you be sure that the two curves intersect in just two places?

6. It can be shown that the graphs of the following two curves on the same rectangular coordinate system do not intersect at all.

$$y = 1 + 2^{x}, \quad x^{2} + y = 1$$

In order to clearly define or specify a "region between them" you must restrict x, say to an interval like $[-2, 1]$.

(a) Make a sketch showing the specified region between the curves with $-2 \le x \le 1$.

(b) How can you be sure that the two curves never intersect?

7. **Tuple**s, or finite sequences, can have more than two components. One use for **tuples** with many components is to represent partitions of an interval. Below is **ISETL** code which constructs a **tuple**, or sequence of numbers, that represents a partition of the interval from -1.5 to 2.02 with 8 subintervals of equal length using nine equally spaced points.

$$[-1.5, \ -1.06 .. 2.02];$$

In **MapleV**, the same sequence, or **tuple**, is obtained using the following code:

$$[\,\mathrm{seq}(-1.5 + 0.44 * (i - 1), \ i = 1..9];$$

Notice that the length of each subinterval of the partition is given by

$$\frac{2.02 - (-1.5)}{8} = 0.44.$$

What do you think the above code in **ISETL** and **MapleV** each represents? Use the above code in **ISETL** or **MapleV**, or the appropriate code in your MPL to check your response. Now, write out by hand the following.

(a) A partition of the interval from 0.05 to 12 into 17 equal subdivisions.

(b) A diagram of a line with the endpoints of the interval and the subdivisions clearly marked.

8. Write **computer function**s that will accept two pairs of points, or two tuples with two components each, representing two points on a rectangular coordinate system and that will return each of the following quantities, respectively. Can you write a single computer function which can return a **tuple** containing all four of the following quantities? If so, do it instead of writing four separate **computer function**s.

(a) The distance between the two points, i.e., the length of the line segment connecting them.

(b) Coordinates of the midpoint of the line segment connecting the two points.

(c) The slope of the line passing through the two points.

(d) The slope of the perpendicular bisector of the line segment connecting the two points.

Test your **computer functions** (or function) on at least two different pairs of points.

9. Use your computer to produce a graph of the following line on a rectangular coordinate system.

   The line determined by the perpendicular bisector of the line segment connecting the two points P and R shown in Figure 1.16.

10. Consider that portion of the curve in Activity 4 that is above the line $y = 0.5$. Make a large picture of it, select a scale and estimate the length of the resulting curve.

11. Make a large picture of the region specified in Activity 5. Estimate the area of this region by placing an appropriate upright box around it. Can you do any better? Explain how.

12. Make a large picture of the region that is "between" the graph of the equation $y = x^3 + 0.4x^2 - 0.55x + 0.05$ and the x-axis for the interval $-1.5 \le x \le 0.7$. Estimate the area of this region by placing an upright box around it. Can you do any better? Explain how.

13. In this activity you are to investigate comparative sizes on the horizontal and vertical dimensions of the window of a graph that will keep all of the graph in a given interval visible. Consider the radio frequency pulse expression that you worked with in Activity 5, Section 1.4, p. 31). Here is what you have to do.

   Graph the expression in the interval for x going from $c - d$ to $c + d$ and y goes from $L - eps$ to $L + eps$. You will be given c, eps and L. Your job is to choose d in such a way that all of the graph over this horizontal interval is in the picture — that is, lies below the upper border and above the lower border.

   (a) $c = 2.5$, $L = \dfrac{\sqrt{2}}{2}$, $eps = 1$ (b) $c = 2.5$, $L = \dfrac{\sqrt{2}}{2}$, $eps = 0.5$

   (c) $c = 1$, $L = 2$, $eps = 2$ (d) $c = 1$, $L = 2$, $eps = 1$

   (e) $c = 1$, $L = 2$, $eps = 0.5$ (f) $c = 2$, $L = 0.5$, $eps = 2$

   (g) $c = 2$, $L = 0.5$, $eps = 1$ (h) $c = 2$, $L = 0.5$, $eps = 0.5$

   (i) $c = 2$, $L = 0.5$, $eps = 0.25$

14. A **tuple** such as $[-1.5, \ -1.06..2.02]$ which you considered in Activity 7, is often expressed in mathematical notation more compactly as

$$-1.5 + 0.44(i - 1), \quad i = 1, \ 2 \ldots 9$$

Notice that this is very similar to the **MapleV** code in Activity 7. In general to express the partition of the interval from a to b into n equal subdivisions, in mathematics we write,

$$a + \left(\frac{b - a}{n} \right)(i - 1), \quad i = 1, \ 2 \ldots (n + 1)$$

A similar notation can be used in some MPL's which is what you saw in the **MapleV** code in Activity 7. The **ISETL** code for the **tuple [−1.5, −1.06..2.02]** is as follows:

$$[-1.5 + 0.44 * (i - 1) : \text{ i in } [1..9]];$$

Such an expression is called a *tuple former*. Use the appropriate code in your MPL to see the above **tuple**, or sequence, written out explicitly. Your instructor will provide you with the code if you MPL is not **ISETL** or **MapleV**. Use your MPL to construct the following.

(a) A tuple former that produces the partition of the interval from 0.05 to 12 into 17 equal subdivisions.

(b) A **computer function**, named **partition**, that will accept two numbers **a**, **b** (with **a** < **b**) and a positive integer **n** and will return the **tuple** of the partition from **a** to **b** into **n** equal subdivisions. Test your **computer function** on the following intervals for the indicated values of n.

      i. $[0, \ 2]$, $n = 8$

      ii. $[-1, \ 7]$, $n = 16$

1.5.3 Discussion

Coordinate Systems and Points   Everyone has in their mind a notion of where something is, its location or position. Sometimes it is necessary and useful to specify a location very precisely. Think about where the mercury line sits on a thermometer, or where Sri Lanka is relative to Chicago, or the point at which the spaceship Columbia can meet a satellite that has gone astray. Position can be an idea in one, two, or three dimensions. It can be on a straight line or a curved surface. It can be relative to a fixed point, or a moving one.

In the seventeenth century, mathematicians like Descartes, Fermat and Mercator came up with the idea of using numbers to make positions precise: one number for one dimension, two numbers for two and three numbers for three dimensions. They realized that making a connection between the geometric (or geographic) intuition of position and the exactness of numbers could be a powerful tool for analyzing the position of an object as it moves in time, or traces a curve or surface, or fills in a three dimensional region.

There are many different ways of making the connection between position and numbers, but they have several things in common. They all have a picture of a reference frame called a *coordinate system*, a fixed position called the *origin* which is the reference point, a choice of unit values for distances and angles, and a description of precisely how the position of a point in the coordinate system is specified. In addition, there are transformation formulas for changing coordinate systems.

In all of this remember that there are two ways of locating a point: your mental image of "where the point is", and a list of numbers that "specify" the location. In one dimension the list has one number, in two dimensions it has two numbers written as an ordered pair such as [3, −2] and in three dimensions it has three numbers which represent its coordinates as an ordered triple.

One Dimension — a Coordinate Line, $\mathcal{R}^1$ Here there is only one way to do it. The frame of reference is a line, called a *coordinate line*, with the origin a fixed point on the line. The unit is the distance between 0 and 1. The position of a point is specified by a single coordinate which represents the distance from the origin, positive if the point is to the right of the origin, negative if the point is to the left of the origin. The coordinate of the origin is 0.

For example, the numbers in the list

$$[-1.5, \ -1.06, \ -0.62, \ -0.18, \ 0.26, \ 0.7, \ 1.14, \ 1.58, \ 2.02];$$

that you considered in Activity 7 are shown in Figure 1.18.

You can see how these points *partition* the interval from −1.5 to 2.02 on the coordinate line in Figure 1.18.

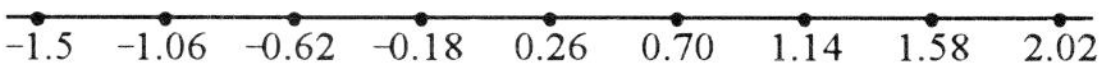

$$\begin{array}{ccccccccc} \overset{\bullet}{-1.5} & \overset{\bullet}{-1.06} & \overset{\bullet}{-0.62} & \overset{\bullet}{-0.18} & \overset{\bullet}{0.26} & \overset{\bullet}{0.70} & \overset{\bullet}{1.14} & \overset{\bullet}{1.58} & \overset{\bullet}{2.02} \end{array}$$

Figure 1.18. A partition of the interval [−1.5, 2.02].

We denote the set of all real numbers by $\mathscr{R}^1$, i.e., all points on a coordinate line.

Two Dimensions — Rectangular Coordinates, $\mathscr{R}^2$

In two dimensions, there are two standard ways of connecting position with numbers. We begin with the *rectangular coordinate system*, often referred to as a *Cartesian coordinate system* after the famous mathematician Rene Decartes. This is a frame of reference consisting of two perpendicular lines, called *axes* as shown in Figure 1.19. Their intersection is the origin. There are two systems of units, one for each axis. The position of any point is specified by two numbers representing, respectively, the distance from the horizontal axis (traditionally the *x*-axis) and the distance from the vertical axis (traditionally the *y*-axis). The first distance is taken to be positive when the point is to the right of the vertical axis and negative otherwise. Similarly, the second distance is taken to be positive when the point is above the horizontal axis and negative otherwise. The coordinates of the origin are the numbers in the pair [0, 0].

Usually (but not always, especially when applications are involved), the horizontal axis is called the *x*-axis and the vertical axis is called the *y*-axis. The entire system is sometimes referred to as an *xy*-coordinate system.

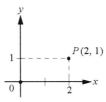

Figure 1.19. A rectangular coordinate system.

We denote the set of all pairs of real numbers by $\mathscr{R}^2$.

Representations of Lines in $\mathcal{R}^2$ If coordinate systems were only used to represent single points, they would not be very important. Things get interesting when they are used to represent sets of points. This is because important properties of a set, for example if those points form a line or circle or parabola, can be better understood when you see a picture of them. Actually, a really exciting thing that mathematics has to offer is the opportunity to make a synthesis in your mind between the pictorial and analytic representations of a set of points.

   We begin with lines. A line is determined by two points. What this means is that if you take any two points A and B on a line, then all possible values (that is, coordinates) of a general point P on the line can be determined by looking at appropriate expressions. Consider for example the figure in which A has coordinates $[-1, 2]$ and B has coordinates $[-4, -2]$. We consider that the general point P has coordinates $[x, y]$. Using similar triangles as in Figure 1.20,

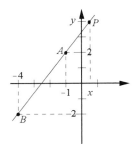

Figure 1.20. A line, similar triangles and proportionality.

you can see that the points on the line connecting a and b are just those points P whose coordinates satisfy the proportionality relation

$$\frac{y-2}{2-(-2)} = \frac{x-(-1)}{-1-(-4)}$$

This equation can be simplified to either of the forms,

$$y = \frac{4}{3}x + \frac{10}{3}$$

or

$$4x - 3y + 10 = 0.$$

   The first is the form that you needed to use in Activity 1 and the second is the form for Activity 2. To summarize, we have seen two ways to represent a line:

- as a function in an expression of the form $y = mx + b$ (Activity 1),
- as an equation in an expression of the form $Ax + By + C = 0$ (Activity 2).

The important issue in both cases is to describe just how the analytic expression "represents" a set of points, in this case, in a two-dimensional rectangular coordinate system. The description must explain just which points are included on the figure, usually by specifying the "acceptable" coordinates.

The graphical interpretation of the form $y = mx + b$ is that the points on the figure are obtained by first taking all real numbers and letting each of them be an x coordinate. For each of these x coordinates, a single y coordinate is obtained by calculating the value $mx + b$ and taking this for y. See Figure 1.21.

Notice in this form that the *slope* of the line which is the ratio of the *rise* (change in y) to the *run* (change in x) can be read off as the value of m. Similarly, this line meets the y-axis at the point $[0, b]$. Can you explain why a non-vertical line in $\mathcal{R}^2$ is completely determined by specifying its slope and y-intercept?

The next form has a completely different interpretation. In the previous form, you might think of the equation as generating *all* the points to be included. The form $Ax + By + C = 0$ is more of a *test* for inclusion. That is, you imagine considering all possible pairs $[x, y]$ of real numbers and plugging them into the equation. If it is satisfied, then the point is included. Otherwise, not. See Figure 1.22.

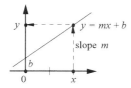

Figure 1.21. A graph of $y = mx + b$.

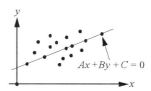

Figure 1.22. A graph of $Ax + By + C = 0$.

Not much can be read off here, but it is not too difficult to get the points where the line meets the two coordinate axes, in this case at the values for the x and y intercepts are, respectively, $-C/A$ and $-C/B$.

To summarize we might refer to these two ways of representing a set as *generative* (all of the points are generated by the equation) and *relational* (the points are represented by an equation which gives the relationship between x and y).

Slope of a Line

A line is determined by two points — any two points. This means that all information about a line should be expressible in terms of two points, and this information should be the same no matter which points on the line you use.

For example, in two dimensions there is the idea of the *slope* of a line. It refers to the slant of the line or the direction you would travel if you moved along the line. It can be positive or negative. The slope is measured by taking (any) two points on the line and forming the ratio of its "rise" (a difference between the y-coordinates) to its "run" (the corresponding difference between the x-coordinates).

Do you see that the slope m of a non-vertical line, such as the one through the points P_1 and P_2 pictured in Figure 1.23 is given as follows?

$$m = \frac{\text{rise}}{\text{run}} = \frac{y_2 - y_1}{x_2 - x_1}$$

provided $x_1 \neq x_2$. Do you see that slope can also be seen as the change in y, $\Delta y = y_2 - y_1$, over the change in x, $\Delta x = x_2 - x_1$?

$$\text{slope} = \frac{\Delta y}{\Delta x} = \frac{y_2 - y_1}{x_2 - x_1}.$$

Using any point on the line, say $[x_1, y_1]$, and the slope m, we can write an equation of the line as

$$m = \frac{y - y_1}{x - x_1}$$

or

$$y - y_1 = m(x - x_1).$$

The latter equation is sometimes referred to as the *point-slope form* of a line.

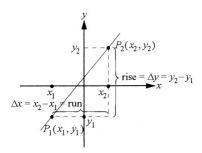

Figure 1.23. slope = rise/run.

Notice that for a line given by an equation of the form $y = mx + b$ the *slope* can be read off as the value of m. Such a line meets the y-axis at the point $[0, b]$. Can you explain why a non-vertical line in $\mathcal{R}^2$ is completely determined by specifying its slope and y-intercept?

How can you easily find the *x-intercept* of a line in the form $y = mx + b$ or $Ay + Bx + C = 0$? That is, how can you find the value of x where a line crosses the x-axis. You will get a chance to explain this in the exercises.

What is the slope of a horizontal line? Any two points on a horizontal line have the same y coordinates, so what can you say about the slope of such a line? What about the equation of a horizontal line? What about the slope of a vertical line? What about the equation of a vertical line?

Comparing Slopes in Two Dimensions   If you have a line in two dimensions, then there are infinitely many lines that are parallel to it. All of them have the same slope as the original line. Can you prove this? What basic properties of parallel lines would you use?

There also are infinitely many lines perpendicular to a given line in two dimensions. They are all parallel to each other and so have the same slope. But what is the relation of this slope to the slope of the original line? Can you guess a relationship? Can you prove it? If you can't think of anything else to do, try some examples.

Do you recall the relationship between the slopes of two perpendicular lines, neither of which is vertical or horizontal? Suppose you have two such lines with slopes m_1 and m_2, respectively. Then the relationship between the slopes is given by

$$m_1 = -\frac{1}{m_2}$$

In other words, *two lines in the plane are perpendicular if and only if their slopes are negative reciprocals of each other.*

Incidentally, the terms *orthogonal* and *normal* are sometimes used as a synonyms for perpendicular. We will use the three terms interchangeably in this book.

Representations of Regions in $\mathcal{R}^1$

In each of $\mathcal{R}^1$ and $\mathcal{R}^2$ (and later in this text $\mathcal{R}^3$), we will be interested in the set of all points *between* two objects. In $\mathcal{R}^1$ it will be the *interval* between two points and in $\mathcal{R}^2$, the region between two curves (and in $\mathcal{R}^3$, the region between two surfaces.) We start with $\mathcal{R}^1$.

Definition 1.1

The closed interval between two points a, b in $\mathcal{R}^1$ is the set of all numbers x in $\mathcal{R}^1$ which satisfy

$$a \leq x \leq b$$

We denote this set by [a, b]. The open interval between two points a, b in $\mathcal{R}^1$ is the set of all numbers x in $\mathcal{R}^1$ which satisfy

$$a < x < b$$

We denote this set by (a, b).

What do you think is meant by the intervals $(a, b]$ and $[a, b)$? What are the sets $[a, a]$, (a, a), $(a, a]$, $[a, a)$? Do you think it is necessary that $a \leq b$ in all of this? Suppose it is not? See the exercises for examples.

There is a notational difficulty here that we should mention. The symbol $[a, b]$ is the closed interval from a to b. The symbol **[a, b]** is the tuple with two elements **a** and **b**. Unfortunately, we are locked into this notation, in the first case because we don't want to do anything different from what most math books use, and the second because that is fixed in the language. You need to use the context to decide which meaning to give to the symbol. In this book, we will use the type font to distinguish between them.

You cannot use a MPL to represent the set of *all* numbers in an interval. However, using partitions as in Activity 14, we can *approximate* this set and you will see as the book proceeds that this can be very useful.

Representations of Regions in $\mathcal{R}^2$

Once you can make pictures of curves and know a little bit about their properties, you can represent entire regions of a plane. For example, if you know that the graph of a curve stays above the x-axis, then there is a region between them — that is, the set of all points below the curve and above the x-axis. Since the curve could extend indefinitely in both horizontal directions, it is necessary to "chop it off" by restricting the values of the domain variable if you wish to deal with a finite region. This is what was done in Activity 4.

Similarly, if the curve stays below the x-axis, then the region between them is the set of points above the curve and below the x-axis as in Figure 1.24. Again, you must chop it off by restricting the x values.

It is important to be able to specify the region not only with a picture, but also by means of inequalities that describe exactly which points are in the region. Using inequalities to specify a region is referred to as an *analytic representation* of the region as in Figure 1.25.

Of course, it is possible for a curve to have points both above and below the x-axis. In this case, we might break up the domain into separate intervals on which the curve is above or below the axis as in Figure 1.26.

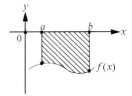

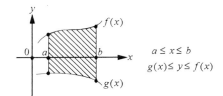

Figure 1.24. A region below the x-axis.   *Figure 1.25. An analytic representation of a*
                                                              region.

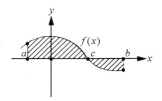

Figure 1.26. A region which is above and below the x-axis.

The most general situation is the region between two curves. You looked at this in Activities 5 and 6. In those activities, the two curves intersected exactly once or not at all. How would you specify the region between them if they intersected several times?

There are many important issues regarding regions between curves and you will learn about them as you study calculus. One of the most important issues is the area of a region. If you have a rectangle and you know its dimensions, it is pretty easy to calculate its area. You also have formulas for the area of a circle and even a circular sector. But what about the shapes that you worked with in the activities? How would you go about finding their areas?

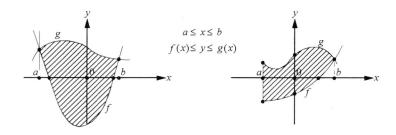

Figure 1.27. Analytic representations of regions between two curves.

Distance and Midpoint   Once we have a method for accurately specifying position of a point in $\mathcal{R}^2$, we can discuss such ideas as the distance between two points and the midpoint of the line segment between two points. Do you recall how to find the distance between two points? The Pythagorean

Theorem gives us a formula for the distance between two points, that is, the length of the line segment connecting them. In a rectangular two dimensional coordinate system, the distance between two points P and Q with coordinates $[x_1, y_1]$, $[x_2, y_2]$, respectively, is given by

$$\sqrt{(x_1 - x_2)^2 + (y_1 - y_2)^2} .$$

The midpoint of a line segment is, essentially, the average of the endpoints. Thus its coordinates are obtained by averaging the coordinates of the endpoints. As you can see in the picture in Figure 1.28, the midpoint R of the line connecting P, Q has coordinates given by

$$\left[\frac{x_1 + x_2}{2}, \ \frac{y_1 + y_2}{2} \right].$$

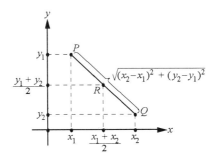

Figure 1.28. Distance and midpoint formulas in two dimensions.

Estimating Lengths of Curves

Many curves have the nice property that if you just look at a very small portion of it, it seems to be a straight line. This suggests that a curve is made up of very "small line segments". That's not exactly true, but if you use very "small line segments" between points on the curve, what you get is a pretty close approximation to the curve.

Such an approximation can be really useful. For instance, what is the length of a curve? You could use the idea in Activity 10, to get at the idea of length and approximate its value. You start with the curve. Pick some points along it that are fairly close together. Measure the distances between these points and add them up. This is an approximation to the length of the curve. See Figure 1.29.

Figure 1.29. An approximation to the length of a curve.

The closer the points are together, the better will be your approximation. You can use the picture here as a first approximation, but to get anything reasonable, you probably have to use a finer subdivision — that is, more points closer together.

Estimating Areas   The same basic idea can be used for areas. What we are doing is taking a problem, chopping it up into little problems which are easy to solve and then combining the solutions to get a final answer. Something is lost in the chopping up, and this is why we only get an approximation.

How does this look in the case of estimating areas? In Activities 11 and 12 you made an estimate by enclosing the region in a rectangle. How can you do better? Look at the pictures in Figure 1.30.

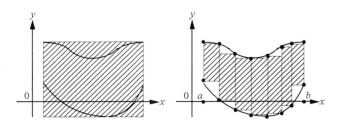

Figure 1.30. Approximations to the area between two curves.

It is easy to figure out the area of the one big rectangle. Notice how much error is introduced because of the large sections in the rectangle that are not in the region, but are counted in the approximation.

Now look at what happens if the big rectangle is subdivided into several smaller ones. It is a little more work since you have to compute several areas and add them up. But it is still areas of rectangles which are easy. Of course if you use all of the rectangles, you get the same answer as if you used the one large rectangle. The point here is that you don't have to use all of the little rectangles. Obviously, a rectangle which does not touch the region should not be counted. Do you think that such judicious omissions will give a better approximation? Why or why not?

Can you do even better than using small rectangles and omitting the ones that don't touch the region?

Curves in windows   Take a look at Activity 13. Think about the work you did and also think about what would happen if, in each of the examples you kept all of data the same but took smaller and smaller values of *eps*. How would this affect your possibility of success or failure in the task?

Is your response to this question the same for all examples? Does it seem to depend on *c*? On *L*? What about if you use a graph other than the one you worked with in Activity 13? What different phenomena are possible?

Partitions   In Activity 14 of this section, you considered the notion of a partition of an interval in $\mathcal{R}^1$. That activity is part of a connected string that we will carry through several chapters as we go along. In the end, you will have built up a powerful structure that you will be able to use in solving several fundamental problems in calculus.

One of these problems has to do with area and the questions you were considering in Activities 11 and 12. Look at these two activities and compare them with Activity 14. Can you guess what ultimately will be the connection between these?

1.5.4 Exercises

1. For each of the following points in rectangular coordinates, plot and label each point on a rectangular coordinate system.

   (a) [3, 0]   (b) [−2, 0]

   (c) [3, 3]   (d) [0, −3]

   (e) $\left[\sqrt{3}, 1\right]$ (f) [2, −1]

   (g) [−3, 2]   (h) [−4, −7]

2. For the following two dimensional reference frames give the rectangular coordinates of the indicated points.

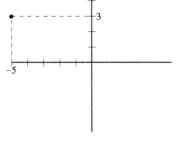

C.

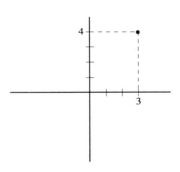

A.

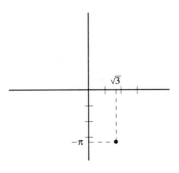

B.

D.

3. In each of the following situations, find an analytic description of the line L in each of the two forms: generative and relational.

A.

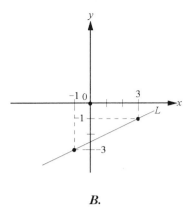

B.

4. What do you think is meant by the intervals $(a, b]$ and $[a, b)$. Give a definition for each half-open interval analogous to Definition 1.1. What are the sets $[a, a]$, (a, a), $(a, a]$, $[a, a)$? Do you think it is necessary that $a \leq b$ in all of this? Suppose it is not? Explain your answer.

5. How would you specify the region between two curves if they intersected several times? Illustrate your answer geometrically using graphs and inequalities.

6. Write a brief essay to explain how you would go about finding the approximate areas of regions like those in Activities 4, 5, and 6? Illustrate what you would do geometrically.

7. Explain why a line in $\mathcal{R}^2$ is completely determined by specifying its slope and y-intercept?

8. A manufacturer's total cost of producing a product is the sum of a fixed cost of $15,000 and a variable cost of $90 per unit. Find the total cost as a function of the number of units produced. Sketch graph of total cost versus the number of units produced. What about the domain?

9. A company builds an office building for $675,000. The company plans to depreciate the cost of the building for tax purposes over a 15 year period using linear depreciation to a value which is 20% of the original value. Find the value of the building as a function of the number of years after the building is completed. Sketch a graph of the value of the building versus time. What about the domain?

10. Sketch and shade the "region between" the indicated curves, clearly labeling each boundary curve.

    (a) $0 \leq x \leq 2$ and $0 \leq y \leq x^2$

    (b) $-1 \leq x \leq 2$ and $0 \leq y \leq x^2 + 4$

    (c) $-1 \leq x \leq 1$ and $x^2 - 3 \leq y \leq x - 1$

    (d) $0 \leq x \leq 3$ and $x - 3 \leq y \leq \sqrt{x}$

    (e) $0 \leq y \leq 2$ and $-y \leq x \leq y^2$

11. Sketch and shade the "region between" the indicated curves, clearly labeling each boundary curve. Find all points of intersection of the curves (if any) and describe the region between them using inequalities with x and y.

    (a) $y = x^2$ and $y = 3x$

    (b) $y = x^2$ and $y = 6 - x$

    (c) $y = 2x + 1$ and $y = x^2 - 2$

    (d) $y = x - 1$ and $y = x^2 - x$

    (e) $y = \sqrt{x}$ and $y = x^2$

    (f) $y = \sqrt{x}$, $y = x + 1$, $x = 0$ and $x = 3$

    (g) $y = 1/x$, $y = x$ and $x = 3$

    (h) $x = y^3$ and $y = x$

    (i) $x = 2 - y^2$ and $x = y$

12. Use inequalities with x and y to describe the "region between" the curves in each situation.

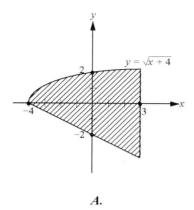

A.

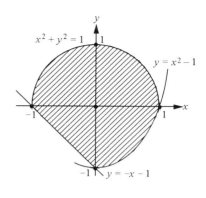

D.

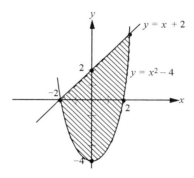

B.

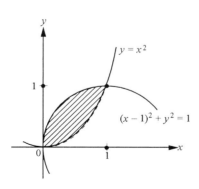

C.

13. Let a be a fixed number. Consider the function f_a given by

$$f_a(x) = a^x$$

Thus, if $a = 2$, then the function is given by $f_2(x) = 2^x$.
Sketch the graph of f_a for $a = 0$, 0.5, 1, 1.5, 2, 2.5, 3. Which values of a give you something special?

14. For the function f_a defined by $f_a(x) = a^x$, write a **computer function** which will accept a number a and return the value of the slope of the secant line to the graph of f_a through the points $[0, 1]$ and $\lfloor h, a^h \rfloor$. Find a value of a to three decimal places so that your **computer function** returns the value 1 by experimenting with values of a where $2.5 \le a \le 3$ and using $h = 0.001$. What do you think this says about the slope of the "tangent" to the graph of f_a at the point $[0, 1]$?

15. Make a sketch of the graphs of the following lines. Find the slope, x-intercept and y-intercept.

   (a) $y = 3x - 2$ (b) $y = 1 - 2x$

   (c) $x + 2y = 1$ (d) $3x + 2y = 6$

   (e) $5y - 3x = 15$ (f) $4x - 5y - 7 = 0$

16. Find an equation of the line L in each of the following situations.

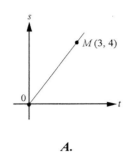

A.

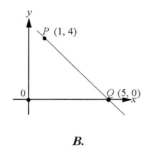

B.

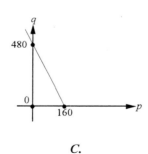

C.

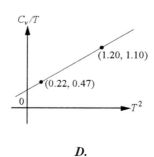

D.

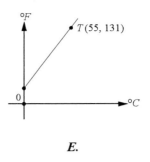

E.

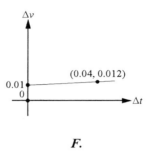

F.

17. Find an equation of the line from the following information. Give your answers in both generative and relational form.

   (a) The line passing through the point $[2, -1]$ with slope 3.

   (b) The line passing through the point $[0, 3]$ with slope $-2/5$.

   (c) The line passing through the points $[1, 4]$ and $[-2, 5]$.

   (d) The line passing through the points $[0, 3]$ and $[2, -1]$.

   (e) The line passing through the points $[1, 4]$ and $[1, -2]$.

   (f) The line passing through the points $[3, -2]$ and $[0, -2]$.

18. Find an equation of the line passing through the indicated point and (i) parallel (ii) perpendicular to the indicated line. Give your answer in the both the generative and relational forms.

   (a) $[3, 2]$, the line in Exercise 17 (a).

(b) $[-1, 4]$, the line in Exercise 17 (b).

(c) $[2, 1]$, the line in Exercise 17 (c).

(d) $[0, 0]$, the line in Exercise 17 (d).

(e) $[5, 1]$, the line in Exercise 17 (e).

(f) $[-2, 1]$, the line in Exercise 17 (f).

(g) $[1, 1]$, the line $y = 2x + 4$.

(h) $[-3, 4]$, the line $2y + 4x = 5$.

(i) $[-1, -2]$, the line $3x - 5y = 11$.

(j) $[0, 1]$, the line $y = 3$.

(k) $[2, 5]$, the line $x = -2$.

19. Are the graphs of the following lines parallel, perpendicular or neither?

    (a) $2y - x = 1$ and $y + 2x + 4 = 0$.

    (b) $y - 3x - 7 = 0$ and $6x - 2y = 1$.

    (c) $4x + 3y = 2$ and $2y - 3x = 6$.

20. A thirty six month investment is made and interest is to be paid on a quarterly basis.

    (a) Partition the time interval in months to indicate when interest is to be paid. Make a sketch showing the partition points.

    (b) Partition the time interval of the investment in years to indicate when interest is to be paid. Make a sketch showing the partition points.

21. Injections of a certain drug are to be administered to a patient ten times over a 3 hour period.

    (a) Partition the time interval in minutes to indicate the times when the injections are to be made. Make a sketch showing the partition points.

    (b) Partition the time interval in hours to indicate the times when the injections are to be made. Make a sketch showing the partition points.

22. Estimate the area between the graph of the function f given by $f(x) = x^2 - 2x$ and the x-axis on the interval $[1, 4]$.

23. Make a picture of the region determined by the curves in Activity 6, p. 46. Estimate the area of this region.

24. A slider moves in a vertical plane along a wire that is bent into the shape of the curve $y = 36/x$. Estimate the distance along the curve that the slider has traveled when it moves from the point $[1, 36]$ to the point $[6, 6]$.

25. A variable force acts on an object moving along a coordinate line. Assume that the coordinate line lies along the x-axis in an xy-coordinate system. The force F, in Newtons, at a point x is given by the function

$$F(x) = \frac{9}{x^2}.$$

The *work* done by the force F on the object from $x = a$ meters to $x = b$ meters is equal to the measure of the area under the graph of F over the interval $[a, b]$. Sketch the graph of F over the interval $[1, 4]$. Estimate the work done by the force on the object as it moves from $x = 1$ meter to $x = 4$ meters. What are the units of work in this situation?

26. The rate R at which a machine generates revenue is given by $R(x) = 30 - 0.3x^2$ thousand dollars per year when the machine is x years old. The total net earnings generated by the machine from $x = a$ to $x = b$ is equal to the measure of the area under the graph of R on the interval $[a, b]$. Sketch the graph of R over the appropriate interval. Estimate the total net earnings during the first six years the machine is in service. What are the units of total net earnings in this situation?

27. An object is 4 units long. A vertical cross-section of the object x units from one end is a circle with radius x^2. The volume of the object is equal to the measure of the area under the graph of the function A, which represents the area of a vertical cross-section in terms of x, from $x = 0$ to $x = 4$ in an xy-coordinate system. Sketch the graph of A for $0 \le x \le 4$. Then estimate the volume of the object.

28. Find an equation which represents the set of all points in two dimensions equidistant from two fixed points (a, b) and (c, d). Describe geometrically the set of all such points.

29. Find an equation which represents the set of all points on the line perpendicular to a line in two dimensions. Describe geometrically the set of all such points.

30. If you have a line in two dimensions, then there are infinitely many lines that are parallel to it. All of them have the same slope as the original line. Can you prove this? What basic properties of parallel lines would you use?

31. Suppose that you estimate the area of a region in two dimensions using "little rectangles" as discussed in this section, p. 56. As stated in the text, obviously a rectangle which does not touch the region should not be counted. Do you think that such judicious omissions will give a better approximation? Why or why not? Explain your answer and illustrate geometrically.

32. Take a look at Activity 13, p. 47. Think about the work you did and also think about what would happen if, in each of the examples you kept all data the same but took smaller and smaller values of *eps*. How would this affect your possibility of success or failure in the task? Is your response to this question the same for all examples? Does it seem to depend on c? On L? What about if you use a graph other than the one you worked with in Activity 13? What different phenomena are possible? Write a brief essay discussing the questions posed in this exercise.

33. What, if any, is the meaning of the slopes of two lines being reciprocals (not negative reciprocals) of each other?

34. Show that the slope of a line in $\mathcal{R}^2$ containing two points is independent of the choice of two points on the line.

35. Consider the following properties of a function. For each property give an analytic description using function notation, inequalities, etc. and then explain in your own words how you think each property is manifested in the graph of the function. You are expected to guess even when you have no idea what the words mean.

    (a) zero(s)   (b) positive values

    (c) negative values

    (d) regions (intervals) where increasing

    (e) regions (intervals) where decreasing

    (f) periodic   (g) even

    (h) odd

    (i) local maximum

    (j) local minimum

    (k) horizontal asymptote

    (l) vertical asymptote

    (m) regions (intervals) where concave up ("graph holds water")

    (n) regions (intervals) where concave down ("graph does not hold water")

36. Use a graph to answer the questions below (approximately) for the function in each of the following situations.

(a) $s(x) = \begin{cases} \frac{2}{x} & \text{if } x < 0 \\ 3|\sin(x)| & \text{if } 0 \leq x < \pi \end{cases}$

(b) $h(x) = \begin{cases} -(x+4)^2 - 2 & \text{if } x < -3 \\ (x+2)^2 - 4 & \text{if } |x+1| < 2 \\ x+4 & \text{if } |x-1.5| < 0.5 \\ -2x+5 & \text{if } x \geq 2 \end{cases}$

(c) You have 3000 feet of fencing to enclose a rectangular field, separated by fencing down the middle so that there are two equal areas. The area of the field as a function of the length of fencing down the middle of the field.

(d) You want to fence in a rectangular garden having an area of 900 square feet. The amount of fencing as a function of the width of the field.

(e) You have a rectangular piece of metal 18 in by 30 in and you wish to cut squares out of each corner and then fold up the sides to make an open box. The volume of the box as a function of the length of a side of each square cut out.

(f) The side of a cube has length x cm. The average change in the volume with respect to a change in the length of a side, x, is the change in the volume over the change in the length of a side. The average change in the volume when each side is decreased by 0.01 cm as a function of x.

(g) A point on the curve $y = x^2 + 3$ and its distance to the point $(-1, 2)$. The distance as a function of x.

(h) You wish to build a closed rectangular box having volume 60 m^3 so that the length of the base is three times the width. The cost of the base is $3 per square meter and the cost for each side is $2 per square meter. The total cost of the box as a function of the width of the base.

(i) You drive your car 5 miles to a train station and then ride a train 150 miles to another town. The train averages 20 miles per hour faster than your average speed. The total time it takes you to make the trip as a function of your average speed.

(j) An island is 0.6 mile from a straight shoreline and there is a village 5 miles down the shoreline. A shipping company can ship goods at a rate of 8 mph on water and truck goods at a rate of 45 mph on shore. The total shipping time between the village and the island as a function of the position of the pier on the shore.

Here are the questions.

   i. For what values of the independent variable, if any, does the function have zeros?

   ii. For what intervals of the independent variable, if any, is the function is positive? negative?

   iii. For what values, if any, of the independent variable does the expression have a local maximum? local minimum?

   iv. For what intervals of the independent variable, if any, is the function is increasing? decreasing?

   v. For what intervals, of the independent variable, if any, is the graph of the function concave up? concave down?

   vi. For what values of the independent variable, if any, is the function undefined?

    vii. What are the equations, if any, of the horizontal and vertical asymptotes?

37. Your instructor will provide you with a file named **signs** containing a **computer function signs** for use in your MPL. Print a copy of **signs**. This file has defined values for a function **G** and a tuple **p** so you can perform the command

$$\text{signs}(G, p);$$

and try to make sense of what happens. Print a copy of the file **signs**, study it to try to guess what is going on and then write a brief essay on what **signs** is doing. Make a graph of the function **G**. Do you see any connection between the graph of **G** and the output of **signs(G, p)**? You may wish to make several graphs of **G** on various intervals.

38. The **computer function signs** which you used in the previous problem can be used to solve inequalities. First you put everything on one side and then inspect the expression to determine the points at which it is 0 or undefined. These are the only points at which the expression can change sign and the **signs** uses them and the expression to determine the regions where the expression is positive or negative. You will need to write a **computer function f** represented by the expression and a **tuple p** listing the points where your **f** is 0 or undefined (in increasing order). Then run **signs** on **f** and **p**. Use **signs** to solve each of the following inequalities. Then solve the inequalities by hand.

(a) $x - 1 > 3x + 4$ (b) $|1 - 3x| \le 2$

(c) $|2x + 5| > 3$ (d) $x^2 - 3x + 2 < 0$

(e) $4x \le x^3$ (f) $\dfrac{x+3}{x-2} < 0$

(g) $\dfrac{x^2 + 3x + 2}{x - 1} \ge 0$ (h) $\dfrac{x+1}{x-1} + 5 \le 0$

(i) $\dfrac{x^3 - x^2}{(x-3)^2} > 0$ (j) $\dfrac{1}{x} > \dfrac{3}{x-1}$

(k) $\dfrac{2}{x^{1/3}} > x^{2/3}$

(l) $\cos(x) - \sin(x) > 0$

(m) $\tan(x) - \cos(x) \le 0$

◼ 1.6 CHAPTER SUMMARY

Computer Components Used in This Text   In this chapter you were introduced to the two computer components, a symbolic computer system, or SCS, and a mathematical programming language, or MPL, that you will use throughout this text.

You should have a good feel for what your SCS can do to the *objects*, i.e., symbolic expressions, it was designed to act upon. At this point, you should be able to do the following with your SCS:

- enter symbolic expressions

- perform arithmetic and algebraic manipulations

- produce graphs

- combine functions with arithmetic operations and composition

- do all of your pre calculus symbol manipulating

You should also have a good feel for what your MPL can do with various *processes* (represented by **computer functions**): compute, evaluate, graph, etc. At this point, you should be able to do the following with your MPL:

- construct and evaluate **computer functions**

- produce a table of values of a function

- construct sets of ordered pairs, **smaps**

- combine computer functions with arithmetic operations and composition

- produce graphs of portions of **computer functions**

In addition, through your experiences with these two computer components, your thinking has been focused on some of the basic problems which you will encounter in your study of Calculus in the remainder of this text.

Since most mathematical ideas can be interpreted in (at least) two ways: first as a collection of mathematical objects and processes that you think about and, second, as a symbol system that you use to work with and communicate the ideas, these are two of the reasons that this text uses two computer components, a SCS and a MPL. A MPL tends to interpret mathematics in the first way and a SCS does it the second way. Knowing how to use these two different components will help you use the right tool for the right job, and each has different and important uses in the learning process. The use of an SCS is a fast and efficient way of investigating the manipulation of powerful mathematical symbols. Programming mathematical ideas or concepts in a MPL provides a very powerful way to construct mathematics for yourself, on the computer and thereby in your mind, and hence learn mathematics. This is the basic philosophy of learning embodied in this text, a philosophy which is supported by current educational research into how people learn mathematics.

Approximate Calculus Ideas
In this chapter you worked with the two major ideas which are the basis of calculus.

Difference Quotients
In this chapter you encountered the idea of a difference quotient of a function f over the interval x to $x+h$, for $h \neq 0$, which is given by

$$\frac{f(x+h) - f(x)}{h}.$$

A difference quotient can be interpreted in the following two ways:

- as an average rate of change of the values of the dependent variable with respect to the independent variable, and

- as the slope of the secant line through the points $(x, f(x))$ and $(x+h, f(x+h))$ on the graph of f.

You began to think about what happens as the value of h is allowed to get closer and closer to zero. In this way you were beginning to think about the powerful concept of a "limit". The limit of a difference quotient, and its interpretation as a slope or average rate of change, provide the basic idea behind one of the two fundamental concepts in calculus.

Approximate Areas and Lengths of Curves

In this chapter, you also worked with partitions of intervals of real numbers on a coordinate line, or $\mathcal{R}^1$. The partition of an interval $[a, b]$ on a coordinate line into n equal subdivisions with endpoints of the n subintervals is given by

$$a = x_1, \ x_2, \ x_3, \ \dots, \ x_i, \ \dots, \ x_{n+1} = b$$

where

$$x_i = a + \left(\frac{b-a}{n}\right)(i-1), \ i = 1, \ 2, \ \dots, \ n+1.$$

In addition, you made approximations of areas of regions and lengths of curves in a coordinate plane, or $\mathcal{R}^2$. The ideas of a partition of an interval and the approximation of quantities (like areas of regions, lengths of curves and many others), linked with the notion of "limit", provide the basic ideas behind the second fundamental concept in calculus.

Lines and Slopes, Distances and Midpoints

Finally, in this chapter, your thinking has been focused on some intuitive notions related to regions in two dimensions, and how to represent various mathematical objects analytically (using equations and inequalities) and pictorially (using graphs) on a rectangular coordinate system. You should have a feeling for the various ways in which mathematical objects like points, lines and regions in two dimensions can be represented using a rectangular coordinate systems. You should be able to find: the equation of a line between two points, or the equation of a line knowing a point on the line and the slope of the line; the distance between two points; and the midpoint of a line segment between two points. If P and Q are two points with coordinates $[x_1, y_1]$, $[x_2, y_2]$, respectively, and $x_1 \neq x_2$, then:

- The slope m of the non-vertical line L containing P and Q is given by

$$m = \frac{\text{rise}}{\text{run}} = \frac{y_2 - y_1}{x_2 - x_1}$$

provided $x_1 \neq x_2$, and the point-slope form of the equation of a line is given by

$$y - y_1 = m(x - x_1).$$

The relational and generative (or slope-intercept) forms are given respectively by

$$Ax + By + C = 0 \text{ and } y = mx + b$$

where the latter equation the point $[0, b]$ is the y-intercept of the line.

- The distance, or length of the line segment, between P and Q is given by

$$\sqrt{(x_1 - x_2)^2 + (y_1 - y_2)^2}.$$

- The midpoint between of the line segment between P and Q is given by

$$\left[\frac{x_1 + x_2}{2}, \ \frac{y_1 + y_2}{2}\right].$$

Conclusion   As you proceed through this text, remember that the computer activities you will work on are designed to help you to think about and begin to understand various mathematical concepts and problem situations you will encounter in calculus, and in other mathematics courses or mathematics related courses. The activities in each section are there to get you to think and lay the groundwork (or shall we say "mind work") for the mathematical ideas, concepts and problem solving situations to be discussed in that section or later sections. You may not leave a given section knowing how to solve every problem situation introduced in that section. However you should be able to recognize which problems you can solve and which you cannot with the mathematical knowledge and tools at your disposal after you finish each section.

The intuition and estimation skills you have worked with in this chapter will provide you with a solid foundation on which to base your study of Calculus. Remember: You have already studied many of the mathematical ideas on which Calculus is based. One of the major ideas you worked with in this chapter was that of a function and one of the primary goals of this chapter was to help you begin to construct lots of functions in your mind! In the next chapter, you will once again focus your thinking on *functions* — the most important *processes* and *objects* in Calculus.

CHAPTER 2

FUNCTIONS

2.1 SOURCES AND REPRESENTATIONS OF FUNCTIONS

2.1.1 Overview

In previous mathematics courses you encountered the notion of a function and you worked with functions in the first chapter of this text in two fundamental ways: as *processes* which transform numbers and as *objects* which can themselves be acted upon. The concept of a function is a fundamental and very powerful tool, not only in calculus, but in all branches of mathematics. In this chapter you will further develop your concept image (i.e., all that you can imagine) of, and intuition about, functions.

This first section is about where functions come from and where they go. You will learn about the various sources of functions and how to construct a function when you are faced with certain kinds of situations. You will learn how to construct a function on a computer using various kinds of representations, how to express a function in mathematical notation and, perhaps the most important, you will make functions in your mind. When you are dealing with a situation that relates to a function, as you are constructing the function on a computer or with paper and pencil, try to be aware of what is going on in your mind. It will be very helpful in this chapter and in all of your mathematical and scientific work if you learn how to construct mental processes that express a function you are working with. Think about the function and what kinds of things are given to it to work with; think about what comes out of the function; and think about how what goes in is converted to what comes out. In other words, think about the process that is connected with a given function, the process which takes a domain value and produces a range (or *co-domain*) value. Once you get good at making such processes in your mind, you will have gone a long way towards understanding functions.

In the remainder of the chapter you will begin to deal with explanations and definitions related to the function concept. In particular, you will learn about the usefulness of multiple representations of functions, interpolation of function values, and properties of functions. You will also learn ways to take one or more functions and combine them to get new functions. That is, you will be acting on functions and you will find yourself thinking about functions, not so much in terms of their processes, but as objects on which other functions can act.

Pay careful attention to the examples in this section because they will be continually revisited throughout the remainder of the book.

2.1.2 Activities

1. Consider the freezing point of a chemical solution formed by two elements A and B. This depends on the relative amount of the two chemicals in the solution. If the relative amount m (in mole fractions) of A is less than or equal to $76\% = 0.76$, then the temperature T in degrees Centigrade at the freezing point is given by

$$T = 5m^4 - 23.6m^3 - 146m^2 + 129m - 17$$

If the amount of A is more than that, then the freezing point is given by

$$T = 45.056(m - 0.76)^{\frac{1}{3}} - 11.98132480.$$

Write a **computer function** that expresses this situation. What is the domain of your function? Use your function to determine the freezing point when the percentage of A in the mixture is 0%, 10%, 50%, 75%, 85% and 100%.

2. The Heat Capacity C_p (in calories per mole) in terms of temperature T (in degrees Kelvin) of a certain material has been experimentally determined and it depends on whether the material is in solid, liquid or gaseous state. If it is solid, then the experimental data is reasonably closely approximated by the formula,

$$C_p = 7 - (2.73 \times 10^{-4})(T - 160)^2$$

If it is in liquid form, then the functional relationship is

$$C_p = 12.7 - 0.016T$$

For gaseous form, this material's heat capacity is independent of temperature and has the approximate value 5.4.

This situation leads to a function, but there is a closely related function that is more useful in certain situations (such as in the calculation of entropy). Rather than take the value of Heat Capacity as the result of applying the function for a given temperature, it is useful to take as the value of the function, the ratio of Heat Capacity to temperature.

Write two **computer functions** that express these functions. You will need to know that this material melts at temperature 160° K and vaporizes (transforms into a gaseous state) at temperature 220° K.

Evaluate your functions at temperatures 0° K, 30° K, 50° K, 100° K, 200° K, 220° K and 300° K.

3. A slender bar of mass m and length L is at an angle of elevation θ in a static equilibrium position. The spring is unstretched in the position where $\theta = 60°$. Suppose that a force is applied vertically downward at the point F as indicated in Figure 2.1.

Write a **computer function** that expresses the amount by which the spring is shortened in terms of the angle of the bar in its new position.

Evaluate your **computer function** at angle positions of 80°, 70°, 60°, 50°, 30°, 10°, 0°, −10°, 375°. Use a mass of 0.25 kg and length of 10 cm.

You should think carefully about whether you wish to use degrees or radians here for the size of an angle.

4. In the context of the previous activity, the stiffness k of the compressed spring is a physical quantity defined in terms of work. It can be calculated to be

$$k = \frac{mg\cos(\theta)}{L\sin(\theta)(2\cos(\theta) - 1)}$$

where m is the mass of the bar, g is the gravitational constant and L is the length of the bar.

Write a **computer function** that expresses the stiffness of the spring in terms of the angle of the bar. Use the values $m = 5$ kilograms, $L = 0.36$ meters and $g = 9.8$ meters/sec^2. What are the units of the stiffness of the spring?

Evaluate your **computer function** at angle positions of $80°$, $70°$, $60°$, $50°$, $30°$, $10°$, $0°$, $-10°$, $375°$.

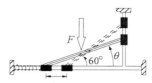

Figure 2.1. A slender bar in static equilibrium.

5. Your instructor can provide you with a computer tool for producing a table of values for any **computer function** you have constructed, over any interval in the domain and at any increment step. We will call this tool **table**. For example, in **ISETL**, the **func table** has the following syntax.

table(f , a, b, n);

where **f** is the name of the function you have constructed, **[a, b]** is the interval over which you wish the table to run and **n** is the number of points, evenly spaced between **a** and **b** at which you wish the function evaluated.

Use **table** for each of the functions you constructed in the previous four problems. Choose your interval to include all values that make sense in terms of the source of the function. Have each of the tables contain 25 points.

6. Your **table** should have a feature which could help prevent trouble. (For example, in Activity 2, if you used the first expression for Heat Capacity with a value of $T = 160$, you might not get any value. What would that mean physically?) A sophisticated version of **table** would have a feature that allowed the user to set conditions under which a value will be ignored.

If any of the functions on which you tried to generate a table gave you any trouble, use this feature to avoid the difficulty.

If you did not run into any trouble, find at least one situation (that means a choice of one of these four functions or perhaps a variation of them, and values for the parameters that *do* lead to trouble. Now use the feature to fix things.

7. Your instructor will provide you with a file **data9** which contains two columns of numbers based on a turbulent flow of a very viscous fluid through a thin pipe. The first column represents the logarithm of the distance from the wall of the pipe and the second column represents the mean velocity of the fluid.

Edit the file **data9** and use your MPL to make a set of ordered pairs (page 35) of this data to represent velocity as a function of distance from the wall. Now graph this function.

The functions which you constructed in Activities 1 and 2 have the property that their values are given by different expressions in different parts of their domains. Would you judge that the function represented by the table in this activity is given by a single expression on all parts of its domain or is more than one expression required?

Relative to this splitting of the domain, the three functions (this one, Activity 1 and Activity 2) differ in certain ways. How many of these differences can you see?

8. In quantum mechanics the energy of a diatomic molecule is approximated by regarding it as a dumbbell connected by a massless bar. The possible energy levels E_J are given by,

$$E_J = \frac{\hbar^2}{2I} J(J+1), \quad J = 0, 1, 2, 3\ldots$$

where $\hbar$ is Planck's constant divided by 2π, I is the moment of inertia and J is the quantum number. Use your MPL to construct a finite sequence or **tuple** (see Activity 14, p. 47) to express E_J as a function giving energy levels corresponding to the first 30 quantum numbers. Make a graph of E_J. Use the value

$$M = \frac{\hbar^2}{2I} = 2.1181 \times 10^{-22}$$

which is appropriate for a certain situation involving a hydrogen chloride molecule.

9. In the context of Activity 8, consider the ratio of the energy levels to the square of the quantum numbers (i.e., E_J/J^2). What would you predict would be the value of this ratio for arbitrarily high quantum numbers? What would you say its ultimate value would be?

One way to think about this question is to write a computer function and evaluate it for several (increasingly large) values of J, make a guess, test it on more values, and try to find a reason why your guess is correct.

10. Look again at Activity 1 and the freezing point function. Consider a new function with the same expressions defining it, but with no restriction on the domain. Draw the graph of this new function and also the graph for the restricted domain $0 \le m \le 1.0$. Why do they look so different?

Investigate different regions of the graph of the freezing point function and make a hand sketch of the *complete graph* that shows all relevant (important) features.

11. In the file **data10** there is a set of ordered pairs representing a function. Use your MPL to plot this function on the domain interval $[-10, 10]$ and co-domain interval $[-5, 5]$ and notice that it appears to have 3 or 4 breaks which could be due to regions on which the function is not defined or inaccuracies of the data points. Investigate by sketching the graph in different regions (note that the zoom feature may not work for a set of ordered pairs — does zoom make sense in this situation?) to decide, for each break, what is the cause. In order to look at different regions blown up, you may have to edit the **data10** file to restrict the domain appropriately.

12. The *ideal gas law* is the relation

$$PV = nRT$$

where P is pressure, V is volume, T is temperature and n, R are numbers giving the number of moles and a universal constant. Choosing units for a particular situation so that $nR = 1$, we may work with the relation,

$$PV = T.$$

Write three **computer functions** giving each of these variables as a function of the other two. Put in appropriate tests to make sure that no errors will occur in running any of your functions.

Write a short statement describing the main differences between the three functions that arise out of this situation.

Write a **computer function** that will accept values for n, R, T and will return a **computer function** that expresses pressure as a function of volume.

2.1.3   Discussion

Where Do Functions Come From?   Functions don't fall from the sky. Human beings make functions. We don't go around looking for functions that have been sitting there waiting for us to discover them. We study a situation, like mixture and freezing point, or temperature and heat capacity and we construct functions that we use as tools to understand what is going on.

This is what we mean by saying that a function comes from a situation. There is really nothing in a chemical compound, its temperature, or its freezing point that *is* a function. All that is there is a situation; a situation in chemistry, for example. There are also situations in physics, in engineering, in economics, in mathematics — really in all walks of life. It takes a human mind (and this is one of the great contributions of our culture) to realize that, in general, the idea of function can often be used to make sense out of these situations.

Constructing Functions   You can construct a function in your mind by thinking, you can construct it on paper by writing, or you can do it on a computer by programming. The best way is to make the construction in your mind — but that's hard. The most efficient way is probably to write it out — but you have to learn how to do it. Constructing functions on a computer is a way of learning how to write down your function and ultimately to develop the ability to make the construction in your mind.

So we begin with the computer. The construction of a function is often called a *representation* of it. You can represent a function on a computer in several ways: as an expression, as a **computer function**, as a set of ordered pairs, as a tuple, or as a graph. Each of these representations have their strengths and weaknesses. The one you should choose in a particular situation depends on the problem, and what you wish to do with your construction.

No matter which representation you use, there are a number of issues that need to be dealt with. For example, you must understand how the representation implements the process of the function. That is, for a given representation, how is the value of the independent variable transformed into the corresponding value of the dependent variable? You must also understand how the particular representation might restrict the domain of the function. By *domain* we mean the set of possible values on which the process of the function can work, to evaluate, to transform.

Advance Organizer

Advance Organizer   The next few paragraphs describe various ways in which functions can be represented on a computer. As you begin reading each paragraph, look at the heading and try to answer the following questions before reading. Then read the paragraph and try to synthesize what you thought with what is written.

- How can this representation be described?
- Which of the activities used this representation?
- How is the process of a function manifested in this representation?
- How is the domain determined in this representation?
- Are there any particular advantages or disadvantages to using this representation?

Functions as Expressions   Any "legal" combination of mathematical symbols involving variables, constants and operations is an expression. The key feature of a symbolic computer system is the possibility of representing any mathematical expression. Thus, for example, in Activity 4, (taking the value of mg/L to be 1 for simplicity), then the expression

$$\frac{\cos(\theta)}{\sin(\theta)(2\cos(\theta)-1)}$$

can be represented in your symbolic computer system. The same is true of the expression you can derive in Activity 3.

The process in an expression is implemented by choosing a value for the variable(s), plugging this into the expression and chugging through the calculations indicated by the expression. This will work for some numbers and not for others. Consider for example, the above expression which you worked with in Activity 4. It will not "work" if $\theta = 60°$. The numbers for which the calculation can be made form the domain of the function. How does this relate to the situation which gave rise to the function?

Representing functions as expressions is very simple and calculating with them can be fast and efficient. Unfortunately, many functions cannot be represented by simple expressions. For example, in Activity 1, the chemical properties of mixtures makes it impossible to use a single expression to represent the function. How many are needed? What about Activity 2? In Activities 7 and 11 there might be an expression lurking in the background for these functions, but the information given in the problem does not provide it.

Functions as Computer Functions   Many functions can be defined by a sequence of manipulations that can be implemented by computer operations and controls that are supported by an **MPL**. The manipulations are organized in a computer procedure with parameter(s) that represent the independent variable(s). The operation of the procedure is the function's process. Again, the domain consists of those numbers which, if given as values of parameters, will cause the procedure to return a value.

The choice of the domain comes from the situation that gave rise to the function. For example, in Activity 1, the situation has to do with the ratio of chemical A to the total number of mole fractions of chemicals A and B. Therefore, it seems reasonable to take the domain to be the set of numbers in the interval from 0 to 1.

The domain can be made explicit in a **computer function** by putting in, at the beginning, a test on the parameter. In this way, values that are not in the domain and would result in errors if given as parameter values can be omitted. Also, the domain can be made to reflect facts that are not contained in the expression. Consider the following situation.

For a fixed number of molecules of an ideal gas, using appropriate units, the quantity $Z = PV/T$, which is called the *compressibility*, is a constant. For real gases, it will vary depending on the nature of the gas and other factors. If pressure and temperature are constant, then one can express Z as a function of the volume V. If the pressure and temperature (and perhaps other features of the gas) are different within different regions of values of the volume variable, then the situation would be described by what we call a *function defined in parts*. That means that the function is given by different expressions in different regions of its domain. You have already worked with examples of functions defined in parts in Activities 1 and 2.

An additional complication can arise from the fact that the expressions that are used do not exactly represent an actual gas in "real-life". They are theoretical approximations that will agree with experimental data for some values of the variables, but not for others.

Consider the situation of a gas for which the compressibility z is given as a function Z of its volume v by three different expressions in three different regions. We write this mathematically in the following way.

$$z = Z(v) = \begin{cases} 1 + \frac{(v-2)^3(v-6)}{54} & \text{if } 0 \le v < 6 \\ \frac{v^3}{3 \times 10^3} - 0.048v^2 + 1.73v - 7.724 & \text{if } 6 \le v < 90 \\ 6.432 - \frac{4v^{\frac{3}{2}}}{5\sqrt{v^3 - 7 \times 10^5}} - \frac{27}{v-200} & \text{if } 90 \le v \end{cases}$$

Since a gas cannot have negative volume, the domain of this function must consist of non-negative numbers. There are more complicated domain issues, however. In the third branch, the quantity $v - 200$ appears in the denominator, so the value $v = 200$ must be ruled out as well. Also in the third branch, there is a denominator with a square root and this means that the quantity $v^3 - 7 \times 10^5$ must be positive or an error will occur. This is not a problem, however, because for this branch, $v \ge 90$. (Do you see why this restriction eliminates the difficulty?)

We take all of this into consideration when writing a **computer function** to express this function. Here is how it might look in **MapleV**.

```
Z := proc(v);
        if v < 0 or v = 200 then
                undefined;
        elif v < 6 then
                1 + ((v-2)**3*(v-6))/54;
        elif v < 90 then
                v**3/(3.0*10*3) - 0.048*v**2 + 1.73*v - 7.724;
        else
                6.432 - 4*v**(3/2)/(5*sqrt(v**3 - 7.0*10**5)) - 27/(v-200);
        fi;
    end;
```

Another example can be taken from Activity 4. Although one can use an expression for this function, it is also possible to use a **computer function**. If that is done, then a test for the value $\theta = 60°$, or more generally, $2\cos(\theta) = 1$ can deal more smoothly with values that would lead to computer errors.

Using a **computer function**, it is usually possible to obtain information, as accurately as you need about many, if not all of the important properties of a function. It is not always easy in practice to extract this information from the representation. One way to have the best of both worlds is to convert a procedure to some other representation appropriate to the particular phenomenon under study.

Functions as Sets of Ordered Pairs — Tables

Two or more columns of data form a table. In order for the table to represent a function, one of the columns must be free of repetitions. This column represents the possible values of the independent variable. In Activities 7 and 11 the functions were given this way. The process applies to a value by looking it up in the column representing the independent variable and finding the corresponding value in the other column. Do you see why one of the columns is not allowed to have any repetitions? Which one?

The domain of a function represented by a table is the set of numbers in the column corresponding to the independent variable. Recall from Activity 6, how you were able to avoid points that gave difficulties in a table.

A function given by a table can be represented in **ISETL** by an **smap**. Recall the skateboard problem in Section 1.3, page 18. In that situation, the following table represents some of the values of the function N as follows.

p	N
45	75
47	72
49	69
51	66
53	63

Table 2.1 Skateboard Sales.

We can represent Table 2.1, and hence the function by the following set of ordered pairs.

sk := {[45, 75], [47, 72], [49, 69], [51, 66], [53, 63]};

It is important to note that this *does not* represent the same function as does the expression (Section 1.3, page 18)

$$N = 142.5 - \frac{3}{2}p$$

(Why would one think that these functions might be the same?)

This expression represents a function which can be applied to any number. Thus its domain is the set of all real numbers. The table and the set of ordered pairs **sk**, however, can only be applied to numbers in the set {45, 47, 49, 51, 53} which is *their* domain.

Because of the heavy domain restriction, tables and sets of ordered pairs, are not very powerful ways of representing functions that are dealt with in calculus. We need them however, because in many scientific situations, the only knowledge of the function that we have comes from experimental data which, at least initially, can only be represented by a table. In the next section you will see how interpolation can make a table more useful.

Functions as Tuples — Sequences   A *sequence* is a function whose domain is the set of positive integers. The situations in Activities 8 and 9 are best described by sequences (assuming that a numerical value is chosen for $\hbar^2/2I$). A finite sequence can be represented by a tuple. The process applied to an integer j in the domain consists of moving along the numbers in the sequence until the jth term is encountered. The value of this term is the value of the function at j.

There is a difficulty with most MPL representations of tuples in that, usually it can only handle a finite number of terms, so you can't really represent the entire sequence. The best you can do is to define a large number of terms in the sequence that is sufficient for a particular consideration. Fortunately, it is usually easy to increase the number of terms represented any time this becomes necessary.

One important thing that is done with sequences is to try to figure out what will be the values of the function for larger and larger values of the independent variable. In Activity 9, you worked with this for the ratio of energy level to the square of the quantum number. This is an introduction to limits of sequences which is an important topic in Calculus.

Functions as Graphs   A graph is a pictorial representation of a function on a coordinate system. In Chapter 1 you looked into the most common way of setting up a coordinate system in $\mathscr{R}^2$, called a rectangular (or Cartesian) coordinate system in which the horizontal or x-axis represents the independent variable and the vertical or y-axis represents the value of the function. You apply the process to a value of the independent variable by locating it on the horizontal axis and then moving vertically until you hit the graph. The distance moved (positive or negative) is the value of the function for that value of the independent variable. The value of the function can be located on the vertical axis by moving horizontally from the point on the graph to the vertical axis.

Do you recall the "vertical line test" from previous math courses? It says that a curve represents a function if each vertical line passes through the graph no more than once. We can make that criterion a little more flexible. Look at Figure 2.2.

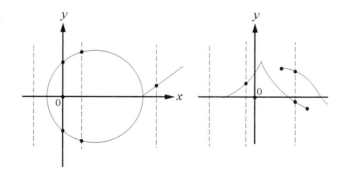

Figure 2.2. The vertical line test.

You will notice that the vertical line test succeeds sometimes and fails other times. That is, depending on where the vertical line crosses the horizontal axis, it may intersect the graph in one point, in several points, or in no points. One way of constructing a function out of this situation is to take for its domain the set of numbers represented as points on the horizontal axis where the vertical line test succeeds. Can you finish a description of a function?

A graph can be a powerful tool for studying functions because it displays the properties of the function visually and for most people this makes it very easy to understand what is going on. On the other hand, the information given by a graph can be incomplete or even misleading. Look at Activity 10. One "branch" of this function is represented by an expression which is a fourth degree polynomial. If you construct a graph for say $-10 \leq m \leq 10$ instead of the proper interval $0 \leq m \leq 1$, you might be misled. Suppose you wanted to know where this freezing point function has the value 0, or where it is increasing or decreasing, or where it peaks. The conclusions you would draw from a single graph are not likely to tell the full story. In Activity 10, you had to look at several regions more closely to get a full picture of what is going on.

The conclusion from this example is not that the graph is better or worse than another kind of representation (such as an expression or a **computer function**). The conclusion is rather that full knowledge about a function is best obtained by considering more than one representation. In Activity 10, for example, you might have used the fact that a fourth degree polynomial has a chance of crossing the horizontal axis at most four times. This is an analytic fact (which you might remember from previous math courses) that comes from its representation as an expression. You can use the graph to look for these four points and decide if the graph of your particular function does cross the horizontal axis four times. You can also use the graph to get approximations of the values of these points. For high degrees of accuracy however, you might have to return to the expression and use a more complex mathematical analysis.

A great deal of calculus is about analyzing one representation of a function to get information about another representation. We will begin to work on this in the next section and continually add information as the course progresses.

Function Notation

In discussing functions, there are three main things we have considered that make up a function. There is the domain which is the set of values that can be assigned to an independent variable; there is the process which acts on the value of the independent variable; and there is the value of the function which is generally assigned to the dependent variable. We have not mentioned it very much, but the set of all potential values of the dependent variable is called the *co-domain*, or range, of the function.

The question arises as to what notation should be used for the independent variable, the process, and the dependent variable. If you are working in straight mathematics, the most common choice is to use x for the independent variable, f for the process and y for the dependent variable. The expression $f(a)$ refers to the value obtained by applying the process of the function f to the value a of the independent variable. Since this value is the value of the dependent variable, we can write $b = f(a)$ for specific values or $y = f(x)$ for the general situation.

Each representation of a function that we have considered has its interpretation of the expression $f(a)$. Here is a tabulation of those meanings.

Representation	Implementation of $f(a)$
Expression	Substitute a in the expression and calculate
computer function	Run the **computer function** with the value a passed as an argument
Set of ordered pairs	Look for a as the first term of a pair and return the second term
tuple	The value of the ath component
Graph	The signed length of the vertical line from a on the horizontal axis to the graph

Of course, if there is more than one function, other letters such as $g, h, \ldots$ must be used instead of just f. Also, if you are working with a function of more than one variable, then x and y could be the names for the independent variables, and the dependent variable would have to be called something else, like z.

Another reason for using other letters is to reflect the original situation from which the function came. Thus in the activities we used variable names such as T for temperature or C_p for heat capacity.

The upshot of all this is that you should not allow yourself to become too attached to a particular naming system. It is true that in mathematics the choice of x, f, y is most common and indeed, we shall use that convention often in this book. We will also use capital letters, especially when they refer to physical quantities. But you should be prepared to work with any notational system as long as the essential ideas: independent variable, process, and dependent variable are clear to you.

There is one ambiguity in that f can be used for the name of the whole function including its process, domain and co-domain, as well as just its process. People working in mathematics usually don't have too much difficulty in living with it. There is another ambiguity, however, which we will try to avoid. It is the confusion that can arise between the name of the function (or its process) and the name of the dependent variable. Thus, for the compressibility function, we write Z for the *function* that gives the compressibility as a function of the volume. The lower case letter z refers to a typical *value* of the compressibility and the letter v refers to a typical value of the volume. Thus, the expression $z = Z(v)$ is interpreted as "the value z, given a volume v of the compressibility is given by applying the (process of the) function Z to v. We will try to be careful about making these distinctions, often (but not always) using a capital letter for the function and lower case letters for the variables.

As a final example, recall that in Section 1.4, p. 34 we asked if there was a function specified by the following equation.

$$x = y^2 + 1$$

What is your answer now? Can you specify the independent and dependent variables? Can you describe the process?

Functions of Several Variables

Most of the discussion in this section has been about functions of one variable. The situation in Activity 12, is best represented by a function of two variables. There is not much change in representing functions of more than one variable as expressions, procedures, or as tables (although for a table, more than two columns would be necessary). They can also be represented as graphs, but the issues can be very different from graphs of functions of one variable. The idea of sequence with its linear order of the integers, does not readily apply to functions of more than one variable.

If you have a situation which leads to a function of several variables, there are a number of ways to deal with it. One way is to treat it directly as in Activity 12. Another approach is to reduce the number of variables by "bundling". For example, in Activity 4, you might consider that stiffness is a function of four variables: mass, gravitation, length of bar, and angle. Another possibility (suggested by the form of the expression) is to bundle the first three into a single variable, say t. Then the expression can be written

$$k = \frac{t \cos(\theta)}{\sin(\theta)(2\cos(\theta) - 1)}$$

so that it is a function of two variables instead of four. A similar technique can be applied in Activity 8.

You can also go all the way down to functions of one variable by introducing the idea of families of function. Consider the function of Activity 8 and suppose we have bundled the square of Planck's constant divided by twice the moment of inertia to obtain the single variable $M = \hbar^2/2I$ so the expression giving the function (which we may call E) is

$$e = E(J) = MJ(J+1).$$

The variable M expresses essentially the moment of the bar and so you might think of several bars with several different moments. Thus for each value of M you have a *function* E that gives the energy e at a given level J. We might say that instead of one function of several variables, we have several functions, or a *family of functions* each a function of one variable.

This interpretation can be implemented directly in **ISETL** as follows.

> **E := func(M);**
> **return func(J);**
> **return M*J*(J+1);**
> **end;**
> **end;**

This representation raises a number of questions. How would you explain the meaning of the code in lines 2-4 of these five lines? What is the meaning of **E(M)(J)**? Once you have run this code in **ISETL** what will be returned by an expression like

E(32.87)(6);

Would it make sense if you interchanged 32.87 and 6 in that expression? What about if you just replaced 6 by 6.3?

The following **MapleV computer function** implements the function E.

> **E := proc(M);**
> **EM := M*J*(J+1);**
> **makeproc(EM, J);**
> **end;**

Then, **E(M)** and **E(M)(J)** have the same meaning as they do in **ISETL**. Notice that the use of **makeproc** is how a **MapleV computer function** can return a **computer function** as an output. To use **makeproc** you must first load the *student package* in **MapleV** by typing the command **with (student):**.

2.1.4 Exercises

1. For the expression representing the stiffness k in Activity 4, page 70, the expression will not "work" if $\theta = 60°$. The numbers for which the calculation can be made form the domain of the function. How does this relate to the situation which gave rise to the function? How do the other values of θ where the expression will not "work" relate to the situation?

2. Is there a function specified by the equation $x = y^2 + 1$? If so, can you specify the independent and dependent variables? Can you describe the process?

3. Consider the energy function E of Activity 8, page 72, given by

   $$e = E(J) = MJ(J+1),$$

   Read through Activity 8 page 72 again and then answer the following.

   (a) How would you explain the meaning of the **ISETL** code in lines 2-4 for the **func** representing E on p. 80?

(b) Once you have run this code in **ISETL** what will be returned by an expression like

$$E(32.87)(6); \ ?$$

(c) Would it make sense if you interchanged 32.87 and 6 in that expression? What about if you just replaced 6 by 6.3?

4. Following are a number of expressions defining transformations of a function f. After that is a list of specific functions. Use each of the specific functions in each of the transformations and simplify the resulting expressions.

Here are the transformations.

(a) $f(x) - 1$ (b) $f(x-1)$

(c) $f(x^2)$ (d) $(f(x))^2$

(e) $f(f(x))$ (f) $f(1/x)$

(g) $1/f(x)$

(h) For $h \neq 0$, $\frac{f(x+h)-f(x)}{h}$

(i) For $x \neq a$, $\frac{f(x)-f(a)}{x-a}$

Here are the functions.

     i. $f(x) = 3x - 2$

     ii. $g(x) = 1 - 2x^2$

     iii. $f(x) = x^3 - 2x$

     iv. $g(x) = \sqrt{x}$

     v. $f(x) = \frac{1}{x}$

     vi. $g(x) = \frac{3}{x^2}$

     vii. $f(x) = \frac{x}{1-x}$

     viii. $f(x) = \begin{cases} -2 & \text{if } x \leq 0 \\ x-2 & \text{if } x > 0 \end{cases}$

     ix. $g(x) = \begin{cases} x^2 & \text{if } x < 1 \\ 2x-1 & \text{if } x \geq 1 \end{cases}$

5. For each of the following functions use your graphing utility to make a complete graph, i.e., a graph which shows all relevant features of each function. Explain why the graph is a complete graph.

(a) $3x^2 + 24x$ (b) $x^3 - 2x + 1$

(c) $3x - 2x^3$ (d) $x^3 - 6x^2$

(e) $15 - 25x - x^3$ (f) $|2x - 3| + 1$

(g) $\left|3x - 2x^3\right|$ (h) $2x + \frac{3}{x} + 5$

(i) $\frac{x^4 - x^2}{x-2}$ (j) $\frac{3}{x^2} + \frac{1}{x}$

(k) $\frac{300}{x^6} - \frac{100}{x^3}$ (l) $\frac{x}{x-2}$

(m) $\frac{1-2x}{x^2+1}$ (n) $\frac{x}{x^2-4}$

(o) $\frac{x^3}{x^2-2x-3}$ (p) $-2\sin(\pi x)$

(q) $\sin^2(2x)$ (r) $x^2 - \sin(x)$

(s) $x\cos(2x)$

6. In each situation listed below, represent the specified function as follows.

(a) An expression(s).

(b) A **computer function**.

(c) A table of values using your **table** with at least 20 points from the domain.

(d) A (complete) graph using your graphing utility. Explain why your graph is a complete graph.

Here are the situations.

     i. You have 2000 feet of fencing to enclose a rectangular field, separated by two lengths of fencing down the middle so that the field is separated into three equal rectangular areas. Express the area of the field as a function of the length of fence down the middle.

ii. You want to screen in a rectangular patio having an area of 200 square feet. Express the amount of fencing as a function of the length of the patio.

iii. You have a rectangular piece of cardboard 20 in by 30 in and you wish to cut squares out of each corner and then fold up the sides to make an open box. Express the volume of the box as a function of the length of a side of a square cut out of the corner.

iv. A triangle has sides of length 10 and 7 inches. Express the area of the triangle as a function of the angle between the two sides given.

v. For the triangle in the previous problem, express the perimeter of the triangle as a function of the angle between the two sides given.

vi. The distance traveled by a falling object is approximately given by $4.9t^2$ meters t seconds after it is dropped from the top of a cliff. Express the average rate (distance over time) of the object when it falls from t seconds to $t + 0.01$ seconds as a function of time, t.

vii. A rectangular box has a square base and its height is twice the base length. Each side is decreased by 0.02 cm. The average change in the volume of the box is the ratio of the

change in the volume to the change in the length of a side of base. Express the average change in the volume as a function of the base length.

viii. A certain number of molecules of an ideal gas satisfies the law $PV = T$ where P is the pressure, V is the volume and T is the temperature. Express the pressure as a function of volume when the temperature is kept at 270 degrees.

ix. A point on the curve $y = x^2 + 1$ and its distance to the point $(1, -1)$. Express the distance as a function of x.

x. You wish to build an open rectangular box having volume 1000 in^3 so that the length of the base is two times the width. The cost of the base is \$0.2 per square inch and the cost for each side is \$0.3 per square inch. Express the total cost of the box as a function of the width of the base.

xi. A manufacturer of tin cans is to make a can holding 80 in^3. Express the surface area of the can as a function of the radius of the top (and bottom) of the can.

xii. For the tin can manufacturer in the previous problem, suppose that the top and bottom cost of the can is 0.03 per in^2 and the sides cost 0.02 per in^2. Find the total cost of a tin can as a function of the radius of the top (and bottom) of the can.

xiii. You drive your car 30 miles to a train station and then ride a train 70 miles to another town. The train averages 15 miles per hour faster than the average speed you drive. Express the total time it took you to make the trip as a function of the your average speed.

xiv. An island is 1 mile from a straight shoreline and there is a village 10 miles down the shoreline. A shipping company can ship goods at a rate of 6 mph on water and truck goods at a rate of 40 mph on shore. The company is about to build a pier. Express the total shipping time between the village and the island as a function of the position of the pier on the shore.

xv. The postal rate for first class letters in 1990 weighing no more than 1 ounce was 25 cents plus 20 cents more for each additional ounce or fraction of an ounce up to 2 pounds. Express the postal rate of a letter as a function of its weight.

xvi. The federal income tax rate in 1986 for a single taxpayer with an income of

over \$16,190 but not over \$19,640 was \$2,160 plus 23% of the excess over \$16,190,

over \$19,640 but not over \$25,360 was \$2,954 plus

26% of the excess over \$19,640,

over \$25,360 but not over \$31,080 was \$4,441 plus 30% of the excess over \$25,360.

Express the federal income tax in terms of income.

xvii. A rectangle with one side lying along the x-axis is to be inscribed in the triangular region enclosed by the curves $y = 0$, $y = 3x$ and $y = 30 - 2x$. Express the area of the rectangular region as a function of the height of the rectangle.

xviii. A rectangle is inscribed in a circle with diameter 20, with one side lying on a diameter. Express the area of the rectangle as a function of the length of a side lying along the diameter.

7. For each of the following representation of a function, express it using: a **computer function**; a table using your **table** with at least 20 points; a computer representation of a set of ordered pairs using at least 20 points; and a complete graph. Explain why your graph is complete.

(a) $g(x) = \begin{cases} 0.75(x+1)^2 & \text{if } x < -3 \\ |x| & \text{if } |x| <= 3 \\ 3 & \text{if } x > 3 \end{cases}$

(b) $f(x) = \begin{cases} x & \text{if } 0 \le x \le 1 \\ 2-x & \text{if } 1 < x \le 2 \\ x-2 & \text{if } 2 < x \le 3 \\ 4-x & \text{if } 3 < x \le 4 \\ x-4 & \text{if } 4 < x \le 5 \\ \cdots & \cdots \\ f(-x) & \text{if } x < 0 \end{cases}$

(c) $s(x) = \begin{cases} \frac{1}{2x} + 0.5 & \text{if } x \le -1 \\ |\sin(\pi x)| & \text{if } -1 < x \end{cases}$

8. For each of the following representation of a function, express it using: an expression; a table using your **table** with at least 20 points; a computer representation of a set of ordered pairs using at least 20 points; and a complete graph. Explain why your graph is complete.

   (a) The function represented by the following **ISETL func**.

   ```
   f := func(y);
       if  y  <  0  then  return
       2*y**2;

       elseif y>= 0 and y < 3 then
       return 3 - y**3;

       elseif  y  >=  3  and  y  <  10
       then return 3*y + 1;

       else return "function not
       defined";

       end;

   end;
   ```

   (b) The function represented by the following **Maple proc**.

   ```
   f := proc(x);
       if x > 0 then x**2 - 1;

       elif x = 0 then 1.0;

       else x + 1;

       end;

   end;
   ```

   (c) The function represented by the following.

   ```
   f := func(t);
       if abs(t) <= 1 then return
       cos(Pi*t);

       else return 3*f(t/2);

       end;

   end;
   ```

9. For each of the following representation of functions, express the function using: a computer representation of a set of ordered pairs; an expression(s); a **computer function**; a complete graph. Explain why your graph is complete.

   (a) The vapor pressure P as a function of temperature T is represented by the following table of data for ethanol.

T	P
25.00	55.9
30.00	70.0
35.00	97.0
40.00	117.5
45.00	154.1
50.00	190.7
55.00	241.9

Table 2.2 Pressure as a function of temperature for ethanol.

   (b) The (constant-pressure) Heat Capacity C_p as a function of temperature T for solid zinc represented by the following data.

   (c) For the table in the previous problem, C_p/T as a function of T.

T	C_p
20	0.406
30	1.187
40	1.953
50	2.671
60	3.250
70	3.687
80	4.031
90	4.328
100	4.578

Table 2.3 Heat Capacity as a function of temperature.

10. Your instructor will provide you with several graphs made using dots on the same sheet and separate definitions of some functions. Your task is to connect the dots by hand and match (when possible) the definition of the function with its graph.

11. A rectangular region is to be fenced in along a building using F feet of fencing with no fencing along the building. For each equation or **computer function** indicated below, put in appropriate tests to make sure that no errors will occur in running them.

    (a) Write an equation which relates the area A of the region, its width W and the amount of fencing F.

    (b) Write three **computer functions** giving each of the variables as a function of the other two.

    (c) Write six **computer functions** giving each variable as a **computer function** representing a function of one variable which returns a **computer function** representing a function of the other.

12. In this problem it is assumed that your **table** tool allows you to write conditions guaranteeing that certain values will be ignored (see Activity 6, p. 71). These conditions must be designed to rule out any values that are not in the domain of the function being considered. In this way, the table can reflect restrictions on the domain of the function.

For each of the following functions, write **computer functions** to implement them and conditions to reflect the specified domain. Then run **table** on the indicated interval with at least 25 points (not all of which will satisfy the domain restrictions.)

- The function g given by $g(x) = \sqrt{(x+3)(x-4)}$. The domain is the set of values for which the function is defined and the interval is $[-5, 5]$.
- The function h given by $h(x) = x^2$. The domain is a set on which no value of the function occurs more than once (that is, the function is 1-1). The interval is any interval including this domain.
- The function is sin and the domain is a set on which no value of the function occurs more than once (that is, the function is 1-1). The interval is any interval including this domain.
- The function is cos and the domain is a set on which no value of the function occurs more than once (that is, the function is 1-1). The interval is any interval including this domain.
- The function is tan and the domain is a set on which the function is defined and no value of the function occurs more than once (that is, the function is 1-1). The interval is any interval including this domain.

■ 2.2 Properties of Functions

2.2.1 Overview

One goal of working with multiple representations of a function, as we discussed in the last section, is for you to be able to construct in your mind a higher order concept of function, one that will be useful in your study of calculus and all other branches of mathematics. Thinking on this lofty height, you can then use your mind to descend at will to a particular representation — the one most useful when you wish to investigate a particular property of a function or problem situation. In this section we will discuss and define some basic properties of functions: including zero, positive and negative values; monotonicity (increasing and decreasing values); rates of change and slopes, extrema (maximum and minimum values); and upper and lower bounds in regions.

2.2.2 Activities

1. In Figure 2.3, several plots are shown of absolute viscosity v as a function of temperature T for several common fluids.

   What is the viscosity at $0°$C for SAE 10W-30 oil, SAE 10W oil, Kerosene, Mercury, and Carbon Dioxide?

   For several of the liquids, Castor oil, SAE 30 oil, water, Carbon Tetrachloride, the domain of the curve appears to be restricted. Do you think the restrictions are due to lack of data? Explain your answer.

   What do you think is going on with Glycerine at low temperatures?

2. Referring to Activity 1 above, consider SAE 10W oil. Construct a computer representation of a set of ordered pairs using 15 values. Use your MPL to plot your set of ordered pairs and compare it with the given graph.

3. Evaluating a continuous function represented as a set of ordered pairs can present a problem if the point at which you wish to evaluate is not in the table. One simple way around this is to use interpolation. Your instructor can provide you with a computer tool we will call **interpolate**. This will be a **computer function** that accepts a set of ordered pairs and returns a **computer function** that represents a function that agrees with the original at the given data points and approximates values for points in between.

   Use **interpolate** on the set of ordered pairs of the previous activity and find a point at which the resulting **computer function** has the value as close to 0.43 as you can get it.

   Now use your graphing utility to graph your **computer function** and compare it with the given graph as well as the graph you obtained in the previous exercise. Which do you think are the best?

4. Answer the following questions relative to the function in Section 2.1, Activity 1, page 69. Give numbers as accurately as you can find them.

   (a) What is the highest freezing point of the solution?

   (b) What is the lowest freezing point in which the solution has at least 50% of chemical A?

   (c) What is the highest freezing point in which the solution has less than 75% of chemical A?

   (d) For what mixture(s) is the freezing point $0°$?

   (e) For what mixtures is the freezing point above, below $0°$?

   (f) For what mixtures does the freezing point increase as the amount of chemical A increases?

   (g) For what mixtures does the freezing point decrease as the amount of chemical B increases?

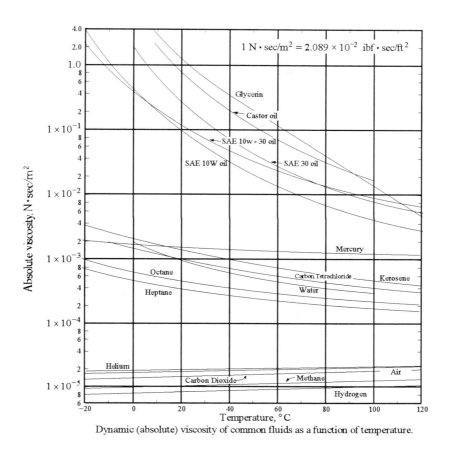

Dynamic (absolute) viscosity of common fluids as a function of temperature.

Figure 2.3. Plots of (absolute) viscosity v versus temperature T for common fluids.

5. Consider the function Z that is discussed in Section 2.1.3, page 75, giving compressibility as a function of volume. Since the values of this function refer to compressibility, the expressions can only be realistic when they give positive values. It may be interesting to know when the compressibility is less than a certain bound, or in between a lower bound and an upper bound. One may also want to know what are the largest and smallest possible values of the function in a region. Answer the following questions relative to the particular function given in Section 2.1.3, page 75.

(a) For what values of the volume is the compressibility positive?

(b) For what values of the volume is the compressibility less than or equal to one?

(c) For what values of the volume is the compressibility between 0.1 and 0.9?

(d) What is the largest value of compressibility? The smallest?

(e) What is the largest value of compressibility when the volume is less than 95?

6. Look again at Figure 2.3 that you worked with in Activity 1.

   (a) What would you say is the rate of change of the viscosity of Hydrogen with respect to temperature? Estimate it as closely as you can.

   (b) Estimate the rate of change of viscosity of Mercury with respect to temperature. How does the rate differ from Hydrogen?

   (c) Estimate the rate of change of viscosity of Heptane with respect to temperature. How does the rate differ from Mercury and Hydrogen?

   (d) Estimate the rate of change of viscosity of SAE 10W oil with respect to temperature. How does the situation differ from the previous three?

7. In this activity, you will need to have a device in your MPL that will allow you to add up all of the components of a tuple. Your instructor will give you information on this, if necessary. In **ISETL** the operation %+ will do this. For example, if P is the partition of the interval from a to b into n equal parts (Chapter 1, page 46) and f is some function whose domain includes this interval then there are many situations in calculus where one would like to consider each subinterval and multiply the value of f at, say, the left endpoint times the length of the subinterval and then add up all of these products. The mathematical notation for this quantity is

$$x = \left(a + \frac{(b-a)}{n}(i-1) \right)_{i=1}^{n+1}$$

$$\sum_{i=1}^{n} f(x_i)(x_{i+1} - x_i)$$

and, using %+, the **ISETL** code for obtaining it is very similar.

```
x := [a+((b-a)/n)*(i-1) : i in [1..n+1]];
%+[f(x(i))*(x(i+1)-x(i)) : i in [1..n]];
```

Compute this quantity for the function in Section 2.1, Activity 2 page 70, that gave C_p/T as a function of T. Use the interval from 100 to 250 degrees, and a partition of the interval into 20 equal parts.

Make a graph of the function and indicate on it an interpretation of the quantity you are computing.

8. In this activity you will look at specific portions of graphs and investigate the possibility of keeping them within a given window. We will select an upper and lower bound for the vertical axis of your coordinate system and you will look for an interval on the horizontal axis. Next you will use these two intervals as parameters in producing a computer graph of a given function. Your task is to choose your interval so that everything that appears on the screen is within the given bounds on the vertical axis.

To make things a little more specific, we will give you, in each case, a value for the domain variable and insist that this value be the center of your interval.

Suppose for example the function was the compressibility function Z of Section 2.1.3, page 75, and we gave you the interval [0.9, 1.1] for Z and the domain value 2. To do this, you must find an interval of the form $[2-d, 2+d]$ so that if you defined the function as a **computer function Z** and gave a command to plot **Z** with domain $[2-d, 2+d]$ and co-domain $[0.9, 1.1]$ then the curve you see would not leave the upper or lower boundaries of your window and you would see all of the curve for this interval of the domain variable.

In fact, for this specific case, you could choose $d = 1$ and use the command to plot **Z** with domain $[1, 3]$ and co-domain $[0.9, 1.1]$.

Now you do some, for the following situations.

(a) The freezing point function from Section 2.1, Activity 1 p. 69, the interval $[-12, -11]$ and the domain value 0.76.

(b) The Heat Capacity function from Section 2.1, Activity 2 p. 70, (C_p, not C_p/T), the interval $[0, 20]$ and the domain value 160.

(c) The Heat Capacity function from Section 2.1, Activity 2, p. 70, (C_p, not C_p/T), the interval $[6, 8]$ and the domain value 160.

2.2.3   Discussion

Representations of Functions and Properties   In the previous section, you learned about where functions come from and the various ways in which they can be represented. The reason you want to represent a function is so that you can find out things about it. What you want to find out are called the function's *properties*. We pointed out in the previous section that some representations are useful in answering some kinds of questions while others are good for other kinds of questions. For those who insist on knowing everything, the thing to do is to switch from one representation to another according to the needs of the moment.

In Activity 1, you read values of a function from its graph as a preliminary to making a table or set of ordered pairs which you did in Activity 2. The individual points on a graph and the pairs in a set of ordered pairs are simply different ways of organizing exactly the same information. Hopefully, the graph you constructed in Activity 2 from the table was not too far from the original graph from which you got the table. What would you do to make it closer? Can you think of any situation in which you were limited in the agreement you could get?

Some functions can be represented by expressions. In this case, the pairs of numbers come from linking a value for the independent variable with the result you get by plugging it into the expression and chugging (calculating). You should be flexible in your thinking about representing functions by graphs, tables, and expressions.

The **computer function** is more complicated and can be used to get more precise information. One thing you can do to get more information about the values of a function is use a **computer function** to evaluate the function at more and more points, or to make a graph that shows more detailed information in a narrow region.

We can say, perhaps, that the **computer function** and the expression are similar in that they carry (potentially by evaluating or calculating) *all* information about the function but you have to work to get this information out, while the set of ordered pairs and graph are similar in that they display all of their information readily, but not everything can be known from them.

Interpolation of Function Values   Because of the completeness of information, going from a **computer function** or expression to a set of ordered pairs or graph is a straightforward technical operation and very useful when you would like a more visual grasp of the data. Going in the other direction is problematic in that something must be done about the missing information.

Consider, for example, the 15 points you selected for Activity 2. If this was the only information you had about the viscosity of SAE 10W oil, you would not know very much. On a graph, all you would have would be something like Figure 2.4.

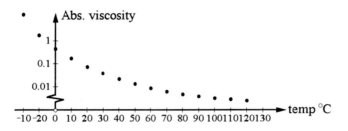

Figure 2.4. Points on a graph of the viscosity of SAE 10W oil.

On the other hand, a great deal of information can be obtained *approximately* by connecting the dots in some reasonable way. For example, you could connect with line segments and it would look like Figure 2.5.

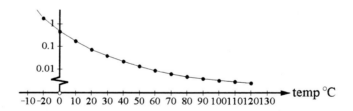

Figure 2.5. A graph of the viscosity v of SAE 10W oil versus temperature T.

Of course, this is based on the assumption that the function does not do anything rash in between the points. By *interpolation* we refer to any scheme for approximating values of a function between data points by making assumptions about what happens between those points. If you assume that the function follows a straight line between the points, as we did in making the graph in Figure 2.5, it is called *linear interpolation*. This is what the tool **interpolate** (from Activity 3) does.

Can you figure out an expression for the function obtained by linear interpolation on a set of data points? If it is available to you, look at how **interpolate** is built and see if you can make any sense out of it.

Suppose you were not satisfied with the accuracy obtained from interpolation with your 15 points. What could you do to make it better?

Can you think of any alternatives to *linear* interpolation? Figure 2.6 shows a graph of the compressibility function from Section 2.1, page 75.

Suppose you had only 15 data points for this function. Depending on your choice, it might look like Figure 2.7 below. Do you think linear interpolation is the best idea for this function? Whether it is or not, what do you think would happen in this case when the domain values are near 200?

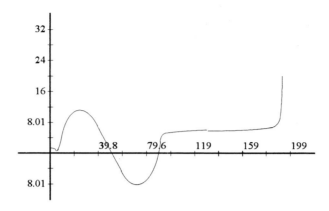

Figure 2.6. A graph of the compressibility function Z.

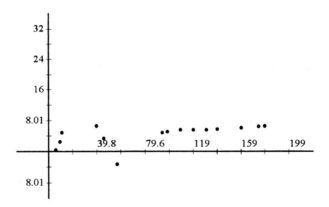

Figure 2.7. Points on the graph of the compressibility function Z.

Properties of Functions

Now we turn to a systematic discussion of properties of functions. Most of calculus can be considered to be a study of the properties described in this section. In this chapter you are learning how the computer with its particular representations of functions can be used to get information about these properties. Here, this information is usually in the form of numerical approximation. In the rest of the book, you will learn to use calculus to obtain information about properties when computers and approximations are either insufficient or inappropriate.

Sometimes what you want to know is not just data about a single property. You might also be interested in the relationship between two or more properties (for example, zeros and extrema or areas and rates of change). Again, calculus is a vehicle for studying relationships between properties, for using information that is relatively easy to obtain about one property to find out something hard to figure out about another property.

There is yet another way in which calculus is useful for properties. Sometimes you have an idea, such as instantaneous rate of change, slope of a curve, area of an irregularly shaped body, or entropy of a substance that makes sense intuitively, but seems to elude a precise definition. That can make it hard to say things with any degree of exactness, such as what direction are you moving when you drive along a sharply curved section of road; or to compute things, such as the area of a saw toothed blade. You will see that calculus concepts like derivative and integral are just the thing for getting a strong hold on certain vague ideas. But that is for later chapters; here you will work on developing intuitive notions about various concepts.

In order to use mathematics to study properties, it often becomes necessary to describe phenomena more formally and less vaguely than we have been doing so far in this book. This means, among other things, precise definitions and even theorems and proofs. They are tools used in mathematics in order to find out things, or to develop confidence about the reliability of information already obtained. We will gradually introduce more formality as the book progresses. For example, later in this section, we will be giving some formal definitions. Hopefully, in each case, the definitions will refer to concepts that are already familiar to you. Although the amount of formality will grow, we will continually return to the informal and intuitive to construct mental processes and objects to use in making sense out of formal statements.

Zero, Positive and Negative Values of Functions

Definition 2.1
A zero of a function f is a value a in the domain of f such that $f(a) = 0$.

In Activity 4(d), you were asked to find a zero of the freezing point function from Section 2.1, Activity 1, page 69. How did you do it? You could have made a graph and read off an approximation. How could you get a more accurate value from the graph? You could also have used **table** by listing values and seeing where they pass from positive to negative. Do you see how to use **table** to get increasingly more accurate approximations? Can you imagine a situation in which using a table will not be helpful?

One thing you will notice about methods for finding zeros is that they all require you to first *isolate* the zero. That means, you must find an interval $[c, d]$ in the domain such that there is exactly one zero a in the interval, that is, $c \leq a \leq d$. Then what most approximation methods do is let you make the interval smaller and smaller to shrink down to the desired value of the zero.

The same methods can be used to find values other than zeros. Suppose you have a function f and you wish to find a 2.45. That is, you wish to find a value in the domain for which the value of the function is 2.45. How can you easily convert this to a problem of finding a zero?

Obviously, zeros are closely connected to regions in the domain in which the value of the function is always positive — or always negative. Is it true that in a region between two zeros of a function the values are always positive, or always negative? In Exercises 37 and 38 in Chapter 1, page 64, when you used the tool **signs**, you made use of this idea to determine regions in which the values of the function were always of the same sign. What assumptions are you making in using this method?

Often, zeros and regions of positive, negative values have physical meaning. What is their meaning in the case of the freezing point function of Section 2.1, Activity 1 page 69?

Monotonicity

In Activity 4, you looked at where a function was increasing or decreasing. A function is said to be increasing if, whenever the values in the domain increase, then the corresponding values of the function also increase. We can state this and related definitions more formally. (The mathematical symbol $\in$ refers to a quantity or quantities being in a set.)

Definition 2.2

A function f is said to be increasing *in a region S of its domain if*

$$(x_1, x_2 \in S \text{ and } x_1 < x_2) \text{ implies that } f(x_1) \le f(x_2)$$

The function is said to be strictly increasing *in S if*

$$(x_1, x_2 \in S \text{ and } x_1 < x_2) \text{ implies that } f(x_1) < f(x_2)$$

The function is said to be decreasing *in S if*

$$(x_1, x_2 \in S \text{ and } x_1 < x_2) \text{ implies that } f(x_1) \ge f(x_2)$$

The function is said to be strictly decreasing *in S if*

$$(x_1, x_2 \in S \text{ and } x_1 < x_2) \text{ implies that } f(x_1) > f(x_2).$$

There is a single generic term for all of these.

Definition 2.3

A function f is said to be monotonic *in a region S of its domain if it is either increasing, strictly increasing, decreasing, or strictly decreasing in S.*

Can you give examples in which a function is monotonic but not strictly increasing nor strictly decreasing? How about functions which are strictly monotonic but not monotonic?

Take a look at the viscosity graphs in Figure 2.3. Do you think that they are all monotonic? What does that mean physically? Notice that Helium, Air, Carbon Dioxide, Methane and Hydrogen are all different from the others. In what way(s)? What does it all mean? What do you think would happen in the graph for water as the temperature goes beyond $100°$?

At first glance, there might seem to be little or no connection between regions in which a function is monotonic and regions in which it is positive or negative. Actually there is, but it is fairly subtle. We will discuss the relationship in depth in Chapter 4. Look at Figure 2.8. Do you see any connection between monotonicity of one function and regions of positive and negative for some other function?

Extrema

There are many problems in which it is important to find a point at which the value of the function is larger (or smaller) than any other value. It can also be important to know that the value at a point is larger (or smaller) than the value at any point in the immediate neighborhood. All of these are referred to as extreme points, or extrema. In Activity 4, you looked for such points.

In the following definitions we suppose that f is a function and D is its domain.

Definition 2.4

A point $(c, f(c))$ *with* $c \in D$ *is said to be a* maximum point *if*

$$\text{for every } x \in D \text{ we have, } f(x) \le f(c)$$

and the value $f(c)$ *is called the* maximum value *of* f. *A point* $(c, f(c))$ *with* $c \in D$ *is said to be a* minimum point *if*

$$\text{for every } x \in D \text{ we have, } f(x) \ge f(c)$$

and the value $f(c)$ *is called the* minimum value *of* f.

A point $(c, f(c))$ *with* $c \in D$ *is said to be a* relative maximum point *if there is an open interval* (a, b) *containing c and contained in D such that*

$$\text{for every } x \in (a, b) \text{ we have, } f(x) \le f(c)$$

and the value $f(c)$ *is called a* relative maximum value *of* f. *A point* $(c, f(c))$ *with* $c \in D$ *is said to be a* relative minimum point *if there is an open interval* (a, b) *containing c and contained in D such that*

$$\text{for every } x \in (a, b) \text{ we have, } f(x) \ge f(c)$$

and the value $f(c)$ *is called a* relative minimum value *of* f.

Definition 2.5

A point $(c, f(c))$ *which is either a maximum, a minimum, a relative maximum, or a relative minimum point for a function f is said to be an* extreme point *of* f. *The extreme points of a function are referred to collectively as its* extrema. *For emphasis, a maximum or minimum is called an* absolute maximum or minimum *— to distinguish it from relative maximum or minimum.*

Note that we sometimes also refer to relative extrema as *local* extrema. Can you give examples of relative maxima, minima that are not absolute? How about absolute maxima, minima that are not relative?

As with zeros, you can use the graph of a function to approximate its extrema. You can make the approximation better by looking at smaller regions of the graph. You will see later that calculus provides you with a precise way of determining, in some cases exactly, where the extrema of a function could occur. The essential ideas are actually contained in the following example, although it may be a little while before you are ready to go into it. Look at Figure 2.8 again. Do you see any connection between zeros of one function and extrema of some other function?

Rates of Change and Slopes

How did you respond to the questions in Activity 6, about rates of change? Did you say for Hydrogen that it was something like 2×10^{-7}? What about SAE 10W? Does your answer depend on the temperature? Is the rate of change of SAE 10W different at $0°$C than it is at $100°$C? In fact, what do you think is meant (exactly — so that you could calculate it) by the rate of change of viscosity with respect to temperature *at a certain temperature*? You could ask the same question about any function. What is meant by the rate of change of the function f with respect to x at the point a in its domain?

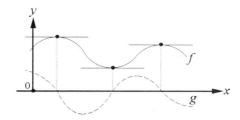

Figure 2.8. *What is the connection between zeros and extrema?*

That's a lot of questions. And the path to answering them is not short. We can only begin in this chapter. One starting point has to do with slope and tangents. Look at Figure 2.9 which is just the viscosity of the SAE 10W oil function pulled out of Figure 2.3.

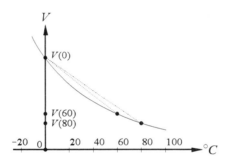

Figure 2.9. *A graph of the viscosity of SAE 10W oil with two secants.*

We have drawn a line connecting the point $[0, V(0)]$ with the point $[80, V(80)]$. This is called a *secant*. Now, keeping the domain value 0 fixed, we moved the point 80 back to 60, evaluated V and drew the secant connecting $[0, V(0)]$ to $[60, V(60)]$. Now imagine that we keep moving the domain point towards 0, each time evaluating V and drawing the secant. What do you think ultimately happens to the secant?

Hold that question for a chapter or two and return to the rate of change of v with respect to T at a point. This is called the *instantaneous rate of change* at the point. Let's consider it at $T = 0$. It is not obvious how to compute it, but at least we can approximate it. Consider the average rate of change over the interval from $T = 0$ to $T = 80$. We can calculate this by taking the difference of the viscosities at the two temperatures and dividing by the difference in temperature. We get,

$$\frac{V(80) - V(0)}{80 - 0} \approx \frac{0.008 - 0.4}{80} = -4.9 \times 10^{-3}$$

We got the approximate numbers by reading off the graph. Can you calculate what would happen if we moved 80 to 60? Suppose we kept going, what would we get? What does this have to do with instantaneous rate of change?

Now, here is the big question. What does this way of approximating instantaneous rate of change have to do with our question about the ultimate position of the secant? Take another look at that question and see if you feel any differently about it.

One thing that should be fairly clear is that the expression we are using to approximate instantaneous rate of change has to do with the slope of the secant line. What does it have to do with the tangent to the graph (as discussed, for example, in Chapter 1, page 55)?

You should also review the material in Chapter 1 (p. 23) where difference quotients were considered.

Upper and Lower Bounds in Regions; Bounds

In the analysis of a function, a first indication of how it behaves can be obtained by trying to find regions in the domain for which the values of the function are less than a given number and/or greater than some number. In the following definitions, f is a function and S is a subset of its domain.

> **Definition 2.6**
> *An* upper bound *for f on S is a number B such that*
>
> $$\text{for every } x \in S \text{ we have, } f(x) \le B.$$
>
> *A* lower bound *for f on S is a number b such that*
>
> $$\text{for every } x \in S \text{ we have, } f(x) \ge b.$$

In these terms, what would you say is the meaning of the expression, $|f(x)| \le M$ for every $x \in S$? What would you call M? The next definition provides an answer, where again f is a function and S is a subset of its domain.

> **Definition 2.7**
> *A* bound *for f on S is a number M such that*
>
> $$\text{for every } x \in S \text{ we have, } |f(x)| \le M.$$

Upper and lower bounds are easily indicated on a graph as in Figure 2.10. How would you indicate a bound for a function on its graph? How are the upper and lower bounds related to a bound of a function?

In Activity 5, you found regions on which various numbers were upper and lower bounds for the compressibility function of Section 2.1, page 75. The answers to parts (a), (b) and (c) were all intervals, or unions of intervals. This is usually the case.

Think about your answer to parts (d) and (e) of Activity 5. What does this have to do with the maximum and minimum of the compressibility function?

Limits of Continuous Predictions

Who can predict the future? It turns out that, using Calculus, you can actually predict what will ultimately happen in certain calculations — provided you are willing to accept certain assumptions.

For example, in Section 2.1, Activity 9, page 72, did you say that ultimately the value of the ratio would be 2.1181×10^{-22}? What is the meaning of such an answer? It certainly depends on the assumption that, as the quantum numbers go on and on, the value of the energy level is still given by the

3written formula. Presumably you calculated values of the ratio for higher and higher quantum numbers and saw that it got closer and closer to 2.1181×10^{-22}.

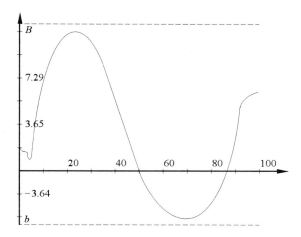

Figure 2.10. Upper and lower bounds of a function.

There is another kind of prediction you can try to make. Look at the freezing point function in Section 2.1, Activity 1, page 69. Suppose you didn't have the formulas but could only make measurements of the freezing point for various mixtures. Suppose also that, for some reason, it was not possible to test a mixture of exactly 76% of Chemical A. Imagine trying to predict the freezing point at this value by testing at mixtures less than 76%. Could you make a prediction? Would you be right? How would your prediction be effected if you tested the freezing point for values larger than 76%?

Now do the same for the Heat Capacity function in Section 2.1, Activity 2, page 70. Assume that you can only test individual values by experimentation, you are interested in the Heat Capacity at $T = 160$, and (possibly because the gas is changing state) you cannot test at that exact value. Make your prediction by testing close to 160 when the gas is solid and then by testing when the gas is liquid. What do you conclude?

Now, again we have a big question. Take a look at Activity 8 in this section. You worked with the same functions and the same values in the domain. Is there any connection with the discussion here and what you were doing in Activity 8?

Finally, there is a situation of this kind that is absolutely critical for just about everything that happens in Calculus. Look at the difference quotient we worked with on page 95 in calculating secants.

$$\frac{V(80) - V(0)}{80 - 0} = -4.9 \times 10^{-3}$$

and recall that we replaced 80 by smaller numbers. Thinking about this in general we could write,

$$\frac{V(h) - V(0)}{h - 0} = \frac{V(h) - V(0)}{h}.$$

Now, the problem is to predict what happens to this quantity as h gets closer and closer to 0. Suppose that, instead of the graph, we had an expression for this particular viscosity function, say,

$$\frac{360}{(1.5h + 30)^2}.$$

Could you use this to make a prediction of what happens as *h* goes to 0?

What happens if you try to do the same thing with the freezing point function of Section 2.1, Activity 1, page 69? Write out the expression with the fixed domain value 0 replaced by $T = 0.76$. Anything strange happen? Now try it with the compressibility function from Section 2.1 for the domain value $v = 6$. What is similar and what is different about these two situations?

Step back and think about all of this "prediction" business. As we said, it requires that you make certain assumptions. What are the assumptions that must be made for the examples in this paragraph? What about interpolation. Does this involve prediction? Is it the same kind of prediction? What kinds of assumptions are made when you interpolate? Are they the same as the ones you are making here? The issues in this paragraph lie at the heart of the foundations of calculus. We will return to them in Chapter 3.

Area Under a Curve   Entropy is a scientific concept that can only be defined precisely using Calculus. We can't do that for quite a while because in order to define the entropy function you need to know about definite integrals, and defining a function by integration. All we can do here is to point out that your calculation in Activity 7, amounted to approximating the change in entropy as the temperature of the gas went from 100 to 250 degrees.

Finally, let's end with a meaty question for you to think about. Suppose we said that the entropy of this material was a function which gave, for each temperature *T* the "change in entropy" as the temperature goes from 0 to *T* degrees. Given what we just said about approximating this change in entropy, can you write a **computer function** that will approximate the entropy function? You will get a chance to do just that in Activity 11 of Section 2.3.

Definition of Function   As a way of summarizing our study of functions, we now give two equivalent definitions of the concept of function. We leave it as an exercise for you to show they are the same.

> **Definition 2.8**
> *A function f is a process which can take any element of a set D, called the* domain *of f, and transform it into a unique element f(a) of a set R, called the* co-domain *(or range) of f.*

> **Definition 2.9**
> *A function f is a set of ordered pairs (x, y) such that no two pairs have the same first element and different second elements. We write* $f = \{(x, y) : x \in D, y = f(x) \in R\}$, *where the set D is called the* domain *of f and the set R is called the* co-domain (or range) *of f.*

2.2.4 Exercises

1. For the function indicated in each of the following situations, answer (approximately) each of the questions below using an appropriate representation of the function.

    (a) Exercise 6 (i) of Section 2.1.

    (b) Exercise 6 (ii) of Section 2.1.

    (c) Exercise 6 (iii) of Section 2.1.

    (d) Exercise 6 (iv) of Section 2.1.

    (e) Exercise 6 (v) of Section 2.1.

    (f) Exercise 6 (vi) of Section 2.1.

    (g) Exercise 6 (vii) of Section 2.1.

(h) Exercise 6 (viii) of Section 2.1.

(i) Exercise 6 (ix) of Section 2.1.

(j) Exercise 6 (x) of Section 2.1.

(k) Exercise 6 (xi) of Section 2.1.

(l) Exercise 6 (xii) of Section 2.1.

(m) Exercise 6 (xiii) of Section 2.1.

(n) Exercise 6 (xiv) of Section 2.1.

(o) Exercise 6 (xv) of Section 2.1.

(p) Exercise 6 (xvi) of Section 2.1.

  i. What are the zero(s)?

  ii. In what domain regions does the function have positive values?

  iii. In what domain regions does the function have negative values?

  iv. In what domain regions is the function increasing?

  v. In what domain regions is the function decreasing?

  vi. What is an upper bound of the function?

  vii. What is a lower bound of the function?

  viii. Is the function periodic?

  ix. Is the function even?

  x. Is the function odd?

  xi. Where in the domain do the extreme points occur? What are the corresponding extreme values of the function?

  xii. What are the horizontal asymptotes?

  xiii. What are the vertical asymptotes?

  xiv. In what regions of the domain is the function concave up ("graph holds water")?

  xv. In what regions of the domain is the functions concave down ("graph does not hold water")?

2. Your instructor will provide you with a file **data11** that contains a table of values of the function f given by

$$f(x) = \frac{1}{(1-x)(x-2)^2}.$$

Use **interpolate** to produce a **computer function** and then graph it. Is there anything wrong? That is, anything other than the fact that producing this graph may take a very long time.

3. Find increasingly accurate approximations (to three decimal places) of all zeros of each of the following functions by using

  (a) the tool **table**.

  (b) the **zoom** feature in your graphics tools.

  (c) Check your answer using your symbolic computer system.

  i. $x^2 - 3x - 1$

  ii. $x^3 - 13.7x^2 + 40.87$

  iii. $x^3 - 2x + 1$

  iv. $3 + 3x^3 - x^4$

  v. $x - 2\sin(x)$

  vi. $x^2 - \cos(x)$

  vii. $x - \tan(x)$ for $-\pi \le x \le \pi$

  viii. $8.5y^3 - 1.8y^2 + 0.026y + 0.001 - 17.6\sin\left(\frac{y}{2.3}\right)$

4. In each of the following situations, find an increasingly accurate approximation (to three decimal places) of the solution to the problem using the tool **table** and the **zoom** feature in your graphics tool.

   (a) Solve for x

   $$\frac{25}{x-20} - \frac{50}{x^2} = 2.$$

   (b) Solve for x

   $$x^{\frac{3}{2}} = 50 - x^{\frac{2}{3}}.$$

   (c) Find all x coordinates for all points of intersection of the graphs of the curves given by $f(x) = 2\cos(x)$ and $g(x) = \tan(x)$ for $0 \le x \le \pi/2$.

   (d) Find all x coordinates for all points of intersection of the graphs of the curves $f(x) = \sin(x)$ and $g(x) = x^2$.

   (e) A container is to be formed by placing a hemisphere with radius r cm on the top of a cylinder with radius r cm and height 6 cm. Find the radius of the cylinder and hemisphere so that the total volume is 200 cm³. (The volume of a sphere with radius r is $4\pi r^3/3$ and the volume of a cylinder with radius r and height h is $\pi r^2 h$.)

   (f) The temperature of the human body is elevated when a dosage d, in grams, of a certain drug is injected into the blood stream and absorbed by the body. The elevation t of the temperature, in degrees Fahrenheit, is given by

   $$t = T(d) = 1.73d^2 - 0.31d^3$$

   where the domain of the dosage is in the interval $0 \le d \le 4$. Find the dosage d required to raise the body temperature 4.5°.

5. For the situation in Section 2.1, Activity 2, page 70, graph

   (a) C_p as a function of T.

   (b) C_p/T as a function of T.

6. Graph the following functions given by the indicated data. Then find an expression representing a function that best seems to "fit" this data, omitting any "bad" points.

   (a) The pressure P as a function of temperature T represented by the following table of data for ethanol.

T	P
25.00	55.9
30.00	70.0
35.00	97.0
40.00	117.5
45.00	154.1
50.00	190.7
55.00	241.9

   Table 2.4 Pressure as a function of temperature for ethanol.

   (b) The (constant-pressure) Heat Capacity C_p as a function of temperature T for a certain material represented by the following data.

T	C_p
20	0.406
30	1.187
40	1.953
50	2.671
60	3.250
70	3.687
80	4.031
90	4.328
100	4.578

   Table 2.5 Heat Capacity as a function of temperature.

(c) For the situation in the previous exercise, C_p/T as a function of T.

(d) For the situation in the previous exercise, C_p/T as a function of T^2.

7. Use the **zoom** feature in your graphics tools to get increasingly accurate values (to three decimal places) of all extreme points in the domain where the function represented by each of the following expressions has extrema (if any) and all corresponding extreme values of the function.

(a) $3x^2 + 24x$

(b) $x^2 - 20x + 215$

(c) $x^3 - 2x + 1$

(d) $3x - 2x^3$

(e) $x^3 - 6x^2$

(f) $15 - 25x - x^3$

(g) $\left|3x - 2x^3\right|$

(h) $2x + \frac{3}{x} + 5$

(i) $\frac{3}{x^2} + \frac{1}{x}$

(j) $\frac{300}{x^6} - \frac{100}{x^3}$

(k) $x\cos(2x)$

(l) $x^2 - \sin(x)$

(m) $x + \cos(3x)$

(n) $x - \tan(x)$

8. Using **zoom** feature in your graphics tool, predict the indicated values in each of the following situations.

(a) At $x = 0$, what is the ultimate value of $\frac{\sin(x)}{x}$.

(b) At $x = 0$, what is the ultimate value of $\frac{2x}{\tan(3x)}$.

(c) At $x = 0$, what is the ultimate value of $\frac{1 - \cos(x)}{x}$.

(d) The distance traveled by a falling object is approximately given by $4.9t^2$ meters t seconds after it is dropped from the top of a cliff. Use a difference quotient to figure out the instantaneous velocity, three seconds after it began its fall. What are the units of velocity?

(e) An ideal gas confined to a piston satisfies the law $PV = 270$ where

P is the pressure and V is the volume of the gas at $270°$ K. The instantaneous rate of change of the pressure, in lbs/in^2, with respect to the volume at the instant the volume is 30 in^3 using a difference quotient. What are the units of the instantaneous rate of change of pressure?

(f) The instantaneous rate of change of pressure with respect to temperature for the data in Exercise 6 (a) using a difference quotient.

(g) The instantaneous rate of change of Heat Capacity with respect to temperature at $T = 40°$ for the data in Exercise 6 (b) using a difference quotient and linear interpolation.

9. For each of the following situations, use sums as in Activity 7 on page 88, to estimate the desired quantity. Make a sketch which shows the geometric meaning of the sum you are using to make the estimate.

(a) For a certain spring which hangs vertically from a ceiling, the force F required to stretch the spring x inches from its normal position ($x = 0$) is given by $f(x) = 3x$ pounds. The *work W* required to stretch the spring 4 inches beyond its normal position is the measure of the area under the graph of F in an xy-coordinate system from $x = 0$ inches to $x = 4$ inches. Estimate W using a partition of the interval $[0, 4]$ into 20 equal subintervals. How can you make your estimate more accurate? What are the units of work in this situation? What is the exact amount of work W? How did you find it?

(b) An object is 6 feet long. A vertical cross-section of the object x feet from one end is a circle with radius

$\sqrt{x}$ feet. The volume of the object is equal to the measure of the area under the graph of the function A, which represents the area of a vertical cross-section in terms of x, from $x = 0$ to $x = 6$ in an xy-coordinate system. Estimate the volume of the object (rounded to three decimal places) using a partition of the interval $[0, 6]$ into 20 equal subintervals. How can you make your estimate more accurate? What are the units of volume in this situation? What is the exact volume of the object? How did you find it?

(c) A variable force acts on an object moving along a coordinate line. Assume that the coordinate line lies along the x-axis in an xy-coordinate system. The force F, in Newtons, at a point x is given by the function

$$F(x) = \frac{9}{x^2}.$$

The *work* W done by the force F on the object moving from $x = 1$ meters to $x = 3$ meters is equal to the measure of the area under the graph of F over the interval $[1, 3]$. Estimate the work done by the force on the object as it moves from $x = 1$ meters to $x = 3$ meters using a partition into 20 equal subintervals. How can you make your estimate more accurate? What are the units of work in this situation?

(d) The rate R at which a machine generates revenue is given by $R(x) = 5 - 0.2x^2$ thousand dollars per year when the machine is x years old.

The total net earnings generated by the machine during the first n years it is in use is equal to the measure of the area under the graph of R on the interval $[0, n]$. Estimate the total net

earnings of the machine during its lifetime (i.e., during the time it generates revenue) using a partition with 20 equal subintervals. How can you make your estimate more accurate? What are the units of total net earnings in this situation?

10. For the function represented by each of the following expressions

(a) Use the **zoom** feature in your graphics tool to get increasingly accurate values (if any) of the upper bound B and lower bound b.

(b) Use the **zoom** feature in your graphics tool to get an increasingly accurate value (if any) of the bound M on the absolute value of the function.

(c) Estimate the bound M analytically using inequalities, their properties, and whichever of the following absolute value properties you need:

$$|f(x) + g(x)| \le |f(x)| + |g(x)|$$

$$|f(x)g(x)| = |f(x)||g(x)|$$

$$\left|\frac{f(x)}{g(x)}\right| = \frac{|f(x)|}{|g(x)|}$$

If $0 < N \le |f(x)|$, then $\dfrac{1}{|f(x)|} \le \dfrac{1}{N}$.

  i. $3.2\sin(x^2)$

  ii. $1 - 2.1\cos(x)$

  iii. $\sin(x) + \cos(x)$

  iv. $3.2\sin(x) - 1.4\cos(2x)$

  v. For $\pi/2 < x < 3\pi/4$, $2.3\sec(x)$

  vi. For $x > -2$, $\dfrac{0.3}{x+2}$.

  vii. For $x > 2$, $\dfrac{3}{x+2}$.

viii. For $x > 3$, $\dfrac{2}{2x-1}$.

ix. For $|x| \leq 2$, $\dfrac{0.5x}{x-3}$.

x. For $|x-2| < 1$, $\dfrac{x+6}{(x+2)^2}$.

xi. For $|x| < 2$, $x\sin(x)$.

xii. For $|x| \leq \pi$, $x^2\cos(x)$.

11. For each of the following pairs of graphs, find a transformation (e.g., $f(ax+b)$, $af(x)+b$, etc.) that makes them the same.

(a)   (b)

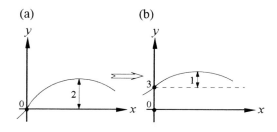

(c)   (d)

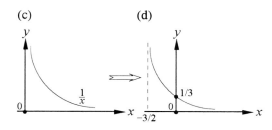

(e)   (f)

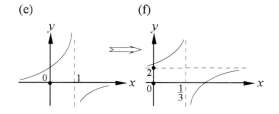

(g)   (h)

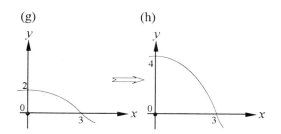

12. A *fixed point* of a function f is a value c in the domain of f such that $f(c) = c$. For each of the following functions, find increasingly accurate approximations (to three decimal places) for all fixed points (if any) for each of the following functions using **table** and the **zoom** feature in your graphics tool.

    (a) $f(x) = \cos(x)$

    (b) $f(x) = \sin(2x)$

    (c) $g(x) = 0.5\sin(x)$

    (d) $h(x) = x^3 - 3x + 1$

13. Let f be the function of Section 2.1, Activity 2, page 70, that gives C_p/T as a function of T. Suppose it is desired to find a function E on the interval [50, 250] and a partition of that interval into 20 equal subintervals, where the requirement on E is that in each subinterval its slope is the value of f at the left endpoint. Derive a formula for the best you can do for E.

14. In Activity 1, page 86, you read values of a function from its graph as a preliminary to making a table or set of ordered pairs which you did in Activity 2. The individual points on a graph and the pairs in an **smap** are simply different ways of organizing exactly the same information. Hopefully, the graph you constructed in Activity 2 from the table was not too far from the original graph from which you got the table. What would you do to make it closer? Give an example, if possible, of a situation in which you were limited in the agreement you could get?

15. The notion of *extrapolation* is the predicting of values beyond those that you know about. What problems could you imagine in using extrapolation to predict function values beyond known values? How might extrapolation differ from interpolation? How is extrapolation related, if at all, to predicting ultimate "limiting values"?

16. Can you think of any alternatives to *linear* interpolation? Explain your answer.

17. Consider the graph of the compressibility function Z in Figure 2.7 page 91. Do you think linear interpolation is the best idea for this function? Whether it is or not, what do you think would happen in this case when the domain values are near 100?

18. In Activity 4 (d), page 86, you were asked to find a zero of the freezing point function from Section 2.2, Activity 1, page 86. Explain how you can use **table** to get increasingly more accurate approximations? Give an example of a situation in which using a table will not be helpful?

19. Suppose you have a function f and you wish to find a "c value". That is, you wish to find a value in the domain for which the value of the function is c. How can you easily convert this to a problem of finding a zero of a function?

20. Zeros are closely connected to regions in the domain in which the value of the function is always positive or always negative. Is it true that in a region between two zeros of a function the values are always positive, or always negative? In the exercises of Chapter 1, p. 62, you made use of this idea to determine regions in which the values of the function were always of the same sign using the tool **signs**. What assumptions are you making in using this method?

21. Often, zeros, regions of positive values and regions of negative values have a physical meaning. What is their meaning in the case of the viscosity functions of Activity 1, page 86?

22. Can you give an example of a function which is monotonic but not strictly increasing nor strictly decreasing? Can you give an example of a function which is strictly monotonic but not monotonic?

23. Take a look at the viscosity graphs in Figure 2.3 page 87. Do you think that they are all monotonic? What does that mean physically? Notice that Helium, Air, Carbon Dioxide, Methane and Hydrogen are all different from the others. In what way(s)? What does it all mean? What do you think would happen in the graph for water as the temperature goes beyond 100°?

24. Look at Figure 2.8, page 95, which shows the graphs of two functions on the same coordinate system. Do you see any connection between monotonicity of one function and regions of positive and negative for the other?

25. If possible, give examples of functions which satisfy the following.

    (a) A function which has a relative maximum value which is not an absolute maximum.

    (b) A function which has a relative minimum value which is not an absolute minimum.

    (c) A function which has an absolute maximum value which is not a relative maximum.

    (d) A function which has an absolute minimum value which is not a relative minimum.

26. Try seeing what happens to a difference quotient

$$dq(x) = \frac{f(x+h) - f(x)}{h}$$

for appropriate interpretations of f and x, as h goes to 0, in the following situations.

(a) The compressibility function from Section 2.1, page 75, for the domain value $v = 6$.

(b) The freezing point function of Activity 1, page 86, at $m = 0.76$. Anything strange happen? What is similar in this situation as compared to the situation in part (a)?

(c) The graph of the viscosity of SAE 10W oil in Figure 2.9, page 95, shows secants connecting the points $[0, V(0)]$ and $[80, V(80)]$ and also the points $[0, V(0)]$ to $[60, V(60)]$. Imagine that you keep moving the domain point towards 0 (from 60), each time evaluating V and drawing the secant.

    i. What do you think ultimately happens to the position of the secant?

    ii. What does this have to do with the notion of instantaneous rate of change of the viscosity?

    iii. Can you explain how the two questions in parts (i) and (ii) are related?

27. For each of the functions described below, you are given a small positive number ε, a value c in the domain of definition of the function and a number L. Your task is to find another positive number δ having the property that if the independent variable is in the interval from $c - \delta$ to $c + \delta$, then the values of the function lie in the interval

from $L - \varepsilon$ to $L + \varepsilon$. Use the method of Activity 8, page 88.

(a) The function is given by the expression $\frac{\sin(x - 0.5)}{x - 0.5}$, $\varepsilon = 0.01$, $c = 0.5$, and $L = 1$.

(b) The function is given by the expression $\frac{\sqrt{2+x} - \sqrt{2}}{x}$, $\varepsilon = 0.01$, $c = 0$, and $L = \frac{\sqrt{2}}{4}$.

28. Write a **computer function** that will accept a **computer function** representing a function and a finite subset of the domain of the function and return a computer representation of a set of ordered pairs that represents the same function with domain restricted to the finite set.

Apply your **computer function** to the two **computer function**s that you are required to write in parts (a) and (b) of Exercise 9, Section 2.1, p. 84, and see if you get the same set of ordered pairs as the one you are to write in the first part of that exercise.

29. A number U is a *least upper bound* (lub) for a set of real numbers S if U is an upper bound (page 96) for S, and $U \le B$ for all upper bounds B of S. In the theory of the branch of mathematics known as *Real Analysis*, a fundamental axiom for the set of real numbers $\mathcal{R}^1$ referred to as the *completeness axiom* is as follows.

If a non-empty set S has an upper bound, it must have a least upper bound.

Note that a lub of a set S need <u>not</u> be a member of the set S. Similar statements can be made for the concept of *greatest lower bound* (glb).

(a) Give a definition for the greatest lower bound of a set of real numbers S.

(b) Find the lub and glb of the set of values of the function f given by

$f(x) = x^3 + 1$ on the interval $2 \le x < 9$.

(c) Find the lub and glb of the set of values of the function f given by $f(x) = 1 - x^2$ on the interval $-1 \le x \le 4$.

(d) Let S be the set of all rational numbers in the interval $0 \le x \le 2$. Where the domain of f is S. Find the lub and glb of the set of values of the function f given by

$$f(x) = -\frac{3}{2}\sqrt{x}$$

2.3 COMBINATIONS TO MAKE NEW FUNCTIONS

2.3.1 Overview

In this section you will investigate various ways in which you can combine functions to make new functions — combinations of functions that will be useful in problem solving. In particular, we will consider combinations of functions using arithmetic (addition, subtraction, multiplication and division); compositions (functions of functions); and the making of new functions from difference quotients and approximate areas.

Once again, much of this material forms the background for things you will be studying in the remainder of this text. Many of the examples will revisited several times.

2.3.2 Activities

1. If you mix two substances, it is not clear how the viscosity of the mixture will be related to the viscosities of the original substances. One possibility that will occur in some cases is that the viscosity of the mixture will be the proportional sum of the two viscosities. Assume this is the case for the substances in Figure 2.3, page 87. Suppose, then, that you mixed 7.32 parts of SAE 10W oil with 385 parts of Kerosene. Construct a graph that approximates the viscosity of the mixture.

   Solve this problem in the following two ways.

   (a) Read some points on the graph for the two components, perform the appropriate calculation by "hand" to get corresponding points for the mixture, plot them, and connect them.

   (b) Read some points on the graph for the two components and enter them into two computer representations of sets of ordered pairs. (You can do a lot more points in less time because you will not be performing any calculations on them.) Now, write code that will construct a new computer representation of a set of ordered pairs by making the appropriate calculations with the two original sets of ordered pairs. Then apply **interpolate** and plot the resulting **computer function**.

   Compare the two graphs you obtained in this way. Which is better?

   Think carefully about the combination formula that you calculated. For example, should the viscosity of the mixture be greater than or less than the viscosity of the two components?

2. The forces on two interacting molecules lead to an energy e that depends on the distance r between the two molecules. Actually, this energy is made up of two parts. One is given (in appropriate units) by r^{-6} and is due to an attractive force. The other is given by r^{-12} and is due to a repulsive force. The energy is the sum or difference of these two energies. Which is it? Explain your choice.

   Write two **computer functions** representing these two energy components and combine them appropriately to get a **computer function** for the total energy. Construct the graph of the energy function.

   Write a short explanation of how you can "see" the effects of the two kinds of forces by looking only at the graph of the total energy. Include in your discussion responses to the following questions.

   (a) What does it mean for the energy function to be 0?

   (b) How would you interpret regions of positive and negative for this function?

3. Look at the two functions in Section 2.1, Activity 2, page 70. Show how the second function (which is the one used in approximating the entropy) can be obtained from the Heat Capacity function by multiplying, or dividing two functions.

4. Look at the compressibility function Z introduced in Section 2.1, page 75. That function can be obtained as an appropriate combination of the following two functions. Find the combination.

$$Z_1(v) = \begin{cases} 0.5 & \text{if } 0 \leq v < 6 \\ \frac{v^3}{6 \times 10^3} - 0.024v^2 + 0.865v - 3.862 & \text{if } 6 \leq v < 90 \\ -\frac{27}{2v-400} + 3.216 & \text{if } 90 \leq v \end{cases}$$

$$Z_2(v) = \begin{cases} \frac{(v-2)^3 (v-6)}{18} & \text{if } 0 \leq v < 6 \\ 0 & \text{if } 6 \leq v < 90 \\ \frac{-12v^{\frac{3}{2}}}{5\sqrt{v^3 - 7 \times 10^5}} & \text{if } 90 \leq v \end{cases}$$

5. In this activity you are to investigate what happens to the property of monotonicity when you add, subtract, or multiply two functions. Look again at the viscosity graphs in Figure 2.3, page 87. Assuming no chemical reactions you can form all possible mixtures to investigate the question of adding two functions. Forgetting about physical interpretations, you can also use these graphs to investigate other combinations.

   For each of following assertions, either show that it is false by giving a counterexample from one of these graphs (or one that you might make up) or conjecture that it might be true. In the latter case, try to give a plausible explanation as to why it should be so.

   (a) If two functions are decreasing over a region, then their sum is also decreasing over that region.

   (b) If two functions are decreasing over a region, then their difference is also decreasing over that region.

(c) If two functions are monotonic over a region, then their product is also monotonic over that region.

6. Return once more to the compressibility function Z, page 75, in terms of volume v. It is possible to imagine an apparatus that varies the volume with time, but keeps the temperature constant so that the expression for Z remains valid. Suppose for example, the volume is varied as follows.

$$v = V(t) = \begin{cases} 1.26t^2 + 1 & \text{if } 0 \le t < 10 \\ (4/3)\pi(t-7)^3 + 2 & \text{if } 10 \le t \end{cases}$$

(a) Write a **computer function** that represents this function.

(b) Construct a **computer function** that represents a function W whose process works like this:

Given t, apply the function V to get the volume v. Then apply the function Z to that value of v to get the compressibility.

We call this *composition*.

(c) Figure out by hand the values for W for the following values of time and then check them with your **computer function**.

$$t = 0.2, 0.5, 0.8, 7, 8, 9, 9.5, 9.8, 10, 11$$

(d) Graph the function W.

(e) Write a mathematical formula for W.

7. For each of the following expressions, figure out two functions whose composition would be given by the expression. What we mean is that if the given function is f then your job is to find two functions g and h such that, for each x in the domain of g for which $g(x)$ is also in the domain of h, it is the case that

$$f(x) = h(g(x)).$$

(a) $\sqrt{1+x}$ (b) $\dfrac{7}{\sqrt[3]{\sin(x)}}$ (c) $\dfrac{1}{(1-4x)^5}$

(d) $2\cos(3x)$ (e) $\sin\left(\sqrt{x^2+3}\right)$ (f) $\tan^3(\pi x)$

Use your symbolic computer system to check your result.

8. In the file **data12** there is a computer representation **H** of a set of ordered pairs giving some experimentally determined data for the Heat Capacity C_v of Potassium as a function of time T. The goal is to obtain a reasonable expression that fits this data. If you just produce the graph, you will see that it is a curve and it is not immediately obvious just what expression will fit this curve.

Experience shows that if you pick the right positive integer n and use the data to graph C_v/T as a function whose domain values are given by T^n, instead of T, then you will get a straight line.

Construct a set of ordered pairs that uses the **H** and a value for n (n a positive integer) to produce a **computer function** that will give C_v/T as a function of T^n. Your construct will work on those numbers T^n which you can get from values of T that are in the domain of the function represented by **H**. It will go from such a number to the corresponding T, apply **H** to get the value of C_v and divide by T. Finally, it will apply **interpolate** to convert your **H** to a **computer function G**.

Use your set of ordered pairs to experiment with n and graphs of the resulting **computer function** to try to find a choice of n that makesyour graph a straight line. After deciding on your choice of the parameters, write the required expression giving C_v as a function of T.

9. In the previous section, you worked with an expression called the difference quotient. For example, beginning with the freezing point function T of Activity 1, page 69, you chose a point in its domain, say 0.76 and wrote

$$\frac{T(0.76+h) - T(0.76)}{h}$$

where h is a very small number.

It turns out that in many situations, the value of h is not very critical, but it is interesting to vary the value chosen in the domain. For this activity, let us fix $h = 0.0001$. Write a **computer function** that will accept a value x in the domain of the freezing point function and return the value of the difference quotient, taken between x and $x+h$.

Graph this **computer function**.

Write down, in your own words, what this graph means. Can you say anything special about the behavior of the graph near the point 0.76?

10. To do this activity you will need to use a computer tool, like **graphR** in **ISETL** or **implicitplot** in **MapleV**, which graphs equations in two variables in $\mathcal{R}^2$. Make a graph of the portion of the curve

$$\left(\sqrt{3y^4 + 4x}\right)e^{\cos y} - x^2 \sin xy - 4 = 0$$

with horizontal interval [0, 4] and vertical interval [−6, 6]. Check with your instructor on how to represent e^x on your computer. One common notation is **exp(x)**. Depending on your computer system, this might take a few minutes, so you should set it running on one computer and do some other work while you are waiting.

Can you think of any sense in which this curve defines a function? Could there be a function, for example, whose value at 3 is about −0.9 ? Try to use this curve to set up a function (in your mind) and, by taking values from the graph, write down a few values of your function.

11. In Section 2.2, Activity 7, page 88, you developed code for approximating the change in entropy from a to b using the function which gave C_p/T as a function of T (See the Section 2.2 discussion: Area under a curve, page 98). Write a **computer function** that will approximate, for any non-negative number T the change in entropy from 1 to T. The function being approximated here is called the *entropy function*.

2.3.3 Discussion

Making New Functions from Arithmetic   What you did in Activity 1, was to combine the viscosity function f for SAE 10W oil and g for Kerosene by multiplying one by a number a, the other by another number, b and adding them to form the function $af + bg$ which gives the viscosity of the mixture as a function of temperature. Let's analyze in more detail how this works in a simpler example.

The first important thing about a function is its process. Think about the energy function for two particles in Activity 2. The energy due to attractive force is given by r^{-6}. This means that lurking somewhere in the background there is a process that takes a distance r (it must be positive — negative makes no sense and zero would mean that there is only one particle), and performs the calculation of raising its reciprocal to the 6th power. That is the process. A similar discussion applies to the repulsive force given by r^{-12}.

So you could imagine that there are two functions, A and R which give the energy due to attractive and repulsive forces respectively as a function of the distance r between two particles. We could write, therefore, that the total energy function E is the difference between these two functions, that is,

$$E = A - R$$

The process of the function E can be described as follows. It takes a distance r, and computes the energy due to attraction by running the process for the function A on it. This gives the value $A(r)$. Then it runs the process for the repulsive energy function R on the same value r to obtain $R(r)$. Finally it subtracts (because they act in opposite directions) the two results to obtain the total energy e. We summarize this entire process by writing,

$$e = A(r) - R(r)$$

This is how you construct a function which is the difference of two functions. In a completely analogous manner, you can do sums, products, quotients and linear combinations of functions. For each of these, you can construct a **computer function** that will take as input two **computer functions** representing functions and produce a **computer function** that represents the combination. Here is how it looks in **ISETL** for the difference of two functions.

```
diff := func(f,g);
    return func(x);
        return f(x) - g(x);
    end;
end;
```

Make sure you understand this **func** because there are a lot of subtle mathematical ideas contained in it. The **func** accepts any two representations **f, g** of functions. Then, there is a **return** statement which takes up three lines. These three lines use the processes of **f** and **g** to construct a new **func** whose process is the difference of the original two processes. This is what is returned for the difference. Such a program can also be implemented in **MapleV** using the **makeproc** command (see the **MapleV** program on page 80).

If you had constructed **diff** in **ISETL**, and **funcs** **A** and **R** were **funcs** representing the attractive and repulsive energies in the above example, you could get the **func** that represented total energy by writing **diff(A,R)**. Once you did that, what would be the meaning of the following expression?

diff(A, R)(64);

There is an interesting variation on notation, available in some MPLs that lets you write this in a language that is closer to standard mathematical usage. In **ISETL**, for example, by placing a period in front of the operator, you can use it between the two things it operates on. Thus, instead of **diff(A,R)** you can write **A .diff R** and the above expression can be written,

(A.diff R)(64);

Can you do something similar with your MPL? What would happen if **f** and/or **g** were not **computer functions** but some other representation of a function, for instance, a tuple or a set of ordered pairs? Suppose that **f** was a set of ordered pairs and **g** a **computer function**?

Functions as Combinations of Functions — Arithmetic

If you do something, and understand what you are doing, then it should be possible to recognize the result — and even to go backwards and see how it was done. This is what we were driving at in Activities 3 and 4. Instead of starting with two functions and combining them to form a new function as in Activities 1 and 2, you are given the resulting function in Activities 3 and 4 and asked to figure out how to combine two functions to get this result.

Look at a function f given by an expression like

$$f(x) = 1 + \frac{(x-2)^3(x-6)}{54}$$

You should have no difficulty in writing it as combination of other functions in several different ways. For example, if we define f_1, f_2 by

$$f_1(x) = 1, \quad f_2(x) = \frac{(x-2)^3(x-6)}{54}$$

then

$$f = f_1 + f_2$$

Or we might define three functions f_1, f_2, f_3 by

$$f_1(x) = 2, \quad f_2(x) = (x-2)^3, \quad f_3(x) = x-6$$

in which case we would have

$$f = \frac{1}{2}f_1 + \frac{1}{54}f_2 \cdot f_3$$

Now look at Activity 4, where you were to express the compressibility function Z given by

$$Z(v) = \begin{cases} 1 + \frac{(v-2)^3(v-6)}{54} & \text{if } 0 \le v < 6 \\ \frac{v^3}{3\times10^3} - 0.048v^2 + 1.73v - 7.724 & \text{if } 6 \le v < 90 \\ 6.432 - \frac{4v^{\frac{3}{2}}}{5\sqrt{v^3 - 7\times10^5}} - \frac{27}{v-200} & \text{if } 90 \le v \end{cases}$$

as a linear combination of the two given functions Z_1 and Z_2 given by

$$Z_1(v) = \begin{cases} 0.5 & \text{if } 0 \le v < 6 \\ \frac{v^3}{6\times10^3} - 0.024v^2 + 0.865v - 3.862 & \text{if } 6 \le v < 90 \\ -\frac{27}{2v-400} + 3.216 & \text{if } 90 \le v \end{cases}$$

$$Z_2(v) = \begin{cases} \frac{(v-2)^3(v-6)}{18} & \text{if } 0 \le v < 6 \\ 0 & \text{if } 6 \le v < 90 \\ \frac{-12v^{\frac{3}{2}}}{5\sqrt{v^3 - 7\times10^5}} & \text{if } 90 \le v \end{cases}$$

You have to begin by considering each of the branches of Z separately. The first branch is the function, say F, given by

$$F(v) = 1 + \frac{(v-2)^3(v-6)}{54}.$$

You have to write this as a linear combination of the two functions given by the first branches of the two functions Z_1 and Z_2. These branches are given by functions, say F_1, F_2 where

$$F_1(v) = 0.5, \quad F_2(v) = \frac{(v-2)^3(v-6)}{18}.$$

You must find a linear combination (a sum of constant multiples) of F_1 and F_2 that gives F and then you must try to do the same thing with the other two branches in such a way that *the same linear combination works for all three branches.*

Properties of Combinations — Arithmetic   When you form combinations of functions, there can be changes in what it means for a property to hold. For example, in Activity 2(a), you might have answered the question by saying that if a function f is equal to the difference of two functions, g and h, then what it means for a point to be a zero of f is that it is a point at which $g - h = 0$, or equivalently, a point where g and h give the same value, that is, $g(x) = h(x)$.

Sometimes, you can make deductions about a property of a combination by looking at properties of the ingredients. Here is a general statement about the kind of situation you were investigating in Activity 5. Looking over the proof will give you some practice with the use of terminology and formal definitions.

Theorem 2.1

If two functions are increasing over a region, then their sum is increasing over the same region.

Proof. First let's make up some labels. Let f and g be the two functions, so their sum can be called $f + g$. Let S be the region.

Next, we use this terminology to recall what it means for a function to be increasing (see Definition 2.2, page 93). It means that, for f, if x and y are in S and $x < y$, then $f(x) \le f(y)$. We know that the same statement is true with f replaced by g and our goal is to show that it is also true if f is replaced by $f + g$.

Thus, we must show that if x and y are in S and $x < y$ then $(f+g)(x) \le (f+g)(y)$. Well, suppose indeed that x and y are in S and $x < y$. Our hypotheses allow us to conclude that

$$f(x) \le f(y)$$

and

$$g(x) \le g(y)$$

Now think for a moment. What do we have to show? It is that $(f+g)(x) \le (f+g)(y)$. We can expand each of these two terms,

$$(f+g)(x) = f(x) + g(x)$$

$$(f+g)(y) = f(y) + g(y)$$

so if we add the two inequalities that we know are true, we may conclude that

$$(f+g)(x) = f(x)+g(x) \le f(y)+g(y) = (f+g)(y)$$

which is just what we wanted.

$\square$

On the other hand, it is sometimes the case that a property satisfied by functions will be lost in the combination. Consider for example, the statement,

> *If two functions are decreasing over a region, then their product is also decreasing over that region.*

This statement has the same form as the one which we just proved but, nevertheless, it is not always true — so we say that it is false. Consider for example the two functions f, g given by,

$$f(x) = -x, \; g(x) = -x^2$$

and let us take the region S to be the set of positive numbers. Then both of these functions are decreasing on S. But what about the product $f \cdot g$? It is given by

$$(f \cdot g)(x) = x^3.$$

Is that decreasing on S? Do you see why the proof that we made for the theorem will not work in this case?

Making New Functions from Composition

In Activity 6, you formed a function W by composing two functions, Z and V. Let's see how we can understand the process of this new function. The process for V is given by the formula,

$$V(t) = \begin{cases} 1.26t^2 & \text{if } 0 \le t < 10 \\ (4/3)\pi(t-7)^3 & \text{if } 10 \le t \end{cases}$$

and the process for Z is given by,

$$Z(v) = \begin{cases} 1 + \frac{(v-2)^3(v-6)}{54} & \text{if } 0 \le v < 6 \\ \frac{v^3}{3 \times 10^3} - 0.048v^2 + 1.73v - 7.724 & \text{if } 6 \le v < 90 \\ 6.432 - \frac{4v^{\frac{3}{2}}}{5\sqrt{v^3 - 7 \times 10^5}} - \frac{27}{v-200} & \text{if } 90 \le v \end{cases}$$

To form the process for W we proceed as follows.

> We start with a given value t in the domain of V, that is, $t \ge 0$. Then we apply the function V. This means that we must first check whether $t < 10$ or $t \ge 10$ and then plug the value for t into the appropriate formula and calculate the answer, $v = V(t)$. Next, we apply Z to this answer. This means that we must first check that it is in the domain of Z, that is, that $v \ge 0$. If so, then we must check which of the inequalities, $v < 6$, $6 \le v < 90$ or $v \ge 90$ is satisfied for this $v = V(t)$. This tells us which of the three expressions to put the value $V(t)$ for v in the definition of Z. We then calculate with this value and the result is the value of $W(t)$.

A general formula for $W(t)$ can be obtained by thinking along these lines. You will have an opportunity to work it out in the exercises.

In Activity 6, you did those calculations several times by hand. It is a lot of thinking and a lot of calculating. All of the thinking is indicated by the simple expression

$$W(t) = Z(V(t))$$

and in your MPL, all of the calculation will be done (provided the **computer functions Z** and **V** have been defined) if you enter something like the corresponding expression

Z(V(t));

Of course, the whole business is a lot simpler if your functions are given by single expressions. In this case, it should be easy for you to substitute in formulas (see Chapter 1, p. 7). You can even do something a little harder like in Activity 7, where you start with the "answer", a composite function, and try to find two functions whose composition will be the one you started with. Being able to do this will be very helpful for you when you come to the *chain rule for differentiation* in Chapter 4. It can also help in understanding how graphs are simplified by transformations. We will look into all of this in the next three paragraphs.

There is a standard notation for a function constructed as the composition of two functions.

Definition 2.10

For two functions g and h, the composition of h with g is a function f defined by the following relationship

$$f(x) = h(g(x))$$

where the domain of f is the set of all x in the domain of g for which $g(x)$ is in the domain of h. We write $f = h \circ g$.

Functions as Combinations of Functions — Composition
If you understand how to form the composition of two functions to construct a new function, then you should be able to reverse the process — that is, you should be able to look at a function and see how to write it as the composition of two functions, or even three functions. This is the same kind of thing as what you were doing a couple of paragraphs above on page 112 with arithmetic. It is exactly what you had to do in Activity 7. This little skill will be very useful to you when you begin to learn how to compute derivatives in Chapter 4.

Consider the function f given by

$$f(x) = \left(1 - x^2\right)^{3.2}$$

We can write this as the composition of two functions, g and h. Here is one way of doing it.

$$g(x) = 1 - x^2, h(x) = x^{3.2}$$

so that

$$f = h \circ g$$

Do you see that with these definitions, it is the case that $f(x) = h(g(x))$? Is this the only way to write f as the composition of two functions?

We can even write f as the composition of three functions. Define g, h, and k by

$$g(x) = \sqrt{x}, h(x) = x^{3.2}, k(x) = 1 - x^4$$

Do see that it then follows that

$$f = h \circ (k \circ g)?$$

Of course there are issues of domains and co-domains here. Can you explain the various domain restrictions that are implicit in the above discussion?

Application of Composition — Translations of Graphs
In Chapter 1, Section 3, Activity 3 (p. 16) you saw how replacing the domain variable, say x, in a function f by a linear expression such as $ax+b$ (where $a \neq 0$ and b are real numbers) effected the graph of f. Multiplication by a has the effect of expanding or contracting (depending on whether $0 < |a| < 1$, or $|a| > 1$) horizontally and addition of the term b has the effect of shifting the graph b/a units left, say if $b > 0$ and $a > 0$, or $|b/a|$ units right, say if $b < 0$ and $a > 0$. Do you see why? Do you see what happens if $b > 0$ and $a < 0$, or $b < 0$ and $a > 0$? If $a < 0$, what happens to the graph of f? Do you see that it is reflected about the y-axis? What about the other sign combinations of a and b? Such transformations effect the domain of the function, whereas a transformation like $af(x)+b$ effects the co-domain of the function. Do you see how?

Consider the function f whose domain is the set D of all non-negative numbers and is given by $f(x) = \sqrt{x}$. Figure 2.11 shows a picture of the graph of f.

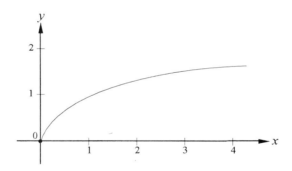

Figure 2.11. A graph of $f(x) = \sqrt{x}$ for $0 \leq x \leq 4$.

If we replace x by $3.7x - 2.5$ we obtain the function g given by

$$g(x) = f(3.7x - 2.5) = \sqrt{3.7x - 2.5}$$

and its domain is the set $E = \left\{ x \in \mathcal{R}^1 | x \geq 0.676 \right\}$ which can be found by solving the inequality $3.7x - 2.5 \geq 0$, since the domain of the square root function is all real numbers ≥ 0. Note that the set builder notation for the set E is read "the set of x in $\mathcal{R}^1$ such that $x \geq 0.676$."

The graph can be obtained by "squeezing" the shape by a factor of 3.7 and then translating it 0.676 units to the right (that is, -0.676 units to the left.) See Figure 2.12.

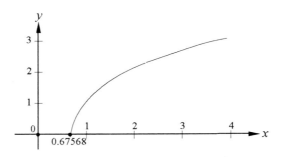

Figure 2.12. *A graph of* $g(x) = \sqrt{3.7x - 2.5}$.

We can interpret this transformation by a composition. If we define the function h by $h(x) = 3.7x - 2.5$ and f by $f(x) = \sqrt{x}$, then we have $g = f \circ h$. Here is how we could have reasoned if the original problem had been to find the graph of g.

> Since g is given by $g(x) = \sqrt{3.7x - 2.5}$, define f and h by $f(x) = \sqrt{x}$ and $h(x) = 3.7x - 2.5$. Then $g = f \circ h$. We know the graph of f (at least it is simpler to figure out than the graph of g) so we can draw that and just perform the changes indicated by composing f with h.

Application of Combinations — A Case Study

We can use composition to explain what is happening in Activity 8. In this situation, there is a function, let us call it H which gives the heat capacity C_v of Potassium as a function of time, T. We don't have complete information about H, but only some data giving several values of $C_v = H(T)$. The problem is to find a reasonable expression for H.

The key points to interpret are contained in the second paragraph of Activity 8. It says, first of all that we should try to graph, not H, but the function obtained by dividing H by the function J where $J(T) = T$. Let's call this K so we have,

$$K = \frac{H}{J}, \ H = K \cdot J,$$

The second thing it says is that, with the positive integer n to be determined, we get a straight line if we use the data to graph C_v/T as a function whose domain values are given by T^n. This means that if we define the function F by $F(T) = T^n$ then there is a *linear* function G such that $K = G \circ F$. The reason this is useful is that we can decide with our eye whether G is a straight line or not and get the parameters of the equation of the line by reading off the graph.

In Activity 8, it turns out that you get something close to a straight line if you choose $n = 2$. This straight line is the graph of the function G. Reading points off the graph you will see that G is given by something like $G(S) = 2.08 - 2.57S$. Hence we may calculate,

$$H(T) = (K \cdot J)(T) = ((G \circ F) \cdot J)(T) = TG(F(T))$$
$$= T \cdot G(T^2) = T(2.08 - 2.57T^2) = 2.08T - 2.57T^3$$

You might try comparing the graph of this expression for H with the original set of ordered pairs **H** (with some graphing utilities, you can graph them on a single window) to see if the result appears to be reasonable.

Preservation of Properties — Rates of Change

Some properties of functions, like monotonicity that we discussed above, are really questions that have yes or no answers. Others, like rate of change, provide numerical information. Look again at the viscosity graphs in Figure 2.3, page 87. Suppose you had a mixture of equal parts Castor Oil and SAE 10W-30 Oil and you wanted to estimate the average rate of change of viscosity with respect to temperature as the temperature went from 75 to 80.

One of the things you had a chance to figure out in Section 2.2 is that you can estimate the average rate of change over an interval when a function is represented by a graph by drawing a secant and measuring its slope. This is what we were referring to on page 96. If we make such an estimate with Castor Oil, for example, calling the function C and the temperature T, we get (measuring values off of the graph),

$$\frac{C(80) - C(75)}{80 - 75} = \frac{C(80) - C(75)}{5} \approx \frac{0.016 - 0.038}{5} = -0.0044$$

Why is this answer negative?

We can make the same calculation for SAE 10W-30 Oil. Let's call the function S so we would calculate,

$$\frac{S(80) - S(75)}{80 - 75} = \frac{S(80) - S(75)}{5} \approx \frac{0.03 - 0.02}{5} = -0.002$$

Here is an interesting side question. Can you relate the fact that the second number has a larger magnitude than the first to what you see on the graph?

But let's return to our main question. We want to make a mixture of equal parts of these two oils and estimate the rate of change of the viscosity of the mixture with respect to temperature over the same interval from 75 to 80, making the same assumption as in Activity 1. First, it is necessary to determine the viscosity function for the mixture. We don't want to use the sum of the two functions because that would mean that the viscosities simply add. A more reasonable function would be a proportional sum. Since we are using equal parts of the two oils, perhaps the best function to work with for the mixture is

$$M = \frac{1}{2}C + \frac{1}{2}S.$$

Now, to estimate the rate of change of M we could do things like in Activity 1, to get a reasonable approximation of the graph of M and then make the same calculation. Actually, a little formula manipulation reveals that we already have made all the estimates necessary and only simple arithmetic is needed. In fact, we wish to compute,

$$
\begin{aligned}
\frac{M(80) - M(75)}{80 - 75} &= \frac{\left(\frac{1}{2}C + \frac{1}{2}S\right)(80) - \left(\frac{1}{2}C + \frac{1}{2}S\right)(75)}{5} \\
&= \frac{\frac{1}{2}C(80) + \frac{1}{2}S(80) - \left(\frac{1}{2}C(75) + \frac{1}{2}S(75)\right)}{5} \\
&= \frac{\frac{1}{2}C(80) - \frac{1}{2}C(75) + \frac{1}{2}S(80) - \frac{1}{2}S(75)}{5} \\
&= \frac{1}{2}\frac{C(80) - C(75)}{5} + \frac{1}{2}\frac{S(80) - S(75)}{5} \\
&\approx \frac{1}{2}(-0.0044) + \frac{1}{2}(-0.002) \\
&= -0.0032.
\end{aligned}
$$

In other words we conclude that the average rate of change of a linear combination of two functions is the same linear combination of the average rates of change of the two functions.

Do you think this is the case for all combinations of functions? What if we had the sum, or the difference of two functions? How about the product, or the composition of two functions. Instead of working this out here, we will put these questions over to the exercises. There you will construct a computer tool that will help you compute average rates of change of functions very quickly.

Implicitly Defined Functions   The curve you worked with in Activity 10 can be used to set up a function. Actually, there are many functions which can be defined using this curve. We chose a particularly complicated example for this activity, but now let's take one which is a little simpler. Figure 2.13 shows the graph of the curve $\left(x^2 + y^2\right)^2 - 4xy = 0$. That is, all pairs (x,y) which satisfy this equation are plotted. This is what the tool **graphR** is supposed to do.

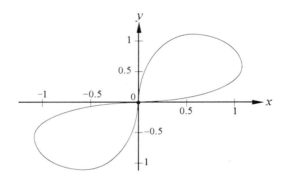

Figure 2.13. Graph of the curve $\left(x^2 + y^2\right)^2 - 4xy = 0$.

Now, how can we use this curve to define a function? If we wish the horizontal axis to represent the domain variable, then we can try to evaluate the function at a point, say $x = 0.8$ by going to that point at the x-axis and moving up with a vertical line until it hits the curve.

But the vertical line hits the curve twice, at about $y = 0.15$ and $y = 1.12$. You might say that this curve fails the vertical line test! Nevertheless, we can still get a function by restricting ourselves to a "branch".

Suppose that we decide that the domain of our function will be the interval [0.6, 0.9] and that the co-domain will be the interval [0, 0.5]. Then the only choice for the value of the function at 0.8 is the smaller one, 0.15 (approximately).

In this way, by cutting down to a branch, we can construct a function — at least in our mind. A function defined in this way is called an *implicitly defined function*. Do you see what the value of this function will be at 0.65? What about the value at 1 or 1.5?

In a situation like this, there is more than one way to define the function. Do you see how we could get a completely different function out of this which still has the domain [0.6, 0.9]?

Now here is an interesting question. How much work would it be to compute the slope of the tangent line at various points on this curve? Could you do it at all? Try. You will see in Chapter 4 that, using Calculus, it is much easier than you might think.

As a final point, compare implicitly defined functions with the situation on p. 77.

Making New Functions From Difference Quotients

Consider now, for a given function f, the difference quotient or average rate of change (by now you should realize that they are the same thing) between the two domain values x and $x+h$. The expression is

$$\frac{f(x+h)-f(x)}{h}$$

Notice that we have here two variables. Do you recall that you can think of fixing one of these variables and thereby create a function of one variable? We have a choice of fixing x and considering this to be a function of the single variable h or fixing h and considering it to be a function of the single variable x. In this situation, each alternative has an interpretation.

If we fix x then we are focusing our attention on a single point of the graph of the function f and the expression gives the slope of the secant connecting the points at x and $x+h$. We can also interpret this as the average rate of change of f with respect to x over the region from x to $x+h$.

If, on the other hand, we fix h, one can say that we are looking at the variation of the average rate of change, or the slope of the secant as x varies through the domain of f. One interesting thing to do is to fix h at a very small value. This means our expression approximates the slope of the tangent to the curve at x or the instantaneous rate of change of f with respect to x as a function of x. Indeed, after we study limits in Chapter 3 we will even be able to remove the "approximation" and speak of the exact value of the slope of the tangent or the instantaneous rate of change.

This is what you did in Activity 9. Reviewing that activity in general, you take a function F and use it to construct a new function dq given by,

$$dq(x)=\frac{F(x+0.0001)-F(x)}{0.0001}$$

so that, starting with F we construct a new function, called the *difference quotient*.

This second alternative is very important in Calculus because what we are actually approximating is the *derivative* of the function f at x. We will introduce this concept more formally in Chapter 4.

There are some questions that arise here that will have to be dealt with sooner or later. Notice that we do all this with the fixed value of $h=0.0001$. What if we made h smaller? Or negative? Does anything depend on the size or sign of h? Is there any way that all effects of h can be eliminated? Try to think about these questions in light of the examples of functions connected with freezing points, viscosity and compressibility.

Making New Functions From Approximate Area

One more way of making new functions from old is very similar in spirit to what we did with difference quotients in the previous paragraph although the details are completely different. This is the idea of area under a curve.

With difference quotients, we took the variation of h and more or less ignored it (or deferred considering it) to obtain a function of one variable. In the case of area under the curve, we again have approximations. One must first select a partition of an interval $[a, b]$. This is written

$$x=(x_i)_i=\left(a+\frac{(b-a)}{n}(i-1)\right)_{i=1}^{n+1}$$

Then the area (see Figure 2.13) under the graph of f from $x=a$ to $x=b$ is approximated by

$$\sum_{i=1}^{n}f(x_i)(x_{i+1}-x_i)=f(x_1)(x_2-x_1)+f(x_2)(x_3-x_2)+\ldots+f(x_n)(x_{n+1}-x_n).$$

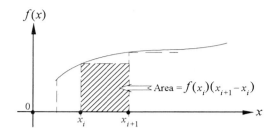

Figure 2.14. An approximation to the area under the graph of f.

(Recall Section 2.2, Activity 7, page 88, where we considered MPL code to compute this expression.)

For now, once more, there are some issues that we will defer until Chapter 5. In analogy with the value of h for difference quotients, we will have to deal with how the approximation depends on the choice of the partition. Suppose that n gets very large. Do all of the subdivisions have to be the same length?

We will ignore this and just think about the entire process of making the computation and obtaining a number which approximates the area. Suppose we changed b? That would give another number. In fact, this defines a function when interpretation of the domain variable is the upper endpoint b of the interval over which the original function is considered as a variable, say x. The process is to take the partition — that is, with a fixed, very large, positive integer n, we divide the interval $[a, x]$ into n equal subdivisions — and compute the approximation of the area from the fixed a to the variable x. See Figure 2.15.

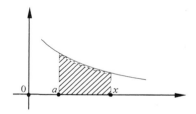

Figure 2.15. An approximation to the area under the graph of f from a to x.

If the original function was the one which gave Heat Capacity divided by temperature, C_p/T of Activity 2, Section 2.1, page 70, then the process we just described is what you were supposed to implement as a **computer function** in Activity 11.

Preview of Coming Attractions   *This paragraph is only for the daring. It is for those of you who are excited by ideas and want a chance to think about them and maybe get some of your own ideas rather than just be spoon-fed fully baked ideas of other people. Don't read further in this paragraph if you think that every question can be answered in 5 minutes — or in a week. Don't read on if you can't stand to live with a problem for a long time without knowing the answer, or even if there is an answer. For the rest of you,...*

Think about the following in connection with the last two paragraphs. What do you think would happen if you started with a function, constructed a new function by approximating the area under its graph from a fixed point a in its domain to a variable point and then took *that* function and formed a function that was an approximation to its instantaneous rate of change? What do you get? Could you write an MPL expression that implements these steps?

Now try it another way. Take your same function, form a function that is an approximation to its instantaneous rate of change and then take *that* function and approximate the area under its graph from a fixed point *a* in its domain to a variable point? What do you get? How does this compare with the first one? Again, could you do it in your MPL?

2.3.4 Exercises

1. What is the difference, if any, between each of the following pairs of expressions? Explain your answer.

   (a) $f(x) + g(x)$ and $(f + g)(x)$

   (b) $f(x)g(x)$ and $fg(x)$

   (c) $f(x)g(x)$ and $f \circ g(x)$

   (d) $\frac{f(x)}{g(x)}$ and $\left(\frac{f}{g}\right)(x)$

   (e) $f(f(x))$ and $f^2(x)$

   (f) $f(g(x))$ and $f \circ g(x)$

   (g) $f(g(x))$ and $g \circ f(x)$

2. Below are the graphs of three functions *f*, *g* and *h* which represent curves giving the Maxwell-Boltzmann distributions of molecular velocities for two chemicals and a mixture of the chemicals. For each of the following combinations of functions, determine which function *f*, *g* or *h* it coincides with.

   (a) $f - g$ (b) $f - h$ (c) $g + h$

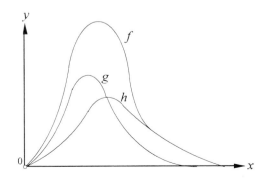

3. For each pair of functions *f* and *g* given below, find and simplify the following combinations of the functions. What is the domain of the resulting function? How does the domain of the simplified combination compare with the domains of *f* and *g*?

   (a) $(f + g)(x) = f(x) + g(x)$

   (b) $(f - g)(x) = f(x) - g(x)$

   (c) $fg(x) = f(x)g(x)$

   (d) $\left(\frac{f}{g}\right)(x) = \frac{f(x)}{g(x)}$

   (e) $f^2(x) = (f(x))^2$

   (f) $f \circ f(x) = f(f(x))$

   (g) $g^2(x) = (g(x))^2$

   (h) $g \circ g(x) = g(g(x))$

   (i) $f \circ g(x) = f(g(x))$

   (j) $g \circ f(x) = g(f(x))$

   Here are the functions.

   i. $f(x) = 3x^2 - x$ and
        $g(x) = 1 + x$

   ii. $f(x) = \sqrt{x^2 - 3}$ and
        $g(x) = 2x - 1$

   iii. $f(x) = 3|1 + 2x| - 1$ and
        $g(x) = 2 + 3x^2$

   iv. $f(x) = \begin{cases} 1 - x & \text{if } x > 0 \\ 1 & \text{if } x \le 0 \end{cases}$ and
        $g(x) = \begin{cases} 1 - x & \text{if } x > 1 \\ x - 1 & \text{if } x \le 1 \end{cases}$

v. $f(x) = \begin{cases} x^2 & \text{if } x < 1 \\ 2x - 1 & \text{if } x \geq 1 \end{cases}$ and

$g(x) = \begin{cases} x & \text{if } x > 0 \\ 3 - x & \text{if } x \leq 0 \end{cases}$

4. For each pair of functions f and g, use your graphing utility to make a sketch of the functions on the same graph. Then draw by hand the indicated combinations of functions. Check your hand drawn sketch using the computer. Hand in both your hand sketches and the sketches you made using the computer.

(a) $f(x) = x^2 + 1$ and $g(x) = 1 + x$

(b) $f(x) = \sqrt{x - 1}$ and $g(x) = 2x - 1$

(c) $f(x) = |1 + x| + 2$ and $g(x) = 1 - x^2$

(d) $f(x) = \begin{cases} 1 - x & \text{if } x > 0 \\ 1 & \text{if } x \leq 0 \end{cases}$ and

$g(x) = \begin{cases} 1 - x & \text{if } x > 1 \\ x - 1 & \text{if } x \leq 1 \end{cases}$

(e) $f(x) = \begin{cases} x^2 & \text{if } x < 1 \\ 2x - 1 & \text{if } x \geq 1 \end{cases}$ and

$g(x) = \begin{cases} x & \text{if } x > 0 \\ 3 - x & \text{if } x \leq 0 \end{cases}$

Here are the combinations.

  i. $(f + g)(x) = f(x) + g(x)$

  ii. $(f - g)(x) = f(x) - g(x)$

  iii. $fg(x) = f(x)g(x)$

  iv. $\left(\frac{f}{g}\right)(x) = \frac{f(x)}{g(x)}$

  v. $f^2(x) = (f(x))^2$

  vi. $f \circ f(x) = f(f(x))$

  vii. $g^2(x) = (g(x))^2$

  viii. $g \circ g(x) = g(g(x))$

  ix. $f \circ g(x) = f(g(x))$

  x. $g \circ f(x) = g(f(x))$

5. For two functions f and g represented as **computer functions**, write **computer functions** to implement each of the following. Be sure to put in an appropriate test for the domain if necessary. Test your **computer function** on the indicated function(s) and the specified values of x. Check your results by hand.

(a) $(f - g)(x) = f(x) - g(x)$;
     $f(x) = x^2 + 1$; $g(x) = 3x + 1$;
     $x = -1, 2$

(b) $fg(x) = f(x)g(x)$; $f(x) = x^2 + 1$;
     $g(x) = 3x + 1$; $x = -1, 2$

(c) $\left(\frac{f}{g}\right)(x) = \frac{f(x)}{g(x)}$; $f(x) = x^2 + 1$;
     $g(x) = x + 1$; $x = -3, -1, 2$

(d) $f^2(x) = (f(x))^2$; $f(x) = \sqrt{2x + 1}$;
     $x = -1, 2, 4$

(e) $f \circ f(x) = f(f(x))$;
     $f(x) = \sqrt{2x + 1}$; $x = -1, 2, 4$

(f) $f \circ g(x) = f(g(x))$; $f(x) = x^2 + 1$;
     $g(x) = 3x + 1$; $x = -1, 2$

(g) $g \circ f(x) = g(f(x))$; $f(x) = 3 - x^2$;
     $g(x) = \sqrt{3x + 1}$; $x = -1, 1, 2$

(h) $f \circ g(x) = f(g(x))$;
     $f(x) = \begin{cases} 1 - x & \text{if } x > 0 \\ 1 & \text{if } x \leq 0 \end{cases}$;
     $g(x) = \begin{cases} 2 - x & \text{if } x < 2 \\ x - 2 & \text{if } x > 2 \end{cases}$; $x = -1, 2, 1, 3$

6. For the function f given by $f(x) = x^2 - x$, find an expression for the function g which satisfies $g(x) = f(x + 1) - f(-x)$ for all real numbers x.

7. For a function f and all real numbers x,

$$f\left(\frac{1 + 3x}{2}\right) = 1 - x.$$

Find

$$f\left(\frac{2 + 5y}{3}\right).$$

8. For $t > 1$ and a function f find an expression for $f(t)$ if

$$f\left(\frac{1}{z+3}\right) = \frac{1}{2-5z}.$$

9. Find $f(f(2/x))$ and simplify your answer for the function f given by

$$f(x) = \frac{3x-4}{2x-1}.$$

10. For all $x \neq 3$, a function f satisfies

$$f(x+2) = \frac{1}{3-x}.$$

Find

$$f\left(\frac{1}{3-x}\right).$$

11. For a function f, $f(x) = 2f(x-1) + f(x-2)$ for all $x > 2$ and $f(1) = f(2) = 2$. Find $f(5)$.

12. For functions f, g, h and k and two real numbers a and b,

$$f(x) = (x^2 - 4)g(x) + ax + b$$
$$= (x-2)k(x)$$
$$= (x+2)h(x) + 7$$

for all real numbers x. Determine the values of a and b.

13. In this exercise you are to construct an MPL tool that will produce difference quotients of functions in general. You are to write a **computer function ad** that will accept any computer representation of a function f and return a **computer function** that gives, for any value x in the domain of the function, the value of the difference quotient (using $h = 0.0001$)

$$dq(x) = \frac{f(x+h) - f(x)}{h}.$$

(a) What does **ad(f)** mean?

(b) What does **ad(f)(x)** mean?

(c) Apply the **computer function ad** to each of the following functions and produce a complete graph of both the original function f and the function represented by **ad(f)** on the same coordinate axes.

   i. $f(x) = x^2$

   ii. $f(x) = x^3$

   iii. $f(x) = 1/x$

   iv. $f(x) = |x|$

   v. $f(x) = 5x^3 + 2x^2 - x + 4$.

   vi. $f(x) = |2x - 1| + 1$.

   vii. $f(x) = \sin(x)$

   viii. $f(x) = \cos(x)$

   ix. $f(x) = \tan(x)$

   x. $f(x) = \begin{cases} 1-x & \text{if } x > 0 \\ 1 & \text{if } x \leq 0 \end{cases}$

   xi. $f(x) = \begin{cases} x^2 & \text{if } x < 1 \\ 2x-1 & \text{if } x \geq 1 \end{cases}$

   xii. $g(x) = \begin{cases} x & \text{if } x > 0 \\ 3-x & \text{if } x \leq 0 \end{cases}$

   xiii. $g(x) = \begin{cases} 1-x & \text{if } x \leq 1 \\ x-1 & \text{if } x > 1 \end{cases}$

   xiv. $f(x) = \begin{cases} 0.75(x+1)^2 & \text{if } x < -3 \\ |x| & \text{if } |x| \leq 3 \\ 3 & \text{if } x > 3 \end{cases}$

   xv. The freezing point function in Section 2.1, Activity 1, page 69.

   xvi. The set of ordered pairs you constructed in Section 2.2, Activity 2, page 86, representing the viscosity function v of SAE 10W.

xvii. The compressibility function Z on page 75.

14. For the **computer function ad** you constructed in Exercise 13, suppose that Z is the compressibility function, page 75. Explain in your own words what is meant by the following expressions.

    (a) **ad(Z)**   (b) **ad(Z)(4)**

15. Decide whether or not the following statements are true or false. If true, give a proof of the statement. If false, give a counterexample, that is, an example which shows it is false.

    (a) If two functions are monotonic over a region, then their sum is also monotonic over that region.

    (b) If two functions are increasing over a region, then their difference is also increasing over that region.

    (c) If two functions are monotonic over a region, then their difference is also monotonic over that region.

    (d) If two functions are increasing over a region, then their product is also increasing over that region.

    (e) If two functions are increasing over a region, then their quotient is also increasing over that region.

16. Find an expression for a function f so that $h = f \circ g$ where g and h are functions given by

    $$g(a) = 2\cos^3(a) \quad \text{and} \quad h(b) = \sqrt{1 + \cos^6(b)}.$$

    Explain why your expression for f works.

17. Find an expression for a function g so that $h = f \circ g$ where f and h are functions given by

    $$f(z) = 2z - 1 \quad \text{and} \quad h(r) = 6r + 7.$$

    Explain why your expression for g works.

18. Suppose that the functions A, B, C are given by the following expressions.

    $$A(x) = \frac{1}{x^2}$$
    $$B(x) = \log(1 + x)$$
    $$C(x) = \log(x^2 + 1) - \log(x^2)$$

    Find all correct formulas of the form $H = F \circ G$ using A, B, and C for H, F, and G in all possible orders.

19. Find two functions f and g so that $f \circ g \neq g \circ f$. Explain why your functions satisfy this situation.

20. For the situation in Activity 2, page 107, there were functions A and R representing the attractive and repulsive energies, respectively. The function representing the total energy can be found by writing **diff(A,R)** using the computer function **diff** on page 110. What is the meaning of the expression,

    diff(A,R)(64);?

    What would happen if **A** and/or **R** were not **computer functions**, but some other representation of a function, for instance, a tuple or a set of ordered pairs? In particular, what would happen if **A** was a set of ordered pairs and **R** was a **computer function**?

21. Consider the two functions f, g given by,

    $$f(x) = -x, \; g(x) = -x^2$$

    and take the region S to be the set of positive numbers. Then both of these functions are decreasing on S. But what about the product $f \cdot g$? It is given by

    $$(f \cdot g)(x) = x^3.$$

    Is that decreasing on S? Explain why the proof for Theorem 2.1, page 112, will not work in this case?

22. In Activity 6, page 108, you formed a function W by composing two functions, Z and V. Using expressions, find a general formula for $W(t)$. (Hint: The formula will be one for a function defined in parts.)

23. Consider the function f given by

$$f(x) = \left(1 - x^2\right)^{3.2}$$

    (a) $f = h \circ g$ where g and h are functions given by $g(x) = 1 - x^2$ and $h(x) = x^{3.2}$. Show that $f(x) = h(g(x))$? Is this the only way to write f as the composition of two functions? Explain.

    (b) $f = h \circ (k \circ g)$ where $g(x) = \sqrt{|x|}$, $h(x) = x^{3.2}$ and $k(x) = 1 - x^4$. Show that $f(x) = h(k(g(x)))$. Is this the only way to write f as the composition of three functions? Explain.

24. The file **data13** contains a set of ordered pairs which gives data for the Heat Capacity C_p of methane as a function of temperature T at low temperatures (the temperature units are K - absolute).

    (a) Make a computer graph.

    (b) Use **interpolate** to obtain a **computer function** to represent this function and plot it. Explain what additional information about the curve this gives you.

    (c) What do you think is happening with Methane at temperature about 20 K?

    (d) How would you explain the fact that, except for a drop at about 20 K, the curve appears to be rising?

    (e) What do you think is happening with Methane at temperature about 90 K?

25. Write a general **computer function** that will approximate the area under a curve with variable endpoint, x, as discussed on page 120. See Figure 2.15.

26. In the discussion of implicitly defined functions (p. 118), consider the example $\left(x^2 + y^2\right)^2 - 4xy = 0$ which was discussed there. Use this curve to define three different functions, all different from the one that was discussed on p. 118. Pick two domain values for each function and figure out an approximation to the value of your function for these values. For example, you could use your graphing facility to locate points on the curve.

27. Repeat the previous exercise for the curve,

$$x^3 + y^3 = 4.5xy$$

28. Write **computer functions** that will accept any computer representation of a function f, or two functions f and g, and return a **computer function** that gives, for any value x in the domain of the function, the value of the indicated combination of functions. Test your **computer functions** on the functions given by $f(x) = x^2$ and $g(x) = x^3$ for $x = -1, 0, 1, 2$ and compare the results to the answers you get by hand.

    (a) $f + 3$ (b) $f + g$

    (c) $f - g$ (d) $3f - 2g$

    (e) fg (f) f / g

    (g) $f \circ g$

29. Show that the two definitions of function given on page 98 are equivalent.

30. In this exercise you are to use the **computer function ad** from Exercise 13. Investigate **ad** acting on functions, combinations of the same functions and their graphical relationships. Use the following functions in your investigations.

$$f(x) = 1, \quad g(x) = x, \quad h(x) = x^2, \quad \text{and}$$
$$k(x) = x^3$$

Sketch each set on the same graphics window. Apply the computer function ad to each of the following sets of functions.

   (a) $f, 2f, -5f$ (b) $g, 2g, -3g$

   (c) $h, 2h, -3h$ (d) $k, 2k, -k$

   (e) $g, h, g+h$ (f) $h, k, h+k$

   (g) $g, h, g-h$ (h) $h, k, k-h$

   (i) $g, h, 3g-2h$ (j) $h, k, 3h+2k$

31. Based on your investigations in the previous exercise, answer the following questions (approximately).

   (a) For a constant function $f(x) = k$, what is **ad(f)**?

   (b) For a function f and constant k, how are **ad(f)** and ad of **kf** related?

   (c) For two functions f and g, how are **ad(f)**, **ad(g)** and **ad** of (**f + g**) related?

   (d) For two functions f and g, how are **ad(f)**, **ad(g)** and **ad** of (**f–g**) related?

   (e) For two functions f and g and constants a and b, how are **ad(f)**, **ad(g)** and **ad** of (**af+bg**) related?

32. Use the **computer function ad** in Exercise 13 to approximate the indicated rates in each of the following situations.

   (a) The volume v, in gallons, of water in a tank after t hours is given by

$$v = V(t) = 600\sin^2\left(\frac{\pi t}{12}\right)$$

for $0 \le t \le 6$. The rate of flow into the tank for $t = 2$. Sketch a computer graph of the functions V and $ad(V)$ on the same graph.

   (b) The position s of a car traveling along a straight highway after the driver applies the brakes is given by $s = S(t) = 30t - t^2$, where s is the number of feet from the point at the brakes are applied and t is the number of seconds after the brakes are applied. The rate at which the car is traveling at $t = 5$. Sketch a graph of the functions S and $ad(S)$ on the same graph.

2.4 Chapter Summary

Formal Definitions   In this chapter you have considered the following concepts, given by their formal definitions which should, at this point, have a rich meaning for you. In looking over these definitions, think about how you might express, informally, in your own words, what the terms mean.

1. A function f is a process which can take any element a of a set D, called the domain of f, and transform it into a unique element $f(a)$ of a set R, called the co-domain (or range) of f.

2. A function f is a set of ordered pairs (x, y) such that no two pairs have the same first element and different second elements. We write $f = \{(x, y): x \in D, y = f(x) \in R\}$, where the set D is called the domain of f and the set R is called the co-domain (or range) of f.

3. A zero of a function f is a value a in the domain of f such that $f(a) = 0$.

4. A function f is said to be increasing in a region S of its domain if

$$(x_1, x_2 \in S \text{ and } x_1 < x_2) \text{ implies } f(x_1) \le f(x_2).$$

The function is said to be strictly increasing in S if

$$(x_1, \ x_2 \ \in S \text{ and } x_1 < x_2) \text{ implies } f(x_1) < f(x_2).$$

The function is said to be decreasing in S if

$$(x_1, \ x_2 \ \in S \text{ and } x_1 < x_2) \text{ implies } f(x_1) \geq f(x_2).$$

The function is said to be strictly decreasing in S if

$$(x_1, \ x_2 \ \in S \text{ and } x_1 < x_2) \text{ implies } f(x_1) > f(x_2).$$

5. A function f is said to be monotonic in a region S of its domain if it is either increasing, strictly increasing, decreasing, or strictly decreasing in S.

6. A point $(c, f(c))$ with $c \in D$ is said to be a maximum point if

$$\text{for every } x \in D \text{ we have, } f(x) \leq f(c)$$

and the value $f(c)$ is called the maximum value of f. A point $(c, f(c))$ with $c \in D$ is said to be a minimum point if

$$\text{for every } x \in D \text{ we have, } f(x) \geq f(c)$$

and the value $f(c)$ is called the minimum value of f.

7. A point $(c, f(c))$ with $c \in D$ is said to be a relative maximum point if there is an open interval (a, b) containing c and contained in D such that

$$\text{for every } x \in (a, b) \text{ we have, } f(x) \leq f(c)$$

and the value $f(c)$ is called a relative maximum value of f. A point $(c, f(c))$ with $c \in D$ is said to be a relative minimum point if there is an open interval (a, b) containing c and contained in D such that

$$\text{for every } x \in (a, b) \text{ we have, } f(x) \geq f(c)$$

and the value $f(c)$ is called a relative minimum value of f.

8. A point $(c, f(c))$ which is either a maximum, a minimum, a relative maximum, or a relative minimum point for a function f is said to be an extreme point of f. The extreme points of a function are referred to collectively as its extrema. For emphasis, a maximum or minimum is called an absolute maximum or minimum — to distinguish it from a relative maximum or minimum.

9. An upper bound for f on S is a number B such that

$$\text{for every } x \in S \text{ we have, } f(x) \leq B.$$

A lower bound for f on S is a number b such that

$$\text{for every } x \in S \text{ we have, } f(x) \geq b.$$

10. A bound for f on S is a number M such that

for every $x \in S$ we have, $|f(x)| \le M$.

Formulas for Combinations of Functions

Let f and g be two functions. In order to form various combinations of these two functions, there sometimes are restrictions on the domains and/or co-domains of the functions. In cases where all required conditions are satisfied, the following formulas hold.

addition. The sum of the two functions f and g is written $f + g$ and it is given by,

$$(f + g)(x) = f(x) + g(x).$$

subtraction. The difference of the two functions f and g is written $f - g$ and it is given by,

$$(f - g)(x) = f(x) - g(x).$$

multiplication. The product of the two functions f and g is written fg and it is given by,

$$(fg)(x) = f(x)g(x).$$

division. The quotient of the two functions f and g is written $\frac{f}{g}$ and it is given by,

$$\left(\frac{f}{g}\right)(x) = \frac{f(x)}{g(x)}.$$

composition. The composition of the two functions f and g is written $f \circ g$ and it is given by,

$$(f \circ g)(x) = f(g(x)).$$

Constructing and Representing Functions

All of these concepts relate, in one way or another, to the function concept. A major goal in this chapter is that you have begun to develop an almost automatic reaction to certain kinds of situations by which you construct a process in your mind. This process, in which something comes in, something is done to it, and something comes out, is your conception of function.

You have studied, in this chapter a number of ways of representing a function. These are:

- expression
- **computer function**
- set of ordered pairs or table
- finite sequence or tuple
- graph

In order to think about functions, to answer questions about them, or to use them to solve problems, you have to work with one or more representations of functions. But as you do this, don't forget about the process idea: in, transform, out. The way in which the transform is seen for various representations is laid out for you in the table on p. 78. The process interpretation is often expressed in what we call function notation:

$$f(x)$$

where f is a name of a function, x is a name of something which could go in, f is the name of the process that transforms x (note the use of the symbol f for two different meanings) and $f(x)$ stands for what comes out.

It is important to realize that, whatever representation of a function you are using, this function notation is the same.

Sometimes, it is necessary or helpful to go back and forth between representations. You should know how to convert one representation of a function into another. Sometimes you do this with a computer program and sometimes you do it in your mind. In either case, interpolation is an important technique for going from representations in which values of the function are given for only a discrete, finite set of points to one in which values are given along a continuum of numbers. The simplest way of doing this is by linear interpolation in which the discrete point are connected by line segments.

The set of points that are inputs for a function are those for which the function has values, i.e., the set where a function is defined which is called the domain of the function. The set of values that can come out of the function's process is called the co-domain of the function. Sometimes the process of a function is specified by an expression. Another possibility is that it be given by several different expressions, depending on the value in the domain on which the process is to act. In the latter case we talk about a function defined in parts or a function with split domain.

Functions can be combined to make new functions. The main ways of combining functions in this chapter are through arithmetic and composition of functions. It is necessary to be able to form these combinations, but it can be even more important to be able to take a given function and see how it might have been obtained by combining simpler functions. Another important question has to do with what happens to the properties of functions when you combine them.

Yet another way to make a function is to define it implicitly. This can be done, for example, by looking at a curve and breaking off a piece on which the vertical line test is satisfied.

Finally, you got a first introduction to functions of more than one variable. These can be treated as they are, or converted to functions whose process takes a value of a single variable and transforms it to a function of one or more variables.

Futures   A number of ideas which you studied in this chapter are meant to introduce you to important notions that will be studied later in this book. These include predictions of values through limits, difference quotients, and area under a curve. In fact, these three concepts will be studied extensively in the next three chapters. The idea is that sometimes a concept has to be looked at more than once in order to understand it. At the very least, when you begin your thorough study of limits, derivatives and integrals, you may perhaps be comforted by a realization that you have seen these things before — here in Chapter 2.

Conclusion   Our hope is that, as a result of studying this chapter, you have climbed higher in your thinking about functions via your exploration of the sources of functions, various ways of representing functions, fundamental properties of functions and ways to combine and make new functions. You are now ready to begin your study of the fundamental notion of *limit* — a concept which provides the link between precalculus concepts, especially those related to functions, and the powerful problem solving concepts of the calculus. In the remaining chapters of this text, you will return to many problem situations you encountered in the first two chapters, where often you were only able to solve a problem approximately or not at all. Using the concepts of calculus, based on the limit concept, you will be able to find better approximations or even exact solutions to many problems . It's time to take you to the limit!

CHAPTER 3

LIMITS OF FUNCTIONS

3.1 LIMITS AND CONTINUITY

3.1.1 Overview

In this chapter you will study the notion of a *limit* — the fundamental link between pre-calculus and calculus concepts. You have already worked with the intuitive notion of limit in previous chapters when you were asked to consider what happened to the difference quotient

$$dq(x) = \frac{f(x+h) - f(x)}{h}$$

as h approached 0. The difference quotient has various interpretations, both geometric (as the slope of the secant line of the graph connecting the points $(x, f(x))$ and $(x+h, f(x+h))$ and physical (as the average rate of change of a function between x and $x+h$). Furthermore, *limits of difference quotients* have analogous (instantaneous) interpretations such as slopes of tangents (limits of slopes of secants) to curves and instantaneous rates of change (limits of average rates of change). Such interpretations of limits of difference quotients are the essence of *differential calculus* — one of the two broad branches of calculus.

   You will begin this chapter by developing in more detail an intuitive understanding of the limit of a function at a point. Using the idea of a limit, you will see that various concepts such as the slope of a curve at a point and instantaneous rates of change can be made precise by making appropriate limit definitions. In Section 2, you will extend the idea of the limit to include a discussion of limits at infinity and infinite limits. Then, in Section 3, you will work with the formal mathematical definition of a limit and prove some of the most important theorems concerning limits. As we proceed throughout the remainder of this text, you will see that the theory of limits provides a foundation for the theoretical aspects of the calculus.

3.1.2 Activities

1. A certain particle in motion is constrained to move in a linear track. Suppose that the function S gives its distance s from some fixed reference point as a function of time t according to the following formula.

$$s = S(t) = \frac{\sqrt{1+t^2}}{1+t}$$

We can use the idea of a limit to find its *instantaneous velocity* at time $t = 0.5$.

To do this, you must first construct a new function V which gives the *average velocity* over a time interval between 0.5 and $0.5+h$ (See Chapter 1, pages 15, 31, and 41). Thus $V(h)$ is the average velocity from $t = 0.5$ to $t = 0.5+h$. Then the instantaneous velocity is given by the limiting value of $V(h)$ "as h goes to 0".

Calculate an expression for V (you may use your symbolic computer system if you like.) Determine the limiting value in three ways.

Warning: Be careful about what happens when $h = 0$.

   (a) Use your MPL and the tool **table** to make a table of values of $V(h)$ for small values of h. Be sure to use both positive and negative values of h.

   (b) Use your graphing utility to plot the graph of V and read its values near $h = 0$.

   (c) Use your SCS to evaluate

$$\lim_{h \to 0} V(h).$$

Explain why the instantaneous velocity cannot be found simply by evaluating the function V at $h = 0$.

2. Look at the picture in Figure 3.1.

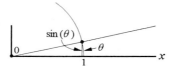

Figure 3.1. A small central angle θ in a circle of radius 1.

You may recall that the radian measure of an angle θ is the length of the arc it subtends in a circle of radius 1. Thus, in the picture, the arc has length θ and the vertical side of the triangle has length $\sin(\theta)$. It appears from the picture that when θ is small, these two lengths are not very different. Use your SCS to evaluate

$$\lim_{\theta \to 0} \frac{\sin(\theta)}{\theta}.$$

3. In Chapter 2 you worked with a function T which gave the freezing point of a mixture of two chemicals A and B as a function of the mole fraction m of chemical A. The expression for this function can be written as,

$$T(m) = \begin{cases} 5m^4 - 23.6m^3 - 146m^2 + 129m - 17 & \text{if } 0 \le m \le 0.76 \\ 45.056(m - 0.76)^{\frac{1}{3}} - 11.98132480 & \text{if } 0.76 < m \le 1 \end{cases}$$

Use your graphing utility to draw a graph of this function. Suppose you would like to compute the slope of the line which is tangent to the curve at the point $P = [0.9, T(0.9)]$.

Do this in the following two ways.

(a) Read points Q off the graph of T with Q close to P. Compute the slope of the secant line connecting P with Q for each point Q you have chosen. Do this for several choices of Q and see what you think the limiting value will be.

(b) Move a little bit away from the point P to another point Q (that is, you replace 0.9 by another number m which is slightly different from 0.9). Next, write an expression for the slope of the line connecting P and Q — as a function of m. You may want to use your SCS to do this. Finally, compute the "limiting value" of this slope as Q approaches P (that is, as m approaches 0.9).

4. To compute the numerical value of a quantity such as

$$3^{\sqrt{2}}$$

you could just enter it into any good calculator and get the answer. But what does the answer mean? You learned in pre-calculus how to compute rational roots of numbers. Thus you should be able to compute

$$3^{\frac{14}{10}} = \sqrt[10]{3^{14}} = 3^{1.4}$$

and interpret what this means in terms of integral powers and roots. Since 1.4 is an approximation to $\sqrt{2}$, the quantity $3^{1.4}$ approximates 3 raised to the power $\sqrt{2}$. Write a **computer function** that will accept two positive integers **p, q** and compute the value of 3 raised to the power $\frac{p}{q}$. Use this to approximate 3 raised to the power $\sqrt{2}$.

You might want to make use of the approximation

$$\sqrt{2} \approx 1.414213562373095049.$$

5. Which side are you on? If a function is given in parts, then it is often necessary to figure out what is happening at the "seam". If you want to get a limit at a seam point, you have to see what it is approaching from the left and from the right and then compare. For each of the following functions and point a, use your symbolic computing system to figure out the two one-sided limits and then decide if you want to say that the function has a limit at a.

(a) The freezing point function f of Activity 1, Chapter 2, Section 1 with $a = 0.76$

(b) The heat capacity function C_p of Activity 2 (not C_p/T) in Section 2.1.2, page 70, with $a = 160$.

(c) The compressibility function Z in Chapter 2, Section 1, page 75, with $a = 90$.

6. Estimate each of the following limits graphically. Then use your SCS to evaluate the limits.

(a) $\displaystyle\lim_{x \to 8} 2x^{\frac{1}{3}} - 1$

(b) $\displaystyle\lim_{x \to 8} 12x^{-\frac{2}{3}}$

(c) $\displaystyle\lim_{x \to 8}\left(2x^{\frac{1}{3}} - 1 + 12x^{-\frac{2}{3}}\right)$

(d) $\displaystyle\lim_{x \to 8}\left(2x^{\frac{1}{3}} - 1 - 12x^{-\frac{2}{3}}\right)$

(e) $\displaystyle\lim_{x \to 8}\left(2x^{\frac{1}{3}} - 1\right)\left(12x^{-\frac{2}{3}}\right)$

(f) $\displaystyle\lim_{x \to 8}\left(\frac{2x^{\frac{1}{3}} - 1}{12x^{-\frac{2}{3}}}\right)$

(g) $\lim\limits_{x \to 8} 17.36\left(2x^{\frac{1}{3}} - 1\right)$ (h) $\lim\limits_{x \to 0} \sin\left(\dfrac{1}{x}\right)$

(i) $\lim\limits_{x \to 0} x$ (j) $\lim\limits_{x \to 0} x \sin\left(\dfrac{1}{x}\right)$

(k) $\lim\limits_{x \to 3} \dfrac{x^2 - 2x - 3}{x - 3}$ (l) $\lim\limits_{x \to 3} \dfrac{x^2 - 2x - 3}{(x - 3)^2}$

Now, let your imagination run free and guess what you can about the following questions.

i. Suppose you know the limits of two functions as the independent variable approaches a certain point. What can you say about the limit of the sum of those two functions as the independent variable approaches the same point?

ii. Suppose you know the limits of two functions as the independent variable approaches a certain point. What can you say about the difference between two functions, the product, the quotient, and the product of a number times a function.

iii. Look at the two limits involving $\sin(\frac{1}{x})$ in the activities above. Can you explain, in general terms, what is going on in this situation?

iv. Suppose you are taking the limit of the quotient of two functions, and both the numerator and denominator have limits zero. What can you say about the limit of the quotient?

7. Write a paragraph or two that discusses the difference between the three functions V_1, V_2, V_3 given by the following formulas.

$$V_1(x) = \frac{3\sqrt{4x^2 + 4x + 5} - 2\sqrt{5}x - 3\sqrt{5}}{6x^2 + 9x}$$

$$V_2(x) = \begin{cases} \frac{3\sqrt{4x^2 + 4x + 5} - 2\sqrt{5}x - 3\sqrt{5}}{6x^2 + 9x} & \text{if } x \neq 0 \\ 0 & \text{if } x = 0 \end{cases}$$

$$V_3(x) = \begin{cases} \frac{3\sqrt{4x^2 + 4x + 5} - 2\sqrt{5}x - 3\sqrt{5}}{6x^2 + 9x} & \text{if } x \neq 0 \\ -\frac{4\sqrt{5}}{45} & \text{if } x = 0 \end{cases}$$

8. For a function f such that $\lim_{x \to a} f(x) = b$, what values can $f(a)$ be? Explain your answer and illustrate your explanation graphically.

9. This problem is a continuation of Activity 8 in Chapter 2, Section 2, page 88, so go back and read through that activity. You will recall that this was a problem to force the graph of a function to be within a window of specified dimensions. More specifically, relative to a particular function, you were given an interval on the vertical axis and the center of an interval on the horizontal axis. You had to find the interval on the horizontal axis to restrict the domain over which you were graphing in such a way that the curve never left the window through the top or bottom.

If you look at Activity 8, page 88, and some of the corresponding exercises in Chapter 2, Section 2, you will notice that sometimes you are able to do this and sometimes you are not. The question for you to investigate in this activity is, for a given function f and center of the horizontal interval c, for which intervals on the vertical axis is this possible?

For each of the following functions f and center c, determine the intervals $[u, v]$ for which it is possible to find an interval of the form $[c-d, c+d]$ such that the graph of f "stays in the window" defined by horizontal span $[c-d, c+d]$ and vertical span $[u,v]$.

(a) The function f given by the expression $f(x) = x^3 + 3x^2 + 13.2$, with the center of the horizontal interval 0.5.

(b) The natural exponential function $f = \exp(x) = e^x$, with the center of the horizontal interval 27.2.

(c) The characteristic function χ of the unit interval defined by

$$\chi(x) = \begin{cases} 1 & \text{if } 0 \le x \le 1 \\ 0 & \text{otherwise} \end{cases}$$

with the center of the horizontal interval 0.5.

(d) The characteristic function χ of the unit interval with center of the horizontal interval 1.0.

(e) The freezing point function from Activity 3 with the center of the horizontal interval 0.76.

(f) The Heat Capacity function of Activity 2, Chapter 2, Section 2, page 70, (C_p, not C_p/T), with the center of the horizontal interval 160.

3.1.3   Discussion

The Idea of a Limit   The assertion that "the limit of a function f at a point a is the number L" is a statement about the values of the function near a. That is, this is a statement about the numbers $f(x)$ when x is close to a. For example, in Activity 1 if you computed the average velocity of the point over the range from $t = 0.5$ to $t = 0.5 + h$, you should have constructed a function, say V, of the independent variable h whose values are given by

$$V(h) = \frac{S(t+h) - S(t)}{h} = \frac{3\sqrt{4h^2 + 4h + 5} - 2\sqrt{5}h - 3\sqrt{5}}{6h^2 + 9h}.$$

Notice that the function V doesn't have a value at $h = 0$, and yet the function V does have a limit at $h = 0$. Do you see why?

As another example, consider Activity 2. Again the expression

$$\frac{\sin(\theta)}{\theta}$$

is not defined when $\theta = 0$ (that is, it has no value at $\theta = 0$), and yet the limit of this function at $\theta = 0$ does exist.

In Activities 1, 3 and 4 you tried several ways to determine the behavior of a function near a specified point. In some cases this point was in the domain of the function, in other cases it was not.

Table.   You can construct a table of values (using the tool **table** in your MPL, for example) of the function f at points x closer and closer to a. By looking at the numbers $f(x)$, you might decide that they seem to be approaching some limit L.

Graph.   You can make a graph of the function and read off the values of $f(x)$ for x close to a and again make a determination of what the limit L seems to be.

Theory.   You used theory when you entered an expression in your SCS and calculated the limit. That is, your SCS performed calculations based on theory. This theory gives rules or algorithms for computing limits exactly. You will investigate some of these rules later in this section and revisit the theory again in the last section of this chapter.

Can the Value of the Function Equal the Limit?   Think about the limit L of a function

f at a point a. We write this in mathematics as follows.

$$\lim_{x \to a} f(x) = L$$

In our discussion above we emphasized that the limit of f at a is all about what happens to the values of $f(x)$ as x gets closer and closer to a, but not about the actual value $f(a)$. Indeed, the function may not even have a value at a. In other words, when talking about limits, we are saying that x approaches a, but never reaches it. We should emphasize here that $\lim_{x \to a} f(x)$ is <u>not</u> the same as $f(x)$ for some specific value of x that is "very close" to a, except when f is constant in an interval about a.

Now, what about the limit itself? That is, if $\lim_{x \to a} f(x) = L$, is it possible that for some x in the domain of f the function value $f(x)$ is equal to L? Think about your answer, and then look at the example, from Activity 2,

$$\lim_{\theta \to 0} \frac{\sin(\theta)}{\theta} = 1.$$

Think some more about your answer and then look at the example, from Activity 6,

$$\lim_{x \to 0} x \sin\left(\frac{1}{x}\right) = 0.$$

Do you see any reason to change your mind? See Figures 3.2 and 3.3.

Two-Sided Limits   In determining $\lim_{x \to a} f(x)$, you must consider the approach of the

independent variable from both sides of a. This means that we are making a determination of what happens to $f(x)$ as x approaches a. In the limit calculations which we have done so far, you might have been thinking that we meant that the value of x is greater than a and it is "coming down" to a. For instance, in our calculations of rates of change and slopes, you might have thought that we were usually taking h to be positive and coming down to 0. We do not mean this to be the case. It is true that often it does not really make a difference. That is, often, whatever happens in coming from one direction is the same as what happens in coming from the other direction. But *often* is not *always*. There are important situations in which you *must* consider both sides separately.

One situation in which seeing both sides of the matter is essential is the case of functions defined in parts when you are looking for the limit as the independent variable approaches the seam. Consider Activity 5 where you investigated limits of the freezing point, heat capacity, and compressibility functions. In each case, the point a was at a seam. In which cases was the limit the same from both sides? Would you say the limit exists if it is not? We would not. We are only prepared to say that a limit exists if the limit from the left exists, the limit from the right exists, and the two limits are the same.

Just to make sure you understand how to compute a limit when a function is defined in parts, look at how you did Activity 3. In part (b) you computed a formula for the function which gives the slope. Did you realize that although the freezing point function is defined in parts, if you want to compute the slope of the curve at $m = 0.9$, then you only have to use the second part? Why is it that here the second part is enough, but for the freezing point function in Activity 5 part (a) you had to use both parts? Which parts of the Heat Capacity function did you have to use in Activity 5 part (b)? What about the compressibility function in Activity 5 part (c)?

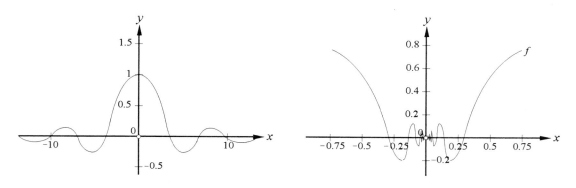

Figure 3.2 A graph of $\frac{sin(x)}{x}$.   *Figure 3.3 A graph of $x \, sin(\frac{1}{x})$.*

Using Limits to Make Definitions — Velocity

Throughout the Activities of this section, we have seen that the notion of a limit can be used to actually define other mathematical concepts. In Activity 4 you saw how to use limits to give meaning to raising 3 to a power $\sqrt{2}$ which is not a rational number. We can also use limits to precisely define physical quantities for which you might have a vague notion, but not a precise understanding.

Consider instantaneous velocity for example. What does it mean? It is not hard to say what is the meaning of average velocity over a period of time. It is the ratio of the distance traveled divided by the length of the time interval. You can't get instantaneous velocity at a point by dividing the distance by the time because both are 0 and you can't divide by 0. What you do is to take the average velocity of an interval containing that point, and then take the limit as the length of the interval goes to 0.

For example, in the situation described in Activity 1 and analyzed at the beginning of this section, the function V gives the average velocity $V(h)$ of the moving point as time ranges between 0.5 and $0.5 + h$. According to Activity 1, we then take the instantaneous velocity at this point to be $\lim_{h \to 0} V(h)$.

It is important to understand that this statement is not a determination of what instantaneous velocity is. It is more complicated. What we have done was first decide that we didn't know what instantaneous velocity really was. Then we realized that the idea of average velocity over a time interval was close to what we wanted, and that it got closer as the interval got smaller. Then we made a decision to formalize this vague notion to *give meaning* to instantaneous velocity when it did not previously have a precise meaning. The following is a definition of instantaneous velocity which depends on the concept of limit.

Definition 3.1

Let S be a function which gives the signed distance (or position) from an initial point as a function of time t for a particle moving in a straight line. The velocity or instantaneous velocity of this point at time t_0 is defined to be

$$\lim_{h \to 0} V(h)$$

where V is the average velocity function near t_0 given by,

$$V(h) = \frac{S(t_0 + h) - S(t_0)}{h}.$$

You may recall that in earlier chapters we referred to expressions like the one for V as the difference quotient. This is a very important formula in calculus.

Using Limits to Make Definitions — Rate of Change

We can find the instantaneous rate of change of any dependent variable with respect to an independent variable exactly as we did with position and time to obtain instantaneous velocity. Look at the freezing point function in Activity 3 and consider the rate of change of the freezing point f with respect to the mole fraction m of chemical A. You might want to know this rate at a certain point m_0. To compute this, you once again compute the average change over an interval from m_0 to $m_0 + h$,

$$\frac{T(m_0 + h) - T(m_0)}{h}$$

and take the limit as h goes to 0.

We formalize this procedure for any function in the following definition.

Definition 3.2

Let f be a function which gives the value of the variable y as function of x, that is, $y = f(x)$. Then the instantaneous rate of change of y with respect to x at a point x_0 in the domain of f is defined to be

$$\lim_{h \to 0} \frac{f(x_0 + h) - f(x_0)}{h}.$$

Notice that again we have the formula for the difference quotient. In many texts, you will find the symbol Δx instead of h. The symbol Δx refers to the "change in x". You will also find the symbol Δy which refers to the change in y or f, $f(x + \Delta x) - f(x)$. Thus you will see the alternative notation,

$$\lim_{\Delta x \to 0} \frac{\Delta y}{\Delta x} = \lim_{\Delta x \to 0} \frac{f(x + \Delta x) - f(x)}{\Delta x}.$$

Using Limits to Make Definitions — Slope of a Curve

One nice thing about mathematics is that once you get a good idea, and understand it clearly, you can use it for many things. Sometimes things which appear quite different are seen to be more similar by virtue of the fact that the same mathematical process can be used for both.

Did you think that rate of change and slope of a curve were completely different? In Activity 3 you used the limit idea to get at the slope of a curve. Let's discuss that in the same way that we analyzed instantaneous velocity and rate of change.

Again, it is easy to see what is meant by the slope of a secant. You take two points on a curve, construct the line connecting them and figure out the slope of that line. But what is the tangent? Well, you begin with a slope of a secant. Let's do it for the curve which is the graph of the function S of Activity 1 and try to figure out what is meant by the tangent to this curve at the point $(0.5, S(0.5))$. See Figure 3.4.

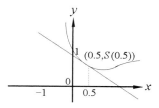

Figure 3.4. A graph of the function S with a tangent at (0.5, S(0.5)).

Now, take a nearby point on the curve. One way to do this is to take a point in the domain near 0.5 and go up to the curve. What do you think we call this nearby point in the domain? That's right, we call it $0.5+h$. Now we consider the two points $P = (0.5, S(0.5))$, $Q = (0.5+h, S(0.5+h))$ and draw the secant connecting them. We do this for several values of h, positive and negative, and with h getting close to 0 as shown in Figure 3.5.

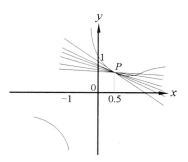

Figure 3.5. A graph of S with several secants through (0.5, S(0.5)).

Do you see clearly what happens to $0.5+h$ as h goes to 0? What happens to Q? Can you describe it carefully in words?

Now the point is to agree that, visually, as h goes to 0, then the secant line seems to approach the tangent line. Since lines passing through P are determined by their slopes, then it makes sense to think of the slope of the secant line approaching the slope of the tangent line. In other words, we compute the slope of the secant line that corresponds to $0.5+h$ and then take the limit as h goes to 0.

So far, none of this has anything to do with rates of change (except that both involve an initial approximation followed by taking a limit). The full connection comes when we write down the formula for the slope of the secant line that corresponds to $0.5 + h$. It is

$$\frac{S(0.5+h) - S(0.5)}{h}.$$

Does this look familiar? Yes, it is the same as the average velocity over the time interval from 0.5 to $0.5 + h$. Furthermore, we are getting the slope of the tangent line by taking the limit as h goes to 0. Thus the slope of the tangent line is

$$\lim_{h \to 0} \frac{S(0.5+h) - S(0.5)}{h}$$

and this is exactly the same as the instantaneous velocity at 0.5. Let us formalize our idea of slope of a curve (remember, we are using the limit to give precise meaning to this notion understood only vaguely until now).

Definition 3.3

Let f be a function and x_0 a number in its domain. Let γ be the curve which is the graph of f over a domain which includes x_0 and let $P = (x_0, f(x_0))$ be the point on γ corresponding to x_0. The tangent to the curve γ at the point P is the line which passes through the point P and has slope given by

$$\lim_{h \to 0} \frac{f(x_0 + h) - f(x_0)}{h}.$$

Thus we can say that *the slope of the graph of a function at a point is the same as the rate of change of the dependent variable with respect to the independent variable at that point.* (Once again you see the difference quotient formula.)

We can also connect this formula with material on lines discussed in Chapter 1. In Chapter 4 you will see that the notation $f'(x_0)$ is commonly used for the slope of the curve at x_0, and using that notation an equation for the tangent line to the curve at the point $P = (x_0, f(x_0))$ is, by the point-slope formula,

$$y - f(x_0) = f'(x_0)(x - x_0), \quad \text{or} \quad y = f(x_0) + f'(x_0)(x - x_0).$$

See Figure 3.6.

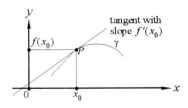

Figure 3.6. A curve with slope $f'(x_0)$ at $(x_0, f(x_0))$.

Using Limits to Make Definitions — Continuity

Let's look at one more important definition which we can make using limits. We'll begin with a question. Do you think that it is always possible to make a prediction for the value of the function f at the point a by considering values $f(x)$ when x is near a? If you *can* make a prediction, will it be correct in some sense? Look at the three functions V_1, V_2, and V_3 of Activity 7 given by the following formulas,

$$V_1(x) = \frac{3\sqrt{4x^2 + 4x + 5} - 2\sqrt{5}x - 3\sqrt{5}}{6x^2 + 9x}$$

$$V_2(x) = \begin{cases} \frac{3\sqrt{4x^2 + 4x + 5} - 2\sqrt{5}x - 3\sqrt{5}}{6x^2 + 9x} & \text{if } x \neq 0 \\ 0 & \text{if } x = 0 \end{cases}$$

$$V_3(x) = \begin{cases} \frac{3\sqrt{4x^2 + 4x + 5} - 2\sqrt{5}x - 3\sqrt{5}}{6x^2 + 9x} & \text{if } x \neq 0 \\ -\frac{4\sqrt{5}}{45} & \text{if } x = 0 \end{cases}$$

Each of these functions is given by exactly the same expression for domain values other than 0, but in each case, something different happens at 0. The first function is not defined at 0 and the other two are defined, but they have different values. Do you see that in spite of using the same humongous expression, these are three different functions?

Now let's look at the limit at 0. The fact that the same expression is used *for domain values other than 0* means that the limit at 0 must be the same for all three functions. Hopefully, by the time you finished Activity 7 you were already bored by the fact that the limit of that expression as x approaches 0 is $-\frac{4\sqrt{5}}{45}$.

But what happens to the value of the function at 0? For V_3 there is a value and we have $V_3(0) = -\frac{4\sqrt{5}}{45}$, the limit. That is, the value of the function at 0 is the same as the limit of the function as x approaches 0. In this case, we say that the function V_3 is *continuous* at 0.

For V_2, however, the situation is different. We have $V_2(0) = 0$ and this value is different from the limit. In this case, we say that V_2 is *discontinuous* at 0.

For V_1 there is no definition of the function at 0 and so there is no issue of continuity. We only say that V_1 has a limit as x approaches 0, but is undefined at 0.

These three examples illustrate several possible relations between the limit of a function at a point and the value of the function at that point. All three functions have a limit at the point 0, and the value of that limit is $-\frac{4\sqrt{5}}{45}$. Now there are three possibilities. The function V_1 is not defined at 0. The function V_2 is defined at 0 but its value is something else, $V_2(0) = 0$. The function V_3 is defined at 0 and its value is the limit, that is $_3(0) = \lim_{x \to 0} V_3(x) = -\frac{4\sqrt{5}}{45}$. We generalize this situation in the following definition.

■Definition 3.4

If a function f has a limit L at a point a in the domain of f, and $f(a) = L$, then we say that f is continuous at a. This means that f is continuous at a when

$$f(a) = \lim_{x \to a} f(x).$$

A function f is continuous on a set S, if it is continuous at each point of S. A function f is said to be continuous, if it is continuous at each point of its domain D.

The property of continuity might not seem very important compared to limit because once you have a limit at a point, it is just a question of how you happen to define the value of the function at the point. You could even redefine the function, if necessary, to guarantee that it is continuous.

It turns out, however, that continuity is very important, even more so than limits. Think about what it means from the visual point of view. How would you describe the property of continuity in terms of the appearance of the graph of the function near the point?

Continuity also has a practical value. If you know that a function is continuous at a point, then you can calculate its limit just by evaluating it once — at the point. A little later in this section we will have some theorems that can be used to tell, almost without thinking, that certain functions are continuous at certain points. Armed with this theory, we will then be able to calculate many limits by simple evaluation. In other words, the theory will permit us to predict the value of a continuous function at a point if we know its values near that point.

Rules for Computing Limits   We come now to the question of how limits can be computed. What is your SCS doing when it computes a limit? How does it determine that

$$\lim_{x \to 8}\left(2x^{\frac{1}{3}} - 1\right) = 3?$$

It does it with rules. One rule is that the limit of x as $x \to a$ is a, no matter what a is. This rule, and an algorithm for using it can be stored in the system. Another rule is that if you have an expression (such as x) raised to a power, then the limit of the expression raised to the power is calculated by taking the limit of the expression and then raising that to the power. For example, $\lim_{x \to 8} x^{\frac{1}{3}}$ is obtained by taking the limit of x which is 8 and then raising that to the power $\frac{1}{3}$. We can write that in mathematics as follows.[1]

$$\lim_{x \to 8} x^{\frac{1}{3}} = \left(\lim_{x \to 8} x\right)^{\frac{1}{3}} = 8^{\frac{1}{3}} = 2$$

In this way, we can compute the limit of an entire expression.

$$\lim_{x \to 8}\left(2x^{\frac{1}{3}} - 1\right) = \lim_{x \to 8}\left(2x^{\frac{1}{3}}\right) - \left(\lim_{x \to 8} 1\right) = 2\left(\lim_{x \to 8} x\right)^{\frac{1}{3}} - 1 = 2(8)^{\frac{1}{3}} - 1 = 2(2) - 1 = 3$$

Do you see what additional rules were used here? Do you see that we had to have a rule about the limit of the difference of two expressions, the limit of a constant times an expression, and the limit of a constant. Do you see how each rule was used in the calculation? The main purpose of the examples you looked at in Activity 6 was to get you thinking about using rules.

We have emphasized here the use of rules for making calculations and that is very important. It shows how a theoretical analysis can be of value in numerical computations. Later in this course you will see some examples of the use of these rules in theoretical situations as well as computational situations.

3.1.4 Exercises

1. In this exercise, you are to create a computer function named **lim** which computes an approximation to a limit of a function at a point. This **computer function** should accept as input any other **computer function f** and a number **a**. Your

---

[1]We ignore here the difficulties that would arise if x had a negative value or the exponent was negative and x was equal to 0.

computer function **lim** should give as output an approximate value of the limit at **a** of the function represented by **f** if this limit exists, or the message "limit not found" if there is no limit (or the computer can't find an appropriate approximation).

You are to test out your **computer function lim** with each of the test cases below to be sure that it gives the correct response. (Of course you should first figure out what the correct response is! If you're not sure, try looking at the graph of the given function or use your SCS.)

(a) The function f given by

$$f(x) = \frac{x^2 + 3x - 10}{x - 2}$$

with $a = 2$ and $a = 3$.

(b) The function f given by

$$f(x) = \frac{x^2 + 3x - 11}{x - 2}$$

with $a = 2$ and $a = 3$.

(c) The function f given by

$$f(x) = \sin\left(\frac{1}{x}\right)$$

with $a = \frac{1}{2\pi}$ and $a = 0$.

(d) The function f given by

$$f(x) = x\sin\left(\frac{1}{x}\right)$$

at $a = 0$.

(e) The **greatest integer function** which accepts any real number a as input and gives the greatest integer less than or equal to a as output at $a = 1.9$, at $a = 2$, and at $a = 2.1$.

(f) The freezing point function in Activity 1, Chapter 2, Section 1, page 69, with **a = 0.76**. Compare

your result with what you got in Activity 5 (a).

2. Explain why the **computer function** of the previous exercise will only give an approximate value for the limit. Give an example in which it gives the wrong answer. What happens if you apply it to the situation in Activity 3?

3. Find the following limits by hand. Which limit rules did you use to evaluate each limit? (Factoring and other algebraic manipulations may be helpful. Also, in simple cases, trying to mentally visualize the graph near the point may also be helpful. For more complicated functions, use your graphing utility to help you to visualize the behavior of the functions near the points in question.) You may wish to use your SCS to check your work.

(a) $\lim_{x \to -1} 5\pi$ (b) $\lim_{y \to 2} (3.1y - 7)$

(c) $\lim_{x \to 0} (x^2 - 3x + 5)$ (d) $\lim_{x \to 9} (5\sqrt{x} - x)$

(e) $\lim_{s \to 2.3} (s^3 + 4s - 1.1)$

(f) $\lim_{x \to 4} x^{\frac{3}{2}} (0.5x^2 + 2)$

(g) $\lim_{t \to 2} \frac{t^5 + 2}{1 - \sqrt{t}}$ (h) $\lim_{y \to 0} \frac{2y^2}{y}$

(i) $\lim_{x \to 1} \frac{x^2 - 2x + 1}{x - 1}$

(j) $\lim_{w \to -3} \frac{w^2 + 2w - 3}{w + 3}$

(k) $\lim_{x \to -1} \frac{x^2 + 2x + 1}{x^2 - 2x - 3}$ (l) $\lim_{y \to 1} \frac{y^2 - 5y + 4}{y^2 + 2y - 3}$

(m) $\lim_{x \to 2} \frac{x^2 - 4}{x - 2}$

(n) $\lim_{a \to -3} \frac{9 - a^2}{a^2 + 4a + 3}$ (o) $\lim_{x \to 5} \frac{\sqrt{x} - \sqrt{5}}{x - 5}$

(p) $\lim\limits_{x \to 3} \dfrac{\frac{1}{3} - \frac{1}{x}}{x - 3}$ (q) $\lim\limits_{y \to 0} \dfrac{y^2 - 1}{1 - \frac{1}{y^2}}$

(r) $\lim\limits_{x \to 0} \dfrac{\sqrt{3 + x} - \sqrt{3}}{x}$

(s) $\lim\limits_{h \to 0} \dfrac{h}{2 - \sqrt{h + 4}}$ (t) $\lim\limits_{x \to 0} \dfrac{\sin(3.2x)}{x}$

(u) $\lim\limits_{x \to 0} \dfrac{x}{\cot(4x)}$ (v) $\lim\limits_{x \to 0} \dfrac{\sin(5x)}{\sin(2x)}$

(w) $\lim\limits_{\theta \to 0} \dfrac{2\theta}{\cos(\theta)}$

(x) $\lim\limits_{x \to 1} \dfrac{7.26 \sin(x - 1)}{x^2 - 1}$

(y) $\lim\limits_{x \to 0} \dfrac{\sin^2(x)}{x}$ (z) $\lim\limits_{h \to 0} \dfrac{2^h - 1}{h}$

4. Approximate each of the following limits in three ways

   (a) Using your MPL to create a table of function values.

   (b) Using your graphing utility.

   (c) Using your **computer function lim** from Exercise 1. Compare each of your approximations to the limit given by your SCS.

   i. $\lim\limits_{x \to 0} \dfrac{\sqrt{x + 1} - 1}{x}$

   ii. $\lim\limits_{x \to 0} \dfrac{2^x - 1}{x}$

   iii. $\lim\limits_{t \to -2} \dfrac{t^4 - t^3 - 24}{t^2 + t - 2}$

   iv. $\lim\limits_{y \to \pi} \dfrac{\cos(y) + 1}{y - \pi}$

5. In each of the following situations find the indicated limit. If not possible, explain why not.

(a) $\lim_{x \to 0} f(x)$, $\lim_{x \to 2} f(x)$

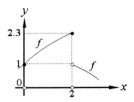

(b) $\lim_{x \to a} f(x)$, $\lim_{x \to b} f(x)$

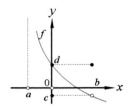

(c) $\lim_{x \to 0^+} f(x)$, $\lim_{x \to 0^-} f(x)$

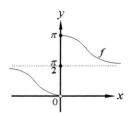

6. What is wrong with the following statement?

$$\lim\nolimits_{x \to 2} \sqrt{x} = 1.414.$$

Explain your answer.

7. What is wrong with the following statement?

$$\lim\nolimits_{x \to \pi} x = 3.14159.$$

Give reasons for your answer.

8. Write two variations of the **computer function lim** in Exercise 1 and call them **leftlim** and **rightlim**. Apply them to the following two situations.

   (a) The heat capacity function in Activity 2, Chapter 2, Section 1, page 70, at the domain point 160.

(b) The compressibility function Z of Chapter 2, Section 1, page 75, at the domain point 6.

Explain what happened.

Compare your answer with what you obtained in Activity 5.

Try to formulate a general statement that relates the existence and value of the three **computer functions**, **lim**, **leftlim** and **rightlim**.

9. Solve the following problems either by hand or by using your SCS.

   (a) A rock is thrown straight down from the top of a cliff so that the distance s traveled by the rock t seconds after it is released is given by $s = 16t^2 + 48t$ feet.

      i. Find an expression for the average velocity $V(h)$ of the rock from $t = 2.3$ seconds to $t = 2.3 + h$ seconds.

      ii. Find the instantaneous velocity of the rock 2.3 seconds after it is released.

      iii. Find the instantaneous velocity of the rock 3.7 seconds after it is released.

   (b) A cylinder with a piston contains a gas such that the pressure p of the confined gas occupying a volume of V in^3 is given by

   $$p = P(V) = \frac{300}{V} \text{ lbs}/\text{in}^2$$

   when the temperature is held constant.

      i. Find an expression for the average rate of change $A(h)$ of the pressure with respect to the volume from $v = 30$ in^3 to $v = 30 + h$ in^3.

      ii. Find the instantaneous rate of change of pressure with respect to volume when $v = 30$ in^3.

      iii. Find the instantaneous rate of change of pressure with respect to volume when $v = 12.7$ in^3.

   (c) For the curve γ which is the graph of the function f given by $f(x) = 0.5x^2 - 4\sqrt{x}$,

      i. Find an expression for the slope of the secant line $S(h)$ to the graph of γ from $x = 4$ to $x = 4 + h$.

      ii. Find an equation of the tangent line to the curve γ at $x = 4$.

      iii. Find an equation of the tangent line to the curve γ at $x = 5$.

   (d) For the curve γ which is the graph of the function g given by

   $$g(t) = \frac{t}{\sqrt{3+t}}.$$

      i. Find an expression for the slope of the secant line $S(h)$ to the graph of γ from $t = 1$ to $t = 1 + h$.

      ii. Find an equation of the normal line (i.e., the line perpendicular to the tangent line) to the curve γ at $t = 1$.

      iii. Find an equation of the normal line (i.e., the line perpendicular to the tangent line) to the curve γ at $t = 0.5$.

10. Find values of a and b so that the line $2x + 3y = a$ is tangent to the graph of the function f given by $f(x) = bx^2$ at the point where $x = 3$.

11. Find values of m and n so that the line $y = mx + 1$ is tangent to the graph of the function g given by $g(x) = nx^3$ at the point where $x = -1$.

12. For each of the following functions, find an expression for the difference quotient

$$dq(x) = \frac{f(x+h) - f(x)}{h}$$

and then use your SCS to find the limit of the difference quotient as $h \to 0$.

(a) $f(x) = x \sin\left(\frac{1}{x}\right)$

(b) $f(x) = x^2 \sin\left(\frac{1}{x}\right)$

13. In Activity 4 you saw that you could use the notion of a limit to extend the definition of rational exponents to irrational exponents. Use this method to approximate the following numbers using the approximation $\pi = 3.141592654$.

(a) 2^π (b) π^π

14. Compute the difference quotient at the "seam point" for each of the following functions and then take the limit of the difference quotient at the "seam point" as h approaches 0.

(a) $f(x) = \begin{cases} x & \text{if } x \le 0 \\ x^2 & \text{if } x > 0 \end{cases}$

(b) $g(x) = \begin{cases} x^2 + 1 & \text{if } x > -1 \\ 1 - x & \text{if } x \le -1 \end{cases}$

(c) $k(x) = \begin{cases} x^2 + 1 & \text{if } x > 1 \\ x^3 & \text{if } x \le 1 \end{cases}$

15. For each of the following functions, find out what happens as the independent variable approaches each point a from the left and from the right, compare the results and then decide if the function has a limit at a. You may wish to use your symbolic computer system to figure out the two one-sided limits.

(a) The compressibility function Z in Chapter 2, Section 1, page 75, with $a = 6$.

(b) The voltage function of Activity 5, Chapter 1, Section 4, page 31, with $a = 1$; $a = 2$; $a = 3$; and $a = 4$.

(c) The Heat Capacity function C_p (not C_p/T) in Activity 2, Chapter 2, Section 1, page 70 with $a = 220$.

16. Do this problem in the spirit of Activity 9. That is, for each of the following functions f and center c, determine the vertical intervals $[u, v]$ for which it is possible to find a horizontal interval of the form $[c-d, c+d]$ such that the portion of the graph of the function "stays in the window" determined by $[c-d, c+d]$ and $[u, v]$.

(a) The compressibility function in Chapter 2, Section 1, page 75, with the center of the horizontal interval equal to 6.

(b) The voltage function in Activity 5 of Section 1.4.2, page 31, for several values of the center of the horizontal interval such as 1, 2, 3, and 4.

(c) The Heat Capacity function in Activity 2, Chapter 2, Section 1, page 70, with the center of the horizontal interval equal to 220.

17. Is there an a so that

$$\lim_{x \to -1} \frac{x^2 + ax + 4}{x^2 + 2x + 1}$$

exists? Explain your answer.

18. Is there an a so that

$$\lim_{x \to 3} \frac{2x^2 - 3ax + x - a - 1}{x^2 - 2x - 3}$$

exists? Explain your answer.

19. If $\lim_{x \to a} f(x) = L$, is it possible that the value $f(x)$ is ever equal to L? Explain your answer.

20. For each of the following functions, find a and b so that the difference quotient at the "seam point" has a limit at the "seam point".

    (a) $f(x) = \begin{cases} ax & \text{if } x \le 1 \\ bx^2 + x + 1 & \text{if } x > 1 \end{cases}$

    (b) $g(x) = \begin{cases} ax^2 & \text{if } x \le 1 \\ bx^3 + 2ax & \text{if } x > 1 \end{cases}$

21. The *acceleration* of a particle at a particular time is the (instantaneous) rate of change of the velocity of the particle with respect to time.

    (a) What are the units of acceleration?

    (b) Using a limit, how would you define the acceleration a of a particle at time t when the particle has velocity given by a function $v = V(t)$?

    (c) For each of the following situations, find the acceleration a at the indicated time t for the indicated velocity functions.

         i. $v = V(t) = t^2 - 2t$; $t = 1$.

         ii. $v = V(t) = t - \sqrt{t}$; $t = 3$.

22. A manufacturer produces x units of a commodity and obtains a total profit p given by the function $p = P(x)$ when the x units are sold. The manufacturer's *marginal profit* is the (instantaneous) rate of change of total profit with respect to the number of units produced. What are the units of marginal profit?

    (a) Using a limit, how would you define the manufacturer's marginal profit mp when x units are produced?

    (b) For the following situation, find the marginal profit when 50 units are produced and the total profit is given by $p = P(x) = 0.03x^2 + 10x$ dollars.

    (c) How do you think you would use a limit to define *marginal cost* when the total cost c is given by a function $c = C(x)$ when x units are produced?

    (d) How do you think you would use a limit to define *marginal revenue* when total revenue r is given by a function $r = R(x)$ when x units are produced?

23. Physicians often prescribe nitroglycerin tablets to heart patients in order to help dilate blood vessels that become constricted. In this situation, the (instantaneous) rate of change of the cross sectional area of a blood vessel with respect to its radius is of interest.

    (a) Using a limit, how would you define the (instantaneous) rate of change R of the cross sectional area a of a blood vessel with respect to its radius r where the cross sectional area a is given by the function $a = A(r)$?

    (b) Find the (instantaneous) rate of change R of the cross sectional area a of a blood vessel with respect to its radius for a blood vessel that has a circular cross sectional area with radius r.

24. In the theory of electricity, the *current i* at a specified time t is the (instantaneous) rate of change of the charge q at that instant. Using a limit, how would you define the current i at time t when the charge q is given by a function $q = Q(t)$ and the current i is given by $i = I(t)$?

25. Using a hand-held calculator find the following limit

$$\lim_{x \to 0} \frac{\sin(x)}{x}$$

in the following two ways.

(a) Using your calculator in radian mode.

(b) Using your calculator on degree mode.

Finally, explain any difference between the two answers you obtained.

26. For each of the following functions, find a value for a (if one exists) so that the function is continuous at the "seam point."

(a) $f(x) = \begin{cases} ax & \text{if } x \le 2 \\ ax^2 + x + 1 & \text{if } x > 2 \end{cases}$

(b) $g(x) = \begin{cases} ax^2 & \text{if } x \le -1 \\ x^3 + ax & \text{if } x > -1 \end{cases}$

(c) $g(x) = \begin{cases} x^2 - 2x & \text{if } x \ne 1 \\ a + 3 & \text{if } x = 1 \end{cases}$

(d) $g(x) = \begin{cases} ax - 4 & \text{if } x > 2 \\ 4 - x^2 & \text{if } x \le 2 \end{cases}$

27. The function q defined below is a *net charge* function of time t from electrical theory. At which "seam points" is the charge continuous? Use your graphics utility to sketch q.

$$q(t) = \begin{cases} 0 & \text{if } t < -4 \\ 4t + 16 & \text{if } -4 \le t \le -2 \\ 2t^2 & \text{if } -2 < t \le 1 \\ 5\sqrt{t+3} - 8 & \text{if } 1 < t \le 6 \\ 7 & \text{if } t > 6 \end{cases}$$

28. Show that the function f given by

$$f(x) = \begin{cases} x \sin(\frac{1}{x}) & \text{if } x \ne 0 \\ 0 & \text{if } x = 0 \end{cases}$$

is continuous at $x = 0$. Use your graphing utility to make a sketch of f.

29. For each of the following functions, determine whether or not the function is continuous at each "seam point".

(a) The voltage function in Activity 5 of Chapter 1, Section 4, page 31.

(b) The temperature function in Activity 1, Chapter 2, Section 1, page 69.

(c) The Heat Capacity function in Activity 2, Chapter 2, Section 1, page 70.

(d) The compressibility function on in Chapter 2, Section 1, page 75.

(e) The **computer function** represented by the following **MapleV proc.**

```
f := proc(x);
    if x > 0 then x**2 - 1;
    elif if x = 0 then 1.0;
    else x + 1
    fi;
end;
```

(f) The computer function represented by the following **ISETL func.**

```
g := func(y);
    if is_number(y) then
        if y >= -5 and y < -3 then
        return -y**2;
        elseif y >= -3 and y < 1
        then return y**3 - 1;
        elseif y >= 1 and y < 2 then
        return y+4;
        else return "function not
        defined";
        end;
    end;
end;
```

■ 3.2 INFINITE LIMITS AND LIMITS AT INFINITY

3.2.1 Overview

The intuitive understanding of limits which you have developed in Section 1 enables you to not only make precise mathematical definitions, but also to make important qualitative analyses concerning the graphs of functions. Armed with the concept of a limit, you should now be able to mentally visualize what it means for a function to be continuous at a point. Your understanding of limits should also enable you to visualize how a function may fail to be continuous at a point.

In this section you will extend the idea of a limit in two ways. You will make a definition involving a limit as the domain variable "tends to infinity" and make a definition involving a limit as the range variable "tends to infinity". Armed with these powerful extensions of the notion of a limit, you will be able to make even more comprehensive qualitative analyses of the behavior of functions and their graphs.

3.2.2 Activities

1. You can use the idea of limit to try to predict the "ultimate value" of a function as the independent variable gets large. Consider the functions given by the following expressions. For each of them, use your graphics utility to make graphs over the domain $[a, b]$ with b a very large positive number. If you run into trouble with certain values in this domain, change the value of a so as to avoid them.

   After looking at the graphs for larger and larger values of b, read off coordinates to see if you can tell what will be the "ultimate value" of the function as the domain variable "goes to infinity". Then use your SCS to compute this limit exactly.

   (a) $g(x) = \dfrac{3x^3 + 57}{2x^3 + 2.8x^2 + 7.4x + 10.4}$

   (b) $h(u) = \dfrac{\sqrt{9u^{\frac{4}{3}} - 2}}{3 - 4u^{\frac{2}{3}}}$

   (c) $k(z) = \dfrac{\sqrt{9z^{\frac{5}{3}} - 2}}{3 - 4z^{\frac{2}{3}}}$

   (d) $f(x) = \sin\left(\dfrac{1}{x}\right)$

   (e) $d_1(t) = \dfrac{1+t}{t}\sin(t)$

   (f) $d(t)_2 = \dfrac{1}{\sqrt{t}}\sin(t)$

   (g) $m(x) = \dfrac{2x^3 + 7x}{x^2}$

   (h) $p(x) = \dfrac{2x^3 + 7}{x}$

2. In this activity, you are to create a computer function named **limatinf** which figures out an approximation to a limit of a function at infinity. This **computer function** should accept as input any other **computer function f** and should give as output an approximate value of the limit at infinity of the function represented by **f** if this limit exists, or the message "limit not found" if the limit does not exist.

   Test out your **computer function limatinf** on each of the functions given in Activity 1 above.

3. Estimate each of the following limits graphically. Then use your **computer function limatinf** to approximate these limits. Finally, check your results with your symbolic computer system.

   (a) $\lim\limits_{x \to \infty} \dfrac{2x^3 - 1.45x^2 + 34.87}{5x^3 + 3.61x - 27.8}$ (b) $\lim\limits_{x \to \infty} \dfrac{2x^3 - 1.45x^2 + 34.87}{5x^4 + 3.61x - 27.8}$

   (c) $\lim\limits_{x \to \infty} \dfrac{2x^3 - 1.45x^2 + 34.87}{5x^2 + 3.61x - 27.8}$

4. Formulate a general rule for

$$\lim_{x \to \infty} \frac{a_0 x^n + a_1 x^{n-1} + \cdots + a_n}{b_0 x^k + b_1 x^{k-1} + \cdots + b_k}$$

5. Modify the **computer function limatinf** from Activity 2 above to create a new **computer function** called **limatminf** which figures out an approximation to a limit of a function at minus infinity. This **computer function** should accept as input any other **computer function f** and should give as output an approximate value of the limit at minus infinity of the function represented by **f** if this limit exists, or the message "limit not found" if the limit does not exist.

   Test out your **computer function limatminf** on each of the functions given in Activity 1 above.

6. A hot piece of steel which is $94°C$ is placed in a room that is kept at a constant temperature of $20°C$. The steel begins to cool, and has its temperature t minutes after being placed in the room given by

$$T(t) = 20 + 74e^{\frac{\ln 0.4t}{40}}$$

   (a) Use your graphing utility to obtain a graph of temperature as a function of time.

   (b) How long will it take the steel to cool down to $22°$?

   (c) How long will it take the steel to cool down to $21°$?

   (d) What will be the ultimate temperature of the steel?

7. Chemists use mathematical formulas to describe the relationships between pressure, volume and temperature of gasses. For an ideal gas, Boyle's Law states that for a fixed number of molecules of gas at a constant temperature, volume is inversely proportional to temperature, that is,

$$V = \frac{k_1}{P}$$

   Charles' Law states that for a fixed number of molecules of gas at a constant pressure, volume is proportional to temperature, that is,

$$V = k_2 T$$

   Avagadro's Law states that at a constant temperature and pressure, volume is proportional to the number of molecules present, that is,

$$V = k_3 n$$

(a) By hand, sketch graphs which illustrate each of these laws.

(b) What does Boyle's Law tell us happens to pressure if a fixed amount of gas is compressed into half of the original volume? To one-fourth of the original volume? What happens to pressure as volume is reduced to near zero?

(c) What does Charles' Law tell us happens to volume if a fixed number of molecules is kept at a constant pressure while the temperature doubles? Triples? What happens as the temperature is increased without bound?

(d) What does Avagadro's Law allow you to predict about the volume an ideal gas will occupy if temperature and pressure are kept constant and the number of molecules are doubled? Tripled?

8. The ideal gas law $PV = nRT$ combines the statements of the three gas laws in the previous activity. If we measure pressure in atmospheres (atm), volume in liters, number of molecules in moles and temperature in degrees Kelvin ($^\circ K$), then the ideal gas constant R is found experimentally to be $R = 0.0821 \frac{liter-atm}{mole\,^\circ K}$.

   (a) What volume will one mole of an ideal gas occupy at $273^\circ K$ and one atmosphere pressure?

   (b) Determine the equation which gives pressure as a function of volume when one mole of an ideal gas is kept at a constant temperature of $273^\circ K$.

   (c) What happens to the pressure of the gas described above as the volume increases without bound?

   (d) What happens to the pressure of the above gas as the volume decreases to near zero?

9. All pairs of particles of matter experience a gravitational force of attraction which depends upon the masses of the particles and the distance which separates them. The formula describing this attraction is

$$F = G\frac{m_1 m_2}{R^2}$$

where m_1 and m_2 are the masses of the objects measured in kilograms (kg), R is the distance separating the particles in meters, and G is the constant given by

$$G = 6.67 \times 10^{-11} \frac{Nm^2}{kg^2}$$

where the unit N stands for Newtons and the unit m for meters. Explain in your own words the change in the gravitation attraction between two particles as the distance between these particles:

   (a) is doubled   (b) is quadrupled

   (c) is halved   (d) is quartered

   (e) approaches infinity   (f) approaches zero

10. A skydiver in freefall has two opposing forces acting on her which affects the speed at which she falls. The force of gravity acts on her to increase this speed, but the faster she falls, the greater wind resistance she encounters which tends to slow her speed. Her approximate vertical speed t seconds after leaving the airplane is given by

$$V(t) = 200\left(1 - \frac{1}{e^{0.2t}}\right) \text{ feet per second.}$$

Find this skydiver's terminal velocity, i.e., the "ultimate value" of her velocity.

3.2.3   Discussion

Limits at ∞   In Section 1, you encountered limits as you were studying what happens to the values $f(x)$ of a function f as the independent variable x gets closer and closer to some particular number. You could also be concerned about what happens to the values of the function as the values of the independent variable x get larger and larger — without bound. In this case we say that "x tends to infinity", and we are considering the "limit of f at infinity". We write the value of this limit (if it exists) as follows.

$$\lim_{x \to \infty} f(x)$$

There are practical situations in which this arises. For example, you might imagine a radio signal putting out a steady beam of constant intensity. Perhaps there is a one-shot input signal that effects the intensity of the output. You might imagine that the intensity of the disturbance is given as a function of time by an expression of one of the following kinds. See Figures 3.7 and 3.8.

$$d_1(t) = \frac{1+t}{t}\sin(t)$$

$$d_2(t) = \frac{1}{\sqrt{t}}\sin(t)$$

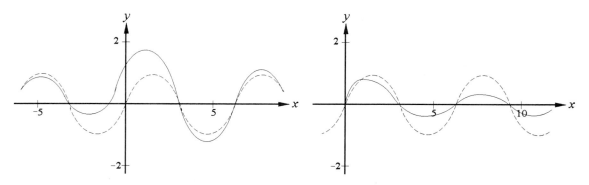

Figure 3.7. A graph of $\frac{1+t}{t}\sin(t)$.   **Figure 3.8. A graph of $\frac{1}{\sqrt{t}}\sin(t)$.**

We can make some qualitative inferences about the effect of the disturbance from the behavior of these two functions which you investigated in Activity 1 parts (e) and (f).

In both cases, the function has no value at $t = 0$, but the limit does exist. This suggests that something violent may have occurred that cannot be described by a simple expression. Now consider what happens as time goes on — that is, as t becomes large. In the first function, the expression $\frac{1+t}{t}$ has limit at infinity equal to 1 (do you see why?), i.e.,

$$\lim_{t \to \infty} \frac{1+t}{t} = 1$$

and so the values of $d(t)$ for large t are very close to $\sin(t)$. You can see this by looking at the graph. Because the values of the function as t increases without bound do not approach a single fixed number L, the function does not have a limit at infinity and we write,

$$\lim_{t \to \infty} d_1(t) \text{ does not exist}$$

With the second function, the oscillation is still present, but this time, it is multiplied by $\frac{1}{\sqrt{t}}$ which is quite small for large t. In fact, for this function, its limit at infinity is 0. We write this as,

$$\lim_{t \to \infty} d_2(t) = 0.$$

It means that the disturbance continues, but becomes smaller and smaller and is "dampened" out, so that after some time, it has no practical effect.

The **computer function limatinf** which you created in Activity 2 helped you to identify whether or not a function possessed a limit at infinity. In Activity 5 you found that only a very slight modification of this **computer function** was necessary to determine whether or not a function possessed a limit at minus infinity.

Throughout the Activity set, you saw that if a function f has a limit at infinity, a very important observation can be made about the graph of f. In general, if either $\lim_{x \to \infty} f(x) = L$ or $\lim_{x \to -\infty} f(x) = L$, or both, then the line $y = L$ is a *horizontal asymptote* of the graph of the function f.

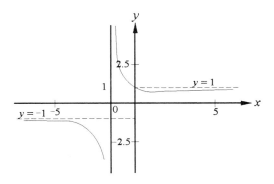

Figure 3.9. Limits at infinity and minus infinity - horizontal asymptotes.

Mental Calculation of Limits at Infinity

There are some expressions for which you can figure out the limit at ∞ by looking at the form of the expression. Consider Activity 1, part (a). Did you determine that the limit was 1.5? Which of the methods, graphs or your symbolic computer system was a better way to get it? You can also do it in your head and some people think that is the best way. Let's look at the expression.

In the numerator there are two terms. One stays at 57. The other gets very large as the values of x get very large. If you think about the numbers that you get by evaluating the numerator for large values of x, you can see that they will not be very much different from what you get if you just neglect 57. So maybe we can calculate the limit at ∞ by just using $3x^3$ in the numerator.

The same analysis holds true for the denominator. The first term $2x^3$ is much larger than the other three terms, the more so as x gets larger. So we might calculate the limit at ∞ by just using $2x^3$ in the denominator.

In other words, as x gets large, the values of this expression are getting closer to

$$\frac{3x^3}{2x^3} = 1.5$$

Can you use the same kind of analysis to do part (b) in Activity 1? What about part (c)? Does this approach help at all with the last three? What about Activity 3?

Infinite Limits   You've seen that through the use of limits at a point and limits at infinity, you are able to say quite a bit about the behavior of a function and the appearance of its graph. The use of limits at infinity allow you to study the behavior of a function as the independent variable increases without bound. There is a dramatic type of behavior of some functions, say f, at some points, say a, where $\lim_{x \to a} f(x)$ does not exist, but rather the values of the dependent variable $y = f(x)$ increases or decreases without bound as x approaches a.

Although no single number exists which the functional values all approach as x approaches a, this situation is so important that we extend the notion of a limit and say that "the limit as x approaches a of $f(x)$ is ∞" or "the limit as x approaches a of $f(x)$ is $-\infty$". In these cases we write, respectively,

$$\lim_{x \to a} f(x) = \infty \text{ or } \lim_{x \to a} f(x) = -\infty.$$

Again, you should try not to be confused by the use of the symbol ∞. This is not a number, but rather it is a way of specifying that the values of $f(x)$ get large without bound as x gets close to a; $-\infty$ is a way of specifying that the values of $f(x)$ get large and negative without bound as x gets close to a. For example, $\lim_{x \to 0} \frac{1}{x^2} = \infty$.

This dramatic "blowing up" (or "down") of the values of a function f near a point a sometimes occurs as the domain values approach a from both sides, but sometimes occurs only as the domain values approach a from the left or from the right. In these cases, we write

$$\lim_{x \to a^-} f(x) = \infty$$

$$\lim_{x \to a^-} f(x) = -\infty$$

$$\lim_{x \to a^+} f(x) = \infty$$

$$\lim_{x \to a^+} f(x) = -\infty$$

A simple example illustrating this behavior is the rational function f given by $f(x) = \frac{1}{x}$. Do you see why $\lim_{x \to 0^-} f(x) = -\infty$ but $\lim_{x \to 0^+} f(x) = \infty$?

In Activities 7 and 8, you studied $\lim_{V \to 0^+} P(V)$ and saw that if a fixed number of gas molecules at a constant temperature was compressed into a smaller and smaller volume, then the pressure would tend to infinity. In Activity 9 you also encountered a limit of this sort, $\lim_{R \to 0^+} F(R)$. Why were right-hand limit used in these situations?

We can summarize this type of behavior of a function f near a point a as follows. If $\lim_{x \to a} f(x) = \infty$ or $\lim_{x \to a} f(x) = -\infty$, then the line $x = a$ is a *vertical asymptote* of the graph of the function f, where $\lim_{x \to a}$ can be replaced by $\lim_{x \to a^+}$ or $\lim_{x \to a^-}$.

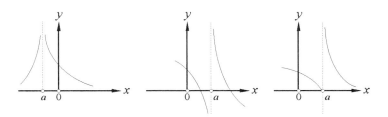

Figure 3.10. Functions which are unbounded near a point a — vertical asymptotes.

Infinite Limits at Infinity

In Activity 3(c) you found that there was no fixed number which the values of the rational function, say f, given by $\frac{2x^3 - 1.45x^2 + 34.87}{5x^2 + 3.61x - 27.8}$ approaches as x approaches infinity. Once again, we can extend the notion of a limit to include cases such as this, that is, cases where the dependent variable increases without bound as the independent variable increases without bound. In such cases, we say that "the limit as x approaches infinity of $f(x)$ is infinity" and we write

$$\lim_{x \to \infty} f(x) = \infty.$$

In a completely analogous manner, we can define the following expressions

$$\lim_{x \to \infty} f(x) = -\infty,$$

$$\lim_{x \to -\infty} f(x) = \infty$$

and

$$\lim_{x \to -\infty} f(x) = -\infty.$$

You should make sure that you understand the different qualitative behavior that is represented by each of these expressions. For each of these four situations, try to mentally visualize the graph of a function which displays this behavior. Is it possible for $\lim_{x \to \infty} f(x) = \infty$ and $\lim_{x \to \infty} f(x) = -\infty$? Is it possible for $\lim_{x \to -\infty} f(x) = \infty$ and $\lim_{x \to -\infty} f(x) = -\infty$?

Oblique Asymptotes   It may seem as if we've exhausted the possibilities for describing the qualitative behavior of a function f as x approaches infinity. We've seen that if $\lim_{x\to\infty} f(x) = L$, then the line $y = L$ is a horizontal asymptote of the graph of f. We've also seen an example in the discussion of limits at ∞ a function $d_1(t)$ which failed to have a limit at infinity. What else could be left? What other meaningful information could we want to convey about the behavior of a function as the domain variable gets larger and larger without bound?

Go back and look at Activity 1(g). By now you should be able to look at the rational expression defining this function, m, and see that $\lim_{x\to\infty} m(x) = \infty$ and $\lim_{x\to-\infty} m(x) = -\infty$. But can you say more? Also look once more at the function p in Activity 1(h). Clearly $\lim_{x\to\infty} p(x) = \infty$ and $\lim_{x\to-\infty} p(x) = \infty$. But can you say more here as well?

There are two good ways to discover what's happening here. If you have your graphing utility handy, look at the graphs of m and p. Do they resemble the graphs of any familiar functions?

Another way to figure out what's happening here is to use long division to simplify the rational expressions defining m and p. Do you see that $m(x) = 2x + \frac{7}{x}$ and $p(x) = 2x^2 + \frac{7}{x}$? By writing the expressions for m and p in this way, you can see that as x gets very large (in either the positive or negative sense) the term $\frac{7}{x}$ approaches zero. Hence the graph of m gets arbitrarily close to the oblique line $y = 2x$ and the graph of p gets arbitrarily close to the parabola $y = 2x^2$. These are specific examples of the following two definitions.

Definition 3.5

A line $y = ax + b$ is called an oblique asymptote of the graph of a function f if $f(x) = q(x) + ax + b$ where $\lim_{x\to\infty} q(x) = 0$ or $\lim_{x\to-\infty} q(x) = 0$ (or both).

Definition 3.6

A curve $y = A(x)$ is called an asymptotic curve of the graph of a function f if $f(x) = q(x) + A(x)$ where $\lim_{x\to\infty} q(x) = 0$ or $\lim_{x\to-\infty} q(x) = 0$ (or both).

Once again, you see the power of using limits (this time the extended notion of limits) to make definitions. Can you find the asymptotic curve to the function $d_1(t)$ which has no limit at infinity?

3.2.4 Exercises

1. Find the following limits at infinity by hand. You may wish to use your SCS or your graphics utility to help you see what is going on or to check some of your answers.

   (a) $\lim_{x\to\infty} \dfrac{x^2 + 2.7x + 6.3}{5x^2 + 3x - 2}$

   (b) $\lim_{y\to\infty} \dfrac{y^3 - 4.21y^2 + 7.1}{5y^4 + y + 7}$

   (c) $\lim_{w\to\infty} \dfrac{w^5 + w - 1}{5w^4 - 3w^3 + w + 13}$

   (d) $\lim_{x\to\infty} \dfrac{x^3 - x + 3}{5x - 2x^3}$

   (e) $\lim_{x\to\infty} \dfrac{\sqrt{x} - x^{\frac{5}{3}}}{x^2 - 23.6x - 11}$

   (f) $\lim_{x\to\infty} \dfrac{2x - x^2}{\sqrt{1 + x^4}}$ (g) $\lim_{x\to\infty} \dfrac{x + 5}{1 + 2x^2}$

   (h) $\lim_{x\to\infty} \cos x^2$ (i) $\lim_{x\to\infty} \dfrac{x \cos\left(\frac{3}{x^2}\right)}{1 - 2x}$

(j) $\lim\limits_{x\to\infty} \dfrac{\sin(2x)}{5\sqrt{x}}$ (k) $\lim\limits_{z\to\infty} \dfrac{3\cos(z)}{\sqrt{1+z^3}}$

(l) $\lim\limits_{x\to\infty} \dfrac{x^2}{\sec\left(\frac{1}{x^2}\right)}$

2. Find the following limits at minus infinity by hand. You may wish to use your SCS or use your graphics utility to help you see what is going on or to check some of your answers.

(a) $\lim\limits_{x\to-\infty} \dfrac{2x^3 - 1.45x^2 + 34.87}{5x^3 + 3.61x - 27.8\sqrt{5}}$

(b) $\lim\limits_{x\to-\infty} \dfrac{2x^3 - 1.45x^2 + 34.87}{5x^4 + 3.61x - 27.8\sqrt{5}}$

(c) $\lim\limits_{x\to-\infty} \dfrac{2x^3 - 1.45x^2 + 34.87}{5x^2 + 3.61x - 27.8\sqrt{5}}$

(d) $\lim\limits_{x\to-\infty} \dfrac{x^3 + 7x^2 + 25}{5x - 3x^2 + 1}$

(e) $\lim\limits_{s\to-\infty} \dfrac{s^5 + 7s^2 - s}{3s^2 - 2s - 7s^5}$

(f) $\lim\limits_{x\to-\infty} \dfrac{x+5}{1+2x^2}$

(g) $\lim\limits_{\theta\to-\infty} \dfrac{\theta^3 + \theta^2 + 5}{5\theta - 3\theta^2 + \theta^4}$

3. Based on your experiences in Exercise 2, formulate a general rule for the following limit situation

$$\lim\limits_{x\to-\infty} \dfrac{a_0x^n + a_1x^{n-1} + \cdots + a_n}{b_0x^k + b_1x^{k-1} + \cdots + b_k}.$$

4. In each of the following situations find the indicated limit. If not possible, explain why not.

(a) $\lim\limits_{x\to-2^+} h(x),\ \lim\limits_{x\to-2^-} h(x)$

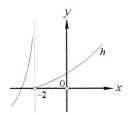

(b) $\lim\limits_{x\to\infty} f(x),\ \lim\limits_{x\to-\infty} f(x)$

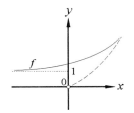

(c) $\lim\limits_{x\to\infty} F(x),\ \lim\limits_{x\to-\infty} F(x)$

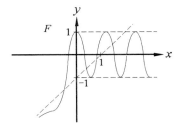

5. A chemical factory which produces quarts of a certain chemical where the average cost a, in dollars per quart, for a batch of q quarts is given as a function of q by

$$a = A(q) = \dfrac{1130 + 7.35q}{q}.$$

(a) Find the limiting average price $\lim\limits_{q\to\infty} A(q)$ and explain the meaning of the limiting average price as the number of quarts q increases without bound.

(b) How many quarts should a batch be so that the average cost per quart is within 10 cents of the limiting average value in part (a)?

(c) Use your graphics utility to make a complete graph of the function A.

6. According to Newton's law of gravity, an object that weighs one pound on the surface of the earth weighs about

$$w = \frac{16}{d^2}$$

pounds when it is d thousand miles from the center of the earth (with a radius of about four thousand miles).

   (a) How many miles from the center of the earth must a 200 pound astronaut be so that her or his weight is less than 100 pounds? less than one pound?

   (b) Using the notion of limit at infinity, give an explanation for the use of the term "zero gravity" to describe the apparent weightlessness of an astronaut in a space vehicle half way to the moon.

7. The value v, in dollars, of an industrial machine t years after it is purchased is given as a function of t by

$$v = V(t) = 1000 + \frac{20{,}000}{3^{0.02t}}.$$

The scrap value of the machine is the limiting value of the machine as the number of years t increases without bound. The scrap value is also referred to as the value of the machine in the "long run."

   (a) What was the value of the machine at the time it was purchased?

   (b) Find the scrap value of the machine in this situation.

   (c) After how many years will the machine be worth less than $10,000?

   (d) After how many years will the value of the machine be within $100 of its scrap value?

   (e) Use your graphics utility to make a complete graph of the function V.

8. Find the following limits by hand. You may wish to use your SCS to help you see what is going on.

   (a) $\lim\limits_{x \to \infty} \left(\sqrt{1+x} - \sqrt{x} \right)$

   (Hint: Multiply the numerator and denominator by some expression.)

   (b) $\lim\limits_{x \to \infty} \left(\sqrt{x + x^2} - \sqrt{x^2} \right)$

   (c) $\lim\limits_{x \to \infty} \left(\sqrt{\dfrac{1}{x} + \dfrac{1}{x^2}} - \sqrt{\dfrac{2}{x^2}} \right)$

   (d) $\lim\limits_{x \to \infty} \dfrac{1 - \cos(x)}{x}$

9. In this exercise you will consider *oblique asymptotes* and *asymptotic curves* of functions.

   (a) Find the oblique asymptotes for each of the following functions. (Hint: First simplify or perform long division.)

   i. $f(x) = \dfrac{x^2 + x - 1}{x}$

   ii. $f(x) = \dfrac{4x^3 - 2x^2 - x + 1}{2x^2}$

   iii. $f(x) = \dfrac{1 + 3x - 2x^3}{1 + x}$

   iv. $f(x) = \dfrac{3x^3 - 4x^2 - 2x + 5}{x^2 - 1}$

   (b) Find asymptotic curves for each of the following functions.

   i. $f(x) = \dfrac{x^3 - 3}{x}$

   ii. $f(x) = \dfrac{1 + 2x - x^4}{x^2}$

   iii. $f(x) = \dfrac{x^3 - x^2 - x + 2}{x + 1}$

   iv. $f(t) = \dfrac{1 + 2t}{t} \sin(t)$

v. $f(y) = \dfrac{\sqrt{y} - y^2}{2y^2}\cos(2y)$

vi. $f(r) = \dfrac{1+r}{r}\sin(r)$

10. Find all horizontal and vertical asymptotes of the graph of the curve $2xy + |x| - y + 1 = 0$.

11. The function h defined by the following **MapleV proc** is called a *recursive* function because its values are defined in terms of previous values.

   h := proc(n)

       if n = 1 then 2;

           elif if n > 1 and

           hastype(n,integer)

           then 0.5*(h(n-1) + 2/h(n-1));

       fi;

       end;

   Look up the word *recursive* in a dictionary and see what it means. What do you think is the limit of h as n increases without bound? You may wish to use the computer to evaluate h for large values of n to investigate $\lim_{n\to\infty} h(n)$ or create a table of values to see what is happening. Can you guess what the *exact* limit is rather than a decimal approximation?

12. Does $\lim_{x\to 0}\frac{|x|}{x}$ exist? Explain your answer.

13. Does $\lim_{x\to 0}\cos\frac{1}{x}$ exist? Explain your answer.

14. Use your **computer function lim** from Exercise 1, page 142, to approximate the following limits and then compare your approximation with the limit obtained using your SCS.

   (a) $\lim_{x\to 0}\dfrac{3^x - 1}{x}$ (b) $\lim_{x\to 1}\dfrac{2 - 2^x}{4^x - 4}$

(c) $\lim_{x\to 0}\dfrac{1 - \cos(x)}{x^2}$ (d) $\lim_{x\to 0}\dfrac{x - \sin(x)}{x^3}$

(e) $\lim_{x\to 0}(1 + x)^{\frac{1}{x}}$

15. For the given function, find the indicated one-sided limits. You may wish to use your SCS or your computer functions **limleft** and **limright** to help you see what is going on or to check your answers.

   (a) $\lim_{x\to 0^-} f(x)$ and $\lim_{x\to 0^+} f(x)$ for
   $$f(x) = \frac{1}{x^3}.$$

   (b) $\lim_{w\to 0^-} f(w)$ and $\lim_{w\to 0^+} f(w)$ for
   $$f(w) = \frac{-5.27}{w}.$$

   (c) $\lim_{x\to 2.4^+} g(x)$ and $\lim_{x\to 2.4^-} g(x)$ for
   $$g(x) = \sqrt{x - 2.4}.$$

   (d) $\lim_{t\to 0^-} h(t)$ and $\lim_{t\to 0^+} h(t)$ for
   $$h(t) = \frac{t - 2}{2}.$$

   (e) $\lim_{x\to 1.6^-} f(x)$, $\qquad \lim_{x\to 1.6^+} f(x)$, $\lim_{x\to -3.1^+} f(x)$ and $\lim_{x\to -3.1^-} f(x)$ for
   $$f(x) = \frac{x}{(x + 3.1)(x - 1.6)}.$$

   (f) $\lim_{x\to 0.48^+} g(x)$, $\qquad \lim_{x\to 0.48^-} g(x)$, $\lim_{x\to 7^+} g(x)$ and $\lim_{x\to 7^-} g(x)$ for
   $$g(x) = \frac{x - 0.48}{x - 7}.$$

   (g) $\lim_{u\to 5^-} s(u)$ and $\lim_{u\to 5^+} s(u)$ for
   $$s(u) = \frac{|u - 5|}{u - 5}.$$

   (h) $\lim_{x\to 2^+} F(x)$ and $\lim_{x\to 2^-} F(x)$ for
   $$F(x) = \frac{|x^2 - 4|}{2 - x}.$$

(i) $\lim_{\theta \to 0^+} h(\theta)$ and $\lim_{\theta \to 0^-} h(\theta)$ for

$$h(\theta) = \frac{\cos(2\theta)}{\sin(\theta)}.$$

16. Find the right and left hand limits at each "seam point" for each of the following functions. Does the limit exist at each "seam point"? Is the function continuous at each "seam point"? Use your graphics utility to make a sketch of each function.

    (a) $f(x) = \begin{cases} x & \text{if } x \le 0 \\ x^2 & \text{if } x > 0 \end{cases}$

    (b) $f(x) = \begin{cases} x & \text{if } x \le -1 \\ x^2 & \text{if } x > -1 \end{cases}$

    (c) $h(x) = \begin{cases} x^2 + 1 & \text{if } x > -1 \\ 1 - x & \text{if } x \le -1 \end{cases}$

    (d) $g(x) = \begin{cases} x^2 + 1 & \text{if } x > 1 \\ x^3 & \text{if } x \le 1 \end{cases}$

    (e) $g(x) = \begin{cases} x & \text{if } |x| > 1 \\ x^3 & \text{if } |x| \le 1 \end{cases}$

17. In this exercise the notation $\lfloor x \rfloor$ represents the greatest integer less than or equal to x. The function f given by $f(x) = \lfloor x \rfloor$ is called the *greatest integer function*. Find the following limits if they exist. If no limit exists explain why the limit does not exist. (This function is represented by the function **floor** in many computer systems.)

    (a) $\lim_{x \to 0^+} \lfloor x \rfloor$, $\lim_{x \to 0^-} \lfloor x \rfloor$, $\lim_{x \to 0} \lfloor x \rfloor$

    (b) $\lim_{x \to 3^+} \lfloor x + 1 \rfloor$, $\lim_{x \to 3^-} \lfloor x + 1 \rfloor$, $\lim_{x \to 3} \lfloor x + 1 \rfloor$

    (c) $\lim_{x \to \sqrt{3}^+} \lfloor x^2 \rfloor$, $\lim_{x \to \sqrt{3}^-} \lfloor x^2 \rfloor$, $\lim_{x \to \sqrt{3}} \lfloor x^2 \rfloor$

    (d) $\lim_{x \to -\frac{1}{3}^+} \lfloor 3x \rfloor$, $\lim_{x \to -\frac{1}{3}^-} \lfloor 3x \rfloor$, $\lim_{x \to -\frac{1}{3}} \lfloor 3x \rfloor$

    (e) $\lim_{x \to 2^-} \lfloor \frac{x}{2} \rfloor$, $\lim_{x \to 2^+} \lfloor \frac{x}{2} \rfloor$, $\lim_{x \to 2} \lfloor \frac{x}{2} \rfloor$

    (f) $\lim_{x \to \frac{\pi}{2}^+} \lfloor \cos(x) \rfloor$, $\lim_{x \to \frac{\pi}{2}^-} \lfloor \cos(x) \rfloor$, $\lim_{x \to \frac{\pi}{2}} \lfloor \cos(x) \rfloor$

    (g) $\lim_{x \to \frac{\pi}{2}^+} \lfloor 3 \sin(x) \rfloor$, $\lim_{x \to \frac{\pi}{2}^-} \lfloor 3 \sin(x) \rfloor$, $\lim_{x \to \frac{\pi}{2}} \lfloor 3 \sin(x) \rfloor$

18. Find $\lim_{x \to 0} f(f(x))$ for the function f given by

    $$f(x) = \begin{cases} x - 3 & \text{if } x > 0 \\ 5 & \text{if } x = 0 \\ x^2 + 4x - 1 & \text{if } x < 0 \end{cases}$$

19. What can you say about the following statement?

    $\lim_{x \to a} f(x) = L$ if and only if $f(x)$ gets close to L whenever x gets close to a.

    Give reasons for your answer.

3.3 THE FORMAL THEORY OF LIMITS

3.3.1 Overview

In this section you will work with a mathematically precise (or formal) definition of the limit of a function at a point. Proofs of some limit statements and theorems will be given. You will also work with other formal definitions of the limit concept which will be used in the remainder of this text. You will begin with some activities that will help you make the mental constructions needed to understand the theoretical notion of limit which underlies the theory of calculus.

3.3.2 Activities

1. This activity is the culmination of the ε-δ window activities that have appeared periodically in this and the previous chapter. The general problem is that you are given a function f and a number a in its domain. You were looking at horizontal intervals of the form $(a - \delta, a + \delta)$, and vertical intervals of the form $(L - \varepsilon, L + \varepsilon)$ where δ, ε are positive numbers. In Section 3.1, Activity 9, page 134, your goal was to figure out when it was possible, given an interval of the form $(L - \varepsilon, L + \varepsilon)$, to find an interval of the form $(a - \delta, a + \delta)$ such that the graph of f over the interval $a - \delta$ to $a + \delta$ stays completely within the window with vertical range from $L - \varepsilon$ to $L + \varepsilon$.

   In this activity, you will take f to be the function given by $f(x) = x^2$, $a = 2$, $L = 4$. We don't mind telling you that in this case, given any $\varepsilon > 0$, that is, given any interval of the form $(4 - \varepsilon, 4 + \varepsilon)$, you can find a $\delta > 0$, that is, an interval $(2 - \delta, 2 + \delta)$. Your problem is to verify this for a few values of ε, but then to figure out analytically a general formula for δ, given ε.

2. Do exactly the same thing as in the previous problem, but this time for two situations. One is the function g, given by $g(x) = 1 + x$ with $a = 2$ and the other is for the function h given by $h(x) = 3$ with $a = -4.5$. In both cases, you should be able to figure out the value of L. You should also be able to completely solve this problem by making a rough sketch of the situation and thinking about it, without necessarily using the computer.

3. Try to do the same thing as in the previous two problems for the signum function, denoted sgn, defined by

$$\mathrm{sgn}(x) = \begin{cases} -1 & \text{if } x < 0 \\ 0 & \text{if } x = 0 \\ 1 & \text{if } x > 0 \end{cases}$$

   Explain what happens if you take $a = -0.5, 0, 2$. Note: The function **sgn** is a predefined function in many computer software systems.

4. In Activities 1 and 2 you looked for relationships that told you how to find δ, given ε so that the graph stayed in the window. In particular, you did this for the function g given by $g(x) = 1 + x$ and the function f given by $f(x) = x^2$. Now do it for the function $g + f$ at the domain value 2.

5. Consider the functions f, g, $f \circ g$ where $f(x) = x^2$ and $g(x) = 1 + x$. Investigate $\lim_{x \to 3} f(x)$, $\lim_{x \to 2} g(x)$, $\lim_{x \to 2} (f \circ g)(x)$.

   (a) Why is the first domain point different from the other two?

   (b) How did we choose the three domain points, 3, 2, 2?

   (c) Guess the value of each of these three limits. If you can't do these in your head, you may wish to use your **computer function lim**.

   (d) Do the ε-δ analysis on the three functions. (You already did it for the first two, except that you did f at 3, not 2, but it will not be much different.)

6. Try to do exactly the same things as the previous problem to investigate $\lim_{x\to 1} f(x)$, $\lim_{x\to 0} g(x)$, $\lim_{x\to 0}(f \circ g)(x)$ where these functions are given by

$$f(x) = \begin{cases} \frac{x^2-1}{x-1} & \text{if } x \neq 1 \\ 0 & \text{if } x = 1 \end{cases}$$

$$g(x) = \begin{cases} 1 + x\sin(\frac{1}{x}) & \text{if } x \neq 0 \\ 1 & \text{if } x = 0 \end{cases}$$

7. A function f is said to be bounded on an interval I if there is a number M such that $|f(x)| \leq M$ for every $x \in I$ provided x is in the domain of f. Use the tools and examples we have been working with along with your intuitive understanding of limits to answer the following questions.

   (a) Can you find an example of a function f, an interval I and a point $a \in I$ such that f is bounded on the interval I but does not have a limit at a? If so, find one, if not, explain why you can't.

   (b) Can you find an example of a function f, a point a such that f has a limit at a but is not bounded on any interval I which contains a? If so, find one, if not, explain why you can't.

8. Use your graphics utility to obtain graphs of

$$f(x) = 2 - \frac{x^2}{3}, \ g(x) = \frac{x\sin(x)}{1-\cos(x)}, \text{ and } h(x) = 2$$

all on the same coordinate axes. Evaluate $\lim_{x\to 0} f(x)$ and $\lim_{x\to 0} h(x)$. Use the picture which you have generated to evaluate $\lim_{x\to 0} g(x)$.

9. Use your **computer function lim** and your computer tool **table** to construct a table with the following information. The first column should consist of twenty points in the domain of the sine function. For each point a in the first column, there should be a corresponding entry of $\sin a$ in the second column. Finally, for each point a, the third column should contain the value of $\lim_{x\to a} \sin(x)$.

   Repeat this procedure of constructing a table for the cosine function. Speculate about a general rule which would explain the results found in your tables.

10. In Activity 7 of Section 3.1, you encountered the three functions V_1, V_2 and V_3. Later in the discussion of that section you saw that

$$\lim_{x\to 0} V_1(x) = \lim_{x\to 0} V_2(x) = \lim_{x\to 0} V_3(x)$$

Consider the following similar situation with the function f given by

$$f(x) = \frac{x^2-9}{x-3} = \frac{(x-3)(x+3)}{x-3}.$$

Explain in your own words why you think the following is true

$$\lim_{x\to 3} f(x) = \lim_{x\to 3}(x+3).$$

Make a general conjecture about limits the following

$$\lim_{x \to a} \frac{(x-a)g(x)}{x-a}$$

where g is a function which is continuous at a.

3.3.3 Discussion

The Formal Definition of Limit   Now it is time to bring together our different ways of thinking about limits to obtain a full understanding of this concept which we express in a formal definition of the limit of a function f at a point a.

In the first place there is an intuitive notion of the process of the independent, or domain, variable x taking on values that get "closer and closer" to a, but not reaching a for which we write $x \to a$. As this happens, the dependent, or co-domain, variable $y = f(x)$ takes on corresponding values which may get "closer and closer" and "stays close" to some value L. In other words, the function f coordinates the domain process of $x \to a$ with the co-domain process of $f(x)$ getting "closer and closer" and "staying close" to some number L. It could happen that $f(x)$ actually equals L for some value(s) of x — or not. That is not the issue with limits. The issue is that $f(x)$ becomes close to L and "stays close" to L as x gets "closer and closer" to a with $x \neq a$.

This intuitive notion of a limit is expressed in the many tables you constructed of values of x approaching a with corresponding values of $f(x)$ approaching L. It is also expressed in a graph when you think about moving along the horizontal axis towards a so that the values of the function (controlled or determined by the values of the independent variable) move along the vertical axis towards L. The **computer function lim** which you wrote in Exercise 1, Section 3.4.1, implements this intuitive point of view. The intuitive notion has a dynamic flavor with points moving around and approaching fixed positions.

Activity 1 is part of a group of connected activities and exercises that we have put in for you to work with and familiarize yourself with the situation. We are almost ready to discuss what is going on and we will do so in the next section. For now, we just want to try to get you to do some hard thinking about it.

Go back, read through Activity 1 and think about how you worked on it. Try to write down in your own words what you did, or should have done in solving this problem. Instead of explaining things in terms of windows, try to formulate the relationships in terms of inequalities and absolute values. For example, can you write a mathematical expression that corresponds to specifying the domain over which you are producing the graph as being the interval $[c-d, c+d]$? What about restricting the range values to be in the interval $[u, v]$?

The ε-δ windows that you worked with in Activities 1, 2, 3 and elsewhere in this book are computer activities that can help you move from an intuitive understanding to a formal specification of a limit. In the ε-δ window, you don't have values moving towards a stationary point. You only have a single window in which the value of the domain variable is restricted to an open interval $(a-\delta, a+\delta)$ and, as a result, the values of the range variable are restricted to some other open interval $(L-\varepsilon, L+\varepsilon)$ and therefore stay in the window. The dynamics of the intuitive notion appear here only in the idea that you are doing this for many windows. That is, in order for the function to have the limit L at a, it is required that *for every open interval $(L-\varepsilon, L+\varepsilon)$, it is possible to find an open interval $(a-\delta, a+\delta)$ so that the graph of f over the interval $a-\delta$ to $a+\delta$ stays completely within the window with vertical range from $L-\varepsilon$ to $L+\varepsilon$.*

If you want to think about it, you might imagine a contract between two persons. The party of the first part agrees to provide a δ (satisfying certain conditions) whenever the party of the second part submits an ε. The contract itself is static. The execution of the contract however, which might involve repeated submissions by the party of the second part followed by responses from the party of the first part, involves considerable motion and activity.

In other words, the idea of $f(x)$ approaching L is embodied in the idea of considering every possible open interval $(L - \varepsilon, L + \varepsilon)$, no matter how small, and the idea of x approaching a is embodied in the idea of being able to take any open interval $(a - \delta, a + \delta)$, no matter how small. What do you think about the fact that in one formulation, we speak of x getting close to a first and then about $f(x)$ getting close to L, whereas in the other formulation, we reverse it and consider the open interval, $(L - \varepsilon, L + \varepsilon)$ first and then try to find an open interval $(a - \delta, a + \delta)$?

Well, there is only one more step required to complete the progress from the intuitive notion of "approaching" through the idea of windows to obtain our formal definition. That step is to be more precise about what we mean when we say x is close to a (but is not equal to a) or $f(x)$ is close to L (and might equal L). This is formulated by saying that x is in $(a - \delta, a + \delta) - \{a\}$ and $f(x)$ is in $(L - \varepsilon, L + \varepsilon)$. We can write these two statements symbolically as follows.

$$0 < |x - a| < \delta, \ |f(x) - L| < \varepsilon$$

Thus, our formalization will read something like a condition that guarantees that for any $\varepsilon > 0$ it is possible to find a $\delta > 0$ such that if $0 < |x - a| < \delta$, then $|f(x) - L| < \varepsilon$. We now formalize this.

Definition 3.7

Let f be a function, a, L numbers. We say the limit of f at a is L provided that for every $\varepsilon > 0$, there exists a number $\delta > 0$ such that if $0 < |x - a| < \delta$, then $|f(x) - L| < \varepsilon$.

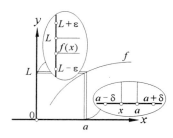

Figure 3.11. A geometrical interpretation of the definition of limit.

You should pore over this definition, the discussion that precedes it and think about all of the activities you have done to illustrate the ideas we are putting forth.

Examples of Limits — Applying the Definition
One of the first important things that you do with a definition in mathematics is to use it with examples to determine definitely whether or not a particular example has the property specified by the definition. What you had to do in Activities 1, 2, and 3 came very close to just this. Finding a formula for δ in terms of ε is just a way of describing how to proceed with any ε to find the appropriate δ.

You might ask how the potential limit L is obtained. This is a very different task from checking the definition once you have L. You get the limit by evaluating the function at nearby domain points. You might construct a table, or look at a graph, or apply a **computer function** like **lim**. Or you might just guess. Somehow or other, you come up with a candidate for the limit. Then you check the definition using this value for L. If it doesn't work, you try another value for L. If it keeps refusing to work, you then consider the possibility that the limit does not exist at that point.

Consider the function f given by $f(x) = x^2$ at the point 2. Do you see immediately that the obvious candidate for the limit is 4? Why?

Now think about applying the definition in this case. The goal is to show that, given any positive number ε, we must find a positive number δ such that if x is in the domain of f and $0 < |x - a| < \delta$, then $|f(x) - L| < \varepsilon$. Let's translate that into this particular situation. Imagine that you are given some $\varepsilon > 0$. You are looking for δ. You don't have it yet, but once you do have it, what must you do with it? You have to take any x in the domain of f. In this case, that just means any x. Then you assume that $0 < |x - a| < \delta$ and try to prove that $|x^2 - 4| < \varepsilon$.

Every limit problem is different at this point. You just have to look at your situation and see what can be done. In this case, we can analyze the inequality we are trying to establish. We write,

$$|x^2 - 4| = |(x - 2)(x + 2)| = |x - 2||x + 2|$$

What happens here is typical of many problems. We have written our expression in such a way that the factor $|x - 2|$ is explicit. This is a quantity that we know something about. We know that it is less than δ. Do you see why? It means that if we use our assumption, we can continue the expression,

$$|x^2 - 4| = |(x - 2)(x + 2)| = |x - 2||x + 2| < \delta|x + 2|$$

This is the quantity that we would like to conclude to be less than ε. Well, we know a little more. Given our assumption that $|x - 2| < \delta$, that is, that x is in the interval $(2 - \delta, 2 + \delta)$, what is the worse case? What is the very largest that $|x + 2|$ can be? Do you see that it can't be as much as $4 + \delta$?

This allows us to continue the expression one more step,

$$|x^2 - 4| = |(x - 2)(x + 2)| = |x - 2||x + 2| < \delta|x + 2| < (4 + \delta)\delta$$

Thinking a bit ahead, since we are just about to choose δ, let us decide beforehand that whatever we decide, we will make sure that $\delta \leq 1$. This will mean that $(4 + \delta)\delta < 5\delta$ and our string of expressions can be summarized as,

$$|x^2 - 4| < 5\delta.$$

Now it is time to review. We are given $\varepsilon > 0$. We want to be able to assume that $0 < |x - 2| < \delta$ and conclude from that the relation $|x^2 - 4| < \varepsilon$. We see that we *can* get the relation $|x^2 - 4| < 5\delta$. But we are free to choose δ! This relation tells us that all we need to do is take $\delta \leq \frac{\varepsilon}{5}$ and everything will work out, provided we remember our earlier decision and take $\delta \leq 1$. In other words, everything will be fine if we choose,

$$\delta = \min\left(\frac{\varepsilon}{5}, 1\right)$$

That was the investigative phase. Now we know what to do so we can summarize it. We are given a positive number, ε. We choose $\delta = \min(\frac{\varepsilon}{5}, 1)$ which is again a positive number. Now, assuming that $0 < |x - 2| < \delta$ we can compute,

$$\left|x^2 - 4\right| = |(x-2)(x+2)| = |x-2||x+2| < \delta|x+2| < 8(4+\delta)\delta \leq 5\delta \leq \varepsilon$$

so it follows that $\left|x^2 - 4\right| < \varepsilon$. This completes the verification of the example and we can say that

$$\lim_{x \to 2} x^2 = 4.$$

Examples of Limits — Eyeballing the Definition
Consider the example in which f is given by $f(x) = x$ and we are checking the limit at 2, again. If you understand what is going on, then some problems are very easy. It should be clear to you that the candidate for a limit is 2. Then you can jump ahead to the critical statement,

$$\text{if } 0 < |x-a| < \delta \text{ then } |f(x) - L| < \varepsilon.$$

and translate it into the specifics of this example,

$$\text{if } 0 < |x-2| < \delta \text{ then } |x-2| < \varepsilon.$$

Well, that's pretty simple. Given $\varepsilon > 0$, how do you choose δ? Do you see that the very obvious thing to do is to choose $\delta = \varepsilon$?

What would you do if you were checking the limit at some point other than 2?

An even simpler example is the case in which $f(x) = 3$, a constant function. No matter where you are checking the limit, since the values of $f(x)$ are always 3, the only thing they can approach is 3, so that is always going to be your candidate for the limit. Here the right thing to do is to just translate part of the general definition for this function, the part that replaces $f(x)$ by 3, and L by 3 as well. We get,

$$\text{if } 0 < |x-a| < \delta \text{ then } |3-3| < \varepsilon.$$

or

$$\text{if } 0 < |x-a| < \delta \text{ then } 0 < \varepsilon.$$

Well, $0 < \varepsilon$ is always true, given a positive number ε, so the definition of limit is satisfied, no matter what positive number you choose for δ.

Examples of Limits — Violating the Definition
If f is a function and a is a real number, the limit of f as x approaches a might not exist. Look at the function sgn in Activity 3. What would be the limit at 0? Intuitively, you can see that if x approaches 0 from the left, then the values of the function will "approach" (actually they will be right there!) -1. If you approach from the right, the values will approach 1. So it doesn't seem possible that the values of the function will approach a single number as x approaches 0. Let's try to see formally how our definition is violated.

Once again, given any $\varepsilon > 0$, we have to find δ. Suppose $\varepsilon = 4$. Would it work? Could you find a δ? The answer is yes. Indeed, any positive number would work for δ. In other words, you can always keep this function inside a window of height 4.

But the definition says that it has to be done for *any* $\varepsilon > 0$. Can you think of a window height that would be too small to contain this function? What about 0.5? Could you fit it in a window of height 0.5? This doesn't appear to be possible. Let's analyze the situation formally.

Consider any choice L for the limit. Suppose you are given $\varepsilon = 0.5$. Try to do it with some positive number δ. Do you think it will always be true for numbers x satisfying $0 < |x| < \delta$ (recall that $a = 0$ here) that $|f(x) - L| < 0.5$? We can see that this is impossible, no matter what was our original choice of L.

The point is that the condition $|x| < \delta$ does not prevent us from taking two values for x, one positive, one negative. This will give two values for $f(x)$ one equal to 1 and the other equal to -1. Now do you think we can have both of the following statements true,

$$|1 - L| < 0.5$$

$$|-1 - L| < 0.5?$$

The first one says that L is within 0.5 of 1 and the other says that L is within 0.5 of -1. Give it some thought and make sure that you understand why this is impossible. Thus the definition is not satisfied and we say that the limit of this function at 0 does not exist.

One-Sided and Two-Sided Limits

The function we just analyzed is an example of a general situation. You can look at what happens to the values of a function as the domain variable takes on values that approach a number a from the left, that is values that are less than a. If the values of the function approach some number, then we say that this number is the *limit from the left*. Similarly we have the idea of a *limit from the right*.

Look at the function sgn that we considered in the previous paragraph. Does it have a limit from the left? What is it? Does it have a limit from the right? What is it? Are the two limits the same?

Now compare this with what you found in Exercise 8, page 144. Sometimes it can happen that a function has a limit from the left at some point and also from the right at that point. It would seem that in this case, having a limit should mean that these two limits exist and are the same.

In Exercise 8, Section 3.1.4, and Activity 3 of this section you saw all of the possibilities when the one-sided limits exist. What about the function whose values are given by $\sin(\frac{1}{x})$? What about its one-sided limits at 0? You see that it can also happen that a function does not have a limit from the left or from the right. In this case it will not have a limit, of course.

Can you think of an example of a function which has a limit from the left but not from the right? How about a limit from the right but not from the left?

We can formalize all of this with a definition and a theorem. Study them carefully and try to make the connection in your mind between the intuitive ideas and their formal specification.

Definition 3.8

Let f be a function and let a and L be numbers. We say the limit from the left of f at a is L provided that for every $\varepsilon > 0$, there exists a number $\delta > 0$ such that if $a - \delta < x < a$, then $|f(x) - L| < \varepsilon$. We write this as

$$\lim_{x \to a^-} f(x) = L.$$

We say that the limit from the right of f at a is L, provided that for every $\varepsilon > 0$, there exists a number $\delta > 0$ such that if $a < x < a + \delta$, then $|f(x) - L| < \varepsilon$. We write this as

$$\lim_{x \to a^+} f(x) = L.$$

Among other things, this definition exemplifies a feature of mathematical discourse. It is very concise and precise. Look at the two definitions of limit from the left and limit from the right and compare them with the definition of limit. All three statements are almost exactly the same except for one small variation. In each definition one has either $a - \delta < x < a$, $a < x < a + \delta$, or $0 < |x - a| < \delta$. Everything else is identical in all three. The first condition says that our only restriction in demanding

Finally we can formalize the idea that a function having a limit at a point is the same as having both a limit from the left and a limit from the right at this point when these two one-sided limits are equal.

Proof. First, let us suppose that $\lim_{x \to a} f(x) = L$. We will show that $\lim_{x \to a^-} f(x) = L$. We are given $\varepsilon > 0$ and we are looking for a δ. Let's just use the same one that is given to us by virtue of the fact that the limit exists. Thus we have $\delta > 0$. What we know is that

$$\text{if } 0 < |x - a| < \delta \text{ then } |f(x) - L| < \varepsilon$$

and what we have to show is that

$$\text{if } a - \delta < x < a \text{ then } |f(x) - L| < \varepsilon$$

Our solution is to compose these two implications. We may assume that $a - \delta < x < a$. It follows that $|x - a| < \delta$ and hence by the implication that we know, $|f(x) - L| < \varepsilon$ which is the desired conclusion. Hence the limit from the left exists.

The proof that the limit from the right exists is done in exactly the same way.

Now let's prove the converse. We suppose that $\lim_{x \to a^-} f(x) = L$, and $\lim_{x \to a^+} f(x) = L$. We must show that $\lim_{x \to a} f(x) = L$. If we are given $\varepsilon > 0$, we know that the existence of the two one-sided limits provides us with two positive numbers, δ_1, δ_2 and we have two implications,

$$\text{if } a - \delta_1 < x < a \text{ then } |f(x) - L| < \varepsilon$$

and

$$\text{if } a < x < a + \delta_2 \text{ then } |f(x) - L| < \varepsilon$$

We want to find a value of δ that will make both of these implications go. We choose $\delta = \min(\delta_1, \delta_2)$. We have to show that

$$\text{if } 0 < |x - a| < \delta \text{ then } |f(x) - L| < \varepsilon$$

To this end, suppose that $0 < |x - a| < \delta$. Then it follows that either $a - \delta < x < a$ or $a < x < a + \delta$. In the first case we have,

$$a - \delta_1 < a - \delta < x < a$$

and in the second,

$$a < x < a + \delta < a + \delta_2$$

In either case we are permitted to conclude that $|f(x) - L| < \varepsilon$ as desired.

$\square$

As we have seen, this theorem can be used to show that a limit does not exist in a particular example. In many situations you can just determine the limit from the left and from the right and see that they are not equal.

Limits at ∞

In Section 3.2 you considered what happened as the value of the independent variable became very large in a positive direction or very large in a negative direction. You saw that if either $\lim_{x \to \infty} f(x) = L$ or $\lim_{x \to -\infty} f(x) = L$, then the line $y = L$ is a horizontal asymptote of the graph of f. We can now give the formal definition for the limit at ∞.

Definition 3.9

Let f be a function on an interval (c, ∞), c and L numbers. We say the limit of f at ∞ is L provided that for every $\varepsilon > 0$, there exists a number $M > 0$ such that if $x > M$, then $|f(x) - L| < \varepsilon$. We write this as

$$\lim_{x \to \infty} f(x) = L.$$

How would you make a formal definition for the limit of a function at $-\infty$, that is $\lim_{x \to -\infty} f(x) = L$? You'll get a chance to do this in the exercises.

Continuity at a Point

Because the definition of continuity of a function depends upon the notion of a limit, we can now give a more technically correct definition of continuity at a point. Here we have another example of the conciseness and preciseness of mathematical discourse. Look at the following alternative definition of continuity which is equivalent to Definition 3.4, and compare it with the definition of limit which we repeat.

Definition 3.10

Let f be a function, a, L numbers with a in the domain of f. We say that f is continuous at a provided that for every $\varepsilon > 0$, there exists a number $\delta > 0$ such that if $|x - a| < \delta$, then $|f(x) - L| < \varepsilon$.

Definition 3.7

Let f be a function, a, L numbers. We say the limit of f at a is L provided that for every $\varepsilon > 0$, there exists a number $\delta > 0$ such that if $0 < |x - a| < \delta$, then $|f(x) - L| < \varepsilon$.

In a moment we will give a detailed proof that Definition 3.10 is an equivalent definition of continuity (Definition 3.4), but first let us consider how this alternate statement captures the idea and how little it differs from the definition of limit.

The only differences between these two statements is that in the first, we have insisted that a is in the domain of f and replaced the inequality, $0 < |x - a| < \delta$ by $|x - a| < \delta$. What do these changes mean? The first means that f is defined at a. The second says that, in addition to everything that is required for the limit, we do not exclude the case, $0 = |x - a|$. That is, we also insist that $|f(x) - L| < \varepsilon$ when $x = a$. In other words, we are saying that for every $\varepsilon > 0$ we have $|f(a) - L| < \varepsilon$. Do you see why the fact that the number $|f(a) - L|$ is smaller than every positive number implies that $f(a) = L$? That is, we are requiring that the limit L at a equal the value of f at a.

Now we will try to express that in mathematical language.

Theorem 3.2

The two definitions, Definition 3.4 and Definition 3.10, of continuity are equivalent.

Proof. Suppose that we have a function f and a number a. First we have to show that if Definition 3.4 is satisfied, then so is Definition 3.10 with $L = f(a)$. Conversely, we must show that if Definition 3.10 is satisfied (for some L), then so is Definition 3.4, and it follows that $L = f(a)$.

So let us begin by assuming (Definition 3.4) that a is in the domain of f and $\lim_{x \to a} f(x) = f(a)$. We have to establish the requirements of Definition 3.10 for $L = f(a)$.

Given any $\varepsilon > 0$, we know from Definition 3.4 that $\lim_{x \to a} f(x) = f(a)$ and so we have some $\delta > 0$ such that

$$\text{if } 0 < |x - a| < \delta, \text{ then } |f(x) - f(a)| < \varepsilon$$

This is all that Definition 3.10 requires except that, in addition, we must know that if $0 = |x - a|$ then in this case as well it follows that $|f(x) - f(a)| < \varepsilon$. That is, we have to know that if $x = a$ then $|f(x) - f(a)| < \varepsilon$. But if $x = a$ then

$$|f(x) - f(a)| = |f(a) - f(a)| = 0 < \varepsilon$$

as desired.

Conversely, let us suppose that Definition 3.10 is satisfied for some L. Then we have the statement that for every $\varepsilon > 0$, there is a $\delta > 0$ such that

$$\text{if } |x - a| < \delta, \text{ then } |f(x) - L| < \varepsilon$$

This is actually saying more than the requirement

$$\text{if } 0 < |x - a| < \delta, \text{ then } |f(x) - f(a)| < \varepsilon$$

because if $0 < |x - a| < \delta$ then it follows that $|x - a| < \delta$ and hence $|f(x) - f(a)| < \varepsilon$.

Finally we observe that the statement in Definition 3.10 says that the number $|f(a) - L|$ is smaller than every positive number. Hence it if it were positive, it would be smaller than itself which is impossible, so it can't be positive and therefore must be 0. Hence $f(a) = L$.

□

Discontinuities In the examples you have worked with so far in this course you have seen most of the types of discontinuities that can occur. A function like V_1 in Section 3.1.2, Activity 7, page 134, is *discontinuous* at 0 because it is not defined at 0. We say that the function V_2 in that activity has a *removable discontinuity* at 0 because you can make it continuous at 0 because it has a limit at 0 and you could redefine it at that single point if you wish. See Figure 3.12. A function like the heat capacity function of Activity 5, Section 3.1.2, page 133, has a *jump discontinuity* at the point 160 because it jumps values as it crosses that point. See Figure 3.13. We say that a function like the compressibility function Z of Chapter 3 has a *"blow-up"*, or *infinite*, discontinuity at the point $V = 200$ because the values of Z go to ∞ as the domain values approach 200. See Figure 3.10, page 155. Finally, we say that a function like $\sin(\frac{1}{x})$ has an *oscillatory discontinuity* at 0 as shown in Figure 3.14.

Think about other possible examples of these types of discontinuites. Some of them are important enough to give formal definitions specifying them. Think about how to do this before reading the next few lines.

Definition 3.11 (Removable discontinuity)

Let f be a function, and a a number. If $\lim_{x \to a} f(x) = L$ but $L \neq f(a)$, then we say that f has a removable discontinuity at a.

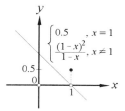

Figure 3.12. A removable discontinuity.

Figure 3.13. A jump discontinuity.

Definition 3.12

Let f be a function, and a a number. If $\lim_{x \to a^+} f(x)$ and $\lim_{x \to a^-} f(x)$ exist, but are not equal, then we say that f has a jump discontinuity at a.

As you saw in section 2, if the function f has no finite limit at $x = a$ the values of the function might increase without bound. In this case we write $\lim_{x \to a} f(x) = \infty$ to indicate the behavior of the values of f as $x \to a$. Such a "limit" is called an *infinite limit*. We formalize this situation in the following definition.

Definition 3.13

Let f be a function, and a a number. If, for every positive number M, there is a positive number δ such that,

$$\text{if } 0 < |x - a| < \delta, \text{ then } f(x) > M,$$

then we say that f is unbounded at a. In this case, we write,

$$\lim_{x \to a} f(x) = \infty.$$

What is the definition of the infinite limit $\lim_{x \to a} f(x) = -\infty$? What about infinite one-sided limits at a? What about infinite limits at ∞ and $-\infty$? You'll get a chance to make these formal mathematical definitions in the exercises.

The next one is a little tough. See if you can understand how it captures the behavior of $\sin(\frac{1}{x})$ as shown in Figure 3.14.

Definition 3.14

Let f be a function, and a a number. If there are at least two different numbers L_1, L_2 such that in every open interval containing a there are at least two numbers x_1, x_2 such that $f(x_1) = L_1$ and $f(x_2) = L_2$, then we say that f has an oscillatory discontinuity at a.

Preservation of Limits Under Arithmetic

In Activity 4 you had a chance to think about the ε-δ window for the sum of two functions in terms of the ε-δ window for the individual functions. What this leads up to is that you can figure out the limit of the sum of two functions by simply adding the limits of the two functions (if they exist). This is a very convenient fact that is true for most arithmetic operations between functions and therefore simplifies a host of computations. These are the rules that you had a chance to figure out from the examples in Activity 6 of Section 3.1.2, page 133. The proofs of these rules for computing limits are not very interesting. Their value is mainly for giving you practice in working with this kind of logic. In this paragraph, we state all the rules and prove one of them. The proofs of the others are left for you in the exercises.

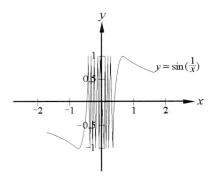

Figure 3.14. $\sin(\frac{1}{x})$ *has an oscillatory discontinuity at x = 0.*

Theorem 3.3

Let f, g be functions and a a number. Let $\lim_{x \to a} f(x) = L$ *and* $\lim_{x \to a} g(x) = K$. *Then the following statements hold.*

1. $\lim_{x \to a} (f(x) + g(x)) = L + K$

2. $\lim_{x \to a} (f(x) - g(x)) = L - K$

3. $\lim_{x \to a} (f(x)g(x)) = LK$

4. *If* $K \neq 0$ *then* $\lim_{x \to a} \dfrac{f(x)}{g(x)} = \dfrac{L}{K}$

5. *If t is any number then* $\lim_{x \to a} tf(x) = tL$

Proof. *(We give, by way of illustration, the proof of the third statement. We are going to present this proof in its final form, without mentioning any of the thinking that went into it. See if you can figure out where it all comes from. We use an idea that is almost exactly the same as what we did in checking that* $\lim_{x \to 2} x^2 = 4$ *(p. 165) and another analysis that will be repeated in the proof of Theorem 3.9.)*

As usual, for a basic limit proof, we begin with an arbitrary positive number ε and we are looking for a positive number δ. The two assumptions, that $\lim_{x \to a} f(x) = L$ and $\lim_{x \to a} g(x) = K$ allow us to choose a number δ for any given positive number ε. We apply this three times, once for f and twice for g.

First, taking $\varepsilon = 1$, choose δ_1 such that

$$\text{if } 0 < |x - a| < \delta_1, \text{ then } |g(x) - K| < 1$$

It follows that for $0 < |x - a| < \delta_1$ we have $|g(x)| < |K| + 1$.

Now we apply the limit definition to f using $\frac{\varepsilon}{2(|K|+1)}$. This gives δ_2 such that

$$\text{if } 0 < |x - a| < \delta_2, \text{ then } |f(x) - L| < \frac{\varepsilon}{2(|K|+1)}$$

Finally, we apply the limit definition to g again with $\frac{\varepsilon}{2|L|}$ to obtain δ_3 satisfying,

$$\text{if } 0 < |x - a| < \delta_3, \text{ then } |g(x) - K| < \frac{\varepsilon}{2|L|}$$

Putting this all together, we can take $\delta = \min(\delta_1, \delta_2, \delta_3)$. It follows that if $0 < |x - a| < \delta$, then all three of our relations will hold and we can calculate,

$$
\begin{aligned}
|(fg)(x) - LK| &= |f(x)g(x) - Lg(x) + Lg(x) - LK| \\
&\leq |f(x)g(x) - Lg(x)| + |+ Lg(x) - LK| \\
&= |f(x) - L||g(x)| + |L||g(x) - K| \\
&< \frac{\varepsilon}{2(|K|+1)}(|K|+1) + |L|\frac{\varepsilon}{2|L|} \\
&= \frac{\varepsilon}{2} + \frac{\varepsilon}{2} = \varepsilon
\end{aligned}
$$

as desired.

$\square$

Theorem 3.3 has an important consequence for continuity. You should see that the following theorem is an immediate consequence of the previous theorem and the relationship between limits and continuity.

Theorem 3.4

Let f, g be functions which are continuous at a. Then the following statements hold.

1. *The function $f + g$ is continuous at a.*

2. *The function $f - g$ is continuous at a.*

3. *The function fg is continuous at a.*

4. *If $g(a) \neq 0$ then the function $\frac{f}{g}$ is continuous at a.*

5. *If t is any number, then the function tf is continuous at a.*

The above two theorems really settle the questions of limit and continuity for many (but not all) of the functions with which you are familiar. Let's see how.

Continuity of Rational Functions

You saw in our discussion of Activity 2 that any constant function is continuous at every point and so is the function whose value at x is $1+x$. Now apply Theorem 3.4. If, from the latter function, you subtract the constant function 1, then you get the identity function, whose value at x is x itself and the theorem tells us that it is continuous at every point. You can apply the theorem again to conclude that you can multiply the identity function by any number and it will still be continuous. So functions like $12x$ are continuous at every point. The theorem also tells you that you can add two continuous functions and the result is continuous. Hence functions like $6+12x$ are continuous at every point.

How about higher powers of x? Well, just use multiplication. Hence functions which are given by expressions like

$$2x^5 - 3.7x^4 + 17.64x + \sqrt{5}$$

are continuous at every point.

We can summarize this in a simple way.

Theorem 3.5

Every function given by a polynomial is continuous at every point.

We get even more by throwing in division. Functions given by expressions like

$$\frac{1-3x^2-6.5x^3-17x^4+\sqrt{3}x^6}{x^2-2x+1}$$

are called *rational* functions because they are the ratio of two polynomials. Such a function will also be continuous because you can divide without losing continuity — but not at all points. You have to avoid dividing by 0.

Theorem 3.6

Every rational function is continuous at every point except those at which the denominator is 0.

These are not all of the functions you know. There are functions like $\sqrt{1+x^4}$, the trigonometric functions (sin, cos, ...) and also the functions defined in parts by piecing together two or more expressions. Nevertheless the rational functions form a big portion of your repertoire. Most of these other functions are also continuous, or it is easy to see where they fail to be so. Thus, it is possible to use mathematics to determine the continuity of many functions — probably almost all of the functions given by single expressions that most of you will ever see. Hence limits very often can be computed just by evaluating the function.

So why did we make such a fuss about limits? Well, one reason is that you can develop your mathematical power by learning how to think in these terms. But there is another reason. Hopefully by now you have come to realize that not every function is given by an expression you can evaluate. For such functions you have to understand about limits. These functions are not rare. They arise almost all the time when you work with what will be one of the most important concepts in calculus — the derivative, which is coming in the next chapter.

Preservation of Limits Under Composition

A couple of paragraphs above we showed

that limits were preserved under arithmetic operations. The limit of a sum is the sum of the limits. Similarly for difference, product and multiplication by a constant. For quotients it is the same, except you have to rule out dividing by 0.

It is not quite so simple with compositions. In Activity 6 you saw how this could break down. In that situation, $\lim_{x \to 0} g(x) = 1$ and $\lim_{x \to 1} f(x) = 0$. However, $\lim_{x \to 0} (f \circ g)(x)$ does not exist. If you have any difficulty seeing what is going on with this example, try your **computer function lim** on it.

Difficulties such as this disappear completely if you are dealing with continuous functions.

Theorem 3.7

Let $\lim_{x \to a} g(x) = L$ and $\lim_{x \to L} f(x) = K$ and assume further that f is continuous at L. Then $\lim_{x \to a} (f \circ g)(x) = K$.

Proof. Given any $\varepsilon > 0$ our task is to find a $\delta > 0$ such that

$$\text{if } 0 < |x - a| < \delta \text{ then } |(f \circ g)(x) - K| < \varepsilon.$$

Given this ε, we apply the definition of the limit for f at L and the continuity to find a $\delta_1 > 0$ such that

$$\text{if } |x - L| < \delta_1 \text{ then } |f(x) - K| < \varepsilon.$$

Now, reversing roles, we use this δ_1 in the definition of limit for g at a to find $\delta > 0$ such that

$$\text{if } 0 < |x - a| < \delta \text{ then } |g(x) - L| < \delta_1.$$

We now summarize. We were given an $\varepsilon > 0$ and have found a $\delta > 0$. If $0 < |x - a| < \delta$, we have $|g(x) - L| < \delta_1$. But then, by the previous statement, $|f(g(x)) - K| < \varepsilon$, as desired.

□

An important way of expressing this result is that, under the assumptions of the theorem, we have,

$$\lim_{x \to a} (f \circ g)(x) = f\left(\lim_{x \to a} g(x) \right) = f(L) = K.$$

In other words, the limit can "pass through" the continuous function f.

Properties of Limits — Uniqueness

From an intuitive point of view, if you think about a limit in terms of the value that a function approaches, then a function cannot have more than one limit at a point. It is an interesting exercise in logic to see that our formal definition captures this observation with little difficulty. The argument actually is very similar to our proof that if a function has a limit at a point, then its left and right limits must be the same. The whole idea rests on the fact that if a non-negative number b is smaller than every positive number, it follows that b must be 0.

Theorem 3.8

Suppose that $\lim_{x \to a} f(x) = L_1$ and $\lim_{x \to a} f(x) = L_2$. Then $L_1 = L_2$.

Proof. We will show that $|L_1 - L_2|$ is smaller than every positive number. Let $\varepsilon > 0$. We apply the definition of the limit to f twice, once for L_1 and once for L_2. We use, however, not ε, but $\frac{\varepsilon}{2}$. This gives us positive numbers δ_1, δ_2 such that,

$$\text{if } 0 < |x - a| < \delta_1 \text{ then } |f(x) - L_1| < \tfrac{\varepsilon}{2}.$$

and,

$$\text{if } 0 < |x - a| < \delta_2 \text{ then } |f(x) - L_2| < \tfrac{\varepsilon}{2}.$$

Now we pick any number x_0 which satisfies both $0 < |x - a| < \delta_1$ and $0 < |x - a| < \delta_2$. We simply choose a point in the interval $(a - \min(\delta_1, \delta_2), a + \min(\delta_1, \delta_2))$ which is different from a but sufficiently close to it.

Now we have both, $|f(x_0) - L_1| < \tfrac{\varepsilon}{2}$ and $|f(x_0) - L_2| < \tfrac{\varepsilon}{2}$ and so, adding, we get,

$$|L_1 - L_2| = |L_1 - f(x_0) + f(x_0) - L_2| \le |L_1 - f(x_0)| + |f(x_0) - L_2| < \frac{\varepsilon}{2} + \frac{\varepsilon}{2} = \varepsilon$$

as desired.

$\square$

Properties of Limits — Boundedness

In Activity 7 you looked for examples to illustrate the relation between boundedness and having a limit. The property of a function being bounded on an interval is less of a requirement than that the function have a limit. For example, the function whose values for $x \ne 0$ are given by $\sin(\tfrac{1}{x})$ is certainly bounded on every interval. As we have seen, it does not have a limit at 0. Can you think of other examples? What about functions defined in parts?

On the other hand, having a limit at a point is enough to guarantee that the function is bounded on some interval containing the point.

Theorem 3.9

Suppose that the function f has a limit at a. Then there is an open interval centered at a on which the function is bounded.

Proof. Let the limit be L. We apply the definition with $\varepsilon = 1$. Then we have $\delta > 0$ satisfying,

$$\text{if } 0 < |x - a| < \delta \text{ then } |f(x) - L| < 1.$$

This says that for x in the interval $(a - \delta, a + \delta)$, $x \ne a$, the values of the function are in the interval $(L - 1, L + 1)$. In other words, on the interval $(a - \delta, a + \delta) - \{a\}$, which is centered at a, the function is bounded by $|L| + 1$. Suppose now that a is in the domain of f. If we also throw in $|f(a)|$, then we can conclude that on the entire interval $(a - \delta, a + \delta)$, $|f(x)|$ is bounded by $\max(|f(a)|, |L| + 1)$.

$\square$

Although boundedness is not enough to guarantee the existence of a limit, there is one special situation in which it is helpful. Consider the limit at 0 of the function whose values for $x \ne 0$ are given by $x \sin(\tfrac{1}{x})$. You have seen that this limit exists and it is equal to 0. The reason is that the bounded function, $\sin(\tfrac{1}{x})$ is multiplied by the function x which goes to 0 as x goes to 0. This is a completely general phenomenon and we formalize it in the following theorem.

Theorem 3.10

Suppose that g is a function which is bounded on some open interval centered at a and $\lim_{x \to a} f(x) = 0$. Then $\lim_{x \to a} (fg)(x) = 0$.

Proof. Since g is bounded, there exists an M and a $\delta_1 > 0$ such that $|g(x)| < M$ on the interval $(a - \delta_1, a + \delta_2)$. To show that $\lim_{x \to a} (fg)(x) = 0$, we must consider an arbitrary $\varepsilon > 0$ and find $\delta > 0$ such that

$$\text{if } 0 < |x - a| < \delta \text{ then } |f(x)g(x)| < \varepsilon.$$

Given ε we apply the definition of limit for f and the number $\frac{\varepsilon}{M}$ which is still positive. This gives a δ_2 such that

$$\text{if } 0 < |x - a| < \delta_2 \text{ then } |f(x)| < \frac{\varepsilon}{M}.$$

Now we choose $\delta = \min(\delta_1, \delta_2)$ so that if $0 < |x - a| < \delta$ then both of our statements will be true and we have,

$$|f(x)g(x)| = |f(x)||g(x)| < \frac{\varepsilon}{M} M = \varepsilon$$

as desired.

$\square$

Properties of Limits — The Squeeze Theorem

You can think about Theorem 3.10 as saying that if you take a function f that has limit 0 at some point a and you multiply it by a function g which doesn't get too large, then the product will also have the limit 0 at the point.

A similar, but more general situation occurs when you squeeze a function between two functions that both approach the same number. Then the squeezed function will be forced to approach that number as well. This is the situation you encountered in Activity 8. See Figure 3.15 below.

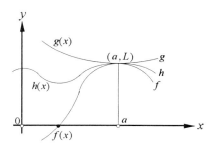

Figure 3.15. The Squeeze theorem.

Theorem 3.11 (Squeeze theorem)

Let f, g, and h be three functions such that $\lim_{x \to a} f(x) = L$, $\lim_{x \to a} g(x) = L$, and for every x in some open interval centered at a, we have $f(x) \le h(x) \le g(x)$. Then $\lim_{x \to a} h(x) = L$.

Proof. As usual, we begin with an arbitrary $\varepsilon > 0$. To make a long story short we repeat considerations we have made several times before to find a $\delta > 0$ such that for $0 < |x - a| < \delta$ we have the following three relations,

$$|f(x) - L| < \varepsilon$$

$$|g(x) - L| < \varepsilon$$

$$f(x) \leq h(x) \leq g(x)$$

Then, using the fact that $|A| < B$ is the same as $-B < A < B$, we have, for $0 < |x - a| < \delta$,

$$-\varepsilon < f(x) - L \leq h(x) - L \leq g(x) - L < \varepsilon$$

so that $|h(x) - L| < \varepsilon$ as desired.

$\square$

This theorem can be used, together with a geometric argument to establish a very important special limit. The fact should not surprise you since you have seen it several times in this chapter, at least in approximation, but the proof is quite different from anything we have used before.

Theorem 3.12

$$\lim_{x \to 0} \sin(x) = 0.$$

Proof. Recall that a central angle of a unit circle has radian measure x if it intercepts an arc of length x. From the picture in Figure 3.16, we can see that the area of the triangular segment is $\frac{1}{2}\sin(x)$, which is smaller than the area of the circular segment, $\frac{x}{2}$.

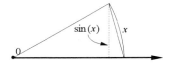

Figure 3.16. A circular angle cut off by a small positive angle of x radians.

This gives,

$$\frac{1}{2}\sin(x) < \frac{x}{2}$$

or, for $x > 0$,

$$0 < \sin(x) < x$$

Since $\sin x$ is negative for small negative x, we may conclude that for small x, positive or negative,

$$\sin(x) < |x|$$

Multiplying this inequality by -1 and using the fact that sin is an odd function, we have, for small x, positive or negative,

$$-|x| < -\sin(x) = \sin(-x)$$

and replacing x by $-x$ gives,

$$-|x| = -|-x| < \sin(x)$$

Hence we have, on some interval centered at 0,

$$-|x| < \sin(x) < |x|.$$

Now you can check that $\lim_{x \to 0} |x| = 0$ by considering the left and right sided limits. Hence, both $|x|$ and $-|x|$ have limit 0 at 0. By the squeeze theorem, the same is true of $\sin x$.

$\square$

Of course it follows immediately from this theorem that the function sin is continuous at 0.

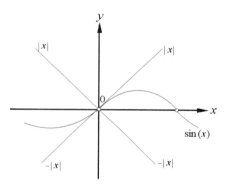

Figure 3.17. $-|x| < \sin(x) < |x|$.

Continuity of Trigonometric Functions
We are now in a position to establish the continuity of all the trigonometric functions at (almost) all points. This result for the sine and cosine functions was previewed in Activity 9. The result for the remaining trigonometric functions follows from Theorem 3.12 and the standard trigonometric identities.

Theorem 3.13

The functions sin and cos are continuous at every point. The functions tan, cot, sec, and csc are continuous at every point except where they are not defined.

Proof. We begin by showing that for any number a we have $\lim_{x \to a} \sin(x) = \sin(a)$. Look at the trigonometric formula

$$\sin(x) - \sin(a) = 2\cos\left(\frac{x+a}{2}\right)\sin\left(\frac{x-a}{2}\right)$$

Consider the function h given by $h(x) = \sin(\frac{x-a}{2})$. We can write this as the composition $f \circ g$ where $f = \sin(x)$ and g is given by $g(x) = \frac{x-a}{2}$. By Theorem 3.5, $\lim_{x \to 0} g(x) = g(0) = 0$ and as we just saw, f is continuous at 0. Hence by Theorem 3.7 (using the formulation just after the theorem), we have

$$\lim_{x \to a} h(x) = \lim_{x \to a} f(g(x)) = f\left(\lim_{x \to a} g(x)\right) = \sin(0) = 0.$$

Finally, using the fact that the limit of a constant is the constant to conclude that $\lim_{x \to a} \sin(a) = \sin(a)$ and the fact that the limit of a sum is the sum of the limits, we conclude that $\lim_{x \to a} \sin(x) = \sin(a)$. Hence sin is continuous at any a.

The rest are easy. We obtain the continuity of cos at any a from the continuity of sin, the relation

$$\cos a = \sin\left(a + \frac{\pi}{2}\right)$$

and similar arguments with the arithmetic and composition rules. The others are done completely with the arithmetic rules.

$\square$

We end this chapter by formally proving a very important special limit through the use of the Squeeze Theorem. The proof makes use of the picture in Figure 3.18.

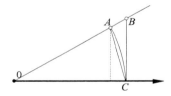

Figure 3.18. Squeezing $\frac{\sin(\theta)}{\theta}$.

Theorem 3.14

$$\lim_{\theta \to 0} \frac{\sin(\theta)}{\theta} = 1$$

Proof. From the picture in Figure 3.18, we see that the area of triangle OAC is less than the area of the circular segment OAC which is less than the area of triangle OBC. This gives us the relation,

$$\frac{\sin(\theta)}{2} < \frac{\theta}{2} < \frac{\tan(\theta)}{2}$$

Simplifying and dividing by $\frac{\sin(\theta)}{2}$ yields

$$1 \le \frac{\theta}{\sin(\theta)} < \frac{1}{\cos(\theta)}$$

By the continuity of cos we have $\lim_{\theta \to 0} \cos(\theta) = 1$ and so, by Theorem 3.3, $\frac{\theta}{\sin(\theta)}$ is squeezed by two functions both of which approach 1. The theorem then follows from the Squeeze Theorem.

□

The Replacement Theorem

In Activity 10 you considered the function f given by

$$f(x) = \frac{x^2 - 9}{x - 3} = \frac{(x-3)(x+3)}{x - 3}.$$

Do you know why the following is true?

$$\lim_{x \to 3} f(x) = \lim_{x \to 3} (x + 3)$$

Did you say that the limit of f at 3 does not depend on the value of f at 3, only on what happens at $x \to 3$ with $x \neq 3$? The following theorem, commonly called the Replacement theorem, characterizes any such situation. The proof is left as an exercise.

Theorem 3.15 (The Replacement Theorem)

If f and g are functions with $f(x) = g(x)$ for $x \neq a$ and $\lim_{x \to a} g(x) = L$, then $\lim_{x \to a} f(x) = L$.

3.3.4 Exercises

1. For the function f shown in Figure 3.19, $\lim_{x \to 0.5} f(x) = 0.75$. For $\varepsilon = \frac{1}{4}$ which of the following values of δ satisfy the definition of limit: $1, \frac{1}{2}, \frac{1}{4}, \frac{1}{8}, \frac{1}{16}, \frac{1}{32}$?

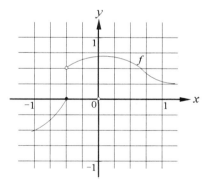

Figure 3.20. $\lim_{x \to -0.5^+} f(x) = 0.5$.

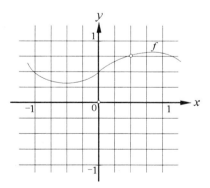

Figure 3.19. $\lim_{x \to 0.5} f(x) = 0.75$.

2. For the function f shown in Figure 3.20, $\lim_{x \to -0.5^+} f(x) = 0.5$. For $\varepsilon = \frac{1}{8}$ which of the following values of δ satisfy the ε-δ definition of limit: $1, \frac{1}{2}, \frac{1}{4}, \frac{1}{8}, \frac{1}{16}$.

3. Explain in your own words why, in the definition of limit, we have $0 < |x - a| < \delta$, but only $|f(x) - L| < \varepsilon$.

4. What is the largest possible value of δ for $\varepsilon = 0.01$ in the definition of limit for $\lim_{x \to 3} x^{\frac{2}{5}} = 8$?

5. Prove that the statement $\lim_{x\to 3} 2x = 5.99$ is false by showing that for some $\varepsilon_0 > 0$ there is no value of δ which will satisfy the definition of limit. Find such an ε_0 and explain why this shows that the statement is false.

6. Use the definition of a limit to establish each of the following limits.

   (a) $\lim_{x\to 2} 3x = 6$
   (b) $\lim_{x\to 1} 2x + 3 = 5$
   (c) $\lim_{x\to -1} 5 - x = 6$
   (d) $\lim_{x\to 9} \sqrt{x} = 3$
   (e) $\lim_{y\to 3} \dfrac{3}{y} = 1$
   (f) $\lim_{x\to 1} 3x^2 + 5 = 8$
   (g) $\lim_{x\to -1} \dfrac{3}{x+4} = 1$
   (h) $\lim_{x\to 2} \dfrac{4}{x^2} = 1$
   (i) $\lim_{x\to 8} \sqrt[3]{x} = 2$
   (j) $\lim_{x\to \pi} \sin(x) = 0 = \sin(\pi)$
   (k) $\lim_{x\to \pi} \cos(x) = -1 = \cos(\pi)$

7. Do each of the following problems in the two ways:

   (a) Using ε-δ windows on a computer.

   (b) Analytically using estimates involving absolute values and inequalities.

      i. A soil sample has an approximate volume of 400 cm^3 and a mass m of exactly 200 grams. Find a positive number δ so that the error in the measurement of the volume V would yield an error less than 10^{-2} grams/cm^3 in the calculated density ρ of the soil. Use $\rho V = m$ where ρ is the density in grams/cm^3, V is the volume in cm^3 and m is the mass in grams. Can you find the largest such δ that will work? Can you find all such δ that will work?

      ii. The area of a square picture is measured to be approximately 4 meters^2. Find a positive number δ such that an error less than δ in this measurement will yield an error less than 10^{-3} meters in the calculated width of the picture. Can you find the largest such δ that will work? Can you find all such δ that will work?

      iii. The radius of a glass cube is measured to be approximately 1 inch. Find a positive number δ so that an error less than δ in the measurement of the radius would yield an error less than 10^{-2} in^3 in the calculated volume of the cube. Can you find the largest such δ that will work? Can you find all such δ that will work?

      iv. The radius r of a steel ball bearing is measured to be 2 cm. Find a positive δ so that an error less than δ in the measurement of the radius would yield an error less than 10^{-2} in^3 in the calculated volume V of the bearing. Can you find the largest such δ that will work? Can you find all such δ that will work? (Use $V = 4\pi R^3/3$)

      v. A 100 yard long football field is to be marked off using a measuring tape that is approximately 10 yards long. How close should the length of the measuring tape be to 10 yards so that the actual length of the field is 100 yards with an error of no more than 0.02 yards?

      vi. A square garden is to have an area of approximately 400 feet^2. How close to 20 feet should the length of each side be so that the actual area of the square is 400 feet^2 with an error of no more than 0.01 feet^2? Assume each side is within 0.1 feet of 20 feet.

vii. How close to 4 must the radius of a circle be so that the area of the circle is within 0.001 of 16π? Assume that the radius is within 0.1 of 4.

viii. An electrical system is to have a voltage of exactly 6 volts. How close to 2 ohms must the resistance be in the system so that the current is within 10^{-2} amps of 3 amps? Use the fact that $I = V/R$ where I is the current in amps, V is the voltage in volts and R is the resistance in ohms.

8. In the ε-δ proof of $\lim_{x\to 2} x^2 = 4$, it was assumed (see page 165) that $|x - 2| < \delta$, that is, that x is in the interval $(2 - \delta, 2 + \delta)$. Show that $|x + 2|$ can't be as much as $4 + \delta$, i.e., $|x + 2| < 4 + \delta$.

9. Use the definition of limit to prove that $\lim_{x\to a} f(x) = L$ is equivalent to
$$\lim_{h\to 0} f(a + h) = L.$$

10. For the following situations, use the definition of limit to prove that
$$\lim_{x\to a} \sqrt[n]{x} = \sqrt[n]{a}$$

   (a) n is an even positive integer and $a > 0$. Why must a be greater than 0 in this case?

   (b) n is an odd positive integer.

11. (a) Prove that if $\lim_{x\to a} f(x) = L$, then $\lim_{x\to a} |f(x)| = |L|$.

(b) Show that the following converse of part (a) is false by giving a counterexample.

If $\lim_{x\to a} |f(x)| = |L|$, then $\lim_{x\to a} f(x) = L$.

12. Complete the proof of Theorem 3.1 (p. 168) by showing that the existence of a limit implies the existence of the limit from the right and then show that the limit and limit from the right are equal.

13. Using the definition of limit at ∞, page 169, verify that the definition is satisfied for the following limit.
$$\lim_{x\to\infty} \frac{2x}{1+x} = 2$$

14. (a) Use the definition of limit at ∞, page 169, to prove the following

   For any real numbers c and n with $n > 0$,
$$\lim_{x\to\infty} \frac{c}{x^n} = 0.$$

   (b) Use the result of part (a) to prove the general rule you found in Activity 4, Section 3.2.2, page 150, for the following limit situation.
$$\lim_{x\to\infty} \frac{a_0 x^n + a_1 x^{n-1} + \cdots + a_n}{b_0 x^k + b_1 x^{k-1} + \cdots + b_k}$$

15. Let f be a function.

   (a) Give definitions for each of the following (general) limit situations.

   (b) Make a sketch illustrating each of the indicated (specific) limit situations.

   (c) Using your definition in part (a), verify that the definition of limit is satisfied for each of the indicated (specific) limit situations.

   Here are the limit situations and limits to verify.

i. $\lim_{x \to -\infty} f(x) = L$;

$\lim_{x \to -\infty} \frac{1}{x-1} = 0$

ii. $\lim_{x \to a^+} f(x) = \infty$;

$\lim_{x \to 1^+} \frac{1}{x-1} = \infty$

iii. $\lim_{x \to a^-} f(x) = \infty$;

$\lim_{x \to 1^-} \frac{1}{1-x} = \infty$

iv. $\lim_{x \to a^+} f(x) = -\infty$;

$\lim_{x \to 1^+} \frac{1}{1-x} = -\infty$

v. $\lim_{x \to a^-} f(x) = -\infty$;

$\lim_{x \to 1^-} \frac{1}{x-1} = -\infty$

vi. $\lim_{x \to \infty} f(x) = \infty$;

$\lim_{x \to \infty} x^2 = \infty$

vii. $\lim_{x \to \infty} f(x) = -\infty$;

$\lim_{x \to \infty} -x^2 = -\infty$

viii. $\lim_{x \to -\infty} f(x) = \infty$;

$\lim_{x \to -\infty} x^2 = \infty$

ix. $\lim_{x \to -\infty} f(x) = -\infty$;

$\lim_{x \to -\infty} x^3 = -\infty$

16. Give an example of a function which satisfies the following.

(a) A function which is defined at 1, discontinuous at 1 and which has right and left hand limits at 1.

(b) A function which is not defined at 3, discontinuous at 3 and which has right and left hand limits at 3.

(c) A function which is not defined at 2, is discontinuous at 2, has a left hand limit at 2, but no right hand limit at 2.

(d) A function which is not defined at 1, is discontinuous at 1, has a right hand limit at 1 and such that the limit from the left at 1 does not exist.

(e) A function defined at 3 and having a jump discontinuity at 3.

(f) A function not defined at −1 and having a removable discontinuity at −1.

(g) A function not defined at 1 and having a non removable discontinuity at 1.

(h) A function discontinuous at 2 and having oscillatory behavior at 2.

17. Show that the function f given by

$$f(x) = \begin{cases} \frac{\sin(2x)}{x} & \text{if } x \neq 0 \\ 1 & \text{if } x = 0 \end{cases}$$

is discontinuous at $x = 0$. Show that the discontinuity at 0 is removable. How would you redefine f so that it is continuous at 0. Use your graphics utility to make a sketch of f.

18. For the function g shown in Figure 3.21, $\lim_{x \to 0.5} g(x)$ does not exist. That is, for each possible choice of a "limit" L for this situation, there is an $\varepsilon_0 > 0$ for which no $\delta > 0$ can be found to satisfy the δ-ε definition of limit. Find such a value for ε_0, given the choices $L = 0, 0.25, 0.5, 1$.

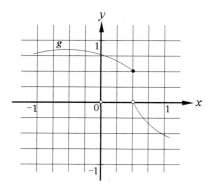

Figure 3.21. g has no limit at 0.5.

19. The function g shown in Figure 3.22 is discontinuous at 0.75. That is, there is an $\varepsilon_0 > 0$ for which no $\delta > 0$ can be found to

satisfy the δ-ε definition of continuity. Find such a value for ε_0.

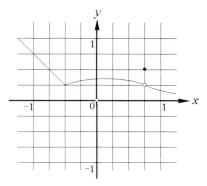

Figure 3.22. g is discontinuous at 0.75.

20. Let f be the *greatest integer function* defined in Exercise 17, page 160.

    (a) Show that f has a jump discontinuity at 0.

    (b) Show that f has a jump discontinuity at every integer value.

    (c) Show that $f(ax+b)$ is discontinuous at $x = \frac{2-b}{a}$ for any real numbers a and b with $a \neq 0$.

21. Use the definition of limit to prove that $\lim_{x \to a} b = b$. What does this say about the constant function $f(x) = b$?

22. Use the definition of limit to prove that $\lim_{x \to a} x = a$. What does this say about the identity function $f(x) = x$?

23. Prove that $\lim_{x \to a} x^n = a^n$ for n a positive integer. What does this say about the power function $f(x) = x^n$ for n a positive integer?

24. Prove that $\lim_{x \to a} (f(x))^n = \lim_{x \to a} (f(x))^n$ where n is a positive integer, provided $\lim_{x \to a} f(x)$ exists.

25. Use the definition of limit to prove that $\lim_{x \to a} |x| = |a|$. What does this say about the absolute value function given by $f(x) = |x|$?

26. Determine the points (if any) at which the following functions are continuous. Use the fact that the rational and irrational numbers are both *dense* in the real numbers, that is, between every pair of real numbers there is a rational number and an irrational number.

    (a) $f(x) = \begin{cases} 0 & \text{if } x \text{ is a rational number} \\ 1 & \text{if } x \text{ is an irrational number} \end{cases}$

    (b) $f(x) = \begin{cases} x & \text{if } x \text{ is a rational number} \\ -x & \text{if } x \text{ is an irrational number} \end{cases}$

27. For the functions represented by each of the following expressions,

    (a) Find all points of discontinuity and classify each as removable, jump, infinite, or oscillatory.

    (b) Find equations of all vertical and horizontal asymptotes, if any.

    (c) Use your graphics utility to make a complete graph. Explain why your graph is complete.

    Note: In the last three parts, $\lfloor x \rfloor$ represents the greatest integer function defined in Exercise 17, page 160.

    i. $\dfrac{x+3}{x^2-4}$

    ii. $\dfrac{x^2-x}{x^2-1}$

    iii. $\dfrac{x^2+x}{x^3-x}$

    iv. $\dfrac{x^2-7x+12}{x^3-8}$

    v. $\dfrac{x^2-6x-16}{x^2-x-6}$

vi. $\dfrac{x - \frac{2}{x}}{x^2 + 3}$

vii. $\dfrac{3x + 1}{\sqrt{1 - x^2}}$

viii. $\dfrac{3|x - 1|}{x + 1} + 8$

ix. $\sqrt{\dfrac{x + 2}{x + 1}}$

x. $\dfrac{x^2 - x - 6}{x^2 - 6x - 16}$

xi. $\dfrac{\sqrt{x} - 1}{x - 1}$

xii. $\sec(2x)$

xiii. $\tan\left(x + \dfrac{\pi}{2}\right)$

xiv. $\dfrac{\sin(2x)}{3x}$

xv. $\dfrac{\tan(3x)}{x}$

xvi. $(x^2 - 1)\sin\left(\dfrac{1}{x^2 - 1}\right)$

xvii. $\cos\left(\dfrac{1}{x}\right)$

xviii. $\lfloor 0.5x \rfloor$

xix. $\lfloor x^{-1} \rfloor$

xx. $\lfloor x^2 \rfloor$

28. For the function represented by the expression

$$\frac{1}{x}\sin\left(\frac{1}{x}\right)$$

(a) Explain why the discontinuity at $x = 0$ does not fit any of the classifications: removable, jump, infinite, or oscillatory.

(b) Give a definition of the type of discontinuity in this situation.

29. Prove the following parts of Theorem 3.3 (p. 172).

(a) Part 1 (b) Part 2 (c) Part 4 (d) Part 5

30. Prove that functions such as the square root function, the cube root function, etc., i.e. functions given by $\sqrt{x}, \sqrt[3]{x}, \ldots$ are continuous for $x > 0$. Functions like this are called *algebraic functions*. Use this to show that combinations of functions obtained by taking roots are continuous.

31. Prove the following parts of Theorem 3.4 (p. 173) using the results of Theorem 3.3 and the definition of continuity.

(a) Part 1 (b) Part 2 (c) Part 3 (d) Part 4 (e) Part 5

32. Prove that the composition of two continuous functions is continuous using Theorem 3.7 and the definition of continuity.

33. Let f be the *greatest integer function* defined in Exercise 17, page 160, and g the function given by $g(x) = x^2$. Show that $\lim_{x \to 2}(f \circ g)(x)$ does not exist, yet $\lim_{x \to 2} g(x) = 4$. Explain why this does not contradict Theorem 3.7.

34. Use the squeeze theorem to find the following limits.

(a) $\lim_{x \to 4} f(x)$, where $x^2 - 4x \le f(x) \le 4x - x^2$ for all $x \ne 2$.

(b) $\lim_{x \to 0} x^N g(x)$, where $-N \le g(x) \le N$ for all $x \ne 0$ and $N > 0$.

(c) $\lim_{x \to 0} x^M \cos(\frac{1}{x})$ where $M > 0$.

35. Prove the *replacement theorem* which is as follows:

    If f and g are functions with $f(x) = g(x)$ for $x \neq a$ and $\lim_{x \to a} g(x) = L$, then $\lim_{x \to a} f(x) = L$.

    (a) Explain why you think the theorem is called the replacement theorem.

    (b) Give an example of a pair of functions f and g which satisfy the theorem.

36. For a function f defined on an interval $[a, b]$ and a number c with $a \leq c \leq b$, f is called *left-continuous* at c, if $\lim_{x \to c^-} f(x) = f(c)$, and f is *right-continuous* at c if $\lim_{x \to c^+} f(x) = f(c)$.

    (a) Show that f is continuous at c if and only if f is both right-continuous and left-continuous at c.

    (b) Give a definition of continuity for a function f on a closed interval $[a, b]$.

    (c) Give a definition of continuity for a function f on a half-open interval $(a, b]$.

    (d) Give a definition of continuity for a function f on a half-open interval $[a, b)$.

37. Prove that the following trigonometric functions are continuous at every point of their respective domains using the continuity of sine and cosine.

    (a) tan   (b) cot   (c) sec   (d) csc

3.4 CHAPTER SUMMARY

Formal Definitions   In this chapter you have considered the concepts whose formal definitions are collected together below. You should try to reconstruct each definition for yourself to test your understanding. As you think about these definitions, try to construct examples in your mind which illustrate each definition, and examples which fail to satisfy each definition.

1. Let f be a function, a, L numbers. We say the limit of f at a is L provided that for every $\varepsilon > 0$, there exists a number $\delta > 0$ such that if $0 < |x - a| < \delta$, then $|f(x) - L| < \varepsilon$.

2. Let S be a function which gives the signed distance (or position) from an initial point as a function of time t for a particle moving in a straight line. The *velocity* or *instantaneous velocity* of this point at time t_0 is defined to be

$$\lim_{h \to 0} V(h)$$

where V is the *average velocity function near* t_0 given by,

$$V(h) = \frac{S(t_0 + h) - S(t_0)}{h}.$$

3. Let f be a function which gives the value of the variable y as function of x, that is, $y = f(x)$. Then the instantaneous rate of change of y with respect to x at a point x_0 in the domain of f is defined to be

$$\lim_{h \to 0} \frac{f(x_0 + h) - f(x_0)}{h}.$$

4. Let f be a function and x_0 a number in its domain. Let γ be the curve which is the graph of f over a domain which includes x_0 and let $P = (x_0, f(x_0))$ be the point on γ corresponding to x_0. The tangent to the curve γ at the point P is the line which passes through the point P and has slope given by

$$\lim_{h \to 0} \frac{f(x_0 + h) - f(x_0)}{h}.$$

5. If a function f has a limit L at a point a in the domain of f, and $f(a) = L$, then we say that f is continuous at a. This means that f is continuous at a when

$$f(a) = \lim_{x \to a} f(x).$$

Alternately, let f be a function, a, L numbers with a in the domain of f. We say that f is continuous at a provided that for every $\varepsilon > 0$, there exists a number $\delta > 0$ such that if $|x - a| < \delta$, then $|f(x) - L| < \varepsilon$. A function f is *continuous on a set* S, if it is continuous at each point of S. A function f is said to be *continuous*, if it is continuous at each point of its domain D.

6. Let f be a function and let a and L be numbers. We say the limit from the left of f at a is L provided that for every $\varepsilon > 0$, there exists a number $\delta > 0$ such that if $a - \delta < x < a$, then $|f(x) - L| < \varepsilon$. We write this as

$$\lim_{x \to a^-} f(x) = L.$$

We say that the limit from the right of f at a is L provided that for every $\varepsilon > 0$, there exists a number $\delta > 0$ such that if $a < x < a + \delta$, then $|f(x) - L| < \varepsilon$. We write this as

$$\lim_{x \to a^+} f(x) = L.$$

7. Let f be a function defined on an interval (c, ∞), c and L numbers. We say the limit of f at ∞ is L provided that for every $\varepsilon > 0$, there exists a number $M > 0$ such that if $x > M$, then $|f(x) - L| < \varepsilon$. We write this as

$$\lim_{x \to \infty} f(x) = L.$$

8. Let f be a function and a a number. If, for every positive number M, there is a positive number δ and we have, if $0 < |x - a| < \delta$, then $f(x) > M$ then we say that f is *unbounded* at a. In this case, we write,

$$\lim_{x \to a} f(x) = \infty.$$

9. The line $y = L$ is called a horizontal asymptote of the graph of f if either $\lim_{x \to \infty} f(x) = L$ or $\lim_{x \to -\infty} f(x) = L$. The line $x = a$ is called a vertical asymptote of the graph of f if either $\lim_{x \to a^+} f(x) = \infty$, $\lim_{x \to a^+} f(x) = -\infty$, $\lim_{x \to a^-} f(x) = \infty$, or $\lim_{x \to a^-} f(x) = -\infty$.

10. A line $y = ax + b$ is called an oblique asymptote of the graph of a function f if $f(x) = q(x) + ax + b$ where $\lim_{x \to \infty} q(x) = 0$ or $\lim_{x \to -\infty} q(x) = 0$ (or both).

11. A curve $y = A(x)$ is called an asymptotic curve of the graph of a function f if $f(x) = q(x) + A(x)$ where $\lim_{x \to \infty} q(x) = 0$ or $\lim_{x \to -\infty} q(x) = 0$ (or both).

12. Let f be a function and a a number. If $\lim_{x \to a} f(x) = L$ but $L \ne f(a)$, then we say that f has a removable discontinuity at a.

13. Let f be a function and a a number. If both $\lim_{x \to a^+} f(x)$ and $\lim_{x \to a^-} f(x)$ exist, but they are not equal, then we say that f has a jump discontinuity at a.

14. Let f be a function and a a number. If there are at least two different numbers L_1, L_2 such that in every open interval containing a there are at least two numbers x_1, x_2 such that $f(x_1) = L_1$ and $f(x_2) = L_2$, then we say that f has an oscillatory discontinuity at a.

The Power of Limits

Each of the above concepts relate directly to a concept of a limit. You began this chapter by developing an intuitive understanding about what it means for a function to have a limit at a point, and how a function may fail to have a limit at a point. You quickly learned to apply this intuitive notion of a limit to define other important concepts: instantaneous velocity, instantaneous rate of change, slope of a curve, and continuity.

In Section 2, you saw that by extending the idea of a limit at a point to include the notions of a limit at infinity and infinite limits, you were able to make important qualitative observations about the graphs of functions in terms of horizontal, vertical and oblique asymptotes as well as asymptotic curves.

Armed with the formal mathematical definitions of the different types of limits you saw in Section 3 how it was possible to rigorously prove theorems about limits and more rigorously define other mathematical ideas such as continuity. Each of the concepts treated earlier in the chapter in an intuitive way was revisited and made more precise using the formal definition of the limit.

Conclusion

Your intuition for limits and your knowledge of the kinds of problems that can be attacked and solved using limits should now be sufficient to construct the problem solving tools of the calculus. Keep in mind as you study the remaining chapters of this text that the calculus concepts and related theoretical constructs are all based on the notion of limit. So now that limits are in your toolbox, let's get on with the construction of the powerful calculus problem solving skills that the rest of this text has waiting for you!

CHAPTER 4

DERIVATIVES

▌4.1 THE CONCEPT OF THE DERIVATIVE

4.1.1 Overview

Calculus is essentially made up of two broad branches known as *differential calculus* and *integral calculus*. This chapter is concerned with the fundamental concepts of *differential calculus*. In the first section you will study the concept of the *derivative* and its related process called *differentiation*. As in all branches of mathematics you will see that one idea, in this case the derivative, has many interpretations and applications which are useful in problem solving. To find a derivative we will combine two basic operations from the previous chapters — the difference quotient and limit. The result is one of the most powerful and fundamental tools in calculus — the derivative. We will make a formal definition of the derivative and then study notations for derivatives, how to calculate derivatives by hand using difference quotients and limits, and various interpretations and uses of the derivative — both geometrical and physical.

In the second section you will learn rules for finding formulas for derivatives of many basic functions. In the third section you will learn about various powerful problem solving ideas using derivatives. Finally, in the last section covers the formal theory of differentiation, including proofs of the basic derivative formulas and many of the important theorem about derivatives.

4.1.2 Activities

1. In this activity you will put together two operations that you worked with in previous chapters: difference quotient and limit. The result is the derivative. The idea begins with a function f and number a in its domain.

    The first step is to write a **computer function dq** that will accept a number h and, if $h \neq 0$, will return the value of the difference quotient of f at a with difference h.

    The function represented by the **computer function dq** is not defined at 0. It might have a limit so apply your **computer function lim** from Exercise 1 of Chapter 3, Section 3.1.4, page 142.

    The number we approximate by these two operations is called the *derivative* of the function f at the point a.

    (a) Apply the above two-step operation to the function whose value at x is given by $|x|$, at the points $a = -3$, $a = 2$ and $a = 0$.

    (b) Apply the above two-step operation to the function whose value at x is given by $\sqrt{|x|}$, first at the point $a = -4$ and then at the point $a = 0$.

(c) Use the operations in this activity to produce a number and use it to write a **computer function** that will represent the equation of the tangent line to the graph of the function in part (b) above at the point −9. That is, find an equation of the tangent line $y = mx + b$, for appropriate values of m and b, and then write a **computer function** which accepts a value of x in the domain of the function in part (b) and returns the y value $mx + b$. Make a sketch of a portion of the function in part (b) and the tangent line to the graph of that function in an interval around the point where $x = -9$ on the same coordinate system.

(d) Heat flow in a one-dimensional situation is defined (in appropriate units) to be the rate of change of the temperature with respect to the position. Consider two concentric cylinders in which the outer radius of the smaller is 1.124m and the inner radius of the larger is 1.128m. The space between is filled with Helium gas at a temperature of 330K. The inner cylinder is maintained at 301K and the outer at 299K. After some time, the temperature at various points between the cylinders is sampled and the data suggests that a reasonable measure of temperature at a point is given by

$$299 + 2\cos\left(\frac{\pi d}{0.008}\right)$$

where d is the distance from the point to the inner cylinder. Use the operations in this activity to estimate the heat flow, i.e., the (instantaneous) rate at which the temperature is changing with respect to distance, at a point which is half way between the two cylinders.

2. The operations of the previous activity are somewhat unsatisfactory because of the fact that, every time you want to change the function or the point, you have to write a new version of the **computer function dq** you wrote in Activity 1. The goal of this activity is to write a single **computer function D** that will accept a **computer function** representing a function f and return a **computer function** which represents a function f' where $f'(x)$ is the derivative of f at x.

Here are some suggestions for writing the **computer function D** but before you read them, try it yourself. Note: In **MapleV** the letter **D** is a reserved name, so you will need to choose another name if you use the MPL which is part of **MapleV**.

   (a) Set up **D** as a **computer function** which accepts a **computer function f** and then returns a **computer function** which itself accepts a number **x**.

   (b) In defining the **computer function** which is to be returned, put a definition of **dq** for **f** at the point **x**.

   (c) Now take the limit of **dq** at **0** using your **computer function lim**. This is what your constructed **computer function** will return for a given **x**.

   Apply **D** to the function **f** in Activity 1, part (a) above. Explain the meaning of **D(f), D(f)(x)**. What happens with **D(f)(0)**?

3. In this activity you are to apply your **computer function D** from the previous activity to two situations: one that arises in studying mechanical equilibrium and one from electrical current.

(a) In a mechanical system with one degree of freedom, the potential energy v is given in terms of the (generalized) coordinate q by the function V where

$$V(q) = \frac{12q^5 - 45q^4 + 40q}{8}.$$

(You can imagine, for example, a particle rolling on a hilly surface. The graph of this function is related to the contour of the surface.)

The *virtual work* of such a system is again a function depending on q given by the derivative of the energy function. The (instantaneous) rate of change of the potential energy with respect to the (generalized) coordinate q.

    i. Use your **computer function D** to estimate the virtual work at $q = 1$.

    ii. Construct a graph of both V and $D(V)$ on the same coordinate axes.

    iii. Produce a table of 50 values of the two **computer functions V** and **D(V)** in the interval $[-1, 3]$. Use the tool **table2** in your MPL to produce a table for **V** and **D(V)**. In **ISETL**, for example, the syntax for **table2** is

table2(f, g, a, b, n);

(b) The electrical current i at a point in a conductor is a function I of time t determined by the total charge q which has flowed past the point from the beginning until time t. In fact, if Q is the function which gives the total charge as a function of time, then I is the derivative of Q. The current is the (instantaneous) rate of change of charge which has flowed past a point with respect to time.

In a certain conductor, the total charge function has been determined to be

$$q = Q(t) = 10(1.2^{-t})\cos(5t).$$

    i. Use your **computer function D** to approximate the current at $t = 2$.

    ii. Construct graphs of both Q and $I = D(Q)$ on the same coordinate axes.

    iii. Use the tool **table** in your MPL to produce a table of 50 values of the charge and current at times ranging from 0 to 10.

4. The heat capacity C of a certain substance is determined by its temperature T according to one of three formulas depending on whether the substance is in solid, liquid or gas form.

If the substance is solid, the formula is

$$C(T) = 214 + 1.2T - 0.01T^2.$$

If it is liquid, the formula is

$$C(T) = 470 - 2T.$$

If it is a gas, the formula is

$$C(T) = 360 - 1.5T.$$

The substance melts at $T = 160$ and boils at $T = 220$.

(a) Use the description of the operations to compute, *by hand*, the (instantaneous) rate of change of C with respect to temperature at the points $T = 100, 160, 200$ and 220.

(b) In at least one of these cases, a problem occurs with the calculation. Explain why.

(c) Use your imagination to make up some meaning to the phrases *left derivative* and *right derivative*.

5. A particle is constrained to move in a straight line. Observations are made of its position p relative to a starting point and its velocity v at various times. Consider the functions f and g given by the following expressions. One is an approximation to the position function and one to the velocity function.

$$f(t) = 2^t (1.443\cos(t) + \sin(t))$$
$$g(t) = 2^t (2\cos(t) - 0.75\sin(t))$$

(a) By making computer graphs of these expressions and reasoning from the graphs, or any other method you can think of, determine which is the position and which is the velocity. Explain your reasoning.

(b) Using your choice for the position function draw a picture which indicates the motion along coordinate line and clearly displays the oscillation of the particle.

6. For each of the following functions, sketch a graph of the function and its derivative on the same axes in the indicated intervals. Then answer the questions on the list at the end of this activity. There are several ways in which you can compute the derivative. You can use your **computer function D** to approximate it, calculate it yourself if you know how to do that, or use your SCS (your instructor will provide you with the necessary syntax). If it is a **data** function, then your only choice is to use your **computer function D** and the **computer function interpolate**.

(a) The function is given by the expression

$$2\sin(x) - \cos(3x)$$

and the interval for x is from -0.1 to 7.

(b) The function **data14** in your MPL on the interval for x from -1 to 4.

For each pair of a function and its derivative, answer the following questions (approximately) by inspecting the graph.

(a) Where are the zeros of the function?

(b) Where are the zeros of the derivative?

(c) Where does the function take a local (relative) maximum value?

(d) Where does the derivative take a local maximum value?

(e) Where does the function take a local minimum value?

(f) Where does the derivative take a local minimum value?

(g) On which region is the function positive?

(h) On which region is the derivative positive?

(i) On which region is the function negative?

(j) On which region is the derivative negative?

(k) On which region is the function increasing?

(l) On which region is the derivative increasing?

(m) On which region is the function decreasing?

(n) On which region is the derivative decreasing?

(o) On which region is the function concave up ("holds water")?

(p) On which region is the derivative concave down ("holds no water")?

(q) On which region(s) is the function concave down?

(r) On which region(s) is the derivative concave down?

7. The purpose of this activity is to investigate graphically whether or not the following three functions have derivatives at the point 0.

$$h(x) = |x| = \text{abs}(x)$$

$$f(x) = \begin{cases} x\sin(\frac{1}{x}) & \text{if } x > 0 \\ 0 & \text{if } x \le 0 \end{cases}$$

$$g(x) = \begin{cases} x^2\sin(\frac{1}{x}) & \text{if } x > 0 \\ 0 & \text{if } x \le 0 \end{cases}$$

For each of these functions, use your graphics utility to zoom in on its graph in extremely small intervals centered at 0. Try to decide from what you see if the function has a derivative or not. Explain how you came to your conclusion.

8. Consider the functions given by the following expressions. In each case, you are to represent the function as a simpler function or as a combination of two or more functions, using the particular combination(s) of function(s) indicated.

(a) $x^7, \dfrac{1}{y^3}, \sqrt{x}, \dfrac{1}{\sqrt{z}}$, power of a variable.

(b) $x^2 + \sin(x), 2x^3 - x + 1$, sum of two functions.

(c) $\tan(x) - \sqrt{x}, 2y^3 - y + 1$, difference of two functions.

(d) $x\sin(x), y^3\tan(y), \sin(a)\sqrt{1 + \cos(a^2)}$, product of two functions.

(e) $\dfrac{\sin(x)}{x^2}, \dfrac{t^2}{\sqrt{1+t}}, x^{-3}\exp(x)$, quotient of two functions.

(f) $\sin(3x), \cos(x^2), (7x-4)^{-\frac{3}{2}}, \sqrt{1 - \tan(x)}$, composition of two functions.

(g) $\sin(v)\sqrt{1 + \cos v^2}$, product of two functions with one composition.

(h) $\sin\left(\sqrt{9-v^2}\right)$, composition of three functions.

NOTE: The **computer function D**, as we have suggested you write it, illustrates the idea of the derivative very nicely and is extremely useful for understanding this concept. However, it might be very slow on some computer systems. This problem will disappear as computers become faster, but for now, the use of **D** in its present form may lead to some long delays. Therefore, once you have completed this set of activities, <u>but</u> not <u>before</u>, change **D** by eliminating the use of the **computer function lim** and replacing h by a small value, such as $h = 0.001$.

4.1.3   Discussion

Definition of the Derivative at a Point
Start with a function f and a point a. The first thing you do is define the difference quotient. You worked with difference quotients throughout the previous two chapters. The *difference quotient* of a function f at a point a is another function which we may call, dq. It is defined by the following formula, if $h \neq 0$,

$$dq(a) = \frac{f(a+h) - f(a)}{h}.$$

If h is a positive or negative number, then dq is defined by this formula. If $h = 0$, then dq is not defined.

In Chapter 3, you worked with functions that were not defined at a single point and saw that one possibility is to take the limit at this point. If it exists, then it is a reasonable choice for the value of the function at that point.

We do this for dq. We take the limit as $h \to 0$. If the limit of dq exists, then we can take this as the value of dq at 0. In fact, this number is actually defined to be the *derivative of f at a*.

To summarize we write that the derivative of f at a is given by the formula,

$$\lim_{h \to 0} dq(a) = \lim_{h \to 0} \frac{f(a+h) - f(a)}{h}$$

We emphasize that, in this statement, the phrase "if it exists" is understood, whether it is explicitly mentioned or not. We formulate all this in the following definition.

Definition 4.1 (Definition of derivative at a point)

If f is a function and a is a point in its domain, the derivative of f at a is,

$$f'(a) = \lim_{h \to 0} \frac{f(a+h) - f(a)}{h}$$

provided this limit exists.

Sometimes the above definition is given using Δx in place of h, where Δx is read "delta x" (not delta times x). The h and Δx both represent the change in x.

A function f is said to be *differentiable at a* if $f'(a)$ exists. We say f is differentiable on a set S, if f is differentiable at each point of S. We say f is *differentiable*, if $f'(x)$ exists at each point x in the domain of f.

Here is another useful and equivalent definition of the derivative. We leave it as an exercise for you to show the two definitions of derivative are equivalent.

> **Definition 4.2 (Alternative definition of derivative at a point)**
>
> *If f is a function and a is a point in its domain, the derivative of f at a is,*
>
> $$\lim_{x \to a} \frac{f(x) - f(a)}{x - a}$$
>
> *provided this limit exists.*

Notation for the Derivative

There are several notations used for the derivative. Some relate to particular interpretations that are intended, some are of historical origin, and some are used just because of personal taste or habit. We will use three notations in this book for the derivative of f at a: $f'(a)$, $D(f)(a)$, and $\frac{dy}{dx}\big|_{x=a}$. Thus we can write,

$$f'(a) = \lim_{h \to 0} \frac{f(a+h) - f(a)}{h}$$

or,

$$D(f)(a) = \lim_{h \to 0} \frac{f(a+h) - f(a)}{h}$$

or,

$$\frac{dy}{dx}\bigg|_{x=a} = \lim_{h \to 0} \frac{f(a+h) - f(a)}{h}.$$

Although formally, these three notations mean exactly the same thing — they have the same *denotation*, the interpretations — their *connotations* are different.

The f' notation is used to emphasize the fact (as you will see in a few pages) that the operations we perform in this definition create a function f' which is intimately related to f and which must be evaluated at a.

The D notation emphasizes the fact that the derivative is an operation on functions. It is also used because it is the easiest to implement in the syntax of an MPL. The $\frac{dy}{dx}$ notation emphasizes the interpretation of a derivative as a rate of change of one variable (y in this case) with respect to another (x in this case). It makes the most sense in conjunction with the common notation, $y = f(x)$, since

$$\frac{dy}{dx} = \lim_{\Delta x \to 0} \frac{\Delta y}{\Delta x} = \lim_{\Delta x \to 0} \frac{f(x + \Delta x) - f(x)}{\Delta x}.$$

Often the notation

$$\frac{d}{dx}$$

(or D_x), synonymous with the notation D, is used to indicate the derivative operation on functions. This notation emphasizes that you are to take the derivative of "the function represented by the expression that follows" $\frac{d}{dx}$ with respect to x. Hence,

$$\frac{d}{dx} f(x)\bigg|_{x=a} = f'(a)$$

is the derivative of the function f with respect to x, evaluated at a.

Calculating Derivatives by Hand   There are actually three ways in which you can compute or approximate the derivative of a function at a point. One is by using the computer as in Activity 1. Another is to build up a theory that provides rules through which you can calculate the derivative in most situations that you will encounter in elementary mathematics and science. A third method, which we will discuss in this paragraph is to apply the definition of derivative to a formula (or formulas) which define the function and then calculate the limit by hand. Although calculating a derivative from the definition can be cumbersome, it can be useful in helping you understand the meaning of the derivative.

Consider the function f given by the formula

$$f(x) = x^2 - 3x.$$

Let us compute the derivative at the point $a = 1$. The first step is to define the difference quotient function and this is not too difficult. We have, for $h \neq 0$,

$$
\begin{aligned}
dq(1) &= \frac{f(1+h) - f(1)}{h} \\
&= \frac{\left((1+h)^2 - 3(1+h)\right) - \left((1)^2 - 31\right)}{h} \\
&= \frac{(1+h)^2 - 3(1+h) - (1)^2 + 3(1)}{h} \\
&= \frac{1 + 2h + h^2 - 3 - 3h - 1 + 3}{h} \\
&= \frac{h^2 - h}{h} \\
&= h - 1.
\end{aligned}
$$

Hence, for $h \neq 0$, $dq(1) = h - 1$. So we can easily conclude from the limit rules derived in Chapter 3 that $\lim_{h \to 0} dq(1) = -1$. Therefore, for this function f, we have shown that $f'(1) = -1$.

The Derivative of a Function is a Function   This is a paragraph for rising above the details and thinking grandiose thoughts. Try to understand the essence of what you are doing to get at the derivative of a function.

First you take a function. Now stop here and set that function as a fixture upon the firmament of your mind. It will not change throughout the rest of the discussion. You can reminisce about all of the functions that you ever "took" — and did something with. Now try to avoid thinking about a particular function but concentrate on a "generic" function.

Now you take a number. Any number — but make sure it is in the domain of the function. You go through some business and if it works (that is, if the limit exists), you get another number. Now this time, don't think so much about these individual numbers, but try to focus on the *process* that you are going through. You start with a number, do something to it and get a number. What is all that? Do you hear the crescendo? That's right, it is a function.

Definition 4.3 (Definition of derivative function)
If f is a function. The derivative f' of f is defined to be the function whose domain is the set of points x in the domain of f at which the derivative exists and, at such a point, the value of f' at x is given by

$$f'(x) = \lim_{h \to 0} \frac{f(x+h) - f(x)}{h}.$$

The derivative of a function is a function. Its process consists of the following two steps which are performed on any number in the domain of the original function.

1. Form the difference quotient.

2. Take the limit of the difference quotient as $h \to 0$.

If the limit exists, you get a number as an output, and the original number input is in the domain of the new function — the derivative. The value of this new function at the original input number is the output number you obtained by first forming the difference quotient and then evaluating the limit — the derivative of the original function at the original number.

We try to express this with the notation. If f is the original function, then the new function, its derivative, is denoted f'. The value of this new function at a point a in its domain is denoted $f'(a)$.

A word about domains. See if you can make some sense out of the following.

> Up until now, the specification of the domain of a function has been fairly arbitrary. Now, with derivative, you see that the domain is not always under your control but can be a consequence of the particular situation you are working with.

To what extent is that statement accurate? Is it an exaggeration? Does it actually say anything?

In light of all this discussion, you should have little trouble with answering the questions at the end of Activity 2. $D(f)$ is the function you get by differentiating f. $D(f)(a)$ is its value at a. The issue with $D(f)(0)$ when f is the function in Activity 1, part (a), has to do with whether or not 0 is in the domain of $D(f)$. Is it? How can you tell?

An Example

Once again, for simple functions, we can compute the derivative by hand directly from the definition, and in this way determine a formula for the function which is the derivative of a given function. Consider the function f, discussed above, given by

$$f(x) = x^2 - 3x.$$

We can go through exactly the same computations that we did to compute its derivative at $a = 1$. The only difference is that, instead of using 1, we use a generic value which we symbolize by x. In other words, to compute the derivative at any point x, we calculate as follows. First we define the difference quotient dq and compute, for $h \neq 0$,

$$
\begin{aligned}
dq(x) &= \frac{f(x+h) - f(x)}{h} \\
&= \frac{(x+h)^2 - 3(x+h) - \left(x^2 - 3x\right)}{h} \\
&= \frac{x^2 + 2xh + h^2 - 3x - 3h - x^2 + 3x}{h} \\
&= \frac{2xh + h^2 - 3h}{h} \\
&= 2x + h - 3.
\end{aligned}
$$

Hence, once again, we can compute the limit to obtain,

$$f'(x) = \lim_{h \to 0} dq(x) = \lim_{h \to 0} (2x + h - 3) = 2x - 3.$$

Therefore, for this f we have shown that f' is the function given by $f'(x) = 2x - 3$. Notice that $f'(1) = -1$ as we saw earlier on page 198.

Derivatives of the Simplest Functions

You can run through calculations just like in the previous example for very simple functions such as the "identity" function f given by $f(x) = x$ and a constant function g given by $g(x) = 1$. If you make such calculations you obtain that g' is given by $g'(x) = 0$ and f' is given by $f'(x) = 1$.

What about the derivative of the constant function h given by $h(x) = c$ where c is any real number? A calculation similar to that for the function g given by $g(x) = 1$ shows that $h'(x) = 0$ also. We leave this as a simple exercise.

We could now proceed in a manner very similar to what we did in Theorem 3.3 in Chapter 3. In fact, in the next section we will lay out some rules for differentiation. Using these and the two facts we just noted about the derivative of x and of 1, it is possible to derive formulas for the derivative of any function which is given by an expression that is a ratio of two polynomials. We turn now, however, to an examination of the basic interpretations of the derivative.

Interpretations of Derivatives — Rates of Change

Each interpretation of the derivative is based on an interpretation of the difference quotient. The limit is the connecting link. You worked with many interpretations of the difference quotient in previous chapters. Consider, for example, heat capacity c given as a function C of temperature T as in Activity 4. If you form the difference quotient, say at $T = 100$, then the difference quotient gives, for a given h the ratio of the change in C to the change in T from 100 to $100 + h$. The derivative at $T = 100$, is then the *rate* of change of C with respect to T at $T = 100$, or

$$\left. \frac{dC}{dT} \right|_{T=100}$$

Interpretations of Derivatives — Slopes

If you think about the graph of a function, then the difference quotient gives the slope of a secant to the graph. For example, in part (c) of Activity 1 you considered the function $\sqrt{|x|}$ at the point where $x = -9$.

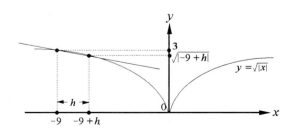

Figure 4.1. A graph of $f(x) = \sqrt{|x|}$.

As in Figure 4.1, you can plot the points $(-9,3)$ and $\left(-9+h, \sqrt{|-9+h|}\right)$. Then draw the line determined by these two points. It is a secant line of the graph and its slope is the value of the difference quotient,

$$\frac{\sqrt{|-9+h|}-3}{-9+h-(-9)} = \frac{\sqrt{|-9+h|}-3}{h}.$$

(Do you see why it is not correct to remove the absolute value sign?)

Now, the limiting value of the different quotient as h goes to zero, corresponds to the limiting value of the slope of the secant line as the point $-9+h$ approaches -9. Do you see what happens to this secant line? Do you agree that it approaches the tangent?

Making use of our intuition we *define* the tangent to a graph of a function f at the point $(a, f(a))$ to be the line passing through the point $(a, f(a))$ with slope given by the derivative of f at a, $f'(a)$. This is expressed formally as follows.

> **Definition 4.4**
> *Let f be a function and a a number in its domain. Let γ be the curve which is the graph of f. The tangent to γ at the point $(a, f(a))$ is the line passing through $(a, f(a))$ with slope given by $f'(a)$.*

Applying the point-slope formula $y-y_0 = m(x-x_0)$ at the point $(a, f(a))$ with $x_0 = a$, $y_0 = f(a)$ and $m = f'(a)$, we obtain the following as an equation of the tangent line to the graph of f at $(a, f(a))$,

$$y-f(a) = f'(a)(x-a) \quad \text{or} \quad y = f(a)+f'(a)(x-a).$$

We refer to this expression as a *linear approximation* to the curve at the point $(a, f(a))$. The reason is that if you look at the picture "near" the point $(a, f(a))$, the graph of the tangent is a straight line segment that "closely" approximates the curve γ "near" $(a, f(a))$. This notion of linear approximation is related to the idea of *local straightness* which will be discussed later in this chapter.

Interpretations of Derivatives — Physical Quantities
Now we come to one of the most profound issues concerning the derivative and its relationship to many aspects of science. The *concept* of derivative provides a means of establishing a logical foundation for a number of scientific concepts. The thing to be realized is that, if you think only intuitively, in terms of everyday discourse, you don't really *know* what is meant by certain concepts such as instantaneous velocity, force, heat flow, current, power, and many others.

Suppose you have a really good grasp of the concept of derivative (which we hope will be the case by the time you finish this chapter). Suppose that this allows you both to make computations concerning the derivative and understand the meaning of those calculations. Then, there is a whole list of scientific concepts for which you have an instantaneous *initial* knowledge and understanding. This is a powerful starting point from which to begin your study of science.

Just to show you how widespread is this feature, we give a (far from exhaustive) list of definitions of scientific concepts based entirely on the derivative. In each case, we express it in terms of the derivative of one quantity (the dependent variable) with respect to another quantity (the independent variable) and leave you to think about the underlying function. Some of these scientific concepts were considered in Activities 3 and 4.

Instantaneous velocity. The derivative of position with respect to time, or the instantaneous rate of change of position with respect to time.

Instantaneous acceleration. The derivative of velocity with respect to time, or the instantaneous rate of change of velocity with respect to time.

Force. The product of mass and acceleration, or the product of the mass times the instantaneous rate of change of velocity with respect to time.

Heat flow. The derivative of temperature with respect to position, or the instantaneous rate of temperature with respect to time.

Current. The derivative of electrical charge with respect to time, or the instantaneous rate of change of electrical charge with respect to time.

Voltage. The derivative of current with respect to time, or the instantaneous rate of change of current flowing past a point with respect to time.

Power. The derivative of amount of work done with respect to time, or the instantaneous rate of change of work with respect to time.

It will be useful for you to think about science in these terms as you study the above topics and others. We are not suggesting that this mathematical foundation is the whole story. Far from it. There is a rich body of knowledge that is derived from a scientific point of view. Our only suggestion is that putting scientific thinking together with a mathematical point of view will put you in a position to master these important ideas.

Interpretations of the Derivative — Marginals

In economics, the rate of change of the total production cost of a commodity (product) with respect to the number of units produced is called the *marginal cost*. By the rate of change interpretation of the derivative, it follows that marginal cost MC is the derivative of the total production cost C with respect to the number of units produced. That is, $MC(x) = C'(x)$. Similarly, the rate of change of the total revenue and the rate of change of the total profit with respect to the number of units produced are called the *marginal revenue* and *marginal profit*, respectively.

In the theory of economics, an analysis of a production situation often involves the notion of a *marginal* — the additional cost (revenue or profit) due to the production of 1 additional unit of a commodity. Consider a situation where C is a total cost function and $C(x)$ represents the total cost of producing x units of a commodity. Then we have,

$$MC(x) = C'(x) \approx \frac{C(x+1) - C(x)}{1}.$$

Hence, when x units have been produced, the marginal cost, $MC(x)$, can be used to approximate the cost of producing 1 additional unit, $C(x+1) - C(x)$. That is,

$$MC(x) = C'(x) \approx C(x+1) - C(x).$$

An analogous discussion can be made for marginals of revenue or profit due to the production of 1 additional unit of a commodity.

Similarly, to produce an additional Δx units, the additional cost $\Delta C = C(x + \Delta x) - C(x)$ can be approximated using the derivative $C'(x)$ as follows:

$$\Delta C = C(x + \Delta x) - C(x) \approx C'(x)\Delta x.$$

This approximation follows by the definition of the derivative, provided Δx is small relative to x. Do you see why? Using derivatives in this way to make approximations is referred to as *marginal analysis* in economics.

Interpretations of the Derivative — Linear Motion

Before ending our discussion of interpretations of the derivative, let's dwell for just a bit longer on one particular example. Think about a particle constrained to move in a straight line, back and forth, with varying speed. Imagine a picture of the path of this particle. Now, consider that its motion is described by a function P which gives the position (signed distance from a fixed point) as a function of time. Imagine a picture, or graph of this function. The two pictures are quite different.

Consider for example an object projected directly upward. Because of gravity, we may consider that it is constrained to move in a straight line. It will move directly upward, and then turn around to complete its journey downward. Suppose that the height of the object t seconds after it is projected is given by the formula

$$p = P(t) = 86.5t - 16t^2.$$

Figure 4.2 shows a graph of the function P.

Now can you imagine a picture that actually shows the particle (with its downward path displaced slightly to distinguish it from the upward path)? You could show a vertical axis with positions marked and write the time at various points on the path. It could look something like that in Figure 4.3. Figure 4.3a shows a graph of the derivative $D(P) = P'$ which gives the (instantaneous) velocity v of the particle at any time t. Hence, $v = V(t) = P'(t)$, the rate of change of position P with respect to time, t.

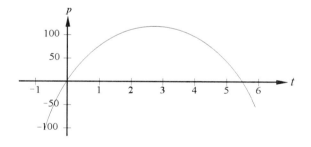

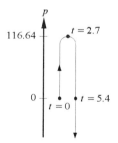

Figure 4.2. A graph of $p = P(t) = 86.5t - 16t^2$.

Figure 4.3. A picture of the path.

What can you say about the initial velocity of this object? How can you decide which direction it is going? How long does it take for the particle to reach its maximum height above the point at which it was released? How long does it take to return to the point at which it was released? How does the graph of the velocity function relate to such questions?

Now let's go the other way. Figure 4.4 shows a picture of a path, this time of an object constrained to move horizontally with position p given by

$$p = P(t) = t^3 - 7t^2 + 12t.$$

Can you make a rough sketch of what the motion function (position as a function of time) would look like? The object begins at time $t = 0$ so the graph starts at the origin. As time increases, the object moves to the right. This means that the position increases, so the graph of P in Figure 4.5 goes up. This continues until about $t = 1$ when it turns around and starts coming back. This means the position value decreases, so the graph of P in Figure 4.5 goes down. The particle passes through the initial point at about $t = 3$ which means that the graph crosses the horizontal axis at about $t = 3$. At about $t = 3.5$ the object turns around again so the graph starts increasing. The next crossing occurs at about $t = 4$ and then the object continues on to the right (or positive direction).

You can graph the position function and its derivative, the velocity function, from this data. The graphs look something like those in Figure 4.5.

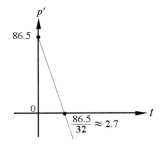

Figure 4.3a. A graph of $D(P) = P'$, or $v = V(t) = P'(t)$.

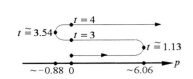

Figure 4.4. The path of an object moving according to $p = P(t) = t^3 - 7t^2 + 12t$.

One-Sided Derivatives

When you studied limits in Chapter 3, you encountered the phenomenon of one-sided limits. A function could have one limit L_1 at a point if the domain variable approached L_1 from one side, and a different limit L_2 if the approach were from the other side. You saw in Theorem 3.1 that the limit exists if and only if both two-sided limits exist and they are equal.

Since the definition of derivative involves a limit, you should expect a similar phenomenon to occur with derivatives. It does. In most cases, the situation we are about to discuss arises when a function is defined in parts.

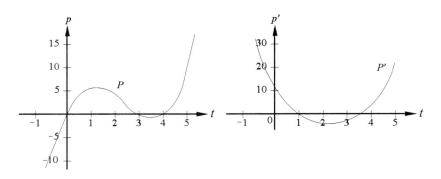

Figure 4.5. Graphs of $p = P(t) = t^3 - 7t^2 + 12t$ and $D(P) = P'$.

In Activity 1(a), you investigated the absolute value function given by the expression $|x|$. There you used **computer functions** to see if the derivative existed at 0.

You may have found that it was more convenient to express this function with a definition in parts. If we call the function f, then we have, using the definition of absolute value,

$$f(x) = \begin{cases} x & \text{if } x > 0 \\ 0 & \text{if } x = 0 \\ -x & \text{if } x < 0. \end{cases}$$

So, to compute the derivative at 0 we define the difference quotient and take its limit. We can set up the operations all at once, using $f(0) = 0$, to obtain

$$f'(0) = \lim_{h \to 0} \frac{f(h) - f(0)}{h} = \lim_{h \to 0} \frac{f(h)}{h}.$$

In order to compute this limit, we have to evaluate $f(h)$ and this is done in two ways. We get one expression if $h > 0$ and another if $h < 0$. This is the source of the two directions. We have,

$$\lim_{h \to 0^+} \frac{f(h)}{h} = \lim_{h \to 0^+} \frac{|h|}{h} = \lim_{h \to 0^+} \frac{h}{h} = \lim_{h \to 0^+} (1) = 1$$

and

$$\lim_{h \to 0^-} \frac{f(h)}{h} = \lim_{h \to 0^-} \frac{|h|}{h} = \lim_{h \to 0^-} \frac{-h}{h} = \lim_{h \to 0^-} (-1) = -1.$$

Hence, the two one-sided limits are different, so the *limit* does not exist and this function does not have a derivative at 0.

The situation is quite different at any point other than 0. Let us check the derivative at some point $a > 0$. Do you understand why we can, in such an investigation take the function as being given by the formula $f(x) = x$? Once you agree to this you conclude that the derivative is equal to 1. What formula can be used if you want to compute the derivative at some negative point? You should run through and check this carefully. The conclusion is that the derivative of this function is given by the following function defined in parts,

$$f(x) = \begin{cases} 1 & \text{if } x > 0 \\ -1 & \text{if } x < 0 \end{cases}$$

You will note that f' is undefined at $x = 0$. See Figure 4.6 for graphs of f and f'.

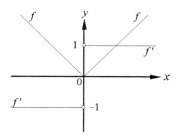

Figure 4.6. Graphs of $f(x) = |x|$ and $f'(x)$.

Definition 4.5

If f is a function and a is a point in its domain and,

$$\lim_{h \to 0^+} \frac{f(a+h) - f(a)}{h}$$

exists, then we say that f has a right *derivative at a and its value is this limit. Similarly, if*

$$\lim_{h \to 0^-} \frac{f(a+h) - f(a)}{h}$$

exists, then we say that f has a left *derivative at a and its value is this limit.*

Now consider another example. Let g be the function defined as follows,

$$g(x) = \begin{cases} x^2 & \text{if } x > 0.5 \\ x - 0.25 & \text{if } x \le 0.5 \end{cases}$$

Let's compute the derivative at the seam, 0.5. Looking ahead we see that it will be necessary to compute the limit from both directions, so let's do that from the outset. From the right we have,

$$\lim_{h \to 0^+}\left(\frac{g(0.5+h) - g(0.5)}{h}\right) = \lim_{h \to 0^+}\left(\frac{(0.5+h)^2 - 0.25}{h}\right)$$

$$= \lim_{h \to 0^+}\left(\frac{0.25 + h + h^2 - 0.25}{h}\right)$$

$$= \lim_{h \to 0^+}\left(\frac{h + h^2}{h}\right) = \lim_{h \to 0^+}(1+h) = 1$$

and from the left,

$$\lim_{h \to 0^-}\left(\frac{g(0.5+h) - g(0.5)}{h}\right) = \lim_{h \to 0^-}\left(\frac{(0.5+h) - 0.25 - 0.25}{h}\right)$$

$$= \lim_{h \to 0^-}\left(\frac{h}{h}\right) = \lim_{h \to 0^+}(1) = 1$$

Hence the two limits are equal so we may conclude that,

$$g(0.5) = \lim_{h \to 0}\left(\frac{g(0.5+h) - g(h)}{h}\right) = 1$$

Now again, as you will see later, there is little difficulty in computing the derivative at any of the other points and we have,

$$g'(x) = \begin{cases} 2x & \text{if } x > 0.5 \\ 1 & \text{if } x = 0.5 \\ 1 & \text{if } x < 0.5 \end{cases}$$

Figure 4.6a shows a graph of the functions g and g'.

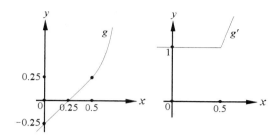

Figure 4.6a. Graphs of g and g′.

Vertical Tangents

In Activity 1(b), you investigated the function given by the expression $\sqrt{|x|}$ using **computer functions** to see if the derivative existed at 0.

We first express this function as a definition in parts using the definition of absolute value. If we call the function f, then we have,

$$f(x) = \begin{cases} \sqrt{x} & \text{if } x > 0 \\ 0 & \text{if } x = 0 \\ \sqrt{-x} & \text{if } x < 0. \end{cases}$$

To compute the derivative at 0 we define the difference quotient and take its limit. We set up the operations all at once, using $f(0) = 0$, to obtain

$$f'(0) = \lim_{h \to 0} \frac{f(h) - f(0)}{h} = \lim_{h \to 0} \frac{f(h)}{h}.$$

As in the example above, in order to compute this limit, we have to evaluate $f(h)$ in two ways. We get one expression if $h > 0$ and another if $h < 0$. We have,

$$\lim_{h \to 0^+} \frac{f(h)}{h} = \lim_{h \to 0^+} \frac{\sqrt{h}}{h} = \lim_{h \to 0^+} \frac{1}{\sqrt{h}} = \infty$$

and

$$\lim_{h \to 0^-} \frac{f(h)}{h} = \lim_{h \to 0^-} \frac{\sqrt{-h}}{h} = \lim_{h \to 0^-} \frac{\sqrt{-h}}{-h} = \lim_{h \to 0^-} -\frac{1}{\sqrt{-h}} = -\infty.$$

Hence, the two one-sided limits do not exist. In fact, they both "blow up" at 0. So the *limit* does not exist and this function does not have a derivative at 0. In this situation, the function f has a *vertical tangent* at 0 as shown in Figure 4.7.

This situation is quite different at any point other than 0. What formula can be used if you want to compute the derivative at some point a other than 0? The answer is that the derivative of this function is given by the following function defined in parts,

$$f'(x) = \begin{cases} \frac{1}{2\sqrt{x}} & \text{if } x > 0 \\ -\frac{1}{2\sqrt{-x}} & \text{if } x < 0. \end{cases}$$

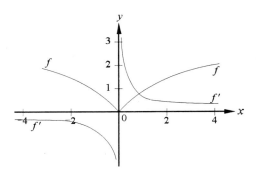

Figure 4.7. Graphs of $f(x) = \sqrt{|x|}$ and $f'(x)$.

We leave the derivation of the above formula for $f'(x)$ as an exercise. Notice that f' is undefined at $x = 0$. In this situation, we have $\lim_{x \to 0^+} f'(x) = \infty$ and $\lim_{x \to 0^-} f'(x) = -\infty$. Hence, using absolute values, we have $\lim_{x \to 0} |f'(x)| = \infty$.

The following definition characterizes situations in which a graph of a function has a vertical tangent.

Definition 4.6 (Definition of vertical tangent)

If f is a continuous function and a is a point in its domain, then f has a vertical tangent at a if,

$$\lim_{x \to a} |f'(x)| = \infty.$$

For functions defined on a closed interval, vertical tangents at an endpoint can be defined as in Definition 4.6 using one-sided limits.

Derivative at an Endpoint   Suppose that you are analyzing a function whose domain is a closed interval. Consider for example, the freezing point function of Chapter 3,

$$F(m) = \begin{cases} 5m^4 - 23.6m^3 - 146m^2 + 129m - 17 & \text{if } 0 \leq m \leq 0.76 \\ 45.056(m - 0.76)^{\frac{1}{3}} - 11.98132480 & \text{if } 0.76 \leq m \leq 1. \end{cases}$$

The domain of this function is the closed interval $[0,1]$.

What would we mean by the derivative of this function at 0? At 1? At $m = 0$ for example, we consider only the right derivative and at $m = 1$ we consider only the left derivative. This convention is used in general.

Definition 4.7

Let f is a function whose domain is a closed interval $[a, b]$. The derivative of f at a is taken to be the right-derivative at a. The derivative of f at b is taken to be the left-derivative at b.

Derivatives and Graphs
In Activity 6, you observed several properties of the graph of a function and the graph of its derivative. One of the most important features of the theory of differentiation is that certain "coincidences" always occur. Certain combinations of properties of the derivative and of the function always occur simultaneously. In the next section of this chapter we will list those connections

explicitly and in the last section we will prove that they *must* occur under appropriate conditions. For the time being we think it is valuable to give you a chance to determine for yourself, from your observations, as many of the connections as you can.

Make a list of all connections which you can possibly imagine. The only rule is that each connection must be consistent with every example that you have observed.

One interpretation of the derivative in which these connections are particularly important has to do with linear motion. Can you think about connections between any of the properties listed in Activity 6 and properties of a particle constrained to move in a straight line according to some position function of time? What does it mean for the particle to be moving to the right? Turning around? Speeding up or slowing down? Turn back and look at the discussion on page 202 as you think about these questions.

Local Straightness

Everybody (or almost everybody) knows that the earth is round. Not too very long ago, however, most people thought that the earth was flat. Well, it is "locally flat". That means, if you take a very small portion of the earth's surface and examine it closely, it is very close to being flat.

The same is true of certain curves. In Activity 7, you "blew up" small portions near the origin of the graphs of two function f and g. How would you describe what you found? We hope that you found that one looked pretty much like a straight line, but the other did not. The difference is that one is differentiable at $x = 0$ and the other is not.

Let's check this out for two functions A, B which are similar to f and g of Activity 7. We define,

$$A(x) = \begin{cases} x \sin(\frac{1}{x}) & \text{if } x \neq 0 \\ 0 & \text{if } x = 0 \end{cases}$$

$$B(x) = \begin{cases} x^2 \sin(\frac{1}{x}) & \text{if } x \neq 0 \\ 0 & \text{if } x = 0 \end{cases}$$

How do these two functions differ from f and g?

We want to check differentiability at 0 for A and B. We begin with A and the start of the calculations are always the same.

$$A'(0) = \lim_{h \to 0} \left(\frac{A(h) - A(0)}{h} \right) = \lim_{h \to 0} \left(\frac{h \sin(\frac{1}{h})}{h} \right)$$
$$= \lim_{h \to 0} \left(\sin\left(\frac{1}{h} \right) \right)$$

and you have seen in Chapter 3 that this limit does not exist. Hence A is not differentiable at 0.

On the other hand, for B we have,

$$B'(0) = \lim_{h \to 0} \left(\frac{B(h) - B(0)}{h} \right) = \lim_{h \to 0} \left(\frac{h^2 \sin(\frac{1}{h})}{h} \right)$$
$$= \lim_{h \to 0} \left(h \sin\left(\frac{1}{h} \right) \right) = 0$$

Hence, B is differentiable at 0 and $B'(0) = 0$ (Why is the last limit is 0?).

If you did the same thing to A and B that you did to f and g in Activity 7, which do you think would look locally straight near 0?

Preparing for Differentiation — a Parthian Shot   Activity 8, which ends the activities, does not relate directly to the concept of derivatives. We put it here to form a bridge to the next section where you will be concerned mainly with computing derivatives. There will be a number of rules pertaining to computing derivatives of functions and combinations of functions. You can only apply these rules to differentiate a particular function if you see it in a simpler form or as an appropriate combination of functions. So Activity 8 is there to give you some practice in seeing a function as a combination of functions. Once you get good at this, calculating derivatives will become a routine mechanical procedure.

4.1.4   Exercises

1. Consider the following statement given on page 199.

   > Up until now, the specification of the domain of a function has been fairly arbitrary. Now, with derivatives, you see that the domain is not always under your control but can be a consequence of the particular situation you are working with.

   To what extent is that statement accurate? Is it an exaggeration? Does it actually say anything? Write a brief essay discussing these questions.

2. For a function f, the function $D(f)$ is what you get by differentiating f. $D(f)(a)$ is its value at a. The issue with $D(f)(0)$ when f is the absolute value function in Activity 1(a), page 191, has to do with whether or not 0 is in the domain of $D(f)$. Is it? How can you tell?

3. For each of the following functions,

   (a) Explain how you can use the slope interpretation of the derivative to find the derivative of the following functions.

   (b) Find the derivative using the definition of the derivative.

   Here are the functions.

      i. The constant function f given by $f(x) = c$ where c is any real number.

      ii. The linear function g given by $g(x) = mx + b$ where m and b are any real numbers.

4. Each of the following situations says something about the derivative of a given function. Do the following for each situation.

   (a) Define an appropriate function and variables which describe the situation.

   (b) Interpret what each situation says about the derivative of your function.

   (c) Sketch a graph of your function and its derivative.

      i. The price of oil decreases steadily as more of it is produced.

      ii. The price of oil increases steadily for the first three months of the year, then the price levels off for the next two months, and finally the price decreases twice as fast as it increased the first three months.

      iii. The cost of health care is rising at an ever increasing rate.

      iv. A car is gradually slowing down to a stop.

v. A person's temperature has been rising for the past four hours, but not so rapidly since the person took an antibiotic an hour ago.

5. A liquid is flowing into a large tank at a constant rate. Let $V(t)$ be the volume of liquid in the tank and $H(t)$ be the height of the liquid in the tank t seconds after it starts to flow into the tank. For each of the following types of tank, answer the questions below.

   (a) A spherical tank.

   (b) A tank in the shape of a right circular cylinder with a circular base.

   Here are the questions.

      i. What physical interpretations can you give for $\frac{dV}{dt}$ and $\frac{dH}{dt}$?

      ii. Is $\frac{dV}{dt}$ positive, negative, or zero when the tank is one-third full? Explain your answer. Is $\frac{dH}{dt}$ positive, negative, or zero when the tank is one-third full? Explain your answer.

      iii. Is $\frac{dV}{dt}$ a constant? Is $\frac{dH}{dt}$ constant? Explain your answers.

6. Find the derivative of each of the following functions using the definition. You may wish to use your SCS for some of the algebra. However, in your solution be sure to show each step in your work.

   (a) $f(x) = 3 - 4x$ (b) $f(x) = 3x^2$

   (c) $g(x) = 5 - x^2$ (d) $h(x) = 2x^3 - 1$

   (e) $h(x) = \sqrt{x}$ (f) $f(x) = \dfrac{4}{x}$

   (g) $f(x) = \dfrac{1}{3x+1}$ (h) $g(x) = \dfrac{x}{x+1}$

   (i) $g(x) = \sqrt[3]{x}$ (j) $f(x) = \sqrt[3]{x^2}$

7. Use the definition of the derivative to find derivatives in each of the following.

   (a) Find an equation of the line tangent to the curve $y = 0.5x^2 - 2x + 3$ at $x = 1$.

   (b) Find an equation of the line normal (perpendicular to the tangent) to the curve $y = 3 - 2x^2$ at $x = 0.5$.

   (c) A ball falls so that its distance s, in feet, at time t, in seconds, is given by $s = 16t^2 + 32t$.

      i. Find a function which gives the instantaneous velocity $v = V(t)$ at time t.

      ii. What is the instantaneous velocity when $t = 3$ seconds?

   (d) The total cost to produce x units of a certain commodity is given by $C(x) = 0.05x^3 - 30x$ hundred dollars.

      i. Find the marginal cost function when x units have been produced.

      ii. Estimate the cost to produce 1 additional unit when the cost level is 100 units.

      iii. What is the actual cost of producing 1 additional unit when 100 units have already been produced?

      iv. Sketch a graph of the marginal cost function and the cost functions on the same graph "near" $x = 100$.

   (e) Find the rate of change of the area of a circle with respect to its radius when the radius is 5 feet.

(f) Show that the rate of change of the area of a circle with respect to its radius is equal to the circumference of the circle.

8. For the function represented by each of the following expressions, use the definition of the derivative to investigate the following. Explain your answers.

   (a) Does the function have a limit at a?

   (b) Is the function continuous at a?

   (c) Use your graphics utility to zoom in on the graph of each of the following functions to investigate the local straightness of each function at the indicated point(s) a. Is the function differentiable at the indicated point(s) a?

   (d) Find the one-sided derivatives at the "seam" point.

   $\quad$ i. $|1+2x|$, $a = -0.5$, $a = 0$, $a = 3$

   $\quad$ ii. sgn, $a = -3$, $a = 0$, $a = 2$

   $\quad$ iii. $x|x|$, $a = -1$, $a = 0$, $a = 1$

   $\quad$ iv. $|1-x^2|$, $a = -1$, $a = 0$, $a = 1$

   $\quad$ v. $|x^3 - 4x|$, $a = -2$, $a = 0$, $a = 2$

   $\quad$ vi. $x^{1/3}$, $a = -2$, $a = 0$, $a = 1$

   $\quad$ vii. $x^{2/3}$, $a = -1$, $a = 0$, $a = 8$

   $\quad$ viii. For $a = 0$, $a = 1$, $a = 3$

   $$f(x) = \begin{cases} x^2 & \text{if } x < 1 \\ 2x-1 & \text{if } x \geq 1 \end{cases}$$

   $\quad$ ix. For $a = 0$, $a = 1$, $a = 3$

   $$g(x) = \begin{cases} 2\sqrt{x} & \text{if } x > 0 \\ 3-x & \text{if } x \leq 0 \end{cases}$$

9. For the function F in the graph shown below, use the graph to determine whether or not the function has a derivative at the indicated points. Give reasons for your answer.

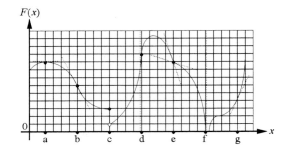

$F(x)$

   (a) $x = a$ (b) $x = b$

   (c) $x = c$ (d) $x = d$

   (e) $x = e$ (f) $x = f$

   (g) $x = g$

10. Use your graphics utility to zoom in on the graph of each of the following functions to investigate the local straightness of the each of the following functions. At what points, if any, is each function not differentiable?

   (a) $f(x) = |\cos(x)|$

   (b) $f(x) = \sin^2(x)$

11. For an object that moves along a line according to

   $$p = P(t) = 86.5t - 16t^2$$

   what can you say about the initial velocity of this object? How can you decide which direction the object is going?

12. What are the connections between the properties listed in Activity 6, page 194, and properties of a particle constrained to move in a straight line according to the position function P given by $p = P(t) = t^3 - 4t^2 - 3t$ where t represents time in seconds and p is in feet? What does it mean for the particle to be moving to the

right? Turning around? Speeding up or slowing down? Write a brief essay discussing these questions. Turn back and look at the discussion on page 203 as you think about these questions.

13. Let A and B be the functions given by

$$A(x) = \begin{cases} x \sin\frac{1}{x} & \text{if } x \neq 0 \\ 0 & \text{if } x = 0 \end{cases}$$

$$B(x) = \begin{cases} x^2 \sin\frac{1}{x} & \text{if } x \neq 0 \\ 0 & \text{if } x = 0 \end{cases}$$

(a) How do the functions A and B differ from the functions f and g, respectively, of Activity 7 on page 195?

(b) For the function B in part (a), $B'(0)$ is given by the following limit.

$$B'(0) = \lim_{h \to 0}\left(\frac{B(h) - B(0)}{h} \right)$$

$$= \lim_{h \to 0}\left(\frac{h^2 \sin(\frac{1}{h})}{h} \right)$$

$$= \lim_{h \to 0}\left(h \sin\left(\frac{1}{h}\right) \right) = 0$$

Hence $B'(0) = 0$. Explain why the last limit is 0?

14. The graph of the function f has a tangent line $y = 3x - 2$ at the point $(x_0, f(x_0))$. Find equations of the tangent and normal lines to the graph of the function $g(x) = f(x+1)$ at the point $(x_0 - 1, g(x_0 - 1))$.

15. For the function given by the each of the following expressions, find an expression for the derivative and state any restrictions on x.

(a) $x|x|$ 　　　　(b) $x\sin(\frac{1}{x})$

(c) $x^2 \sin(\frac{1}{x})$

16. Use the definition of the derivative to check if the derivative of the function f given by $f(x) = \sqrt{|x|}$ is given by

$$f'(x) = \begin{cases} \frac{1}{2\sqrt{x}} & \text{if } x > 0 \\ -\frac{1}{2\sqrt{-x}} & \text{if } x < 0 \end{cases}$$

Give a single expression which represents $f'(x)$ and state any restrictions on x that are necessary.

17. For the graph of each function, make a sketch by hand of the derivative function.

(a)

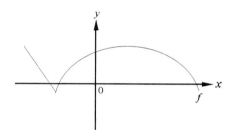

(b)

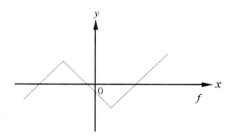

(c)

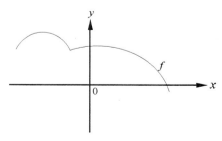

(d)

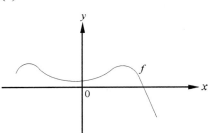

(d)

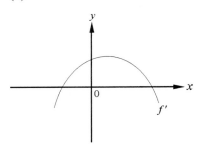

18. The graphs shown below are the graphs of the derivative of a function. Make a sketch by hand of the function. Can there be more than one such function? Give reasons for your answer.

19. In each of the following sketches there are two or more functions. One is a function and another is its derivative. Which is the function and which is its derivative? Give reasons for your answer.

(a)

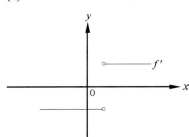

(a)

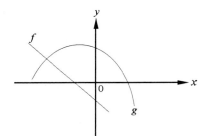

(b)

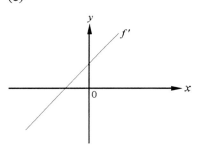

(b)

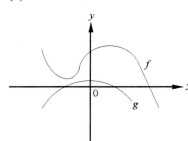

(c)

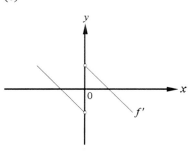

(c)

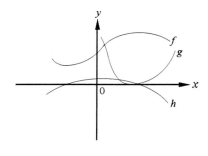

(d)

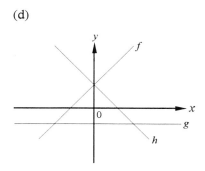

20. For each pair of data functions, one is the derivative of the other. Sketch each pair of functions on the same graph using your MPL. Which is the function and which is the derivative? Give reasons for your answer.

   (a) **data19** and **data20**

   (b) **data21** and **data22**

   (c) **data23** and **data24**

   (d) **data25** and **data23**

   (e) **data26** and **data27**

21. Use the definition of the derivative to show that the following function is differentiable at each point of its domain. Make a sketch of the function.

$$f(x) = \begin{cases} 3 & \text{if } x \leq 0 \\ \sqrt{9 - x^2} & \text{if } 0 < x \leq 3 \end{cases}$$

22. Show that each of the following functions has a vertical tangent at the indicated point. Make a sketch of each function showing the vertical tangent.

   (a) $x = 0$

$$f(x) = x^{\frac{2}{3}}$$

   (b) $x = -1$

$$g(x) = \sqrt[3]{x + 1}$$

23. (a) For a function defined on the closed interval $[a, b]$, what is the meaning for the derivative at a? Give appropriate limit definitions for the derivatives at a and b.

   (b) On the indicated interval, find the derivatives at the endpoints of the interval using limits of difference quotients.

      i. $[1, 4]$, $f(x) = x^2 - 3x$.

      ii. $[-1, 3]$,
$$g(x) = \begin{cases} 1 - x^2 & \text{if } -1 \leq x \leq 0 \\ \sqrt{x + 1} & \text{if } 0 < x \leq 3 \end{cases}$$

      iii. $[-1, 0]$, g is the function in part (ii).

■ 4.2 CALCULATING DERIVATIVES

4.2.1 Overview

Computing derivatives using difference quotients and limits is cumbersome, but not always necessary. Since derivatives are of such fundamental importance in using calculus to solve problems, it is valuable to have efficient procedures for calculating them. In this section we will investigate rules for finding derivatives of functions. After studying this section you will be able to find formulas for derivatives of functions which are constant, powers of a variable, the basic trigonometric functions the natural exponential function, functions defined in parts and implicitly defined functions. You will also be able to find derivatives of various combinations of functions, including: sums, differences, constant multiples, products, quotients and compositions. We will also consider derivatives of derivatives and you will see how derivatives are related to finding extrema of functions.

4.2.2 Activities

1. Consider the functions represented by the following expressions.

   (a) $2, -7, \pi$

   (b) $x^7, \dfrac{1}{y^3}, \sqrt{x}, \dfrac{1}{\sqrt{z}}$

   (c) $\sin(x), \cos(t), \tan(y), \sec(x), \csc(v), \cot(z)$

   (d) $\exp(x), \ln(x)$

   (e) $2x^5, 10y^{-2}, -3t^{4/3}, 3\exp(x), 5\cos(t)$

   (f) $x^2 + \sin(x), 2x^3 - x + 1$

   (g) $\tan(x) - \sqrt{x}, 2y^3 - y + 1$

   (h) $x\sin(x), y^3\tan(y), x\exp(x)$

   (i) $\dfrac{\sin(x)}{x^2}, \dfrac{t^2}{\sqrt{1+t}}, \dfrac{x}{\exp(2x)}$

   (j) $\sin(3x), \cos(x^2), \exp(3x), \ln(1+x^2), (7x-4)^{\frac{3}{2}}$

   (k) $\sin(a)\sqrt{1+\cos(a^2)},$

   (l) $\sin\left(\sqrt{9-v^2}\right), \exp(\sin^2(3t))$

   Calculate their derivatives by hand if you can or guess. Then use your SCS to calculate their derivatives and see how close you came.

   After doing this, see if you can make up rules for differentiating the following functions and combinations of functions.

   (a) The function whose value at x is x^a where a is fixed real number.

   (b) The six basic trigonometric functions, sin, cos, tan, cot, sec and csc.

   (c) The exponential and logarithmic functions, exp and ln.

   (d) For each of the following, assume f and g are differentiable functions. Make up a rule in terms of f, g, f', and g'. Make up your own examples to test your rule in each part.

        i. the sum $f+g$,

        ii. the difference $f-g$,

        iii. the constant multiple cg where c is a constant,

        iv. the product fg (Hint: try some products of the form $x^n\sin(x)$ and $x^n\cos(x)$ where n is a positive integer.),

        v. the quotient $\dfrac{f}{g}$ (Hint: try some expressions of the form

   $$\frac{x^n}{\sin(x)}, \frac{x^n}{\cos(x)}, \text{ and } \frac{x^n}{\exp(x)}$$

        where n is a positive integer.),

        vi. the composition $f \circ g$. (Hint: try some expressions like $\sin(ax), \sin(x^n), \cos(ax), \cos(x^n), \exp(ax), \exp(x^n)$ where a is a constant.)

2. What is the derivative of a function defined in parts? Suppose that you have a function f defined in three parts, so that there are two seams. Suppose the seams are at $x = a_1$ and $x = a_2$, and the individual parts are given by expressions $p_1(x)$, $p_2(x)$ and $p_3(x)$. In other words, the definition of the function might look something like,

$$f(x) = \begin{cases} p_1(x) & \text{if } x < a_1 \\ p_2(x) & \text{if } a_1 \leq x < a_2 \\ p_3(x) & \text{if } a_1 \leq x \end{cases}$$

Suppose that you have found expressions $q_1(x)$, $q_2(x)$ and $q_3(x)$ to give the derivatives of $p_1(x)$, $p_2(x)$ and $p_3(x)$, respectively. Suppose moreover that you have checked the derivatives at the seams, found them to exist and that their values are the numbers b_1 and b_2, respectively. Write a formula for the derivative of f.

3.   (a) Pick a number a with $2 \leq a \leq 3$ and let f be the exponential function given by $f(x) = a^x$. (Notice that this is not the same as x^a.) Make a graph of the functions f and $D(f)$ where D is an approximate derivative **computer function**. **ISETL** code for such a function is as follows:

```
D := func(f);
       return func(x);
                  return (f(x+0.0001)-f(x))/0.0001;
            end;
    end;
```

Notice that, depending on the value of a, the two curves may be quite close to each other. Try to find a value for a such that the graph of f and $D(f)$ are as close to each other as you can make them.

   (b) Now consider once again the exponential function f with $f(x) = a^x$ where $2 \leq a \leq 3$. For $h \neq 0$ the difference quotient $dq(x)$ is given by

$$dq(x) = \frac{a^{x+h} - a^x}{h} = a^x \left(\frac{a^h - 1}{h} \right).$$

Do you see why? Find a number a with $2 \leq a \leq 3$ so that

$$\lim_{h \to 0} \frac{a^h - 1}{h}$$

is as close to 1 as you can make it. To do so, you may wish to use your **computer function lim** that you used in Chapter 3, or you can write a **computer function dq** which accepts a number a and returns the value of

$$\frac{a^h - 1}{h}$$

with $h = 0.0001$. You may also wish to try some smaller values of h. How does the number you found for a in this instance compare to the number you found so that the graph of f and

$D(f)$ were as close as you could make them above? Are you surprised by the comparison? For the number a that you found, what is the relationship between $f(x)$, $D(f)(x)$, and $\lim_{h\to 0}\frac{a^{x+h}-a^x}{h}$?

(c) It can be shown that there is an exact number, customarily denoted by e, such that the graph of the function given by e^x is the same as the graph of its derivative and so that

$$\lim_{h\to 0}\frac{e^h-1}{h}=1.$$

In other words, we have

$$\left(e^x\right)'=e^x.$$

    i. What do you think is the derivative of the function given by

$$e^{2x^3-4.7}?$$

Try to figure it out on your own and then use your SCS to find it.

    ii. Suppose that f is a function whose derivative you know and let g be the function given by

$$g(x)=e^{f(x)}.$$

Write a formula for the derivative of g in terms of the derivative of f.

4. Use your SCS to find the first four derivatives of the following functions.

(a) The function given by the expression

$$\cos(2x).$$

(b) The function given by the expression

$$\frac{1}{1+y}.$$

5. Consider the following equation.

$$x^3+y^3=4xy$$

Use your SCS to find a formula for $\frac{dy}{dx}$. This is essentially done in two steps. Your instructor will provide you with directions on how to complete this exercise.

(a) Differentiate the entire equation with respect to x.

(b) Solve the resulting equation for $\frac{dy}{dx}$.

6. The **computer function bis** implements the *Bisection Method* for finding zeros of a continuous function on an interval $[u, v]$ which contains exactly one point x such that the value of the function at x is 0. The following is **ISETL** code for a **computer function bis**:

```
bis := func(f, a, b, eps);
        if f(a) = 0 then return a; end;
        if f(b) = 0 then return b; end;
        c := (a + b)/2;
        if abs(b-a) < eps then return c; end;
        if f(a)*f(c) <= 0 then return bis(f, a, b, eps); end;
        if f(c)*f(b) <= then return bis(f, a, b, eps); end;
        return "cannot find any zeros";
    end;
```

Study the code for **bis**. Explain what you think **bis** is doing. Your instructor will provide you with the proper code, or a computer file you can access, if your MPL is not **ISETL**. Apply this to the freezing point function F of Activity 1, Chapter 2, Section 2.1.2, page 69. Recall that the value of the temperature T from that activity is given by

$$T = F(m) = \begin{cases} 5m^4 - 23.6m^3 - 146m^2 + 129m - 17 & \text{if } 0 \le m \le 0.76 \\ 45.056(m - 0.76)^{\frac{1}{3}} - 11.981 & \text{if } 0.76 < m \le 1 \end{cases}$$

Use the graph of F to find intervals $[u, v]$ which isolate the zeros and apply **bis** for each interval to find an approximation for that zero.

7. You can put several things together to find extreme points of a function on an interval. If we consider the domain to be the interval $[0, 0.76]$, then the freezing point function is given by the formula

$$T = F(m) = 5m^4 - 23.6m^3 - 146m^2 + 129m - 17.$$

In Chapter 2, Section 2, page 94 you considered relative and absolute extrema of a function on an interval. You can use your SCS to find these.

Apply the following steps to the function F on the interval $[0, 0.76]$.

   (a) Find the derivative F' of F.

   (b) Draw the graphs of F and F' on the same coordinate axes.

   (c) Find intervals $[u, v]$ in which there is exactly one point x for which $F'(x) = 0$.

   (d) For each interval $[u, v]$ use **bis** to find the zero for F'.

   (e) The zeros that you found are the possible extreme points of F. Look at the graph of F to determine which is which.

   (f) Compare the extrema that you found with $F(0)$, $F(0.76)$ to find the absolute maximum and minimum values of F on the interval $[0, 0.76]$.

8. Your instructor will provide you with a **computer function** named **signs** which determines all regions in which a given function is positive, negative or undefined. In **ISETL** the syntax for such a function is as follows,

signs(f, p);

where **f** is a **computer function** and **p** is a tuple of numbers in increasing order which includes every point at which the function is zero or at which it changes its state from one of being defined, undefined, positive or negative to any other of these states.

Apply the **computer function signs** to the function f given by

$$\frac{7(x-1)^2(x+1)^5}{x(x+2)(x-3)^5(3x^2-1)}.$$

Of course you must think about this expression in order to determine the list of points to give **signs**.

Look at the listing of the code for the **computer function signs** provided by your instructor and write a detailed verbal description of how it works.

9. In Chapter 2, Section 2, Activity 7 (page 88), you computed the quantity

$$\sum_{i=1}^{n} f(x_i)(x_{i+1}-x_i)$$

where f is a function and $\{x_i : i = 1, \ldots, n+1\}$ is a partition of the interval from a to b into n subintervals. In this problem you are to put it all together and write an **computer function** which accepts a **computer function**, say **f**, which represents a function, the endpoints **a**, **b** of an interval contained in the domain of **f**, and a positive integer **n**. Your **computer function** is to construct the partition of the interval from **a** to **b** into **n** equal subintervals and then return the value of the above expression. Call your **computer function** by the name **RiemLeft**. Apply your **RiemLeft** in the following situations. Choose your own value of n.

(a) The function is **data28**, $a = -1$, $b = 4$.

(b) The function is the absolute value function, $a = -1$, $b = 2$.

(c) The function is f' where f is the absolute value function, $a = -1$, $b = 2$.

4.2.3   Discussion

Rules for Derivatives — Constant Functions
In Activity 1, you investigated the derivatives of functions with constant values, i.e., constant functions. Since, the graph of a constant function is a horizontal line (or line segment) with slope 0, its derivative is simply zero as stated in the following rule.

The constant function rule
If f is a function, i.e., $f(x) = c$ for some constant c for all x, then $f' = 0$, or $f'(x) = 0$.

Rules for Derivatives — Sums, Differences and Multiples
Suppose that you can view a function as the sum of two functions. Then you just differentiate right term-by-term. That is, you look at the first term and take its derivative; look at the next term and take its derivative; and then add the two derivatives. This is a way of making a differentiation task simpler by breaking it into smaller tasks — the individual terms of a sum. The rule is as follows.

The sum rule

If f is a function which can be written as the sum of two functions, that is,
f = g + h, then f' = g' + h', or f'(x) = g'(x) + h'(x).

Of course you can repeat this rule to take the derivative of a sum of three or more functions.

For example, in the previous section, page 200, we observed that the derivative of x is 1 and the derivative of 1 is 0. Can you now see what must be the derivative of the function given by $1 + x$?

The situation is the same for differences.

The difference rule

If f is a function which can be written as the difference of two functions, that is,
f = g − h, then f' = g' − h', or f'(x) = g'(x) − h'(x).

and very similar for a number times a function.

The constant multiple rule

If f is a function which can be written as the product of a number c and
function g, i.e., f = cg, then f' = cg', or f'(x) = cg'(x).

Do you see how to use this to get the derivative of the function given by $7x - 4$?

Again from page 200 we know that the derivative of x is 1 and the derivative of (the constant function) −4 is 0. Now, using the constant multiple rule you get the derivative of $7x$ is 7 times the derivative of x, or 7 times 1, which is 7. Also, by the constant function rule, we know that the derivative of 4 is 0. Putting the two together with the difference rule leads to the conclusion that the derivative of the function given by $7x - 4$ is 7.

Rules for Derivatives — Powers of Variables

The power rule

If a is a number and f is a function given by $f(x) = x^a$ then f' is the function
given by

$$f'(x) = ax^{a-1}$$

provided this expression is defined.

The requirement that the expression be defined is needed in case a is negative in order to avoid division by 0 and also in case a is other than an even integer in order to avoid taking roots of negative numbers. Of course in the case $a = 0$, we assume that $x \neq 0$, since 0^0 is not defined.

We now have enough rules to take the derivative of $2x^3 - x + 1$. By the power rule, the derivative of x^3 is $3x^2$. By the rule for multiplying by a constant, the derivative of $2x^3$ is therefore $6x^2$. The derivative of x is 1 and the derivative of 1 is 0. Applying the addition and subtraction rules we conclude that the derivative of $2x^3 - x + 1$ is $6x^2 - 1$.

Rules for Derivatives — Products   At this point we actually hope you are beginning to find this a little boring. Why should we be making all of this fuss instead of just saying something general. We could simply say that just as with limits and continuity, to take the derivative of a combination, you take the derivative of the individual parts and put them back according to the same combination. The reason we don't jump to such a simple statement is that it is not true!

Consider the function f given by $f(x) = x\sin(x)$. Now you know that the derivative of x is 1 and by now you may have come to know that the derivative of the function sin is the function cos. So is the derivative of f the product of the derivatives $1 \cdot \cos(x) = \cos(x)$? In fact, it is not. The formula is

$$f'(x) = \sin(x) + x\cos(x)$$

which can be found by using the following general rule for products.

The product rule
If f is a function which can be written as the product of two functions g and h,
that is, $f = gh$, then $f' = g'h + gh'$, or $f'(x) = g'(x)h(x) + g(x)h'(x)$.

In other words, to take the derivative of a product of two functions, you take the derivative of the first factor and multiply it by the second, then multiply the first function times the derivative of the second and finally add the two products. Try the rule to find $f'(x)$ for the function $f(x) = x\sin(x)$ and see if you get $f'(x) = \sin(x) + x\cos(x)$.

Rules for Derivatives — Quotients   The quotient rule is even more complicated. You start out just like with the product rule except that you subtract instead of add. Then when you are all done, you divide by the square of the denominator function.

The quotient rule
If f is a function which can be written as the quotient of two functions g and h,
that is, $f = \frac{g}{h}$, then

$$f' = \frac{hg' - gh'}{h^2},$$

or

$$f'(x) = \frac{h(x)g'(x) - g(x)h'(x)}{(h(x))^2}$$

where this formula holds only at points at which h is not 0, i.e., $h(x) \neq 0$.

In other words what we have is that f' is a function whose domain is the set of points x which are in the domain of f for which $g'(x)$ exists, $h'(x)$ exists, and $h(x) \neq 0$. For such a point, we have,

$$f'(x) = \frac{h(x)g'(x) - g(x)h'(x)}{(h(x))^2}.$$

You can now use this formula to derive,

$$\left(\frac{x}{1+x}\right)' = \frac{1 \cdot (1+x) - x \cdot 1}{(1+x)^2} = \frac{1}{x^2 + 2x + 1}.$$

Rules for Derivatives — Compositions   The rule for compositions is as follows.

The chain rule
If h is a function which can be written as the composition of two functions f and
g, that is, $h = f \circ g$, then

$$h' = (f' \circ g)g'$$

or, written out in terms of individual points,

$$h'(x) = f'(g(x))g'(x).$$

In other words, you take the derivative of the first function f — but evaluate it at $g(x)$ — and multiply it by the derivative of the second function, g, evaluated at x. This rule is called the *chain rule*. We can see that this rule makes sense as follows (a rigorous proof will be given later in Section 4.4). Suppose $y = f(g(x))$ and we let $u = g(x)$. Then we have, $y = f(u)$ and $u = g(x)$. Then, for a "small change" Δx in x, there is a corresponding "change" $\Delta u = g(x + \Delta x) - g(x)$ in u, and we have (since as $\Delta x \to 0$, $\Delta u \to 0$)

$$\frac{\Delta y}{\Delta x} = \lim_{\Delta x \to 0} \frac{\Delta y}{\Delta u} \cdot \frac{\Delta u}{\Delta x} = \lim_{\Delta u \to 0} \frac{\Delta y}{\Delta u} \cdot \lim_{\Delta x \to 0} \frac{\Delta u}{\Delta x} = \frac{dy}{du} \cdot \frac{du}{dx}.$$

However, since

$$\frac{dy}{du} = f'(u) \text{ and } \frac{du}{dx} = g'(x)$$

it follows that,

$$\frac{d}{dx} f(g(x)) = (f(g(x)))' = f'(g(x))g'(x).$$

Do you see the problem with the above derivation of the chain rule? Well, it may happen that for a "small change" Δx in x, the corresponding change $\Delta u = g(x + \Delta x) - g(x)$ may be zero. It turns out that this is not really a big problem, as you will see in the rigorous proof of the chain rule in Section 4. Hence, we see that the chain rule is a product of (instantaneous) "rates", or derivatives.

It is really important in computing derivatives, to be able to recognize a given function as some combination of simpler functions. For instance, if want to differentiate the function h given by $h(x) = \cos(x^2)$, then the first step is to recognize this as a composition — of cos and the squaring function. In other words, $h = f \circ g$ where $f(x) = \cos(x)$ and $g(x) = x^2$. We will use the fact that the derivative of cos is $-\sin$ and the derivative of x^2 is $2x$.

Then $f'(x) = -\sin(x)$ so $f'(g(x)) = -\sin(g(x)) = -\sin(x^2)$. Also, $g'(x) = 2x$. Hence we have,

$$\left(\cos(x^2)\right)' = \left(-\sin(x^2)\right)(2x) = -2x\sin(x^2).$$

Rules for Derivatives — Functions Defined in Parts   A function is defined in parts if its domain is divided into intervals and the function is defined separately on each interval.

If a function is defined in parts, then you have to compute its derivative separately on each part and deal with the seams separately as well. The derivative will also be a function defined in parts. If each part of the original function is a function whose derivative you can compute, then do that. This gives the parts for the derivative. There is no simple way to compute the derivative at the seam. You just have to go back to defining the difference quotient and taking the limit. This may or may not exist and therefore, the derivative function may or may not be defined at the seams.

For example, in Section 1, page 191, we investigated the derivative of the absolute value function f given by

$$f(x) = \begin{cases} x & \text{if } x > 0 \\ 0 & \text{if } x = 0 \\ -x & \text{if } x < 0 \end{cases}$$

There are actually only two parts here because a part must extend over an interval. We could define this function a little more compactly as

$$f(x) = \begin{cases} x & \text{if } x \geq 0 \\ -x & \text{if } x < 0 \end{cases}$$

Now, you can take the derivative of the two parts x and $-x$ to get 1 and -1 respectively. Thus we have, for the derivative,

$$f'(x) = \begin{cases} 1 & \text{if } x > 0 \\ -1 & \text{if } x < 0 \end{cases}$$

and so far, the value of the derivative at 0 has not been determined. In fact, this was investigated in Section 1, and it turned out that the derivative did not exist at 0, so we have the complete definition of f'.

Rules for Derivatives — Trigonometric Functions   Here is a list of the derivatives of the six trigonometric functions.

$$
\begin{aligned}
(\sin)'(x) &= \cos(x) \\
(\cos)'(x) &= -\sin(x) \\
(\tan)'(x) &= \sec^2(x) \\
(\cot)'(x) &= -\csc^2(x) \\
(\sec)'(x) &= \tan(x)\sec(x) \\
(\csc)'(x) &= -\cot(x)\csc(x)
\end{aligned}
$$

You will see how to derive these formulas in the last section of this chapter. For now, it is important that you are able to apply all of the rules to compute derivatives of functions such as the one given by the expression,

$$\sin(a)\sqrt{1+\cos(a^2)}.$$

By the product rule and the rule for sin, we have,

$$\frac{d}{da}\left(\sin(a)\sqrt{1+\cos(a^2)}\right) = \left(\sin(a)\sqrt{1+\cos(a^2)}\right)' = \cos(a)\sqrt{1+\cos(a^2)} + \sin(a)\left(\sqrt{1+\cos(a^2)}\right)'$$

Now, applying the power rule and the chain rule,

$$\left(\sqrt{1+\cos(a^2)}\right)' = \frac{1}{2}(1+\cos(a^2))^{-\frac{1}{2}}(1+\cos(a^2))'$$

and applying, the sum rule, the rule for cos, the chain rule and the power rule, we have,

$$\left(\sin(a)\sqrt{1+\cos(a^2)}\right)' = \cos(a)\sqrt{1+\cos(a^2)} - 2a\sin(a)\sin(a^2)(1+\cos(a^2))^{-\frac{1}{2}}.$$

The Euler Constant e If a is any positive number then there is a function given by a^x. There is a very special number e named after the mathematician L. Euler who first worked with it. The function givenby e^x is very important in mathematics. Sometimes it is called the *exponential function* and given the name exp. The first interesting thing you know about this function is that its derivative is equal to itself. This is what you used in Activity 3, to estimate e. In other words,

$$\left(e^x\right)' = \lim_{h \to 0} \frac{e^{x+h} - e^x}{h} = e^x \lim_{h \to 0} \frac{e^h - 1}{h} = e^x$$

using the fact that the (natural base) e (whose value is approximately 2.71828) is that real number with the property that

$$\lim_{h \to 0} \frac{e^h - 1}{h} = 1.$$

The proof of the existence of such a number is beyond the scope of this book.

What about a function h given by an expression like

$$h(x) = e^{2x^3 - 4.7} \text{ ?}$$

It is not hard to calculate its derivative if you recognize it as a composition of two functions f with g. Here, we have $f = \exp$ and we can let g be the function given by

$$g(x) = 2x^3 - 4.7$$

Can you see how h is the composition of these two functions? We have

$$h = f \circ g = \exp \circ g$$

Hence, by the chain rule for compositions, we have,

$$h'(x) = (\exp)'(g(x))g'(x) = (\exp)'(2x^3 - 4.7)(6x^2) = 6x^2 e^{2x^3 - 4.7}$$

Can you write out in words what is the derivative of any function given by an expression of the form $e^{g(x)}$?

Higher Derivatives

Once you understand that the derivative of a function is a function, then there is nothing more to understand about the concept of higher derivatives. If you can do something once, you can try to do it again. Thus, the first derivative of a function is a function. The derivative of the first derivative is also a function and it is called the *second derivative* of the original function.

Here are some notations that are commonly used for higher derivatives.

First Derivative f', $D(f)$, $\frac{dy}{dx}$

Second Derivative f'', $D^2(f)$, $\frac{d^2y}{dx^2}$

Third Derivative $f^{(3)}$, $D^3(f)$, $\frac{d^3y}{dx^3}$

Let's see what happens with products. If you have the product fg, then the product rule says,

$$(fg)' = f'g + fg'$$

We can differentiate this,

$$\begin{aligned}(fg)'' &= (f'g)' + (fg')' \\ &= f''g + f'g' + f'g' + fg'' \\ &= f''g + 2f'g' + fg''\end{aligned}$$

Can you see a general rule for the higher derivatives? Compute a few more by hand or use your SCS and see if you can discern a pattern.

Higher derivatives have geometrical and physical interpretations. In particular, consider the situation where a particle is moving along a line with its position p given by a function P, where $p = P(t)$ at time t. You have seen that $P'(t)$ represents the velocity v of the particle at time t at the position $P(t)$ and v is given by the function V where $V(t) = P'(t)$. What do you think $V'(t) = P''(t)$ represents? Of course, it's the acceleration a of the particle at time t. That is,

$$a = A(t) = V'(t) = P''(t).$$

What are the units of acceleration?

Derivatives of Implicitly Defined Functions

Again, with implicit differentiation the problem is simple once you understand the situation. Consider an expression like

$$x^3 + y^3 = 4xy$$

which you worked with in Activity 5. If you can accept the idea that in this equation, y represents (implicitly) a differentiable function f whose independent variable is called x, then it is perhaps more informative (but less convenient) to write

$$x^3 + (f(x))^3 = 4xf(x)$$

This is a technique we use often in mathematics. The first expression is easier to work with so that is what we use. What we actually mean is embodied in the second expression. This we keep in our mind and only trudge it out if we want to contemplate the meaning — or explain it to someone else.

So, looking at the longer expression, the point is that this statement is understood to be true for every x in the domain of the function y. Thus it is an equation about *functions*, not just numbers. The equality asserts that two functions are equal. Therefore their derivatives are equal. This means that we can differentiate both sides of the equation and we still have an equation.

The actual computation can be done by hand, remembering things like the derivative of $(f(x))^3$ is, using the chain rule, $3(f(x))^2 f'(x)$ and the derivative of $4xf(x)$, using the product rule, is $4xf'(x) + 4f(x)$. Or you can use a SCS to find the derivatives.

The important point is that the resulting equation *can always be solved for* $y' = f'(x)$. This is because y' always appears as a first order term. Hence you can collect the terms containing y', factor it out, put everything else on the other side and divide to get an expression for y'. Or you can take the lazy way and find the derivative y' using your SCS.

There is one point we have not mentioned here. Starting with this equation in x and y, where does the function y come from? Does it fall from the sky? How do we even know that there is such a function y? In the last section of this chapter you will see how this equation can actually be used to *define* y as a function of x (or x as a function of y).

Extrema

In Chapter 2, page 94, we gave a detailed discussion of extreme points — that is, relative and absolute maxima and minima — of a function defined on an interval. We will return to a formal discussion of these concepts in the last section of this chapter. Here is a simple, but very important rule that is a very helpful in finding extreme points. The problem of finding extrema usually arises in the context of a function defined on a closed interval and therefore in considering its derivative at each endpoint we must use the left and right derivatives as discussed above (page 208).

> *If f is a function defined on an interval $[a, b]$, then the only place that extreme points can occur is at points x where $f'(x) = 0$, where $f'(x)$ does not exist, or at one of the endpoints of the interval, a and b.*

What this means computationally, and what you did in Activity 7, is to compute the derivative of f, find the points at which the derivative is zero, and then compare these with the endpoints and points at which the derivative is not defined to check for the absolute maxima or minima of a function f. By looking at the graph, or checking values of the derivative at nearby points, you can make a determination as well of the relative maxima and minima.

In Activity 7 you worked with one of the parts of the freezing point function. Let's look at the whole thing. It is given by,

$$F(m) = \begin{cases} 5m^4 - 23.6m^3 - 146m^2 + 129m - 17 & \text{if } 0 \le m \le 0.76 \\ 45.056(m-0.76)^{\frac{1}{3}} - 11.981 & \text{if } 0.76 < m \le 1. \end{cases}$$

Use the rule for parts to check that the following is the derivative of this function. We get, initially, that

$$F'(m) = \begin{cases} 20m^3 - 70.8m^2 - 292m + 129 & \text{if } 0 \le m < 0.76 \\ \frac{45.056}{3}(m-0.76)^{-\frac{2}{3}} & \text{if } 0.76 < m \le 1. \end{cases}$$

It is then necessary to check the derivative at the seam, 0.76. Can you determine that the derivative does not, in fact, exist at this point? It does not. Nevertheless, the function has a relative minimum at this point. Is it an absolute minimum?

4.2.4 Exercises

1. Find the derivative of each of the following functions.

   (a) $f(x) = \sqrt{2} - 5x$

   (b) $g(x) = 0.5x - x^4 + 3$

   (c) $h(t) = 2\sqrt[3]{t} - t^{-3} + 1$

   (d) $f(s) = 4\sqrt{s}$

   (e) $g(x) = \dfrac{x}{2} - \cos(x)$

   (f) $G(x) = \dfrac{-3}{x^2}$

   (g) $f(x) = \dfrac{3}{2}x^5 - \dfrac{2}{3x^{\frac{3}{4}}}$

   (h) $s(t) = \sqrt{t} - 2\tan(t)$

   (i) $f(x) = \dfrac{3x}{5} + \dfrac{1}{x^{2.3}} - \pi$

   (j) $f(x) = \dfrac{1}{3x+1}$

   (k) $f(x) = \dfrac{\sin(5x)}{-3}$

   (l) $h(x) = \sec\left(1 - \sqrt[3]{x}\right)$

   (m) $g(x) = \dfrac{\cos(2x)}{\sqrt[3]{x}}$

   (n) $f(x) = e^{-2x}\sin(\pi x)$

   (o) $f(x) = \sqrt{x}\cos(1-3x)$

   (p) $f(x) = 2x - \dfrac{3}{x} + 5$

   (q) $F(x) = x^2 e^{1-3x}$

   (r) $g(x) = \dfrac{3x}{x+1}$

   (s) $h(x) = \dfrac{5+x^2}{1-x}$

   (t) $f(x) = \tan^3(x^2 - x)$

   (u) $h(x) = 2x - \sin^2(3x)$

   (v) $f(x) = \dfrac{\tan(x^2)}{x+\cos(x)}$

   (w) $g(x) = 3\cos\left(\sqrt{1-2x}\right)$

(x) $f(x) = \sqrt{\dfrac{1-x}{x+2}}$

(y) $H(x) = 2\sqrt{1+\sqrt[3]{2-x}}$

2. Find the first two derivatives of each of the following functions.

   (a) $f(x) = 4x^5 - 2x^3 - x^2 - x + 1$

   (b) $g(t) = t^{\frac{2}{3}} - \dfrac{2}{t^3}$

   (c) $g(x) = \sqrt{x} - \sqrt[3]{x}$

   (d) $h(x) = \sqrt{1-x^2}$

   (e) $f(x) = x\sqrt{x+2}$

   (f) $h(x) = \sin^3(2x)$

   (g) $F(x) = \dfrac{2x}{1-x}$

   (h) $g(t) = \sec(2t)$

   (i) $h(x) = \dfrac{\cos(x)}{x}$

   (j) $f(x) = \dfrac{2}{1-x}$

   (k) $g(v) = 3\sin(\sqrt{v})$

   (l) $H(z) = \tan(\tfrac{1}{z})$

   (m) $f(x) = x^2 \sin(3x)$

   (n) $g(w) = w^2 \sqrt{1-w}$

3. Find an equation of the tangent line and an equation of the normal line to the graphs of each curve at the indicated point. Sketch the curve, the tangent line and the normal line at the indicated point on the same graph.

   (a) $x = 2,\ f(x) = 3x - x^2$

   (b) $x = 1,\ g(x) = x^3 - \sqrt{x}$

   (c) $x = \dfrac{\pi}{6},\ h(x) = 1 + 3\cos(x)$

(d) $x = 0,\ f(x) = \dfrac{3+x}{1-x^2}$

4. Find the derivatives of the following functions.

   (a) $g(x) = \begin{cases} -2 & \text{if } x > 0 \\ 3 - x & \text{if } x \le 0 \end{cases}$

   (b) $f(x) = \begin{cases} x^2 & \text{if } x < 1 \\ 2x - 1 & \text{if } x \ge 1 \end{cases}$

   (c) $f(x) = \begin{cases} x^3 & \text{if } x < -1 \\ 1 + 2x^2 & \text{if } -1 \le x < 2 \\ 4x + 1 & \text{if } x \ge 2 \end{cases}$

5. The following functions give the position p at time t of a particle moving along a coordinate line.

   (a) Find the functions giving velocity and acceleration functions at any time t.

   (b) On what intervals for t is the particle moving in the positive direction? the negative direction?

   (c) On what intervals for t is the particle slowing down? speeding up?

   (d) Make a sketch showing the path of the particle.

   (e) Make a sketch of the position function, the velocity function and the acceleration function on the same graph.

   i. $p = P(t) = 3t^2 + t^3$

   ii. $p = P(t) = t^3 - 6t^2 + 9t$

   iii. $p = P(t) = 5t^3 - 3t^5$

   iv. $p = P(t) = t^4 - 4t^2 + 4$

6. Assuming y is defined to be a differentiable function of x in each of the following equations, find $y' = \frac{dy}{dx}$.

(a) $3x^2 - y^2 = 3$

(b) $y^3 + x^2 y = x + 1$

(c) $\dfrac{1}{x} - \dfrac{2}{y^2} = 1$ (d) $2y = x - \sqrt{y}$

(e) $\dfrac{y}{x} + \dfrac{1}{\sqrt[3]{y}} = 5$

(f) $(1 + xy)^4 = x - y$

(g) $\cos(y) = y + 2x$

(h) $\sin(2y) = x^2 + y$

(i) $\tan\left(\dfrac{x}{y}\right) = x - 1$ (j) $\sqrt{\sin(xy^2)} = y$

(k) $x^3 - y = \sqrt[3]{\sin(y)}$

(l) $x - 1 = \sqrt{2x - xy^3}$

7. Assuming that y is defined to be a twice differentiable function of x, find y''.

   (a) $x^2 - y^3 = 1$ (b) $\sin(y) = 2x + y$

   (c) $x^2 y = \tan(xy^2)$ (d) $\dfrac{x^2 y}{y - x}$

8. Find the indicated derivative assuming the equation defines an appropriate differentiable function. Assume a and b are constants.

   (a) $ay^2 - x = bx^3 + \sin(y), \dfrac{dx}{dy}$

   (b) $x^2 t + a3t^2 = bx^3 + a^2, \dfrac{dt}{dx}$

9. The volume v, in gallons, of water in a tank after t hours is given by

$$v + V(t) = 300 \sin^2\left(\frac{\pi t}{6}\right)$$

for $0 \le t \le 3$.

   (a) Find a function which gives the rate of flow of water into the tank for any time t.

   (b) At what rate is the water flowing into the tank after one hour?

10. The position s of a car traveling along a straight highway after the driver applies the brakes is given by $s = S(t) = 30t - t^2$, where s is the number of feet from the point at the brakes are applied and t is the number of seconds after the brakes are applied.

   (a) Find the rate at which the car is traveling at any time t.

   (b) At what rate is the car traveling at $t = 5$?

   (c) Find the deceleration of the car.

   (d) How long does it take the car to stop?

   (e) How many feet does it take the car to stop?

11. A particle moves along a line with its position s given by $s = S(t) = 0.5t^2 - 2\sqrt{t}$ feet after t seconds.

   (a) Find the velocity and acceleration functions at any time t.

   (b) When is the velocity zero, positive, negative?

   (c) Find the acceleration when the velocity is zero.

12. For the functions represented by each of the following expressions, find the extrema and tell whether each is a relative extremum or an absolute extremum. Sketch of the complete graph, i.e., a graph showing all relevant features of the graph.

   (a) $3x^2 + 24x$ (b) $x^3 - 2x^2 + 1$

   (c) $\sqrt[3]{x^2} - 3$ (d) $2\sqrt{x} - x$

   (e) $3x - x^3$ (f) $x^3 - 6x^2$

   (g) $15 - 25x - x^3$ (h) $|2x - 3| + 1$

   (i) $|3x - 2x^3|$ (j) $2x + \frac{3}{x} + 5$

(k) $\frac{3}{x^2}+\frac{1}{x}$ (l) $\frac{x}{x-2}$

(m) $\frac{1-2x}{x^2+1}$ (n) $\frac{x}{x^2-4}$

13. Find all extrema of the function on the indicated interval. Tell whether each of the extrema is relative or absolute. Make a sketch of the situation.

   (a) $f(x)=x^3-3x^2$, $[-1, 4]$

   (b) $g(x)=x^2-2x$, $[-2, 0]$

   (c) $h(x)=2x^3-3x^2-12x$, $[-4, 4]$

   (d) $f(x)=9x^2-2x^3-12x$, $[0, 5]$

   (e) $G(x)=\sin(x)-\dfrac{x}{2}$, $[0, \pi]$

   (f) $f(x)=\begin{cases} x^2 & \text{if } -1\le x<1 \\ 2x-1 & \text{if } 1\le x\le 3 \end{cases}$

   (g) $g(x)=\begin{cases} (x+2)^2-1 & \text{if } -3\le x\le -1 \\ x^3+1 & \text{if } -1<x<1 \\ 1+(x-2)^2 & \text{if } 1\le x\le 2 \end{cases}$

14. Write out in both words and symbols a rule for finding the derivative of any function given by an expression of the form $e^{g(x)}$, where g is a differentiable function of x.

15. For a particle moving along a line with its position p given by a function of time P, that is $p=P(t)$,

   $$a=A(t)=V'(t)=P''.$$

   What are the units of acceleration?

16. For the **computer function bis** in Activity 6 on page 218, explain in your own words what you think **bis** is doing.

17. In Activity 7, page 219 you worked with the freezing point function F given by

$$F(m)=\begin{cases} 5m^4-23.6m^3-146m^2+129m-17 \\ \qquad \text{if } 0\le m\le 0.76 \\ 45.056(m-0.76)^{\frac{1}{3}}-11.981 \\ \qquad \text{if } 0.76<m\le 1 \end{cases}$$

Show that F is not differentiable at the seam 0.76. Use your computer graphing system to show that the function does <u>not</u> have a relative minimum at this point. Is it an absolute minimum? How can you tell?

18. Let f be the absolute value function given by $f(x)|x|$.

   (a) Show that f' is given by

   $$f'(x)=\frac{x}{|x|}$$

   What is the domain of f'?

   (b) For a differentiable function g, find a formula for the derivative of $|g|$. What is the domain of the derivative of $|g|$?

   (c) Using your formula in part (b), find the derivative of each function and give the domain of each derivative.

      i. $f(x)=|1+2x|$

      ii. $f(x)=\left|1-x^2\right|$

      iii. $g(x)=\left|x^3-4x\right|$

      iv. $h(x)=|\sin(2x)|$

      v. $f(x)=x|x|$

19. (a) Find a formula for the 3rd derivative of fg.

   (b) Find a formula for the kth derivative of fg where k is a positive integer.

20. Find a formula for the second derivative of the function h defined by $h(x)=f(g(x))$. What assumptions must be made about the functions involved, their derivatives, and the domains of the functions and their derivatives?

4.3 PROBLEM SOLVING WITH DERIVATIVES

4.3.1 Overview

This is the first of an occasional section which is a departure from the standard organization we are using in this book. Since Chapter 1, the material has been divided into sections, each of which consists of a collection of computer activities, a discussion and a set of exercises. In general, the discussion is closely related to the activities. The activities are designed to set your mind working in ways that will make the mathematical material in the discussion more accessible to you. Our goal is that the explanations and analyses in the discussion section is always about things you are familiar with — because you worked with them concretely when you did the activities.

Occasionally, as in the present section, there are no activities related to many of the discussion sections. The reason for this is that because of the work you have done up to this point, you are already familiar with all of the mathematical objects and processes, and relationships that make up the mathematics we would like you to learn in studying this section. In a sense, this is pay-off time. You have done a lot of work and now you can put things together and learn a great deal of mathematics in a very short time. You will learn to solve many problems, some of which you have may have encountered earlier in this text, using derivatives and the ideas of calculus.

The activities section is not completely empty. There is one set of computer tasks that you should do relative to graphs before studying the methods of drawing graphs by hand presented in this section. In addition, there are a few activities which deal with a method of applying derivatives to find limits that are of the *indeterminate form* $\frac{0}{0}$. We also include an activity related to the **computer function RiemLeft** that you constructed in the previous section. You may have noticed that we have been giving a connected set of activities on this topic running through several chapters. This is a build-up to a major piece of mathematical analysis. The climax will occur in the following chapter.

4.3.2 Activities

1. The task of this activity is to make as careful and informative as possible graphs of each of the functions f, g, h defined as follows.

$$f(x) = x^3 - x^2$$

$$g(x) = 10x^{2/3} - x^{5/3}$$

$$h(x) = \frac{x^3 - 7x}{x^3 + 3x - 3}$$

For each of these, use a computer, or graphics calculator, to make a graph to investigate how the curve looks in various regions. Sketch each graph carefully on a large sheet of paper. Use labels and write neatly on the sheet to indicate the best answers you can give to the following questions.

(a) In what regions is the function increasing?

(b) In what regions is the function decreasing?

(c) At what points does the function have a relative maximum and what is it?

(d) At what points does the function have a relative minimum and what is it?

(e) In what regions is the graph concave up ("holds water")?

(f) In what regions is the graph concave down ("does not hold water")?

(g) What are the vertical asymptotes of the graph?

(h) What is the behavior of the function at ∞ and $-\infty$? What are the horizontal asymptotes of the graph? If there are none, indicate whether the function goes to ∞, $-\infty$, or oscillates.

2. For the (linear) functions f and g given by

$$f(x) = 6 - 2x \text{ and } g(x) = x - 3,$$

consider the following limit.

$$\lim_{x \to 3} \frac{f(x)}{g(x)}$$

Can you evaluate this limit by simply substituting 3 for x? What happens if you try to do so? Such an expression is said to be "indeterminate" at 3 and is called an *indeterminate form* at 3 of the type $\frac{0}{0}$.

(a) Use your computer tool **table2** to make a table of values for f and g on the interval $[2.99, 3.01]$ with $n = 20$. Then make a table showing values of both of the functions

$$h = \frac{f}{g} \text{ and } k = \frac{f'}{g'}$$

on the interval $[2.99, 3.01]$ with $n = 20$ (be sure to take care of any problems at $x = 3$ for the functions h and k).

(b) Make an sketch of f and g on the interval $[2, 4]$ on the same coordinate axes. Then make a sketch of the function $h = f/g$ on the interval $[2, 4]$. By inspecting your sketch for h, what do you think is the value of the limit of $h = f/g$ as $x \to 3$? How does the value of the limit you found relate to the ratio of the slopes of the lines $y = f(x) = 6 - 2x$ and $y = g(x) = x - 3$? Do you see how to find the ratio of the slopes of the these two linear functions from the graph of f and g on the same coordinate axes? Explain.

(c) Using your results in parts (a) and (b), what do you think is the relationship (if any) between the following two limits?

$$\lim_{x \to 3} \frac{f(x)}{g(x)} \text{ and } \lim_{x \to 3} \frac{f'(x)}{g'(x)}$$

3. For the functions f and g given by

$$f(x) = x\left(1 - e^{x-1}\right) \text{ and } g(x) = 1 - e^{2(1-x)},$$

consider the following limit.

$$\lim_{x \to 1} \frac{f(x)}{g(x)}$$

Is the ratio $\frac{f(x)}{g(x)}$ of the indeterminate form $\frac{0}{0}$?

(a) Use your computer tool **table2** to make a table of showing values of both f and g on the interval $[0.99, 1.01]$ with $n = 20$. Now use the tool **table2** for the functions

$$h = \frac{f}{g} \text{ and } k = \frac{f'}{g'}$$

(make sure you take care of the problem at 1 for $h = f/g$) on the interval $[0.99, 1.01]$ with $n = 20$.

(b) Make a sketch of f and g on the interval $[0, 2]$ on the same coordinate system. Now make a sketch of the function $h = f/g$. By inspecting your sketch for $h = f/g$, what do you think is the value of the following limit?

$$\lim_{x \to 1} \frac{f(x)}{g(x)}$$

(c) Make a sketch of both of the functions f and g on the interval $[0, 2]$ on the same coordinate axes. Now use your graphics utility to zoom on your sketch so that it looks similar to the graphs of the two linear functions in Activity 1. Are you surprised? Explain your answer. What do you think is the value of the following limit?

$$\lim_{x \to 1} \frac{f'(x)}{g'(x)}$$

(d) Using the results parts (a)-(c), what do you think is the relationship (if any) between the following two limits in this activity?

$$\lim_{x \to 1} \frac{f(x)}{g(x)} \text{ and } \lim_{x \to 1} \frac{f'(x)}{g'(x)}$$

4. In this activity you will make conjectures (guesses) about limits of ratios which are of the indeterminate form $\frac{0}{0}$.

(a) Suppose that the ratio of functions in the following limit

$$\lim_{x \to a} \frac{f(x)}{g(x)}$$

is of the indeterminate form $\frac{0}{0}$. Based on your investigations in the previous two activities, what do you think is the relationship between the following two limits?

$$\lim_{x \to a} \frac{f(x)}{g(x)} \text{ and } \lim_{x \to a} \frac{f'(x)}{g'(x)}$$

Give an explanation for your answer.

(b) Suppose that the ratios of functions in the following limits

$$\lim_{x \to a} \frac{f(x)}{g(x)} \text{ and } \lim_{x \to a} \frac{f'(x)}{g'(x)}$$

are both of the indeterminate form $\frac{0}{0}$, what do you think is the relationship between the following three limits

$$\lim_{x \to a} \frac{f(x)}{g(x)}, \ \lim_{x \to a} \frac{f'(x)}{g'(x)} \text{ and } \lim_{x \to a} \frac{f''(x)}{g''(x)}$$

Give an explanation for your answer.

(c) For differentiable functions f and g, what do you think could be a general rule for finding limits like

$$\lim_{x \to a} \frac{f(x)}{g(x)}$$

which involve the indeterminate form $\frac{0}{0}$? Give two versions for your general rule, a symbolic one and a verbal one in your own words.

Such a rule gives a another "analytical" way of possibly finding limits of the indeterminate form $\frac{0}{0}$ using derivatives. The rule is called *L'Hôpital's rule*. As you will see in the discussion section after these activities, L'Hôpital's rule can also be used for limits from the right and left of a number a and for limits at infinity and minus infinity.

5. Think about your experiences in the previous three activities. Do you think *L'Hôpital's rule* can be applied to limit situations, involving quotients of two functions, that are of the indeterminate form $\frac{\infty}{\infty}$? That is, if f and g are differentiable functions such that

$$\lim_{x \to \infty} f(x) = \infty \text{ or } -\infty \text{ and } \lim_{x \to \infty} g(x) = \infty \text{ or } -\infty$$

do you think *L'Hôpital's rule* can be extended to such a situation? If so, give a brief explanation for your answer and also state what you think might be a general rule for finding limits like

$$\lim_{x \to a} \frac{f(x)}{g(x)}$$

you encounter the indeterminate form $\frac{\infty}{\infty}$. Give two versions for your general rule, a symbolic one and a verbal one in your own words.

6. Return to Section 2, Activity 9, page 220, and your **computer function RiemLeft**. Run **RiemLeft** with n sufficiently large, on the same three examples that you used then except reverse a and b. Compare your answers with the answers you got originally.

7. Return again to Section 2, Activity 9, page 220, and consider the first function, **data28** on the interval $[-1, 4]$. Sketch the graph of this function and, using a value of $n = 10$, shade in the quantity on the graph that is represented by the calculation that **RiemLeft** would make.

Repeat the process except this time, use the calculation that **RiemRight** would make. Of course you have to guess what **RiemRight** would be!

Repeat the process one more time, using the calculation that **RiemMid** would make. Again you have to guess what **RiemMid** would be! (How would you find the midpoint of an interval of real numbers $[x_i, x_{i+1}]$? Now do you see what to do?)

4.3.3 Discussion

Using Calculus in Situations — General The first step in using mathematics to analyze a situation and solve a problem is to decide what mathematics to use. The next step is to give the problem a mathematical formulation. This involves a translation from the language of the problem to the language, mathematics, of the solution. In the situations considered in most applications of mathematics, the translation will include a definition of one or more functions. This will certainly be the case in this section. We will also be calculating derivatives.

So here is a very brief recipe for solving the problems in this section.

1. Define a function.

2. Lay out the mathematical steps required in solving the problem.

3. Perform the mathematical steps and required calculations — by hand or with the computer.

4. Translate the mathematical solution back to the language of the problem. Make any appropriate interpretations.

In the next few paragraphs we will illustrate how these steps are applied to solve what have become some of the "standard problems" of differential calculus.

Using Calculus in Situations — Max/Min You have worked with problems concerning maximum and minimum values throughout this book.

Consider the following situation.

> A farmer wishes to use some spare fencing to build a rectangular pen for goats. It is possible to use one side of a 100 ft. long barn as all or part of one side of the pen. If there is 180ft. of fencing available, what dimensions should be used to get the maximum area?

Before starting the solution it is sometimes worthwhile to give some thought to the situation. One question that needs to be settled is whether the barn wall should form all of one side of the pen, or just a part of a side. Common sense tells us that we should use as much of the barn wall as possible for one side so that more fencing is available for the three remaining sides. This suggests that the wall should form all of one side. Nevertheless, we will check this "intuition" by considering the other possibility as well.

1. It can help in seeing how to define a function to draw a picture. For this problem, we might sketch something like the pictures shown in Figures 4.8a and 4.8b.
   The function we must define has to give the area as a function of some variable connected with the dimensions of the pen. At first thought, this means that there are two independent variables, but if we remember that the total amount of fencing is fixed, then we can realize that this will restrict one of the variables. That is, if we take l to be the length of the pen, then we can compute the width from the length and the perimeter. The area will then be the length times the expression for the width. So our first task is to determine the width as a function of the length.

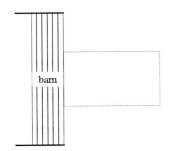

Figure 4.8a. A picture with wall length = side length.   **Figure 4.8b. A picture with wall length > side length.**

We will have to consider two cases corresponding to how much of the wall is used for a side of the pen. If the length parallel to the barn is less than or equal to 100 ft. then the wall will be all of one side. Otherwise, the wall will only take up a part of one side.

In the first case, we only have three sides of fencing and the width is half of the amount of fencing left after l ft. is used for one side parallel to the barn. Thus, the width is given by

$$\frac{1}{2}(180-l) = 90-\frac{l}{2}.$$

In the other case, the width is half of the amount of fencing left after l ft. is used for one side and $l-100$ ft. is used for the other side. Thus, the width is given by,

$$\frac{1}{2}(180-l-(l-100)) = 140-l.$$

Therefore, remembering that we use the first case if the length is less than or equal to 100 ft. and the second case otherwise, and the length must be positive and cannot exceed 140 ft., we have,

$$a = A(l) = \begin{cases} (90-\frac{1}{2})l & \text{if } 0 \le l \le 100 \\ (140-l)l & \text{if } 100 < l \le 140. \end{cases}$$

This is the function whose value must be maximized.

2. According to the discussion of extrema in the previous section, the mathematical steps in determining the maximum value of this function consist of determining the extreme points by finding out where the derivative vanishes, i.e., has a value of zero, and to compare the values of the function at the extreme points with its value at the endpoints, and at the seam. (It is not necessary to compute the derivative at the seam since it is just as easy to check the value.) The biggest value that we get from evaluating the function at these critical points is the maximum value of the function on the closed interval.

3. The derivative A' at points other than $l = 100$ is given by

$$a' = A'(l) = \begin{cases} 90-l & \text{if } 0 \le l < 100 \\ 140-2l & \text{if } 100 < l \le 140 \end{cases}$$

which is zero when $l = 90$. (Do you see why the second branch does not give a zero of the derivative?) Hence we must check the values $A(0)$, $A(90)$, $A(100)$, $A(140)$. We get, respectively, 0, 4050, 4000, 0. Therefore the maximum occurs at $l = 90$ so the dimensions giving maximum area are 90×45.

4. The maximum occurred in the first case, where less than all of the barn wall was used, so the original "common sense" analysis turns out to be incorrect. What happens is that using the wall saves fencing, but it also affects the shape of the pen. If all of the wall is used for one side or part of a side, then the pen has to be too long and narrow to achieve maximum area.

Using Calculus in Situations — Velocity and Acceleration   You have seen that the derivative of a position function for an object moving along a coordinate line represents the velocity of the object and the second derivative of the position function (the derivative of the velocity function) represents the acceleration of the object. Consider the following situation.

A weight is hung from a spring as shown in Figure 4.9. The weight is pulled vertically downward to a position 5 centimeters(cm) below its resting position and then released. As a result, the weight oscillates vertically up and down in such a way that the position s of the top of the weight from the resting position after t seconds is given by the formula,

$$s = S(t) = -5\cos\left(\frac{\pi}{3}t\right).$$

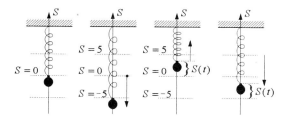

Figure 4.9. A weight hung from a spring oscillates along a vertical line.

At what times, during the first 10 seconds, is the distance of the weight from resting position the greatest? the least? When is its velocity increasing? Decreasing?

1. For this problem, there are actually three functions to define. There is the position function $s = S(t)$ given by the stated formula. Because there are questions about the velocity and also about the extreme points of the position, we must define the velocity function V, where $v = V(t)$. This is the derivative of the position function, that is, $V = S'$. Finally, in order to find out where the velocity is increasing we might look at the acceleration function A, where $a = A(t)$. This is the derivative of the velocity function, $A = V' = S''$.

2. Of course the first mathematical steps are to compute the two derivatives to find formulas for the functions V and A.

Next, in order to find the extreme points of S we find out where the velocity is 0. You could do this according to the method described in Section 2, Activity 6 (page 218). That is, a sketch of the graph of V will allow us to pick intervals that isolate zeros. It is necessary to blow up, i.e., examine more closely, appropriate intervals so that nothing is missed. Then we can apply **bis** to estimate the zeros of V if necessary. In this example, however, you should be able to solve for the zeros of the functions S, V and A by hand. Finally, a glance at the graph of S will tell us if these represent points of relative maxima or minima or neither.

Now we can turn to the points of increasing and decreasing velocity. These correspond to regions in which the acceleration is positive or negative. One way to do this is to approximate, or solve for, the zeros of the acceleration function A. Then a glance at the velocity graph will tell us whether those are regions of positive or negative values of the velocity. An alternative would be to use the **computer function signs** (page 64) to find out where the functions V and A are positive and negative.

3. Although the actual calculations for this problem can be done using your SCS, you should be able to do them by hand. The following formulas are obtained immediately by differentiation.

$$V(t) = \frac{5\pi}{3}\sin\left(\frac{\pi}{3}t\right)$$

$$A(t) = \frac{5\pi^2}{9}\cos\left(\frac{\pi}{3}t\right)$$

From the graph of V, or from the expression representing $V(t)$, you can determine that V has zeros at 0, 3, 6, and 9 seconds. The velocity is positive between 0 and 3 seconds and 6 and 9 seconds, and it is negative elsewhere in the first 10 seconds. Similarly, the acceleration has zeros at 1.5, 4.5, and 7.5 seconds. The acceleration is negative in the intervals from 0 to 1.5 seconds and between 4.5 and 7.5 seconds, and it is positive elsewhere in the first 10 seconds.

4. The relative maxima of the position function represents those times when the object has been rising and then stops rising, and begins to descend. Similarly, the relative minima correspond to times at which the object stops descending and starts to rise. The velocity is increasing when the object is accelerating and decreasing when it is decelerating.

Notice that exactly when the object begins to rise it starts accelerating. Sometime before it stops rising, it stops accelerating and then begins to slow down until eventually it stops and begins to descend. Then it speeds up again, and so on. This is why the zeros of the velocity and acceleration function alternate in time.

Using Calculus in Situations — Tangent/Normal Lines   You have seen that the derivative of a function at a point represents the slope of the tangent line to the graph of the function at that point. We illustrate this with the following problem.

Consider the curve in the first quadrant described by the equation

$$x^{\frac{2}{3}} + y^{\frac{2}{3}} = 1.$$

If P is a fixed point on this curve (see Figure 4.10), find a formula for the length of the segment of the tangent line to the curve at that point P which lies in the first quadrant.

1. In this problem there is more than one function that we have to define. In the first place, since everything is restricted to the first quadrant, we can solve the given equation for y and so we have a function f which give the y-coordinate of a point on the curve as a function of its x-coordinate. That function is given by

$$y = f(x) = \left(1 - x^{\frac{2}{3}}\right)^{\frac{3}{2}}$$

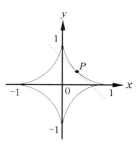

Figure 4.10. *A graph of the curve* $x^{\frac{2}{3}} + y^{\frac{2}{3}} = 1$.

Next, we have to define a function which is the equation of the tangent line at the point P. We will need the coordinates of P. If its x coordinate is a, then, by what we have just defined, its y coordinate is $f(a)$. The equation of the tangent line will also use the slope which is given by the derivative at a. This is $f'(a)$. Using the point slope formula, we can determine our function to be,

$$y = f(a) + f'(a)(x - a)$$

$$= \left(1 - a^{\frac{2}{3}}\right)^{\frac{3}{2}} - \left(\frac{\left(1 - a^{\frac{2}{3}}\right)^{\frac{1}{2}}}{a^{\frac{1}{3}}}\right)(x - a).$$

2. Here are the mathematical steps we will have to take to solve our problem.

   (a) Find the y intercept of the tangent line by setting x equal to 0 in the equation and computing y.

   (b) Find the x intercept of the tangent line by setting y equal to 0 in the equation and solving for x.

   (c) Apply the distance formula to these two intercepts to find the desired length.

3. The calculations are quite tedious and you might want to use your SCS to do them (don't forget to simplify). We list the results of these three steps.

   (a) The y intercept is

$$\left(1 - a^{\frac{2}{3}}\right)^{\frac{3}{2}} + \left(1 - a^{\frac{2}{3}}\right)^{\frac{1}{2}} a^{\frac{2}{3}}.$$

(b) The x intercept is

$$\frac{\left(\left(1-a^{\frac{2}{3}}\right)^{\frac{3}{2}}+\left(1-a^{\frac{2}{3}}\right)^{\frac{1}{2}}a^{\frac{2}{3}}\right)a^{\frac{1}{3}}}{\left(1-a^{\frac{2}{3}}\right)^{\frac{1}{2}}}.$$

(c) The distance turns out to be 1.

4. The interesting interpretation of this result has to do with the fact that the coordinates of P do not appear in the solution. In other words, this curve has the property that the length of the specified line segment is a constant.

Using Calculus in Situations — Implicit Differentiation   Consider the following equation in x and y.

$$\left(x^2+8y-16\right)^2 = y^2\left(16-x^2\right)$$

We know that this defines a curve in the xy-plane. Suppose that it also defines y as a function of x. Recall that this means that there is some function f which satisfies,

$$\left(x^2+8f(x)-16\right)^2 = (f(x))^2\left(16-x^2\right)$$

for points x in its domain.

Answer the following questions about this function.

What is the equation of the line normal to this curve at the point $\left(\sqrt{7},\frac{9}{5}\right)$?

What can be said about the tangent lines to this curve at the points on the curve which intersect the y-axis?

We will try to apply our general outline of a solution to both problems simultaneously.

1. In both cases, the function to define is the function f assumed to exist in the statement of the problem. We don't have any explicit definition for this function and that is why this kind of problem is referred to as an *implicit differentiation* problem.

2. The mathematical steps consist of implicit differentiation and plugging in values to obtain the slopes of the tangent lines. The slope of the normal line is obtained by computing the negative reciprocal. Then the point-slope formula for the equation of a line is used.

3. In both problems, we must first deal with the points in question. In the first problem, it is only a matter of checking that $\left(\sqrt{7},\frac{9}{5}\right)$ is a point on the curve, that is, it satisfies the equation. You can do this by hand or with your SCS.

For the second problem, we must compute the y-intercepts, that is, the points at which $x=0$. In other words, we set $x=0$ in the equation and solve for y. This gives,

$$16y^2 = (8y-16)^2$$

or

$$48y^2 - 256y + 256 = 0$$

or

$$3y^2 - 16y + 16 = 0.$$

Again you can solve this by hand or with your symbolic computer system. There are two solutions, $y = 4$ and $y = \frac{4}{3}$.

To summarize, then, we have three points, $\left(\sqrt{7}, \frac{9}{5}\right)$, $(0, 4)$, and $\left(0, \frac{4}{3}\right)$. We must find the equation of the normal line at the first point and the tangent lines at the other two points.

In all cases, we need a formula for the derivative of our implicit function f. We do this by differentiating the equation which defines f with respect to x. We can actually work more efficiently if we use the original equation in terms of x and y, but it is important to remember that y represents a function and so its derivative y' is the derivative of our function f. You can do the following calculations with your symbolic computer system, but you should also make sure you follow the steps written here.

We have,

$$\left(x^2 + 8y - 16\right)^2 = y^2\left(16 - x^2\right)$$

and differentiating on both sides, using the power rule and chain rule on the left-hand side and the product rule and chain rule on the right-hand side, gives,

$$2\left(x^2 + 8y - 16\right)(2x + 8y') = 2yy'\left(16 - x^2\right) + y^2(-2x).$$

Collecting terms in y' gives, after some manipulations,

$$y'\left(16x^2 + 2x^2y + 96y - 256\right) = -4x^3 - 2xy^2 + 32xy + 64x$$

and so,

$$y' = -\frac{4x^3 + 2xy^2 - 32xy - 64x}{16x^2 + 2x^2y + 96y - 256}$$

Finally, this must be evaluated at the points in question. We can see immediately that for the second and third points, since $x = 0$, the derivative is 0. This means that the two tangent lines are,

$$y = \frac{9}{5}$$

and

$$y = \frac{9}{11}.$$

For the first point, we must compute the normal line. Its slope is the negative reciprocal of the derivative. Plugging $x = \sqrt{7}$ and $y = \frac{9}{5}$ in our expression for y' and calculating (by hand or with your SCS), we obtain

$$y' = -\frac{39\sqrt{7}}{115}$$

and so the slope is

$$\frac{115}{39\sqrt{7}}.$$

Since the point it must pass through is $\left(\sqrt{7}, \frac{9}{5}\right)$, we can use the point slope formula to obtain, for the equation of the normal line,

$$y - \frac{9}{5} = \frac{115}{39\sqrt{7}}\left(x - \sqrt{7}\right).$$

4. Finally, we must interpret these results. The main point to observe is that the tangents at the second and third points are both parallel to the horizontal axis. Moreover, these two points both have the same x-coordinate. What does that tell you about the position of the points? Does the position of the tangent say anything about the curve itself?

Using Calculus in Situations — Related Rates

A ship on the ocean is traveling due south towards port at a rate of 21 mph. A second ship is traveling due west away from port at a rate of 17 mph. At noon, the first ship is 93 miles from port and the second is 13 miles from port. At what rate is the distance between the ships changing. Is it increasing or decreasing?

1. The function which must be defined for this problem will give the distance d between the two ships as a function of time t. We may call it D. We define this function as a combination of two other functions. Our thinking goes like this. The distance between the two ships can be computed, using the Pythagorean theorem if we know the distance of each ship from port. See Figure 4.11.

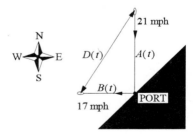

Figure 4.11. The distance D between two ships.

This suggests that we begin with two functions, A and B which give the distances from port of the two ships as functions of time. Then the function D which gives the distance between the two ships as a function of time can be defined as a combination of these two functions, as follows,

$$d = D(t) = \sqrt{A(t)^2 + B(t)^2} = \left((A(t))^2 + (B(t))^2\right)^{\frac{1}{2}}.$$

2. The mathematics that must be done in order to solve this problem is to compute $D'(12)$.

3. Assuming A and B are differentiable functions of t and using the rules for differentiation, we have,

$$D'(t) = \frac{\frac{1}{2}(2A(t)A'(t) + 2B(t)B'(t))}{\left((A(t))^2 + (B(t))^2\right)^{\frac{1}{2}}}$$

$$= \frac{A(t)A'(t) + B(t)B'(t)}{\left((A(t))^2 + (B(t))^2\right)^{\frac{1}{2}}}.$$

Fortunately, for the evaluation at $t = 12$, we know the values of all of the quantities on the right hand side of this equation. From the information given in the problem we have, $A(12) = 93$, $B(12) = 13$, $A'(t) = -21$ and $B'(t) = 17$. Why did we take one rate to be negative and the other positive? Plugging these values in gives,

$$D'(12) = \frac{(-93)(21) + (13)(17)}{\sqrt{93^2 + 13^2}} = -\frac{1732}{93.9} \approx -18.44.$$

4. The interpretation of our solution is that the ships are getting closer together at a rate of approximately 18.44 mph.

Using Calculus in Situations — Marginals in Economics

We previously discussed the notion of marginal analysis in economics on page 202. Consider the following situation.

The total cost C, in thousands of dollars, to produce x units of a certain commodity is given by

$$C(x) = 1 + 0.3x + 0.1x^2$$

for $0 \leq x \leq 50$. Use the marginal cost function MC to estimate the additional cost that will result from the production of 1 additional unit when 30 units have already been produced. Compare your estimate to the actual additional cost to produce 1 more unit.

1. From our previous discussion on page 202, the marginal cost is the derivative of the cost function C. Hence,

$$MC(x) = C'(x) = 0.3 + 0.2x$$

2. Recall that after x units have been produced,

$$MC(x) = C'(x) \approx C(x+1) - C(x)$$

where the cost difference $C(x+1) - C(x)$ represents the cost to produce one additional unit. So when $x = 30$, the cost of producing one additional unit is about

$$MC(30) = C'(30) = \$6300.$$

3. The actual cost of producing one more unit, when 30 units have been produced, is given by the difference

$$C(31) - C(30) = \$6400.$$

Hence, we see that the marginal cost $MC(30)$ is a pretty good estimate to the actual cost of producing one more unit.

Figure 4.12 shows a graph of the total cost function C near $x = 30$. The slope of the tangent line to the graph of C, $C'(x) = MC(x)$, approximates the slope $C(x+1) - C(x)$ of the secant line through $(x, C(x))$ and $(x+1, C(x+1))$. You can also see from Figure 4.12 that the tangent line at $(30, C(30))$ provides a "linear approximation" to the graph of C for values of x "near" 30. Hence, the values of the function represented by the graph of the tangent line approximate the values of the function C near 30.

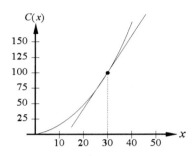

Figure 4.12. A graph of C near $x = 30$ and the tangent line at $(30, C(30))$.

Using Calculus in Situations — Errors in Measurement

You are managing the unloading and storage of grain that has just arrived on a cargo ship. The grain is to be stored in bins which you know to be in the shape of a cube. You also know that there are 100,000 cubic feet of grain. Except for the fact that they are almost perfect cubes and identical, you don't know the size of the bins. Unfortunately, you are caught down at the docks without calculator, or any writing material. It is raining heavily and you need to know how many bins will be necessary. You do have a tape measure in your possession. You make several measurements of the length of the edge of a bin and you get the following values 9'11", 10', 10'1", where ' and " represent feet and inches, respectively. Make calculations in your head to determine how far off you will be in estimating the volume of a bin if you take the length of a side to be 10'? Should you go back and measure more accurately, or is this good enough?

1. The function is represented by the formula for the volume v of a cube with side s. We have,

$$v = V(s) = s^3.$$

2. The problem is to estimate the error in the volume that would result if 10 were used for s with an error of one inch or $\frac{1}{12}$ ft. We can use the *idea* of the derivative at the point 10 to develop a method for making such a mental estimate.

The derivative is given by,

$$V'(s) = \lim_{h \to 0} \frac{V(s+h) - V(s)}{h} = 3s^2.$$

The idea that we take from this is that, for h small, we have

$$\frac{V(s+h) - V(s)}{h} \approx 3s^2$$

or

$$V(s+h) - V(s) \approx 3s^2 h.$$

Hence, for the purpose of estimation, we can replace the difference $\Delta V = V(s+h) - V(s)$ by $3s^2 h = V'(s)h$. Note that sometimes Δs is used in place of h to indicate the error (or change) in the measurement s.

3. If we do this for $s = 10$ we conclude that the difference we get for the volume by using a side length that differs from 10 by $h = \frac{1}{12}$ is about $3(10)^2 \left(\frac{1}{12}\right) = \frac{300}{12} = 25$ ft^3. This is referred to as the *absolute error*.

   If the absolute error is divided by the value we get using the approximation 10, we obtain the *relative error* which in this case is $\frac{25}{1000} = 0.025$. We get the *percent error* by multiplying this by 100. Thus the error is about 2.5%.

4. Translating this back into the language of the problem, we conclude that using $10'$ as a measurement of the side would give us a volume error of about 2.5%. It also means that we would use 1000 ft^3 as our estimate of the volume of a bin so that 100 bins is what would be needed. The error of 2.5% suggests that we could be off by as much as 3 bins. If this is unacceptable, then it would be necessary to take out the tape measure and try to make more careful measurements of the side of a bin. What percent error in side measurement would make it reasonably certain that your estimate would get the right number of bins?

   There is another interpretation of this problem that we would like to add. When we replaced $\Delta V = V(s + \Delta s) - V(s)$ by $3s^2 h = V'(x)\Delta s$ we were considering s to be a constant (10 in this case) and the error function is a function of Δs which gives the measurement error. Thus, in considering the variation of the *cubic* function given by s^3 near the value $s = 10$, we are looking at the variation of the *linear* function given by $300\Delta s$. This is called a *linear approximation* of V near $s = 10$.

Linear Approximations — Differentials   For a differentiable function f, suppose that the domain variable changes from a particular value x to another domain value $x + \Delta x$, where Δx could be positive or negative. Then the corresponding change Δy, or Δf, in the values of f is given by

$$\Delta y = f(x + \Delta x) - f(x).$$

Figure 4.13 suggests that the actual change Δy can be approximated by $f'(x)\Delta x$. As we have seen previously, the tangent line provides a "linear approximation" to the curve $y = f(x)$ at the point $(x, f(x))$.

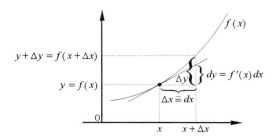

Figure 4.13. Linear approximation and differentials.

Based on the situation in Figure 4.13, we make the following definition.

> **Definition 4.8**
> *Let f be a differentiable function of x.* The differential of the independent variable x, *which we denote by dx or* Δx *is defined to be the change in x,* Δx. The differential of the dependent variable y, *which we denote by dy is defined to be* $f'(x)dx = f'(x)\Delta x$.

From the previous definition, we see that $dx = \Delta x$ and $dy \approx \Delta y$, where x and y are the independent and dependent variables, respectively. Moreover, as in the discussion of marginals on page 202 and the situation involving the bin problem (where $f = V$ and $x = s$), we have

$$\frac{dy}{dx} = f'(x) \approx \frac{f(x+\Delta x)-f(x)}{\Delta x} = \frac{\Delta y}{\Delta x},$$

or

$$dy = f'(x)\Delta x \approx f(x+\Delta x) - f(x) = \Delta y.$$

(Notice that in the bin situation, we used h in place of Δs to indicate the error in the measurement of a side s.)

Hence, the *relative error* in y is given by

$$\frac{\Delta y}{y} = \frac{f(x+\Delta x)-f(x)}{f(x)}$$

and the relative error in y can be approximated by

$$\frac{dy}{y} = \frac{f'(x)\Delta x}{f(x)} \approx \frac{\Delta y}{y}.$$

The *percentage relative error* is then given by

$$100\frac{\Delta y}{y} = 100\frac{f(x+\Delta x)-f(x)}{f(x)}$$

and it is approximated by

$$100\frac{dy}{y} \approx 100\frac{\Delta y}{y}.$$

You will get a chance to investigate applications of differentials in the exercises.

The Properties of a Graph — Drawing Complete Graphs
The following issues must be considered in using the graph of a function to display as much information as possible.

Intercepts. These are points at which the graph intersects the x-axis and the y-axis. They are found by setting one of the variables equal to zero and solving for the other.

Domain. This is the set of values at which the function is defined. For example, points at which even or non-fractional roots are applied to negative numbers, or which lead to division by zero are not in the domain and will not be part of the graph.

Vertical asymptotes. If there is a point a at which evaluation of an expression defining a function leads to division by zero, then near this point the graph may become arbitrarily large in the positive or negative direction (or both) near a. If the graph becomes arbitrarily large near a, i.e., the limit from the right and/or left at a approach ∞ or $-\infty$, then the line $x = a$ is a vertical asymptote.

Horizontal asymptotes. If the limit L of the function as the independent variable goes to ∞ or $-\infty$ exists, then the horizontal line $y = L$ is a horizontal asymptote to the graph of f.

Extreme points. The only points at which the graph of a function will reach a relative maximum or minimum are points in the domain of the function at which the value of its derivative is 0 or undefined.

Monotonicity. The graph of a function whose derivative exists at every point on an interval will be increasing on this interval if the values of the derivative are positive at every point on the interval. Can you formulate a similar statement for decreasing?

Is there any general relation you can imagine that connects regions where a function is increasing or decreasing with points at which the function has maximum(s) and minimum(s)?

Shape. Speaking somewhat loosely, we say that the graph of a function is *concave up* on an interval if it is shaped in such a way that "it will hold water". Again there is a similar statement for concave down. What is it?

The following test is a simple test for concavity which involves the second derivative.

> If a function has a first and a second derivative on an interval, then, if the second derivative is positive (negative) at every point on the interval, the graph is concave up (down) on that interval.

Related to concavity are points at which a graph of a function changes concavity. A point at which the graph of a function changes from one kind of concavity to the other is called an *inflection point*. Such points in the domain of a function can only occur where the values of the second derivative are 0 or undefined.

Symmetry. If a function f has the property that $f(x) = f(-x)$ for all x in its domain, then the graph is symmetric with respect to the y-axis and it is referred to as an *even function*. If it satisfies, $f(-x) = -f(x)$ for all x in its domain, then it is symmetric with respect to the origin and it is referred to as an *odd function*.

What exactly does it mean for a graph to be symmetric with respect to something?

We will illustrate the above discussion using two of the examples you considered in Activity 1. First, the function f given by $f(x) = x^3 - x^2 = x^2(x - 1)$.

Intercepts. Clearly the graph intersects the y-axis at the origin only and the x-axis additionally at $x = 1$.

Domain. This expression defines the function for all values of x.

Vertical asymptotes. There are no vertical asymptotes.

Horizontal asymptotes. As $x \to \infty$ the value of the function goes to ∞ and as $x \to -\infty$ it goes to $-\infty$. Hence there are no horizontal asymptotes.

We can draw a partial graph using only the information obtained so far. It looks like that in Figure 4.14.

Figure 4.14. A partial graph of $f(x) = x^3 - x^2$.

Extreme points. The derivative f' of f is given by $f'(x) = 3x^2 - 2x = x(3x - 2)$. Hence the derivative is zero when $x = 0, \frac{2}{3}$. Looking at the expression for f, we see that for very small values of x, the quantity $x^3 - x^2$ is negative (why?) and since $f(0) = 0$, it follows that 0 is a relative maximum. Can you make a similar analysis for the other extreme point? We will return to this issue.

Monotonicity. Since $f'(x) = 3x^2 - 2x$, we can solve the inequality $3x^2 - 2x > 0$ (or use the **computer function signs**) to determine that $f'(x)$ is positive when $x < 0$ and when $x > \frac{2}{3}$. Similarly, $f'(x)$ is negative on the open interval $\left(0, \frac{2}{3}\right)$.

Hence, the function is increasing for $x < 0$. It then begins to decrease. This is another indication that 0 is a relative maximum. The function continues decreasing until it reaches $x = \frac{2}{3}$ when it begins to increase. Thus, the point $x = \frac{2}{3}$ is a relative minimum occurs at $x = \frac{2}{3}$.

Again we can draw a partial graph that uses the information we have, as shown in Figure 4.15.

Shape. The second derivative f'' is given by $f''(x) = 6x - 2$. Hence the graph is concave down for $x < \frac{1}{3}$ and concave up for $x > \frac{1}{3}$. An inflection point occurs at $x = \frac{1}{3}$.

Symmetry. This graph has no symmetry.

Using all of this information, we obtain a complete graph, i.e., a graph showing all relevant features, like the one in Figure 4.16.

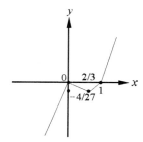

Figure 4.15. A partial graph of $f(x) = x^3 - x^2$ without concavity indicated.

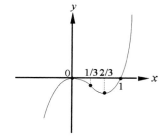

Figure 4.16. A complete graph of $f(x) = x^3 - x^2$.

We can also consider the function g from Activity 1, given by,

$$g(x) = 10x^{2/3} - x^{5/3} = x^{2/3}(10-x).$$

Intercepts. Clearly the graph intersects the y-axis at the origin only and the x-axis additionally at $x = 10$.

Domain. This expression defines the function for all values of x.

Vertical asymptotes. There are no vertical asymptotes.

Horizontal asymptotes. As $x \to \infty$ the value of the function goes to ∞ and as $x \to -\infty$ it goes to $-\infty$. Hence there are no horizontal asymptotes.

We can draw a partial graph using only the information obtained so far. It looks like that in Figure 4.17.

Extreme points. The derivative f' of f is given by

$$f'(x) = \frac{20}{3}x^{-1/3} - \frac{5}{3}x^{2/3} = \frac{5(4-x)}{3x^{1/3}}.$$

Hence the derivative is zero at $x = 4$ and it is undefined at $x = 0$. Looking at the expression for f, we see that for very small values of x, the quantity

$$f(x) = 10x^{2/3} - x^{5/3} = x^{2/3}(10-x).$$

is positive (why?) and since $f(0) = 0$, it follows that 0 is a relative minimum. We can also see this by using the first derivative test. That is, $f'(x) < 0$ for "near" zero and $x < 0$ and $f'(x) > 0$ for x "near" zero and $x > 0$. Can you make a similar analysis for the other extreme point?

Monotonicity. Since

$$f'(x) = \frac{5(4-x)}{3x^{1/3}}$$

we need to determine when this expression for f' is positive and negative. Do you see that this boils down to finding out when

$$\frac{4-x}{3x^{1/3}}$$

is positive and negative on the intervals determined by the points 4 and 0? Why? We can do so by hand (or use the **func signs**) to determine the following.

$$f'(x) > 0 \text{ for } 0 < x < 4 \text{ and } f'(x) < 0 \text{ for } x < 0 \text{ or } x > 4.$$

Do you see why? Test some numbers in the intervals $(-\infty, 0)$, $(0, 4)$, and $(4, \infty)$ for yourself and you will see.

Therefore, the function is increasing on the interval $(0, 4)$ and it is decreasing on the intervals $(-\infty, 0)$ and $(4, \infty)$. That is, f is decreasing to the left of 0 and increasing to the right of 0, another indication that f has a relative minimum at 0. Similarly, we see that the function has a relative maximum at 4, since the function increases just to the left of 4 and it decreases just to the right of 4. The points on the graph corresponding to 0 and 4 are $(0, 0)$ and $\left(4, 6(2)^{2/3}\right) \approx (4, 15.12)$.

Again we can draw a partial graph that uses the information we have, as shown in Figure 4.18.

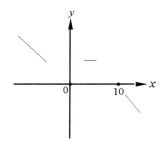

Figure 4.17. A partial graph of
$f(x) = 10x^{2/3} - x^{5/3}.$

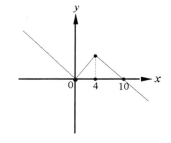

Figure 4.18. A partial graph of
$f(x) = 10x^{2/3} - x^{5/3}$ **without concavity indicated.**

Shape — Concavity. The second derivative f'' is given by

$$f''(x) = -\frac{20}{3}x^{-4/3} - \frac{5}{3}x^{-1/3} = -\frac{10(2+x)}{9x^{4/3}}.$$

The points where the second derivative are zero and undefined are −2 and 0. So to determine the concavity, or shape of the curve, we need to find out the sign of the second derivative on the intervals $(-\infty, -2)$, $(-2, 0)$, and $(0, \infty)$. Since the denominator $9x^{4/3}$ is always positive for $x \neq 0$ (why?), we need only determine the sign of the numerator $-(x+2)$ to determine the sign of f''. It is easy to see that the numerator, and hence f'', is positive in the interval $(-\infty, -2)$ and negative in the intervals $(-2, 0)$ and $(0, \infty)$. Hence the graph is concave down for $-2 < x < 0$ and $x > 0$ and it is concave up for $x < -2$. An inflection point occurs at $x = -2$, since the concavity changes at this point. The coordinates of the point on the graph where the inflection changes is $(-2, 12(4)^{1/3}) \approx (-2, 19.05)$.

Using all of this information, we obtain a complete graph, i.e., a graph showing all relevant features, like the one in Figure 4.19. Notice that the graph has a "cusp" at 0 — a point at which the derivative of the function is undefined.

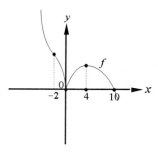

Figure 4.19. A complete graph of $f(x) = 10x^{2/3} - x^{5/3}.$

Finally, we consider the function h from Activity 1, given by,

$$h(x) = \frac{x^3 - 7x}{x^3 + 3x - 3} = \frac{x(x^2 - 7)}{x^3 + 3x - 3}.$$

We will not give a complete analysis, but consider only asymptotes since none occurred appear in the previous two examples.

For horizontal asymptotes, we compute the limits,

$$\lim_{x\to\infty} h(x) = \lim_{x\to-\infty} h(x) = 1$$

You can use the methods of Chapter 3 to evaluate this, or use your SCS, or just reason it out. In any case, this implies that we have a horizontal asymptote at $x = 1$ in both directions.

The question arises as to whether the graph stays above the asymptote, below it, or oscillates back and forth. Do you see how to use the analysis of monotonicity to determine this?

For vertical asymptotes, we must see if there are any solutions to the equation $x^3 + 3x - 3 = 0$. We can actually apply our general analysis to determine this. If you take the derivative, you get $3x^2 + 3$. This is always positive, so the values of the expression are increasing. Moreover, for very large positive x this expression is positive and for very large negative x it is negative. Hence, it is negative and increases to being positive. It must therefore cross the x-axis once and only once. You can use any of our previous methods to determine that this happens at about $x = 0.818$.

We know that there is a vertical asymptote at about $x = 0.818$, that is, the function is going to ∞ or $-\infty$ as x approaches this value. But which is it? Must it be the same on both sides, or can it be different?

The behavior near a vertical symptote can be obtained as a consequence of the rest of the analysis — mainly from whether the function is increasing or decreasing near the asymptote. You can do that, but it is also possible to just reason about the situation and draw conclusions.

The only question is whether $h(x)$ is positive or negative when x is close to 0.818 on either side. We see that the numerator is definitely going to be negative in this region (is that clear?) and so our answer is the opposite of what happens to the denominator.

Now we can use our previous analysis that got us this asymptote in the first place. We determined that the values of the expression for the denominator were negative for large negative values of x and increased steadily until the value of the expression was zero, and then the values became positive and continued to increase. Thus, if x is to the left of the asymptote, the denominator is negative so $h(x)$ is positive. If x is to the right, then $h(x)$ is negative. Hence the asymptotic behavior is to go to ∞ on the left and $-\infty$ on the right of the asymptote.

The results of our analysis so far show that the graph looks something like the one shown in Figure 4.20.

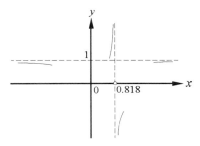

Figure 4.20. A partial graph of $h(x) = \frac{x^3 - 7x}{x^3 + 3x - 3}$.

We leave it for you to finish this example and make a complete graph of

$$h(x) = \frac{x^3 - 7x}{x^3 + 3x - 3}.$$

What else needs to be determined? You should check the list of properties of graphs which begins on page 246 to be sure that you include all relevant features.

Indeterminate Forms — L'Hôpital's Rule

We now consider a useful technique for calculating limits of certain ratios of functions as you did in Activities 2, 3 and 4. The method is commonly known as *L'Hôpital's rule*. This method uses the ratio of the derivatives of the functions in the numerator and denominator (NOT the quotient rule!). Hence it is an application of derivatives to solve limit problems. The rule can be used for many situations you will encounter in the remainder of this text. It is as follows. A proof will have to wait until Chapter 6.

L'Hôpital's Rule Let f and g be two functions which are differentiable in an open interval (a, b) containing the point c.

1. If $\lim_{x \to c} f(x) = 0$ and $\lim_{x \to c} g(x) = 0$, i.e., $\frac{f(x)}{g(x)}$ is of the indeterminate form $\frac{0}{0}$ at c, then if $g'(x) \neq 0$, except possibly at c,

$$\lim_{x \to c} \frac{f(x)}{g(x)} = \lim_{x \to c} \frac{f'(x)}{g'(x)}$$

provided the latter limit exists.

2. If $\lim_{x \to c} f(x) = \infty$ and $\lim_{x \to c} g(x) = \infty$, i.e., $\frac{f(x)}{g(x)}$ is of the indeterminate form $\frac{\infty}{\infty}$ as x approaches c, then

$$\lim_{x \to c} \frac{f(x)}{g(x)} = \lim_{x \to c} \frac{f'(x)}{g'(x)}$$

provided the latter limit exists.

Note that to use the above rule, or any of the following generalizations, you must first check that the limit of the ratio of the two functions is one of the indeterminate forms $\frac{0}{0}$ or $\frac{\infty}{\infty}$ in order to use the rule and make valid conclusions.

The above rules also hold for one-sided limits at c, and limits at ∞ and $-\infty$. A ratio of two functions in which the limits of both the functions in the denominator and numerator are 0 is called an *indeterminate form of type* $\frac{0}{0}$. Similarly, for a ratio of functions such that the limit of both functions in the denominator and numerator are either ∞ or $-\infty$, we say that the expression is an *indeterminate form of the type* $\frac{\infty}{\infty}$. In Activities 2-4 you only worked with the indeterminate form $\frac{0}{0}$. Do you see why it makes sense to extend L'Hôpital's rule from indeterminate forms of the type $\frac{0}{0}$ to indeterminate forms of the type $\frac{\infty}{\infty}$? Think about taking reciprocals of differentiable functions which have limit zero. Now do you see why?

Can you think of some situations you have seen in this text which are indeterminate forms? Well, if you think about it every derivative problem is really of the form $\frac{0}{0}$. Why? Also finding horizontal asymptotes for the graph of many rational functions is of the form $\frac{\infty}{\infty}$. For rational functions we saw in

Section 3.2 that you can reason with your mind to find the limits of some rational functions as the independent variable tends to infinity or minus infinity. However, some indeterminate form situations are not so easy to determine or reason out in your mind. It is the latter situations in which L'Hôpital's Rule is often useful.

For example, consider the following limit problem.

$$\lim_{x \to \infty} \frac{3x}{e^{2x}}.$$

It is not too hard to see that the functions in both the numerator and denominator tend to ∞ as $x \to \infty$. If you think about it for a moment, it appears that the denominator is growing much faster than the numerator, so maybe the ratio is approaching 0. You can check this out with your computer or a calculator. Since it iseasy to find the derivatives of the two functions, this is a certainly situation where you can easily apply L'Hôpital's Rule. Taking the derivative of the numerator and denominator, and then taking the limit of each we have,

$$\lim_{x \to \infty} \frac{3}{2e^{2x}} = 0$$

so by L'Hôpital's Rule it follows that

$$\lim_{x \to \infty} \frac{3x}{e^{2x}} = 0.$$

It may happen that you will have to apply L'Hôpital's Rule repeatedly to solve some limit problems. Consider the following limit problem.

$$\lim_{x \to 0} \frac{1 + x - e^x}{x^2}$$

The functions in the numerator and denominator both approach 0 as $x \to 0$ so we can try to apply L'Hôpital's Rule. Doing so, we get

$$\lim_{x \to 0} \frac{1 - e^x}{2x}$$

which is again of the form $\frac{0}{0}$, so we can try to apply L'Hôpital's Rule once again. This time we get,

$$\lim_{x \to 0} -\frac{e^x}{2} = -\frac{1}{2}.$$

Hence, we can conclude that

$$\lim_{x \to 0} \frac{1 + x - e^x}{x^2} = -\frac{1}{2}$$

As can happen with any good thing, L'Hôpital's Rule may fail to work in some situations. Try to find the following limit using L'Hôpital's Rule.

$$\lim_{x \to \infty} \frac{\sqrt{1 + x^2}}{x}$$

What happened? Can you think of another way to find the limit? You will get a chance to discuss this example as an exercise.

L'Hôpital's Rule can be helpful in some limit situations with the help of a little algebra. Two such limit situations are the following indeterminate forms,

$$0 \cdot \infty \qquad \infty - \infty.$$

Can you give limit interpretation of these two types of indeterminate forms in terms of functions and limits? The indeterminate form $0 \cdot \infty$ occurs when taking a limit of a product of two functions, where one function approaches 0 and the other tends to ∞ (or $-\infty$). The form $\infty-\infty$ occurs when taking the limit of the sum (or difference) of two functions, where one tends to ∞ and the other tends to $-\infty$ (or both tend to ∞, hence the difference is of the form $\infty-\infty$). These two limit situations can usually be handled by using the algebraic manipulations to change the indeterminate form into one of the indeterminate forms $\frac{0}{0}$ or $\frac{\infty}{\infty}$.

Consider, for example, the following limit.

$$\lim_{x \to \infty} x\left(1 - e^{1/x}\right)$$

Do you see that this problem is of the indeterminate form $0 \cdot \infty$? Do you see how you could manipulate the expression in the following limit problem so that you can apply L'Hôpital's Rule? Try rewriting the expression $x\left(1 - e^{1/x}\right)$ as follows.

$$\frac{1 - e^{1/x}}{\frac{1}{x}}.$$

Now what do you think? You'll get a chance to finish this problem and to do others like it in the exercises.

What do you think you could do with an indeterminate limit of the form $\infty-\infty$? Be careful, $\infty-\infty$ is not necessarily 0. Why? What if one function grew faster than the other? What do you think you might do in the following limit situation?

$$\lim_{x \to 0^+} \left(\frac{1}{x} - \frac{1}{x^2}\right)$$

Yes, that's right, combine the two functions into one function by getting a common denominator. Try it! Other indeterminate limit situations can also be handled using L'Hôpital's Rule as you will see in the next section.

Some Determinate Forms   What about limit situations like ∞^∞, $\frac{\infty}{0}$, and $\frac{0}{\infty}$?

Do you see why these are not too hard to "determine"? Such limit situations are often referred to as *determinate forms*. Do you see that the third one has value 0 and the other two have value ∞ (or possibly $-\infty$)? Can you give examples of each situation? You will get a chance to do so in the exercises.

4.3.4 Exercises

1. You have 2000 feet of fencing to enclose a rectangular field, separated by two lengths of fencing down the middle so that the field is separated into three equal rectangular areas. What is the largest such rectangular area you can enclose?

2. You have a rectangular piece of cardboard 20 in by 30 in and you wish to cut squares out of each corner and then fold up the sides to make an open box. Answer the following questions for this situation.

(a) What is the length of the side of a square cut out so that the volume of the box is a maximum?

(b) What is the maximum volume of the box?

3. Find the minimum distance from a point on the curve $y = x^2 + 1$ to the point $(1, -1)$. Is there a point where the distance is a maximum? If so, what are the coordinates of the point?

4. A cattle rancher wants to enclose a rectangular region along a straight river bank so that the region is separated into two equal rectangular regions by a fence down the middle of the region that is perpendicular to the straight riverbank. The rancher intends to graze horses in one region and cattle in the other. The rancher has 720 feet of fencing available. Answer the following questions for this situation.

(a) What is the largest amount of grazing land that can be fenced as described?

(b) What are the dimensions of the region which give the greatest amount of grazing land enclosed by the fence?

5. An appliance manufacturer produces washing machines and determines that to sell x machines its unit selling prices must be $s = S(x) = 300 - 0.3x$ dollars per machine. The manufacturer has also determined that the its total cost of producing x washing machines is given by $0.7x^2 + 5600$ dollars. Answer the following questions for this situation.

(a) What is the manufacturer's maximum total profit?

(b) How many washing machines produced gives the greatest total profit?

(c) What is the price of a washing machine so that the profit is the largest?

6. The postal service restricts the size of packages which can be accepted for delivery. The restriction is that the length of a package plus the girth (perimeter around a cross section perpendicular to the length) cannot exceed 108 inches. Answer the following questions for this situation.

(a) Find the volume of the largest box package?

(b) What are the dimensions of the box package with the largest volume?

7. A manufacturer of tin cans is to make a can holding 80 in^3. Answer the following questions for this situation.

(a) What is the least amount of tin needed to make the can?

(b) What is the radius and height of the can requiring the least amount of tin?

8. Consider the situation in the previous exercise where the top and bottom of the tin can cost $0.02 per in^2 and the sides cost $0.03 per in^2. Answer the following questions for this situation.

(a) What is the least cost of such a can?

(b) What is the radius and height of the most economical can?

9. A triangle has two sides of length 5 and 8. Answer the following questions for this situation.

(a) What is the angle between the two sides so that the area is the largest?

(b) What is the perimeter of the triangle when its area is the largest?

10. For the triangle in the previous exercise, do the following.

(a) Find the angle between the sides so that the perimeter is the largest.

(b) Find the angle between the sides so that the perimeter is the smallest.

(c) Find the triangle so that the perimeter is the largest.

(d) Find the triangle so that the perimeter is the smallest.

11. A charter airline runs a promotion on its late night flight to Reno with a maximum capacity of 120 passengers. The fare per person is $100 with a reduced fare of $2 per person for every person over the 75th person. Each $2 reduction is for every person on the plane. Do the following for this situation.

(a) Find the number of tickets sold so that the revenue is the greatest.

(b) Find the largest revenue the airline can expect.

12. You wish to construct an open rectangular tank with the length of the base three times the width. The volume of the tank is to be 100 m^3. The cost of material for the base is $3 per square meter and the cost of material for sides is $2 per square meter. Do the following for this situation.

(a) Find the minimum cost of the tank.

(b) Find the dimensions of the tank so that the cost is a minimum.

13. A woman is in a boat, 3000 meters from her house on the shore on a line perpendicular to the (straight) shoreline. A lighthouse is 2000 meters downshore from her position. The woman intends to get to the lighthouse by rowing towards the shore at an angle so as to arrive at a point between her house and the lighthouse. She will walk the rest of the way. She can row at 2 km/hr (kilometers/hour) and walk at 4 km/hr. Do the following for this situation.

(a) Find the maximum time it takes the woman to make the trip.

(b) Find the minimum time it takes her to make the trip.

14. A worker arrives on the job at 7:00 AM and gets a thirty minute lunch break at 11:00 AM. The worker can produce p units of a product t hours after the lunch break where

$$p = P(t) = -\frac{t^3}{2} + 5.5t^2 + 20t.$$

Do the following for this situation.

(a) At what time after the lunch break does the worker reach peak efficiency?

(b) What is the worker's peak efficiency?

15. Due to environmental factors, there is an upper bound, N, on the wildlife population in a forest preserve. In addition, the wildlife population grows at a rate proportional to the product of the current number of wildlife and the difference between the population upper bound and the current size of the population. Find the size of the wildlife population at which the growth rate of the wildlife population is the greatest.

16. An island is 1 mile from a straight shoreline and there is a village 10 miles down the shoreline. A shipping company can ship goods at a rate of 6 mph on water and truck goods at a rate of 40 mph on shore. The company is about to build a pier. Find the minimum total shipping time between the village and the island for a shipment of goods.

17. The power of an electrical circuit is the (instantaneous) rate of change of work with respect to time. Suppose that the work

being done in an electrical circuit is given by $W = 15\cos(2t)$. Find an expression for the power in the circuit at any time t.

18. The voltage v in an electrical circuit with an inductor is given by

$$v = V(t) = L\frac{di}{dt} \text{ volts}$$

where L is the inductance (a constant) measured in Henrys, i is the current in amperes, and t is the time in seconds. Find an expression for the voltage in an electrical circuit with a 2 henry inductor and current given by

$$i = I(t) = 20\cos\left(60\pi t + \frac{\pi}{3}\right) \text{ amperes}$$

19. A particle is constrained to move in a straight line. Observations are made of its position p relative to a starting point at various times. The (approximate) position function is given by

$$p = P(t) = e^{-2t}(\cos(t) + \sin(t))(t) \text{ cm}.$$

Find an expression for the velocity as a function of time. Give correct units with your answer. Make a sketch of the function P on the interval $[0, 10]$.

20. The light intensity distribution i in the theory of making holographic images is given by $i = I(\theta) = cE_o^2 \cos^2(0.5\theta)$ where c and E_o^2 are constants and θ is the phase angle between two light waves. Find an expression for the rate of change of the light intensity with respect to the phase angle.

21. Plants usually do not grow at a constant rate each day (due to the need for sunlight). The amount of growth g of a certain type of plant is given by

$$g = G(t) = 0.15t + 0.05\sin(2\pi t) \text{ cm}$$

where t is the number of hours after midnight. Do the following for this situation.

(a) At what time of day is the growth rate of the plant the greatest? the least?

(b) Sketch of the function g and the line given by $y = 0.15t$ on the same coordinate system over the interval $[0, 24]$. Use your graphics utility to zoom in on a portion of the graph to take a closer look at a small portion of your graph. The line $y = 0.15t$ is called a *trend line*. Plant growth often follows a cyclic pattern about a trend line.

22. The electrical current i at a point in a conductor is given by a function I of time t determined by the total charge q which has flowed past the point from the time when the current begins to flow until time t in seconds. In fact, if Q is the function which gives the total charge as a function of time, then I is the derivative of Q.

In a certain conductor, the total charge function has been determined to be

$$q = Q(t) = 100e^{-t}\cos(5t) \text{ coulombs}.$$

Find the current as a function of time. Give correct units with your answer. Use your computer to make a sketch of the function Q on the interval $[0, 10]$.

23. A woman's blood pressure p ranges from a high (the systolic pressure) to a low (the diastolic pressure) so that at time t in seconds, p is given (approximately) by

$$p = P(t) = 100 + 20\sin(5t).$$

The time between two successive high pressures is the time needed for one heartbeat. Do the following for this situation.

(a) Find an expression for the rate at which the woman's blood pressure changing at time t.

(b) At what time t is the woman's blood pressure a maximum? a minimum?

(c) What is the woman's maximum blood pressure? minimum blood pressure?

(d) How many times per minute does the woman's heart beat?

24. A ferris wheel has a radius of 20 feet and the height h of a seat above the ground at the loading point is located three feet above ground level. During a ride on the ferris wheel your height above the level ground is given by $h = H(t) = 23 - 20\cos(2\pi t)$ feet where t is the time, in minutes, after you leave (or pass) the loading point. Use calculus to find the following.

(a) The speed of the ferris wheel in feet per minute.

(b) Assuming your ride begins when you leave the loading point, at what times during a two minute ride is your height above the ground a maximum? a minimum?

25. The current i in an electric circuit is given by

$$i = I(t) = \sin^2(\pi t) \text{ amperes}$$

where t is in seconds. Use calculus to find all times at which the current is a maximum. Can you solve this exercise without calculus? How?

26. A model for the number of hours of sunlight s per day in a non-leap year in Lafayette, Indiana is given (approximately) by

$$s = s(t) = 12 + 3\sin\left(\frac{2\pi}{365}(t-80)\right)$$
hours per day

where t the number of days after January 1. Answer the following questions for this situation.

(a) At what rate is the number of hours of daily sunlight changing with respect to time on the first day of Spring (March 21)?

(b) On what day(s) is the number of hours of daily sunlight the most? the least?

27. Based on a nationwide survey, a discount chain store determines that the sales s of a certain product are given (approximately) by

$$s = S(t) = 50\left(1 - \cos\left(\frac{\pi t}{13}\right)\right) \text{ thousands}$$
of dollars per week

where t is the number of weeks after the first of the year. What is the total amount of sales predicted by this sales model during the first half of the year? the entire year?

28. A rock is thrown into a calm lake causing concentric waves about the point of impact on the lake. The radius of the wave is increasing at 6 in/sec. At what rate is the area of the circular region inside the waves increasing when the radius is 70 in?

29. The radius of a spherical soap bubble is increasing at a rate of 0.02 in/sec. How fast is the volume of the bubble increasing when the radius is 3 in? (Use $V = 4\pi r^3/3$.)

30. The radius of a spherical soap bubble is increasing at a rate of 0.02 in/sec. How fast is the volume of the bubble increasing when the volume is 288π in³? (Use $V = 4\pi r^3/3$.)

31. A rocket is launched vertically and its height h, in feet, is given by $h = H(t) = 40t^2$ where t is the number of seconds after the launch. An observer is located 3000 ft from the launch pad. At what rate is the angle of elevation changing 10 seconds after the launch?

32. A 10 meter ladder is leaning against a vertical building. The foot of the ladder is slipping away from the building at a rate of 0.5 meters/sec. At what rate is the top of the ladder moving downward when the foot of the ladder is 4 meters from the base of the building?

33. A person is standing at the end of a pier 12 feet above the water and is pulling in a rope attached to a raft (at water level) at a rate of 4 feet of rope per minute. At what rate is the angle between the rope and the surface of the water changing with respect to time when the raft is 16 feet from the pier?

34. A light on the ground is shining on the wall of a barn which is 100 ft away. A boy 5 feet tall runs at a rate of 10 ft/sec away from the light directly toward the barn. How fast is the boy's shadow on the barn decreasing when the boy is 20 ft from the barn?

35. A load of grain is being poured at a rate of 200 ft^3/min onto the floor of a storage bin. The grain forms a conical pile so that the height is always two-thirds the radius of the base. At what rate is the radius of the base of the pile changing when the volume is 20,000 ft^3? (Hint: $V = \pi r^2 h/3$.)

36. A trough has cross sections which are equilateral triangles (tips down) and it is 10 ft long and 6 ft wide. The trough is filled with water at a rate of 0.2 ft^3/min. At what rate is the water level rising when the water level is 2 ft?

37. A mass is attached to one end of a spring. The other end of the spring is attached to a ceiling. If the mass is pulled downward and then released, then the position s of the weight with respect to its resting position is given by

$$s = S(t) = 0.5\cos(\pi t) - 0.25\sin(\pi t)$$

where s is in inches and t is in seconds after the weight passes the resting position on the way up.

   (a) Find the velocity and acceleration functions at any time t.

   (b) Find the position of the mass, the velocity of the weight and the acceleration of the mass when $t = 1$ sec and $t = 3$ sec.

   (c) Sketch a graph of the position, velocity and acceleration functions on the same graph on the interval $[0, 4]$.

38. A triangle is formed in the first quadrant by a tangent line to the graph of the curve $y = e^{-x}$ and the coordinate axes. What is the area of the largest such triangle? Explain why your answer gives the largest area.

39. A cone is to be formed by revolving a right triangle with a hypotenuse of length 3 feet about one of its legs. What is the volume of the largest cone that can be formed in this manner? Explain why your answer gives the largest volume.

40. The reaction r to a drug dose x administered to a patient is the value at x of the function R given by

$$R(x) = kmx^2 - kx^3$$

where p and m are constants with $k > 0$, $m > 0$, and $r > 0$. The sensitivity s of a patient's body to the dose of the drug of size x is defined to be the value of the derivative of R at x, that is, $s = S(x) = R'(x)$. Answer the following questions for this situation.

   (a) What is the domain of the reaction function?

   (b) What physical interpretations can you give for the constants m and k?

   (c) For what dose of the drug is a patient's reaction the greatest?

Explain why your answer gives the greatest reaction.

(d) For what dose of the drug is a patient the most sensitive? Explain why your answer gives the most sensitivity.

(e) Why do you think $R'(x)$ is defined to be the sensitivity of a patient?

41. A projectile whose muzzle velocity in meters per second is v (a constant) and whose angle of elevation in radians is θ has a range r which is the value at θ of a function R given by

$$r = R(\theta) = \frac{v^2 \sin(2\theta)}{g}$$

where g is a gravitational constant (the magnitude of the constant acceleration of gravity). At what angle of elevation is the range of the projectile the greatest? Why does your answer give the greatest range? What is the greatest range?

42. A storage building in the shape of a rectangular box is to have a square floor. Twice as much heat is lost per square foot through the roof as through the exterior walls, and no heat is to be lost through the floor (i.e., a negligible amount of heat is lost through the floor). What are the dimensions of such a building which encloses 12 thousand cubic feet and has the least heat loss? Why does your answer describe the desired storage building?

43. A truck travels over a flat interstate at a constant rate of 50 miles per hour and its (average) fuel mileage is five miles per gallon. Fuel costs 95 cents per gallon and for each mile per hour increase in the speed, the trucks mileage decreases by one-tenth mile per gallon. The truck driver gets $30 per hour in wages and the fixed cost for the use of the truck is $12 per hour. If possible, find a constant speed between 50 mph and 70 mph that the driver should maintain on 100 miles of straight interstate for the most economical total operating cost. Explain why your answer gives the most economical total operating cost.

44. For circle of radius r, a sector of the circle subtended by an angle θ is cut out to form a pie shaped wedge. The two cut sides of length r are then joined together to form a cone. What angle θ produces the cone with the greatest volume? Explain why your answer gives the greatest volume.

45. A right triangle is inscribed in a semicircle with radius r so that one of its legs is on the diameter and two of its vertices are on the circumference of the semicircle. What are the dimensions and the area of the largest right triangle inscribed in the semicircle in this manner? Explain why your dimensions give the greatest area.

46. A pizza box is formed from a rectangular piece of cardboard 16 in by 32 in by cutting out six squares of equal size, three from each of the 32 in sides with one square from each corner and one from the middle of each side. The result is then folded to form the square box. Explain how to make the box with the greatest volume. Why does your method give the greatest volume?

47. What are the dimensions and area of the largest rectangle that can be circumscribed around a given rectangle so that all four vertices of the given rectangle lie on the circumscribed rectangle? Why does your answer give the dimensions of the largest area?

48. What is the volume of the largest cylinder that can be formed by revolving a given rectangle whose perimeter is fixed about one of its sides? Why does your answer give the largest volume?

49. A rectangular piece of metal is to be made into the bottom of a trough by bending the piece of metal into a circular arc. Explain how to bend the metal so that the trough will have the greatest volume. Why does your method give the greatest volume?

50. Find the equation of the line (a) tangent and (b) normal to each of the following curves at the specified point.

    (a) $x^2 + y^3 = 3$, $(2, -1)$

    (b) $2x + \sin(y) = 1$, $(0.5, 0)$

    (c) $x^2 + 1 = 2\cos(x + y)$, $(0, \pi/3)$

    (d) $x^2 - \sqrt{xy} + y = 1$, $(-1, -1)$

51. Sketch a complete graph (i.e., a graph showing all relevant features) of the function represented by each of the following expressions. Be sure to consider all the issues about graphing in the list beginning on page 246.

    (a) $f(x) = x^3 - 6x^2$

    (b) $f(x) = x^3 - 6x^2 + 9x$

    (c) $g(x) = x^3 - 2x^2 + x + 3$

    (d) $h(x) = 8x^2 - x^4$

    (e) $f(x) = 9x^3 - x^5$

    (f) $f(x) = x\sin(x)$

    (g) $f(t) = t^2\cos(t)$

    (h) $h(x) = 0.5x + \sin(x)$

    (i) $f(x) = x - \sqrt{x}$ (j) $g(t) = (t - 1)^{\frac{3}{5}}$

    (k) $g(x) = \dfrac{x}{4 - x^2}$ (l) $f(x) = \dfrac{x^2 - 2}{x}$

    (m) $f(t) = \dfrac{2x}{1 - x^2}$ (n) $h(x) = \dfrac{3 + x^2}{x}$

    (o) $g(x) = \dfrac{3 + x^2}{x^2}$ (p) $g(x) = \sqrt[3]{x} + x$

    (q) $f(x) = 2x\sqrt[3]{x - 4}$

    (r) $g(t) = \dfrac{t^2 - 2t}{t}$ (s) $f(x) = \dfrac{1 - x^2}{x - 3}$

    (t) $g(x) = \dfrac{4 - x^2}{x^3}$ (u) $h(x) = \dfrac{2 - 3x}{x^2 + 1}$

52. In each of the following situations, sketch a complete graph (i.e., a graph showing all relevant features) of a function with the indicated properties.

    (a) $f'(x) > 0$ for $-1 < x < 2$ and $3 < x < 4$; $f'(x) < 0$ for $x < -1$, $2 < x < 3$ and $x > 4$; $f'(-1)$ does not exist; $f'(2) = 0 = f'(3)$; f has a vertical asymptote at $x = 4$.

    (b) $g'(x) < 0$ and $g''(x) > 0$ for $|x| < 1$; $g'(x) > 0$ and $g'' < 0$ for $x > 1$; $g'(x) = 0$ for $x \le -1$; $g(1) = 2$; $g'(1)$ does not exist.

    (c) $f(0) = 0$, $f(3) = 4$, and $f(7) = 0$; $f'(x) > 0$ for $x < 2$, $2 < x < 3$, $4 < x < 5$ and $x > 5$; $f'(x) < 0$ for $3 < x < 4$; $f'(3)$ is not defined; $f''(x) > 0$ for $2 < x < 3$, $3 < x < 5$; $f''(x) < 0$ for $x < 2$ and $x > 5$; $\lim_{x \to 5^+} f(x) = -\infty$ and $\lim_{x \to 5^-} f(x) = \infty$; $\lim_{x \to \infty} f(x) = 1$ and $\lim_{x \to -\infty} f(x) = -\infty$.

53. Find values of a and b so that the line $2x + 3y = a$ is tangent to the graph of the function f given by $f(x) = bx^2$ at the point where $x = 3$.

54. Let f be the function defined by

    $$f(x) = \begin{cases} ax & \text{if } x \le 1 \\ bx^2 + x + 1 & \text{if } x > 1 \end{cases}$$

    Find a and b so that f is differentiable for all x.

55. In the previous problem, find a and b so that the second derivative of f exists for all x.

56. What exactly does it mean for a graph to be symmetric with respect to something?

57. For the functions represented by each of the following equations and the indicated values of x and Δx, evaluate the differential dy and the actual change Δy. Sketch each function, the tangent line at $(x, f(x))$ and indicate x, Δx, dy and Δy on your sketch.

    (a) $y = 4x^2$, $x = 1$, $\Delta x = 0.02$

    (b) $y = 2\sin(x)$, $x = \pi/6$, $\Delta x = -0.03$

58. For a differentiable function f and "small" values of Δx, derive the following approximation formula

    $$f(x + \Delta x) \approx f(x) + f'(x)\Delta x$$

    where x and $x + \Delta x$ are in the domain of f. Explain the steps in your derivation. Using this approximation formula and appropriate choices of f, x and Δx, approximate the following by hand.

    (a) $\sqrt{65}$ (b) $\sqrt[3]{7.8}$

    (c) $\sin(31°)$

59. Using the approximation formula in the previous exercise, show that $\sin(h) \approx h$ for "small" values of h.

60. You wish to construct a square concrete basketball court so that the length of each side is 30 feet. When the measurement of a side has an error of at most 1 inch, use differentials to approximate the

    (a) (actual) error ΔA, in the calculated area, A.

    (b) relative error in the calculated area, A.

    (c) percentage relative error in the calculated area, A.

61. The speed s of blood flowing in an artery with radius r is given by the function S where

    $$s = S(r) = 1620r^2 \text{ cm/sec}.$$

    Assume that he measurement of the radius has an error of at most 0.02 cm. For the calculated speed S, use differentials to approximate

    (a) (actual) error, ΔS, in the calculated speed, S.

    (b) relative error in the calculated speed, S.

    (c) percentage relative error in the calculated speed, S.

62. A cylinder with a piston contains a gas with volume v given by the function

    $$v = V(p) = \frac{300}{p} \text{ in}^3$$

    where p is the pressure in lb/in^2. Suppose that the pressure decreases from 30 lb/in^2 to 29 lb/in^2. Use differentials to approximate the

    (a) (actual) change ΔV, in the calculated volume.

    (b) relative change in the calculated volume, V.

    (c) percentage relative error in the calculated volume, V.

63. The total cost C of producing x units of a certain product is given by

    $$C(x) = 0.15x^3 - 0.5x^2 + 300x + 1000 \text{ dollars}.$$

    When 50 units have been produced, use differentials to approximate the cost of producing one more unit.

64. For the situation in the previous exercise, use differentials to approximate the additional cost of producing another 5 units when 100 units have already been produced.

65. For each of the following curves, find the extreme values (if any) of y and x. Use your computer to make a sketch of each equation.

(a) $\left(x^2 + 8y - 16\right)^2 = y^2\left(16 - x^2\right)$

(b) $\left(x^2 + y^2\right)^2 = 10\left(x^2 - y^2\right)$

66. A function f given by $y = f(x)$ is defined implicitly by the equation

$$1 - y^5 = x^2 - y.$$

When f satisfies

(a) $f(1) = -1$

(b) $f(1) = 0$

    i. Use your computer to make a sketch of the curve given by $1 - y^5 = x^2 - y$.

    ii. Use your sketch in part i to make a sketch of f.

    iii. Find an equation of the tangent line to the graph of f at $x = 1$.

    iv. Find the extreme points and extreme values of f.

67. Find the following limits. Explain the reasoning you used to obtain each limit. You may wish to use your computer to see what is happening.

(a) $\lim\limits_{x \to 0} \dfrac{\sin(3x)}{2x}$ (b) $\lim\limits_{\theta \to 0} \dfrac{1 - \cos(2\theta)}{\theta}$

(c) $\lim\limits_{x \to 0} \dfrac{1 - e^x}{\sin(x)}$

(d) $\lim\limits_{x \to 0}\left(\dfrac{1}{x} - \dfrac{1}{e^x - 1}\right)$

(e) $\lim\limits_{x \to 0^+} x \ln(x)$

(f) $\lim\limits_{x \to 0}\left(\dfrac{1}{\sin(x)} - \dfrac{1}{x}\right)$

(g) $\lim\limits_{y \to 0} \dfrac{y\left(1 - e^y\right)}{1 - \cos(y)}$ (h) $\lim\limits_{x \to \pi} \dfrac{1 + \cos(x)}{(x - \pi)^2}$

(i) $\lim\limits_{x \to 0} \dfrac{3 - 3\cos(x)}{x^2}$

(j) $\lim\limits_{x \to 1}\left(\dfrac{1}{x - 1} - \dfrac{1}{\ln(x)}\right)$

(k) $\lim\limits_{x \to \infty} \dfrac{\sin(x)}{x}$ (l) $\lim\limits_{x \to \infty} x\left(1 - e^{1/x}\right)$

(m) $\lim\limits_{x \to \infty} \dfrac{1 - e^x}{1 - \cos(x)}$ (n) $\lim\limits_{x \to \infty} x^2 \sin\left(\dfrac{1}{x}\right)$

(o) $\lim\limits_{x \to 0^+}\left(\dfrac{2}{x} - \dfrac{1}{x^2}\right)$ (p) $\lim\limits_{x \to 0} \dfrac{\cos(x)}{x}$

(q) $\lim\limits_{x \to 0} \dfrac{1 - e^x}{x}$ (r) $\lim\limits_{u \to 0} \dfrac{1 - u - e^u}{5u^2}$

(s) $\lim\limits_{x \to 0} \dfrac{1 - x - e^{-x}}{x^2}$ (t) $\lim\limits_{x \to 0^+} xe^{1/x}$

68. Find the following limits. Explain the reasoning you used to obtain each limit. You may wish to use your computer to see what is happening.

(a) $\lim\limits_{t \to 0} \dfrac{e^{2t}}{1 - e^{-t}}$ (b) $\lim\limits_{x \to \infty} \dfrac{e^{-2x}}{1 + e^x}$

(c) $\lim\limits_{x \to \infty} \dfrac{e^{2x}}{x^3}$ (d) $\lim\limits_{x \to \infty} \dfrac{e^{-2x}}{x^3}$

(e) $\lim\limits_{x \to \infty} \dfrac{1 + x^2}{e^{-x}}$ (f) $\lim\limits_{x \to \infty} \dfrac{x^n}{e^x}, n > 0$

(g) $\lim\limits_{x \to \infty} \dfrac{x^n}{e^{-px}}, n > 0, p > 0$

(h) $\lim\limits_{x \to \infty} \dfrac{e^{px}}{x^n}, n > 0, p > 0$

(i) $\lim\limits_{x \to \infty} 20\left(5 + 2e^{-3x}\right)$

(j) $\lim\limits_{x \to \infty} \sqrt{x}e^{-x}$ (k) $\lim\limits_{x \to \infty} e^{-2x} \ln(x)$

(l) $\lim\limits_{x \to \infty} 3 + 2e^{-x}$ (m) $\lim\limits_{y \to 0} \dfrac{y\left(1 - e^y\right)}{1 - e^{-y}}$

(n) $\lim\limits_{t \to -\infty} \dfrac{1 + e^{-2t}}{t}$ (o) $\lim\limits_{x \to 0} \dfrac{1 - e^{2x}}{x^2}$

(p) $\lim\limits_{x\to\infty} \dfrac{x^2}{e^x}$ (q) $\lim\limits_{x\to\infty} 1 - e^{3x}$

69. Explain what happens when you try to use *l'Hôpital's rule* to find the following limit

$$\lim_{x\to\infty} \frac{\sqrt{1+x^2}}{x}.$$

What is the value of the limit? Why?

70. Prove the following form of L'Hôpital's rule.

Suppose that f and g are functions which are differentiable at a point c such that $f(c) = g(c) = 0$ and $g'(c) \neq 0$. Then

$$\lim_{x\to c} \frac{f(x)}{g(x)} = \frac{f'(a)}{g'(a)}.$$

71. In this exercise, you will consider the following limit situations.

$$\infty^\infty, \quad \frac{0}{\infty} \quad \text{and} \quad \frac{\infty}{0}$$

Explain why these three limit situations are known as *determinate forms* of limits. Do you see that the third one has value 0 and the other two have value ∞ (or possibly $-\infty$)? Give examples of each situation.

72. Find the following limits. Assume that a, b, and c are constants, and that f is a differentiable function.

(a) $\lim\limits_{x\to c} \dfrac{\sin^3(x) - \sin^3(c)}{x - c}$

(b) $\lim\limits_{x\to a} \dfrac{f^2(x) - f^2(a)}{x - a}$

(c) $\lim\limits_{x\to b} \dfrac{f(f(x)) - f(f(b))}{x - b}$

(d) $\lim\limits_{x\to b} \dfrac{e^{f(x)} - e^{f(b)}}{x - b}$

4.4 THE THEORY OF DIFFERENTIATION

4.4.1 Overview

In this section you will consider some of the theoretical aspects of differentiation. You will study the relationship between differentiability and continuity, proofs of some differentiation rules, and implicit differentiation. Next, you will work with some theorems which are important in problem solving, including the Intermediate Value Theorem, the Extreme Value Theorem, Rolle's Theorem and the Mean Value Theorem. We will also give proofs of some of the basic notions relating derivatives, monotonicity and extreme points, and derivatives and concavity. You will also study inverse functions and their derivatives from which the differentiation rules for the natural logarithm function and the inverse trigonometric functions are easily derived. You will also learn to approximate zeros of a function using the concept of derivative and a method often referred to as the *Newton-Raphson* method. You will get a chance to do some proofs of your own in the exercises. In addition, the exercises will provide you further opportunities to: calculate derivatives of more functions; find limits of indeterminate forms not considered in the previous section; and to do more problem solving using derivatives.

4.4.2 Activities

1. Use an implicit function grapher, such as **graphR** in **ISETL** or **implicitplot** in **MapleV**, to sketch a graph of each the following equations. Your instructor will provide you with the necessary information to make such computer graphs.

$$x^2 + y^2 = 1$$

$$x^3 + y^3 - 4.5xy = 0$$

$$|x|^{2/5} + |y|^{2/3} = 1$$

2. In this problem you will use the equation $x^3 + y^3 - 4.5xy = 0$ and your computer to construct an implicity defined function and graph its (approximate) derivative. Your instructor will tell you whether to use your MPL or your SCS. If you are to use your MPL, then you will use a **computer function** called **implicit** like the one shown below in **ISETL** code.

```
implicit := func(rel, I0, J0);
     return func(x);
          local f;
          if not ((I0(1)<=x) and (x<=I0(2))) then return; end;
          f := func(y);
                return rel(x, y);
             end;
          return bis(f, J0(1), J0(2), 0.001);
       end;
  end;
```

The **computer function implicit** accepts a **computer function rel** which represents the values of the expression $rel(x, y)$ which is obtained from an equation in x and y by rewriting the equation in the form "an expression in x and $y = 0$", and two **tuples I0 and J0** that are the endpoints of intervals that restrict x and y respectively. **I0** and **J0** are assumed to be chosen so that in these intervals, for each x, there is exactly one pair **[x, y]** which has **rel** return the value **0**. The choice of the intervals **I0** and **J0** can be made by inspecting the graph of the equation $x^3 + y^3 - 4.5xy = 0$.

Note that **implicit** uses the **computer function bis**, so you will need to be sure to have it defined in your MPL.

Make a sketch of the graph of the function represented by the **computer function** which **implicit** produces for the equation for two different choices of **I0, J0** in different quadrants. Sketch the graphs of the resulting implicitly defined functions. The make a sketch using your MPL of the approximate derivatives of your two implicit functions using your modification of the **computer function D** with the **computer function lim** left out (see the NOTE at the end of the Activities in Section 4.1.2 on page 196).

On the other hand, if you are instructed to use your SCS, then do as follows.

Make a graph of the equation $x^3 + y^3 - 4.5xy = 0$ using a graphics command in your SCS which plots an equation in two variables. Then make two different choices of **I0, J0** in different quadrants which produce two implicit functions. Sketch the graphs of the two resulting implicitly defined functions on the two choices of intervals **I0** and **J0**, respectively. Then make a sketch of the derivative of each implicit function using your computer, or make rough sketches (each in the proper quadrant) of the two derivatives by hand.

3. Return to the compressibility function Z of Chapter 2, Section 1, page 75. We wish to use the computer function **bis** to estimate zeros of this function. Recall that **bis** requires as input a **computer function**, the endpoints of an interval which contain only one zero of the **computer function** and a small value for **eps**. Use **eps** = 0.001. Give it Z and the following intervals,

$$(5, 7), [0, 100], [80, 100], [196, 210].$$

Which intervals lead to success and which do not? Explain what happened.

4. An equation which is of the form $y = $ an expression in x can be used to define a function whose domain variable is called x and co-domain variable is called y. In this activity, you will try to use such an equation to define a function whose domain variable is called y and co-domain variable is called x.

   You can do it by solving for x in terms of y (for instance using your SCS), drawing the graph and inverting it, or any other method that occurs to you. Try it with the following two equations.

   $$y = x^3 - 4x + 1$$
   $$y = x^3 + 4x + 1$$

5. The following **computer function newt** written in **ISETL** code implements a method for approximating a zero a of a function f, i.e., $f(a) = 0$, using what is called the Newton-Raphson Method.

```
newt := func(f, f ', x0, eps);
        local x1;
        print x0;
        if f(x0) < 0.001 then return x0;  end;
        if f '(x0) = 0 then return "Method fails";  end;
        x1 := x0-f(x0)/f '(x0);
        if abs(x1-x0) < eps then return x1;
        else return newt(f, f ', x1, eps);
        end;
    end;
```

   If your MPL is not **ISETL**, you should be able to write a similar **computer function** for your MPL or your instructor will provide you with the necessary code.

   (a) Explain in your own words what is going on in the computer function **newt** (ignore the **local** statement in line two).

   (b) How could you think you could find a starting point x_0 which is "near" to a zero a of a function f?

   (c) The temperature of the human body is elevated when a dosage d of a certain drug is injected into the blood stream and absorbed by the body. The elevation t of the temperature, in degrees Fahrenheit, is given by

   $$t = T(d) = 1.80d^2 - 0.33d^3$$

   where the dosage, in grams, is in the interval $0 \le d \le 4$. Use **newt** to estimate the dosage d required to raise the body temperature $3.4°$. Use **eps** = 10^{-6}.

(d) For the function f given by $f(x) = x^{1/3}$, use **newt** to approximate the zero of f at a using any non-zero approximation x_0 near 0 as a starting point. Explain what happens. Make a sketch of the function and explain by drawing on your sketch what you think is happening.

(e) In the Newton-Raphson method, explain what happens if $f'(x_n) = 0$ for some value of n? Make a graph and give a geometric explanation related to the behavior of such a function f.

6. Write **computer function RiemMax** which will be the same as the others except for the following.

(a) **RiemMax** will accept an additional parameter **crit** which is a **tuple** consisting of a list of all points at which the function in question could have a relative maximum.

(b) Instead of computing

$$\sum_{i=1}^{n} f(x_i)(x_{i+1} - x_i)$$

the quantity $f(x_i)$ will be replaced by an estimate of the maximum value of f on the interval $[x_i, x_{i+1}]$. In **ISETL**, the code is as follows.

%max [f(t) : t in T]

where **T** is a tuple of all possible values in the interval $[x_i, x_{i+1}]$ at which f could attain its maximum.

HINT: If **S** *is a tuple,* **p** *and* **q** *numbers, then the tuple of all numbers in* **S** *that are in the interval* $[p, q]$ *together with these two endpoints is given in* **ISETL** *by the following code*

[p] + [s : s in S | p < s and s < q] + [q];

As part of your **computer function**, you will need to have a construction of a tuple of tuples, corresponding to your partition that give the possible points at which the maximum could occur.

Run your **computer function** on **data29** with the three following choices of the interval $[a, b]$. Remember that you will have to use the **computer function interpolate** on **data29** in order to complete this activity.

(a) $[-1, 4]$

(b) $[-1, 0]$

(c) $[0, 4]$

Use $n = 20$ in all cases.

Of course, you must figure out the tuple **crit** by hand.

Think about the following questions.

> What property of the function on the interval will guarantee that the results of **RiemMax** and **RiemLeft** are the same? What will guarantee that the results of **RiemMax** and **RiemRight** are the same? That the results of **RiemRight** and **RiemLeft** are the same?

7. Make up your own function and an interval on which it has at least 3 points on which there is a relative maximum and apply **RiemMax** to it. (Note that this will force your tuple **crit** to have at least 5 components).

4.4.3   Discussion

Differentiability Implies Continuity   Consider the basic statement defining the derivative of a function f at a point a.

$$f'(a) = \lim_{h \to 0}\left(\frac{f(a+h)-f(a)}{h}\right)$$

Looking at this expression, we see the limit of a fraction in which the denominator goes to zero. If this limit exists, then the numerator must also go to zero.

Put another way, the existence of the derivative can be expressed as saying that $f(a+h)-f(a)$ approaches zero at least as fast as h approaches 0.

In particular, this means that if the derivative exists, that is, if the limit exists, then the function must also be continuous at a because continuity means nothing other than that the limit of a function at a point equals the value of the function at that point (see the definition of continuity on page 141).

We can formalize this in a theorem.

Theorem 4.1 (Differentiability implies continuity)
If a function f is differentiable at the point a, then it is continuous at a.

Proof. We have,

$$f'(a) = \lim_{h \to 0}\left(\frac{f(a+h)-f(a)}{h}\right).$$

We also have,

$$0 = \lim_{h \to 0} h$$

Multiplying the two and applying the product rule for limits, we obtain,

$$0 = f'(a)\cdot 0 = \lim_{h \to 0}\left(\frac{f(a+h)-f(a)}{h}\right)\lim_{h \to 0} h$$

$$= \lim_{h \to 0}\left(\frac{f(a+h)-f(a)}{h}\right)h = \lim_{h \to 0}(f(a+h)-f(a))$$

Or, applying the sum rule for limits, we have

$$\lim_{h \to 0} f(a+h) = f(a).$$

□

We will make use of this theorem several times in the remaining proofs of this section.

Is the converse true? That is, does the fact that a function is continuous at a point necessarily imply that it is differentiable at that point?

Rules for Differentiation — Arithmetic   Some of the rules for differentiation are immediate consequences of corresponding rules for limits.

Theorem 4.2 (Derivative of a sum)

If f and g are two functions which are differentiable at a then the sum f + g is differentiable at a and

$$(f+g)'(a) = f'(a) + g'(a).$$

Proof. We simply calculate from the definition with the difference quotient.

$$
\begin{aligned}
(f+g)'(a) &= \lim_{h \to 0}\left(\frac{(f+g)(a+h) - (f+g)(a)}{h} \right) \\
&= \lim_{h \to 0}\left(\frac{f(a+h) + g(a+h) - f(a) - g(a)}{h} \right) \\
&= \lim_{h \to 0}\left(\frac{f(a+h) - f(a)}{h} + \frac{g(a+h) - g(a)}{h} \right) \\
&= \lim_{h \to 0}\left(\frac{f(a+h) - f(a)}{h} \right) + \lim_{h \to 0}\left(\frac{g(a+h) - g(a)}{h} \right) \\
&= f'(a) + g'(a)
\end{aligned}
$$

Others require some additional calculation.

Theorem 4.3 (Derivative of a product)

If f and g are two functions which are differentiable at a then the product fg is differentiable at a and

$$(fg)'(a) = f'(a)g(a) + f(a)g'(a).$$

Proof.

$$
\begin{aligned}
(fg)'(a) &= \lim_{h \to 0}\left(\frac{(fg)(a+h) - (fg)(a)}{h} \right) \\
&= \lim_{h \to 0}\left(\frac{f(a+h)g(a+h) - f(a)g(a)}{h} \right) \\
&= \lim_{h \to 0}\left(\frac{f(a+h)g(a+h) - f(a)g(a+h) + f(a)g(a+h) - f(a)g(a)}{h} \right) \\
&= \lim_{h \to 0}\left(\left(\frac{f(a+h) - f(a)}{h} \right)g(a+h) + f(a)\left(\frac{g(a+h) - g(a)}{h} \right) \right) \\
&= \lim_{h \to 0}\left(\frac{f(a+h) - f(a)}{h} \right)g(a+h) + \lim_{h \to 0} f(a)\left(\frac{g(a+h) - g(a)}{h} \right) \\
&= \lim_{h \to 0}\left(\frac{f(a+h) - f(a)}{h} \right)\lim_{h \to 0} g(a+h) + \lim_{h \to 0} f(a)\lim_{h \to 0}\left(\frac{g(a+h) - g(a)}{h} \right) \\
&= f'(a)g(a) + f(a)g'(a).
\end{aligned}
$$

Note that in taking the last step, we used the fact (Theorem 4.1), that a function which is differentiable at a point is also continuous at that point. Hence, since g is assumed to be differentiable at a it is continuous there and its value is equal to its limit, so

$$\lim_{h \to 0} g(a+h) = g(a).$$

Rules for Differentiation — Chain Rule

By now you should not have too much difficulty applying the chain rule to take the derivative of the composition of two functions. The basic idea of the proof is very similar to the proof of the product rule, but there are some technical difficulties that must be dealt with.

Theorem 4.4 (Chain Rule (composition))

Let f, g be functions and a a point in the domain of g such that $g(a)$ is in the domain of f. Suppose further that g is differentiable at a and f is differentiable at $g(a)$. Then $f \circ g$ is differentiable at a and

$$(f \circ g)'(a) = f'(g(a))g(a).$$

Proof. We begin by presenting a "proof" which is really the idea of what we want to do, but is not quite correct. Our argument is similar in spirit to what we did for the product rule. In that situation we added and subtracted a term in the basic definition. For the chain rule, we divide and multiply by a term. Thus we have,

$$
\begin{aligned}
(f \circ g)'(a) &= \lim_{h \to 0}\left(\frac{(f \circ g)(a+h) - (f \circ g)(a)}{h} \right) \\
&= \lim_{h \to 0}\left(\frac{f(g(a+h)) - f(g(a))}{h} \right) \\
&= \lim_{h \to 0}\left(\left(\frac{f(g(a+h)) - f(g(a))}{g(a+h) - g(a)} \right)\left(\frac{g(a+h) - g(a)}{h} \right) \right) \\
&= \lim_{h \to 0}\left(\frac{f(g(a+h)) - f(g(a))}{g(a+h) - g(a)} \right)\lim_{h \to 0}\left(\frac{g(a+h) - g(a)}{h} \right) \\
&= f'(g(a))g'(a).
\end{aligned}
$$

First, let us consider what is correct about this argument. The main point to consider is how we obtain the last line. The second factor in the previous line is precisely the definition of $g'(a)$.

The first factor amounts to taking the limit as $g(a+h)$ approaches $g(h)$ of the difference between f evaluated at $g(a+h)$ and at $g(a)$, all divided by $g(a+h) - g(a)$, which goes to 0 (because g is continuous at a.) This is the same as taking the limit as k goes to 0 of the difference between f evaluated at $g(a) + k$ and f evaluated at $g(a)$, all divided by k. We are just replacing $g(a+h) - g(a)$ by k.

In other words, the first factor is just,

$$\lim_{k \to 0}\left(\frac{f(g(a) + k) - f(g(a))}{k} \right) = f'(g(a)).$$

All well and good, except for one sticky point. We have divided by $g(a+h) - g(a)$. We know very little about the value of this expression. It might be 0. Where would we be then? We have to give some serious consideration to this possibility.

Its not too bad if $g'(a) \neq 0$. In this case, we can be sure that for h very small, $g(a+h) - g(a) \neq 0$. Indeed, if $g(a+h) - g(a) = 0$ for values of h arbitrarily close to 0 then we would have,

$$g'(a) = \lim_{h \to 0}\left(\frac{g(a+h) - g(a)}{h} \right) = 0.$$

So the only thing to worry about is the case in which $g'(a) = 0$. Here we go back to our original calculation and consider two possibilities for the points h. For those values of h for which $g(a+h) - g(a) \neq 0$, our argument works. If we have values of h arbitrarily close to 0 for which $g(a+h) - g(a) = 0$, that is, $g(a+h) = g(a)$, considering only such points, we have,

$$(f \circ g)'(a) = \lim_{h \to 0} \left(\frac{(f \circ g)(a+h) - (f \circ g)(a)}{h} \right)$$
$$= \lim_{h \to 0} \left(\frac{(f \circ g)(a) - (f \circ g)(a)}{h} \right) = 0$$
$$= f'(g(a))g'(a)$$

where we used the fact that $g(a+h) = g(a)$ in the second line above.

$\square$

Rules for Differentiation — Trigonometric Functions
The rules for differentiating the six trigonometric functions are all derived from the standard trigonometric identities and the following fact about limits that we obtained in Chapter 3, Theorem 3.14 (page 180.)

$$\lim_{\theta \to 0} \frac{\sin(\theta)}{\theta} = 1$$

The main consequence of this limit is the following.

> **Theorem 4.5 (Derivative of sin)**
> *The derivative of the sine function is the cosine function.*

Proof. We can compute, from the definition of derivative at any point and using the trigonometric identity,

$$\sin(A) - \sin(B) = 2 \sin\left(\frac{A-B}{2} \right) \cos\left(\frac{A+B}{2} \right)$$

that

$$(\sin)'(x) = \lim_{h \to 0} \left(\frac{\sin(x+h) - \sin(x)}{h} \right)$$
$$= \lim_{h \to 0} 2 \left(\frac{\sin(\frac{h}{2})}{h} \right) \cos\left(\frac{2x+h}{2} \right)$$
$$= \lim_{h \to 0} \left(\frac{\sin(\frac{h}{2})}{\frac{h}{2}} \right) \lim_{h \to 0} \cos\left(x + \frac{h}{2} \right)$$
$$= \cos(x)$$

where we used the continuity of the cos function in the last step.

$\square$

Now the formulas for differentiating the other five trigonometric functions are easily derived. For example, we can use the identity

$$\cos(x) = \sin\left(x + \frac{\pi}{2}\right)$$

and then apply the chain rule to obtain the formula,

$$(\cos)'(x) = \cos\left(x + \frac{\pi}{2}\right).$$

Now using the trigonometric identity

$$\cos\left(x + \frac{\pi}{2}\right) = -\sin x$$

we conclude that

$$(\cos)'(x) = -\sin(x).$$

You will have a chance to work with the other trigonometric functions in the exercises.

Rules for Differentiation — Functions Defined in Parts   Consideration of functions defined in parts allows us to make an important remark about the derivative at a point. It is a local property. This means that if two functions f and g have the same value at a point a and at an interval containing this point in its interior, then all questions about the derivative of f at a and g at a have the same answer. That is, if one derivative exists then so does the other and they are equal in value. We formalize this in a theorem.

Theorem 4.6

Let I be an interval (open or closed) with endpoints a, b, and let c be an interior point in this interval, that is $a < c < b$. Let f and g be two functions such that $f(x) = g(x)$ for every $x \in I$. Then, if one of these two functions is differentiable at c, then so is the other. Moreover, in this case,

$$f'(c) = g'(c).$$

Proof. It is only a question of looking at the definition of the derivative and considering the meaning of the limit. We have, for example,

$$f'(c) = \lim_{h \to 0}\left(\frac{f(c+h) - f(c)}{h}\right).$$

Now, in this limit, the only thing that matters is what happens when h is small, that is, when $c+h$ is close to c. As soon as $c+h$ is close enough to c, it will follow that $c+h \in I$. When this happens, nothing changes in the limit if f is replaced by g. Hence the result of taking the limit will be the same in both cases.

□

The rule for differentiating a function f defined in parts follows immediately from this theorem. The derivative of f at a point c away from a seam can be taken as if the expression defining that part defines the function everywhere, since c will be in an interval where the values of f will be the same as if f were the function defined by that expression.

The derivative of *f* at a seam can only be computed by checking the definition at that point and taking the left and right handed limits. Because of Theorem 3.1 (page 168), this is the same as saying that the derivative at a seam exists if and only if the left and right derivatives exist and are equal.

Implicit Functions and Their Derivatives

Let's begin with the standard formulation of an implicit differentiation problem. We will use the second equation that you worked with in Activity 1.

Assume that the equation

$$x^3 + y^3 - 4.5xy = 0$$

defines *y* as a function of *x*. Find $\frac{dy}{dx}$.

There are three issues that must be dealt with in such a situation.

- How do we know that this equation does determine a function?

- Does the derivative of this function exist?

- How do we compute the derivative?

We will discuss the first and last of these. The second is very complicated and uses something called the *Implicit Function Theorem*, which is a topic in more advanced math courses.

Addressing these in reverse order, we begin with the last issue which leads to a computation that you have already performed. Let us assume that the first issue has been settled and we do have a function, say *h*. Assuming also that the second issue has been settled, we may suppose that h' exists. The question is, can we find some expression for $h'(x)$ in terms of *x*, and also, perhaps, in terms of *y*.

The key is to understand what is really meant by saying that the equation

$$x^3 + y^3 - 4.5xy = 0$$

"defines" the function *h*. It means that there is a function *h* such that the following equation is satisfied for all values of *x* in the domain of *h*.

$$x^3 + (h(x))^3 - 4.5xh(x) = 0$$

If you look at that expression, how many functions do you see? One? Two? Three? Or more?

The assertion is that the left hand side of that equation is an expression which defines a function. The domain variable of this function is *x*. What is more, since this equation is to be satisfied for all *x* in the domain of *h*, the assertion also is that this function is 0. That is, it is the function which returns, for every *x* in its domain (which is the domain of *h*) the value 0. Therefore its derivative is also the 0 function.

This last sentence is what does the trick for us. On the one hand, we have an expression, $x^3 + (h(x))^3 - 4.5xh(x)$, defining a function. This means we can compute its derivative. On the other hand, we *know* that this function is the 0 function. Hence this derivative we can compute is actually 0. Therefore, we compute the derivative and set it equal to 0.

This gives,

$$3x^2 + 3h^2(x)h'(x) - 4.5h(x) - 4.5xh'(x) = 0$$

which we can solve for $h'(x)$ to obtain,

$$h'(x) = \frac{4.5h(x) - 3x^2}{3h^2(x) - 4.5x}$$

or, translating back to the language of x and y,

$$y' = \frac{4.5y - 3x^2}{3y^2 - 4.5x}$$

so that unless we are at a point (x, y) on the graph where $3y^2 - 4.5x = 0$ or $y^2 = 1.5x$, we have a formula for the derivative.

Now we are ready for the first issue, the existence of the function h. Where does it come from? Perhaps we can get an idea by looking at the graph shown in Figure 4.21.

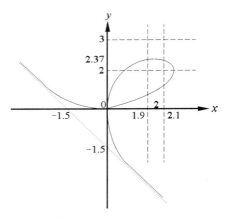

Figure 4.21. A graph of $x^3 + y^3 - 4.5xy = 0$ ***from Activity 1.***

Does this define y as a function of x? In fact, it does *not*! If you think about the vertical line test, you can see that for some values of x, the vertical line passes through the graph, not once, but three times! Or, if you prefer numbers, take $x = 2$. Substitute 2 for x in the equation and use your SCS or hand calculations to solve for y. There are three solutions for y which are approximately, -3.37, 1, 2.37 for y. That is, each of the points, $(2, -3.37)$, $(2, 1)$, $(2, 2.37)$ satisfies the equation. So where is the function? What is the value of y if $x = 2$?

Part of the problem is taken care of by restricting the domain of h. You can see from the graph that if you take $x < 0$ or x greater than about 3, then there will be only one y that gives a point on the curve. Hence, in this domain, we have a function. But what about small positive values of x? Do we just say there is no function here and leave it at that?

No, what we do is restrict not only the value of x, but also the value of y. This is what you did in Activity 2. Consider for example, the value $x = 2$. You might take $I_0 = (1.9, 2.1)$ as the domain interval. Then you choose an interval J_0 which restricts the values of y to make sure that for each $x \in I_0$, there is only one $y \in J_0$ such that this (x, y) satisfies the equation. You might choose $J_0 = (2, 3)$. This would give you a branch of the function h at the top of the curve as shown in the picture. Or you might choose $J_0 = (-4, -3)$. This would give you a branch that is part of the wing flying below the loop.

In other words, there are actually many such functions h. Or, if you prefer, one function which has several branches, depending on how you choose I_0, J_0.

No matter how you make this choice of the intervals, you get a function (the first issue) so this provides the theoretical basis for saying that there is a function and, if it has a derivative (the second issue), then its value is given by

$$h'(x) = \frac{4.5h(x) - 3x^2}{3h^2(x) - 4.5x}$$

where x and y are not only points on the curve, but also x and y restricted to a particular pair of intervals I_0 and J_0, respectively. We note that the expression for $y' = h'(x)$ represents the derivative for all branches of the curve $y = h(x)$. What happens at the points where the denominator is zero, i.e., where $3y^2 - 4.5x = 0$? Are there tangent lines at points which are solutions $3y^2 - 4.5x = 0$?

Once again (as in derivatives of functions defined in parts) we see that the derivative is a local property.

Special Values of a Function

In Activity 3, you applied the **computer function bis** to the compressibility function Z to get estimates of zeros contained in various intervals. Sometimes it worked and sometimes it didn't. Can you explain what went wrong in each case that it failed?

The point of **bis** is that if a continuous function has a negative value at one point and a positive value at another, then in going from positive to negative, the graph must somehow cross the x-axis, that is, there must be a point in between at which the function has a zero.

So one issue affecting whether a function has a zero in an interval $[a, b]$ is whether or not $f(a)$ and $f(b)$ have different signs.

What about the last interval, $[196, 210]$? It is certainly the case that $Z(196) < 0$ while $Z(210) > 0$. Did you find a zero of Z in between 196 and 210? Why not?

The reason is that Z is not continuous everywhere in the interval $[196, 210]$. Indeed, what it does is go down to $-\infty$ as the independent variable v approaches 200 from the left. Then, on the other side of 200 it "jumps" up to the positive side of the x-axis, without ever passing through 0.

This can happen with very simple functions. Look at

$$f(x) = \begin{cases} 1 & \text{if } x < 0 \\ -1 & \text{if } x \geq 0. \end{cases}$$

Again, this function is negative at one point, say $x = 0.5$ and positive at another, say $x = -2$, but it jumps from one to the other without ever touching the x-axis. It has no zero in between the two points, even though the values of the function have opposite signs at these two points.

This anomaly occurs because the function is not continuous at every point. It can never happen if the function is continuous (see Figure 4.22). Indeed, if a function is continuous on a closed interval then it takes on every value in between its values at the endpoints. This is a very important fact that represents a deep property of numbers and the concept of continuity. We will state it formally, but the proof is too advanced for this book.

Theorem 4.7 (Intermediate value theorem)

Let f be a function which is continuous on the closed interval $[a, b]$. Then for every value d which is between $f(a)$ and $f(b)$, there is a number $c \in [a, b]$ such that $f(c) = d$.

The Intermediate Value Theorem is the basis of methods which are used to approximate zeros of functions. If a function f is defined on an interval $[a, b]$ with $f(a) > 0$ and $f(b) < 0$, or $f(a) < 0$ and $f(b) > 0$, what can you say about f? Is there a number $c \in (a, b)$ with $f(c) = 0$?

Actually, more is true. Not only does a continuous function on a closed interval take on every intermediate value between $f(a)$ and $f(b)$, it *must* also take on a maximum and a minimum value. See Figure 4.23. This is another property that we can state, but not prove at this time.

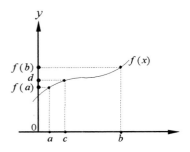

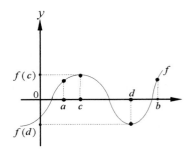

Figure 4.22. The Intermediate Value Theorem.   **Figure 4.23. The Extreme Value Theorem.**

Theorem 4.8 (Extreme value theorem)

Let f be a function which is continuous on the closed interval $[a, b]$. Then there are numbers c, $d \in [a, b]$ such that $f(c)$ is a maximum value, that is $f(c) \geq f(x)$ for all $x \in [a, b]$ and $f(d)$ is a minimum value, that is $f(d) \leq f(x)$ for all $x \in [a, b]$.

Although we cannot prove these two theorems, you should see that the following statement is a summary and reformulation of what these two theorems are saying.

Theorem 4.9

Let f be a function which is continuous on the closed interval $[a, b]$. Then the co-domain of f, that is, the set of values $\{f(x) : x \in [a, b]\}$ is an interval $[m, M]$. The value m is a minimum and the value M is a maximum of f on the interval $[a, b]$. The function f takes on every value from m to M, including m and M.

The Mean Value Theorem   The facts of the preceding paragraph, together with some computational tricks, lead to a very powerful result concerning values of a function which is differentiable. We begin with a simple consequence of the extreme value theorem. This is the first of several theoretical results which amount to formalizing ideas that should be thoroughly familiar to you by now. See Figure 4.24.

Theorem 4.10

Let f be a function that is differentiable on an open interval (a, b) and let $c \in (a, b)$. Then if $(c, f(c))$ is a relative maximum or relative minimum point of f, it follows that $f'(c) = 0$.

Proof. Consider the statement that defines the derivative of f at c.

$$f'(c) = \lim_{h \to 0} \frac{f(c+h) - f(c)}{h}$$

If $f(c)$ is a relative maximum, for example, then for small values of h, the numerator will be non-positive. If $h < 0$ then the fraction will therefore be non-negative. On the other hand, if $h > 0$ then the fraction will be non-positive.

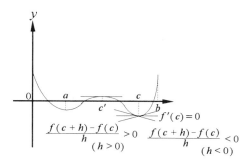

Figure 4.24. The relative maximum and minimum theorem.

Thus it follows that the limit from one side will be a limit of numbers which are not negative, so this limit cannot be negative. (Why?) On the other hand, the limit from the other side will be a limit of numbers which are not positive so the limit cannot be positive. Hence, by the "Goldilocks Principle" (because it's trapped), the limit can only be 0.

□

The converse of this theorem is not true. Can you think of an example of a function whose derivative is 0 at a point c, but $f(c)$ is not a relative maximum or minimum?

We can put some of these results together to prove what seems completely obvious if you look at a nice picture like the one shown in Figure 4.25.

Figure 4.25 shows a function which has the same value at two points. What can be observed is that *somewhere* in between these two points, there is a point at which the tangent is parallel to the x-axis, that is, its slope, or the derivative at that point, is 0.

Hopefully by now you have looked at enough weird situations that can arise in mathematics that you are not completely satisfied that something is true just by looking at a picture. We can give a very simple logical proof of this observation.

Theorem 4.11 (Rolle's Theorem)

Let the function f be continuous on the closed interval $[a, b]$ and differentiable at every point on the open interval (a, b). If $f(a) = f(b)$ then there is a point $c \in (a, b)$ such that $f'(c) = 0$.

Proof. One possibility is that the function has this common value for all $x \in [a, b]$. That is, it is constant, in which case its derivative is 0 at *all* points in (a, b).

If the function is not constant, then it must have at least one point at which it is either larger or smaller than the common value of $f(a)$ and $f(b)$. This means that it must have a relative maximum or relative minimum at c which is neither a nor b, that is, $c \in (a, b)$.

Now, $f(c)$ is a relative maximum or minimum so by the previous theorem, $f'(c) = 0$.

□

Now we come to one of the most important theorems in differential calculus. A geometrical interpretation of the theorem is shown in Figure 4.26. Its proof is a combination of Rolle's theorem and a construction of the "right function".

Theorem 4.12 (Mean Value Theorem)

Let the function f be continuous on the closed interval $[a, b]$ and differentiable at every point on the open interval (a, b). Then there is a point $c \in (a, b)$ such that

$$f'(c) = \frac{f(b) - f(a)}{b - a}.$$

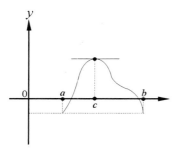

Figure 4.25. Rolle's theorem.   **Figure 4.26. The Mean Value Theorem.**

Proof. The proof is based on applying Rolle's Theorem to the function g constructed out of f as follows,

$$g(x) = f(x) - f(a) - \left(\frac{f(b) - f(a)}{b - a} \right)(x - a).$$

You will get a chance to figure out the "meaning" of g as an exercise.

The definition of g amounts to performing arithmetic operations on f with no division by 0 and so the continuity and differentiability of g is the same as that of f. In particular, g satisfies the first conditions of Rolle's Theorem.

Moreover, we can check that $g(a) = g(b) = 0$. Hence there is a point $c \in (a, b)$ such that $g'(c) = 0$. But we can also compute this derivative. We have,

$$g'(x) = f'(x) - \frac{f(b) - f(a)}{b - a}.$$

Replacing x by c and setting the expression equal to 0 gives,

$$f'(c) = \frac{f(b) - f(a)}{b - a}.$$

$\square$

Applications of the Mean Value Theorem

This very important theorem has many applications, both practical and theoretical. We will look at some of the theoretical applications in the following paragraphs and you will have a chance to make some practical applications in the exercises.

Regarding the latter, here is one example that you might find useful, for example, if you are driving home for vacation.

Part of your trip is by turnpike and you enter at 10:06 am as stamped on the ticket they give you. Down the road about 225 miles, you leave the turnpike and the attendant at the exit stamps your ticket at 1:12 pm. Imagine your consternation when she announces that you have exceeded the speed limit during your trip and will have to pay a fine. Apparently your radar detector did not work at the critical moment!

Actually, there is nothing wrong with your radar detector. It is just not strong enough to counteract the Mean Value Theorem. Can you figure out what happened? The speed limit on the turnpike is 65 mph. You made a trip of 225 miles in 3 hours and 6 minutes. How can the authorities be sure that you must have exceeded the speed limit somewhere in between.

Here is the analysis. Define a function s which gives the distance you traveled from where you entered the turnpike as a function of the time t in hours. The derivative of this function gives your speed at any given time. You started at $t = 10.1$ and ended at $t = 13.2$ According to the Mean Value Theorem, there was at least one time, t_0 at which your speed, $s'(t_0)$ was equal to,

$$\frac{s(13.2) - s(10.1)}{13.2 - 10.1} = \frac{225}{3.1} = 72.58 \, \text{mph}$$

which means that at least one time you were traveling in excess of 72 mph. And anybody who has had a little calculus knows it! Too bad.

Derivatives, Monotonicity and Extreme Points

A function having a positive (negative) derivative is the same as the function being monotonically increasing (decreasing). The latter two terms were defined and exemplified in Chapter 3. This relationship between the derivative and monotonicity should come as no surprise to you at this point since you have been observing and working with this important practical application of a theoretical notion. Now we will formulate the statement in precise mathematical language and prove it. Actually there are four statements. We will state them all and prove two, leaving the others for the exercises. Some parts of the proof rely heavily on the Mean Value Theorem.

Theorem 4.13 (Monotonicity Theorem)

Let f be a function that is differentiable on an open interval (a, b)

1. *If $f'(x) > 0$ for all $x \in (a, b)$ then f is strictly increasing on (a, b).*

2. *If $f'(x) < 0$ for all $x \in (a, b)$ then f is strictly decreasing on (a, b).*

3. *If f is increasing on (a, b) then $f'(x) \geq 0$ for all $x \in (a, b)$.*

4. *If f is decreasing on (a, b) then $f'(x) \leq 0$ for all $x \in (a, b)$.*

Proof.

1. Suppose that c and d are two points in (a, b) with $c < d$. Then we can apply the Mean Value Theorem to the closed interval $c \leq p \leq d$ (Why?) and

$$\frac{f(d) - f(c)}{d - c} = f'(p)$$

CHAPTER 4: DERIVATIVES

for some number $p \in (c, d)$. It follows that $p \in (a, b)$ and so by our hypothesis, $f'(p) > 0$. Hence we have,

$$f(d) - f(c) = f'(p)(d - c) > 0$$

so that $f(d) > f(c)$.

To summarize, we have shown that if $d > c$ then $f(d) > f(c)$. In other words, f is strictly increasing on (a, b).

2. We leave this as an exercise for you.

3. We leave this as an exercise for you.

4. Suppose that f is decreasing on (a, b) and let $c \in (a, b)$. Consider the statement that defines the derivative of f at c.

$$f'(c) = \lim_{h \to 0} \frac{f(c+h) - f(c)}{h}$$

We can see that the values of the fraction are never positive. Indeed, if the denominator $h > 0$, then $c + h > c$ and since f is decreasing, it follows that the numerator $f(c+h) - f(c) \le 0$. Hence the ratio is less than or equal to 0.

Similarly, if $h < 0$ then $c + h < c$ and $f(c+h) - f(c) \ge 0$ so again, the ratio will not be positive. If you take the limit of quantities that are all negative or 0, then the limiting value cannot be positive. (Why?)

$\square$

As a simple consequence of the monotonicity theorem, we can prove a test for monotonicity which you have been using for some time now.

Theorem 4.14 (First derivative test)

Let f be a function that is differentiable on an open interval (a, b), and let $c \in (a, b)$ with $f'(c) = 0$.

1. *If $f'(x) > 0$ for $a < x < c$ and $f'(x) < 0$ for $c < x < b$, then $f(c)$ is a relative maximum.*

2. *If $f'(x) < 0$ for $a < x < c$ and $f'(x) > 0$ for $c < x < b$, then $f(c)$ is a relative minimum.*

Proof. Both of these statements follow from the monotonicity theorem. Indeed, in the first case, that theorem says that f is increasing on the left of c and decreasing on the right of c. Thus c must be a point at which f has a relative maximum. The second statement follows in a similar manner.

$\square$

Second Derivatives and Concavity   In order to establish some precise information about concavity, we need to make the idea itself more precise. It is not quite enough for mathematical analysis to say that a graph will or will not "hold water". This is fine for intuition, but it is also important to relate the intuition to calculations. See Figure 4.27.

See if the following statement agrees with your ideas of concavity.

Definition 4.9

Let f be a function whose graph has a tangent at a point $(c, f(c))$. Then f is said to be concave up *at $(c, f(c))$ if there is an interval centered at c on which the graph is above the tangent line. The function f is said to be* concave down *at the point if there is an interval centered at c on which the graph is below the tangent line.*

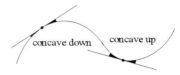

Figure 4.27. Concavity.

You should note that if $y = \alpha x + \beta$ is the equation of the tangent line at the point, then the condition (for concave up, for example) is that

$$f(x) > \alpha x + \beta$$

for all x in the interval, except at $x = c$ where they are equal.

Now look at the following graphs in Figure 4.28a and 4.28b.

Figure 4.28a. f is concave up if $f'' > 0$.   *Figure 4.28b. f is concave down if $f'' < 0$.*

Based on what we can see in Figures 4.28a and 4.28b, we have the following test for concavity which will be proved rigorously below.

Theorem 4.15 (Test for concavity)

Let f be a function which is twice differentiable on an open interval (a, b).

1. If $f''(x) > 0$ for all $x \in (a, b)$ then f is concave up at every point in (a, b).

2. If $f''(x) < 0$ for all $x \in (a, b)$ then f is concave down at every point in (a, b).

Proof.

1. Let $c \in (a, b)$. First we note that the equation of the tangent line at $(c, f(c))$ is $y = f(c) + (x - c)f'(c)$. Hence we must show that for points x sufficiently close to c, but not $x = c$, we have,

$$f(x) > f(c) + (x - c)f'(c).$$

Since the second derivative is positive on (a, b) it follows that the first derivative is strictly increasing on this interval. Hence, if we consider first the case $x > c$ we can apply the Mean Value Theorem to find a point $d \in (x, c)$ with,

$$\frac{f(x) - f(c)}{x - c} = f'(d) > f'(c)$$

and multiplying by (the positive quantity) $x - c$ gives the desired inequality.

A similar argument is used on the other side, if $x < c$.

2. We leave this part as an exercise.

$\square$

A point c in the domain of a function f which has first and second derivatives, except possibly at c, gives rise to a *point of inflection*, or sometimes referred to as an *inflection point* (or simply a "flex" point), $(c, f(c))$ for f if the graph of f changes concavity at c. Sometimes such a point c in the domain is referred to as "an inflection point" of f when we actually mean the inflection point $(c, f(c))$. By Theorem 4.15, it follows that to find the points of inflection c of a function f, we simply find all values of c in the domain of f at which f'' changes sign. To do so, we first find all values of c in the domain of f at which $f''(x) = 0$ or $f''(x)$ is undefined. Then determine where f'' is positive and negative to see where the inflection points are located. For example, consider the function f given by

$$f(x) = x^4 - 4x^3.$$

The second derivative of f is given by

$$f''(x) = 12x^2 - 24x.$$

In this case, f'' is always defined. Factoring, we have

$$f''(x) = 12x(x - 2).$$

and solving for x we get $x = 0$ and $x = 2$. Are either of these inflection points for f? Do you see that $f''(x) > 0$ for $x < 0$ and $x > 2$, and $f''(x) < 0$ for $0 < x < 2$? It follows, that f is concave up for $x < 0$, f is concave down for $0 < x < 2$, and f is concave up for $x > 2$. So, at $x = 0$ and $x = 2$ the function f has points of inflection with coordinates $(0, 0)$ and $(2, -16)$, respectively.

The following test for extrema uses the second derivative. We leave the proof as an exercise.

Theorem 4.16 (Second derivative test)

Let f be a function such that f'' exists on an open interval (a, b), and let $c \in (a, b)$ with $f'(c) = 0$.

1. If $f''(c) > 0$ then $f(c)$ is a relative minimum.

2. If $f''(c) < 0$ then $f(c)$ is a relative maximum.

3. If $f''(c) = 0$, then the test fails and you will have to use the first derivative test or other means to determine whether f has a relative extrema at c.

Inverse Functions — Existence and Derivative

In Activity 4, page 266, you tried to use equations of the form

$$y = x^3 - 4x + 1$$

$$y = x^3 + 4x + 1$$

to define x as a function of y. You may or may not have succeeded. In the given form, these equations can be used without difficulty to define y as a function of x. (Why?) When you try to define x as a function of y, you are looking for the *inverse* of the original function.

Some a functions have inverses and some do not. If you look at the graph shown in Figure 4.29, perhaps you can see why.

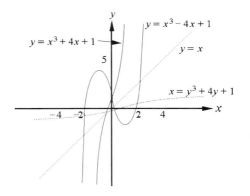

Figure 4.29. Graphs of two cubics with inverse of the first in dotted lines.

Looking at the graph of the second function, you can see that by just reversing your notion of domain and co-domain variables you get another function. You can get an idea of its graph just by reflecting it over the line $y = x$.

But with the first equation, trouble occurs. Because for some values of y there is more than one value of x, you can't tell, given a value of y, what value of x it came from. You can't go back and so you don't have a function. This is the problem of knowing which of several possibilities to go back to. There is also the problem of knowing that there is anything to go back to at all.

We will now try to formulate all of this in general terms and use some of the theory we have developed to get information about a situation in which we might not be able to look at a graph or solve an equation explicitly. To do this, we need the following definitions

---

Definition 4.10

A function f is said to be **one-to-one** *(sometimes written, 1-1) on a set S if, whenever x and y are in S and $x \neq y$, it is the case that $f(x) \neq f(y)$.*

Definition 4.11

A function f is said to be **onto** *a set S if, for every $y \in S$ there is at least one x in the domain of f such that $f(x) = y$.*

It follows from Theorem 4.9 that if f is a continuous function on the closed interval $[a, b]$, then f is onto the interval $[m, M]$ where m is the minimum and M is the maximum of f on $[a, b]$.

Definition 4.12 (Definition of an inverse function)

Let f be a function, D its domain and S its co-domain. We say that a function g with domain S is the **inverse** *of f if $f \circ g$ is the identity function on S and $g \circ f$ is the identity function on D. This means that*

$$f(g(x)) = x \ \text{for every } x \in S,$$

$$g(f(x)) = x \ \text{for every } x \in D.$$

The inverse function g is written f^{-1}.

Warning: Don't confuse the notation f^{-1} for the inverse function of a function f is not to be confused with $\frac{1}{f}$ which represents the reciprocal of f.

A function has an inverse if and only if the function is both 1-1 and onto. We leave the proof as an exercise.

The main tool that we will use to construct inverses of certain functions is the following theorem.

Theorem 4.17 (Existence of inverse functions)

If f is a continuous function on $[a, b]$ and f' exists and never changes sign on $[a, b]$, then f has an inverse which is a function.

Proof. We have already observed that f will be onto an interval $[m, M]$. By Theorem 4.13, if its derivative is always positive it is strictly increasing and if its derivative is always negative it is strictly decreasing. In either case, it cannot come back and repeat any value, so it must be 1-1. $\square$

We can apply this theorem to the second of the two functions in Activity 4. If you calculate the derivative of the function given by $x^3 + 4x + 1$, you get $3x^2 + 4$ which is always positive and hence the original function is (strictly) increasing. That is why the equation $y = x^3 + 4x + 1$ represents a function which has an inverse function. The first, given by $x^3 - 4x + 1$ does not. Its derivative is given by $3x^2 - 4$ which is an expression that will yield both positive and negative values.

The final formula we need for some specific examples is a rule for the derivative of the inverse of a given function. First there is the question of existence. A function satisfying the conditions of the previous theorem will, in fact, have a derivative. Like the Implicit Function Theorem, the proof of this fact is too advanced for the level of this book. We can, however, figure out a formula for the derivative, assuming that it exists.

Theorem 4.18 (Derivatives of inverse functions)

If f is a differentiable function with a differentiable inverse function f^{-1}, then

$$(f^{-1})'(x) = \frac{1}{f'(f^{-1}(x))}.$$

Proof. Let I be the identity function, that is, I is defined by the formula, $I(x) = x$. Then we can write,

$$f \circ f^{-1} = I$$

and, applying the chain rule and the fact that I' is the constant function represented by 1, we have, for any x in the domain of f^{-1},

$$(f \circ f^{-1})'(x) = f'(f^{-1}(x))(f^{-1})'(x) = 1$$

so,

$$(f^{-1})'(x) = \frac{1}{f'(f^{-1}(x))}.$$

$\square$

Natural Exponential and Logarithm Functions

You will recall (page 224) that we somewhat vaguely defined the number e and the natural exponential function exp given by $\exp(x) = e^x$, that is, e, raised to the power x. The number e is approximately equal to 2.71828.

We will use three facts about the natural exponential function. One is that its derivative is equal to itself. The second is that, being a power of a positive number, $\exp(x) > 0$ for all real numbers x. Finally, exp is continuous. Hence, we have,

$$(\exp)'(x) = \exp(x) > 0.$$

Therefore, the function exp is increasing and hence it has an inverse. We call this inverse function the natural logarithm and write ln.

Using the formula of Theorem 4.18, we can compute the derivative of the natural log function. We have,

$$(\ln)'(x) = \frac{1}{\exp'(\ln(x))} = \frac{1}{\exp(\ln(x))} = \frac{1}{x}.$$

Hence,

$$(\ln)'(x) = \frac{1}{x}.$$

Graphs of the natural logarithm function and the natural exponential function are shown in Figure 4.30a. The function represented by $\exp(x) = e^x$ has domain all real numbers, co-domain all positive real numbers, and it is strictly increasing and concave up since $\exp'(x) = \exp''(x) = \exp(x) > 0$ for all x; hence, it follows that its inverse function represented by $\ln x$ has domain all positive real numbers, co-domain all real numbers, and it is strictly increasing and concave down since the graph of $y = \ln x$ is the reflection of the graph of $y = \exp(x)$ about the line $y = x$. Note that the monotonicity and concavity of the function $\ln$ also follow from the fact that the domain of $\ln(x)$ is the set of all real numbers x satisfying $x > 0$, and the first and second derivatives of $\ln$,

$$(\ln)'(x) = \frac{1}{x}$$

and

$$(\ln)''(x) = -\frac{1}{x^2}.$$

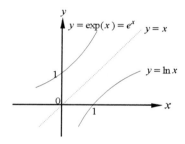

Figure 4.30. Graphs of $y = \ln x$ and $y = \exp(x) = e^x$.

Since the natural logarithm and natural exponential functions are inverses of each other, it follows that the equation $y = \ln x$ is equivalent to the equation $x = \exp(y) = e^y$. Hence, it follows that

$$\ln(\exp(y)) = \ln e^y = y \text{ for all real } y \text{ and } \exp(\ln(x)) = e^{\ln x} = x \text{ for all real } x > 0.$$

The natural exponential function exp has the following properties, which are analogous to the laws of exponents from elementary algebra.

$$\exp(x + y) = \exp(x)\exp(y), \text{ or } e^{x+y} = e^x e^y$$

$$\exp(xy) = (\exp(x))^y = (\exp(y))^x, \text{ or } e^{xy} = \left(e^x\right)^y = \left(e^y\right)^x$$

$$\exp(-x) = \frac{1}{\exp(x)} = (\exp(x))^{-1}, \text{ or } e^{-x} = \frac{1}{e^x} = \left(e^x\right)^{-1}$$

$$\frac{\exp(x)}{\exp(y)} = \exp(x)\exp(-y) = \exp(x - y), \text{ or } \frac{e^x}{e^y} = e^x e^{-y} = e^{x-y}$$

$$\exp(0) = 1 \text{ or } e^0 = 1$$

The natural logarithm function $\ln$ has the following properties which follow from the above properties for the natural exponential function exp and the inverse relationship between the functions exp and $\ln$.

$$\ln(xy) = \ln(x) + \ln(y)$$

$$\ln\left(\frac{1}{x}\right) = \ln(x^{-1}) = -\ln(x)$$

$$\ln\left(\frac{x}{y}\right) = \ln(x) - \ln(y)$$

$$\ln(x^n) = n\ln(x)$$

$$\ln(1) = 0$$

For example, let's show that

$$\ln\left(\frac{1}{x}\right) = \ln(x^{-1}) = -\ln(x).$$

By the inverse relationship between exp and ln, we have $\exp(\ln(x)) = x$ for $x > 0$. It follows that $x^{-1} = (\exp(\ln(x)))^{-1} = \exp(-\ln(x))$ by the third property of exp above. Hence, $x^{-1} = \exp(-\ln(x))$ and using the inverse relationship between ln and exp, we have $\ln(x^{-1}) = \ln(\exp(-\ln(x))) = -\ln(x)$ as desired.

We leave the verification of other properties for the function ln as an exercise.

Inverse Trigonometric Functions   Can we do the same thing about inverses with the trigonometric functions? A quick glance at the graph of the sin function shown in Figure 4.31, for example, shows that it *is* very far from being 1-1, although it is onto the interval $[-1, 1]$. Does the failure to be 1-1 rule out doing *anything*?

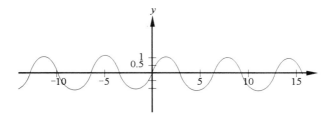

Figure 4.31. A graph of sin(x) with several periods.

Not exactly. Just as we dealt with multiple values in implicit functions by cutting off a branch of the function, we can restrict our attention to just a portion of the sin function. Consider this function, for example, on the interval $\left[-\frac{\pi}{2}, \frac{\pi}{2}\right]$ as shown in Figure 4.32.

The graph certainly suggests that it has an inverse. Let's check Theorem 4.17. We have seen previously that the sin function is continuous and we know that its derivative is the cos function. Now on the interval $\left[-\frac{\pi}{2}, \frac{\pi}{2}\right]$, that is, the first and fourth quadrant, the cos function is positive, except at the endpoints. This means that we only have to check the two endpoints, $-\frac{\pi}{2}$ and $\frac{\pi}{2}$, note that they are the only places where the sin function takes the values -1 and 1 respectively and we can conclude that, on the interval, $\left[-\frac{\pi}{2}, \frac{\pi}{2}\right]$ the sin function has an inverse.

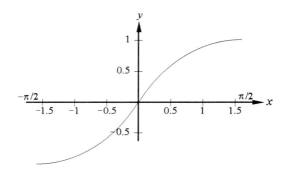

Figure 4.32. A graph of sin(x) on $[-\pi/2, \pi/2]$.

We denote this inverse function arcsin, or $\sin^{-1}$, and refer to the choice of the interval on which it exists as the *principal branch of the* arcsin *function*. Hence, we have that

$$y = \arcsin(x) \Leftrightarrow x = \sin(y) \text{ where } -\frac{\pi}{2} \le y \le \frac{\pi}{2}$$

and the domain of the inverse sine function is the interval $[-1, 1]$. Note that the notation $\Leftrightarrow$ means "is equivalent to". A similar equivalence can be written for each of the other five inverse trigonometric functions with a proper choice of a co-domain for each function. Table 4.1 gives the usual co-domains which have been chosen as convention in mathematics for the inverse trigonometric functions. For a sketch of each inverse trigonometric function use your computer.

Inverse Function Relationship	Co-domain	Domain
$y = \arcsin(x) \Leftrightarrow \sin(y) = x$	$-\frac{\pi}{2} \le y \le \frac{\pi}{2}$	$-1 \le x \le 1$
$y = \arccos(x) \Leftrightarrow \cos(y) = x$	$0 \le y \le \pi$	$-1 \le x \le 1$
$y = \arctan(x) \Leftrightarrow \tan(y) = x$	$-\frac{\pi}{2} < y < \frac{\pi}{2}$	$-\infty < x < \infty$
$y = \text{arcsec}(x) \Leftrightarrow \sec(y) = x$	$0 \le y < \frac{\pi}{2}$ or $\frac{\pi}{2} < y \le \pi$	$-\infty < x \le -1$ or $1 \le x < \infty$
$y = \text{arccsc}(x) \Leftrightarrow \csc(y) = x$	$-\frac{\pi}{2} \le y < 0$ or $0 < y \le \frac{\pi}{2}$	$-\infty < x \le -1$ or $1 \le x < \infty$
$y = \text{arccot}(x) \Leftrightarrow \cot(y) = x$	$0 < y < \pi$	$-\infty < x < \infty$

Table 4.1. The inverse trigonometric functions.

Do you think there are other possible choices of an interval for a co-domain that will give an inverse of the sine function? Can you make other choices of a co-domain that will give an inverse for each of the other five trigonometric functions?

Note that the notation $\sin^{-1}(x)$ for the inverse sine function should <u>not</u> be confused with the fraction $\frac{1}{\sin(x)}$. A similar remark holds for the other five inverse trigonometric functions as well.

We can use Theorem 4.18 to calculate the derivative of the inverse sine function as follows.

$$\arcsin'(x) = \frac{1}{\sin'(\arcsin(x))} = \frac{1}{\cos(\arcsin(x))}.$$

The denominator in the above expression can be described as the cos of an angle (in the first or fourth quadrant, since the co-domain of the inverse sine function is $-\frac{\pi}{2} \le x \le \frac{\pi}{2}$ where cos is positive) whose sin is x. See Figure 4.33. That is,

$$\cos(\arcsin(x)) = \sqrt{1-x^2}$$

and so we can complete the calculation to obtain,

$$\arcsin'(x) = \frac{1}{\sqrt{1-x^2}}.$$

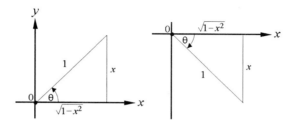

Figure 4.33. $\theta = \arcsin(x)$ *and* $\cos(\arcsin(x)) = \sqrt{1-x^2}$.

In a similar manner you can pick a branch of each trigonometric function on which an inverse exists (such as those in Table 5.1) and compute the derivatives of the inverses of the other five trigonometric functions. We summarize the results here and leave the calculations for the exercises.

$$(\sin^{-1})'(x) = \frac{1}{\sqrt{1-x^2}}$$

$$(\cos^{-1})'(x) = -\frac{1}{\sqrt{1-x^2}}$$

$$(\tan^{-1})'(x) = \frac{1}{1+x^2}$$

$$(\cot^{-1})'(x) = -\frac{1}{1+x^2}$$

$$(\sec^{-1})'(x) = \frac{1}{|x|\sqrt{x^2-1}}$$

$$(\csc^{-1})'(x) = -\frac{1}{|x|\sqrt{x^2-1}}$$

Other Indeterminate Forms and L'Hôpital's Rule

Now that we can calculate the derivatives of many more interesting functions, we will investigate some indeterminate forms not considered in the previous section.

Let's take a look at the following limit problem.

$$\lim_{x \to 0^+} (1+x)^{1/x}$$

This expression is of the indeterminate form 1^∞. What do you intuitively think that the value of the limit is? Did you say it is 1? Some people might think so, since 1 raised to any power is 1. But in this situation we are dealing with "the infinite" which is really a limit process — in this case as x tends to 0 from the right and so the exponent $1/x$ tends to infinity. It turns out that in this situation we can use Theorem 3.7, page 175, from Section 3.3. Theorem 3.7 is as follows.

Let $\lim_{x \to a} g(x) = L$ *and* $\lim_{x \to L} f(x) = K$ *and assume further that* f *is continuous at* L. *Then* $\lim_{x \to a} (f \circ g)(x) = f(\lim_{x \to a} g(x)) = f(L) = K$.

To evaluate the following limit

$$\lim_{x \to 0^+} (1+x)^{1/x}$$

we first need to take care of the problem caused by the infinite limit in the exponent of the expression $(1+x)^{1/x}$. Can you think of a way to do so? How about using the natural logarithm function? Do you see how? Well, we can take care of the infinite limit problem in the exponent by using the natural logarithm function ln, its inverse relationship to exp, and other properties of ln as you will see. Let's let

$$L = \lim_{x \to 0^+} (1+x)^{1/x}$$

and then take the natural logarithm of both sides of the equation to obtain

$$\ln L = \ln \left(\lim_{x \to 0^+} (1+x)^{1/x} \right) = \lim_{x \to 0^+} \ln \left((1+x)^{1/x} \right).$$

We passed the limit through the continuous function ln using Theorem 3.7 above. Using a property of ln (do you see which one?), we obtain

$$\ln L = \lim_{x \to 0^+} \frac{1}{x} \ln(1+x) = \lim_{x \to 0^+} \frac{\ln(1+x)}{x}.$$

Notice that the latter limit is of the form $\frac{0}{0}$ since $\ln(1+x) \to 0$ as $x \to 0^+$. Do you see that the latter limit is 1 by applying L'Hôpital's rule? It is not hard to calculate the latter limit, so we leave it for you. Hence, we conclude that $\ln L = 1$. Do you see how we can get the value of the original limit? Yes, we can use the inverse property of exp and ln once again, and we see that $L = e^1 = e$. Therefore, we have

$$\lim_{x \to 0^+} (1+x)^{1/x} = e.$$

We end this section by making the remark that a method entirely similar to the one above for the indeterminate form 1^∞ can be used to calculate limits of expressions that are of the *indeterminate forms* 0^0 and ∞^0. How would you describe these indeterminate forms in the sense of limits? Can you give a verbal description of such forms? You will get a chance to work out some limits like these in the exercises.

Newton-Raphson Method

We close this section with a brief discussion about a well known and efficient method for approximating zeros of a differentiable function, or equivalently, roots or solutions of some equations, which uses the concept of the derivative.

Figure 4.34 illustrates geometrically a method for finding an approximation to a zero a for a differentiable function f by starting at an initial guess $x_0 \approx a$, where $f'(x_0) \neq 0$. The method is based on the idea of taking the value of x_1, that is, the point where the tangent line at $(x_0, f(x_0))$ crosses the x-axis, to get a "closer" approximation to a. The point x_1 is then used in place of x_0, in a similar manner, to find another approximation x_2 to a. And so on. This *iterative* method is known as the *Newton-Raphson Method*, but is referred to as *Newton's Method*. Do you understand how the method works?

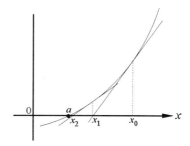

Figure 4.34. Newton-Raphson Method.

Do you see why the point x_1 is given by the following formula?

$$x_1 = x_0 - \frac{f(x_0)}{f'(x_0)}$$

Try the following. Find the value of x where the equation of the tangent line at $(x_0, f(x_0))$ crosses the x-axis, that is, the x-intercept of the tangent line. You should recall that the equation of this tangent line is given by the point-slope formula as follows.

$$y - f(x_0) = f'(x_0)(x - x_0)$$

Did you get the following for the value of the x-intercept of the tangent line?

$$x = x_0 - \frac{f(x_0)}{f'(x_0)}$$

Now we can just rename the x-intercept to be x_1 and we're on our way. What should be true about value of f' at the point x_0? Where can you run into trouble? It is such cases where Newton-Raphson method may fail. Or perhaps another choice of the initial guess might be in order. You will investigate situations where Newton-Raphson method fails in the exercises.

Do you see that a formula for the nth approximation x_{n+1} in terms of the nth approximation x_n is given by the following formula?

$$x_{n+1} = x_n - \frac{f(x_n)}{f'(x_n)}$$

This latter formula is referred to as Newton-Raphson method which was implemented in the **computer function newt** which you used in Activity 5.

Consider the following situation. Suppose that you wanted to find the x-coordinate of the point of intersection (or the coordinates of the point(s) of intersection) of the curve $y = e^{-x}$ and the line $y = x$. See Figure 4.35.

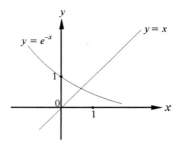

Figure 4.35. Graphs of portions of the curves $y = e^{-x}$ **and** $y = x$.

Do you see how we can find a function for which we can find a zero to solve this problem? Do you see that what we really want to do is find where the graphs of the functions represented by the expressions x and e^{-x} have the same y-coordinate? If you think about it a bit, this is the same as finding the value of x so that

$$e^{-x} = x, \text{ or equivalently } e^{-x} - x = 0.$$

Thus, we want to find (or approximate) the zero of the function f given by

$$f(x) = e^{-x} - x.$$

We now apply Newton-Raphson method. To do so we need an expression for $f'(x)$ which is given by

$$f'(x) = -e^{-x} - 1.$$

From Figure 4.35 we see that the zero for f is in the interval $[0, 1]$ and a reasonable first approximation is $x_0 = 0.5$. Applying Newton-Raphson method with this value of x_0, we get

$$x_1 = 0.5 - \frac{e^{-0.5} - 0.5}{-e^{-0.5} - 1} = 0.5663.$$

Then, using $x_1 = 0.5663$, we have

$$x_2 = 0.5663 - \frac{e^{-0.5663} - 0.5663}{-e^{-0.5663} - 1} = 0.5671.$$

In a similar manner, we can get $x_3 = 0.5671$. Hence, we can conclude that the x-coordinate of the point of intersection of the curves $y = e^{-x}$ and the line $y = x$ is 0.5671 to four decimal places.

We should remark that in order to apply Newton-Raphson method we need to have an interval which contains a zero of a given function. This can usually be determined by using the Intermediate Value Theorem, that is, by finding an interval with endpoints of the opposite sign which isolate the zero as is done in using the **computer function bis** (see Activity 6 on page 267).

4.4.4   Exercises

1. Find the first and second derivatives of the functions represented by each of the following expressions. You may wish to check your answers on your symbolic computer system.

    (a) e^{1-3x}
    (b) e^{-x^2}

(c) $4e^{\sqrt{x}}$

(d) xe^{-x}

(e) $e^{\sin(x)}$

(f) $e^{\tan(3x)}$

(g) $x^2 e^{\cos(x^2)}$

(h) $\cos(x)e^{-3x}$

(i) $\frac{e^{3x}}{x}$

(j) $e^{\sqrt{\sin^2(x)}}$

(k) $\ln(5x)$

(l) $\ln(1-3x^2)$

(m) $x\ln(x)$

(n) $3x^2\sqrt{\ln(x)}$

(o) $\ln(e^{2x})$

(p) $e^{2x}\ln(x^3)$

(q) $\ln(3+2e^x)$

(r) $\ln(\sin(2x))$

(s) $(\arcsin(x^2))^2$

(t) $\arctan(\sqrt{x})$

(u) $x\cos^{-1}(3x)$

(v) $\sin^{-1}(\ln(x))$

(w) $\arctan(e^{2x})$

(x) $\ln(\tan^{-1}(\sin(x)))$

(y) $\frac{\arcsin(x^2)}{x}$

(z) $x\arcsin(\frac{x}{2})+\sqrt{4-x^2}$

2. Assuming that each of the following equations defined a differentiable function of x, find the derivative $y' = \frac{dy}{dx}$ of each implicitly defined function.

(a) $xe^{2y} - 5xy = y^3 + \pi$

(b) $\ln(xy^2)+\dfrac{x}{y} = \sin(y^2)$

(c) $\arctan\left(\dfrac{x}{y}\right) = y^2 + xy - 1$

(d) $2\arctan\left(\dfrac{y}{x}\right) - \ln(x^2+y^2) = 0$

3. In this exercise you will learn how to differentiate functions of the form $f(x)^{g(x)}$ where f and g are differentiable functions of x with $f(x) > 0$.

(a) Using the identity $a^b = e^{\ln a^b}$ for $a > 0$, show that for two functions f and g, wherever $f(x) > 0$,

$$f(x)^{g(x)} = e^{g(x)\ln f(x)}.$$

(b) Show that the derivative of the function h given by $h(x) = f(x)^{g(x)}$ is given by

$$h'(x) = f(x)^{g(x)}\left(\frac{g(x)f'(x)}{f(x)} + g'(x)\ln f(x)\right).$$

(c) Find the derivative of the following functions using the formula in part (b). Check your answer using your SCS.

i. x^x

ii. $x^{\ln(x)}$

iii. $x^{2\sin(3x)}$

iv. $(\cos(x))^{\tan(2x)}$

v. $(\sec(x^2))^{3x}$

4. For each function f,

(a) Sketch a graph of f.

(b) Determine whether or not an inverse function f^{-1} exists. Give reasons for your answer.

(c) If f^{-1} exists, find a formula for $f^{-1}(x)$ and sketch its graph.

(d) Find a formula for $(f^{-1})'$ using Theorem 4.18

i. $f(x) = 2 + 3x$

ii. $f(x) = \frac{5-x}{3x}$

iii. $f(x) = x^2 - 3$

iv. $f(x) = 1 + \sqrt[3]{x+2}$

v. $f(x) = 3\sqrt{2-x}$

vi. $f(x) = 2x^3 + 1$

vii. $f(x) = \sqrt{1-x^2}$, $0 \le x \le 1$

viii. $f(x) = 2x + x^2$

5. Prove the following arithmetic rules for derivatives.

    (a) $(f(x) - g(x))' = f'(x) - g'(x)$

    (b) $(cf(x))' = cf'(x)$

6. Prove the quotient rule for derivatives, assuming f' and g' exist and $g(x) \neq 0$,

$$\left(\frac{f}{g}\right)'(x) = \frac{g(x)f'(x) - f(x)g'(x)}{(g(x))^2}.$$

7. Prove the following derivative rules.

    (a) $\sec'(x) = \sec(x)\tan(x)$

    (b) $\tan'(x) = \sec^2(x)$

    (c) $\cot'(x) = -\csc^2(x)$

    (d) $\csc'(x) = -\csc(x)\cot(x)$

8. Prove the Second Derivative Test for extrema on page 283.

9. Do the following for the monotonicity theorem, Theorem 4.13, on page 279.

    (a) Answer the parenthetical question in the proof of part 1 of Theorem 5.13, i.e., why can we apply the Mean Value Theorem to the interval $[c, d]$?

    (b) Prove part 2 of Theorem 5.13.

    (c) Prove part 3 of Theorem 5.13.

    (d) Explain carefully why the following statement is true.

        If you take the limit of quantities that are all negative (positive) or 0, then the limiting value cannot be positive (negative).

10. Give an example of a function with derivative zero at a point c, but the function does not have a maximum or minimum at c.

11. Prove the second part of Theorem 4.15: Let f be a function that is twice differentiable on an open interval (a, b). If $f''(x) < 0$ for all $x \in (a, b)$, then f is concave down at every point in (a, b).

12. Let f be a function which is continuous on the interval $[a, b]$ and differentiable on (a, b). Let h be the function represented by the graph of the line through the points $(a, f(a))$ and $(b, f(b))$. Show that the function used in the proof of the Mean Value Theorem is $g = f - h$.

13. Prove that for a function f, the existence of an inverse function $g = f^{-1}$ is equivalent to f being 1-1 and onto.

14. Derive the derivative rules on page 289 for the following inverse trigonometric functions.

    (a) $\cos^{-1}$ (b) $\tan^{-1}$ (c) $\sec^{-1}$

    (d) $\cot^{-1}$ (e) $\csc^{-1}$

15. Let h be a continuous function on the interval $[2, 5]$ with $h'(x) \leq 0$ for $2 < x < 5$. Show that $h(5) \leq h(2)$.

16. Let g be a continuous function on the interval $[-1, 3]$ with $g'(x) > 0$ for $-1 < x < 3$. Show that $g(-1) < g(3)$.

17. Use the Mean Value Theorem to derive the following inequalities.

    (a) $|\cos(a) - \cos(b)| \leq |a - b|$ for any real numbers a and b.

    (b) For a differentiable function f with $|f'(x)| \leq M$ for all x,

        $$|f(x) - f(y)| \leq M|x - y|$$

    (c) $|\tan(x) - \tan(y)| \geq |x - y|$ for any real numbers x and y.

(d) For a differentiable function f with $|f'(x)| \geq m$ for all x,

$$|f(x) - f(y)| \geq m|x - y|$$

for any real numbers x and y.

18. Let g and h be differentiable functions for all x, where $g'(x) \leq h'(x)$ for $x > 0$ and $g(0) \leq h(0)$. Show that $g(x) \leq h(x)$ for $x > 0$.

19. Let f be a function and m a real number such that $f'(x) = m$ for all x. Show that f is a linear function, that is, $f(x) = mx + b$ for some real number b.

20. The absolute value function f given by $f(x) = |x|$ is continuous on $[-a, a]$ for any real number $a > 0$ and satisfies

$$\frac{f(a) - f(-a)}{a - (-a)} = 0,$$

yet there is no point $c \in (-a, a)$ where $f'(c) = 0$. Explain why the Mean Value Theorem does not apply in this situation.

21. Use the Mean Value Theorem to show that

(a) If f is a function such that $f'(x) = 0$ for all x in an interval, then f is constant on the interval.

(b) If f and g are functions such that $f'(x) = g'(x)$ for all x in an interval, then there is some constant C such that $f = g + C$ on the interval.

22. Give an example of a function f on an interval $[a, b]$ such that $f(a) = f(b)$, but the Mean Value Theorem (or Rolle's Theorem) does not apply to f.

23. On page 273 you saw that for the equation $x^3 + y^3 - 4.5xy = 0$, the derivative of y with respect to x is given by

$$y' = \frac{dy}{dx} = \frac{4.5y - 3x^2}{3y^2 - 4.5x}$$

where the expression for y' represents the derivative for all branches of the curve which is the graph of the above equation. At what points on the graph of the equation is the denominator of the derivative y' equal to zero? Are there tangent lines at points which are solutions $3y^2 - 4.5x = 0$? Explain what happens at these points where the denominator is zero. Illustrate your answers to these questions on a graph of the equation.

24. Are the following true or false? If true give a proof. If false give a counterexample which shows it is false.

(a) If f and g are 1-1 functions, then $f + g$ is 1-1.

(b) If f and g are onto functions, then $f + g$ is onto.

(c) If f and g are 1-1 functions, then fg is 1-1.

(d) If f and g are onto functions, then fg is onto.

(e) If f and g are 1-1 functions, then $f \circ g$ is 1-1.

(f) If f and g are onto functions, then $f \circ g$ is onto.

25. Use the following properties of the natural exponential function exp to prove the indicated properties of the natural logarithm function.

$$\exp(x + y) = \exp(x)\exp(y)$$
$$\exp(xy) = (\exp(x))^y = (\exp(y))^x$$
$$\exp(-x) = \frac{1}{\exp(x)} = (\exp(x))^{-1}$$
$$\frac{\exp(x)}{\exp(y)} = \exp(x)\exp(-y) = \exp(x - y)$$
$$\exp(0) = 1$$

(a) $\ln(xy) = \ln(x) + \ln(y)$

(b) $\ln\left(\dfrac{x}{y}\right) = \ln(x) - \ln(y)$

(c) $\ln(x^n) = n\ln(x)$

(d) $\ln(1) = 0$

26. Let f be a function which is continuous on a closed interval $[a, b]$ where $f(a) > 0$ and $f(b) < 0$, or $f(a) < 0$ and $f(b) > 0$. Explain why f has at least one zero c with $a < c < b$, that is, $f(c) = 0$.

27. A function f has a fixed point c if $f(c) = c$ for some $c \in [a, b]$. Let f be continuous on an interval $[a, b]$ with co-domain $[a, b]$. Use the Intermediate Value Theorem to prove that f has at least one fixed point in $[a, b]$. (Hint: Consider the function $g(x) = f(x) - x$ and Exercise 25.) Give an example of a function with more than one fixed point on a closed interval.

28. In Activity 3, page 266, you applied a **computer function bis** to the compressibility function Z to get estimates of zeros contained in various intervals. Sometimes **bis** worked and sometimes it didn't. Explain what went wrong in each case that **bis** failed to find a zero in an interval.

29. Is the following statement true or false? If it is true, prove it. If not, then give a counterexample.

Let f be a function that is differentiable on an open interval (a, b), and let $c \in (a, b)$. Then if $f'(c) = 0$, the point $(c, f(c))$ is a relative maximum or minimum point of f.

30. What possible choices of intervals, other than $\left[-\frac{\pi}{2}, \frac{\pi}{2}\right]$, can be used to define an inverse function for the sin function? Give one choice where the interval is the union of two intervals, that is, the interval is disconnected.

31. Show that the following statement is false.

If a function f is continuous at a point, then it is differentiable at the point.

Hence, continuity does not imply differentiability.

32. Determine the continuity and differentiability of the function f given by the following expression,

$$f(x) = \begin{cases} -x^3 & \text{if } x < -1 \\ x^2 & \text{if } -1 \le x < 1 \\ 2x - 1 & \text{if } x \ge 1 \end{cases}$$

33. Make a complete graph of the functions represented by each of the following expressions. Give the xy-coordinates of all extrema and points of inflection, if any.

(a) $x^2 e^{-2x}$ (b) $x\ln(2x)$

(c) $\ln(x^2 + 1)$

(d) $x\arcsin(x) + \sqrt{1 - x^2}$

34. For any real constant $a > 0$, show that

$$\frac{d}{dx}\ln(ax) = \frac{1}{x}.$$

(a) What can you say about the functions $\ln x$ and $\ln(ax)$ for any real constant $a > 0$? Explain your answer.

(b) Use the result of part (a) and the fact that $\ln 1 = 0$ to show that

$$\ln(ab) = \ln a + \ln b$$

for any positive real numbers a and b.

35. The number of bacteria n in a culture at time t is given approximately by the values of the function N where

$n = N(t) = 2000\left(15 + te^{-0.04t}\right)$ for $0 \le t \le 60$.

(a) Find the largest and smallest number of bacteria in the culture during the indicated time interval.

(b) Find the times at which the rate of change of the number of bacteria in the culture is the least and the most.

36. A drug is injected into the bloodstream of a patient and the concentration, c, of the drug in the bloodstream t minutes later is given by the value of the function C where

$$c = C(t) = k\left(e^{-at} - e^{-bt}\right)$$

milligrams per milliliter

and k, a, and b are constants with $k > 0$ and $b > a > 0$. At what time after an injection is the concentration of the drug the greatest? Explain why your answer gives the greatest concentration.

37. Two points P and Q are located opposite each other on the shore of a circular man-made lake with a diameter of two miles. A swimmer wishes to go from P to Q by swimming in a straight line from P to a point R on the shoreline, and then walking from R along the shoreline to Q. The swimmer can swim at a rate of two miles per hour and walk at a rate of five miles per hour. How fast can the swimmer make the trip? Explain why your answer gives the fastest trip.

38. The intensity of a light source located at a point P above the level ground to a point Q on the ground is directly proportional to the sine of the angle of elevation and inversely proportional to the square of the distance from the light source to Q. At what angle of elevation is the intensity of the light the greatest? Explain why your answer gives the greatest intensity.

39. Blood flows into a straight blood vessel of radius r branching off another straight blood vessel with a larger radius R at an angle θ. According to Poiseuille's Law, the total resistance t of the blood in the branching vessel is given by

$$t = T(\theta) = c\left(\frac{a - b\cot(\theta)}{R^4} + \frac{b\csc(\theta)}{r^4}\right)$$

where a, b, and c are constants. Find the angle θ so that the total resistance is the least. Explain why your answer gives the minimum resistance.

40. A rectangular strip of metal 20 feet long is bent to form a gutter by bending up the left and right thirds so that each forms an angle θ with the middle third. Find the angle θ that gives the gutter with the greatest cross-sectional area, that is, find θ so that the gutter flow is optimal. Explain why your answer gives the greatest gutter flow.

41. A plot of land has the shape of a circular section (like a wedge of pie). Flowers are to be planted along the straight line borders of the plot and shrubs are to be planted along the circular arc. The area of the plot is to be 1000 square feet, the flowers cost $6 per linear foot and the shrubs cost $4.50 per linear foot. What is the minimum total cost of the flowers and shrubs? Explain why your answer gives the minimum cost.

42. A fence with height h feet is L feet away from a vertical wall. Find the angle of elevation θ that an adjustable ladder should be propped against the fence so that the length of the ladder from the level ground to the wall is the least possible. Explain why your answer gives the least length of the ladder.

43. For a function p, the values $p(t)$ represent the number of people unemployed t months after the election of a new President. Translate each of the following statements

about the values of the function p and its derivatives into brief written statements which describe the unemployment situation.

(a) $p(0) = 11,000,000$.

(b) $p'(12) = 10,000$.

(c) $p'(24) = -20,000$.

(d) $p'(36) = 0$ and $p''(36) > 0$

44. Find the following limits. Show how you arrived at your answer.

(a) $\lim\limits_{x \to -1} \dfrac{\ln(x^2)}{x+1}$

(b) $\lim\limits_{u \to 1^-} (1-u)^{\ln(u)}$

(c) $\lim\limits_{x \to \infty} (\ln x)^{1/x}$

(d) $\lim\limits_{x \to 0} \dfrac{\ln(1+3x)}{\ln(2x+1)}$

(e) $\lim\limits_{x \to \infty} \dfrac{\ln(x^2+1)}{\ln(x-2)}$

(f) $\lim\limits_{y \to \infty} \dfrac{\ln(1+5y)}{\ln(y^2+3)}$

(g) $\lim\limits_{x \to 0^+} x^x$

(h) $\lim\limits_{x \to \infty} x^{2x}$

(i) $\lim\limits_{x \to \infty} x^{1/x}$

(j) $\lim\limits_{n \to \infty} \left(1 + \dfrac{1}{\sqrt{n}}\right)^n$

(k) $\lim\limits_{x \to \infty} \left(1 + \dfrac{3}{x}\right)^{2x}$

(l) $\lim\limits_{x \to \infty} \left(1 + \dfrac{a}{x}\right)^{bx}$

(m) $\lim\limits_{x \to 0^+} (1+2x)^{1/x}$

(n) $\lim\limits_{x \to \infty} (1+x^2)^{1/x}$

(o) $\lim\limits_{y \to \infty} \sqrt{\left(1 + \dfrac{1}{2y}\right)^y}$

(p) $\lim\limits_{v \to \infty} (e^v + v)^{1/v}$

(q) $\lim\limits_{x \to \infty} (1 - e^{-x})^{e^x}$

45. Suppose that P dollars is invested at an annual interest rate of r percent and it is compounded n times per year. The amount a, in dollars, that the investment of P dollars is worth after t years when interest is compounded n times per year is given by the *compound interest* formula

$$a = A(t) = P\left(1 + \dfrac{r}{n}\right)^{nt}.$$

(a) Use L'Hôpital's rule to find a formula which gives the amount a that an investment of P dollars is worth after t years if interest is compounded continuously. That is, find the *continuous interest* formula which is obtained by taking the limit as $n \to \infty$ of the compound interest formula. Explain why you think this gives the continuous compound interest formula.

(b) How much is an investment of $10,000 worth after five years if interest is compounded monthly?

(c) How much is an investment of $10,000 worth after five years if interest is compounded continuously? Compare your answer with your answer in part (b).

46. Run the **computer function newt** as in Activity 5 on the function represented by each of the following expressions with the given starting point **x0** and **eps** $= 10^{-6}$. Evaluate the function at your final estimate for the zero. What should be true about the value of each function at your final estimate? How should this value compare to $f(x_0)$?

(a) $\sin(2x) - \tan(x)$, **x0** $= -9.5$.

(b) $e^{-x} + \dfrac{x^2}{2} - \dfrac{x^3}{6}$, **x0** $= 4$. (Note that for e^x you must use exp(x) in most MPL's.)

47. Run the **computer function newt** as in Activity 5 on the function given by the expression xe^{-x} with initial point **x0** = **2** and **eps** $= 10^{-6}$. Try to figure out what caused the result you obtained. If necessary, sketch the graph of the function and use your graphics utility to zoom in appropriately on the graph to investigate what is happening. Explain what you think is going on.

(a) $x^2 - 3$

(b) $x - \sin(x)$

(c) The function given by the expression $e^x - 3.62x^3 - 4.13x^2 + 13.24x + 13.11$. (Note that for e^x you must use $\exp(x)$ in most MPL's.)

(d) $\sin(x) - \cos(x)$ on the interval $[-\pi, \pi]$.

(e) $\cos(2x) - x^2$

(f) The data function **data30**.

49. In each of the following situations explain what happens if you try to use Newton-Raphson method to solve for the indicated zero a using the initial guess x_0. Make a sketch and show on your sketch what is happening. Indicate on your sketch the point farthest from a that you can start so that Newton-Raphson method fails to converge a and the point farthest from a so that the method will converge to a.

(a)

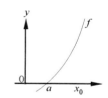

(b)

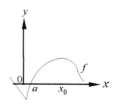

50. For the function represented by each of the following expressions, find all local maxima and minima on the given intervals by approximating the zeros of the derivatives and checking the graphs or nearby points to decide if they are a maximum or a minimum or neither. Use a

computer function newt as in Activity 5 to make your approximations with **eps** $= 10^{-6}$.

(a) $e^x - 3.62x^3 - 4.13x^2 + 13.24x + 13.11$, $[-3, 7.5]$

(b) $\sin(x) - \cos(x)$, $[0, 2\pi]$

(c) The data function **data30**, $[-1.1, 1.1]$.

51. A container is to be formed by placing a hemisphere with radius r inches on the top of a cylinder with the same radius and a height of 5 inches. Find the radius so that the total volume is 150 in^3. (The volume of a sphere with radius r is $4\pi r^3/3$ and the volume of a cylinder with radius r and height h is $\pi r^2 h$.) Approximate the zero of an appropriate equation to solve this problem in the following ways.

(a) The **computer function bis** from Activity 6, Section 2, page 218 with **eps** $= 10^{-3}$.

(b) The **computer function newt** as in Activity 5 with **eps** $= 10^{-3}$.

52. Find all (x, y) coordinates for all points of intersection on the graphs of each of the following pairs of curves using

(a) The **computer function bis** from Activity 6, Section 2, page 218 with **eps** $= 10^{-6}$.

(b) Newton-Raphson method by hand using a calculator to two decimal places.

(c) A **computer function newt** as in Activity 5 with **eps** $= 10^{-6}$.

    i. $f(x) = 2\cos(x)$ and $g(x) = \sin(x)$ for $0 \le x \le \pi$.

    ii. $f(x) = \cos(x)$ and $g(x) = x^2$.

53. A *fixed point* of a function f is a value x in the domain of f such that $f(x) = x$. For each of the following functions approximate

all fixed points by solving the equation $f(x) = x$ for x in each of the following ways.

(a) Newton-Raphson method by hand using a calculator to two decimal places.

(b) The **computer function bis** from Activity 6, Section 2, page 218 with $\text{eps} = 10^{-4}$.

(c) The **computer function newt** as in Activity 5 with $\text{eps} = 10^{-4}$.

   i. $f(x) = 3\sin(x)$

   ii. $f(x) = \cos(2x)$

   iii. $h(x) = x^3 - x + 2$

▎4.5 CHAPTER SUMMARY

In this chapter you have considered numerous concepts related to derivatives and problem solving. The following is a summary of the main concepts and problem solving tools you have studied in this chapter.

Formal Definitions and More

Derivative at a point. If f is a function and a is a point in its domain, the *derivative* of f at a is,

$$f'(a) = \lim_{h \to 0} \frac{f(a+h) - f(a)}{h}$$

provided this limit exists.

In alternative notations, we have

$$f'(a) = \lim_{\Delta x \to 0} \frac{f(a + \Delta x) - f(a)}{\Delta x} \quad \text{and} \quad f'(a) = \lim_{x \to a} \frac{f(x) - f(a)}{x - a}.$$

A function f is said to be *differentiable at* a if $f'(a)$ exists. A function f is differentiable on a set S, if f is differentiable at each point of S. A function f is *differentiable*, if $f'(x)$ exists at each point x in the domain of f.

The derivative of a function is a function. Let f be a function. The *derivative* f' of f is defined to be the function whose domain is the set of points x in the domain of f at which the derivative exists and, at such a point, the value of f' at x is given by

$$f'(x) = \lim_{h \to 0} \frac{f(x+h) - f(x)}{h}.$$

Definition of vertical tangent. If f is a continuous function and a is a point in its domain, then f has a *vertical tangent* at a if,

$$\lim_{x \to a} |f'(x)| = \infty.$$

Definition of an inverse function. Let f be a function, D its domain and S its co-domain. We say that a function g with domain S is the *inverse* of f if $f \circ g$ is the identity function on S and $g \circ f$ is the identity function on D. This means that

$$f(g(x)) = x \text{ for every } x \in S \text{ and}$$

$$g(f(x)) = x \text{ for every } x \in D.$$

The inverse function g is written f^{-1}.

Warning: The notation f^{-1} for the inverse function of a function f is <u>not</u> to be confused with $\frac{1}{f}$ which represents the reciprocal of f.

Natural exponential and logarithm functions. The functions exp, given by e^x, and $\ln = \log_e$, given by $\ln(x) = \log_e(x)$ are inverses of each other, i.e.,

$$\exp(\ln(x)) = x \text{ for } x > 0$$

$$\ln(\exp(x)) = x \text{ for and real number}$$

$$y = \ln(x) = \log_e(x) \text{ is equivalent to } e^y = x$$

The natural exponential function exp has the following properties, which are analogous to the laws of exponents from elementary algebra.

$$\exp(x + y) = \exp(x)\exp(y), \text{ or } e^{x+y} = e^x e^y$$

$$\exp(xy) = (\exp(x))^y = (\exp(y))^x, \text{ or } e^{xy} = \left(e^x\right)^y = \left(e^y\right)^x$$

$$\exp(-x) = \frac{1}{\exp(x)} = (\exp(x))^{-1}, \text{ or } e^{-x} = \frac{1}{e^x} = \left(e^x\right)^{-1}$$

$$\frac{\exp(x)}{\exp(y)} = \exp(x)\exp(-y) = \exp(x-y), \text{ or } \frac{e^x}{e^y} = e^x e^{-y} = e^{x-y}$$

$$\exp(0) = 1 \text{ or } e^0 = 1$$

The natural logarithm function ln has the following properties which follow from the above properties for the natural exponential function exp and the inverse relationship between the functions exp and ln.

$$\ln(xy) = \ln(x) + \ln(y)$$

$$\ln\left(\frac{1}{x}\right) = \ln\left(x^{-1}\right) = -\ln(x)$$

$$\ln\left(\frac{x}{y}\right) = \ln(x) - \ln(y)$$

$$\ln\left(x^n\right) = n\ln(x)$$

$$\ln(1) = 0$$

Definitions of the inverse trigonometric functions.

Inverse Function Relationship	Co-domain	Domain
$y = \arcsin(x) \Leftrightarrow \sin(y) = x$	$-\frac{\pi}{2} \leq y \leq \frac{\pi}{2}$	$-1 \leq x \leq 1$
$y = \arccos(x) \Leftrightarrow \cos(y) = x$	$0 \leq y \leq \pi$	$-1 \leq x \leq 1$
$y = \arctan(x) \Leftrightarrow \tan(y) = x$	$-\frac{\pi}{2} < y < \frac{\pi}{2}$	$-\infty < x < \infty$
$y = \text{arcsec}(x) \Leftrightarrow \sec(y) = x$	$0 \leq y < \frac{\pi}{2}$ or $\frac{\pi}{2} < y \leq \pi$	$-\infty < x \leq -1$ or $1 \leq x < \infty$
$y = \text{arccsc}(x) \Leftrightarrow \csc(y) = x$	$-\frac{\pi}{2} \leq y < 0$ or $0 < y \leq \frac{\pi}{2}$	$-\infty < x \leq -1$ or $1 \leq x < \infty$
$y = \text{arccot}(x) \Leftrightarrow \cot(y) = x$	$0 < y < \pi$	$-\infty < x < \infty$

Basic Derivative Rules

Constant rule.

$$\frac{d}{dx} c = 0.$$

If f is a function with constant values, i.e., $f(x) = c$ for some constant c, then $f'(x) = 0$.

Power of a variable rule.

$$\frac{d}{dx} x^n = nx^{n-1}, \; n \text{ a constant}$$

Sum and difference rules.

$$\frac{d}{dx}(f(x) \pm g(x)) = (f \pm g)'(x) = f'(x) \pm g'(x)$$

Constant multiple rule.

$$\frac{d}{dx}(cf(x)) = (cf)'(x) = cf'(x)$$

Product rule.

$$\frac{d}{dx}(f(x)g(x)) = (fg)'(x) = f(x)g'(x) + f'(x)g(x)$$

Quotient rule.

$$\frac{d}{dx}\left(\frac{f(x)}{g(x)}\right) = \left(\frac{f}{g}\right)'(x) = \frac{g(x)f'(x) - f(x)g'(x)}{(g(x))^2}$$

Chain rule (composition rule).

$$\frac{d}{dx} f(g(x)) = (f \circ g)'(x) = f'(g(x))g'(x)$$

Basic trigonometric function rules.

$$\frac{d}{dx}\sin(x) = \sin'(x) = \cos(x)$$

$$\frac{d}{dx}\cos(x) = \cos'(x) = -\sin(x)$$

$$\frac{d}{dx}\tan(x) = \tan'(x) = \sec^2(x)$$

$$\frac{d}{dx}\sec(x) = \sec'(x) = \sec(x)\tan(x)$$

$$\frac{d}{dx}\csc(x) = \csc'(x) = -\csc(x)\cot(x)$$

$$\frac{d}{dx}\cot(x) = \cot'(x) = -\csc^2(x)$$

Natural exponential, exp, and logarithm, $\ln = \log_e$, function rules.

$$\frac{d}{dx}\exp(x) = \exp'(x) = \exp(x)$$

$$\frac{d}{dx}\ln(x) = \ln'(x) = \frac{1}{x}$$

Inverse trigonometric function rules.

$$\frac{d}{dx}\arcsin(x) = \arcsin'(x) = \frac{1}{\sqrt{1-x^2}}$$

$$\frac{d}{dx}\arccos(x) = \arccos'(x) = -\frac{1}{\sqrt{1-x^2}}$$

$$\frac{d}{dx}\arctan(x) = \arctan'(x) = \frac{1}{1+x^2}$$

$$\frac{d}{dx}\operatorname{arccot}(x) = \operatorname{arccot}'(x) = -\frac{1}{1+x^2}$$

$$\frac{d}{dx}\operatorname{arcsec}(x) = \operatorname{arcsec}'(x) = \frac{1}{|x|\sqrt{x^2-1}}$$

$$\frac{d}{dx}\operatorname{arccsc}(x) = \operatorname{arccsc}'(x) = -\frac{1}{|x|\sqrt{x^2-1}}$$

Note: the notation $\sin^{-1}$ is sometimes used for the inverse sine function arcsin should <u>not</u> be confused with the fraction $\frac{1}{\sin(x)}$. A similar remark holds for the other five inverse trigonometric functions as well.

Implicit differentiation. For an equation in two variables, say x and y, which defines a differentiable function y of x, to find the derivative of y with respect to x implicit differentiation, do the following:

1. Differentiate the entire equation with respect to x using basic differentiation rules.

2. Solve the resulting equation for $\frac{dy}{dx}$.

Functions defined in parts. If a function is defined in parts, then you have to compute its derivative separately on each part and deal with the seams separately. To compute the derivative at the seam use the limit definition of the derivative, i.e., form the difference quotient and take its limit. This may or may not exist and therefore, the derivative function may or may not be defined at the seams. The derivative will also be a function defined in parts.

Derivatives and Graphing
The following issues must be considered in using the graph of a function to display as much information as possible.

Intercepts. Points at which the graph intersects the x-axis and the y-axis. They are found by setting one of the variables equal to zero and solving for the other.

Domain. This is the set of values at which the function is defined. For example, points at which even or non-fractional roots are applied to negative numbers, or which lead to division by zero are not in the domain of the function and hence there will be no points on the graph corresponding to those values.

Vertical asymptotes. If there is a point a at which evaluation of an expression defining a function f leads to division by zero, then near this point the graph may become arbitrarily large in the positive or negative direction (or both) near a. If the graph of f becomes arbitrarily large near a, i.e., the limit from the right and/or left at a approach ∞ or $-\infty$, then the line $x = a$ is a vertical asymptote of the graph of f.

Horizontal asymptotes. If the limit L of the function as the independent variable goes to ∞ or $-\infty$ exists, then the horizontal line $y = L$ is a horizontal asymptote to the graph of f.

Extreme points. The only points at which the graph of a function will reach a relative maximum or minimum are points in the domain of the function at which the value of its derivative is 0 or undefined. Such points are called *critical points*.

Monotonicity. A function whose derivative exists at every point on an interval will be increasing (decreasing) on this interval if the values of the derivative are positive (negative) at every point on the interval.

Shape. Speaking somewhat loosely, we say that the graph of a function is *concave up* on an interval if it is shaped in such a way that "it will hold water" and it is *concave down* on an interval if it is shaped in such a way that "it will not hold water". The following test is a simple test for concavity which involves the second derivative.

> If a function has a first and a second derivative on an interval, then, if the
> second derivative is positive (negative) at every point on the interval, the graph
> is concave up (down) on that interval.

Related to concavity are points at which a graph of a function changes concavity. A point at which the graph of a function changes from one kind of concavity to the other is called an *inflection point*. Such points in the domain of a function can only occur where the values of the second derivative are 0 or undefined.

Symmetry. If a function f has the property that $f(x) = f(-x)$ for all x in its domain, then the graph is symmetric with respect to the y-axis and it is referred to as an *even function*. If it satisfies, $f(-x) = -f(x)$ for all x in its domain, then it is symmetric with respect to the origin and it is referred to as an *odd function*.

L'Hôpital's Rule. Let f and g be two functions which are differentiable in an open interval (a, b) containing the point c.

1. If $\lim_{x \to c} f(x) = 0$ and $\lim_{x \to c} g(x) = 0$, i.e., $\frac{f(x)}{g(x)}$ is of the indeterminate form $\frac{0}{0}$ at c, then if $g'(x) \neq 0$, except possibly at c,

$$\lim_{x \to c} \frac{f(x)}{g(x)} = \lim_{x \to c} \frac{f'(x)}{g'(x)}$$

provided the latter limit exists.

2. If $\lim_{x \to c} f(x) = \infty$ and $\lim_{x \to c} g(x) = \infty$, i.e., $\frac{f(x)}{g(x)}$ is of the indeterminate form $\frac{\infty}{\infty}$ as x approaches c, then

$$\lim_{x \to c} \frac{f'(x)}{g'(x)} = \lim_{x \to c} \frac{f(x)}{g(x)}.$$

Newton-Raphson Method. The Newton-Raphson iterative method for approximating zeros of a differentiable function f is as follows.

$$x_{n+1} = x_n - \frac{f(x_n)}{f'(x_n)}$$

where an initial approximation x_0 is a number such that $f(x_0) \approx 0$.

Important Theorems on Differentiability and Continuity

Differentiability implies continuity. *If a function f is differentiable at the point a, then it is continuous at a.*

Intermediate value theorem. *Let f be a function which is continuous on the closed interval $[a, b]$. Then for every value d which is between $f(a)$ and $f(b)$, there is a number $c \in [a, b]$ such that $f(c) = d$.*

Extreme value theorem. *Let f be a function which is continuous on the closed interval $[a, b]$. Then there are numbers $c, d \in [a, b]$ such that $f(c)$ is a maximum value, that is $f(c) \geq f(x)$ for all $x \in [a, b]$ and $f(d)$ is a minimum value, that is $f(d) \leq f(x)$ for all $x \in [a, b]$.*

Note that the preceding two theorems imply the following:

Let f be a function which is continuous on the closed interval $[a, b]$. Then the co-domain of f, that is, the set of values $\{f(x) : x \in [a, b]\}$ is an interval $[m, M]$. The value m is a minimum and the value M is a maximum of f on the interval $[a, b]$. The function f takes on every value from m to M, including m and M.

Rolle's theorem. *Let the function f be continuous on the closed interval $[a, b]$ and differentiable at every point on the open interval (a, b). If $f(a) = f(b)$ then there is a point $c \in (a, b)$ such that $f'(c) = 0$.*

Mean value theorem. *Let the function f be continuous on the closed interval* $[a, b]$ *and differentiable at every point on the open interval* (a, b). *Then there is a point* $c \in (a, b)$ *such that*

$$f'(c) = \frac{f(b) - f(a)}{b - a}.$$

Monotonicity — increasing and decreasing functions. *Let f be a function that is differentiable on an open interval* (a, b)

1. *If* $f'(x) > 0$ *for all* $x \in (a, b)$ *then f is strictly increasing on* (a, b).

2. *If* $f'(x) < 0$ *for all* $x \in (a, b)$ *then f is strictly decreasing on* (a, b).

3. *If f is increasing on* (a, b) *then* $f'(x) \geq 0$ *for all* $x \in (a, b)$.

4. *If f is decreasing on* (a, b) *then* $f'(x) \leq 0$ *for all* $x \in (a, b)$.

First derivative test for extrema. *Let f be a function that is differentiable on an open interval* (a, b), *and let* $c \in (a, b)$ *with* $f'(c) = 0$.

1. *If* $f'(x) > 0$ *for* $a < x < c$ *and* $f'(x) < 0$ *for* $c < x < b$, *then* $f(c)$ *is a relative maximum.*

2. *If* $f'(x) < 0$ *for* $a < x < c$ *and* $f'(x) > 0$ *for* $c < x < b$, *then* $f(c)$ *is a relative minimum.*

Second derivative test for concavity. *Let f be a function which is twice differentiable on an open interval* (a, b).

1. *If* $f''(x) > 0$ *for all* $x \in (a, b)$ *then f is concave up at every point in* (a, b).

2. *If* $f''(x) < 0$ *for all* $x \in (a, b)$ *then f is concave down at every point in* (a, b).

Second derivative test for extrema. *Let f be a function such that* f'' *exists on an open interval* (a, b), *and let* $c \in (a, b)$ *with* $f'(c) = 0$.

1. *If* $f''(c) > 0$ *then* $f(c)$ *is a relative minimum.*

2. *If* $f''(c) < 0$ *then* $f(c)$ *is a relative maximum.*

3. *If* $f''(c) = 0$, *then the test fails and you will have to use the first derivative test or other means to determine whether f has a relative extrema at c.*

Existence of derivatives of inverse functions. *If f is a continuous function on* $[a, b]$ *and* f' *exists and never changes sign on* $[a, b]$, *then f has an inverse which is a function.*

Derivatives of inverse functions. *If f is a differentiable function with a differentiable inverse function f^{-1}, then*

$$(f^{-1})'(x) = \frac{1}{f'(f^{-1}(x))}.$$

Conclusion   You now have the basic concepts and computational skills of differential calculus at your disposal. You should be able to easily find the derivative of just about *any* function by hand or by using a symbolic computer system. The geometrical and physical interpretations of the derivative, and related concepts, provide you with many powerful tools for problem solving with derivatives. We now move on to Chapter 5 to study the second branch of the calculus called *integral calculus*. The primary operation used in that branch of calculus is called *integration*. Don't think that you can forget about derivatives however — there is a subtle, yet beautiful and powerful relationship between the operations of differentiation and integration. That relationship is one of the most fundamental in all of mathematics.

CHAPTER
5

INTEGRALS

▌5.1 INDEFINITE INTEGRALS

5.1.1 Overview

In this chapter you will study the basic concepts of the second broad branch of calculus known as *integral calculus*. The primary process of integral calculus is called *integration*.

In Chapter 4, a typical problem required you to analyze a situation, come up with a function which described the situation and then analyze the function and/or its derivatives to solve the problem. In a typical problem situation in this chapter, you will know something about the derivative of a function, and you will want to find out about the properties of the function. Moreover, in many problems you will want to "recover" a function from its derivative. You will see that, in a certain sense, the process of integration is a *reversal* of the process of differentiation, and vice-versa. As you will learn, reversals of these two processes (as for many processes in mathematics) turn out to be very powerful problem solving tools. In fact, the process of finding a function from its derivative is often referred to as *anti-differentiation*, or *indefinite integration*.

In Section 1 you will concentrate on indefinite integration as the reversal of the differentiation process. You will learn basic rules for integration (in a sense reversals of the rules you learned for differentiation) and various properties of *indefinite integrals*.

5.1.2 Activities

1. A curve has slope $2x+1$ for each x. Find the equation of the curve in the form $y = f(x)$. Use your SCS or do this problem by hand if you can. Is this solution unique? That is, is there only one function $y = f(x)$ which has slope $2x+1$ for each x? If there are more solutions find them. If the one solution is unique, explain why.

2. If wind resistance and other complicating factors are ignored, an object in freefall has a constant acceleration of $32 \, \mathrm{ft}/\mathrm{sec}^2$. If we define the positive direction for height to be up, and the negative direction down, this tells us that acceleration is a constant function of time, namely $a(t) = -32$.

   (a) Find the velocity of the object as a function of time.

   (b) Find the position of the object as a function of time.

3. According to the Fermi-Dirac distribution, under certain conditions, the number of electrons at a certain energy level is a function N of the energy level E. According to one version of this model, N is a function whose derivative is $E^{\frac{1}{2}}$. We write such a function as

$$N = \int E^{\frac{1}{2}} dE = \int \sqrt{E}\, dE.$$

This is called the indefinite integral of the function represented by $\sqrt{E}$ and it represents (the family of) all functions N for which

$$N' = \frac{dN}{dE} = E^{\frac{1}{2}}.$$

Use your SCS to find such a function. Check your answer by differentiating. Note: in some SCSs (including **MapleV**) the letter E represents the natural base $e = 2.71828...$, so you may need to choose a different name for the independent variable.

Find at least 5 such functions and graph them on a single coordinate system.

4. Just as with differentiation, there are rules for indefinite integration. Each of the following items contains one or more expressions representing functions for which indefinite integrals can be found by applying specific rules. For each item, use your SCS to find anti-derivatives, that is, functions whose derivatives are given by the expressions in the item. Then try to formulate a rule suggested by that item. As you will see, an SCS will give you only one anti-derivative for each function. Do you see how to find all anti-derivatives for a given function?

   (a) 0

   (b) $1, -3, 162, \pi$

   (c) $x, x^2, \sqrt{x}, x^{-3}, \sqrt[3]{x^5}, x^{-1.4}$

   (d) $\sin(x), \cos(y), \sec^2(t), \sec(w)\tan(w)$

   (e) $3u^2 \sqrt{u^3 - 1}, y\cos(y^2), \frac{\cos(x)}{\sqrt{\sin x + 17}}$

   (f) $\frac{1}{x}, (1+3r)^{-1}, \frac{\sin(x)}{\cos(x)}, e^x, e^{2x}, 2xe^{-x^2}$

   (g) $x\sin(x), ue^u, r\ln(r), \arcsin(t)$

5. In addition to the rules for integration, there are a number of properties that hold for indefinite integrals. They can also be used as rules to help you compute integrals. Each of the following items illustrates a particular property. Use your SCS to compute the indicated integrals and then try to guess the property. Try to imagine why your property should hold.

   (a) $\int 5x\, dx, 5\int x\, dx, 3\int x^2\, dx, \int 3x^2\, dx$ (b) $\int x^{\frac{3}{2}}\, dx, \int x^{-1.5}\, dx, \int \left(x^{\frac{3}{2}} + x^{-1.5}\right) dx$

   (c) $\int (1+x)\, dx, \int x^2\, dx, \int \left(3 + 3x + 4.2x^2\right) dx$

   (d) $\int x^{\frac{3}{2}}\, dx, \int \sqrt{x}\, dx, \int \left(2x^{\frac{3}{2}} - 0.005\sqrt{x}\right) dx$

6. In Economics, the notion of present value is used to analyze decisions in which it is necessary to choose between spending money immediately, or over a long period of time. For example: Should you tear down a building and replace it with a new one or should you maintain it over several decades? The immediate cost of demolition and construction must be compared with some concept of the "present value" of the money to be spent over time. Of course, the interest accrued over time must also be considered.

If M is a function of time which gives the amount of money spent from the beginning to time t, and interest is compounded continuously at the rate r, then a reasonable model for the present value V of the money spent from the beginning to time t is a function whose derivative is given by

$$M'(t)e^{-rt}.$$

We denote such a function by

$$V = \int M'(t)e^{-rt}\, dt.$$

This is called the indefinite integral of the function represented by $M'(t)e^{-rt}$. It represents (the family of) all functions V for which

$$V' = \frac{dV}{dt} = M'(t)e^{-rt}.$$

Use your SCS to find an expression for such a function, for a given function M with interest rate r and also for the special case in which $M(t) = t^2$, $r = 12\% = 0.12$.

Check your answer by differentiating.

In fact, there is more than one function which satisfies the condition of this problem. Find at least 5 such functions for the special case ($M(t) = t^2$, $r = 12\% = 0.12$) and graph them on a single coordinate system.

7. Suppose you wish to anti-differentiate the function f given by $f(x) = 2\sin(x)\cos(x)$. On the one hand, you feel that the anti-derivative should be $\sin^2(x)$, since the derivative of the function represented by $\sin^2(x)$ is the function represented by $2\sin(x)\cos(x)$. On the other hand, you feel that the anti-derivative should be $-\cos^2(x)$, since the derivative of this is also $2\sin(x)\cos(x)$. Which is correct? Explain what is going on here.

8. Suppose f_1 and f_2 are two functions defined on the interval $[a, b]$ such that f_1 and f_2 have identical derivatives on $[a, b]$.

   (a) How are f_1 and f_2 related?

   (b) Further suppose that $f_1(c) = f_2(c)$ for some c in $[a, b]$. Can you now say more about the relationship between f_1 and f_2?

9. Consider the current function given by

$$i(t) = \begin{cases} a & \text{if } 0 \le t < b \\ 0 & \text{if } t \ge b \end{cases}$$

where a is a constant.

Show that any voltage function whose derivative is the current function i is given by

$$v(t) = \begin{cases} at + k_1 & \text{if } 0 \le t < b \\ k_2 & \text{if } t \ge b \end{cases}$$

where k_1 and k_2 are constants.

For what values of the constants k_1 and k_2, if any, is the voltage function continuous at $t = b$? For what values of k_1 and k_2, if any, is v differentiable at $t = b$? Give reasons for your answer.

5.1.3   Discussion

Finding Functions with Given Derivatives   Look at the first three activities, step back and ask yourself, what do they have in common? What differences can you see? Of course they are all different in that one is about the slope of a curve, one is about motion of an object, and one is about electrons. These are the areas in which the mathematics is being applied. Try to look beyond the applications and see what mathematics is involved. What single mathematical object is common to all three of these activities?

If you said *function* that would be good. If you said *unknown function* it would be better, and if you said *unknown functions* (and were conscious of the difference) it would suggest that you might have seen this before — or thought ahead about things.

In each one of these activities, you are looking for unknown functions. Of course, they are not completely unknown — otherwise, how would you find them? In fact, in each case, the function has been specified, not by giving a description of its process, but by giving full information about its derivative.

Indeed, let's formulate each of the problems in the first three activities in terms of mathematics alone, stripping away the context of the application. For this purpose we will call the unknown function f, its independent variable x and its dependent variable y. Here they are, but not in the same order as the activities. Can you make the correspondence?

1. $f'(x) = -32$ 2. $f'(x) = \sqrt{x}$ 3. $f'(x) = 2x + 1$

So you can see that, in each of the first three activities, there is a formula, not for the function, but for its derivative. The problem, which is really the issue for this entire chapter is to use the information given about the derivative of a function to find the function itself.

You will notice that in the right hand sides of the three equations above, the independent variable x is the only variable involved. A problem of this kind in which the right hand side depends only on the independent variable is called a problem of *indefinite integration* or *anti-differentiation*. A problem which involves an equation containing an unknown function and/or its derivative(s), we call a *differential equation* problem. Problems like the following: Find a function f so that $f' = g$, where g is a known function, are the simplest type of differential equations. More complicated types of differential equations along with the applications which give rise to them will be considered in Section 5.6.

Anti-Derivatives and the Indefinite Integral   Consider once again the three anti-derivative problems from the activities.

1. $f'(x) = -32$ 2. $f'(x) = \sqrt{x}$ 3. $f'(x) = 2x + 1$

We call such a problem by the name *anti-derivative* because it is the reverse of differentiation. To solve it, you have to ask yourself, what function can I differentiate to get the function represented by -32? or $\sqrt{x}$ or $2x+1$? This is analogous to subtraction, or division. You might want to call subtraction *anti-addition* and division *anti-multiplication*.

Your SCS might have given you, as an answer to $f'(x) = 2x+1$ in Activity 1, the following specification of the process of the unknown function f,

$$f(x) = x^2 + x$$

You can take the derivative of this function and see right away that it certainly has the specified derivative. So it is a solution to the problem. Is it the only solution? What about

$$f(x) = x^2 + x + 37$$

Can you see that this is a solution as well? Can you think of others? Can you determine what all the solutions must be?

The truth is that once you have a function which solves an anti-differentiation problem, then adding any constant whatsoever to this function gives another solution. Moreover, adding constants gives all possible solutions.

One of those statements should be easy to explain (that is, prove) given what you know about derivatives. Can you do it? The other is harder and we will work through it later in this chapter.

Since anti-derivatives are not unique, when we find one we refer to it as *an anti-derivative* of the function. The collection of all anti-derivatives of a given function is referred to as *the indefinite integral* of the function.

Families of Functions — Indefinite Integrals

The set of all solutions to an indefinite integral problem or a differential equation problem is sometimes called a *family of functions*. As we have indicated, there is an arbitrary constant that distinguishes one member of the family from another.

Consider, for example, the simplified description of an object in free-fall described in Activity 2. You can imagine an object moving under the influence of gravity with a constant acceleration, of say -32. The velocity of this particle is then a function V of time. If the particle has an initial velocity of 1135, then one model for V would be that it is the initial velocity plus the acceleration multiplied by time. Thus, the velocity v is given by,

$$v = V(t) = 1135 - 32t.$$

What is a function which gives the position of this object above ground level? You have seen in Chapter 4 that the derivative of the distance function is the velocity function. Hence, the distance function is a function D whose derivative is V. You can determine that the distance function is given by

$$d = D(t) = c + 1135t - 16t^2.$$

Here c is the arbitrary constant. The solution to our problem is the family of functions obtained by taking all possible values of c. Figure 5.1 shows a picture of the graphs of some of the functions in the family.

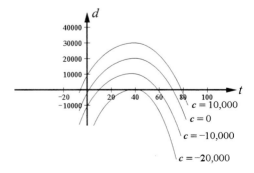

Figure 5.1. A family of functions given by $D(t) = c + 1135t - 16t^2$.

Notice how these curves are related to each other and what the effect is of changing c. You should have no difficulty figuring out the physical interpretation of c in this situation. It is simply the initial height of the particle above ground level.

Can you say anything about relationships between tangents to these curves?

In Activity 7, you saw two different anti-derivatives of the function f given by $f(x) = 2\sin(x)\cos(x)$. You should now be able to see that these are not the only two anti-derivatives of f, but that there is an entire family of functions all having the same derivative. The requirement that a function have a derivative represented by $2\sin(x)\cos(x)$ did not completely specify the function. In other words, the indefinite integral of f is the family of functions which has the functions represented by $\sin^2(x)$ and $-\cos^2(x)$ as two of its members. How do the two members of the family differ? That is, what is the difference between $\sin^2(x)$ and $-\cos^2(x)$? Do you recall that $\sin^2(x) = 1 - \cos^2(x)$? Now can you say how they differ? Did you say by a constant (in this case 1)? As you will see in Section 3, this is no accident. In general, any two anti-derivatives of a given function must differ by a constant. What does this say about the graphs of two functions which have the same derivative?

Rules for Integration

Later in this chapter we will carefully state and indicate proofs of a number of rules for computing indefinite integrals. Right now we want you to try to guess what these rules are. Activity 4 is supposed to help you do that. Another possible help is to give names for the rules. Here is a list of names. See if you can match them up with the examples in Activity 4.

1. The indefinite integral of the zero function.   2. The indefinite integral of a constant function.

3. Power rule.   4. Integrals that lead to the logarithm function.

5. Integral of trigonometric functions.   6. Chain rule (in reverse), or substitution.

7. Integration by parts or the product rule in reverse.

If you recall that the indefinite integral of a function g is the family of functions obtained by finding a single function f such that $f' = g$ and adding an arbitrary constant, then the proofs of these rules are actually quite simple. Try your hand at some of them.

In Activity 4 you may have guessed some of the following integration rules where c and a represent any constants. A more comprehensive list will be given in Section 5.3.

Some Basic Integration Rules

1. $\int 0\, dx = c$

2. $\int a x\, dx = ax + c$

3. $\int x^n\, dx = \dfrac{x^{n+1}}{n+1} + c$, provided $n \neq -1$

4. $\int x^{-1}\, dx = \int \dfrac{1}{x}\, dx = \ln(x) + c$

5. $\int \sin(x)\, dx = -\cos(x) + c$

6. $\int \cos(x)\, dx = \sin(x) + c$

7. $\int \sec^2(x)\, dx = \tan(x) + c$

8. $\int \sec(x)\tan(x)\, dx = \sec(x) + c$

9. $\int e^x\, dx = e^x + c$

Properties of Integrals   We treat the properties at this time in the same way as the rules. Here is a list of names of properties. See if you can match them up with the examples in Activity 5.

1. The integral of the sum of two functions.

2. The integral of the difference of two functions.

3. The integral of a function which is given as a number times a function.

4. The integral of a linear combination $sf + tg$, where f, g are functions and s, t are numbers.

Again you might be able to figure out proofs of some of these properties. Just try to relate the property to some property of differentiation and reverse your thinking.

Integrating Functions Defined in Parts   The question of integrating a function defined in parts which you encountered in Activity 9 raises some additional difficulties. You might have some trouble getting your SCS to deal with this problem. What would you do if you weren't given this anti-derivative?

Here is one point of view. Recall how we handled the derivative of a function defined in parts. The idea was to look at each part individually and then worry separately about the seam. We can take the same point of view for the integral.

Consider the case where the current function from Activity 9 is defined as follows.

$$i(t) = \begin{cases} a & \text{if } 0 \le t < b \\ 0 & \text{if } t \ge b \end{cases}$$

where a is a constant.

How can we find a voltage function v if we know that it is a function whose derivative is i? Since the domain of i is the set of non-negative numbers, we can restrict to that set for v as well.

If we look at each part separately, then we see that on the interval $[0, b)$, the function i is the constant function a. Its anti-derivative is represented by $at + k_1$. On the other branch, i is the 0 function, so its anti-derivative is a constant function k_2. Figure 5.2 shows a graph of what we get for particular choices of k_1 and k_2.

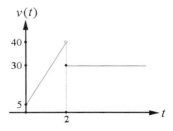

Figure 5.2. A graph of the voltage function v with $k_1 = 5$ and $k_2 = 30$.

Do you think there is much chance of choosing values for k_1 and k_2 that will make the voltage function continuous at $t = b$? Can the voltage function be differentiable at $t = b$? Why or why not?

Perhaps it would be better to use some sense of the situation to decide on these values. At time $t = 0$ it is reasonable to take the voltage to be 0 since we have not been told about any power surge. This suggests a value of $k_1 = 0$. Again, since there is nothing in the situation to cause a jump, we might consider that the current should be continuous at $t = b$. What value of k_2 does this suggest?

Here is a definition of one voltage function whose derivative is the given current function when $a = 4$, $b = 2$, $k_1 = 0$ and $k_2 = 8$.

$$i(t) = \begin{cases} 4t & \text{if } 0 \le t < 2 \\ 8 & \text{if } t \ge 2 \end{cases}$$

Figure 5.3 shows a graph of v.

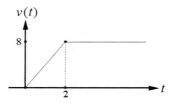

Figure 5.3. *A graph of the voltage function v with $k_1 = 0$ and $k_2 = 40$.*

Do you think that the same rule for all solutions holds for such a situation? What are all possible voltage functions that would give this current?

5.1.4 Exercises

1. This exercise is a continuation of Activity 1. A curve has slope $2x + 1$ for each x and the curve passes through the point $(-1, 0)$. Find the equation of the curve in the form $y = f(x)$. Explain how the additional requirement that the curve pass through the given point makes this problem different from Activity 1.

2. The marginal cost of a certain commodity is the derivative of the total cost function C. Find the total cost function C for a commodity which has a marginal cost of $0.45 + 3x^2$ when x units have been produced and the cost to start production is $7500 (the fixed cost). Use your SCS or do this problem by hand if you can.

3. The acceleration function of an object is the derivative of the velocity function. Find the velocity function V for a mass suspended by a spring, where the mass has an acceleration of $-\cos(t) - 0.5\sin(t)$ feet/sec^2 at any time t and an initial velocity of -0.5 feet/second. Use your SCS or do this problem by hand if you can.

4. A particle starts at the origin and moves along a coordinate line with an initial velocity of -1 meter/min and it has an acceleration of $1 - 6t + 6t^2$ meters/min^2. Find the position of the particle when its acceleration is 1 meters/min^2. Use the fact that acceleration is the derivative of velocity and velocity is the derivative of position. Use your SCS or do this problem by hand if you can.

5. Consider the following anti-derivative problem:

$$f'(x) = 2xe^{-0.02x}.$$

(a) Show by using differentiation that the function f given by

$$f(x) = -(100x + 5000)e^{-0.02x}$$

is a solution to the problem.

(b) Show by using differentiation that the function f given by

$$f(x) = 14.23 - (100x + 5000)e^{-0.02x}$$

is a solution of the problem.

(c) Based on your experience in part (a) and (b), for functions f and g, what can you say about the uniqueness of a solution to a problem like $f = g$? Explain your answer.

6. Consider the following two statements:

Once you have a function which solves an anti-differentiation problem, then adding any constant whatsoever to this function gives another solution.

Moreover, adding constants gives all possible solutions.

One of the above two statements should be easy to prove given what you know about derivatives. Which one is it? Give a proof of the statement you chose.

7. For the family of functions given by $D(t) = c + 1135t - 16t^2$ (see Figure 5.1), can you say anything about relationships between the tangents to these curves? Explain your answer. Illustrate your answer geometrically.

8. For the integration rules given below, perform the indicated tasks.

(a) Match the rules with the examples in Activity 4.

(b) Give a proof using the fact that the indefinite integral of a function g is the family of functions obtained by finding (if possible) a single function f such that $f' = g$ and adding an arbitrary constant.

(c) State each rule in a general way by using the notation $\int f(x)\,dx$ for the indefinite integral of a function f.

Here are the rules:

i. The integral of the zero function.

ii. The integral of a constant function.

iii. Power rule.

iv. Integrals that lead to the logarithm function.

v. Integrals of trigonometric functions.

vi. Chain rule (in reverse), or substitution.

vii. Integration by parts, or the product rule in reverse.

9. For the properties of integrals given below

(a) Match each property with the examples in Activity 5.

(b) State a general rule using the notation $\int f(x)\,dx$ for the indefinite integral of a function f.

(c) Relate a given property about indefinite integrals to some property of differentiation and reverse your thinking to give an explanation of why you think each indefinite integral property is true.

(d) Here are the properties:

i. The indefinite integral of the sum of two functions.

ii. The indefinite integral of the difference of two functions.

iii. The indefinite integral of a function which is given as a number times a function.

iv. The indefinite integral of a linear combination $sf + tg$, where f, g are functions and s, t are numbers.

10. Calculate the following indefinite integrals. Check your answers by differentiation.

(a) $\int x^5\, dx$ 　　　(b) $\int 13\, dw$

(c) $\int (2t^3 + t + 1)dt$ 　(d) $\int \dfrac{1}{z^3}dz$

(e) $\int -3\cos(x)\, dx$ 　(f) $\int 2\sin(w)\, dw$

(g) $\int dx$

(h) $\int (2 + 5\cos(v))dv$

(i) $\int (3\sin(x) - x)dx$

(j) $\int (x - \sec(x)\tan(x))dx$

(k) $\int 2\sec^2(t)dt$

(l) $\int (2 - 3\sec^2(x) + \cos(x))dx$

(m) $\int \dfrac{3}{x}dx$ 　　　(n) $\int 2e^y\, dy$

(o) $\int \cos(2x)\, dx$ 　(p) $\int e^{-3x}\, dx$

(q) $\int \sin(5x)\, dx$

▮ 5.2 DEFINITE INTEGRALS

5.2.1 Overview

In Section 1, you concentrated on one type of integration — *indefinite integration*. Indefinite integration is a process which takes a function as an input and returns a function as an output (or more precisely — a family of functions). In this Section we will study definite integration, a process which takes a function as an input and returns a real number as an output.

In this section, you will again investigate the **Riem computer functions** which you have used previously. In addition, you will consider the notion of taking *limits of Riemann sums*. Such limits are the key to solving various problems in calculus and its applications in Engineering, Science, Mathematics, Management and numerous other fields. You will work with a **computer function** called **DefInt** which approximates the definite integral process.

Once again you will find (or approximate) solutions to certain problems which involve the simplest type of differential equations. You will investigate the properties of the **computer function DefInt** and hence you will be (approximately) studying the properties of definite integrals and why they are valid. You will learn basic rules which apply to definite integrals and various properties of *definite integrals*. In addition, you will use various **Riem computer function**s to approximate *definite integrals* (usually when no exact solution can be found). The notion of the *average value* of a function will also be investigated.

Applications of integrals are plentiful in Engineering, Science, Management and numerous other disciplines. Applications of integration include: area, volume and surface area of irregularly shaped bodies; average values; lengths of curves; work; force; moments of inertia; flow of fluids; centers of mass; force due to fluid pressure; compound interest; electron distribution; motion; electrical current; entropy; chemical reactions; and distribution of gas molecules. As you will see, problems related to such applications can be solved by taking limits of Riemann sums.

5.2.2 Activities

1. The change in entropy associated with heating a substance from temperature T_1 to temperature T_2 is the difference $S(T_2) - S(T_1)$ of values at T_1, T_2 of a function S whose derivative is the function $\overline{C}_p$ given by $\frac{C_p(T)}{T}$ where C_p is the heat capacity of the substance. We denote this number by

$$S(T_2) - S(T_1) = \int_{T_1}^{T_2} \overline{C}_p(T)\, dx = \int_{T_1}^{T_2} \frac{C_p(T)}{T}\, dT$$

and call it the definite integral of C_p from T_1 to T_2.

For $CO_2\,(g)$ the heat capacity in the temperature range from 300 to 400 K is given approximately by

$$C_p(T) = 25.999T + 0.0435T^2 - \left(148.312 \times 10^{-6}\right)T^3$$

Use your SCS to find the change in entropy associated with raising the temperature of $CO_2\,(g)$ from 300 to 350 K.

2. According to circuit analysis, the voltage, as a function v of time t across a capacitor, is the definite integral from 0 to t of the conduction current i divided by the capacitance C. (This assumes that the current and voltage are 0 at time 0). If the conduction current is given by

$$i(t) = \begin{cases} 20 & \text{if } 0 \le t < 2 \\ 0 & \text{if } t \ge 2 \end{cases}$$

and the capacitance is 5, find the voltage function v.

3. Write a **computer function** named **RiemTrap** which is the same as your **RiemLeft computer function** except that the quantity $f(x_i)$ is replaced by $\frac{f(x_i) + f(x_i + 1)}{2}$.

   (a) Apply **RiemTrap** to the function $\overline{C}_p$ of Activity 1 on the domain [300, 350] to estimate the change in entropy. Use 10 subintervals.

   (b) Make a graph of $\overline{C}_p$ in this range and indicate on the picture the quantity that is being computed by **RiemTrap**.

   (c) Explain in your own words the corresponding quantity that would be computed if you used **RiemLeft**, **RiemRight**, and **RiemMid**.

   (d) Apply each of **RiemLeft**, **RiemRight**, **RiemMid**, to the same function $\overline{C}_p$ on the same domain [300, 350], using the same number of intervals, 10. Which answer do you think is the most accurate?

4. In this activity you will use the **computer function DefInt** which approximates the idea of calculating with your **Riem computer functions** and letting $n \to \infty$. Below is the code for **DefInt** in **ISETL**. If your MPL is not **ISETL** your instructor will provide you with the necessary code or a computer file which contains the necessary code. You may need to modify the **0.0001** in the sixth line of code to avoid an endless loop.

```
DefInt := func(f, a, b);

        local previous, current, n;

        n := 50;

        previous := ((f(a)+f(b))/2)*(b–a);

        current := RiemTrap(f,a,b,n);

        while abs(current-previous)/current >= 0.0001 do

            n := n+25;

            previous := current;

            current := RiemTrap(f,a,b,n);

        end;

        return current;

    end;
```

Write out a detailed explanation of what the **computer function DefInt** is doing. (You may ignore the **local** declaration in the second line.)

Apply **DefInt** to the function $\overline{C}_p$ of Activity 1. Compare your answer with what you got in Activity 1. Explain in your own words what is going on.

5. In the previous activity, **RiemTrap** was used. Write a short essay about using **RiemTrap**, as compared to using any other version of the **Riem computer function**.

6. In this activity you are to use **DefInt** to experiment with examples and try to discover various properties that it satisfies. In each case, perform the indicated tasks and try to formulate a general rule. It can be helpful, in looking for general rules, to draw a picture and try to derive the rule from the geometrical quantities that are being computed.

    (a) Using the interval [300, 400] throughout, apply **DefInt** to the function $\overline{C}_p$, the constant function 25.999, the function whose value at T is 0.0435 T, and the function whose value at T is $(148.312 \times 10^{-6})T^2$.

    (b) Apply **DefInt** to $\overline{C}_p$ on the intervals, [300 ,400], [300, 320], [350 , 400], [320, 400].

    (c) Apply **DefInt** to $\overline{C}_p$ on the interval, [400, 300].

    (d) Apply **DefInt** to $\overline{C}_p$ on the interval, [300, 300].

    (e) Apply **DefInt** to the function f given by $f(x) = x^7$, and to the function g given by $g(x) = x^{10}$, both on the interval [0, 1]. Compare the two answers. Now do it again on the interval [1, 2]. Compare the two answers again. What determines which of the two answers is larger?

    (f) Apply **DefInt** to the function whose value at x is $\sin x$ on the interval, $[-\pi, 2\pi]$, and also apply it, on the same interval, to the function whose value at x is $|\sin(x)|$. Compare the two answers.

(g) Let m be the minimum and M the maximum values of the function $\overline{C}_p$ on the interval $[300, 350]$. Compute these numbers, the length of this interval and the number **DefInt (Cp, 300, 350)** (which you have already done). Find some relationship between these numbers.

(h) Can you imagine any meaning for the idea of the "average value of $\overline{C}_p$ on the interval $[300, 350]$"? If you come up with a possible meaning, do you think there is a T_{avg} in this interval such that $C_p(T_{avg})$ is equal to this average value?

(i) Use each of your **computer functions RiemLeft, RiemRight**, and **RiemMid** to estimate the area between the two curves which are graphs of the functions f and g given by $f(x) = x$ and $g(x) = x^2$.

(j) There are a number of properties that hold for definite integrals. They can also be used as rules to help you compute integrals. Each of the following items illustrates a particular property. Use your SCS to compute the indicated integrals and then try to guess the property. Try to imagine why your property should hold.

    i. $\int_0^3 e^x\,dx$, $\int_3^4 e^x\,dx$, $\int_0^4 e^x\,dx$

    ii. $\int_{-1}^2 (1+2x)\,dx$, $\int_2^{-1}(1+2x)\,dx$, $\int_2^2(1+2x)\,dx$

    iii. $\int_1^5 \ln(x)\,dx$, $\int_5^2\ln(x)\,dx$, $\int_1^2 \ln(x)\,dx$

7. For the properties of integrals given below, state a general rule using the notation $\int_a^b f(x)\,dx$ for the definite integral of a function f on an interval [a, b].

    (a) The definite integral from a to b of a function in termsof the definite integral from b to a of the same function.

    (b) The definite integral from a to b of a function when $a = b$.

    (c) The definite integral from a to b of a function f in terms of the definite integral from a to c of f and from c to b of f.

8. Later in Section 6.2 you will see that if a definite integral $\int_a^b f(x)\,dx$ exists, it has the same numerical value as that approximated by **RiemLeft(f,a,b,n)** as $n \to \infty$. Use this fact and your experience from the activity above to give an explanation of why you think each definite integral property listed in the previous activity is true.

5.2.3 Discussion

There is a vast number of topics in mathematics, science, economics and other subjects where you will find applications of integration. We have considered some of them in the activities and will consider several others in Section 5.5, as well as throughout the rest of this book.

Just to give you a feeling for how widespread is the applicability of integration, here is a list of topics where we have or will have considered the integral: area, volume and surface area of irregularly shaped bodies; average values; lengths of curves; work; force; moments of inertia; flow of fluids; centers of mass; force due to fluid pressure; compound interest; electron distribution; motion; electrical current; entropy; chemical reactions; and distribution of gas molecules.

As was the case with differentiation, the definite integration processis a combination of two operations, the operation of limit and another basic operation you have been exposed to in the previous chapters — the formation of Riemann sums. Let's begin by looking at the **Riem computer function**s which you have written so far.

Different Versions of Riem
If you draw a picture, then you can see right away that all of the versions of the **Riem computer function**s calculate approximations to the (net) area between the graph of a given function and the x-axis, over a domain specified by the endpoints **a** and **b**. See Figure 5.4.

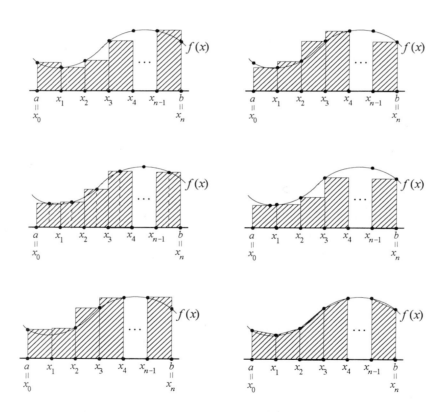

Figure 5.4. RiemLeft, RiemRight, RiemMid, RiemMin, RiemMax and RiemTrap

As you can see from the pictures, most of the **Riem computer function**s approximate the (net) area by using rectangles. The base of the rectangle is the length of the subinterval in the partition and its height is determined by the particular version of the **Riemfunc**. For example, **RiemLeft** uses the value of the functionat the left endpoint of each subinterval, **RiemMax** uses thepoint at which the maximum value of the function occurs in each subinterval, and so on.

The one exception is **RiemTrap** which uses a trapezoid. The base is the same length of the subinterval, but two heights are used — the value of the function at the two endpoints. The trapezoid is completed by connecting the two points on the curve with a straight line. Approximating the (net) area in this way is called the *trapezoid method*.

In the next section you will study a method of computing such areas exactly, at least for some problems. Using this method, you will be able to compare the approximation in Activity 3 with the exact answer. You will see that relative accuracy of the method depends to some extent on the function, but that some methods are better than others in most cases. Can you imagine why?

There are a number of reasons why it is plausible that the trapezoid method is the most accurate. If you look at the formula, you will see that the basic quantity, $\frac{f(x_1)+f(x_1+1)}{2}$, is the average of the value of the function at the two endpoints. In general, when you have two estimates of a quantity and no special reason to pick one or the other, their average is usually more accurate than either.

There is another, perhaps deeper reason. Look at the picture and imagine the broken line obtained by putting together all of the tops of the trapezoid. This is a pretty good approximation to the curve. It should be, because it is the secant connecting two points that are close together. However, it can be shown that through theory, **RiemMid** is more accurate than **RiemTrap** or any other **Riem computer functions** you have used. More precisely, let $|E_n|$ be the *absolute (discretization) error* made in using any one of the **Riem computer functions** you have used or **RiemTrap** to estimate $\int_a^b f(x)\,dx$ for a continuous function f on $[a, b]$. Then,

$$|E_n| = \left| \int_a^b f(x)\,dx - Riem(f,\,a,\,b,\,n) \right|$$

where $Riem(f,\,a,\,b,\,n)$ denotes any **Riem computer function** approximation you have previously used or **RiemTrap**. It can be shown that for the trapezoid method

$$|E_n| \le \frac{(b-a)^3}{12n^2} \max_{a\le x\le b} |f''(x)|$$

and for the midpoint method

$$|E_n| \le \frac{(b-a)^3}{24n^2} \max_{a\le x\le b} |f''(x)|.$$

These error estimates provide an upper bound for the absolute error. Do you see that the estimate of the error for **RiemMid** is sharper than that of the estimate of error for **RiemTrap**? We should note that the absolute error $|E_n|$ is the error due to the method used and not the error made due to round-off or computer arithmetic. Proofs of these error estimates can found in a textbook on Numerical Analysis. You will get a chance to derive an error estimate and you will also work with the error estimates given above in the exercises.

We should note that the trapezoid method uses line segments which are close to the tangent at one of the endpoints of the subintervals. It seems (as we have seen in Chapter 4, for example) that drawing line segments lying on tangents is a reasonable way to approximate a curve as shown in Figure 5.5.

Figure 5.5. *Tangent line segments along a curve.*

Passing to the Limit with Riem

Let's return now to the **Riem computer functions** and consider how to improve the approximation. This is done by using a partition with smaller subintervals. That is, the basic interval $[a, b]$ is divided into n subintervals, each of length $\frac{b-a}{n}$. Increasing n will decrease the length of the subintervals and increase their number. Thus we will be adding up more and more quantities that are smaller and smaller. Do you see any difficulty in getting as accurate an answer as you would like in this way?

The basic quantity we are working with, for a function f on an interval $[a, b]$, is

$$\sum_{i=1}^{n} f\left(x_i^*\right)(x_i - x_{i-1})$$

Here, $f\left(x_i^*\right)$ refers to the value of the function at *some* point x_i^* in the subinterval $[x_{i-1}, x_i]$. It could be one of the endpoints, the midpoint, or just any point in the interval $[x_{i-1}, x_i]$. In the case of **RiemTrap** the quantity $f\left(x_i^*\right)$ is replaced by the average value of f at the two endpoints, $\frac{f(x_i)+f(x_i+1)}{2}$.

Ignoring for the moment which method we use, keeping the function f and using the interval $[1, 50]$, we see that, for any choice of n we go through the calculation and get a number. Thus, if **f** is a **computer function** that implements C_p, in Activity 2 of Section 2.1.2 then calculating **RiemTrap** of f over the interval $[1, 50]$ using $n = 10$ gives the number, 1322.095552, increasing n to 250 gives the number, 1322.124586, increasing n to 600 gives the number, 1322.124625, and so on. For each positive integer n, we get a number. This process defines is a function. A function whose domain is the set of positive integers. Such a function is called a *sequence*.

In the next chapter, you will study sequences and their limits. Just as with limits of functions, the limit of a sequence is a number that the values of the sequence approaches, as the value of n becomes arbitrarily large. The **computer function DefInt** which you used in Activity 4, approximates the idea of calculating with one of your **Riem computer functions** and letting $n \to \infty$. It is only an approximation, of course, since, although n becomes large, it does not become *arbitrarily* large. (The computer, and probably us as well, are all finite.) Nevertheless, it is a useful approximation.

Notice that when we talk about the limit, the question of which version of **Riem** that we use becomes unimportant. Why is that? In Activity 3 how different were the various answers you got by using different versions and taking the limit? Is that what is always going to happen? Can you imagine a situation in which something very different will happen if you use different versions?

DefInt and the Definite Integral

Once you pass to the limit, you get a single number called the definite integral. In other words, we have a process, implemented (approximately) by **DefInt**, whereby you begin with a function f and an interval $[a, b]$ and come up with a number. The purpose of this chapter is to study that process thoroughly.

Although the process is lengthy and complicated, the meaning of what is being computed is fairly simple. It is just the area under the curve given by the graph of the function over the domain [a, b], provided the function is non-negative on the interval [a, b]. We will see in Section 5.5 that the same process can be interpreted in a number of other ways. Each interpretation leads to an application.

We will also see that this process is very intimately related to the other problem we have been talking about in this chapter — finding a function when you know its derivative.

Properties of DefInt

We begin by taking apart the process **DefInt** in order to see what makes it tick, by looking at its properties. In Activity 6 you had a chance to discover some of these properties. Now we will begin to describe them. In each case, you should be thinking about what the property actually is and how it might be used (or what it means for the various interpretations).

We will give some explanations of how these properties can be proved. In every case, the situation is the same. The basic idea of the property, working with the **Riem computer function** is usually quite straightforward. To give a completely rigorous proof, however, requires an analysis of the role played by the limit that is taken of the sequence obtained from applying the **Riem computer function** for various values of n, the number of subintervals. This will be made more precise in Chapter 6, so for now we discuss this aspect only in an informal way.

Before going into the properties, let us review the **Riem** process. We begin with a function f and an interval [a, b] in its domain. Here is a listing of the steps that are taken.

1. Choose a positive integer n and divide the interval [a, b] into a partition P of n equal subintervals. Thus we have,

$$P = \left(a + \frac{i}{n}(b-a) \right)_{i=0}^{n} = \left(a, \; a + \frac{1}{n}(b-a), \; \frac{2}{n}(b-a), \; \ldots, \; \frac{n-1}{n}(b-a), \; b \right).$$

2. Make a decision about a point x_i^* in the subinterval $[x_{i-1}, x_i]$ at which the function is evaluated to obtain $f(x_i^*)$. Alternatively, take for $f(x_i^*)$ the average $\frac{f(x_{i-1}) + f(x_i)}{2}$.

3. Form the sum

$$\sum_{i=1}^{n} f(x_i^*)(x_i - x_{i-1}).$$

4. Pass to the limit as n goes to ∞.

For convenience of calculation, we will write the result of this process as $Riem(f, a, b)$ and call it the *Riemann integral of f on* [a, b]. That is, $Riem(f, a, b)$ is the limit of <u>any</u> Riemann sum as $n \to \infty$. Can you explain the use of the term *any*? An alternate notation for $Riem(f, a, b)$ is the *definite integral* notation

$$\int_a^b f(x)\, dx$$

which you have already encountered.

Properties of DefInt — Linearity   *The Riemann integral of a linear combination of*
continuous functions is the same linear combination of the Riemann integrals of the individual functions.
   If *f* and *g* are functions and *s* and *t* are numbers, then

$$Riem(sf + tg,\ a,\ b) = sRiem(f,\ a,\ b) + tRiem(g,\ a,\ b)$$

or

$$\int_a^b (s(f(x) + tg(x)))\,dx = s\int_a^b f(x)\,dx + t\int_a^b g(x)\,dx.$$

You can see the reason for this by looking at the calculations,

$$\sum_{i=1}^n (sf(x_i^*) + tg(x_i^*))(x_i - x_{i-1}) = s\sum_{i=1}^n f(x_i^*)(x_i - x_{i-1}) + t\sum_{i=1}^n g(x_i^*)(x_i - x_{i-1})$$

If you then take the limit of both sides, you get the desired relation.

Properties of DefInt — Union of Intervals   *If an interval is broken up into two*
subintervals, then the Riemann integral of a continuous function over the original interval is the sum of
the Riemann integrals of the function over the two subintervals.
   If $c \in [a,\ b]$, then

$$Riem(f,\ a,\ b) = Riem(f,\ a,\ c) + Riem(f,\ c,\ b)$$

or

$$\int_a^b f(x)\,dx = s\int_a^c f(x)\,dx + \int_c^b f(x)\,dx.$$

   Using the computations for this one actually introduces so many technicalities that, for some people, it will probably make things even less clear. If you think about area under a curve, then it is reasonable that the area under the curve from *a* to *b* is the sum of the areas under the curve from *a* to *c* and from *c* to *b*. See Figure 5.6.
   Before leaving this property, look again at the examples in Activity 7(c). What do you think happens to the formula if *c* is some other point, not in the interval [*a*, *b*]?

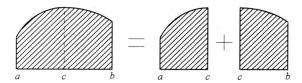

Figure 5.6. $Riem(f,\ a,\ b) = Riem(f,\ a,\ c) + Riem(f,\ c,\ b)$

Properties of DefInt — The Case *a > b*   *The Riemann integral of a continuous function on*
[b, a] is the negative of the Riemann integral of the same function on [a, b].

$$Riem(f,\ b,\ a) = -Riem(f,\ a,\ b)$$

or

$$\int_a^b f(x)\,dx = -\int_b^a f(x)\,dx.$$

Look at the basic formula

$$\sum_{i=1}^{n} f\left(x_i^*\right)(x_i - x_{i-1})$$

and replace the symbols for the endpoints by their actual value. That is, $x_0 = a$, $x_1 = a + \frac{1}{n}(b-a)$, $x_1 = a + \frac{2}{n}(b-a)$, ..., $x_{n-1} = a + \frac{n-1}{n}(b-a)$, $x_n = b$. In general,

$$x_i = a + \frac{i}{n}(b-a)$$

so that,

$$x_i - x_{i-1} = a + \frac{i}{n}(b-a) - \left(a + \frac{i-1}{n}(b-a)\right) = \frac{b-a}{n}$$

and a simpler form of the formula is

$$\left(\frac{b-a}{n}\right)\sum_{i=1}^{n} f\left(x_i^*\right).$$

Now, if we interchange a and b, we get,

$$\left(\frac{b-a}{n}\right)\sum_{i=1}^{n} f\left(x_i^*\right) = -\left(\frac{a-b}{n}\right)\sum_{i=1}^{n} f\left(x_i^*\right)$$

and then you can pass to the limit to obtain the desired relation.

Properties of DefInt — The Case $a = b$
The Riemann integral of a function defined over an interval [a, a] which is degenerated to a single point has the value 0.

$$Riem(f, a, a) = 0$$

or

$$\int_b^a f(x)\, dx = 0.$$

You can see this from the same calculation that we just did. If you take the formula

$$\left(\frac{b-a}{n}\right)\sum_{i=1}^{n} f\left(x_i^*\right)$$

and set $a = b$, then obviously the value is 0. Passing to the limit retains this constant value.

Properties of DefInt — Monotonicity
If the value of a continuous function f at every point of the interval [a, b] is less than or equal to the value of a continuous function g at the corresponding point, then the Riemann integral of f over [a, b] is less than or equal to the Riemann integral of g over [a, b].

If $f(x) \le g(x)$ for every $x \in [a, b]$, then

$$Riem(f, a, b) \le Riem(g, a, b)$$

or

$$\int_a^b f(x)\, dx \le \int_a^b g(x)\, dx.$$

You can see the reason for this by looking at the formula,

$$\sum_{i=1}^{n} f(x_i^*)(x_i - x_{i-1})$$

If you replace f by g, then since $(x_i - x_{i-1}) > 0$ and $f(x_i^*) \le g(x_i^*)$, we have

$$f(x_i^*)(x_i - x_{i-1}) \le g(x_i^*)(x_i - x_{i-1})$$

and adding them up retains the inequality. Passing to the limit also retains the inequality.

Here is a very important formula that can be derived from this relation,

$$|Riem(f, a, b)| \le Riem(|f|, a, b)$$

or

$$\left| \int_b^a f(x)\, dx \right| \le \int_b^a |f(x)|\, dx.$$

The proof of this is obtained by comparing the three functions, f, $|f|$, and $|-f|$. You will have a chance to work out the details in the exercises.

Properties of DefInt — Bounds
If m and M are the minimum and maximum values, respectively, of a continuous function f on the interval [a, b], then the Riemann integral of f on [a, b] is greater than or equal to m times the length of the interval and less than or equal to M times the length of the interval.

If $m \le f(x) \le M$ for every $x \in [a, b]$ then

$$m(b - a) \le Riem(f, a, b) \le M(b - a)$$

or

$$m(b - a) \le \int_a^b f(x)\, dx \le M(b - a).$$

There are a couple of ways to see why this must be so. You can reason just as we did in the previous paragraph, by looking at the basic formula,

$$\sum_{i=1}^{n} f(x_i^*)(x_i - x_{i-1}).$$

If each $f(x_i^*)$ is replaced by M, then the result is not smaller, so we have,

$$\sum_{i=1}^{n} f(x_i^*)(x_i - x_{i-1}) \le \sum_{i=1}^{n} M(x_i - x_{i-1}) = M\sum_{i=1}^{n}(x_i - x_{i-1}) = M(b - a)$$

and, once again, passing to the limit does not change this relation. A similar argument can be made for the inequality on the left.

Another approach is to apply the monotonicity property of the previous paragraph to the three functions, m, f and M.

Can you draw a picture that gives a geometrical interpretation of this property?

DefInt and Piecewise Continuous Functions

In Activity 2, you saw a situation where it was necessary to integrate a function which was defined in parts.

If a function is defined in parts and each part is a continuous function on that subinterval, then we can define the Riemann integral of the function over each part. We add them up to get the Riemann integral over the entire interval.

The following definition formalizes this for a function defined in two parts.

Definition 5.1

Let f be a function defined on [a, b] in two parts. That is, we have $c \in [a, b]$ and,

$$f(x) = \begin{cases} g(x) & \text{if } a \le x \le c \\ h(x) & \text{if } c < x \le b. \end{cases}$$

Suppose further that g is continuous on [a, c] and h is continuous on [c, b]. Then we define

$$Riem(f, a, b) = Riem(g, a, c) + Riem(h, c, b)$$

or

$$\int_a^b f(x)\,dx = \int_a^c g(x)\,dx + h(x)\,dx.$$

We say that a function defined in parts is *piecewise continuous* if each part is continuous. Hence, the above definition gives a way to integrate piecewise continuous functions.

The definition is analogous for the case of functions defined in three or more parts.

Average Value of a Function

We end this section by making an important definition using the definite integral.

Definition 5.2

The average value, f_{avg}, of a continuous function f on the interval [a, b] is given by

$$f_{avg} = \frac{1}{b-a} Riem(f, a, b)$$

or

$$f_{avg} = \frac{1}{b-a} \int_a^b f(x)\,dx.$$

Think about the inequality of the previous paragraph.

$$m(b-a) \le Riem(f, a, b) \le M(b-a)$$

Given this, try to say something about the number which we have called the average value of f. Can you give an interpretation for it? Is the average value of a function f equal to the value of f at any point in $[a, b]$? Can you apply the intermediate value theorem here?

5.2.4   Exercises

1. By interpreting each of the following definite integrals as an area under the graph of a function f over an interval $[a, b]$, find the value of the integral formulas you know from geometry. Specify the function f, the interval $[a, b]$, and make a sketch showing the area you found. In parts (i) and (j), the notation $\lfloor x \rfloor$ represents the greatest integer less than or equal to x.

   (a) $\int_0^1 3\, dx$ (b) $\int_2^5 4\, dx$

   (c) $\int_0^1 2x\, dx$ (d) $\int_1^4 2x\, dx$

   (e) $\int_{-2}^1 |x|\, dx$ (f) $\int_0^2 \sqrt{4 - x^2}\, dx$

   (g) $\int_{-1}^1 2 + \sqrt{1 - x^2}\, dx$

   (h) $\int_{-a}^a \sqrt{a^2 - x^2}\, dx$, $a > 0$

   (i) $\int_{-1}^3 \lfloor x \rfloor\, dx$ (j) $\int_0^2 \lfloor x^2 \rfloor\, dx$

2. By interpreting each of the following definite integrals as an area under the graph of the indicated function over an interval $[a, b]$, find the value of the integral formulas you know from geometry. Make a sketch showing the area you found.

   (a) $\int_0^2 f(x)\, dx$

   where

   $$f(x) = \begin{cases} 3x & \text{if } x \geq 1 \\ 3 & \text{if } x < 1 \end{cases}$$

   (b) $\int_{-1}^3 h(y)\, dy$

   where

   $$h(y) = \begin{cases} |2y| & \text{if } y < 2 \\ 6 - y & \text{if } y > 2 \end{cases}$$

   (c) $\int_{-1}^3 F(v)\, dv$

   where

   $$F(v) = \begin{cases} \sqrt{3 - v^2} & \text{if } -\sqrt{3} \leq v \leq 0 \\ 3 & \text{if } v \geq 0 \end{cases}$$

3. Write a mathematical expression for a Riemann sum

   $$\sum_{i=1}^n f(x_i^*)(x_i - x_{i-1})$$

   which approximates each of the following definite integrals, where x_i^* is any point in the ith subinterval of the interval of integration.

   (a) The integral in part (c) of Exercise 1.

   (b) The integral in part (d) of Exercise 1.

   (c) The integral in part (e) of Exercise 1.

   (d) The integral in part (f) of Exercise 1.

   (e) The integral in part (a) of Exercise 2.

   (f) The integral in part (b) of Exercise 2.

   (g) The integral in part (c) of Exercise 2.

4. In this exercise you are to use the fact that the positions of an object moving with velocity on a coordinate line is given by $s = v \cdot t$ where v is the velocity of the object and t is the time it travels. Do the following for each of the velocity functions for an object moving on a line over the indicated time interval. Suppose an object is moving on a coordinate line with velocity v given by a function V of t, i.e., $v = V(t)$.

   (a) Write an expression for a Riemann sum

   $$\sum_{i=1}^n V(t_i^*)(t_i - t_{i-1})$$

   which approximates the *net* distance traveled by the object on the interval.

(b) Write an expression for a definite integral which represents the *net* distance traveled on the indicated time interval.

(c) Write an expression for a Riemann sum

$$\sum_{i=1}^{n} V\left(t_i^*\right)(t_i - t_{i-1})$$

which approximates the total distance traveled by the object on the indicated time interval.

(d) Write an expression for a definite integral which represents the total distance traveled on the indicated time interval.

(e) Explain in your own words the difference, if any, between parts (a), (b) and parts (c), (d). Make sketches illustrating your explanation geometrically in terms of an area or *net* area.

    i. $v = V(t) = 9.8t$ during the first five seconds.

    ii. $v = V(t) = 32t$ for $1 \le t \le 4$.

    iii. $v = V(t) = 3 + \sqrt{t}$ during the first second.

    iv. $v = V(t) = 2\sin(\pi t)$, $-0.5 \le t \le 1$.

5. A manufacturer predicts that the price per unit of a certain commodity over the next few months will be given by $P'(t) = 0.5t - 0.2$ hundreds of dollars per unit t months from now and 100 units of the commodity are sold each month.

(a) Write a Riemann sum

$$\sum_{i=1}^{n} f\left(t_i^*\right)(t_i - t_{i-1})$$

which approximates the predicted total revenue over the next four months for an appropriate function f.

(b) Write a definite integral which represents the predicted total revenue earned over the next four months.

(c) Find the predicted total revenue over the next four months by interpreting your integral in part (b) as a *net* area.

6. For the situation in the Exercise 5, do the following.

(a) Write a Riemann sum

$$\sum_{i=1}^{n} f\left(t_i^*\right)(t_i - t_{i-1})$$

for an appropriate function f which approximates the predicted total revenue earned between the first and third months from now.

(b) Write a definite integral which represents the predicted total revenue earned between the first and third months from now.

(c) Find the predicted total revenue earned between the first and third months from now by interpreting your integral in part (b) as a *net* area.

7. For the situation in Exercise 5, suppose that the manufacturer sold $20t + 30t^2$ units per month. Use your **computer functions RiemLeft**, **RiemRight**, **RiemMid**, and **RiemTrap** to estimate to the nearest dollar the total profit earned during the first four months from now. What was the smallest value of n you used in each **computer function** to get your answer? Which method appears to be the most efficient and why?

8. Use your **computer functions RiemLeft**, **RiemRight**, **RiemMid** and **RiemTrap** to estimate to five decimal places the area under the graph of the function given by

$$e^{-x^2}$$

over the interval $[-1, 1]$. What was the smallest value of n you used in each case to get your answer? Which method appears to be the most efficient and why?

9. Use your **computer functions RiemLeft, RiemRight, RiemMid,** and **RiemTrap** to estimate to five decimal places the area bounded by the graph of the function given by

$$\sqrt{1+x^3}$$

over the interval $[-1, 2]$. What was the smallest value of n you used in each case to get your answer? Which method appears to be the most efficient and why?

10. Explain why you think that the trapezoidal method is usually the most accurate method for approximating a definite integral. Explain your answer. Illustrate your answer geometrically.

11. In this section we said that $Riem(f, a, b)$ is the limit of any **Riem computer function** as $n \to \infty$. Give an explanation of why you think the use of the term **any** makes sense. That is, explain why the version of **Riem** you used to approximate a definite integral becomes unimportant as you let the number of subintervals n get larger and larger, that is, as you take the limit as $n \to \infty$.

12. Consider the following property of *Riem*:

$$Riem(f, a, b) = Riem(f, a, c) \\ + Riem(f, c, b).$$

Explain what you think happens to the above formula if c is some point <u>not</u> in the interval $[a, b]$? Give a proof of your formula using *Riem*. Illustrate your answer geometrically.

13. Suppose that f and g are continuous functions defined on the interval $[a, b]$ such that $f(x) \geq g(x)$ for every $x \in [a, b]$. Show that

$$Riem(f - g, a, b) \geq 0.$$

14. Suppose that f is a continuous function such that $f(x) \leq 0$ for all x in $[a, b]$. Show that
$$Riem(f, a, b) \leq 0.$$

15. Suppose that g is the continuous function given by $f - k$ where f is a continuous function and k is a constant, i.e., $(f - k)(x) = f(x) - k$. What can you say about f if $Riem(g, a, b) = 0$? Explain your answer.

16. Let x be a number such that $a \leq x \leq b$ and let f be a continuous function of x.

    (a) Is **RiemLeft(f, x, b, n)** $=$ **RiemLeft(f, a, x, n)**? Give an explanation for your answer.

    (b) Could the equality in part (a) be true for any other **Riem computer function**? Explain your answer.

17. Let f and g be the functions given by

$$f(x) = 1 - 3x^2 \text{ and } g(x) = 3 + x - x^2.$$

Show that $Riem(f, a, b) \leq Riem(g, a, b)$ where $a \leq b$.

18. Let g and h be the functions given by

$$g(x) = M^2 \text{ and } h(x) = 2M\,Nt^k - N^2 x^{2k}.$$

Show that $Riem(g, a, b) \leq Riem(h, a, b)$ where $a \leq b$.

19. Prove the following formula for a continuous function f

$$|Riem(f, a, b)| \leq Riem(|f|, a, b)$$

by comparing the three functions, f, $|f|$, and $-|f|$.

20. A chemist produces a gas at a continuous rate during a research experiment. The rate at which the gas is produced, in ml/min, is measured in half-minute intervals and according to the chemist's data a set of

ordered pairs representing the rate at which the gas is produced as a function of time is as follows:

{[0, 0], [0.5, 1.54], [1, 2.06], [1.5, 2.76], [2, 3.14], [2.5, 2.63], [3, 2.48]}.

Estimate the amount of the gas produced during the first three minutes of the experiment using your **computer functions RiemTrap** and **RiemMid**.

21. Use **RiemMid** to estimate each of the following definite integrals to five decimal places. What is the exact value of each integral?

    (a) $\int_1^2 \frac{1}{x}\, dx$. (b) $\int_{0.5}^1 \frac{1}{x}\, dx$.

    (c) $\int_0^1 \frac{4}{1+x^2}\, dx$.

22. The *Bessel function* J_1 is defined by the following integral

    $$J_1(x) = \frac{1}{\pi}\int_0^\pi \cos(\theta - x\sin(\theta))\, d\theta$$

    and is used in several physical applications.

    (a) What is $J_1(0)$?

    (b) Approximate $J_1(2)$ to five decimal places using **RiemMid**. Use your SCS to check your answer.

23. The *Bessel function* $J_1(0)$ is defined by the following integral

    $$J_0(x) = \frac{1}{\pi}\int_0^\pi \cos(x\sin(\theta))\, d\theta$$

    and is used in some physical applications.

    (a) What is $J_0(0)$?

    (b) Approximate $J_0(2.6)$ to five decimal places using **RiemMid**. Use your SCS to check your answer.

24. Show that for any continuous function f on $[a, b]$,

    RiemTrap(f, a, b, n) =

    $$\frac{\textbf{RiemLeft(f, a, b, n)} + \textbf{RiemRight(f, a, b, n)}}{2}.$$

25. Show that any **Riem computer function** gives the exact answer for $Riem(f, a, b)$ when f is a constant function on the interval $[a, b]$.

26. Show that **RiemTrap** gives the exact answer for $Riem(f, a, b)$ when f is a linear function given by $f(x) = mx + d$ on $[a, b]$ for some constants m and d.

27. Suppose that f and g are continuous functions on $[a, b]$ such that each of the following limits exist. Do the following for each of the limits of Riemann sums over the interval $[a, b]$.

    (a) Write a definite integral that is equivalent to the limit.

    (b) Using the indicated conditions on the function(s), give a geometric or physical interpretation for each limit as an (net) area, volume, or length of a curve. Explain your answer using appropriate sketches.

    i. $\lim\limits_{n\to\infty} \sum\limits_{i=1}^{n} f(x_i^*)(x_i - x_{i-1})$,
        $f(x) \geq 0$ on $[a, b]$

    ii. $\lim\limits_{n\to\infty} \sum\limits_{i=1}^{n} (f(x_i^*) - g(x_i^*))(x_i - x_{i-1})$,
        $f(x) \geq g(x)$ on $[a, b]$

    iii. $\lim\limits_{n\to\infty} \sum\limits_{i=1}^{n} \pi(f(x_i^*))^2 (x_i - x_{i-1})$,
        $f(x) \geq 0$ on $[a, b]$

    iv. $\lim\limits_{n\to\infty} \sum\limits_{i=1}^{n} 2\pi x_i^* f(x_i^*)(x_i - x_{i-1})$,
        $f(x) \geq 0$ on $[a, b]$

v. $\displaystyle\lim_{n\to\infty}\sum_{i=1}^{n}\sqrt{1+\left(f'(x_i^*)\right)^2}\,(x_i-x_{i-1})$

28. For each of the definite integrals below, do the following.

(a) Write a mathematical expression for the Riemann sum corresponding to what **RiemLeft** would give as an approximation to the integral when the interval of integration is divided into n equal subintervals.

(b) Use the formulas below to simplify the Riemann sum you found in part (a) to obtain a sequence u, which is a function of n, that represents an approximation to the integral. In the formulas below, a_i and b_i are real numbers, and c is a constant.

$$\sum_{i=1}^{n}ca_i=c\cdot\sum_{i=1}^{n}a_i,$$

$$\sum_{i=1}^{n}(a_i+b_i)=\sum_{i=1}^{n}a_i+\sum_{i=1}^{n}b_i$$

$$\sum_{i=1}^{n}c=n\cdot c,\quad \sum_{i=1}^{n}i=\frac{n(n+1)}{2}$$

$$\sum_{i=1}^{n}i^2=\frac{n(n+1)(2n+1)}{6},$$

$$\sum_{i=1}^{n}i^3=\frac{n^2(n+1)^2}{4}$$

(c) Find the value of the definite integral by taking the limit as $n\to\infty$ of the sequence you found in part (b).

(d) Make a sketch showing a geometric interpretation of each definite integral.

Here are the definite integrals.

i. $\int_0^1 3\,dx$ ii. $\int_0^1 2x\,dx$

iii. $\int_0^3 x^2\,dx$ iv. $\int_0^2 x^3\,dx$

v. $\int_0^1 4-x^2\,dx$

vi. $\int_1^4 2x\,dx$ vii. $\int_1^2 3x^2\,dx$

29. When a Riemann sum

$$\sum_{i=1}^{n}f(x_i^*)(x_i-x_{i-1})$$

is used to approximate the definite integral

$$\int_a^b f(x)\,dx,$$

the *absolute (discretization) error* E_n is defined by

$$E_n=\left|\int_a^b f(x)\,dx-\sum_{i=1}^{n}f(x_i^*)(x_i-x_{i-1})\right|.$$

(a) Show that if the function f is continuous and increasing on $[a,b]$, then the error E_n satisfies the following inequality:

$$E_n\le\frac{b-a}{n}|f(b)-f(a)|$$

when **RiemLeft** is used to estimate the definite integral of f on $[a,b]$.

(b) Show that if the function f is continuous and decreasing on $[a,b]$, then E_n satisfies the inequality inpart (a) when **RiemLeft** is used to estimate the definite integral of f on $[a,b]$.

(c) Show that if the function f is continuous and decreasing on $[a,b]$, then E_n satisfies the inequality in part (a) when **RiemRight** is used to estimate the definite integral of f on $[a,b]$.

(d) Show that if the function f is continuous and decreasing on $[a,b]$, then E_n satisfies the inequality in part (a) when **RiemRight** is used to estimate the definite integral of f on $[a,b]$.

30. Recall the following bounds property of *Riem*: If m and M are lower and upper bounds respectively for the function f on the interval $[a,b]$, then

$$m(b-a) \le Riem(f,a,b) \le M(b-a).$$

(a) Give an alternate proof of the bounds property for $Riem$ by applying the monotonicity property of $Riem$ to the functions represented by the constants m and M, and the function f.

(b) Make a sketch that gives a geometrical interpretation of the bounds property.

(c) Use the bounds property to give an interpretation of the number represented by

$$\frac{1}{b-a} Riem(f,a,b).$$

(d) For a continuous function f on $[a, b]$, use the Intermediate Value Theorem, page 275, and the bounds property of $Riem$ to show that the number represented by $\frac{1}{b-a} Riem(f,a,b)$ is the value of f for some point c in $[a, b]$.

31. (a) Show that

$$\lim_{n \to \infty} \sum_{i=1}^{n} \frac{f(x_i^*)}{n} = \frac{1}{b-a} Riem(f,a,b)$$

using the fact that

$$x_i - x_{i-1} = \frac{b-a}{n}.$$

(b) Based on the result of part (a), give an interpretation of the number represented b

$$\frac{1}{b-a} Riem(f,a,b).$$

32. Write a single equivalent integral using the properties of definite integrals.

(a) $\int_{-1}^{3} f(x)\,dx + \int_{-1}^{3} f(x)\,dx$

(b) $\int_{2}^{5} g(v)\,dv + \int_{0}^{2} g(v)\,dv$

(c) $2\int_{1}^{2} f(x)\,dx + \int_{1}^{2} g(t)\,dt$

(d) $-3\int_{-1}^{3} f(y)\,dy + 5\int_{-1}^{3} f(x)\,dx$

(e) $\int_{1}^{2} f(t)\,dt + \int_{5}^{9} f(v)\,dv + \int_{2}^{5} f(x)\,dx$

(f) $\int_{-1}^{3} g(x)\,dx - \int_{3}^{-1} h(x)\,dx$

33. The average value, f_{avg}, of a function f on the interval $[a, b]$ is given by

$$f_{avg} = \frac{1}{b-a} \int_{a}^{b} f(x)\,dx.$$

Solve the following problems.

(a) Find the average value of the function f given by $f(x) = x^2$ on the interval $[0, 3]$.

(b) Find the average value of the function h given by

$$h(x) = \frac{2}{x}$$

on the interval $[1, 3]$.

(c) Find the average value of the function g given by $g(x) = 2\sqrt{x}$ on the interval $[1, 9]$.

(d) The voltage v in an electrical circuit is given by $v = V(t) = 220\sin(t)$. Find the average voltage on the interval $0 \le t \le \pi$.

(e) The height s of a ball thrown straight up from ground level is given by $s = S(t) = 48t - 16t^2$ feet after t seconds. Find the average height of the ball above the ground during the first two seconds after it is thrown up.

(f) For the ball in part (d), find the average velocity of the ball during the first second after it is thrown up.

34. In this exercise you will use the error estimates for the trapezoid rule given by

$$|E_n| \le \frac{(b-a)^3}{12n^2} \max_{a \le x \le b} |f''(x)|$$

and for the midpoint method given by

$$|E_n| \le \frac{(b-a)^3}{24n^2} \max_{a \le x \le b} |f''(x)|.$$

(a) Estimate the absolute error, $|E_{10}|$, when you use the trapezoid method for computing the following definite integral

$$\int_1^2 \frac{1}{x}\, dx.$$

(b) Estimate the absolute error, $|E_{10}|$, when you use the midpoint method for computing the following definite integral

$$\int_1^2 \frac{1}{x}\, dx.$$

Compare your answer with your answer in part (a).

(c) What value of n will guarantee that the absolute error E_n in using the trapezoid method to approximate the integral inpart (a) will be no more than 0.001?

(d) What value of n will guarantee that the absolute error E_n inusing the midpoint method to approximate the integral in part (a) will be no more than 0.001? Compare your answer with your answer in part (c).

(e) What is the exact number that you are approximating in parts (a) and (b)?

(f) What does your experience with parts (a)-(d) tell you about the use of the trapezoid method versus the use of the midpoint method to approximate a definite integral?

35. Below is **ISETL** code for a **computer function** named **simp** that implements an approximation rule know as *Simpson's Rule*. As for **RiemLeft**, **RiemRight**, **RiemMid**, **RiemMax**, **RiemMin**, and **RiemTrap**, Simpson's Rule uses a partition of an interval $[a, b]$ into n subintervals of equal length and it is usually more accurate than the **Riem computer functions** you have used previously. The rule is derived by using quadratic functions to approximate a given function f on pairs of successive subintervals using three successive partition points as shown in Figure 5.7.
Study the following code for Simpson's Rule.

```
simp := func(f, a, b, n);
  x := func(i);
      return a+(b–a)*(i/n);
    end;
  c := func(i);
      if i in {0, n} then return 1;
      elseif odd(i) then return 4;
      else return 2;
    end;
  end;
  return (b–a)/(3*n)*%+[c(i)*f(x(i)) : i
  in [0..n]];
end;
```

(a) Write a mathematical expression for Simpson's Rule.

(b) Why does the value of n have to be even to use Simpson's Rule?

(c) Explain why you think Simpson's Rule might be more accurate than the **Riem computer functions** you used previously. Make sketches to illustrate your explanation.

(d) Use **RiemTrap** and **Simp** to approximate the following definite integral to four decimal places.

$$\int_0^2 \sqrt{4-x^2}\, dx$$

What is the smallest value of n you can use for each rule to obtain the desired accuracy? Which method is in your estimation better?

36. Give a definition for the definite integral of a function defined in three parts. What about a definition of the general case for the definite integral of a function with $n > 2$ parts?

37. Evaluate the following definite integrals. The notation $\lfloor x \rfloor$ represents the greatest integer less than or equal to x.

   (a) $\int_{-3}^{2} \lfloor 0.5x \rfloor \, dx$ (b) $\int_{-1}^{3} x \lfloor x \rfloor \, dx$

   (c) $\int_{0}^{2\pi} \lfloor 2\sin(x) \rfloor \, dx$

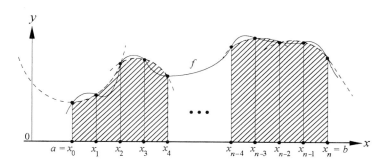

Figure 5.7. Simpson's rule

■ 5.3 THE FUNDAMENTAL THEOREM OF CALCULUS

5.3.1 Overview

In this section you will continue your study of the relationship between the two processes of differentiation and integration. You will learn one of the most important and beautiful theorems in all of mathematics. That theorem, known as *The Fundamental Theorem of Calculus*, characterizes the reversal relationship between differentiation and integration, and it relates limits of Riemann sums (the definite integral) to the indefinite integral.

In this section you will study the Fundamental Theorem of Calculus (Theorem 5.1) and consider some of its consequences. This will be a formal discussion of ideas which, hopefully you already understand, to some extent. We will work entirely from Definition 5.3 and hence no interpretation as anti-derivative can be used until the theorem is proved. The proof of the theorem will rely to a large extent on the properties of *Riem* that were discussed in Section 5.2.

Finally, we will consider the general problem of defining a function using an integral. In particular, the natural logarithm function will be defined using an integral with a variable upper limit. You will see how the various properties of the natural logarithm and exponential functions are consequences of the integral definition of the natural logarithm function and the inverse properties of these functions. The general problem of differentiating functions defined by integrals with variable limits of integration will also be solved.

5.3.2 Activities

NOTE: The **computer function DefInt** which you used in the activities of Section 5.2, illustrates the idea of the definite integral very nicely and is therefore extremely useful for understanding this concept. However, it can be very slow in some cases. This problem will disappear as computers become faster, but for now, the use of **DefInt** in its present form can lead to some long delays. Therefore, in this set of activities and all of the activities and exercises in the remainder of this text which call for you to use **DefInt**, write a new **computer function DefInt** which accepts a **computer function f**, and numbers a and b and returns the value of **RiemMid(f, a, b, 25)**. If any of the following activities takes too much time on your system, then replace 25 by a smaller number.

1. Write a **computer function Int** which accepts a **computer function f** and two numbers **a** and **b** defining an interval $[a, b]$ as input and returns a **computer function** whose domain is $[a, b]$ and whose value at any $x \in [a, b]$ is the result of applying **DefInt** to **f** with endpoints **a** and **x**.

   Apply **Int** to the function represented by $2x$ on the interval $[0, 10]$. The result is a function you have worked with before. Use **table**, graphing, or any other method to discover what that function is.

2. Write a **computer function L** which applies the **computer function DefInt** to the function given by the expression $1/t$ with lower limit of integration $a = 1$ and variable upper limit x, so that **L** has domain all positive real numbers.

   Make a table of twenty values for x and $L(x)$ using the tool **table** on the interval $[1, 21]$. By examining your table of values, experimenting with values of x and $L(x)$, looking at the defining expression for $L(x)$, and thinking about properties of integrals, do the following:

   (a) Determine when the **computer function L** is zero.

   (b) Determine relationships between

      i. $L(a)$, $L(b)$, and $L(ab)$

      ii. $L(a)$, $L(b)$, and $L\left(\dfrac{a}{b}\right)$

      iii. b, $L(a)$, and $L\left(a^b\right)$

   (c) Give a geometric interpretation for the values of $L(x)$ as an area under a curve. Discuss the two cases $0 < x < 1$ and $x > 1$ by making appropriate sketches.

   (d) Use the **computer function bis** to approximate the value of x where $L(x) = 1$ to four decimal places. What value of x are you approximating? Explain your answer. Recall that **bis** is a **computer function** representing the bisection method for approximating zeros of a function.

3. Let **f** be a **computer function** representing the function determined by $2x \sin(x^2)$. Apply **Int** to **f** with endpoints 0, π. Now apply the **computer function D** to the latter result. Call the resulting **computer function g**. Can you say what function g represents?

   Do the same thing again, except in reverse order, i.e., apply **D** to **f** and then apply **Int** to the result, using the interval $[0, \pi]$. Call the resulting **func h**.

Use your MPL tool **table** to make a table off our columns representing the independent variable, and corresponding values of **f**, **g**, and **h**.

WARNING: If you use the full interval $[0, \pi]$ for your table, then one of the three functions gives some trouble. Write a brief explanation as to why.

4. Repeat the previous activity for the function represented by $2x + 1$ on $[0, \pi]$. What is different about your results? Write a brief explanation of why.

5. Let f_1 be the function defined in parts by the following expression.

$$f_1(x) = \begin{cases} x^2 & \text{if } 0 \le x < 1 \\ 2 - x & \text{if } 1 \le x \le 2 \end{cases}$$

and let f_2 be the function defined by

$$f_2(x) = \begin{cases} x^2 & \text{if } 0 \le x < 1 \\ 4 - 2x & \text{if } 1 \le x \le 2 \end{cases}$$

Perform all of the tasks of Activity 3 on these two functions. Write a brief statement describing how your results compare with each other and with the results of Activity 3.

6. Apply your **computer function Int** to a **computer function** representing the cosine function, cos, on the interval $[0, 2\pi]$. What is the derivative of the resulting function?

Now define a **computer function** whose value at x is the result of applying **DefInt** to the cosine function on the interval $[0, x^2]$. Try to find the simplest possible code for doing this. What is the derivative of the function represented by the resulting **computer function**?

7. This activity is a continuation of Activity 1, Section 5.2.2, where we began with an indefinite integral problem and then passed on to a more definite situation by "evaluation at the endpoints". Observe that when you are evaluating at two endpoints and taking the difference, the indefiniteness of the anti-derivative does not matter. Thus, in Activity 1, Section 5.2.2, another function whose derivative is $\overline{C}_p$, where $\overline{C}_p(T) = \frac{C_p T}{T}$, could be

$$S(T) = 25.999T + 0.02175T^2 - \left(49.43733\ldots \times 10^{-6}\right)T^3$$

and the answer to the problem is obtained by substituting in this expression, first 350 for T, then 300 for T and then subtracting the second from the first.

Check out this general observation by making some hand computations. Figure out $\int_1^2 x\, dx$ by first finding a function whose derivative is given by x then evaluating at the endpoints and subtracting. Check your result with your SCS.

5.3.3 Discussion

Area Over an Interval with Variable Endpoint   The **computer function Int** transforms the calculation of a number in **DefInt** to the construction of a function. In **DefInt** the job is to compute a number which estimates the (net) area under a curve. See Figure 5.8.

Figure 5.8. The area under a curve over the interval [a, b].

The (net) area is taken between very definite endpoints. Now suppose one endpoint is varied. For every number x that you pick, you can calculate the (net) area under the curve on the interval $[a, x]$ and you get a number. This process is a function and it can be approximated by the **computer function** produced by **DefInt**. See Figure 5.9.

The question is, "what is the function?" Can we know anything about the function that **DefInt** approximates? If we start with the function represented by $2t$ and construct the function that gives, for each x, the area under $2t$ from 0 to x, what function have we constructed?

In Activity 1, you made a table of this function's values. There is a reasonable guess for what function this is. Try integer values for x and see what you get.

Its not as easy to guess what function gives for each x, the area under the graph of $1/t$ from 1 to x. Can you guess what this function is by its properties which you deduced in Activity 2?

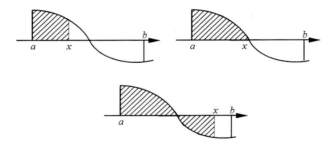

Figure 5.9. The area under a curve over the interval [a, x] with $a \le x \le b$.

Differentiation and Integration   In Activity 3, you applied **Int** to the function given by $2x\sin(x^2)$. If you draw the graph of the function produced by **Int**, it should look familiar.

One thing you can do to find out something about a function is to take its derivative. This is what **D** does (approximately). Are you surprised by the result? The third column of the table you constructed in Activity 3 is the result of applying first **Int** and then **D** to the function given by $2x\sin(x^2)$. Are you surprised by the relationship between the first and third columns?

The same thing happens if the order is reversed. If you apply first **D** and then **Int** to the function given by $2x\sin(x^2)$, then the resulting function is pretty close to the one you started with.

It is not an accident. So what is different about the results obtained in the example in Activity 4? Are the first and third columns of your table the same? How different are they? Do you see a pattern? Is this an accident?

Altogether, in Activities 3, 4 and 5 you made two tables for each of four functions. Which of these resulted in the first and third columns being the same? Can you state a general rule about what happens when you apply **Int** and **D** successively to a function?

The Fundamental Theorem of Calculus — Informal

At this point, you may have begun to realize that there is an amazing connection between computing the (net) area under the graph of a given function — what we have called *Riem*, and finding a function whose derivative is the given function. Indeed all of the facts about this very deep result of calculus are now available to you in approximate form. The only step that remains is to pass to the limit as the length of subintervals in a partition goes to zero. This gives you the Fundamental Theorem of Calculus discovered simultaneously and independently by Gottfried Leibniz and Sir Isaac Newton.

We will review the facts in their approximate form, and then give precise formulations and proofs of these profound relations.

Let's begin with summarizing the facts.

- If you start with a function f and a point a in its domain and construct a new function g which gives for each x the (net) area under the graph of f, then the derivative of g is the function f that you started with.

In mathematical notation, we write,

$$g(x) = Riem(f, a, x) = \int_a^x f(t)\,dt$$

and

$$g'(x) = f(x)$$

This is the basic relation that was one of the things exemplified in Activities 3 and 4.

It says that the problem of finding the (net) area under the graph of f is the same as the problem of finding a function whose derivative is f. This means that what we have been calling the anti-derivative is intimately related to the (net) area.

- Now let's reverse the order. If you start with a function f, differentiate it, and then compute the (net) area function, you don't quite get f back. Your result might be off by a constant. The reason for this is that if you construct a function by computing the (net) area from a to x, then the value of this function at a must be 0 (see page 327). Hence, in order to get back to f you have to add the value of f at a.

Thus we have the following relation. If you start with a function f, differentiate it, compute the (net) area function on the interval $[a, x]$ and add $f(a)$, then the resulting function is the same as f. In mathematical notation, using the integral notation instead of *Riem*,

$$\int_a^x f'(t)\,dt + f(a) = f(x)$$

or, in more standard form,

$$f(x) = f(a) + \int_a^x f'(t)\,dt.$$

The Fundamental Theorem of Calculus — Formal

In order to give precise proofs of relationships it is necessary to work with precise definitions. In this paragraph we formalize not only definitions of concepts you have been working with, but also the statement of the main theorem that makes it all work.

Definition 5.3

Let f be a function which is continuous on the interval $[a, b]$. *For each positive integer n let* P_n *be the partition of* $[a, b]$ *into n equal subintervals. Thus*

$$P_n = (x_i)_{i=0}^n = \left(a = x_0, \, x_0 + \frac{1}{n}(b-a), \ldots, x_0 + \frac{n-1}{n}(b-a), \, x_n = b \right)$$

Let $Riem(f, a, b, n)$ *be the number* R_n *given by*

$$R_n = \sum_{i=1}^n f(x_{i-1})(x_i - x_{i-1}).$$

We define the Riemann integral of f over $[a, b]$ *to be the number* $\lim_{n \to \infty} R_n$ *and we denote it by*

$$\int_a^b f$$

or

$$\int_a^b f(x)dx.$$

You will note that the process of computing R_n is same as that of **RiemLeft(f, a, b, n)**, so this is just a change in notation. We should note that the exact value of R_n and the value of **RiemLeft(f, a, b, n)** may not be the same. Do you see why? Also, after passing to the limit, we switch to the integral notation. It cannot, therefore, refer to the use of an anti-derivative as well, until we have proved they are the same. This will be accomplished as soon as we complete the proof of the Fundamental Theorem of Calculus.

We also note that in Definition 5.3, we could have replaced the function values $f(x_{i-1})$, for $i = 1, 2, \ldots n$, by $f(x_i^*)$ where x_i^* is any point in the ith subinterval $[x_{i-1}, x_i]$.

There are a number of issues raised by this definition. The first is the question of the existence of this integral. That is, if f is continuous on $[a, b]$, how do we know that the limit in the definition actually exists? The next issue has to do with all we have said about the integral being the solution to the problem of finding a function whose derivative is given. Finally, we must show that all of the properties of the integral which we have been working with are actually true, given this definition.

The proof of the existence of the limit used in defining the Riemann integral follows from deep properties of numbers and is beyond the scope of this text. We will deal with the other issues in the following pages. For now, we will concentrate on this definition which formalizes the concept of the definite integral of f over $[a, b]$ as the area under the graph of f from a to b, provided f is non-negative. We have been working with this since the beginning of the book. The expression whose limit is taken in the definition is precisely the number which is the value of the actual mathematical process of **RiemLeft(f, a, b, n)** if f is a **computer function** that represents the function f.

For the remainder of this chapter we will take this definition as the meaning of the integral. Any other meaning which we would like to use (such as anti-derivative) will have to be proved before it is used.

The following theorem is a formalization of the phenomena that you observed in Activities 3, 4 and 5. The theorem is actually in two parts. The first part corresponds to integrating a function and then taking the derivative, and the second part reverses the process by differentiating first and then integrating.

> **Theorem 5.1 (The Fundamental Theorem of Calculus)**
>
> 1. *Let f be a continuous function on [a, b] and for $x \in [a, b]$, let G be the function defined by*
>
> $$G(x) = \int_a^x f(t)\, dt.$$
>
> *Then,*
>
> $$G'(x) = f(x).$$
>
> 2. *Let f be a function whose derivative exists and is continuous on [a, b]. Then*
>
> $$\int_a^b f'(x)\, dx = f(b) - f(a).$$

The second part of this theorem can be related to anti-derivatives in the following manner:

Let F be any anti-derivative of a function f on the interval $[a, b]$, that is, $F' = f$. Then

$$\int_a^b f(x)dx = F(b) - F(a).$$

Proof of the Fundamental Theorem. Following are the properties of *Riem* that will be used in the proof. Look back in Section 5.2.3 for a discussion of how they are proved. We will write them using the definite integral notation in placeof *Riem*.

P1: If f and g are functions on $[a, b]$ and s and t are numbers, then

$$\int_a^b (sf + tg) = s\int_a^b f + t\int_a^b g.$$

P2: If f is a continuous function on $[a, b]$ and $c \in [a, b]$, then

$$\int_a^b f = \int_a^c f + \int_c^b f.$$

More generally, property **P2** is true when c is any real number: that is, if c is not in $[a, b]$. You will have an opportunity to prove the general property in the exercises.

P3: If f is defined at a, then

$$\int_a^a f = 0.$$

P4: If f is continuous on $[a, b]$, then

$$\left|\int_a^b f\right| \le \int_a^b |f|.$$

P5: If f is continuous on $[a, b]$ and $m \le f \le M$ for every $x \in [a, b]$, then

$$m(b-a) \le \int_a^b f \le M(b-a).$$

□

There are two additional facts that we will need to use. They are important in their own right and so we present them as independent propositions. The first has nothing to do with integrals and is a consequence of the Mean Value Theorem for derivatives discussed in Chapter 4. It is, in fact, a formal statement and proof of something you have known and used for some time — any two anti-derivatives of a function differ from each other by a constant function.

Proposition 5.1

Let f and g be two functions which are differentiable on the interval [a, b] and suppose that $f'(x) = g'(x)$ for all x in [a, b].

   Then the function $f - g$ is a constant on [a, b], that is, there is a number c such that

$$f(x) = g(x) + c, \quad x \in [a, b].$$

Proof. Let $h = f - g$. We will show that $h(x) = h(a)$ for all $x \in [a, b]$ which will establish that f and g differ by a constant.

   Note that since $f'(x) = g'(x)$, it follows that h' is the zero function. Using the Mean Value Theorem on page 278, we can write, for any $x \in [a, b]$,

$$h(x) - h(a) = h'(x^*)(x - a)$$

where x^* is some number in (a, b). It doesn't matter which, because $h'(x^*) = 0$ for any such number and so we have, $h(x) = h(a)$, as desired.

$\square$

A geometric interpretation of Proposition 1 is shown in Figure 5.10.

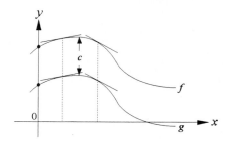

Figure 5.10. *If $f'(x) = g'(x)$, then $f(x) = g(x) + c$.*

Another way of stating this fact is that if the derivative of a function is the zero function, then the original function must be a constant.

   The next property requires us to return to the original definition of the integral (as a limit of Riemann sums, page 342) to evaluate one simple definite integral.

Proposition 5.2

If f is the constant function given by $f(x) = 1$, *then*

$$\int_a^b f(x)\, dx = \int_a^b dx = b - a.$$

Proof. Referring to Definition 6.1 we must take the limit as $n \to \infty$ of

$$R_n = \sum_{i=1}^{n} f(x_{i-1})(x_i - x_{i-1})$$

But before taking the limit, we can substitute 1 for $f(x_{i-1})$ and so we have, for this function,

$$
\begin{aligned}
R_n &= \sum_{i=1}^{n} (x_i - x_{i-1}) \\
&= (x_1 - a) + (x_2 - x_1) + (x_3 - x_2) + \ \ldots \ + (x_{n-1} - x_{n-2}) + (b - x_{n-1}) \\
&= b - a
\end{aligned}
$$

Hence, taking the limit just gives the value of this sum, $b - a$.

$\square$

Now we are ready to prove the theorem. For completeness, we restate it first.

Theorem 5.1 (The Fundamental Theorem of Calculus)

1. *Let f be a continuous function on [a, b] and for* $x \in [a, b]$, *let G be the function defined by*

$$G(x) = \int_a^x f(t)\, dt.$$

    Then,

$$G'(x) = f(x).$$

2. *Let f be a function whose derivative exists and is continuous on [a, b]. Then*

$$\int_a^b f'(x)\, dx = f(b) - f(a).$$

Proof. For the first part, we must determine the derivative G' of G. Using property **P2** we may compute the difference quotient,

$$
\begin{aligned}
\frac{G(x+h) - G(x)}{h} &= \frac{1}{h}\left(\int_a^{x+h} f(t)\, dt - \int_a^x f(t)\, dt \right) \\
&= \frac{1}{h} \int_a^{x+h} f(t)\, dt.
\end{aligned}
$$

Now, we have to take the limit as $h \to 0$ and show that the result is $f(x)$. What we will do is to manipulate $f(x)$ a bit, subtract it from the above expression and show that the limit of the difference is 0. Using Proposition 5.2 and property **P1**, we have for $h \neq 0$,

$$f(x) = f(x) \cdot 1 = f(x) \cdot \frac{1}{h} \int_x^{x+h} 1 \, dt$$

$$= \frac{1}{h} \int_x^{x+h} f(x) \, dt.$$

Hence we may compute, using properties **P1**, **P4**, and **P5**,

$$\left| \frac{G(x+h) - G(x)}{h} - f(x) \right| = \left| \frac{1}{h} \int_x^{x+h} f(t) \, dt - \frac{1}{h} \int_x^{x+h} f(x) \, dx \right|$$

$$= \left| \frac{1}{h} \int_x^{x+h} (f(t) - f(x)) \, dt \right|$$

$$\leq \left| \frac{1}{h} \right| \int_x^{x+h} |f(t) - f(x)| \, dt$$

$$\leq \frac{1}{|h|} M |h| = M$$

where M is the maximum value of $|f(t) - f(x)|$ for t between x and $x+h$.

Now, because we have assumed that f is continuous, this maximum value goes to 0 as $h \to 0$. Hence we have,

$$\lim_{h \to 0} \frac{G(x+h) - G(x)}{h} - f(x) = 0$$

and so,

$$G'(x) = \lim_{h \to 0} \frac{G(x+h) - G(x)}{h} = f(x).$$

This proves the first part and we may turn to the second.

In order to compute $\int_a^b f'$ we define the function G on the interval $[a, b]$ by

$$G(x) = \int_a^x f'(t) \, dt$$

According to the first part of this theorem, which we have already proved, the derivative of G is f'. That is,

$$G' = f'$$

It follows from Proposition 5.1 that $G - f$ is a constant. We can actually determine the numerical value of this constant by evaluating $G - f$ at a single point. We choose the point a because we know from **P3** that $G(a) = \int_a^a f' = 0$. Thus we have,

$$(G - f)(a) = G(a) - f(a) = -f(a)$$

and so, for all $x \in [a, b]$,

$$(G - f)(x) = G(x) - f(x) = -f(a).$$

In particular for $x = b$, we have

$$G(b) - f(b) = -f(a)$$

so

$$\int_a^b f'(x)\, dx = G(b) = f(b) - f(a).$$

☐

Rules for Integration

Rules for Integration   It is now a very easy matter to state and prove the rules for integration listed below. They all follow immediately from the interpretation of the integral as an anti-derivative and they will be stated here in the traditional form of an anti-derivative plus an arbitrary constant. We could have done all of this at the very beginning of the chapter when we were using the term integral solely for its meaning as an anti-derivative. Now, however, you can see that these rules have much more meaning since they can also refer to the (net) area under a curve, as well as to all of the other applications of integrals that will be discussed in the next section.

In the following list, the symbol c refers to an arbitrary constant and a refers to a fixed constant. In each case the statement is followed by an explanation of why it is true. This generally takes the form of differentiating the right hand side to see that it gives the *integrand* (the function which is being integrated).

1. $\int 0\, dx = c$

    The derivative of a constant function is the zero function.

2. $\int a\, dx = ax + c$

    The derivative of the function given by ax is a.

3. $\int x^a\, dx = \dfrac{x^{a+1}}{a+1} + c,\ (a \neq -1)$

    The derivative of the function given by $\frac{x^{a+1}}{a+1}$ is x^a. If $a = -1$, then $\frac{x^{a+1}}{a+1}$ does not define a function.

4. There are two cases of integrals leading to natural logarithms.

    (a) $\int \frac{1}{x}\, dx = \ln x + c,\ x > 0$

    The derivative of the function given by $\ln(x)$, with domain $(0, \infty)$ is $\frac{1}{x}$.

    (b) $\int \frac{1}{x}\, dx = \ln(-x) + c,\ x < 0$

    The derivative of the function given by $\ln(-x)$, with domain $(-\infty, 0)$ is $\frac{1}{x}$.

    Notice that by the chain rule, we have

    $$(\ln(-x))' = -\frac{1}{-x} = \frac{1}{x}$$

    A shorthand version of the above rule can be expressed as,

    $$\int \frac{1}{x}\, dx = \ln(|x|) + c,\ x \neq 0.$$

(Actually when using the shorthand version, you should remember that since $\ln(|x|)$ has two branches and is discontinuous at 0, this rule is really two statements: $\int \frac{1}{x}\, dx = \ln(x) + c_1$ for $x > 0$ and $\frac{1}{x}\, dx = \ln(-x) + c_2$ for $x < 0$, where c_1 and c_2 may be different constants.)

The derivative of the function given by $\ln(|x|)$ is $\frac{1}{x}$.

5. Each of the following integral rules follow immediately by taking the derivative of the function on the right hand side and verifying that the derivative is the integrand function.

(a) $\int \sin(x)\, dx = -\cos(x) + c$ (b) $\int \cos(x)\, dx = \sin(x) + c$

(c) $\int \tan(x)\, dx = -\ln(|\cos(x)|) + c = \ln(|\sec(x)|) + c$

(d) $\int \cot(x)\, dx = \ln(|\sin(x)|) + c$ (e) $\int \sec(x)\, dx = \ln(|\sec(x) + \tan(x)|) + c$

(f) $\int \csc(x)\, dx = \ln(|\csc(x) - \cot(x)|) + c$ (g) $\int \sec^2(x)\, dx = \tan(x) + c$

(h) $\int \csc^2(x)\, dx = -\cot(x) + c$ (i) $\int \sec(x)\tan(x)\, dx = \sec(x) + c$

(j) $\int \csc(x)\cot(x)\, dx = -\csc(x) + c$ (k) $\int \frac{1}{1+x^2} = \arctan(x) + c$

(l) $\int \frac{1}{\sqrt{1+x^2}} = \arcsin(x) + c$ (m) $\int e^x\, dx = e^x + c$

6. $\int f'(g(x))g'(x)\, dx = f(g(x)) + c$

Using the chain rule, we have

$$(f \circ g)'(x) = f'(g(x))g'(x)$$

7. $\int f(x)g'(x)\, dx = f(x)g(x) - \int f'(x)g(x) + c$

According to the product rule for differentiation,

$$(f(x)g(x))'(x) = f'(x)g(x) + f(x)g'(x)$$

and so,

$$f(x)g'(x) = (f'(x)g(x))' - f'(x)g'(x).$$

The technique of integration described in Rule 7 is known as *integration by parts*.

Except for the last two, each of these rules is quite straightforward. If you can recognize an expression as being in the form given on the left hand side of the equation, then its integral is given on the right hand side.

The last two rules work the same way, except that it can be tricky to recognize a given expression as being of the form given on the left hand side. You will see some examples and have a chance to practice with these rules in the exercises.

Of course, the material in this paragraph amounts to just finding specific anti-derivatives. You can always use your SCS for that. Aside from that, the rules do have some value in theoretical analyses. You will see a little of this in the next paragraphs.

In all of these integrals, one can obtain corresponding rules for definite integrals by evaluating at the two endpoints and subtracting. This is straightforward in all cases except the one that uses the chain rule. Here we have,

$$\int_a^b f'(g(x))g'(x)\,dx = f(g(b)) - f(g(a))$$

or, if the function g has an inverse, we write $y = g(x)$, $dy = g'(x)\,dx$, $u = g(a)$, $v = g(b)$, and

$$\int_{g^{-1}(v)}^{g^{-1}(v)} f'(y)\,dy = f(v) - f(u).$$

This is referred to as the *method of integration by substitution*.

Defining Functions by an Integral

Let's begin this paragraph with a little mathematical problem.

What would it mean if you were told that a certain function is given by the expression,

$$\int_{\pi}^{x^2} \cos(t^2)\,dt\,?$$

In the first place there is the question of defining a function in this way. It can be shown that in this case that there is no simpler definition, no expression other than the integral which gives the function. We have to live with that. It certainly produces a function because we can define a function by using a definite integral in which one of the endpoints is a function. This is essentially what the **computer function Int** does.

To say that a function is given by an expression of the form

$$\int_a^{b(x)} f(t)\,dt$$

means that the process of the function works like this. You start with an x, plug it into the expression $b(x)$ and calculate. This gives you a number, say b. Then you have the number $\int_a^b f$ and that is the value of the function.

How would you describe the processes of functions given by expressions of the following two forms?

$$\int_{a(x)}^b f(t)\,dt$$

$$\int_{a(x)}^{b(x)} f(t)\,dt$$

The Natural Logarithm and Exponential Functions

The idea of defining functions by integrals with variable endpoints can be used to put functions like ln and exp on a solid logical foundation. Do you think that it is important to do so? Of any value at all? A waste of time? Suppose that you knew absolutely nothing about these functions, or the number e, but you did have at your disposal all of the material about derivatives and integrals of functions in general. You will need to remember the material about inverses of functions from Chapter 4.

You will also need one fact about differential equations and initial value problems. A problem of the form, given a function F, find a function y which satisfies

$$f' = F(x), \quad f(a) = c$$

has exactly one solution, i.e., one function f which satisfies the equation and has the value c at a.

That last statement is not completely true. It requires that the right hand side F satisfies some technical conditions, but we will not dwell on that. This point will only be used to suggest that the functions we define are, in fact the familiar logarithm and exponential functions.

We begin by recalling the function L defined on the domain $(0, \infty)$ by

$$L(x) = \int_1^x \frac{1}{t}\, dt$$

which you worked with in Activity 2. Is L defined for all positive real numbers? What about $L(0.5)$?

The function L is going to turn out to be the natural log function (no secrets in this book!). We might convince ourselves of the plausibility of that by noticing that L satisfies many of the same properties which ln has. We list those in the following proposition, but we won't spoil your fun by proving them. The proofs are all obtained by using the appropriate properties of integrals, derivatives, and the material in Chapter 4, Section 4.

Proposition 5.3

1. $L(ab) = L(a) + L(b)$

2. $L\left(\dfrac{a}{b}\right) = L(a) - L(b)$

3. $L(a^{-1}) = -L(a)$

4. $L(a^b) = bL(a)$

5. $L'(x) = \frac{1}{x}$

6. L is strictly increasing

7. L has an inverse function E defined on $(-\infty, \infty)$.

8. $E'(x) = E(x)$

9. $E(0) = 1$

The last two items on the list tell us that E is our familiar exponential function. In fact, these two statements say that E is a solution of the initial value problem,

$$y' = y,\ y(0) = 1$$

But the exponential function is also a solution of this problem, so by the uniqueness property, the two are in fact the same function.

Since E is the inverse of L, it follows that L is the inverse of E, i.e., "the inverse of the exponential function". Hence, the logarithm function.

Finally, the number e can be defined to be $E(1) = \exp(1)$.

Derivatives of Integrals with Variable Endpoints

The Fundamental Theorem of Calculus tells us that the derivative of the function given by $\int_a^x f$ is f. So what is the derivative of the function given by

$$\int_\pi^{x^2} \cos(t^2)\,dt\,?$$

Before answering this question, let us first note that by now you should be able to compute (either by hand, or with the help of a SCS) the derivative of any *elementary function*. By an elementary function we mean all functions which can be obtained by making combinations using arithmetic (sums, differences, powers, products, quotients, etc.) and composition of all the familiar *algebraic* and *transcendental* functions. By an algebraic function we mean any function which is explicitly or implicitly defined by a polynomial equation in two variables, including polynomial functions, rational functions (ratios of polynomials) and any function which is a combination obtained by combining polynomials, powers, and roots. Any function which is not algebraic is said to be *transcendental*. Among the transcendental functions are the trigonometric, logarithmic and exponential functions.

On the other hand, the process of integration is not so nice! Some functions have no elementary function as an anti-derivative. This was shown by Joseph Liouville in 1835 when he proved that the function F given by

$$\int_o^x \sin(t^2)\,dt$$

is not an elementary function, hence the function represented by the expression $\sin x^2$ has no elementary function as an anti-derivative. Hence, part (2) of the Fundamental Theorem of Calculus (see page 343) isn't very helpful. But, part (1) says that any function has an anti-derivative theoretically! As you have seen in some of the exercises, some important functions for applications are defined by definite integrals with variable limits of integration. We can differentiate such functions by part (1) of the Fundamental Theorem of Calculus.

In order to differentiate the function represented by $\int_\pi^{x^2} \cos(t^2)\,dt$, what you have to do is put together the Fundamental Theorem of Calculus and the Chain Rule. Remember that in order to apply the Chain Rule, you have to be able to express a function as the composition of two functions. If we write

$$f(x) = \int_\pi^{x^2} \cos(t^2)\,dt\,.$$

can you express f as the composition of two functions g and h, so that $f = g \circ h$?

Here is how you do it. Recall the explanation of how the expression for f works. It requires two steps. First apply the function given by x^2, which in this case is h. We will take

$$h(x) = x^2\,.$$

Then integrate from π to the value given by $h(x)$. That will be our function g, so

$$g(x) = \int_\pi^x \cos(t^2)\,dt\,.$$

Do you see that with these definitions we have $f = g \circ h$?

We can then apply the Chain Rule to get,

$$f'(x) = g'(h(x))h'(x)\,.$$

This is a formula for the derivative we are looking for, so we just have to calculate the parts and plug them in.

Do you have any difficulty with the following?

$$h'(x) = 2x$$

$$g'(x) = \cos(x^2)$$

so

$$g'(h(x)) = \cos(x^4)$$

so

$$f'(x) = (\cos(x^4))(2x) = 2x\cos(x^4).$$

Can you write a general formula for the derivative of the function given by $\int_a^{b(x)} f(t)dt$? Is this problem related to Activity 6?

5.3.4 Exercises

In Exercises 1 - 40, using your SCS, or by hand, find an anti-derivative of the integrand function for each of the following definite integrals, if possible. If so, evaluate the definite integral by hand using the Fundamental Theorem of Calculus. Check your answer by using your SCS to evaluate the definite integral. If you, or your SCS, cannot find an anti-derivative of the integrand function, then use your SCS to estimate the integral.

1. $\int_0^3 -0.5dx$

2. $\int_1^4 2xdx$

3. $\int_{-2}^1 \frac{t}{5}dt$

4. $\int_{-1}^1 \theta d\theta$

5. $\int_0^2 1-vdv$

6. $\int_1^2 3-2ydy$

7. $\int_1^3 3x^2-2x+5dx$

8. $\int_0^1 t^{\frac{2}{3}}dt$

9. $\int_0^{\pi/4} \sec^2(x)dx$

10. $\int_0^{\pi/3} \sin(\theta)d\theta$

11. $\int_{-1}^{0.5} 3\sin(\pi\alpha)d\alpha$

12. $\int_0^1 \sqrt[3]{t^2}dt$

13. $\int_1^9 x-\sqrt{x}dx$

14. $\int_1^2 \frac{3-x-x^2}{x^4}dx$

15. $\int_0^\pi \cos(\theta)d\theta$

16. $\int_0^\pi \sin(2x)dx$

17. $\int_0^1 e^x dx$

18. $\int_0^1 e^{-2x}dx$

19. $\int_0^1 2(\sqrt[3]{y}-1)dy$

20. $\int_0^9 \sqrt{x}(3+x)dx$

21. $\int_1^4 \frac{1-\theta}{\sqrt{\theta}}d\theta$

22. $\int_1^9 1-\sqrt[4]{x}+\frac{1}{x^2}dx$

23. $\int_0^{0.25\pi} \sec(\theta)d\theta$

24. $\int_{0.25}^{0.5} \cot(\pi\alpha)d\alpha$

25. $\int_0^\pi 3\cos(0.5\alpha)d\alpha$

26. $\int_0^{\pi/3} 2\sec(x)\tan(x)dx$

27. $\int_0^{0.25\pi} \tan(m)dm$

28. $\int_2^5 \frac{1}{x}dx$

29. $\int_0^1 \frac{2x}{1+x^2}dx$

30. $\int_0^{0.5} \sqrt{1-x^2}dx$

31. $\int_{-2}^{1} |x| \, dx$

32. $\int_{0}^{2} \sqrt{4 - x^2} \, dx$

33. $\int_{0}^{1} x \sqrt{1 + x^2} \, dx$

34. $\int_{0}^{a} \sqrt{a^2 - x^2} \, dx$

35. $\int_{0}^{1} \sin(x^2) \, dx$

36. $\int_{0}^{1} e^{x^2} \, dx$

37. $\int_{0}^{1} \frac{1}{x^2 + 1} \, dx$

38. $\int_{0}^{0.05\pi} \frac{1}{\sqrt{1 - x^2}} \, dx$

39. $\int_{-2}^{1} |1 - x^2| \, dx$

40. $\int_{0}^{1.5\pi} |\sin(x)| \, dx$

41. An object moves along a coordinate line on the indicated time interval with the specified velocity at any time t. Find (i) the total distance traveled, and (ii) the net distance traveled by the object.

    (a) $v = V(t) = 32t + 48$ during the first four seconds.

    (b) $v = V(t) = 2t + \sqrt{t}$ during the first and fourth seconds.

    (c) $v = V(t) = 2\sin(\pi t)$ during the first 1.5 seconds.

42. A manufacturer predicts that the price p per unit of a certain commodity over the next few months will be given by $p = P(t) = 3t - 1$ thousand dollars per unit t months from now, and 1000 units of the commodity are sold each month. Find the predicted total revenue earned over (i) the next three months and (ii) the second quarter of a year from now.

43. For the situation in the previous Exercise, assume that the manufacturer estimates that the price per unit will be given by $1 + 30t + 50t^2$ thousand dollars per unit and 2500 units of the commodity are sold each month. Now repeat Exercise 42.

44. For the function given by $2t$, construct the function that gives, for each x, the (net) area under $2t$ from 0 to x. In Activity 1, page 338, you made a table of its values. There is a reasonable guess for what function this is. Try integer values for x and see what you get. What function have you constructed? Give an explanation for your answer.

45. Consider the function f given by $f(t) = \sqrt{1 - t^2}$.

    (a) Use your MPL to make a sketch of the function represented by **DefInt** applied to f on the interval $[0, x]$. It should look familiar.

    (b) What expression do you think represents **DefInt(f,0,x)**? Explain your answer.

    (c) Use your MPL to make sketches on the same coordinate system of the functions **DefInt(D(f),0,x)** and **D(DefInt(f,0,x))**. Do you observe anything interesting? Give a brief explanation for what you observe.

46. In Activity 3, page 338, you applied first the **computer function DefInt** to the function given by $2x \sin x^2$ on the interval $[0, x]$ and then the **computer function D** to the resulting function. Are you surprised by the relationship between the first and third columns? Give an explanation for your answer.

47. If you apply first the **computer function D** and then the **computer function DefInt** on the interval $[0, x]$ to the function f, then the resulting function is pretty close to the one you started with. It is not an accident. What is it with the example in Activity 3, page 338? Are the first and third columns of your table the same? How different are they? Do you see a pattern? Is this an accident? Write a brief essay explaining for your answer.

48. Altogether, in Activities 3, 4 and 5 you made tables for each of four functions. Which of these resulted in the first and third columns being the same? Can you state a general rule about what happens if you apply **DefInt** and **D** successively to a function? Give explanations for your answers.

49. Let f be the following function:

$$f(t) = \begin{cases} \frac{1-\cos(t)}{t} & \text{if } t \neq 0 \\ 1 & \text{if } t = 0 \end{cases}$$

(a) The function Cin which is useful in the mathematical analysis of the theory of wave form propagation is defined by the following definite integral with variable upper limit:

$$\text{Cin}(x) = \int_0^x f(t)dt$$

It can be shown that the function f has no elementary antiderivative, so the integral must be evaluated by numerical approximation techniques. Use RiemTrap to estimate Cin(0.65) to five decimal places.

(b) The function Ci which is also used in the theory of wave propagation is the function whose value at x is defined by:

$$\text{Ci}(x) = \gamma + \ln(x) - \int_0^x f(t)dt$$

where γ is the *Euler number*, and to seven decimal places $\gamma = 0.5772157$. Use RiemTrap to estimate Ci(0.65) to six decimal places.

50. Let g be the following function:

$$g(t) = \begin{cases} \frac{\sin(t)}{t} & \text{if } t \neq 0 \\ 0 & \text{if } t = 0 \end{cases}$$

The function Si which is useful in the mathematical analysis of the theory of wave form propagation is defined by the following definite integral with variable upper limit:

$$\text{Si}(x) = \int_0^x g(t)dt.$$

This integral is called the *Sine integral*. It can be shown that the function g has no elementary antiderivative, so the integral must be evaluated by numerical approximation techniques. Use RiemTrap to estimate Si(1.24) to five decimal places.

51. Do you think that the Fundamental Theorem of Calculus holds for functions defined in parts which are piecewise continuous? If so, state and prove a version of the Fundamental Theorem for a piecewise continuous function. If not, give a counterexample.

52. (a) Use your SCS to try to evaluate the integral $\int_{-1}^{1} \frac{1}{x} dx$.

(b) Do you think that the Fundamental Theorem of Calculus holds for improper integrals with an integrand function that has a point in its domain where the function is unbounded? Give an explanation for your answer.

53. Derive the following integration formula

$$\int \sec(x)dx = \ln|\sec(x) + \tan(x)| + c$$

by first multiplying the integrand $\sec(x)$ of the original integral by

$$\frac{\sec(x) + \tan(x)}{\sec(x) + \tan(x)}$$

and then choose an appropriate substitution after verifying that

$$\frac{d}{dx}(\sec(x) + \tan(x)) = \sec(x)\tan(x) + \sec^2(x).$$

54. Prove the following part of the Fundamental Theorem of Calculus:

If F is any anti-derivative of the function f on the interval $[a, b]$, then

$$\int_a^b f(x)dx = F(b) - F(a).$$

Do so as follows:

(a) First subdivide the interval $[a, b]$ into n equal parts by the points

$$x_0 = a, \quad x_1 = a + \frac{b-a}{n},$$

$$x_2 = a + 2\left(\frac{b-a}{n}\right), \ldots,$$

$$x_i = a + i\left(\frac{b-a}{n}\right), \ldots, x_n = b$$

(b) Next write $F(b) - F(a)$ as the following *telescoping sum*

$$F(b) - F(a) = F(x_n) - F(x_{n-1})$$
$$+ F(x_{n-1}) - \ldots - F(x_1)$$
$$+ F(x_1) - F(x_0)$$

(c) Use the fact that F is an antiderivative of f, i.e., $F' = f$ on $[a, b]$ and the Mean Value Theorem (see page 278) for derivatives to show that there is a point c_i with $x_{i-1} \le c_i \le x_i$ so that

$$F(x_i) - F(x_{i-1}) = f(c_i)(x_i - x_{i-1}).$$

(d) Show that

$$\sum_{i=1}^n f(c_i)(x_i - x_{i-1}) = F(b) - F(a)$$

and then take the limit as $n \to \infty$ to finish the proof.

55. For each of the following, evaluate each definite integral and find each indefinite integral by hand using the basic rules on page 347. You may wish to check your answer by using your SCS.

(a) $\int_0^2 (3x - \sqrt{x})dx$ (b) $\int_{-1}^3 (3 - 2x)dx$

(c) $\int_{-2}^1 \frac{3t}{7}dt$

(d) $\int_{-1}^1 (10\theta^4 - \theta^2)d\theta$

(e) $\int_0^4 (11 - 2\sqrt[4]{w})dw$

(f) $\int_1^4 \frac{3}{\sqrt{t}}dt$ (g) $\int_0^{3\pi} \sin(\theta)d\theta$

(h) $\int_{-2}^{0.5} \pi\cos(\pi\beta)d\beta$

(i) $\int_0^{0.5\pi} (\theta - \sin(2\theta))d\theta$

(j) $\int_1^3 \frac{1 + 2x - x^2}{x^3}dx$

(k) $\int_0^\pi \sin(3x)dx$ (l) $\int_0^2 3x - e^x dx$

(m) $\int_0^{10} e^{0.01x}dx$

(n) $\int_0^1 e^{-y} \sin(e^{-y})dy$

(o) $\int_0^{0.25\pi} (1 - \sec^2(x))dx$

(p) $\int_0^2 e^{\ln(t)}dt$

(q) $\int_0^\pi \sin(x)\sin(\cos(x))dx$

(r) $\int d\theta$

(s) $\int (\sqrt[3]{v} - \cot(v))dv$

(t) $\int 2\sec(\theta)d\theta$

(u) $\int \cot(\alpha)\csc^2(\alpha)d\alpha$

(v) $\int 3\sin(\pi\alpha)d\alpha$

(w) $\int \tan(2y)\sec(2y)dy$

(x) $\int \sin(x)\cos^2(x)dx$

(y) $\int \sqrt{\cos(\theta)}\sin(\theta)d\theta$

(z) $\int \cot(\pi\alpha)d\alpha$

56. For each of the following, evaluate each definite integral and find each indefinite integral by hand using the basic rules.

(a) $\int 3\sin(\pi\alpha)d\alpha$ (b) $\int \left(4\theta - \frac{2}{\theta}\right)d\theta$

(c) $\int \left(1 - \sqrt[4]{x} + \frac{1}{x^2}\right)dx$

(d) $\int \tan^2(x)dx$

(e) $\int \left(\sin^2(x) + \cos^2(x)\right)dx$

(f) $\int \csc(2r)dr$ (g) $\int \dfrac{\sin(x)}{1 + \cos(x)}dx$

(h) $\int \dfrac{\sec^2(t)}{(1 - \tan(t))^2}dt$

(i) $\int x\sin(x)dx$ (j) $\int 2xe^x dx$

57. In each of the following, find the derivative of the function defined by a definite integral with a variable upper limit. Explain why the Fundamental Theorem of Calculus can be used to differentiate each function. Note that for each function, it can be shown that the integrand has no antiderivative which is an elementary function. Hence, the values of each function must be approximated by some numerical approximation method like the trapezoid method.

(a) The *Fresnel integral*, which is useful in the analysis of diffraction in optics, is defined by

$$C(x) = \int_0^x \cos\left(0.5\pi t^2\right)dt.$$

(b) In *probability theory* the following definite integral

$$\int_a^x \frac{1}{\sqrt{2\pi}} e^{-0.5t^2}dt$$

gives the *probability* that a *normally distributed* random variable t lies in the interval $[a, x]$, and its value is denoted $P(a \le t \le x)$.

(c) Let f be the following function:

$$f(t) = \begin{cases} \frac{1-\cos(t)}{t} & \text{if } t \neq 0 \\ 1 & \text{if } t = 0 \end{cases}$$

   i. The function Cin which is useful in the mathematical analysis of the theory of wave form propagation is defined by the following definite integral with variable upper limit:

$$\text{Cin}(x) = \int_0^x f(t)dt$$

   ii. The function Ci which is also used in the theory of wave propagation is the function whose value at x is represented by the expression

$$\text{Ci}(x) = \gamma + \ln(x) - \int_0^x f(t)dt$$

   where γ is the *Euler number*, or *Euler's constant*, and to seven decimal places $\gamma = 0.5772157$.

(d) Let g be the following function:

$$g(t) = \begin{cases} \frac{\sin(t)}{t} & \text{if } t \neq 0 \\ 0 & \text{if } t = 0 \end{cases}$$

The function Si which is useful in the mathematical analysis of the theory of wave form propagation is defined by the following definite integral with variable upper limit:

$$\text{Si}(x) = \int_0^x g(t)dt.$$

(e) Let g be the following function:

$$g(t) = \begin{cases} \frac{1-e^{-t}}{t} & \text{if } t \neq 0 \\ 1 & \text{if } t = 0 \end{cases}$$

The function Ein is defined by the following definite integral with variable upper limit:

$$\text{Ein}(x) = \int_0^x g(t)dt$$

The function Ein is called the *exponential integral*.

(f) Let g be the following function:

$$g(t) = \begin{cases} \frac{1-e^{-t}}{t} & \text{if } t \neq 0 \\ 1 & \text{if } t = 0 \end{cases}$$

The function Ei which is defined by $Ei(x) = \int_0^x g(t)dt - \ln x - \gamma$ where γ is *Euler's constant* and $\gamma = 0.5772157$ to seven decimal places.

(g) The *error function* erf, which is useful in probability and statistics applications, is defined by the following definite integral with variable upper limit:

$$\text{erf}(x) = \int_0^x \frac{2}{\sqrt{\pi}} e^{-t^2} dt.$$

This integral is also often referred to as the *probability integral*.

58. Find the derivative of each of the following functions. What function does each of the following integrals represent? Explain your answer.

(a) $\int_0^x \frac{1}{1+t^2} dt$

(b) $\int_0^x \frac{1}{\sqrt{1-t^2}} dt$

(c) $\int_0^x \frac{-1}{\sqrt{1-t^2}} dt$

59. Find the derivative of each of the following functions.

(a) $\int_0^x \frac{t}{1+t^3} dt$

(b) $\int_0^y \sin(\pi t) dt$

(c) $\int_0^{3x} \frac{v^2}{\sqrt{1-v}} dv$

(d) $\int_x^1 \frac{1}{1+w^2} dw$

(e) $\int_{-2x}^{x^2} \tan(t^2) dt$

(f) $\int_{\sin(2x)}^{\ln(x)} \sqrt{1+t} dt$

60. Using the appropriate properties of integrals, derivatives, and the material in Chapter 4, Section 4, prove each of the following properties of the natural logarithm and exponential functions ln and exp using the integral definition of the natural logarithm function

$$\int_1^x \frac{1}{t} dt.$$

(a) $\ln(1) = 0$

(b) $\ln'(x) = \frac{1}{x}$

(c) ln is strictly increasing

(d) $\ln(ab) = \ln(a) + \ln(b)$

(e) $\ln(\frac{a}{b}) = \ln(a) - \ln(b)$

(f) $\ln(a^b) = b \ln(a)$

(g) ln has an inverse function exp defined on $(-\infty, \infty)$.

(h) $\exp'(x) = \exp(x)$

(i) $\exp(0) = 1$

(j) exp is strictly increasing

(k) $\exp(a+b) = \exp(a)\exp(b)$

(l) $\exp(a-b) = \frac{\exp(a)}{\exp(b)}$

(m) $\exp(a)^n = \exp(na)$

61. Consider the following equation

$$\int_1^x \frac{dt}{t} = 1.$$

(a) Show that the equation has one solution and then approximate the zero to six decimal places using the **computer functions bis** and **RiemTrap**.

(b) Do you recognize the number that you are approximating? What is it? Explain your answer.

62. Use property **P2**, page 343, to prove the following extended version of **P2** which allows points c to be outside the interval of integration $[a, b]$ for a function f.

If c is any real number and f is a continuous function on an interval containing a, b, and c, then

$$\int_a^b f(x)dx = \int_a^c f(x)dx + \int_c^b f(x)dx.$$

63. Let f be a continuous function on an interval $[a, b]$ with m and M the absolute minimum and maximum values of f on $[a, b]$, respectively. Prove the following.

(a) $\int_a^b (f_{avg} - f(x))dx = 0$

   where f_{avg} is the average value of f given by

   $$\frac{1}{b-a}\int_a^b f(x)dx.$$

(b) For at least one c with $a < c < b$,

   $$f(c) = \frac{1}{b-a}\int_a^b f(x)dx.$$

(c) $\left|\int_a^b f(x)dx\right| \le \int_a^b |f(x)|dx.$

64. Using the method of proof used to prove Proposition 5.2, page 345, show that

   $$\int_a^b x\,dx = \frac{b^2 - a^2}{2}.$$

65. Write a general formula for the derivative of the function given by

$$\int_a^{b(x)} f(t)dt\,?$$

Is this problem related to Activity 6? Give an explanation for your answer.

66. Write a general formula for the derivative of the function given by

$$\int_{a(x)}^{b(x)} f(t)dt\,?$$

The general formula you are writing is sometimes called *Leibniz's rule* for finding derivatives of integrals with variable limits of integration. Is this problem related to Activity 6? How is the Fundamental Theorem of Calculus related? Give an explanation for your answer.

5.4 INTEGRATION METHODS AND IMPROPER INTEGRALS

5.4.1 Overview

In this section, you will learn some basic methods of integration which can be carried out by hand. In particular, you will study the method of substitution and various related methods, integration by parts and the method of partial fractions. These techniques are useful when you don't recognize the integrand as the derivative of any familiar function.

Although SCSs can be used to find most integrals, a knowledge of methods of integration is useful in reading theoretical arguments in the various disciplines that use calculus and so that one can quickly do simple integrals by hand rather than resorting to the use of a computer. Who knows, maybe one day you will help write a SCS to integrate — maybe one that is better than your current SCS!

Additionally, you will learn how to extend the idea of the definite integral in two different ways. These two extended types of integrals, called **improper integrals**, allow us to extend the notion of the definite integral to cases where the interval of integration is infinite and to cases where the integrand becomes unbounded.

5.4.2 Activities

1. In this activity, you will investigate how some integrals can be transformed into a more basic integral form. The objective is to take a given integral and reduce it to another integral, or constant multiple of another integral, which is more basic than the original integral.

(a) For each of the functions represented by the following expressions, use your SCS to find an anti-derivative.

    i. $2x\left(1+x^2\right)^{\frac{1}{3}}$
             ii. $\int \dfrac{6x}{\sqrt{1+3x^2}}\,dx$

    iii. $\dfrac{5x^4}{2+x^5}$
           iv. $-3x^2e^{-x^3} = -3x^2\exp\left(-x^3\right)$

Carefully look over the form of the expression you obtained for each anti-derivative. Now, in each expression listed above, identify a function u, given by $u = g(x)$, so its derivative u', given by $u' = g'(x)$, and rewrite each expression in terms of u and u'. Can you state a general rule which does the job for all four expressions?

(b) For each of the following integrals, make an appropriate substitution, or change of variable, $u = g(x)$, to find an equivalent integral by hand. See your instructor about the use of your SCS to investigate substitutions in integrals.

    i. $\int x\left(1+x^2\right)^{\frac{1}{3}}dx,\ \int_0^2 x\left(1+x^2\right)^{\frac{1}{3}}dx$
        ii. $\int \dfrac{x}{\sqrt{1+3x^2}}\,dx,\ \int_0^1 \dfrac{x}{\sqrt{1+3x^2}}\,dx$

    iii. $\int \dfrac{x^4}{2+x^5}\,dx,\ \int_{-1}^0 \dfrac{x^4}{2+x^5}\,dx$
        iv. $\int x^2e^{-x^3}dx,\ \int_0^1 x^2e^{-x^3}dx$

2. In this activity you will investigate a way to reduce a given integral to a more manageable integral. For each of the following indefinite integrals, use your SCS to compute each integral.

    (a) $\int x\sin(x)dx$
        (b) $\int xe^{-x}dx$
        (c) $\int \ln(t)dt$

Carefully look over the integrand function and the general form of the expression you obtained for each integral. Can you state a general rule which works for all three integrals? See your instructor about the use of your SCS to investigate integrals like those listed above.

3. Compute the following integrals using your SCS. Carefully look over the answers you obtained. Can you think of a way that you could do each integral by hand with the use of basic trigonometric identities and integration rules.

    (a) $\int \sin^2(x)\cos(x)dx$
        (b) $\int \cos^3(x)dx$
        (c) $\int \sec^2(x)\tan^5(x)dx$

    (d) $\int \sec^4(x)\tan(x)dx$
        (e) $\int \sin^2(x)\cos^3(x)dx$

4. Use the indicated substitution to transform each integral into an equivalent integral in the new variable t. Then use your SCS to compute the original integral and the transformed integral. Do you see how to make the connection between the answer you obtained for the original integral and the answer for the transformed integral?

    (a) $\int \dfrac{\sqrt{4-x^2}}{x}\,dx,\ x = 2\sin(t)$
        (b) $\int \dfrac{1}{\left(9+x^2\right)^2}\,dx,\ x = 3\tan(t)$

    (c) $\int \dfrac{1}{\sqrt{x^2-1}}\,dx,\ x = \sec(t)$

5. Use your SCS to compute the following integrals of rational functions, i.e., functions which are ratios of polynomial functions. Can you think of a way to make a connection between the integrand in each integral and the answer you obtained using your SCS? See your instructor about using your SCS to reduce rational functions into "partial fractions".

    (a) $\int \dfrac{2}{x^2 - x} dx$

    (b) $\int \dfrac{1}{2y + y^2} dy$

    (c) $\int \dfrac{x}{x + 1} dx$

    (d) $\int \dfrac{t^3}{1 - t} dt$

    (e) $\int \dfrac{w}{(w - 1)^2} dw$

    (f) $\int \dfrac{2x^4 + 2x^3 + x^2 + 3x}{(x^2 + 1)(x + 1)} dx$

6. In consideration of gas molecules restricted to one-dimensional motion, the velocity distribution of the fraction of molecules moving at a given velocity is a function f of velocity v given by

    $$f(v) = Ae^{-v^2}$$

    where units have been chosen to normalize various constants. The value of the constant A is given by

    $$A = \frac{1}{\displaystyle\int_{-\infty}^{\infty} e^{-v^2} dv}$$

    Use your SCS to find A.

7. Use your SCS (or do it by hand) to evaluate the definite integral

    $$\int_0^\infty e^{-v} dv$$

    and also the definite integrals

    $$\int_0^{50} e^{-v} dv, \int_0^{100} e^{-v} dv, \text{ and } \int_0^{200} e^{-v} dv.$$

    (a) How would you define

    $$\int_0^\infty e^{-v} dv$$

    using a limit?

    (b) How would you define

    $$\int_0^\infty f(x) dx$$

    using a limit if you knew that

    $$\int_0^b f(x) dx$$

    was defined for each $b > 0$?

8. The drag coefficient for a flat plate parallel to the flow of a fluid is given by

$$\int_0^1 \frac{1}{\sqrt{x}}\,dx$$

(Again, various constants have been normalized). Use your SCS to compute this coefficient.

9. In the previous activity, you used your SCS to evaluate the definite integral

$$\int_0^1 \frac{1}{\sqrt{x}}\,dx\,.$$

Now, using your SCS (or by hand) evaluate the definite integrals

$$\int_{0.1}^1 \frac{1}{\sqrt{x}}\,dx,\ \int_{0.01}^1 \frac{1}{\sqrt{x}}\,dx,\ \text{and}\ \int_{0.001}^1 \frac{1}{\sqrt{x}}\,dx.$$

(a) How would you define

$$\int_0^1 \frac{1}{\sqrt{x}}\,dx$$

using a limit?

(b) Suppose that f is a function which is unbounded at b. How would you define

$$\int_a^b f(x)\,dx$$

using a limit if you knew that

$$\int_a^c f(x)\,dx$$

was defined for each number c with $a \le c < b$?

10. In order to work with situations in which the interval of integration is infinite, we have to take a second limit. Here is how we integrate the function given by e^{-x} on the interval $[0, \infty)$. First, construct a **computer function** to represent a function of integers which, given an integer m as input returns the value of applying **DefInt** to e^{-x} on the interval $[0, m]$ as output. Then we take the limit as $m \to \infty$. You can estimate this limit by evaluating for some large values of m.

What would you do for the function given by e^{-x^2} on the interval $(-\infty, \infty)$?

11. A similar method works for functions which are undefined at one or more points in the given interval. Figure out a method for the following situations.

(a) The situation in Activity 8 above.

(b) The function given by $\frac{1}{\sqrt{|x|}}$ on the interval $[-1, 1]$.

5.4.3 Discussion

Integration Methods — Substitution   In Activity 1 you investigated integrals which can be calculated using a substitution. The *method of substitution*, or *change of variable method*, is based on using the chain rule in reverse. The method is used to transform an integral into another integral, constant times an integral, or sum of integrals which can be found using the integration rules from Section 5.3. The method of substitution is based on the following rules from Section 5.3:

$$\int f'(g(x))g'(x)dx = f(g(x))+c$$

and

$$\int_a^b f'(g(x))g'(x)dx = \int_a^b f'(u)du = f(g(b))-f(g(a)).$$

By the first rule above, we see that we need to look for a function g which is the derivative of the function composed with f. It turns out that we can't always find such a function, but often we can find a function which is "essentially" the derivative of g, usually a constant multiple of g.

An Example

Consider the following integral.

$$\int \sin(3x)dx$$

Can you think of a substitution, or change of variable, which reduces the integral to a simpler integral? Think about the chain rule in reverse. The $3x$ is what distinguishes the integral from being one which can be found by one of the basic integration rules. How about letting $u = 3x$? Then calculating the derivative of u with respect to x, we obtain

$$\frac{du}{dx} = 3.$$

Solving for dx, we get

$$dx = \frac{1}{3}du.$$

Substituting for dx and $3x$, we have

$$\int \sin(3x)dx = \int \sin(u)\frac{1}{3}du = \frac{1}{3}\int \sin(u)du = \frac{1}{3}(-\cos(u))+c = -\frac{1}{3}\cos(3x)+c.$$

In Activity 2, you investigated the following definite integral which can be calculated using a substitution.

$$\int_0^1 \frac{x}{\sqrt{1+3x^2}}dx$$

Can you think of a substitution to make? Think about the chain rule in reverse again. Since the x in the numerator is a multiple of the derivative of $1+3x^2$, i.e., $6x$, let's try $u = 1+3x^2$. Calculating the derivative of u with respect to x, we obtain

$$\frac{du}{dx} = 6x.$$

Solving for dx, we get

$$dx = \frac{du}{6x}.$$

Since we have a definite integral we can change the limits of integration as follows:

When $x = 0$, $u = 1$ and when $x = 1$, $u = 4$. Substituting for dx, $1 + 3x^2$ and the limits of integration and simplifying, we obtain

$$\int_0^1 \frac{x}{\sqrt{1+3x^2}}\,dx = \int_1^4 \frac{1}{6}\frac{1}{\sqrt{u}}\,du = \frac{1}{6}\int_1^4 u^{-1/2}\,du.$$

Using the power rule to find an antiderivative and then the Fundamental Theorem of Calculus, we have

$$\int_0^1 \frac{x}{\sqrt{1+3x^2}}\,dx = \frac{1}{6}\int_1^4 u^{-1/2}\,du = \frac{1}{3}\sqrt{u}\,\Big|_1^4 = \frac{1}{3}.$$

Integration Methods — Integration by Parts

In Activity 2, you investigated integrals which can be calculated using the reversal of the product rule, or what is known as the *method of integration by parts*. This method is based on the following integration rule from Section 5.3 which comes from the product rule for differentiation.

$$\int f(x)g'(x)\,dx = f(x)g(x) - \int g(x)f'(x)\,dx$$

The following shorthand version of the above formula is useful:

$$\int u\,dv = uv - \int v\,du$$

where $u = f(x)$ and $dv = g'(x)\,dx$. Then $du = f'(x)\,dx$ and $v = g(x)$ is any antiderivative of g.

The point of using the rule is to transform an integral into another integral which is easier to evaluate, or on which integration by parts can be used again, finally obtaining an integral which can be found using basic rules. Integration by parts is especially useful for products of different types of functions.

In Activity 2 you investigated the following integral.

$$\int x \sin(x)\,dx$$

If the $\sin(x)$ part of the integrand were $\sin(x^2)$ notice that we could have used the method of substitution. As it stands, however, there are several choices we could make for the parts u and dv. Can you think of some? Here are some choices for the "parts" u and dv: $u = x$ and $dv = \sin(x)\,dx$, or $u = \sin(x)$ and $dv = x\,dx$. Let's try the latter choice of u and dv first. Hence, we have

$$u = \sin(x) \text{ so } du = \cos(x)\,dx \text{ and } dv = x\,dx \text{ so } v = \frac{x^2}{2}.$$

Substituting for u, dv, v and du into the integration by parts formula, we obtain

$$\int x \sin(x)\,dx = (\sin(x))\left(\frac{1}{2}x^2\right) - \int \frac{1}{2}x^2 \cos(x)\,dx.$$

Notice that the latter integral is "worse" than the original integral we wanted to find. So now let's try $u = x$ and $dv = \sin(x)\,dx$. Then we have

$$u = x \text{ so } du = dx \text{ and } dv = \sin(x)\,dx \text{ so } v = -\cos(x).$$

Substituting into the parts formula, we obtain

$$\int x\sin(x)dx = x(-\cos(x)) - \int -\cos(x)dx = -x\cos(x) + \int \cos(x)dx.$$

We now see that the last integral is easy to evaluate, and we have

$$\int x\sin(x)dx = -x\cos(x) + \sin(x) + c.$$

As stated above the method is best suited for products of functions of different types. However, it can be used in other instances too. Consider the following integral

$$\int \ln(2x)dx$$

Letting $u = \ln(2x)$ and $dv = dx$, we have $du = \frac{1}{x}dx$ and $v = x$. Substituting into the parts formula, we obtain

$$\int \ln(2x)dx = x\ln(2x) - \int x\frac{1}{x}dx = x\ln(2x) - x + c.$$

Integration Methods — Trigonometric Integrals

In Activity 3 you investigated integrals involving trigonometric functions, often called trigonometric integrals. We now briefly consider how to handle some trigonometric integrals, i.e., integrals with integrands that are products of trigonometric functions. We shall also consider some algebraic integrands which can be reduced to trigonometric integrals.

Consider the following integral.

$$\int \sin^2(x)\cos(x)dx$$

It is not hard to see that if you let $u = \sin(x)$, then $du = \cos(x)dx$ and this substitution transforms the integral into a simple integral in terms of u

$$\int u^2 du.$$

Hence, after integrating, we obtain

$$\int \sin^2(x)\cos(x)dx = \frac{1}{3}\sin^3(x) + c.$$

How about the following integral?

$$\int \cos^3(x)dx$$

Try to manipulate the integrand using trigonometric identities to see if you can reduce it to a combination of basic integrals and/or some integrals which are easier to work with. Think about the first integral we did above. Any ideas? How about using the trigonometric identity $1 - \sin^2(x) = \cos^2(x)$? We can do so as follows:

$$\int \cos^3(x)dx = \int (\cos^2(x))(\cos(x))dx = \int (1 - \sin^2(x))(\cos(x))dx = \int (\cos(x) - \sin^2(x)\cos(x))dx$$

Hence, by the integral we did earlier and the basic rule for integrating $\cos(x)$, we obtain

$$\int \cos^3(x)dx = \sin(x) - \frac{1}{3}\sin^3(x) + c.$$

The latter integral is typical of the type of manipulation you can do using trigonometric identities and basic algebra to transform an integrand into a more manageable function (or combination of functions) to integrate.

Consider the following integral.

$$\int \sin^2(3\theta)d\theta$$

Any ideas? How about using the identity $1 - \cos^2(3\theta) = \sin^2(3\theta)$? Well that just gets us to $\cos^2(3\theta)$ which is no better than $\sin^2(3\theta)$! What shall we do? Do you see what's causing difficulty? The power of two is causing the difficulty. How about using the double angle identity which gives

$$\sin^2(3\theta) = \frac{1 - \cos(6\theta)}{2} = \frac{1}{2}(1 - \cos(6\theta))?$$

Do you see what the identity does for us? It eliminates the power of 2. Using the identity, we obtain

$$\int \sin^2(3\theta)d\theta = \int \frac{1}{2}(1 - \cos(6\theta))d\theta = \frac{1}{2}\left(\theta - \frac{1}{6}\sin(6\theta)\right) + c.$$

Simplifying, we obtain

$$\int \sin^2(3\theta)d\theta = \frac{\theta}{6} - \frac{\sin(6\theta)}{12} + c.$$

Integrals involving sec, tan, csc and cot are handled in a similar manner using appropriate trigonometric identities.

Integration Methods — Trigonometric Substitution

In Activity 4 you worked with integrals and substitutions involving trigonometric functions. We now discuss how some algebraic integrals can be transformed into an equivalent trigonometric integral. Consider the following integral.

$$\int \frac{\sqrt{4 - x^2}}{x} dx$$

Does the term $4 - x^2$ bring any trigonometric identity to your mind? How about $1 - \sin^2(\theta)$? This identity brings to mind a substitution involving $\sin\theta$. Let $x = 2\sin(\theta)$ where $-\pi/2 \le \theta \le \pi/2$. That is, we are making the substitution $\theta = \arcsin(x/2)$. We transform the x-integral into a θ-integral. We have

$$4 - x^2 = 4 - 4\sin^2(\theta) = 4(1 - \sin^2(\theta)) = 4\cos^2(\theta)$$

and so

$$\sqrt{4 - x^2} = \sqrt{4\cos^2(\theta)} = 2|\cos(\theta)| = 2\cos(\theta),$$

since $\cos(\theta) \ge 0$ for $-\pi/2 \le \theta \le \pi/2$. Computing the differential dx, we get $dx = 2\cos(\theta)d\theta$. Substituting for x, dx and $\sqrt{4 - x^2}$, we obtain

$$\int \frac{\sqrt{4 - x^2}}{x} dx = \int \frac{2\cos(\theta)}{2\sin(\theta)} 2\cos(\theta)d\theta = \int \frac{2\cos^2(\theta)}{\sin(\theta)} d\theta.$$

Using trigonometric identities and the basic integration rules, we have

$$\int \frac{2\cos^2(\theta)}{\sin(\theta)}d\theta = \int \frac{2(1 - \sin^2(\theta))}{\sin(\theta)}d\theta = \int 2(\csc(\theta) - \sin(\theta))d\theta = 2\ln(|\csc(\theta) - \cot(\theta)| + 2\cos(\theta) + c).$$

We now transform our answer to the original variable x using the substitution $\theta = \arcsin(x/2)$ to obtain

$$\int \frac{\sqrt{4-x^2}}{x}\,dx = 2\ln\left|\csc\left(\arcsin\left(\frac{x}{2}\right)\right) - \cot\left(\arcsin\left(\frac{x}{2}\right)\right)\right| + 2\cos\left(\arcsin\left(\frac{x}{2}\right)\right) + c.$$

We can simplify the latter integral using the fact that $\theta = \arcsin(x/2)$ means that θ is an angle whose sine is $x/2$. Hence, by drawing an appropriate right triangle, as shown in Figure 5.11, and then reading off the triangle, we find that

$$\csc(\theta) = \frac{2}{x}, \ \cot(\theta) = \frac{\sqrt{4-x^2}}{x} \ \text{and} \ \cos(\theta) = \frac{\sqrt{4-x^2}}{2}.$$

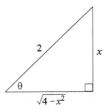

Figure 5.11. Right triangles with an angle $\theta = \arcsin\left(\dfrac{x}{2}\right)$

After making appropriate substitutions and then simplifying, we obtain

$$\int \frac{\sqrt{4-x^2}}{x}\,dx = 2\ln\left|\frac{2-\sqrt{4-x^2}}{x}\right| + \sqrt{4-x^2} + c.$$

Notice that if our integral were a definite integral, then we would not have to change back to the original variable. That's because we could also change the limits of integration.

Integrals involving expressions of the form $a^2 - u^2$, $a^2 + u^2$ and $u^2 - a^2$ where a is a constant and u represents a variable expression can often be transformed using a trigonometric substitution. Think of the identities, $1 + \tan^2(\theta) = \sec^2(\theta)$, $\sec^2(\theta) - 1 = \tan(\theta)$, and the one we used above, $1 - \sin^2(\theta) = \cos^2(\theta)$, when trying to decide what substitution to make. It is not hard to see that for $a^2 + u^2$, you should use $u = a\tan(\theta)$, i.e., $\theta = \arctan(u/a)$. For $u^2 - a^2$ you should use $u = a\sec(\theta)$, i.e, $\theta = \text{arcsec}(u/a)$ and for $a^2 - u^2$ you should use $u = a\sin(\theta)$, i.e., $\theta = \arcsin(u/a)$.

Integration Methods — Partial Fractions   In Activity 5 you computed integrals of rational functions. In order to integrate a *rational function*, i.e, a function which is the ratio of two polynomial functions, it is sometimes necessary to "decompose" the rational function into a sum of *partial fractions*. Consider the following integral from part (a) of Activity 5.

$$\int \frac{2}{x^2-1}\,dx$$

Use your SCS to find the integral. Do you see any clue as to how to proceed? Can you see that the following identity is true?

$$\frac{2}{x^2-1} = \frac{1}{x-1} - \frac{1}{x+1}$$

Just add the fractions on the right-hand side to verify the identity. Hence, we have

$$\int \frac{2}{x^2-1}\,dx = \int \frac{1}{x-1} - \frac{1}{x+1}\,dx = \ln(|x-1|) - \ln(|x+1|) + c = \ln\left(\left|\frac{x-1}{x+1}\right|\right) + c.$$

You can see that by writing a rational function as a sum of its partial fractions, you may be able to transform an integral problem into a sum of more manageable integrals.

The *method of partial fractions* is based on the following theorem from algebra which tells us that we can always decompose a rational function into a sum of partial fractions. Since this is a purely algebraic result, we will omit the proof of the theorem.

Theorem 5.2 (Partial Fraction Decompositions)

Let R be a rational function given by $P(x)/Q(x)$, where the degree of $P(x)$ is less than that of $Q(x)$. Then $R(x)$ can be decomposed into a sum of partial fractions according to the following two cases:

1. *To each linear factor $ax+b$ in the factorization of $Q(x)$ there corresponds a fraction $A/(ax+b)$ in the decomposition of $R(x)$. To each linear factor $ax+b$ occurring n times in $Q(x)$ there corresponds a sum of n fractions*

$$\frac{A_1}{ax+b} + \frac{A_2}{(ax+b)^2} + \cdots + \frac{A_n}{(ax+b)^n}$$

in the decomposition of $R(x)$ for some constants $A_1, A_2, \ldots, A_n$.

2. *To each factor ax^2+bx+c which cannot be factored into linear factors there corresponds a factor of $(Ax+b)/(ax^2+bx+c)$. To each quadratic factor ax^2+bx+c occurring n times in $Q(x)$ there corresponds a sum of n fractions*

$$\frac{A_1x+B_1}{ax^2+bx+c} + \frac{A_2x+B_2}{(ax^2+bx+c)^2} + \cdots + \frac{A_nx+B_n}{(ax^2+bx+c)^n}$$

in the decomposition of $R(x)$ for some constants $A_1, A_2, \ldots, A_n, B_1, B_2, \ldots, B_n$.

Theorem 5.2 covers all possible rational functions because of the following two facts from the theory of algebra:

1. Any rational function can be decomposed by long division into a sum of a polynomial plus a ratio of two polynomials with the degree of the numerator less than that of the denominator.

2. Any polynomial $Q(x)$ with real coefficients can be factored into a product of linear and quadratic factors, where the quadratic factors cannot be factored into linear factors (such quadratic factors are said to be *irreducible*.)

Let's consider the rational function in the indefinite integral we first considered. By Theorem 5.2, we know that there are coefficients A and B so that

$$\frac{2}{x^2-1} = \frac{2}{(x-1)(x+1)} = \frac{A}{x-1} - \frac{B}{x+1}$$

How do you think you could find the coefficients A and B? Let's try clearing fractions by multiplying both sides of the equation by $(x-1)(x+1)$. We get

$$2 = A(x+1) + B(x-1).$$

Now lets pick x so that we can easily solve for A. What do you think would be a good value of x to pick? How about $x=1$? Substituting 1 for x, the last equation reduces to $2 = 2A$ and hence $A = 1$. Now choosing $x = -1$ we see that $B = -1$, and the partial fraction decomposition is as we saw earlier.

Now consider the following integral from part (f) of Activity 5.

$$\int \frac{2x^4 + 2x^3 + x^2 + 3x}{\left(x^2+1\right)(x+1)} \, dx$$

Since the degree of the numerator is not less than that of the denominator, we first multiply out in the denominator and divide to obtain

$$\frac{2x^4 + 2x^3 + x^2 + 3x}{\left(x^2+1\right)(x+1)} = 2x - \frac{x - x^2}{\left(x^2+1\right)(x+1)}.$$

By Theorem 6.2 we know that

$$\frac{x - x^2}{\left(x^2+1\right)(x+1)} = \frac{Ax + b}{x^2+1} + \frac{C}{x+1}$$

for some coefficients A, B and C.

Again we clear fractions to obtain

$$x - x^2 = (Ax+B)(x+1) + C\left(x^2+1\right).$$

Choosing x to be -1 and substituting into the latter equation, we obtain $2C = -2$ and hence $C = -1$. Substituting -1 for C in the equation above, we obtain

$$x - x^2 = (Ax+B)(x+1) - \left(x^2+1\right), \text{ or } x+1 = (Ax+B)(x+1).$$

Choosing $x = 0$, we see that $B = 1$. Hence, we have $x + 1 = (Ax+B)(x+1)$. Finally, choosing $x = 1$ we see that $A = 0$. Hence, the integral reduces to the following integral

$$\int 2x + \frac{1}{x+1} - \frac{1}{x^2+1} \, dx$$

and we can integrate term-by-term to obtain

$$\int \frac{2x^4 + 2x^3 + x^2 + 3x}{\left(x^2+1\right)(x+1)} \, dx = x^2 + \ln|x+1| - \arctan(x) + c.$$

The following theorem characterizes all anti-derivatives of rational functions.

Theorem 5.3 (The Rational Function Theorem)
The anti-derivative of any rational function is a combination of algebraic, inverse trigonometric, and logarithmic functions.

Another way of stating the result of the above theorem is as follows:

> The anti-derivative of any rational function is a combination of elementary functions and hence an elementary function.

What Method Do You Use to Integrate?

The goal in using the basic methods of integration is to transform indefinite integrals into integrals or combinations of integrals that can be found using a basic set of integration rules. In order to decide on which method to use we now suggest a basic strategy for integration based on the article "Presenting a Strategy for Indefinite Integration" by Alan Schoenfeld (see The American Mathematical Monthly, volume 85, 1978, pp. 673-678). Modify our strategy and make one of your own, if you wish.

A Suggested Strategy for Indefinite Integration

Ask yourself the following questions in the indicated order:

1. Can you use a basic integration rule, or combination of basic rules after manipulating the integrand using algebra and/or trigonometry, to find the integral?

2. Can a simple substitution be used to transform the integral into one for which a basic integration rule can be used? In particular look for a substitution which transforms the integral into one for which the power rule, basic trigonometric rules, natural logarithm rule, or exponential rule is applicable.

3. Is the integrand a product of different types of functions? If so, then the method of integration by parts may be applicable. Schoenfeld suggests that the acronym **LIATE** (**L**ogarithmic, **I**nverse trigonometric, **A**lgebraic, **T**rigonometric and **E**xponential functions) be used in selecting the parts u and dv for use in the integration by parts formula. That is, select u to be the first type of function that appears in **LIATE**.

4. Is the integrand function a power or product of trigonometric functions? If so, you may be able to use trigonometric identities and algebra to reduce the integrand to a combination of functions for which basic integration rules are applicable.

5. Does the integrand contain an expression of the form $a^2 - u^2$, $a^2 + u^2$, or $u^2 - a^2$? If so, try a trigonometric substitution.

6. Is the integrand a rational function? Is so, divide first if necessary and then use the method of partial fractions.

If you cannot find an indefinite integral after answering the above questions, try making a substitution to transform the integral into an equivalent integral and then ask yourself questions 1-6 again. Of course, you could just find a SCS to help you out!

Improper Integrals — Infinite Limits of Integration   We have said that the definite

integral, an expression like

$$\int_a^b f(x)dx,$$

can be evaluated as follows:

First, find a function F whose derivative is f and then evaluate

$$F(b) - F(a).$$

If you try this in Activity 2, with the expression,

$$\int_{-\infty}^{\infty} e^{-v^2} dv$$

there are two difficulties. First of all, there is no simple function whose derivative is given by e^{-v^2}. (Try it with your SCS.) The other difficulty is that, even if you had such a function, what would be the meaning of evaluating it at ∞ and $-\infty$?

We are interested in the second problem, so let's look at another example. Consider the problem,

$$\int_0^{\infty} e^{-v} dv$$

This time we *can* find an anti-derivative of e^{-v}. One is given by $-e^{-v}$. Now we can evaluate this at the lower limit, 0, but what do we do about the upper limit, ∞?

In Activity 2 you used your SCS to evaluate the definite integral we are after and also,

$$\int_0^{100} e^{-v} dv$$

Were you able to make any guesses as to how to proceed?

If the interval over which you need to compute the Riemann integral is infinite, then an additional step must be added to the four steps on page 325. For example, if you wish to compute the Riemann integral of a function, say e^{-x} over the interval $[0, \infty)$ as in Activity 2, then you have to run through an additional limit process. Compare the explanation of this process, which we now give, with the work you did on Activity 2.

You have to begin with a preliminary step of replacing the unbounded interval $[0, \infty)$ with a bounded interval $[0, m]$, where m is a positive integer. Then you compute the Riemann integral of your function over $[0, m]$. This will give you a number, but only if you are working with a definite value for m. Every time you change m you get a new number. Thus you have a sequence. Sequences such as this will be examined in detail in Chapter 6. You take the limitof this sequence to get the Riemann integral of your function over the infinite interval $[0, \infty)$.

There is nothing surprising about what needs to be done if your unbounded interval is $(-\infty, b]$, where b is a finite number. You just calculate the Riemann integral of your function over the interval $[-m, b]$ where m is, again, a positive integer. Make the calculation on this bounded interval and then pass to the limit as m goes to infinity.

The situation with the interval $(-\infty, \infty)$ which is unbounded in both directions is a little different. There are actually two ways to proceed. They give the same answer in some cases, but not all. Can you think of two different ways of calculating a Riemann integral of a function over the interval $(-\infty, \infty)$? See what they give you if the function is given by $\sin(\pi x)$.

Improper Integrals — Unbounded Functions

The example in Activity 8 is one of those problems that sometimes make students think that teachers are just trying to make trouble for them. We have a method for evaluating definite integrals and it seems to work for a problem like

$$\int_0^1 \frac{1}{\sqrt{x}}\,dv$$

We just find an anti-derivative for the integrand (that is what we call the function to be integrated), such as the function given by $2\sqrt{x}$, evaluate it at 0 and 1 and then subtract to get the answer 2.

Why should there be any problem here? The explanation is actually quite analogous to what was said in the previous paragraph for integrals with infinite limits. With all these hints, are you ready to try to guess what is going on here?

Suppose that your interval of integration is finite but there are some points on it where the integrand is undefined or discontinuous. This can be because there is real trouble at such a point such as dividing by 0 or taking the square root of a negative number. Or it could be because there is a jump in the integrand.

In any case, the principle is the same as for unbounded intervals. You replace the interval you are working with by one or more intervals whose endpoints are close to the troublesome point, and calculate the Riemann integral. Then you pass to the limit as the endpoint goes to the original point. Can you explain that in detail for an example, say like the situation of Activity 8?

In Activity 8, you were to compute the Riemann integral of the function given by $\frac{1}{\sqrt{x}}$ over the interval $[0, 1]$. The point here is that you cannot let the value 0 appear as a domain value in your calculations of this function because it is undefined at that point. So what you do is calculate the Riemann sum of this function over the interval $\left[\frac{1}{m}, 1\right]$ where, again, m is a positive integer. As with unbounded intervals, you get a number but only if you use a specific number for m. Changing m gives you a new calculation and a new number. Once more you have a sequence. The answer to the problem is the number you get by taking the limit of this sequence as m goes to infinity.

Once again you should have no trouble with a similar situation in which the troublesome value is at the right endpoint. What should you do if the problem occurs in the interior of the integral? For example, how would you compute the Riemann integral of the function in Activity 9(b)?

Improper Integrals — Formal Definitions

We conclude this section by giving formal definitions for integrals over unbounded intervals and for functions which are discontinuous at one or both of the endpoints of the interval. These are the *improper integrals* which you have been working with in an informal way in the previous pages.

In all of the definitions given in this paragraph, the integral may or may not exist. If the integral has a finite value, then it is said to *converge*. If the integral does not converge, it is said to *diverge*.

Definition 5.4

Let f be a function continuous on $[a, \infty)$. We define the improper integral of f over $[a, \infty)$ by

$$\int_a^\infty f(x)\,dx = \lim_{m\to\infty} \int_a^m f(x)\,dx.$$

An analogous definition can be given for a function on $(-\infty, b]$. What is it?

Definition 5.5

Let f be a function continuous on (a, b], but not necessarily at a. We define the improper integral of f over [a, b] by

$$\int_a^b f(x)dx = \lim_{c \to a^+} \int_c^b f(x)dx.$$

An analogous definition is given for a function continuous on [a, b) but not necessarily at b.

Definition 5.6

Let f be a function continuous at every point in [a, b] except at one point $c \in (a, b)$. We define the improper integral of f over [a, b] by

$$\int_a^b f(x)dx = \int_a^c f(x)dx + \int_c^b f(x)dx.$$

If both improper integrals on the right-hand side converge, then the improper integral $\int_a^b f(x)dx$ is said to converge to the sum of the two integrals. If one or both of the integrals on the right-hand side diverges, then the improper integral

$$\int_a^b f(x)dx$$

is said to diverge.

For example, consider the following improper integral

$$\int_{-1}^2 \frac{1}{x^2} dx$$

for which the integrand is unbounded at $x = 0$. Then by Definition 5.6, we have

$$\int_{-1}^2 \frac{1}{x^2} dx = \int_{-1}^0 \frac{1}{x^2} dx + \int_0^2 \frac{1}{x^2} dx$$

provided both integrals on the right-hand side converge. Since, $-\frac{1}{x}$ is an anti-derivative of $\frac{1}{x^2}$, we have

$$\int_{-1}^0 \frac{1}{x^2} dx = \lim_{b \to 0^-} \int_{-1}^b \frac{1}{x^2} dx = \lim_{b \to 0^-} -\frac{1}{b} - 1 = \infty.$$

Hence, the improper integral

$$\int_{-1}^2 \frac{1}{x^2} dx$$

diverges.

5.4.4   Exercises

In Exercises 1-75, evaluate each definite integral and find each indefinite integral by hand using the basic methods of integration. You may wish to check your answer by using your SCS.

1. $\int \dfrac{x}{x^2-1}dx$

2. $\int \dfrac{1}{x^2-1}dx$

3. $\int \dfrac{\sec^2(2y)}{1+\tan(2y)}dy$

4. $\int 3xe^{-x^2}dx$

5. $\int \theta\sin(2\theta)d\theta$

6. $\int \sec^2(x)\tan(x)dx$

7. $\int \dfrac{1}{x^2+9}dx$

8. $\int \dfrac{\sqrt{4-x^2}}{x}dx$

9. $\int_0^1 \dfrac{x^2-1}{1+x}dx$

10. $\int \dfrac{\cos(\sqrt{t})}{\sqrt{t}}dt$

11. $\int \dfrac{x}{\sqrt{2+3x}}dx$

12. $\int \sqrt{4-x^2}\,dx$

13. $\int_0^1 \arctan(x)dx$

14. $\int \sin^2(\pi t)dt$

15. $\int \theta^2\sqrt{2-\theta^3}\,d\theta$

16. $\int_0^{\pi/6} \cos^3(x)dx$

17. $\int \dfrac{\sec(\ln(x))}{x}dx$

18. $\int \dfrac{e^x}{\sqrt{1+e^{2x}}}dx$

19. $\int \dfrac{x}{(x-1)^2}dx$

20. $\int (\ln(x))^2\,dx$

21. $\int \dfrac{2x-1}{\sqrt{x}}dx$

22. $\int \tan^3(2x)dx$

23. $\int_{-1}^0 \ln(1+y^2)dy$

24. $\int \cos^4(\alpha)\sin^3(\alpha)d\alpha$

25. $\int 3\sin(\tan(x))\sec^2(x)dx$

26. $\int \dfrac{e^{0.5x}}{\sqrt[3]{1+e^{0.5x}}}dx$

27. $\int \dfrac{\sqrt{\ln(\sin(t))}}{\tan(t)}dt$

28. $\int \dfrac{1}{e^{-x}+e^x}dx$

29. $\int \dfrac{1}{2x+x^2}dx$

30. $\int \dfrac{1}{(x-1)(x^2+4)}dx$

31. $\int \dfrac{x}{x+1}dx$

32. $\int \dfrac{\sqrt{x^2-1}}{x}dx$

33. $\int \dfrac{d\theta}{1+\cos(\theta)}$

34. $\int x^2\sqrt{1+x}\,dx$

35. $\int \dfrac{w^2}{w^3-1}dw$

36. $\int \sin(\sqrt{x})dx$

37. $\int \dfrac{dy}{2y-y^2}$

38. $\int \dfrac{1}{x^4-1}dx$

39. $\int \dfrac{dx}{\sqrt{16-x^2}}$

40. $\int_{-1}^2 (9-x^2)^{\frac{3}{2}}dx$

41. $\int_0^1 \dfrac{x^2}{\sqrt{9-x^2}}dx$

42. $\int \dfrac{t^2-t}{t+1}dt$

43. $\int \dfrac{dx}{\sqrt{9-4x^2}}$

44. $\int \dfrac{x^3}{1-x}dx$

45. $\int \dfrac{\sqrt{1-4y^2}}{y^2}dy$

46. $\int \dfrac{e^x}{e^{2x}-2e^x-3}dx$

47. $\int \ln(3x)dx$

48. $\int \dfrac{1}{1+\sqrt{x}}dx$

49. $\int \ln\sqrt{1-y^2}\,dy$

50. $\int \dfrac{v^2}{\cot(v^3)}dv$

51. $\int \dfrac{3x^2}{(1+x)^3}dx$

52. $\int \dfrac{2-x^2}{x^3}dx$

53. $\int x^2e^{-x}dx$

54. $\int \dfrac{y}{1+y^4}dy$

55. $\int (1+x^2)e^x dx$

56. $\int \dfrac{5x+4x^2}{(x-1)(2x+1)^2}dx$

57. $\int \dfrac{1}{1+e^x}dx$

58. $\int \sin^5(2\theta)d\theta$

59. $\int \dfrac{y^3 - 20y^2 + 4}{y^3 - 1} dy$

60. $\int_0^1 \ln(e^{2x}) dx$

61. $\int_0^1 \dfrac{x^5}{(x+4)^2} dx$

62. $\int \dfrac{\cos(x)}{1 - \sin^2(x)} dx$

63. $\int \sin(\ln(t)) dt$

64. $\int \dfrac{1 + y^4}{y^2 + 4} dy$

65. $\int x \arcsin(x) dx$

66. $\int \dfrac{y^2 - y + 3}{(y^2 + 3)^2} dy$

67. $\int \sec^2(x) \tan^2(x) dx$

68. $\int \dfrac{5x^2}{4 + x^2} dx$

69. $\int x \sec\left(\dfrac{x}{3}\right) \tan\left(\dfrac{x}{3}\right) dx$

70. $\int \dfrac{w^4}{3 - w^2} dw$

71. $\int \dfrac{5x^2 + 7x - 40}{x^3 + 2x^2 - 8x} dx$

72. $\int x^2 \ln(2x) dx$

73. $\int \dfrac{x - 1}{x^2(x-2)^2} dx$

74. $\int \sqrt{1 - \sin(v)} dv$

75. $\int \dfrac{3 - x^5}{x^3 - x^2} dx$

76. Heat capacities of substances are not independent of temperature, and the heat capacities of thousands of substances have been measured over a wide range of temperatures. For many substances, experimental data for heat capacities versus temperature indicate that the data are often "fitted" by polynomial curves of the form

$$C_p = A + BT + CT^2 + DT^3 + ET^4$$

where A, B, C, D, and E are constants which are chosen to provide the best agreement with experimental data over the relevant temperature range. For one mole of a substance, the heat ΔH needed to raise the temperature of one mole of a substance from T_1 degrees Kelvin to T_2 degrees Kelvin at constant pressure is given by

$$\Delta H = \int_{T_1}^{T_2} C_p dT$$

$$= \int_{T_1}^{T_2} \left(A + BT + CT^2 + DT^3 + ET^4\right) dT \text{ Joules}$$

where Joules is the measure (unit) of heat.

In each of the following situations, find the heat ΔH required to raise one mole of the indicated substance over the indicated range of temperature.

(a) Copper is extracted from CuS ores by "roasting" the ore at $1000°$K. How much heat (in Joules) is required to raise the temperature of one mole of CuS from $300°$K to $1000°$K at a constant pressure for which $A = 44.35$, $B = 0.0111$, and $C = D = E = 0$?

(b) How much heat (in Joules) is required to raise the temperature of one mole of hydrogen bromide gas, HBr, from $500°$K to $1000°$K at a constant pressure for which $A = 27.521$, $B = 0.00400$, $C = 0.000000661$, and $D = E = 0$?

(c) How much heat (in Joules) is required to raise the temperature of one mole of graphite from $300°$ K to $600°$ K at a constant pressure when the heat capacity data for graphite is "fitted" by the curve given by

$$C_p = A + BT - \dfrac{C}{T^2}$$

where $A = 11.18$, $B = 0.0109$, and $C = 489109$?

77. A model for the number of hours of sunlight s per day in a non-leap year in Lafayette, Indiana is given by

$$s = s(t) = 12 + 3 \sin\left(\dfrac{2\pi}{365}(t - 80)\right) \text{ hours}$$

where t the number of days after January 1. Do the following for this situation.

(a) What is the average number of sunlight hours in a day in Lafayette, Indiana?

(b) What is the total number of hours of sunlight in Lafayette, Indiana in a non-leap year?

78. The root-mean square value of a function f on an interval $[a, b]$ is given by

$$f_{\text{rms}} = \sqrt{\frac{1}{b-a} \int_a^b (f(x))^2 \, dx}.$$

What is the root-mean square voltage, v_{rms}, in the circuit over the time interval $[0, 2]$?

79. The current i in an electrical circuit is given by $i = I(t) = 10\cos(\pi t)$ amperes. What is the root-mean square current, i_{rms}, for one period of the current? See the definition of the root-mean square value of a function in part (b) of the previous exercise.

80. In gas kinetic theory the mean speed $\bar{v}$ of a gas molecule is given by

$$\bar{v} = 4\pi \left(\frac{m}{2\pi kT}\right)^{3/2} \int_0^\infty v^3 \exp\left(-\frac{mv^2}{2kT}\right) dv$$

where m, k, and T are constants; m is the mass of the molecule, k is Boltzmann's constant, T is the temperature of the molecule in degrees Kelvin, and $\exp(x) = e^x$. Use integration by parts to find a formula for $\bar{v}$. You may wish to use your SCS to check your answer.

81. Below there are two improper integrals.

(a) Evaluate the following limit for each improper integral

$$\lim_{N \to \infty} \int_{-N}^N f(x)\,dx$$

for an appropriate integrand function.

(b) Use your SCS to find an anti-derivative of the integrand function

in each of the following improper integrals. Then evaluate each integral by hand. Check your answer by using your SCS to evaluate each improper integral.

(c) Explain the differences, if any, between the answers you obtained in parts(a) and (b).

Here are the improper integrals.

i. $\int_{-\infty}^\infty \cos(3x)\,dx$

ii. $\int_{-\infty}^\infty \frac{1}{x}\,dx$

82. Determine whether or not the following improper integrals converge or diverge. If the integral converges, give its value. You may wish to check your answers using your SCS.

(a) $\int_0^4 \frac{3}{\sqrt{x}}\,dx$ (b) $\int_{-1}^4 \frac{1}{\sqrt[3]{x}}\,dx$

(c) $\int_1^\infty \frac{1}{x^2}\,dx$ (d) $\int_1^\infty \sqrt{x}\,dx$

(e) $\int_1^\infty \cos(2x)\,dx$ (f) $\int_{-\infty}^0 \frac{1}{\sqrt[3]{x}}\,dx$

(g) $\int_0^\infty \frac{1}{x^2+1}\,dx$ (h) $\int_0^1 \frac{1}{\sqrt{1-x^2}}\,dx$

(i) $\int_0^{0.5\pi} \sec(x)\,dx$ (j) $\int_0^\pi \frac{\sin(x)}{1-\cos(x)}\,dx$

83. An improper integral of the form

$$\int_{-\infty}^\infty f(x)\,dx$$

can be evaluated as follows:

$$\int_{-\infty}^\infty f(x)\,dx = \int_{-\infty}^c f(x)\,dx + \int_c^\infty f(x)\,dx$$

for any number c, provided each of the improper integrals on the right hand side converges. If one or both of the integrals on the right hand side diverges, then the improper integral

$$\int_{-\infty}^{\infty} f(x)dx$$

is said to diverge.

Determine whether or not the following improper integrals converge or diverge. If the integral converges give its value. You may wish to check your answers using your SCS.

(a) $\int_{-\infty}^{\infty} \frac{1}{x^2+1}dx$ (b) $\int_{-\infty}^{\infty} \frac{2v}{1+v^2}dx$

(c) $\int_{-\infty}^{\infty} \frac{x}{\left(1+x^2\right)^2}dx$ (d) $\int_{-\infty}^{\infty} \sin(\theta)d\theta$

84. Determine whether or not the following improper integrals converge or diverge. If an integral converges give its value. You may wish to check your answers using your SCS.

(a) $\int_{-1}^{2} \frac{1}{\sqrt[3]{x}}dx$ (b) $\int_{0}^{\pi} \sec^2(t)dt$

(c) $\int_{0}^{9} \frac{1}{\sqrt[3]{(x-1)^2}}dx$

85. Evaluate the following integrals.

(a) $\int_{-1}^{2} |1-x^2|dx$ (b) $\int_{0}^{2\pi} |\sin(t)|dt$

(c) $\int_{-1}^{3} f(x)dx$ where
$$f(x) = \begin{cases} 2x-1 & \text{if } x \le 0 \\ 1 & \text{if } x > 0 \end{cases}$$

(d) $\int_{-2}^{5} g(t)dt$ where
$$g(t) = \begin{cases} 1+t^2 & \text{if } t < 0 \\ -1 & \text{if } 0 \le t \le 1 \\ 2-t & \text{if } t > 1 \end{cases}$$

86. Show that the following inequalities are true.

(a) $\int_{0}^{2} \sqrt{1+x^3}dx \ge 2$

(b) $\int_{0}^{\pi} \cos^2(x)dx \le \int_{0}^{\pi} 3-\sin^2(x)dx$

87. An object moves along a line on the indicated time interval with the specified velocity at any time t. Find (i) the total distance traveled and (ii) the net distance traveled by the object.

(a) $v = V(t) = t - \sqrt{t}$ during the first four seconds.

(b) $v = V(t) = 0.5 - \cos(\pi t)$ during the first one and one-half seconds.

88. Determine whether or not the following improper integrals converge or diverge. If the integral converges, find its value.

(a) $\int_{0}^{\infty} xe^{-x^2}dx$ (b) $\int_{-\infty}^{1} e^{2x}dx$

(c) $\int_{1}^{2} \frac{dx}{x^2-2x}$ (d) $\int_{0}^{\infty} 2te^{-t}dt$

(e) $\int_{2}^{\infty} \frac{1}{x^3-2x^2-x}dx$

(f) $\int_{0}^{\infty} \frac{y^3}{1+y^2}dy$ (g) $\int_{-\infty}^{\infty} \frac{2xdx}{\left(1-x^2\right)^3}$

(h) $\int_{0}^{2} x\ln\left(\frac{2}{x}\right)dx$ (i) $\int_{-\infty}^{\infty} \frac{2xdx}{1-x^2}$

89. Show that the improper integral
$$\int_{1}^{\infty} \frac{1}{x^p}dx$$
(a) converges if $p > 1$.
(b) diverges if $p \le 1$.

90. Show that the improper integral
$$\int_{0}^{1} \frac{1}{x^p}dx$$
(a) converges if $p < 1$.
(b) diverges if $p \ge 1$.

91. Find an equation of a curve which has slope given by the indicated expression and that passes through the specified point.

(a) $\sqrt{4+3x}$, $(4, -2)$

(b) $\cos(0.5x)$, $(\pi, 1)$

(c) xe^{3x}, $(0, -1)$ (d) $\dfrac{x}{4 - x^2}$, $(0, 5)$

92. In *probability theory* the following definite integral

$$\int_a^b \frac{1}{\sqrt{2\pi}} e^{-0.5x^2} dx$$

gives the *probability* that a *normally distributed* random variable x, with mean zero, lies in the interval $[a, b]$, and its value is denoted $P(a \le x \le b)$.

(a) Use sufficiently large values of n and **RiemTrap** to estimate to five decimal places

$$\int_0^\infty \frac{1}{\sqrt{2\pi}} e^{-0.5x^2} dx.$$

Use your SCS to show that

$$\int_0^\infty \frac{1}{\sqrt{2\pi}} e^{-0.5x^2} dx = 0.5.$$

(b) Use the result of part (a) to show that

$$\int_{-\infty}^\infty \frac{1}{\sqrt{2\pi}} e^{-0.5x^2} dx = 1.$$

Explain why the above statement is reasonable. Evaluate the integral using your SCS.

(c) Using the results of parts (a) and (b), and properties of *Riem*, prove the following

$$P(x < a \text{ or } x > b) = 1 - P(a \le x \le b).$$

(d) Estimate the following probabilities to five decimal places using **RiemTrap**. You may need to use some of the properties of the probability function P in parts (a)-(c). Evaluate the probabilities by using your SCS to evaluate the definite integral

$$\int_a^b \frac{1}{\sqrt{2\pi}} e^{-0.5x^2} dx$$

for appropriate limits of integration, a and b.

i. The probability that x lies in the interval $[0, 3]$.

ii. The probability that x lies in the interval $[-1, 2]$.

iii. The probability that x lies outside the interval $[1, 4]$.

iv. The probability that $x > 2$.

93. Consider the improper integral for the function f given by

$$f(x) = \sin(\pi x),$$

$$Riem(f, -\infty, \infty) = \int_{-\infty}^\infty \sin(\pi x) dx.$$

(a) Try to find the value of integral by finding the following limit:

$$\lim_{N \to \infty} \int_{-N}^N \sin(\pi x) dx.$$

(b) Try to find the value of the integral by interpreting the integral as the *net* area between the x-axis and the graph of the function represented by $\sin(\pi x)$.

(c) Explain the difference in your answers in parts (a) and (b), if any. Which do you think makes more sense to take as the value of the improper integral and why? What value does your SCS give you for the improper integral?

94. Show that for positive integers m and n,

(a) $\int_0^{2\pi} \sin(m\theta) \sin(n\theta) d\theta = \pi$ if $m = n$.

(b) $\int_0^{2\pi} \sin(m\theta) \sin(n\theta) d\theta = 0$ if $m \ne n$.

(c) $\int_0^{2\pi} \sin(m\theta) \cos(n\theta) d\theta = 0$.

95. In *probability theory*, a function f is said to be a *probability density function* provided

$$\int_{-\infty}^{\infty} f(x)dx = 1.$$

The *expected value* $E(x)$ of the random variable x is given by

$$E(x) = \int_{-\infty}^{\infty} xf(x)dx.$$

Show that each of the following functions is a probability density function and find the expected value of the random variable x.

(a) $f(x) = \begin{cases} \frac{8}{x^3} & \text{if } x \geq 2 \\ 0 & \text{if } x < 2 \end{cases}$

(b) $f(x) = \begin{cases} \frac{12x-3x^2}{32} & \text{if } 0 \leq x \leq 4 \\ 0 & \text{if } x < 0 \text{ or } x > 4 \end{cases}$

(c) $f(x) = \begin{cases} 0.5e^{-0.5x} & \text{if } x \geq 0 \\ 0 & \text{if } x < 0 \end{cases}$

96. Find the following definite integrals and indefinite integrals using the indicated substitution.

(a) $\int \frac{1}{1+\sqrt{x}} dx, u = \sqrt{x}$

(b) $\int_0^9 \sqrt{1+\sqrt{x}} dx, u = \sqrt{x}$

(c) $\int_{-1}^7 x\sqrt[3]{1+x} dx, u = \sqrt[3]{1+x}$

(d) $\int \frac{3x}{\sqrt{1-2x}} dx, u = \sqrt{1-2x}$

97. Use the substitution $u = 1+e^x$ to find the following integral.

$$\int \frac{1}{1+e^x} dx$$

98. Suppose that f is a continuous odd function, i.e., $f(-x) = -f(x)$, for all real numbers x. Is it true that

$$\int_{-\infty}^{\infty} f(x)dx = 0?$$

If so, give a proof. If not, give a counterexample.

99. Find the following integral

$$\int \sin(x)\cos(x)dx$$

in the following two ways: (i) using the subsitution $u = \sin(x)$ and (ii) using the substitution $u = \cos(x)$. Explain the difference between your two answers.

100. Use integration by parts to derive the following integration formula.

$$\int_a^b xf''(x)dx = bf'(b) - af'(a)f(b) + f(a).$$

101. Use integration by parts to derive the following integration formula where a, b and c are arbitrary constants.

$$\int e^{ax}\sin(bx)dx = \frac{e^{ax}(a\sin(bx)-b\cos(bx))}{a^2+b^2} + c$$

5.5 APPLICATIONS OF INTEGRATION

5.5.1 Overview

In this section you will investigate various applications of integration. In particular, you will consider problems which can be solved by thinking of a desired quantity as a limit of a Riemann sum. We will discuss the following applications in the text discussion: areas, volumes of solids of revolution, arc lengths, areas of surfaces of revolution, centers of mass, work and some problems which can be solved using anti-derivatives (such as linear motion, the present value of money, distribution of electrons, electrical circuits and entropy). Applications involving integrals are plentiful and you will see others in the exercises, as well as in courses in other areas. There is no need for further activities in this section, since the concept of a definite integral as a limit of a Riemann sum has been developed in previous sections.

5.5.2 Discussion

We consider the following three categories of applications:

1. Those based on the interpretation of a definite integral as a limit of a Riemann sum.

2. Those based on the interpretation of an indefinite integral as an anti-derivative.

3. Those based on the use of a definite integral with variable endpoints (or limits of integration) as a way of defining a function.

The Integral as a (Limit of a Riemann) Sum   The root of the concept of integration lies in the notion of Riemann sum which you have been working with from the beginning of this book. The idea is that, given a function on an interval, you partition the interval into small subintervals, make the desired calculation on each interval with the function replaced by a constant (but a different constant on each subinterval), and then sum over all the subintervals. The result is an approximation to the quantity you are looking for. The approximation is improved by making the subintervals smaller.

We can use some fancy words to describe this general recipe. When working in the subinterval, we can say we are making *local* or *micro* calculations and when we add it all up, this is a *global* or *macro* calculation. The interplay between local and global, micro and macro is a very important part not only of mathematics, but of all areas in science.

The applications in this category follow the general method of making a local calculation using a constant for the function and adding up to get the global result. The main difference that distinguishes one application from another is the particular calculation you make on each subinterval. For area, you calculate the area of a rectangle, for work you calculate the work done in applying a constant force, for a volume of revolution you calculate the volume of a cylinder. The crux of our discussion of each application will therefore be a description of the local calculation.

The Integral as a Sum — Area   As we mentioned above, the original Riemann sums lead to the area under a curve. For example, if you consider the function f given by

$$f(x) = \begin{cases} 2x+1 & \text{if } -1 \le x < 0 \\ x^2+1 & \text{if } 0 \le x \le 1 \end{cases}$$

and wish to compute the (net) area "under the curve" of this function (or more precisely, the area between the curve and the x-axis), then you begin with a partition, $x_0, x_1, x_2, \ldots, x_n$ of the interval $[-1, 1]$, and calculate the sum used in **RiemLeft**,

$$\sum_{i=1}^{n} f(x_{i-1})(x_i - x_{i-1})$$

As you have seen, this approximates the (net) area (taking area below the x-axis as negative) of each little section by the area of a rectangle and then adds them up. This approximates the (net) area.

Passing to the limit gives the exact area. See Figure 5.12.

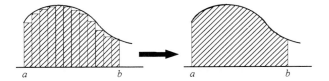

Figure 5.12. An area as the limit of a Riemann sum.

Instead of the area under the curve of one function, you can do the same thing for the area between the curves of two functions f and g. When $f > g$, you just use $f - g$ instead of f. Can you explain why? See Activity 6(i), page 321.

Can you formulate the idea of area under a single curve as a special case of the area between two curves?

The Integral as a Sum — Volumes of Revolution   Consider that the graph of the function given by $x^{1.5}$ on the domain $[0, 4]$ is rotated in three dimensions around the x-axis as shown in Figure 5.13.

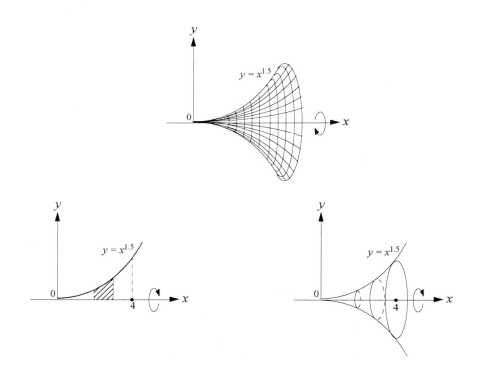

Figure 5.13. A "volume of revolution".

We will find the volume of this "volume of revolution". As usual, we begin by partitioning the interval $[0, 4]$ on the x-axis and replacing the function by a constant on each interval $[x_{i-1}, x_i]$, say the value of the function at the left endpoint, $f(x_{i-1}) = (x_{i-1})^{1.5}$. What is the volume of revolution that we get?

The volume of this cylinder is

$$\pi\left((x_{i-1})^{1.5}\right)^2 (x_i - x_{i-1}) = \pi(x_{i-1})^3 (x_i - x_{i-1})$$

and so, adding them up and factoring out π, we obtain

$$\pi \sum_{i=1}^{n} (x_{i-1})^3 (x_i - x_{i-1})$$

Figure 5.14 shows a thin circular cylinder with radius $(x_{i-1})^{1.5}$ and width $x_i - x_{i-1}$.

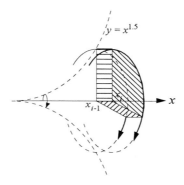

Figure 5.14. A cylindrical approximation to a slice of the solid.

Now, here is a step that must be taken often in these applications. We must recognize the above expression as a certain Riemann sum. Do you see which one?

This is the Riemann sum that you would get by applying **RiemLeft** to the function given by x^3. Hence, passing to the limit we conclude that the volume is given by

$$\int_0^4 \pi x^3 dx = 64\,\pi.$$

The Integral as a Sum — Arc Length

Now let's take the same function, given by $x^{1.5}$ on the domain $[0, 4]$ and try to figure out the length of the curve.

We partition the interval $[0, 4]$ and consider an approximation to the length of the arc on a subinterval. There are several possibilities. We could take the horizontal distance, the vertical distance, the length of the secant, or the length of the tangent at the left endpoint. See Figure 5.15.

It turns out that the first three alternatives give three different answers, and so we have to base our choice on which seems to be the most reasonable way to define arc length. We think it is the third, the secant. Do you agree?

Actually, the fourth choice gives the same answer as the third and the computations are simpler, so we will use it. What we obtain is a small right triangle with one side $x_i - x_{i-1}$. The other side is the rise that you get with a run of $x_i - x_{i-1}$ and a slope of $f'(x_{i-1}) = 1.5\sqrt{x_{i-1}}$, that is $f'(x_{i-1})(x_i - x_{i-1}) = 1.5\sqrt{x_{i-1}}(x_i - x_{i-1})$. See Figure 5.16.

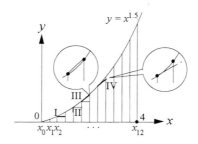

Figure 5.15. *An approximation to the length of a curve.*

The approximation that we want is the hypotenuse of this triangle which is given by

$$\sqrt{(x_i - x_{i-1})^2 + \left(1.5\sqrt{x_{i-1}}(x_i - x_{i-1})\right)^2} = \sqrt{1 + 2.25x_{i-1}}(x_i - x_{i-1})$$

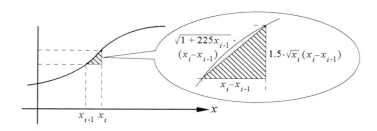

Figure 5.16. *A right triangle with sides $x_i - x_{i-1}$ and $f'(x_{i-1})(x_i - x_{i-1})$.*

If we add these up, we get

$$\sum_{i=1}^{n} \sqrt{1 + 2.25x_{i-1}}(x_i - x_{i-1})$$

Once again, we must look at this expression and ask, of what function is this a Riemann sum? It is the function $\sqrt{1 + 2.25x}$. Hence, the length of the curve is given by

$$\int_0^4 \sqrt{1 + 2.25x}\,dx = \frac{8}{27}\left(10^{\frac{3}{4}} - 1\right) \approx 1.3699.$$

Look again at the specific calculations for the length of the arc given above. Do you see how this specific case can be generalized to the formula

$$L = \int_a^b \sqrt{1 + [f'(x)]^2}\,dx$$

for the length of the arc of f as x goes from a to b? Is there any restriction placed on f' to ensure that this integral exists?

The Integral as a Sum — Surfaces of Revolution

We can put the ideas of the last two paragraphs together. Consider a general function f on the domain $[a, b]$. Draw the graph of this function and rotate it in three dimensions around the x-axis. The problem is to find the surface area of the solid.

It will be very good practice for you to draw a picture of what this looks like. Indicate a partition of the interval and show what the surface would be on a subinterval. Do you see that it is the surface of a cylinder? What is the radius of this cylinder and what is its width? Indicate all of these on your picture.

Now you can calculate the surface on a subinterval and then add them up. You get the following expression

$$2\pi \sum_{i=1}^{n} f(x_{i-1})\sqrt{1+(f'(x_{i-1}))^2}\,(x_i - x_{i-1})$$

and this is the Riemann sum for the function given by

$$2\pi f(x)\sqrt{1+(f'(x))^2}$$

so a formula for area of a surface of revolution is

$$2\pi \int_a^b f(x)\sqrt{1+(f'(x))^2}\,dx.$$

The Integral as a Sum — Center of Mass

We turn now to some concepts from other fields in science, beginning with the center of mass of an object.

Suppose you have a system of finitely many particles located on a line. Let the particles be at (signed) distances x_i, $i = 1\ldots n$, relative to an origin and suppose that the masses of the particles are m_i, $i = 1\ldots n$. Then the center of mass of this system is the point on the line located at $\bar{x}$ where

$$\bar{x} = \frac{\sum_{i=1}^{n} m_i x_i}{\sum_{i=1}^{n} m_i}.$$

The sum

$$\sum_{i=1}^{n} m_i x_i$$

is called the *moment* of the system of particles about the origin of the line.

If a system of particles is spread around a planar area, then one computes the x-coordinate, $\bar{x}$, and the y-coordinate, $\bar{y}$, of the center of mass according to a formula, analogous to the one above, twice. Once with the x-coordinates of the individual particles and once with the y-coordinates. The point $P(\bar{x}, \bar{y})$ so obtained is called the *center of mass* of the system. For the particle with mass m_i, located at a point (x_i, y_i), the x_i in the above formula corresponds to the (signed) distance from the particle to the y-axis. Similarly, y_i corresponds to the (signed) distance from the particle to the x-axis.

Now consider a general planar area, for example the region between two curves given by functions f and g on the domain $[a, b]$. Suppose that a sheet of thin material in the shape of the region is homogeneous in the vertical direction, but not in the horizontal. Thus its density, mass per unit of area, is a function ρ with independent variable x. Our problem is to find the center of mass of this sheet. For such a sheet, the center of mass is a "balance point", that is, a point at which the sheet would balance on the tip of a rod when the tip is placed directly under it.

To find the x-coordinate of the center of mass, we partition the interval $[a, b]$ into n equal subdivisions and assume that the density is constant on the subinterval $[x_{i-1}, x_i]$, equal to, $\rho(x_i^*)$, its value at the midpoint x_i^* of thesubinterval $[x_{i-1}, x_i]$. Moreover, we replace this piece of the region by a rectangle with width $x_i - x_{i-1}$ and height $f(x_i^*) - g(x_i^*)$. What we are actually assuming here, from a physical point of view, is that the entire segment of our region determined by this subinterval is replaced by a single point mass at the center of the rectangle with coordinates

$$\left(x_i^*, \frac{f(x_i^*) + g(x_i^*)}{2} \right).$$

The mass concentrated at this point is taken to be *the density at the point times the area of the rectangle,* or

$$\rho(x_i^*)(f(x_i^*) - g(x_i^*))(x_i - x_{i-1}).$$

See Figure 5.17.

The moment of the point mass located at the center of the (approximating) rectangle about the y-axis is the product of the (signed) distance x_i^* from the y-axis to the mass times the mass, $\rho(x_i^*)(f(x_i^*) - g(x_i^*))(x_i - x_{i-1})$, and is given (approximately) by

$$x_i^* \rho(x_i^*)(f(x_i^*) - g(x_i^*))(x_i - x_{i-1}).$$

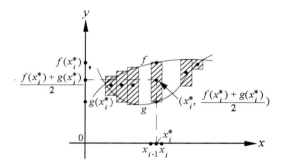

Figure 5.17. $\mathbf{mass} = \rho(x_i^*)(f(x_i^*) - g(x_i^*))(x_i - x_{i-1}).$

Using the above approximation for each point mass, we can apply the formula for a finite system of n point masses, located at the centers of n approximating rectangles, to obtain

$$\frac{\sum_{i=1}^{n} x_i^* \rho(x_i^*)(f(x_i^*) - g(x_i^*))(x_i - x_{i-1})}{\sum_{i=1}^{n} \rho(x_i^*)(f(x_i^*) - g(x_i^*))(x_i - x_{i-1})}$$

and passing to the limit, we obtain, for the x-coordinate of the center of mass,

$$\bar{x} = \frac{\int_a^b x \rho(x)(f(x) - g(x)) dx}{\int_a^b \rho(x)(f(x) - g(x)) dx}.$$

In a similar manner, since the (signed) distance from the x-axis to the center of an approximating rectangle is given by

$$\frac{f\left(x_i^*\right)+g\left(x_i^*\right)}{2}$$

the y-coordinate of the center of mass is given by

$$\overline{y}=\frac{\int_a^b\left(\frac{f(x)+g(x)}{2}\right)\rho(x)(f(x)-g(x))dx}{\int_a^b\rho(x)(f(x)-g(x))dx}$$

or

$$\overline{y}=\frac{\int_a^b\frac{1}{2}\rho(x)\left((f(x))^2-(g(x))^2\right)dx}{\int_a^b\rho(x)(f(x)-g(x))dx}.$$

In the exercises you will deal with the case where the density function depends on y. In Chapter 10 we will describe a more symmetric method that does not require a separate treatment for the x and y coordinates of the center of mass, and that also can handle a density function that depends on both x and y.

If the density is constant, that is, if the material is homogeneous, we call the object a *lamina* and refer to its center of mass as its *centroid*.

The Integral as a Sum — Work

The work W done by applying a constant force F to move an object a distance d in a fixed direction is

$$W=Fd.$$

We will treat the problem of determining the work done in case the force depends on the position. We still assume that it is applied in a constant direction.

We may take the direction of the force to be the x-axis of a coordinate system and the position of the object to be its x-coordinate, with starting position a and ending position b. Hence, the magnitude of the force is a function F. See Figure 5.18.

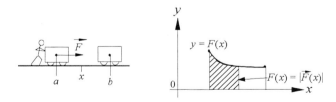

Figure 5.18. A variable force F acting on an object moving along a line from a to b.

Our principle for solving this problem is the same as what we have used before. We partition the interval $[a,b]$ and assume that the force is constant on each subinterval, $[x_{i-1},x_i]$, given by its value at the left endpoint, $F(x_{i-1})$. Hence the work done to move the object from x_{i-1} to x_i is approximated by

$$F(x_{i-1})(x_i - x_{i-1})$$

adding these up we get an approximation of the work done in moving the object from a to b,

$$\sum_{i=1}^{n} F(x_{i-1})(x_i - x_{i-1})$$

and passing to the limit gives the exact value,

$$W = \int_a^b F(x)dx.$$

The Integral as an Anti-derivative
You have already seen and worked with several situations in which the anti-derivative was used to express and compute various quantities. Here we will only summarize them and refer to the locations where they were considered.

Linear Motion. If a function gives the acceleration as a function of time for a particle moving in a straight line, then the anti-derivative of that function gives the velocity. The anti-derivative of the velocity is the position as a function of time. Linear motion was discussed on page 131 and page 203.

Distribution of electrons. According to the Fermi-Dirac distribution, under certain conditions, the number of electrons at a certain energy level is a function N of the energy level E. In various versions of this model, it is possible to derive a function of the energy level which is the derivative of N so that N is the anti-derivative of this function. One such example was considered in Activity 3, page 309.

Present Value. If M is a function of time which gives the amount of money spent from the beginning to time t, and interest is compounded continuously at the rate r, then the present value of the money spent from the beginning to time t is the anti-derivative of the function given by

$$M'(t)e^{-rt}.$$

This is discussed in Activity 6, Section 5.1.2.

Electrical circuits. If i is a function of time giving the conduction current across a capacitor, then the anti-derivative of i divided by the capacitance C is the voltage function. This was discussed in Activity 2, Section 5.2.2.

The Integral as a Method of Defining Functions
The last category of applications of the definite integral is also something that we previously considered at length and so it is only necessary to summarize the ideas.

In several places in this text we have considered the physical quantity called *entropy*. You may well ask, what, exactly is entropy. It is not really possible to give a precise answer outside of calculus. However, entropy provides a measure of unavailable energy in a thermodynamic system — a concept of great importance to physics, chemistry, mechanical and chemical engineering, and many other engineering and physical science fields. You can make some general comments, and develop a vague, intuitive idea of what entropy is. If you wish, however, to have a powerful concept that can be used in scientific analysis, then you must turn to integration. Entropy is the anti-derivative of the function given by heat capacity divided by time. Or, one usually can avoid talking about entropy and refer only to the *change in entropy* which is the definite integral of this function between two points of temperature. Rather than try to build on vague, hand waving arguments, some people find it easier to begin with such analytic definitions and to build their intuition this way, rather than the other way around.

There are other scientific concepts that can be captured in this way, such as *enthalpy* — a measure of heat capacity which is important to engineers and scientists.

Finally we mention our rigorous definition of the natural logarithm and exponential functions in Section 5.3. You also saw in the exercises how to get some of the inverse trigonometric functions in the same way. This is not the only way to construct these functions, but it is a particularly neat, rigorous and powerful method.

5.5.3 Exercises

1. In each of the following exercises, write a definite integral which represents the area of the finite portion of the region bounded by the indicated curves. Then find the area.

   (a) $f(x) = 4 - x^2$ and the x-axis

   (b) $f(x) = 1 - x^2$ and $g(x) = x^2 - 1$

   (c) $y = 3x + 2x^2$ and $y = x^3$

   (d) $y = x^2 - 6$ and $y = x$

   (e) $y = x^2 + 1$, $y = 2x - 2$, $x = -1$ and $x = 2$

   (f) $f(x) = x^2 - 2x - 8$ and the x-axis

   (g) $xy = 1$, $y = x$, $x = 3$ and the x-axis

   (h) $f(x) = \cos(x)$, $g(x) = \sin(x)$, $x = 0$ and $x = \pi/2$

   (i) $x = y^2$ and $y = x^2$

   (j) $y = |x|$ and $x^2 - 4$

   (k) $x = 1 + y^2$ and $x = 2$

   (l) $y = \sin(2x)$, $y = 2\cos(x)$, $x = 0$ and $x = \pi$

   (m) The larger region bounded by the curves $y = 2 + x$, $y = x^2$, and $y = 6 - x$

   (n) $y = \frac{x}{\sqrt{1+x^2}}$, $x = 1$ and the x-axis

   (o) $y^2 + x^2 = x^4$

2. Write a general formula for the area bounded by two functions f and g and the lines $x = a$ and $x = b$ when $f(x) \geq g(x)$ for $a \leq x \leq b$.

3. Write a definite integral which represents the volume of revolution obtained by revolving the finite portion of the region bounded by the indicated curves around the specified line. Then find the volume. In some cases you may wish to use the fact the volume of a "thin" cylindrical shell is given by

$$2\pi RHT$$

where R is the radius from the line about which a region is revolved, H is the height of a typical shell and T is the thickness of a "thin shell".

   (a) $y = \sqrt{x}$, $x = 9$ and the x-axis about the

      i. x-axis   ii. y-axis

      iii. line $x = 9$

      iv. line $x = -1$ v. line $y = 3$

      vi. line $y = 4$

   (b) $y = x^2$, the x-axis and $x = 2$ about the

      i. x-axis   ii. y-axis

      iii. line $x = 2$ iv. line $x = 3$

      v. line $y = 4$

      vi. line $y = -1$

   (c) $yx = 1$, the x-axis, $x = 1$ and $x = 3$ about the

      i. x-axis   ii. y-axis

      iii. line $x = 3$

      iv. line $x = -1$

      v. line $y = -1$

      vi. line $y = 2$

(d) the third quadrant portion of the region bounded by $y = x^2 - 4$, the y-axis, and the x-axis about the

      i. x-axis   ii. y-axis

      iii. line $x = 2$ iv. line $x = 3$

      v. line $y = 4$

      vi. line $y = -1$

(e) $y = x^2$ and $y = x^3$ about the

      i. x-axis   ii. y-axis

      iii. line $x = 2$

      iv. line $x = -1$ v. line $y = 3$

      vi. line $y = -2$

(f) $y = 3\sqrt{x}$ and $y = x$ about the

      i. x-axis   ii. y-axis

      iii. line $x = 9$

      iv. line $x = -2$ v. line $y = 10$

      vi. line $y = -1$

(g) $x = 4 - y^2$ and $y = x - 2$ about the

      i. y-axis   ii. line $x = 4$

      iii. line $x = -2$ iv. line $y = 4$

      v. line $y = -5$

(h) $y = 1 + x^2$, $y = 2x$ and the y-axis about the

      i. x-axis   ii. y-axis

      iii. line $x = 1$

      iv. line $x = -2$ v. line $y = 3$

      vi. line $y = -1$

(i) $y = \sec(x)$, $y = 0$, $x = 0$ and $x = \pi/4$ about the x-axis.

(j) $y = \tan(x)$, $y = 0$ and $x = \pi/6$ about the x-axis.

(k) $y = \sec(x)$, $x = y$ and the y-axis about the x-axis

4. For the region in the first quadrant bounded by the graph of the function f given by

$$f(x) = \frac{1}{1 + x^2},$$

the line $x = 1$ and the coordinate axes, find the

(a) area

(b) volume by revolving the region about the x-axis

(c) volume by revolving the region about the y-axis

(d) volume by revolving the region about the line $x = 1$

(e) volume by revolving the region about the line $y = 1$

(f) volume by revolving the region about the line $x = 2$

(g) volume by revolving the region about the line $y = -1$

5. Find the area of the region bounded by the graph of the ellipse

$$\frac{x^2}{a^2} + \frac{y^2}{b^2} = 1.$$

6. Find the length of each of the following curves on the indicated interval.

(a) $y = 0.5x^2$; $0 \le x \le 3$.

(b) $y = x^2$; $-2 \le x \le 2$.

7. A storage tank is in the shape of a hemisphere with radius 10 feet and its circular top horizontal. Write a definite integral that represents the volume of water in the tank when the tank is filled to a depth of 4 feet. Then find the volume.

8. Suppose that the tank in the previous exercise is filled to a depth of 3 feet with a liquid weighing 60 pounds per cubic foot.

   (a) Write a definite integral which represents the weight of the liquid.

   (b) What is the weight of the liquid in the tank?

   (c) What percentage of the tank's total volume is filled with the liquid?

9. Write a definite integral which represents the volume obtained by revolving the region bounded by two functions f and g where $f(x) \geq g(x) \geq 0$ for $0 < a \leq b$ and the lines $x = a$ and $x = b$ about the (i) x-axis and (ii) y-axis.

10. Write a definite integral which represents the lengths of the following curves on the indicated intervals. Then find the length of the curve by hand and check your answer by using your SCS.

    (a) $y = 2x + 1$ on $[-1, 2]$

    (b) $y = 2\sqrt{x^3}$ on $[0, 1]$

    (c) $y = 1 - 3\sqrt[3]{x^2}$ on $[1, 8]$

11. Write a definite integral which represents the length of a plane curve which is the graph of a function f whose derivative is continuous on an interval $[a, b]$.

12. Write a definite integral which represents the area of the surface of revolution obtained by revolving each curve on the indicated interval about the specified axis. Then find the area by hand and check your answer by using your SCS.

    (a) $y = 2x$ for $0 \leq x \leq 2$ about the (i) x-axis and (ii) y-axis.

    (b) $3y = x^3$ for $0 \leq x \leq 1$ about the x-axis.

    (c) $y = \frac{x^2}{2}$ for $0 \leq x \leq 2$ about the y-axis.

   (d) $y = 1 + \sqrt[3]{x}$ for $0 \leq x \leq 8$ about the y-axis.

   (e) $\sqrt{x} + \sqrt{y} = 1$ about the y-axis.

13. Write a definite integral which represents the area of the surface obtained by revolving about the y-axis a plane curve that does not cross the y-axis and which is the graph of a function f, with independent variable x, whose derivative is continuouson an interval $[a, b]$.

14. Write a definite integral which represents the length of the curve over the indicated interval. Then use **RiemTrap** to estimate the length of each curve on the indicated interval to five decimal places.

    (a) $3y = x^3$ for $-1 \leq x \leq 2$

    (b) $y = \sqrt{4 - x^2}$ for $0 \leq x \leq 1$.

    (c) $y = \sin(x)$ for $0 \leq x \leq 2$.

15. A slider moves in a vertical plane along a wire that is bent into the shape of the curve $y = \frac{36}{x}$. Write a definite integral which represents the distance along the curve that the slider has traveled when it moves from the point $[1, 36]$ to the point $[6, 6]$. Then find or approximate the distance.

16. Write a definite integral which represents the area of the surface obtained by revolving the curve on the indicated interval about the indicated axis. Use **RiemTrap** to estimate the area of the surface of revolution to five decimal places. Finally, find this surface area exactly.

    (a) $3y = x^3$ for $0 \leq x \leq 1$ about the y-axis.

    (b) $y = x^2/3$ for about the x-axis.

    (c) $y = 1 + \sqrt[3]{x}$ for $0 \leq x \leq 8$ about the x-axis.

17. Write a definite integral which represents the area of the surface curves on the indicated intervals. Then find the length of the curve by hand or by using your SCS.

18. The (constant-pressure) Heat Capacity C_p as a function of temperature T for one mole of solid zinc is represented by the following table of data.

T	C_p
20	0.406
30	1.187
40	1.953
50	2.671
60	3.250
70	3.687
80	4.031
90	4.328
100	4.578

Table 5.1. Heat Capacity as a function of temperature.

(a) The change in heat ΔH needed to raise the temperature of one mole of solid zinc from T_1 degrees Kelvin to T_2 degrees Kelvin at constant pressure is given by

$$\Delta H = \int_{T_1}^{T_2} C_p \, dT.$$

When the temperature changes from $T = 20°$ Kelvin to $T = 100°$ Kelvin, use **RiemTrap** to estimate the change in heat, ΔH.

(b) The change in entropy ΔS resulting from a temperature change from T_1 degrees Kelvin to T_2 degrees Kelvin for one mole of solid zinc at constant pressure is given by

$$\Delta S = \int_{T_1}^{T_2} \frac{C_p}{T} \, dT.$$

When the temperature changes from $T = 20°$ Kelvin to $T = 100°$ Kelvin, use **RiemTrap** to estimate the change in entropy, ΔS.

19. In each of the following situations, write the coordinates of the center of mass of a thin sheet in the shape of the finite portion of the plane region bounded by the indicated curves, with the given density function ρ, as a ratio of two definite integrals. Then find the center of mass.

(a) $y = x$, the x-axis and the line $x = 4$, $\rho(x) = 2x$

(b) $y = x^2$ and $y = x$, $\rho(x) = \sqrt{x}$

(c) $y = 1 - x$ and $y = x^2 - 1$, $\rho(x) = x^2$

(d) $y = \sqrt{x}$ the x-axis and the line $x = 9$, $\rho(x) = x + 1$

(e) $y = x^3$, the y-axis and the line $y = 8$, $\rho(y) = y$

(f) $y = 2x - x^2$ and $y = x$, $\rho(x) = 0.5x$

(g) $y = \sqrt{4 - x^2}$ and the x-axis, $\rho(y) = \sqrt{y}$

(h) $x = 6 - y^2$ and the line $x = 2$, $\rho(y) = 3y$

(i) $y = x^2 + 2x - 3$ and the x-axis, $\rho(x) = 2$

20. The *centroid* of a plane region is the center of mass of a "thin" lamina in the shape of the region with a density function that is the constant function 1. The mass is then replaced by the area in computing the centroid. In each of the following situations, write the centroid as a ratio of definite integrals for the finite portion of each plane region bounded by the indicated curves. Then find the centroid.

(a) $y = x$, the x-axis and the line $x = 4$

(b) $y = x^2$ and $y = x$

(c) $y = 1 - x$ and $y = x^2 - 1$

(d) $y = \sqrt{x}$ the x-axis and the line $x = 9$

(e) $y = x^3$, the y-axis and the line $y = 8$

(f) $y = 2x - x^2$ and $y = x$

(g) $y = \sqrt{4 - x^2}$ and the x-axis.

(h) $x = 6 - y^2$ and the line $x = 2$

(i) $y = x^2 + 2x - 3$ and the x-axis

(j) The first quadrant portion of the circle with radius r and center at the origin.

(k) The region bounded by the curve $\sqrt{y} + \sqrt{x} = 4$ and the coordinate axes.

21. The roof of a basketball arena is that portion of a sphere with radius 50 meters cut off by a plane 40 meters from the center of the sphere. Write a definite integral which represents the area of the roof. Then find the area of the roof.

22. A suspension bridge has a length of 4 kilometers with supports 100 meters high. The cables on each side of the bridge hang in the shape of a parabola and each touches the roadway. Write a definite integral which represents the length of each cable. Then use **RiemMid** to estimate the length of each cable.

23. A gas is confined to a cylinder in which there is a piston. The pressure p and volume v of the system are related by the equation

$$pv^{1.4} = 40$$

where the volume v is in cubic inches (in^3) and the pressure p is in pounds per square inch (lbs/in^2). Write a definite integral which represents the work done by the force acting perpendicular to the piston face. Then find the work done by the force against the face as the volume of the gas increases from 10 in^3 to 15 in^3. Use the fact that the force acting perpendicular to the face of the piston head is the pressure times the area of the face of the piston, and the volume displaced by the piston is the product of the cross-sectional area of the piston face times the distance the piston moves.

24. A variable force acts on an object moving along a coordinate line. Assume that the coordinate line lies along the x-axis in an xy-coordinate system. The force F, in Newtons, at a point x is given by the function

$$F(x) = \frac{9}{x^2}.$$

Write a definite integral which represents the work done by the force F acting on the object as it moves from $x = 1$ meter to $x = 4$ meters. Then find the work done.

25. For a spring which hangs vertically from a ceiling, the force f exerted by the spring when it is stretched x inches beyond its normal length is given by $f = F(x) = 3x$ pounds. Write a definite integral which represents the work done in stretching the spring 3 inches beyond its normal length. Then find the work done. Write a definite integral which represents the work done in stretching the spring an additional 2 inches.

26. In this exercise, use *Hooke's Law* which is as follows:

A spring stretched x units beyond its normal length exerts an (elastic) force f that is proportional to x, i.e.,

$$f = F(x) = kx$$

where k is a constant called the *spring constant* which depends on the spring and on the units used.

A force of 30 lbs is exerted by a spring when the spring is stretched two inches beyond its normal length. Write a definite integral which represents the work done in stretching the spring (i) five inches beyond its normal length and (ii) from one inch to three inches beyond its normal length. Then find the work done in stretching the spring in each situation.

27. A 50 foot chain weighing three pounds per linear foot lies coiled on a level concrete slab. Write a definite integral (or an expression involving a definite integral) which represents the work done in raising one end of the chain vertically to

    (a) a level of 20 feet above the slab. Then find the work done.

    (b) so that the bottom of the chain is resting on the slab. Then find the work done.

    (c) so that the bottom of the chain is 20 feet above the slab. Then find the work done.

28. In each of the following exercises, think of lifting "small slabs" of the liquid out of the tank one at a time.

    (a) A conical water tank 6 feet across the top and 10 feet high is full of a liquid weighing 50 lbs/ft^3. Write a definite integral which represents the work done in pumping out

        i. all of the liquid over the edge of the tank. Then find the work done.

        ii. the top half of tank over the edge of the tank. Then find the work done.

        iii. all of the liquid to a storage tank 10 feet above the top of the tank. Then find the work done.

        iv. the top half of tank to a storage tank 10 feet above the top of the tank. Then find the work done.

    (b) A hemispherical tank has a circular base with a radius of 4 feet and it is filled with a liquid weighing 60 lbs/ft^3. Write a definite integral which represents the work done in pumping out

        i. all of the liquid over the edge of the tank. Then find the work done.

        ii. the top half of tank over the edge of the tank. Then find the work done.

        iii. all of the liquid to a storage tank 10 feet above the top of the tank. Then find the work done.

        iv. the top half of tank to a storage tank 10 feet above the top of the tank. Then find the work done.

29. Two electrons of the same charge repel each other with a force that is inversely proportional to the square of the distance between the electrons. The force is 3.2 dynes when the electrons are 10 cm apart.

    (a) Write a definite integral which represents the work done when one electron remains stationary and the other moves from a distance of 15 cm away to a distance of 10 cm away from the stationary electron. Then find the work done.

    (b) Suppose that the distance between the electrons is increased from 15 cm between the electrons to r cm where $r > 15$. Write a definite integral which represents the work done. Then find the work done as a function of r. What is the limiting value of the work done as r increases without bound? That is, what work is done in moving one electron from 15 cm away from the other to "infinity"?

30. A 100 foot cable weighing 3 pounds per linear foot hangs vertically supporting a 100 pound object. Write a definite integral which represents the work done in winding 30 feet of the cable around a drum. Then find the work done.

31. The rate R at which a machine generates revenue is given by $R(x) = 3 - 0.3x^2$ thousand dollars per year when the machine is x years old. Find the total net earnings during the first six years the machine is in service.

32. The rate R at which a machine generates revenue is given by $R(x) = 5 - 0.2x^2$ thousand dollars per year when the machine is x years old. Find the total net earnings of the machine during its lifetime, i.e., during the time it generates revenue.

33. In the following exercises, use the fact that for a solid with known cross-sectional areas, a "thin slab" of volume is the product of the cross-sectional area and the thickness of the thin slab.

    (a) In each of the following, write a definite integral which represents the volume of a solid that has a circular base with radius 4 feet and cross sections perpendicular to a fixed diameter of the base that are

        i. rectangles with height 2 feet. Then find the volume.

        ii. squares. Then find the volume.

        iii. semicircles. Then find the volume.

    iv. isosceles right triangles with hypotenuse in the plane of the base. Then find the volume.

    (b) In each of the following, write a definite integral which represents the volume of a solid that has a base in the shape of an equilateral triangle with sides 3 feet and cross sections perpendicular to the plane of the base are

        i. rectangles with height 3 feet. Then find the volume.

        ii. squares. Then find the volume.

        iii. semicircles. Then find the volume.

        iv. isosceles right triangles with height 2 feet. Then find the volume.

    (c) An object is 6 feet long. A vertical cross-section of the object x feet from one end is a circle with radius $\sqrt{x}$ feet. Write a definite integral which represents the volume of the object. Find the volume of the object.

    (d) An object is 4 meters long. A vertical cross-section of the object x meters from one end is a circle with radius x^2. Write a definite integral which represents the volume of the object. Then find the volume of the object.

    (e) Write a definite integral which represents the volume common to two straight pipes each with a radius of r inches when the axes of the pipes meet at right angles to each other. Then find the volume.

    (f) A solid has as its base the finite portion of the plane region bounded by the curves $y = x^2$ and $y = 8 - x^2$, and it has cross sections perpendicular to the x-axis which are semicircles. Write a definite integral which represents the volume of the solid. Then find the volume of the solid.

(g) A logger cuts a wedge from a tree with a diameter of 6 inches by making two cuts, one parallel to the ground and one at an angle of 30°. Write a definite integral which represents the volume of the wedge removed by the logger. Then find the volume of the wedge.

34. In the following situations, write a definite integral which represents the force due to liquid pressure in each situation by using the following facts.

The pressure p in a liquid at a depth h is the product of the weight *density* ρ (weight per unit of volume) of the liquid times the depth, that is, $p = \rho h$.

Pascal's law of physics states that the pressure at any depth is exerted equally in all directions.

The force f along a "thin strip" of surface area on a vertical plate is the product of the area a of the strip times the pressure p along the strip, that is, $f = ap = a\rho h$.

Assume that the density of water is given by $\rho = 62.4$ lbs/ft^3.

(a) A swimming pool with a horizontal rectangular bottom 30 feet by 20 feet and vertical sides six feet high is filled with water. Find the total force due to the pressure of the water on each side of the pool.

(b) For the swimming pool in part (a), find the total force due to the pressure of the water on the bottom half of each side of the pool.

(c) A vertical rectangular floodgate in a dam is 10 feet high and 20 feet wide with the top parallel to the surface of the water. Find the total force due to the pressure of the water in the dam on the surface of the floodgate when the top of the floodgate is 50 feet under the water.

(d) Suppose that the floodgate in part (c) is in the shape of an isosceles trapezoid with bases parallel to the surface of the water. The gate is 6 feet across the top, 4 feet high and 10 feet across the bottom. Find the total force due to the water pressure on the surface of the floodgate when the bottom of the floodgate is 10 feet below the water level.

(e) A dam in a river has a rectangular face which is 200 feet long and 30 feet wide. Find the total force due to water pressure on the dam face when

    i. the face is vertical and the water level is 20 feet.

    ii. the face is inclined 30° from the horizontal bottom of the river and the water level is 10 feet.

(f) A horizontal cylindrical tank has a diameter of 4 feet and is filled with a liquid weighing 50 lbs/ft^3. Find the total force due to the pressure of the liquid

    i. on one end of the tank.

    ii. on the bottom half of one end of the tank.

    iii. on the top half of one end of the tank.

35. An ellipse with major axis of length 10 and minor axis of length 6 is revolved about its major axis. Write a definite integral which represents the volume of revolution. Then

find the volume. Such a three dimensional figure, like a football or rugby ball, is called a *prolate spheroid*.

36. An oil company owns a certain oil well that produces 300 barrels of light sweet crude oil a month and will run dry in 4 years. The company estimates that t months from now the price of sweet crude oil will be $p = P(t) = 20 + 3\sqrt{t}$ dollars per barrel. Assuming that the crude oil is sold as it is extracted from the well, write a definite integral which represents the total future revenue from the well. Then find the future revenue of the well.

37. An investment is transferred continuously into an account at the constant rate of 1500 dollars per year. The account earns interest at an annual rate of 6 percent compounded continuously. Write a definite integral which represents the value of the account at the end of 3 years. Then find the value of the account. Use the fact that p dollars invested at r percent compounded continuously is worth pe^{rt} dollars after t years.

38. A metal ball with radius 6 inches has a hole with radius 2 inches drilled through it along a diameter. Write a definite integral which represents the volume of the ball drilled. Then find the volume and the percentage of the ball drilled out.

39. A statistical study conducted for a physician who has just opened a medical practice suggests that the fraction of patients who will still be receiving treatment from the physician t months after their initial visit is $e^{-0.03t}$. The physician accepts 50 people for treatment and plans to accept new patients at the rate of 20 per month. Write a definite integral which estimates how many people will be receiving treatment from the physician two years from now. Then find the estimate of how many people will be receiving treatment from the physician two years from now.

40. Blood flows through a circular artery with radius R so that the speed of the blood r centimeters from the central axis of the artery is given by $c(R^2 - r^2)$ where c is a constant. Write a definite integral which represents the rate of flow in centimeters per second at which blood flows through the artery. Then find the rate of flow of blood through the artery.

41. Prove the following *Theorem of Pappus*:

    If a plane region with finite area is revolved around a coplanar axis not crossing the region, then the volume generated is equal to the product of the area of the region and the circumference of the circle described by the centroid of the region.

5.6 ELEMENTARY DIFFERENTIAL EQUATIONS

5.6.1 Overview

This final section of the chapter represents the culmination of your experience with integral calculus. In this section you will formulate and solve certain types of *differential equations* (equations that contain an unknown function and its derivatives).

Differential equations are extremely important in mathematics and the sciences. In fact, differential equations are even more important than integrals in applying mathematics. To a large extent, the fields of Applied Mathematics and Mathematical Physics consist of deriving differential equations that model important physical phenomena and developing methods to solve these equations.

Traditionally, differential equations have been solved by applying a repertoire of specific methods. Most colleges have an entire course in which little else is done but train students to use these methods. Today, most differential equations problems that can be solved using such methods can be solved with a SCS. In light of this reality, it will be interesting to see what happens to the traditional "cookbook" differential equations course.

In this book, we will give only a brief introduction to the methods used in solving differential equations. We will consider two methods for exactly solving some large classes of elementary differential equations, and one numerical method for approximating solutions when an exact solution cannot be found.

5.6.2 Activities

1. Below is a list of six differential equations and a list of ten functions f, given in the form $y = f(x)$. Match each function to the corresponding differential equation for which it is a solution.

   Here are the derivatives.

   (a) $y' + y - 2 = 0$ (b) $y'' + 4y = 0$ (c) $y' + y = 1$

   (d) $y'' + y' - 2y = 0$ (e) $y' - 2y = 3e^{2x}$ (f) $xy' - 2y = -x$

   Here are the definitions of the functions.

   (a) $y = \sin(2x)$ (b) $y = 1 + 2e^{-x}$ (c) $y = 1 + 9e^{-x}$

   (d) $y = 2 + e^{-x}$ (e) $y = 5e^{2x} + 2e^{-x}$ (f) $y = 2e^{2x} + 7e^{-x}$

   (g) $y = c_1 e^{2x} + c_2 e^{-x}$ (h) $y = (3x + 37)e^{2x}$ (i) $y = x + cx^2$

   (j) $y = (3x + c)e^{2x}$

2. When the function that gives the derivative involves both the independent and dependent variable, there are many methods for finding the unknown function. In this activity we give several examples, all of which can be solved by the same method. Use your SCS to find the unknown function and then try to figure out the method.

   (a) $y' = \dfrac{y}{(x+1)^2}$ (b) $y' = (3 - 2y)^2$ (c) $xy' + y = 0$

3. In the spirit of the previous activity, we give here examples of the same kind of problem all of which can be solved by a single method. Use your SCS to solve for the unknown function y. After you have done them all, try to figure out the method.

   (a) $y' - 2y = e^{2x}$ (b) $y' + \dfrac{y}{x} = \dfrac{1}{x^2}$ (c) $y' + xy = 0$

   (d) $y' + xy = 1$ (e) $y' + xy = x$

   (f) $y' + y\tan(x) = \ln(\sec(x))$ (Caution: Some systems crash on this one.)

4. In this activity you will write a **computer function** to represent an approximate solution of the following problem.

> Given a function F of two variables called x and y and a point in the plane (x_0, y_0), suppose that the function F specifies, at each point (x, y) in its domain, the slope of a function. Find a function f of one variable which has the value y_0 at x_0 and whose slope at any point x in its domain is $F(x, f(x))$.

If we write $y = f(x)$, then the conditions of this problems can be expressed as,

$$y' = F(x, y), \quad y(x_0) = y_0$$

Hence you will construct a **computer function** which gives an approximate solution to this *differential equation with initial condition*, that is, to this *initial value problem*.

Your task is to write a **computer function** named **Euler** that accepts a "right hand side" which is a function **F** of two variables, endpoints **a**, **b** of an interval which is to be the domain of the unknown function, a number **c** which is to be the value of the unknown function at **a**, and a positive integer **n** which represents the "mesh size" for your approximation.

Your **computer function** will return a function whose value at a point **x** in **[a, b]** will be **c** if **x = a** and otherwise will be obtained by running through the following steps.

(a) Construct a partition **P** of the interval **[a, x]** into **n** equal subintervals.

(b) Use a loop to construct a **tuple E** of values as follows.

   i. The first component of **E** is **c**.

   ii. Assuming that the value y_i of **E** has been determined, the next value y_{i+1} is given by

$$y_{i+1} = y_i + F(x_i, y_i)(x_{i+1} - x_i).$$

(c) Return the last component of **E**, which is the value at **x** of the function being constructed.

> Run **Euler** for the function F given by $F(x, y) = 3\sin(x) - 27y - 14$, on the interval $[-1, 1]$ with $c = 5.7$. Draw a graph of the function.

5. Compute by hand the nth approximation obtained by **Euler's Method** (see Activity 4 above) for a solution to the problem:

$$y' = \frac{dy}{dx} = F(x), \quad y_0 = c$$

where c is a constant. Do so as follows: Write out the first three approximations y_1, y_2 and y_3, and then look for a pattern to write out the nth approximation y_n.

6. Apply your **computer function Euler** from Activity 4 to the right hand side F given by $F(x, y) = 2x$ on the interval $[0, b]$ and initial condition 0. You should use several values of b. Use a mesh size of $n = 50$. Compare your results with the results obtained by applying **RiemLeft** to the function represented by $2x$ on the same interval. Write a brief explanation of what you observe.

7. Work out the calculations of the previous activity by hand. As much as possible use general symbols instead of specific values for the endpoints, functions, mesh size, etc. What do you conclude about **Euler** and **RiemLeft**?

8. According to certain models, the growth of human population occurs at a rate which is proportional to the population raised to the power 1.5. If P is a function which gives the population at time t, then write an equation which expresses this model. Use your SCS to solve for the function P. Introduce any factual information you like to make your problem easier.

9. According to *Newton's Law of Cooling*, the rate at which a body cools is directly proportional to the difference between its temperature T, at any time t in minutes, and the temperature of its surroundings. A certain liquid cools from 160 degrees to 100 degrees in 30 minutes while in a room kept at a temperature of 70 degrees.

    (a) Write a differential equation describing this situation.

    (b) Solve this differential equation to find the temperature T as a function of t.

    (c) Sketch a graph of T. Such a curve is called a *cooling curve*.

    (d) Find the time required for the liquid to cool to 90 degrees.

10. A vane is attached at a certain angle to a cart. The vane is hit with a jet of water and moves in a straight line at velocity given by a function U of time t. Analysis of momentum shows that the velocity function satisfies the following differential equation,

$$\frac{dU}{dt} = A(B - U)^2$$

    where A and B are constants.

    Use your SCS to find a function U which satisfies this equation. Check your answer using differentiation.

11. According to Kirchoff's voltage law for an RL circuit, the current i is a function of time t which satisfies the following differential equation,

$$\frac{di}{dt} = Vu(t) - Ri(t) - L$$

    where V is a constant voltage, R a constant resistance, L a constant inductance, and u is a given forcing function.

    Take u to be the function given by $u(t) = \sin(t)$ and use your SCS to find a function i which satisfies this equation. Check your answer using differentiation.

    Now do it again with $V = 3$, $R = 27$, and $L = 14$.

    Find at least 5 such functions and graph them on a single coordinate system.

    Finally, compare your solution to that obtained in Activity 4.

5.6.3 Discussion

Differential Equations   In the activities of this section you have gained considerable experience in working with several different types of differential equations and their solutions. In Activity 1, for example, you encountered some differential equations which involved only a first derivative and some which involved a second derivative. Differential equations can be classified in several ways, one of which is according to the highest order derivative which occurs in the equation. Hence you have already seen examples of *first-order differential equations* and *second-order differential equations.*

One observation which we hope you made in Activity 1 is that the solution of a differential equation is not unique. Any one solution to a given differential equation is called a *particular solution,* while the family of solutions to a given differential equation is called the *general solution* of the equation. This is really just an extension of the idea which you are already familiar with in finding anti-derivatives of a given function. Let's see the relationship among particular solutions and the general solution of a differential equation which is slightly more complicated than an anti-derivative problem.

Consider a certain kind of chemical reaction in which a substance A is gradually converted to a substance B. Suppose that the concentration of A is a function C of time and the governing fact is that the rate of change in C is proportional to the concentration of B, that is, the initial amount less the concentration of A. This situation gives rise to the differential equation (with appropriately chosen units),

$$\frac{dC}{dt} = 1 - C.$$

Your SCS should tell you that the general solution to this differential equation is given by

$$C(t) = 1 - e^{k-t}$$

or

$$C(t) = 1 - ke^{-t}$$

where this time, k is the arbitrary constant.

As is the case with anti-derivatives, a family of functions constitutes the general solution of this differential equation. Figure 5.19 shows a picture of some of the members of this family obtained by taking several different specific values of k.

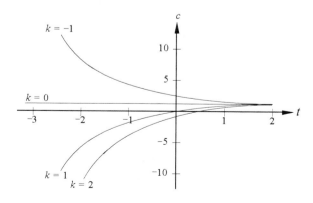

Figure 5.19. A family of functions $C(t) = 1 - ke^{-t}$.

What is the meaning of k in this case? What is the relation between tangents of various curves in this case?

We now turn our attention to several methods which will allow us to produce the general solution when we are faced with a differential equation.

Separation of Variables Look back at Activity 8. Recall that the problem in this activity was to find a function P which gives population p as a function of time t. The assumption that was made is that the rate of change of p is proportional to p raised to the power 1.5. Modeling consists of translating this statement into mathematics. We can write, for example,

$$\frac{dp}{dt} = kp^{1.5}$$

where k is a constant.

You can check by differentiating to see that

$$p = P(t) = \left(-\frac{1}{2}kt + c\right)^{-2}$$

(where c is an arbitrary constant) is a solution to the differential equation. But how was this solution obtained?

The technique used to solve the differential equation $\frac{dp}{dt} = kp^{1.5}$ is the *method of separation of variables*. Activity 8, is an example of a mathematical model which gives rise to a separable differential equation. This method is applicable when the variables, in this case p and t can be "separated" along with their respective differentials. The solution proceeds as follows:

$$\frac{dp}{dt} = kp^{1.5}$$

$$\frac{dp}{p^{1.5}} = kdt$$

$$p^{-1.5}dp = kdt.$$

Integrating both sides of this equation yields

$$\int p^{-1.5}dp = \int kdt$$

$$-2p^{-\frac{1}{2}} = kt + c$$

thus

$$p = \left(-\frac{1}{2}kt + c\right)^{-2}.$$

In general, the *method of separation of variables* illustrated above is useful for solving differential equations of the form

$$y' = \frac{dy}{dx} = F(x, y)$$

where the variables can be separated. We seek a function f so that $y = f(x)$, hence we want to find a solution of the differential equation

$$f'(x) = F(x, f(x))$$

where f is an unknown function of x. When the variables can be separated, we can write the differential equation in the equivalent form

$$h(y)dy = g(x)dx$$

for some functions h and g. Since one side of the resulting equation contains only a function of y and the differential dy and the other side contains only a function of x and the differential dx, it is easy to see why this method is called separation of variables! Differential equations which can be separated in this way are called *separable equations* and as seen above, such equations can be solved by integrating.

Consider the following differential equation problem.

$$yy' = 2x$$

In other notation the equation can be written as

$$y\frac{dy}{dx} = 2x.$$

Separating the variables, we obtain the following differential equation

$$ydy = 2xdx.$$

Integrating each side of the equation and adding a constant of integration on one side only (since if we added a constant on each side we could combine the two constants into one!), we get

$$\frac{y^2}{2} = x^2 + c.$$

Notice that the latter equation can be rewritten as

$$y^2 = 2x^2 + 2c$$

or simply

$$y^2 = 2x^2 + c$$

since $2c$ is just an arbitrary constant also, so we can simply rename it c. Do you understand why we can do this?

When a solution is written in this form, rather than $y = f(x)$, we call the solution an *implicit solution*.

Consider the following initial value problem.

$$yy' = 2x, \; y = 2 \text{ when } x = -1.$$

Using our solution above for the differential equation, we need only solve for the constant c. Substituting 2 for y and -1 for x, we obtain

$$2^2 = 2(-1)^2 + c$$

and hence $c = 2$. Therefore, an implicit solution for our problem is

$$y^2 = 2x^2 + 2.$$

The differential equation which resulted from Newton's Law of Cooling in Activity 9 is another example of a separable differential equation. Go back and look through the activities and try to determine which of the differential equations you encountered there are separable.

Integrating Factors   Consider the following differential equation.

$$y' - 2y = 8e^x$$

It is easily seen that this equation is not separable. Try to separate the variables! Just like separation of variables is used to reduce a differential equation to a one which can be solved by integrating, the *method of integrating factors* also reduces a differential equation to an equation which can be solved by integrating. The method is based on finding a factor which can be used as a multiplier on each side of a differential equation, so that the resulting equation can be integrated. It is not hard to show that any equation of the form

$$y' + f(x)y = g(x)$$

where *f* and *g* are integrable functions, can be solved by integrating after multiplying by the factor

$$e^{\int f(x)dx}$$

where $\int f(x)dx$ is any anti-derivative of *f*. Hence, we usually take the simplest expression we can for an anti-derivative of *f* when computing $e^{\int f(x)dx}$.

Let's try the above method for the equation

$$y' - 2y = 8e^x.$$

Integrating –2 we obtain

$$e^{\int -2\,dx} = e^{-2x}.$$

Now notice that if we multiply the left-hand side of the differential equation by e^{-2x}, we obtain

$$e^{-2x}(y' - 2y) = e^{-2x}y' - 2e^{-2x}y = \left(ye^{-2x}\right)'.$$

Hence, multiplying both sides of the original differential equation by the factor e^{-2x}, the equation is equivalent to the following differential equation

$$\left(ye^{-2x}\right)' = 8e^x e^{-2x} \text{ or } \left(ye^{-2x}\right)' = 8e^{-x}$$

which can be solved by simply integrating each side of the differential equation. Doing so, we obtain

$$ye^{-2x} = \int 8e^{-x}dx = -8e^{-x} + c.$$

Hence,

$$ye^{-2x} = -8e^{-x} + c.$$

Multiplying both sides of the latter equation by e^{2x} and simplifying, we obtain the solution

$$y = -8e^x + ce^{2x}.$$

You should attempt to convince yourself by trying several cases that other integrating factors arising from the expression $e^{\int f(x)dx}$ would have worked equally well to solve this differential equation.

Kirchoff's Law in Activity 11 also gives rise to a differential equation which is not separable but which can be solved through the use of an integrating factor. The terms of the differential equation

$$\frac{di}{dt} = 3\sin(t) - 27i - 14$$

can easily be rearranged so that the equation

$$\frac{di}{dt} + 27i = 3\sin(t) - 14$$

is in the form that we recognize can be solved by using the integrating factor $e^{\int 27\,dt}$. Multiplying both sides of the equation by this integrating factor yields

$$e^{27t}i' + e^{27t}27i = 3e^{27t}\sin(t) - 14e^{27t}.$$

Our motivation for using the integrating factor can now be seen, since the left hand side of the equation simplifies to the derivative of e^{27t}. That is our equation becomes

$$\left(ie^{27t}\right)' = 3e^{27t}\sin(t) - 14e^{27t}.$$

Integrating both sides yields

$$ie^{27t} = \int\left(3e^{27t}\sin(t) - 14e^{27t}\right)dt.$$

Notice that the right hand side can be decomposed into two integrals, one of which you can solve in one step, and the other which requires integration by parts and a little patience. Computing these integrals and finally dividing both sides of the equation by the quantity e^{27t} gives the solution

$$i(t) = -\frac{3}{730}\cos(t) + \frac{81}{730}\sin(t) - \frac{14}{27} + ce^{-27t}$$

where c is an arbitrary constant.

Looking back at the activities of this section, can you now identify those differential equations which can be solved by using an integrating factor? Which are they?

Numerical Approximations to ODE's — Euler's Method

Separation of variables and the use of integrating factors are only two of a myriad of methods for solving differential equations of the form $y' = F(x, y)$ exactly. It can be shown, however, that some differential equations, for example $y' + y^2 = e^{x^2}$, cannot be solved exactly by any method. In cases such as this, we approach the problem numerically and produce an approximate solution.

In this text we will consider only one numerical method, known as **Euler's Method**, to approximate solutions to differential equation problems of the form $y' = F(x, y)$, $y = y_0$ when $x = a$. Look back at Activity 4 where you were given an initial value $y_0 = f(a)$ and a function F which gives, at each point (x, y) in the plane, a number $F(x, y)$ which represents the slope of a curve through that point, or the derivative of a function whose value at x is y. The purpose of the **computer function Euler**, named after the famous mathematician Leonard Euler, is to approximate a function f given by $y = f(x)$ with the initial value $y = y_0 = f(a)$ and with the specified derivative $y' = F(x, y)$.

A differential equation problem of the form $y' = F(x, y)$ can be interpreted as a specification, via the function F on the right hand side, of the slope y' of an unknown curve. In other words, for each x, y in the domain of F the quantity $F(x, y)$ gives a slope or direction. We can display these directions with tangent line segments as in Figure 5.20 for the equation, $y' = 3\sin(x) - 2y + 1$. Such a collection of line segments (which are portions of tangents) is called a *direction field*.

Obviously, this collection of tangents does not specify a particular curve. It actually leads to the family of all curves obtained as graphs of the functions which solve this particular differential equation. If, in addition to the differential equation, we are given the initial value which the solution function must assume, then we are said to have an **initial value** problem. You can see from the above figure that only the initial value of the solution function is needed in order to move from the general solution of this differential equation to a complete specification of a particular solution.

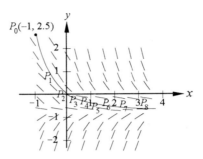

Figure 5.20. Tangent line segments along the curve $y' = 3\sin(x) - 2y + 1$.

The method that **Euler** uses is to construct a function whose value for a given x is obtained as follows: First the interval $[a, x]$ is partitioned into a sequence of points

$$[x_0 = a, x_1, x_2, \ldots, x_{n-1}, x_n = x].$$

Corresponding to each of these points x_i a value y_i is computed. The last value y_n is taken to be the value y at the point x of the function being computed. So our goal is to approximate this value y.

The formula that **Euler** uses is to take $y_0 = c$, where c is the given initial value, and then to successively compute $y_1, y_2, \ldots$ by the recursive relation

$$y_{i+1} = y_i + F(x_i, y_i)(x_{i+1} - x_i).$$

The reasoning behind this is that, if $F(x_i, y_i)$ is the value of the slope of the curve which is the graph of the function we are looking for, then a reasonable approximation for the value of the function at x_{i+1} is obtained by moving up (or down) an amount equal to $F(x_i, y_i)(x_{i+1} - x_i)$. See Figure 5.21.

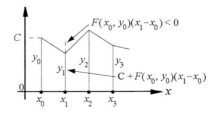

Figure 5.21. A geometric interpretation of Euler's method.

In order to pull out a particular solution, it is necessary to have a starting point. Using the same data as in Activity 4, we look for the solution to the differential equation $y' = 3\sin(x) - 2y + 1$ which passes through the point $P_0 = (-1, 2.5)$.

To approximate the graph of this solution, begin at the point P_0, evaluate f at P_0 to get the slope of the tangent, move along this tangent a little way (more precisely, we move until the x-coordinate has traversed the first subinterval) and stop to get the point P_1. Then repeat the process. This is done as many times as necessary to get the domain value as far to the right as we wish.

It is quite easy to calculate one of the points, say, $P_{i+1} = (x_{i+1}, y_{i+1})$ in terms of the previous point $P_i = (x_i, y_i)$. It is just the sum of the height at the first point y_i plus the rise obtained by moving a horizontal distance of $x_{i+1} - x_i$ along a line of slope $F(x_i, y_i)$. This is given by

$$y_{i+1} = y_i + F(x_i, y_i)(x_{i+1} - x_i)$$

which you can recognize as precisely the formula used in Activity 4.

Once again, the accuracy of this approximation can be improved by using a larger mesh size (the value of n), i.e., by reducing the length, $x_{i+1} - x_i$ of the subinterval $[x_{i+1}, x_i]$.

Thus we can write

$$
\begin{aligned}
y_0 &= c \\
y_1 &= c + F(x_0, y_0)(x_1 - x_0) \\
y_2 &= y_1 + F(x_1, y_1)(x_2 - x_1) \\
&= c + F(x_0, y_0)(x_1 - x_0) + F(x_1, y_1)(x_2 - x_1) \\
y_3 &= y_2 + F(x_2, y_2)(x_3 - x_2) \\
&= c + F(x_0, y_0)(x_1 - x_0) + F(x_1, y_1)(x_2 - x_1) + F(x_2, y_2)(x_3 - x_2) \\
&\vdots \\
t_n &= c + \sum_{i=1}^{n} F(x_{i-1}, y_{i-1})(x_i - x_{i-1}) \\
&\vdots
\end{aligned}
$$

and from the above list, we obtain the formula,

$$y \approx y_n = y_0 + \sum_{i=1}^{n} F(x_{i-1}, y_{i-1})(x_i - x_{i-1}).$$

This value y is the value at x of the function f we are constructing. The idea is that, by using the slope at each point and by using the line determined by this slope to approximate the change in f as the value of the independent variable changes from x_i to x_{i+1}, it is reasonable to expect that the resulting function at x_i will have its derivative close to $F(x_i, y_i)$. In this way we obtain an approximation to a function f which satisfies

$$f'(x) = F(x, y), \ x \in [a, b].$$

A very interesting point becomes evident when we think about the meaning of this calculation and the details of the formula in the special case in which the function F does not depend on y but only on x. In this special case, you are looking for a solution f to the equation,

$$f'(x) = F(x).$$

In other words, you are looking for a function f whose derivative is F. What we are saying here is that the solution to this problem is the (net) area function. Thus, it is no surprise that the formula we obtained in the previous paragraph,

$$y \approx y_n = c + \sum_{i=1}^{n} F(x_{i-1}, y_{i-1})(x_i - x_{i-1})$$

reduces in the case in which F depends only on x to

$$y \approx y_n = c + \sum_{i=1}^{n} F(x_{i-1})(x_i - x_{i-1})$$

or, since $c = f(x_0) = f(a)$ and $f' = F$, we have

$$y = f(x) \approx f(a) + \sum_{i=1}^{n} f'(x_{i-1})(x_i - x_{i-1})$$

Do you recognize the sum

$$\sum_{i=1}^{n} f'(x_{i-1})(x_i - x_{i-1})?$$

It is precisely the formula for left Riemann sum which approximates (net) area under the graph of a function. No wonder area and anti-derivatives are related.

5.6.4 Exercises

1. Solve the following differential equation problems.

   (a) $y^2 \cos(x) + y' = 0$

   (b) $y(1 + x^3)y' - 3x^2 = 0$, $\quad y = 0$ when $x = 2$

   (c) $(3 - y)y' = xy \sin(x^2)$

   (d) $y^2 e^x + y' = 0$

   (e) $2y' + x - xy = 0$

   (f) $\sec(y^3) + xy^2 y' = 0$

   (g) $x^2 y' - y - xy = 0$

   (h) $(\cos(x))y' - y = 0$, $y = 0$ when $x = e$

   (i) $\sqrt{1 - x^2}\, y' - x \csc(y) = 0$

   (j) $y' - 3y = 2e^{-x}$

   (k) $y' + 2xy = x$

   (l) $xy' - 3y + x^3 = 0$

   (m) $x^3 y' + x^2 y = 2$, $y = 1$ when $x = -2$

   (n) $(\cos(x))y' - (\sin(x))y = \cos^3(x)$

   (o) $3\dfrac{di}{dt} + 2i = t$, where $i = 3$ when $t = 0$

   (p) $\dfrac{di}{dt} + 4i = 3\sin(t)$, where $i = 1$ when $t = 0$

   (q) $2\dfrac{di}{dt} + 3i = 12e^{-t}$, where $i = 2$ when $t = 0$

2. Find an equation of a curve which has slope given by the indicated expression and that passes through the specified point.

   (a) xy^2, $(-2, 0.5)$

   (b) $\dfrac{e^{-0.5x}}{2y}$, $(0, -1)$

3. Solve the following *initial value* problems.

   (a) $L\dfrac{di}{dt} + Ri = 0$, where $i = I_0$ when $t = 0$ and L and R are constants.

   (b) $L\dfrac{di}{dt} + Ri = E$, where $i = I_0$ when $t = 0$ and L, R, and E are constants.

4. The current i in an electrical circuit is given by

$$i = I(t) = 5\cos\left(120\pi t + \frac{\pi}{3}\right) \text{ amperes}$$

where t isthe time in seconds. The charge in an electrical circuit is the instantaneous rate of change of current with respect to time. Find an expression for the charge q in the circuit as a function of time when the initial charge is zero.

5. A company purchases a new piece of machinery which decreases in value $v = V(t)$, where V is a function of the number of years t after the machinery is purchased, at a rate of

$$V'(t) = -750e^{-0.2t} \text{ dollars per year.}$$

   (a) Find an expression for the value V of the piece of machinery if the original price was $8,000.

   (b) What is the value of the piece of machinery after 15 years?

6. A manufacturer purchases a new piece of machinery for $100,000 and estimates that the marginal profit from the machinery t years after its purchase is given by

$$P'(t) = 1,200t + 10,000e^{0.08t} \text{ dollars per year}$$

   (a) How much profit does the machine generate for the manufacturer for the first twenty years after it was purchased?

   (b) Find a formula for the total profit that the machine generates for the manufacturer after t_0 years.

7. Solve the following *initial value* problem.

$$y' + 2xy = 1. \ y = 0 \text{ when } x = 1.$$

8. The population P for a certain country grows at a rate given by

$$\frac{dP}{dt} = kP(N - P)$$

where k is a positive constant and N is the population beyond which growth does not occur. This type of problem is called a *logistic model* of population growth.

   (a) Find the population as a function of time t.

   (b) Sketch a graph of P. Such a curve is called a *logistic curve*.

   (c) Show that the population growth rate is a maximum when $P = 0.5N$.

9. Two chemicals are mixed so that the rate at which the amount C of the mixture is formed is given by

$$\frac{dC}{dt} = k(C - C_1)(C - C_2)$$

where k is a constant, and C_1 and C_2 are constants which represent the initial amounts of the two chemicals.

   (a) Find C as a function of t when $C_1 = C_2$. Sketch a graph of C.

   (b) Find C as a function of t when $C_1 \neq C_2$. Sketch a graph of C.

10. Consider the differential equation

$$\frac{dy}{dt} = ky.$$

   (a) Show by using differentiation that the function $y = Y(t) = ce^{kt}$ is a solution to this differential equation.

   (b) Show that if $Y(0) = y_0$ for some real number y_0, then a solution to the problem

$$\frac{dy}{dt} = ky, \ Y(0) = y_0 > 0$$

   is given by

$$y = Y(t) = y_0 e^{kt}$$

(c) When $k > 0$, a problem like that in part (b) is referred to as an *exponential growth problem*. Explain in your own words the use of the term exponential growth. Illustrate your response graphically.

(d) When $k < 0$, a problem like that in part (b) is referred to as an *exponential decay problem*. Explain in your own words the use of the term exponential decay. Illustrate your response graphically.

11. The number n of bacteria in a bacteria colony grows at a rate proportional to its population at any time. That is,

$$\frac{dn}{dt} = kn .$$

Assume that time t is in days. Such a problem is referred to as an *exponential growth* problem.

(a) Use the result of the part (a) of Exercise 10 above to find the number of bacteria n as a function of time, $n = N(t)$.

(b) At 6 A.M. on Tuesday the number of bacteria in a petri dish is 2000 and at 6 A.M. on the following Friday there were 10,000 bacteria in the same petri dish. Write these two conditions in function notation and then using your solution in part (a), find the number of bacteria as a function of time.

(c) For the situation in part (b), how many bacteria are in the petri dish at 6 A.M. on the following Saturday?

(d) At what time will there be 3500 bacteria in the petri dish?

(e) After how many days will the number of bacteria initially in the dish double?

12. The amount of money a in a savings account grows at a rate proportional to the amount of money in the account at any time. That is,

$$\frac{da}{dt} = ka .$$

Assume that time t is in years. Such a problem is referred to as an *exponential growth* problem.

(a) Use part (a) of Exercise 10 above to find the amount a of money as a function of time, $a = A(t)$.

(b) Suppose that at any time t the relative rate of growth of the amount of money, i.e.,

$$\frac{\frac{da}{dt}}{a} ,$$

is 11% = 0.11 and the initial amount of money deposited in the account is 2000 dollars.

(b) Interest that compounds in this way is said to *compound continuously*. Give an explanation of why you think interest is said to compound continuously for such a situation.

(d) For the situation in part (b), find the amount of money in the account five years after the initial deposit.

(e) After how many years will the initial deposit be 4500 dollars?

(f) After how many years will the initial deposit triple, that is, be three times what it was initially?

13. The amount of a radioactive substance a decays at a rate proportional to the amount present at any time. That is,

$$\frac{da}{dt} = ka .$$

Assume that time t is in years. Such a problem is referred to as an *exponential decay* problem.

(a) Use part (a) of Exercise 10 above to find the amount a of such a substance as a function of time, $a = A(t)$.

(b) After 100 years, a 50 gram sample of a certain type of uranium will decay to 45 grams. Write this information as two conditions in function notation, and then using your solution in part (a), find the amount of uranium as a function of time.

(c) For the situation in part (b), find the amount of the 50 gram uranium sample present after 200 years?

(d) After how many years will 10 grams of the uranium decay away?

(e) For the situation in part (b), find the *half-life* of the uranium, that is the number of years it takes for one-half of a sample to decay.

14. When two chemicals A and B are mixed, the rate of change with respect to time t, in hours, of the concentration $[A]$ of A, in moles per liter, is directly proportional to the product of the concentrations of A and B.

(a) Write a differential equation which describes this situation using k as your constant of proportionality.

(b) Suppose that the concentrations of the chemicals are the same throughout the reaction, the initial concentrations of the chemicals are each 5 moles/liter and that after 30 minutes the concentration of each chemical is 3 moles/liter. Write a differential equation problem which describes this situation.

(c) For the situation in part (b), use your SCS (or do it by hand) to find the

concentration of the chemicals as a function of time t in hours.

(d) For the situation in part (c), find the concentration of the chemicals after one hour.

(e) When will the concentration of the chemicals be 2.4 moles/liter?

(f) For the situation in part (c), find the number of hours after the chemicals are combined so that the concentration of the two chemicals is 2 moles/liter.

15. Assuming that there is air resistance, the velocity v of a falling object is a solution to the following differential equation problem.

$$\frac{dv}{dt} = 32 - 0.1v, \ v(0) = 0.$$

Find an expression for the velocity function, $v = V(t)$, at any time t.

16. For a family of curves which is a solution of the differential equation

$$y' = \frac{dy}{dt} = F(x, y),$$

the family of *orthogonal trajectories* of that family is the family of curves which are a solution to the differential equation

$$y' = \frac{dy}{dt} = \frac{1}{F(x, y)}.$$

That is, the tangent lines at the same point on the solution curves of the two differential equations are perpendicular to each other. Such curves are used in applications in thermodynamics (isothermal curves and heat flow curves), electrostatics (equipotential curves and force curves) and hydrodynamics (velocity potential curves and stream, or flow, curves).

For each of the following differential equations, use your SCS (or do it by hand) to do the following.

(a) Find the family of curves which is a solution to the differential equation.

(b) Find the family of curves which are the orthogonal trajectories of the family of solutions to the differential equation.

(c) Use your graphing utility to sketch three solution curves to each family of solutions to the differential equation and three corresponding orthogonal trajectories on the same graph.

    i. $\dfrac{dy}{dx} = -\dfrac{x}{y}$

    ii. $xy' + y = 0$

    iii. $\dfrac{dy}{dx} = \dfrac{3y}{x}$

    iv. $2xy\dfrac{dy}{dx} = y^2 - x^2$

17. Using the solutions you found for the differential equations in Activities 1 and 2 of this section, find solutions to the following *initial value* problems.

(a) $xy' + y = 0$, $y = 0$ when $x = 1$

(b) $y' - 2y = e^{2x}$, $y = 1$ when $x = 0$

(c) $y' + \frac{y}{x} = \frac{1}{x^2}$, $y = -1$ when $x = 1$

(d) $y' + xy = 0$, $y = 5$ when $x = 0$

18. Compute by hand and simplify the expression you get in Euler's Method (see page 397) when the right hand side is independent of y, that is what do you get for the problem

$$y' = F(y)?$$

19. Consider the differential equation

$$\frac{dp}{dt} = kp^{1.5}$$

where k is a constant.

(a) Show that a solution is given by

$$p = P(t) = \left(-\frac{1}{2}kt + c\right)^{-2}$$

where c is an arbitrary constant.

(b) Find a solution of the problem

$$\frac{dp}{dt} = kp^{1.5},\ k < 0,\ P(0) = \frac{1}{4}\ \text{and}$$
$$P'(1) = 2.$$

20. Consider the differential equation

$$f'(x) = (2 - f(x))^2.$$

(a) Show that a solution is given by the expression,

$$f(x) = \frac{2x + 2c - 1}{x + c}$$

where c is an "arbitrary constant".

(b) Find a solution to the problem

$$f'(x) = (2 - f(x))^2,\ f(0) = 1.$$

(c) Find a solution to the problem

$$f'(x) = (2 - f(x))^2,\ f(0.5) = 1.$$

(d) Based on your experience in parts (b) and (c), for functions f and F, do you think a problem like

$$f'(x) = F(f(x)),\ f(x_0) = y_0$$

has a unique (only one) solution? Explain your answer.

21. For the family of functions given by $C(t) = 1 - ke^{-t}$ (see Figure 5.19), what is the meaning of the constant k in this case? What is the relation between tangents of various curves in this case? Explain your answers. Illustrate your answers geometrically.

5.7 CHAPTER SUMMARY

Formal Definitions   In this chapter you have applied your knowledge of derivatives from Chapter 4 to the problem of integration. Along the way, you have encountered new mathematical objects which have strong relationships to derivatives. As you read the summary of definitions below, you should try to reconstruct for yourself not only the individual definitions, but the connections among the defined quantities as well.

1. Let f be a continuous function of an independent variable x. A function F is called an *anti-derivative* of f if $F'(x) = f(x)$ for all x in the domain of f.

2. Let f be a function of the independent variable x which has an anti-derivative. The set of all anti-derivatives of f, i.e., the family of functions with each function having derivative f, is called the *indefinite integral* of f and is denoted by $\int f(x)dx$.

3. Let F be a function which is continuous on the interval $[a, b]$. For each positive integer n let P_n be the partition of $[a, b]$ into n equal subintervals. Thus

$$P_n = (x_i)_{i=1}^n = \left(a = x_0, \; x_0 + \frac{1}{n}(b-a), \; \ldots, \; x_0 + \frac{n-1}{n}(b-a), \; x_n = b \right)$$

Let R_n be the number

$$R_n = \sum_{i=1}^n f(x_{i-1})(x_i - x_{i-1}).$$

We define the *definite integral*, or *Riemann integral*, of f over $[a, b]$ to be the number $\lim_{n \to \infty} R_n$ and we denote it by

$$\int_a^b f$$

or

$$\int_a^b f(x)dx.$$

4. Let f be a function continuous on $[a, \infty)$. We define the improper integral of f over $[a, \infty)$ by

$$\int_a^\infty f(x)dx = \lim_{m \to \infty} \int_a^m f(x)dx.$$

5. Let f be a function continuous on $(-\infty, b]$. We define the improper integral of f over $(-\infty, b]$ by

$$\int_{-\infty}^b f(x)dx = \lim_{m \to -\infty} \int_m^b f(x)dx.$$

6. Let f be a function continuous on $(a, b]$, but not necessarily at a. We define the improper integral of f over $[a, b]$ by

$$\int_a^b f(x)dx = c \lim_{m \to a^+} \int_c^b f(x)dx.$$

7. Let f be a function continuous on $[a, b)$, but not necessarily at a. We define the improper integral of f over $[a, b]$ by

$$\int_a^b f(x)dx = \lim_{c \to b^-} \int_a^c f(x)dx.$$

8. The average value, f_{avg}, of a function f on the interval $[a, b]$ is

$$f_{avg} = \frac{1}{b-a}\int_a^b f(x)dx$$

provided this integral exists.

9. Let f be a piecewise continuous function defined on $[a, b]$ by

$$f(x) = \begin{cases} g(x) & \text{if } a \leq x < c \\ h(x) & \text{if } c < x \leq b. \end{cases}$$

Then the integral of f over $[a, b]$ is defined by

$$\int_a^b f(x)dx = \int_a^c g(x)dx + \int_c^b h(x)dx$$

provided both integrals exist.

10. Equations which contain derivatives are called differential equations; the order of a differential equation is the order of the highest derivative.

11. A function f is called a *particular* solution to a given differential equation if the differential equation reduces to an identity when f and its requisite derivatives are substituted into the differential equation.

12. The family of all solutions to a differential equation is called the *general* solution of the differential equation.

In addition to the formal definitions, this chapter also contained the central theorem of all of calculus, the Fundamental Theorem of Calculus.

1. Let f be a continuous function on $[a, b]$. Then, for $x \in [a, b]$,

$$\frac{d}{dx}\left(\int_a^x f(t)dt\right) = f(x).$$

2. Let f be a function whose derivative exists and is continuous on $[a, b]$. Then

$$\int_a^b f'(x)dx = f(b) - f(a).$$

The second part above can be restated as follows: If F is any anti-derivative of a function f which is continuous on $[a, b]$, then

$$\int_a^b f(x)dx = F(b) - F(a).$$

Basic Integration Formulas   In the following list, the symbol c refers to an arbitrary constant and a refers to a fixed constant.

1. $\int 0\,dx = c$

2. $\int a\,dx = ax + c$

3. $\int x^a\,dx = \frac{x^{a+1}}{a+1} + c, \ (a \neq -1)$

4. $\int \frac{1}{x}\,dx = \ln(|x|) + c, \ x > 0$

(Actually, you should remember that since $\ln(|x|)$ has two branches and it is discontinuous at 0, this rule is really two statements: $\int \frac{1}{x}\,dx = \ln(x) + c_1$ for $x > 0$ and $\int \frac{1}{x}\,dx = \ln(-x) + c_2$ for $x < 0$, where c_1 and c_2 may be different constants.)

5. $\int \sin(x)\,dx = -\cos(x) + c$

6. $\int \cos(x)\,dx = \sin(x) + c$

7. $\int \tan(x)\,dx = -\ln(|\cos(x)|) + c = \ln(|\sec(x)|) + c$

8. $\int \cot(x)\,dx = \ln(|\sin(x)|) + c$

9. $\int \sec(x)\,dx = \ln(|\sec(x) + \tan(x)|) + c$

10. $\int \csc(x)\,dx = \ln(|\csc(x)| - \cot(x)) + c$

11. $\int \sec^2(x)\,dx = \tan(x) + c$

12. $\int \csc^2(x)\,dx = -\cot(x) + c$

13. $\int \sec(x)\tan(x)\,dx = \sec(x) + c$

14. $\int \csc(x)\cot(x)\,dx = -\csc(x) + c$

15. $\int \frac{1}{1+x^2}\,dx = \arctan(x) + c$

16. $\int \frac{1}{\sqrt{1-x^2}}\,dx = \arcsin(x) + c$

17. $\int e^x\,dx = e^x + c$

18. $\int f'(g(x))g'(x)\,dx = f(g(x)) + c$

19. $\int_a^b f'(g(x))g'(x)\,dx = f(g(b)) - f(g(a))$

20. $\int f(x)g'(x)\,dx = f(x)g(x) - \int f'(x)g(x) + c$

Properties of Definite Integrals   If f is a continuous function on an interval $[a, b]$, and m and M are, respectively, the maximum and minimum values of f on $[a, b]$. Then,

1. $\int_a^b s(f(x) + tg(x))\,dx = s\int_a^b f(x)\,dx + t\int_a^b g(x)\,dx$

2. $\int_a^b f(x)\,dx = \int_a^c f(x)\,dx + \int_b^c f(x)\,dx$

3. $\int_a^b f(x)\,dx = -\int_b^a f(x)\,dx$

4. $\int_a^a f(x)\,dx = 0$

5. $\int_a^b f(x)\,dx \leq \int_a^b g(x)\,dx$

6. $m(b-a) \leq \int_a^b f(x)\,dx \leq M(b-a)$

7. $\left| \int_a^b f(x)\,dx \right| \leq \int_a^b |f(x)|\,dx$

A Suggested Strategy for Indefinite Integration   Ask yourself the following
questions in the indicated order:

1. Can you use a basic integration rule, or combination of basic rules after manipulating the integrand using algebra and/or trigonometry, to find the integral?

2. Can a simple substitution be used to transform the integral into one for which a basic integration rule can be used? In particular look for a substitution which transforms the integral into one for which the power rule, basic trigonometric rules, natural logarithm rule, or exponential rule is applicable.

3. Is the integrand a product of different types of functions? If so, then the method of integration by parts may be applicable. Schoenfeld suggests that the acronym **LIATE** (**L**ogarithmic, **I**nverse trigonometric, **A**lgebraic, **T**rigonometric and **E**xponential functions) be used in selecting the parts u and dv for use in the integration by parts formula. That is, select u to be the first type of function that appears in **LIATE**.

4. Is the integrand function a power or product of trigonometric functions? If so, you may be able to use trigonometric identities and algebra to reduce the integrand to a combination of functions for which basic integration rules are applicable.

5. Does the integrand contain an expression of the form $a^2 - u^2$, $a^2 + u^2$, or $u^2 - a^2$? If so, try a trigonometric substitution.

6. Is the integrand a rational function? Is so, divide first if necessary and then use the method of partial fractions.

If you cannot find an indefinite integral after answering the above questions, try making a substitution to transform the integral into an equivalent integral and then ask yourself questions 1-6 again. Of course, you could just find a SCS to help you out!

Some Basic Formulas for Applications   In this chapter you encountered numerous
applications of integration which gave rise to the following basic formulas. We believe that it is best to think through the basic strategy outlined below for solving applications using definite integrals for each particular application. However, for some of the most basic applications which you may frequently encounter, the following list of formulas may be useful.

1. Suppose f and g are two continuous functions such that $f(x) \geq g(x)$ for all $x \in [a, b]$. Then the area A of the region bounded by the graph of f, the graph of g, and the lines $x = a$ and $x = b$ is given by

$$A = \int_a^b (f(x) - g(x))dx.$$

2. Suppose f is a continuous function such that $f(x) > 0$ for all $x \in [a, b]$. If the portion of the graph of f which lies between the lines $x = a$ and $x = b$ is rotated about the x-axis, then the volume V of the resulting solid of revolution is given by

$$V = \int_a^b \pi(f(x))^2\, dx.$$

3. Suppose that f is a differentiable function such that f' is continuous on $[a, b]$. The length L of that portion of the graph of f between the points $(a, f(a))$ and $(b, f(b))$ is given by

$$L = \int_a^b \sqrt{1 + (f'(x))^2}\, dx.$$

4. Suppose f is a continuous function such that $f(x) > 0$ for all $x \in [a, b]$. If the portion of the graph of f which lies between the lines $x = a$ and $x = b$ is rotated about the x-axis, then the surface area SA of the resulting solid of revolution is given by

$$SA = 2\pi \int_a^b f(x)\sqrt{1 + (f'(x))^2}\, dx.$$

5. Suppose f and g are two continuous functions such that $f(x) \geq g(x)$ for all $x \in [a, b]$. Suppose that a lamina has the shape of the region R bounded by the graphs of f and g, and the lines $x = a$ and $x = b$, and also has a density function ρ which is a function of x defined on the region R. Then the mass m and the center of mass $(\bar{x}, \bar{y})$ of the lamina are given by

$$m = \int_a^b \rho(x)(f(x) - g(x))dx$$

$$\bar{x} = \frac{\int_a^b x\rho(x)(f(x) - g(x))dx}{\int_a^b \rho(x)(f(x) - g(x))dx}$$

$$\bar{y} = \frac{\int_a^b \left(\frac{f(x)+g(x)}{2}\right)\rho(x)(f(x) - g(x))dx}{\int_a^b \rho(x)(f(x) - g(x))dx}$$

or

$$\bar{y} = \frac{\int_a^b \frac{1}{2}\rho(x)\left((f(x))^2 - (g(x))^2\right)dx}{\int_a^b \rho(x)(f(x) - g(x))dx}$$

6. Suppose that the continuous function F represents a variable force acting on an object moving along a coordinate line from $x = a$ to $x = b$. Then the work done W by the force acting on the object is given by

$$W = \int_a^b F(x)dx.$$

Applications of Definite Integrals   The fundamental process for doing applications with
definite integrals can be summarized as follows.

Applications of Limits of Riemann Sums

For a quantity Q associated with an appropriate interval $[a, b]$, the following steps will usually lead to a definite integral which represents the value of Q.

1. Partition $[a, b]$ into n subintervals of the same length, where the length of the ith subinterval, $[x_i, x_{i-1}]$, is given by $\Delta x = x_i - x_{i-1} = \frac{b-a}{n}$, $i = 1, 2, ..., n$.

2. Approximate the contribution Q_i to Q over the ith subinterval by forming an appropriate product $f(x_i^*)(x_i - x_{i-1})$ to obtain,

$$Q_i \approx f(x_i^*)(x_i - x_{i-1})$$

   where x_i^* is any point in the ith subinterval $[x_{i-1}, x_i]$.

3. Add up the approximations Q_i to obtain a Riemann sum approximation to Q,

$$Q \approx \sum_{i=1}^{n} f(x_i^*)(x_i - x_{i-1}).$$

4. Take the limit as $n \to \infty$ of the Riemann sum which approximates Q to obtain a definite integral which represents Q,

$$Q = \lim_{n \to \infty} \sum_{i=1}^{n} f(x_i^*)(x_i - x_{i-1}) = \lim_{n \to \infty} \sum_{i=1}^{n} f(x_i^*)\Delta x = \int_a^b f(x)dx.$$

5. Use the Fundamental Theorem of Calculus to obtain the exact value of Q, or approximate the value of Q using a **Riem computer function**.

Elementary Differential Equations.   In this chapter you learned to formulate and solve
certain types of elementary *differential equations* (equations that contain an unknown function and its derivatives).

   We now summarize the methods you learned in this chapter for solving differential equations: two methods for exactly solving some large classes of elementary differential equations and one numerical method for approximating solutions when an exact solution cannot be found.

The Method of Separation of Variables.   In general, the method of separation of
variables is useful for solving differential equations of the form

$$y' = \frac{dy}{dx} = F(x, y)$$

where the variables can be separated. We seek a function f so that $y = f(x)$, hence we want to find a solution of the differential equation

$$f'(x) = F(x, f(x))$$

where f is an unknown function of x. When the variables can be separated, we can write the differential equation in the equivalent form

$$h(y)dy = g(x)dx$$

for some functions h and g with $\frac{g(x)}{h(y)} = F(x, y)$. If the variables can be separated in this way, then you integrate to obtain

$$\int h(y)dy = \int g(x)dx.$$

Then you try to solve for y in terms of x to obtain $y = f(x)$. When you cannot solve the resulting equation for y in terms of x, then the solution is an implicit solution.

Integrating Factors. Just like separation of variables is used to reduce a differential equation to a one which can be solved by integrating, the method of integrating factors also reduces a differential equation to an equation which can be solved by integrating. The method is based on finding a factor which can be used as a multiplier on each side of a differential equation, so that the resulting equation can be integrated. Any equation of the form

$$y' + f(x)y = g(x)$$

where f and g are integrable functions, can be solved by integrating after multiplying by the factor

$$e^{\int f(x)dx}$$

where $\int f(x)dx$ is any anti-derivative of the function f. We usually take the simplest expression we can for an anti-derivative of f when computing $e^{\int f(x)dx}$. After multiplying by the integrating factor $e^{\int f(x)dx}$ on both sides of the differential equation we obtain,

$$e^{\int f(x)dx}(y' + f(x)y) = \left(e^{\int f(x)dx}\right)g(x) \text{ or } \left(y\left(e^{\int f(x)dx}\right)\right)' = \left(e^{\int f(x)dx}\right)g(x).$$

Integrating on both sides, we obtain

$$y\left(e^{\int f(x)dx}\right) = \int \left(e^{\int f(x)dx}\right)g(x)dx.$$

Numerical Approximations — Euler's Method. In this chapter you were introduced to Euler's method for approximating a solution to a differential equation problem of the form

$$y' = F(x, y), \ y = y_0 \text{ when } x = a.$$

That is, you learned to approximate a function f given by $y = f(x)$ with the initial value $y = y_0 = f(a)$ and with the specified derivative $y' = F(x, y)$.

The process that Euler's method uses (which is the process of the **computer function Euler**) to construct a function whose value for a given x is obtained as follows: first, the interval $[a, x]$ is partitioned into a sequence of points

$$[x_0 = a, x_1, x_2, \ldots, x_{n-1}, x_n = x];$$

then corresponding to each of these points x_i a value y_i is computed; and finally the last value y_n is taken to be the value y at the point x of the function being computed. Hence, Euler's method value, and then to successively compute

$$y_{i+1} = y_i + F(x_i, y_i)(x_{i+1} - x_i)$$

where $F(x_i, y_i)$ is the value of the slope $\frac{dy}{dx}$ of the curve which is the graph of the unknown function f, then a reasonable approximation for the value of the function at x_{i+1} is obtained by moving up (or down) an amount equal to $F(x_i, y_i)(x_{i+1} - x_i)$. See Figure 5.21, page 404.

The point $P_{i+1} = (x_{i+1}, y_{i+1})$ is found in terms of the previous point $P_i = (x_i, y_i)$ by adding the height at the first point y_i to the rise obtained by moving a horizontal distance of $x_{i+1} - x_i$ along a line of slope $F(x_i, y_i)$. This is,

$$y_{i+1} = y_i + F(x_i, y_i)(x_{i+1} - x_i).$$

The accuracy of such an approximation can be improved by increasing the mesh size (the value of n), i.e., by reducing the length, $x_{i+1} - x_i$ of the subinterval $[x_{i+1}, x_i]$.

Euler's method can be summarized as follows:

$$
\begin{aligned}
y_0 &= c \\
y_1 &= c + F(x_0, y_0)(x_1 - x_0) \\
y_2 &= y_1 + F(x_1, y_1)(x_2 - x_1) \\
&= c + F(x_0, y_0)(x_1 - x_0) + F(x_1, y_1)(x_2 - x_1) \\
y_3 &= y_2 + F(x_2, y_2)(x_3 - x_2) \\
&= c + F(x_0, y_0)(x_1 - x_0) + F(x_1, y_1)(x_2 - x_1) + F(x_2, y_2)(x_3 - x_2) \\
&\vdots \\
t_n &= c + \sum_{i=1}^{n} F(x_{i-1}, y_{i-1})(x_i - x_{i-1}) \\
&\vdots
\end{aligned}
$$

From the above list, we obtain the approximation formula,

$$y \approx y_n = y_0 + \sum_{i=1}^{n} F(x_{i-1}, y_{i-1})(x_i - x_{i-1}).$$

When the above function F depends only on x, the formula reduces to

$$y \approx y_n = c + \sum_{i=1}^{n} F(x_{i-1})(x_i - x_{i-1})$$

Since $c = f(x_0) = f(a)$ and $f' = F$, it follows that

$$y = f(x) \approx f(a) + \sum_{i=1}^{n} f'(x_{i-1})(x_i - x_{i-1}) \approx \int_a^b f(x)dx.$$

Hence, Euler's formula in this special case is precisely the formula, or process, for **RiemLeft** which approximates the definite integral of f on the interval $[a, b]$. No wonder area and anti-derivatives are related.

Integration Summed-up.

By now you should have an appreciation for the enormous number of problems which can be solved using indefinite and definite integrals. As you saw in this chapter, the process of indefinite integration (anti-differentiation) and the process of definite integration both act on functions as objects. On the one hand, indefinite integration takes a function as an input and returns a family of functions as an output. On the other hand, definite integration takes a function on a specified interval as an input and returns a number as an output.

The central theorem of this chapter, and indeed the central theorem of all calculus, known as the Fundamental Theorem of Calculus detailed the all-important connection between differentiation and integration. As a consequence of this theorem, the problem of evaluating definite integrals is reduced to the problem of indefinite integration. Methods for evaluating definite integrals were then discussed, including the method of substitution, integration by parts, partial fraction decomposition, and trigonometric substitution.

In this chapter you also encountered the powerful method of defining a function using a definite integral with a variable upper limit — a method of defining functions which is very elegant mathematically and which is also a very powerful tool for problem solving in many applications. Finally, you applied your knowledge of integration to the problem of solving differential equations. Three methods for solving differential equations were considered in detail. The method of separation of variables and the use of integrating factors allowed the exact solution of certain types of first order differential equations. When an exact solution cannot be found by these or other methods, you saw that Euler's method provided a technique for numerical approximation of a particular solution.

Conclusion. The integration process is one of the most powerful tools for problem solving ever discovered by mathematicians. You will use the integration process often throughout the remainder of this text to solve problems which require you to find a function (or family of functions) as a solution. In addition, you will be able to solve problems by first dividing a quantity into "small" pieces, approximating each piece appropriately, then forming a Riemann sum which approximates the quantity, and finally taking the limit of the Riemann sum to find the quantity.

CHAPTER 6

SEQUENCES AND SERIES

6.1 SEQUENCES OF NUMBERS

6.1.1 Overview

In this chapter you will study *sequences* and *series*. In Chapter 2, a sequence was defined as a function whose domain is the set of positive integers, or more generally a set of integers greater than or equal to some integer. You encountered the concept of a sequence in Chapter 2, Section 1, Activity 8, page 72, where the possible energy levels of electrons orbiting a nucleus in an atom of a diatomic molecule were interpreted as the terms of a sequence. Also, the **Riem computer functions** you used in previous chapters are examples of sequences. For example, given a function f and an interval $[a, b]$ in the domain of f, one can think of a function defined on the positive integers in which the value at the integer n (number of subintervals) is given by **RiemLeft**(f, a, b, n). The concept of a limit at infinity is used to investigate *infinite* sequences and series.

In Section 1, we begin with a study of sequences and series of numbers. The section includes a formal definition of the limit of a sequence. As you will see, the "sum" of an infinite series is a limit (at infinity) of the sequence of "partial sums" of the sequence used to establish the series.

Your investigations of sequences will include the sources of sequences, various combinations of sequences using arithmetic and composition, properties of sequences, and the relationships between various properties as with functions in Chapter 2.

In the remaining sections of this chapter, you will study infinite series of numbers, combinations of infinite series, the powerful notion of defining functions using *power series* which are generalizations of the idea of polynomials, and representations of functions as *power series*.

6.1.2 Activities

1. Each of the following situations contains within it a sequence of numbers. Do the following for each of them.

    (a) Write out the first 5 terms of the sequence.

    (b) Write a **computer function** that will accept a positive integer and return the value of the term of the sequence indicated by the integer. Use your function to check the first 5 terms.

    (c) Make a table of the first 50 terms of the sequence. Because the domain of a sequence is a set of integers, we use a variation of the tool **table** called **tableint**. This will be provided by your instructor with instructions for its use.

    (d) Write code in your **MPL** that will produce a tuple of the first 50 terms of the sequence.

(e) Finally, produce a computer graph of the sequence represented by your tuple (or **computer function**).

Here are the situations.

    i. The nth power of the number r where $r = -\frac{3}{\pi}$.

    ii. The ratio of $n+1$ to n if n is odd and otherwise the number 0.

    iii. The sum of the reciprocals of the squares of the first n positive integers. (Your **MPL** might have a feature for doing this conveniently. For example, in **ISETL** there is the operation %+ applied to a tuple.)

    iv. The value of **RiemLeft** with n subintervals (you may recall this **computer function** that you used often in the first five chapters of this book) applied to the function given by the expression $3x^2$ on the interval $[1, 2]$.

    v. The situation in Chapter 2, Section 1, Activity 8, p. 72.

2. For each of the following sequences, write code on your **MPL** that will produce a tuple, or write a **computer function** which returns the value of the nth term. To do so you must guess a formula for the nth term of the sequence.

(a) $-1, \dfrac{1}{2}, -\dfrac{1}{3}, \dfrac{1}{4}, -\dfrac{1}{5}, \dfrac{1}{6}, \ldots$ (b) $1, \dfrac{1}{2}, \dfrac{1}{4}, \dfrac{1}{8}, \ldots$

(c) $1, 0, 2, 0, 3, 0, 4, \ldots$ (d) $2, 0, \dfrac{3}{2}, 0, \dfrac{4}{3}, 0, \dfrac{5}{4}, \ldots$

(e) $5, 5, 5, 5, 5, \ldots$ (f) $-4, -1, 0, \dfrac{1}{4}, \dfrac{2}{8}, \dfrac{3}{16}, \dfrac{4}{32}, \ldots$

(g) $2, -\dfrac{1}{2}, \dfrac{3}{2}, -\dfrac{2}{3}, \dfrac{4}{3}, -\dfrac{3}{4}, \dfrac{5}{4}, -\dfrac{4}{5}, \ldots$ (h) $1, 1+\dfrac{1}{2}, 1+\dfrac{1}{2}+\dfrac{1}{4}, 1+\dfrac{1}{2}+\dfrac{1}{4}+\dfrac{1}{8}, \ldots$

(i) $1, 1+\dfrac{1}{2}, 1+\dfrac{1}{2}+\dfrac{1}{3}, 1+\dfrac{1}{2}+\dfrac{1}{3}+\dfrac{1}{4}, \ldots$ (j) $1, 1-\dfrac{1}{2}, 1-\dfrac{1}{2}+\dfrac{1}{3}, 1-\dfrac{1}{2}+\dfrac{1}{3}-\dfrac{1}{4}, \ldots$

3. For each of the sequences in the previous activity, produce a graph of the sequence that you can use to study its behavior. For each of the following terms, find one (or more) of these sequences which the term describes: monotone (increasing, decreasing, or both), convergent, oscillating, unbounded, bounded.

4. Consider the terms which you investigated in the previous activity. Do you see any necessary relations (one implies the other) or incompatibilities (if one occurs, the other can't) between pairs of these terms?

5. Consider the sequences in the following list and try to rank them in terms of which appear to be converging fastest. Guess some meaning for this and list the sequences in increasing "rate of convergence".

(a) $a = (a_n)_{n=1}^{\infty}, \quad a_n = \dfrac{1}{n^2}$ (b) $b = (b_n)_{n=1}^{\infty}, \quad b_n = \dfrac{(-1)^n}{n^2}$

(c) $c = (c_n)_{n=1}^{\infty}, \quad c_n = \dfrac{1}{\ln(n)}$ 　　　　(d) $d = (d_n)_{n=1}^{\infty}, \quad d_n = \dfrac{\ln(n)}{n}$

(e) $e = (e_n)_{n=1}^{\infty}, \quad e_n = \dfrac{n}{n-2}$ 　　　　(f) $f = (f_n)_{n=1}^{\infty}, \quad f_n = \dfrac{1}{n}$

(g) $g = (g_n)_{n=1}^{\infty}, \quad g_n = \left(1 + \dfrac{1}{n}\right)^n$ 　　　　(h) $h = (h_n)_{n=1}^{\infty}, \quad h_n = \dfrac{1}{\ln(\ln(n))}$

6. In this activity you are to write **computer functions** that accept two **computer functions** representing sequences u and v and use them to construct a new sequence w whose n^{th} term is given by the indicated expression. Apply your **computer function** to the pair indicated.

   (a) $w_n = u_n + v_n$ 　　The pair from Activity 2(b) and Activity 2(c).

   (b) $w_n = u_n - v_n$ 　　The pair from Activity 2(a) and Activity 2(c).

   (c) $w_n = u_n \cdot v_n$ 　　The pair from Activity 2(a) and Activity 2(e).

   (d) $w_n = u_n / v_n$ 　　The pair from Activity 1(a) and Activity 1(b).

   (e) $w_n = \displaystyle\sum_{i=1}^{n} u_i \cdot v_i$ 　　The pair from Activity 2(b) and Activity 2(e).

7. This activity is similar to the previous one except that your **computer function** is to take a single sequence x and a number t and return a sequence whose n^{th} term is $t \cdot x_n$.

   Apply your **computer function** to some examples with the sequences taken from Activity 2 and numbers t of your choice. Compute the first few terms of the resulting sequences.

8. Write a **computer function PS** which takes a single sequence a (represented as a **computer function**) and returns a **computer function** representing a sequence s whose n^{th} term is given by

$$s_n = \sum_{i=1}^{n} a_i = a_1 + a_2 + \ \ldots \ + a_n.$$

   The sequence s is called the sequence of nth *partial sums* of the sequence a. Run your **PS** on the following sequences for several values of n.

   (a) $a_n = n$ 　　　　(b) $b_n = 2^n$ 　　　　(c) $c_n = \frac{1}{n(\ln(n))}$

   Try to figure out "closed-form" expressions for each of these, i.e., a "simple" expression which represents the nth partial sum.

9. The following expressions are *recursive* definitions of sequences. For each of them, write a recursive **computer function** to represent them. Try to find a closed form definition of the sequence and try to decide if you think it converges or diverges. (For an example of a recursive **func** in **ISETL** see Chapter 1, Section 4, Exercise 1(b), page 39.)

   (a) The sequence $a = (a_n)$ is given by $a_n = \frac{a_{n-1}}{2}$ for $n > 0$ and $a_0 = 3$.

   (b) The sequence $x = (x_i)$ is given by $x_{i+2} = i x_i$ for $i \geq 0$ and $x_0 = 2$, $x_1 = 1$.

   (c) The sequence $t = (t_k)$ is given by $t_{k+1} = 2 t_k$ for $k \geq 1$ and $t_1 = 1$.

10. Here is an example of a fairly complicated recursive **computer function** in **ISETL** which we will call **taylco**. It will be referred to later in this chapter. If your **MPL** is something other than **ISETL** then you can construct **taylco** in your **MPL**.

Within the **func taylco**, a **func D** is used to approximate the derivative of a function. Below is one possible version of **D**.

```
h := 0.0001;
D := func(f);
        return func(x);
                return (f(x+h) – f(x))/h;
        end;
    end;
```

Note that we have chosen an explicit, fixed value of **h**. One part of this activity will be to see what is the effect of changing this value.

Now here is the code for **taylco**.

```
taylco := func(f,n);
            if n = 0 then return f(0);
            else return (taylco(D(f), n–1))/n;
            end;
         end;
```

The **func taylco** accepts a **func f** which represents a function and a non-negative integer **n**. It returns a number.

(a) Write out a "closed form" expression for **taylco(f,n)**.

(b) Apply **taylco** to the function represented by the expression $1+2x+3x^2+4x^3+5x^4+6x^5$ for **n=0..6**.

(c) Try to guess what the exact answers are.

(d) Now change the value of **h** to 0.00001. Are your results any more accurate?

(e) Now change the value of **h** to 0.001.

(f) Explain what are the effects of changing **h** and make a choice of which you think is best and use it in the future. Why did you make this choice?

11. For each of the sequences listed below, you are given L, a candidate for the limit of the sequence, and a measure ε of closeness to the limit. By playing with various values or inspecting the expression or whatever, make an estimate of from what point on (that means from what index on) the terms of the sequence are of distance less than or equal to ε from the candidate. It may be that there is no such point.

(a) The general term is given by

$$\frac{3n-1}{2n+1}, \quad n = 0, 1, 2, \ldots$$

$L = 3/2, \quad \varepsilon = 0.001$

(b) 2, 1, 1/2, 2/3, 3/4, 4/5, 5/6, ... $\quad L = 1, \quad \varepsilon = 0.01$

(c) 2, 1, 1/2, 0, 2/3, 0, 3/4, 0, 4/5, ... $L = 1$, $\varepsilon = 1$

(d) 2, 1, 1/2, 0, 2/3, 0, 3/4, 0, 4/5, ... $L = 1$, $\varepsilon = 0.99$

(e) 2, 1, 1/2, 0, 2/3, 0, 3/4, 0, 4/5, ... $L = 0$, $\varepsilon = 0.01$

(f) 2, 1, 1/3, 0, 1/5, 0, 1/7, 0, 1/9, ... $L = 0$, $\varepsilon = 0.01$

(g) The sequence obtained by applying the **func PS** you wrote for Activity 8 to the sequence a, given by $a_n = 2^{-n}$, $n = 0, 1, 2, \ldots$ with $L = 2$, $\varepsilon = 0.01$.

12. You can use your symbolic computer system to calculate limits of sequences. Your instructor will provide you with the syntax. Use it to check convergence of those sequences in Activity 2 which you can express in your symbolic computer system.

13. You now have several methods to determine the limit of a sequence: reasoning from the expression, evaluating terms for very large indices, guessing, and applying your symbolic computer system. Use whichever of these methods you like (or better, use combinations of them) to analyze whether or not the sequence given by r^n converges. Here r is a fixed number and you must determine what happens for the different sequences resulting from every possible value of r.

6.1.3 Discussion

Sources and Representations of Sequences

The most important thing for you to realize as you begin thinking about sequences is that a sequence is a special kind of function. So if you are pretty comfortable with functions (and you are, if this course has been any kind of a success for you) then it will be helpful to keep in mind the fact that a sequence is a function whose domain is the set of positive integers, or some subset of the positive integers, or the positive integers with a few additional integer values thrown in. In other words, you put something in (a value of the index) and something comes out (the value of the term at that index.) At this point, you might want to have another look at Chapter 2.

Sequences are found all over the place; they can come from anywhere. A mathematical expression such as $\frac{n}{n+1}$, in which the values of the variable are understood to be positive integers, can generate a sequence. You have seen some examples of sequences earlier in this book. For example, evaluating Riemann sums as approximations to a definite integral with smaller and smaller mesh sizes gives rise to a sequence of numerical approximations. Many physical situations, such as the possible energy levels of electrons orbiting a nucleus in an atom can be interpreted in terms of sequences (see Chapter 2, Section 1, Activity 8, page 72).

Can you think of some situations in mathematics, or science or any other walks of life that lead to sequences?

Once you have a sequence and understand what it is, there is the question of how to represent it. A sequence can be represented by writing out a few terms to establish a pattern, by constructing a function that will give the value of the term for a given index, by constructing a tuple whose components consist of enough of the first few term to make a pattern clear, or in standard mathematical notation.

For example, we can write down various ways of representing the sequence which has zero for all its even indexed terms and, for an odd index, the term is the reciprocal of the square of the index. We use the computer to represent the function and the tuple.

The first few terms.

$$st = 1, \ 0, \ \frac{1}{9}, \ 0, \ \frac{1}{25}, \ 0, \ \frac{1}{49}, \ 0, \ \frac{1}{81}, \ 0, \ \ldots$$

An MPL **computer function** *like the one below in* **ISETL**.

```
sf  := func(n);
          if is_integer(n) and n>0 then
             if even(n) then return 0;
             else return 1/n**2;
             end;
          end;
       end;
```

An **ISETL tuple**.

$$st \ := [sf(n) \ : n \ in \ [1..N]];$$

Of course, in order to evaluate this **tuple**, you have to give a value (presumably a large positive integer) to **N**.

Standard mathematical notation.

$$s = \left(\left(\frac{1-(-1)^n}{2} \right) \frac{1}{n^2} \right)_{n=1}^{\infty}$$

In these notations we write, **sf(i)**, **st(i)** or s_i for the ith term of the sequence s.

Just to make sure you understand, what is the fourth term of this sequence? The fifth term?

Alternating Sequences   A very common type of sequence has terms which alternate between positive and negative values. You can see several of these in Activity 2. Sometimes, as in Activity 2(j) the signs don't alternate, but an alternating plus/minus does appear within successive terms. There is a simple notational trick that uses the fact that the value of $(-1)^n$ is 1 if n is even and -1 if n is odd. Using this we can write the sequence in Activity 2(a) as

$$\left(\frac{(-1)^n}{n} \right)_{n=1}^{\infty}$$

Can you see how to use this trick to the write the sequence in Activity 2(g)? How about Activity 2(j)?

Behaviors and Properties of Sequences   Sequences are special kinds of functions. So, just as with functions, there are a number of interesting and important behaviors and properties that a sequence might have. These have to do with how the values of the terms of a sequence change as the indices increase. If you think of a sequence as a list of values strung out in a long (actually, infinite) row, then you are looking at how the values change as you go out along the row.

We give some definitions, informal and formal, for the most important behaviors. Examples of all these can be found in Activity 2.

A sequence is monotone increasing if the values of the terms increase as you go out along the row and monotone decreasing if the values of the terms decrease. We can express this idea formally as follows.

> **Definition 6.1**
>
> *A sequence* $s = (s_n)_{n=1}^{\infty}$ *is* monotone increasing *if* $s_n \leq s_{n+1}$ *for all* $n = 1, 2, \dots$ *. It is* monotone decreasing *if* $s_n \geq s_{n+1}$ *for all* $n = 1, 2, \dots$ *. The adjective* strictly *is used if the inequalities* $\leq$, $\geq$ *are replaced by* $<$, $>$ *respectively.*

Note that this is analogous to the idea of monotonicity discussed in Chapter 2 (see page 93) for functions whose domains and codomains were sets of real numbers. This should not be surprising since, after all, a sequence is a function with a certain kind of domain!

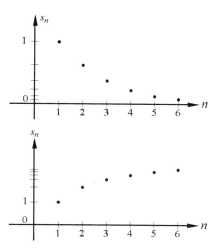

Figure 6.1. Increasing and decreasing sequences.

A sequence is said to be oscillating if the values of the terms alternate between going up and going down as you move along the row. Again we have a formal description.

> **Definition 6.2**
>
> *A sequence* $s = (s_n)_{n=1}^{\infty}$ *is* oscillating *if* $s_n \leq s_{n+1}$ *and* $s_{n+1} \geq s_{n+2}$ *for all* $n = 1, 2, \dots$, *or if* $s_n \geq s_{n+1}$ *and* $s_{n+1} \leq s_{n+2}$ *for all* $n = 1, 2, \dots$ *.*

Can you see an analogy here with properties of functions discussed in Chapter 2?

A sequence is bounded above (respectively below) if there is a number which is greater (respectively less) than every term of the sequence. It is bounded if it is bounded both above and below. The formal version is almost exactly the same except that the bound is named.

> **Definition 6.3**
>
> *A sequence* $s = (s_n)_{n=1}^{\infty}$ *is said to be* bounded above *if there is a number B such that* $s_n \leq B$ *for all* $n = 1, 2, \dots$ *. It is said to be* bounded below *if there is a number b such that* $s_n \geq b$ *for all* $n = 1, 2, \dots$ *. A sequence is said to be* bounded *if it is bounded above and bounded below.*

How does compare with a similar discussion on functions in Chapter 2?

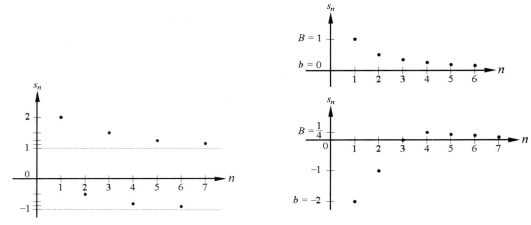

Figure 6.2. *An oscillating sequence.*

Figure 6.3. *Bounded sequences.*

Can you explain, in your own words, exactly what it means for a sequence to be unbounded?

The most important property that a sequence can have is to converge. Everything in the rest of this chapter has to do with convergent sequences (and series). Can you tell if a given sequence converges? What can you do with a sequence once you know that it converges? Are there different "kinds of convergence"? Are there different "rates of convergence"? Try to keep these questions in mind as you go through the many examples and phenomena which follow in the text and exercises.

For a sequence to converge means that there is a number such that, as you go further and further out in the sequence, the values of the terms become closer and closer to this number and, moreover, there is no restriction as to how close the values become. We express this informally by saying: *The terms of the sequence become arbitrarily close to the number as the indices become arbitrarily large.* This is a very profound notion so we will work with it for a while on an intuitive basis before considering a formal definition later (see page 432). You might want to look again at some of the material in Chapter 3 to refresh your memory about limit ideas. If a sequence does not converge, it is said to *diverge*.

Rates of Convergence of a Sequence

In Activity 5 you were asked to construct some meaning for the term "rate of convergence". The basic idea is to look more closely at the criteria for convergence. As you will see, convergence has to do with being able to guarantee that the values of the terms of a sequence will satisfy a particular criterion of closeness to a particular number provided that you go out far enough, that is, take sufficiently large values of the index.

Rate of convergence has to do with just how far you have to go out. The farther you must go, the slower is the convergence. In Activity 5, you were to investigate the values of the terms of each of several sequences, and, hopefully, you saw that in every case, they appear to be approaching a particular number. The rates of approach, however, are different in the sense that in one case you have to go much further out before getting to a particular degree of closeness than in another case. Your ranking in this activity was to compare these rates of approach.

The First Few Terms of a Sequence

The behavior of a sequence is mainly about what happens "ultimately", as you go farther and farther out, as the index increases to infinity. What happens with the first few terms is often unimportant and might be ignored. (How does this compare with the situation in which f is a function whose domain is the set of real numbers? What effect do the values of $f(1)$, $f(2)$, or $f(10^{200})$ have on $\lim_{x \to \infty} f(x)$?) Thus, in Activity 5, we used the same general notation for all of the sequences even though the expression for the general term makes no sense if $n = 1$ in one of them and for $n = 1, 2$ in another. Do you see which these are? We sometimes write (a_n) instead of $(a_n)_{n=1}^{\infty}$.

Also, it is often the case that a sequence will begin with the index $n = 0$ or some other integer value of n. This happens in Activities 9 (a), for example.

Relations Between Properties of Sequences

You have looked at properties of a sequence being monotone, oscillating, bounded and convergent. Perhaps you were able to notice in the examples some apparent relations and incompatibilities. For example, can a sequence be both monotone and oscillating? It should be clear to you from the definitions that this is impossible, so the two properties are incompatible.

What about convergence and bounded above? If a sequence converges, must it be bounded above? If you think about the values, it stands to reason. If the terms of the sequence are getting closer and closer to a single number, then how could the values of these terms go "out of bounds"? Is that reasoning enough for you or does it need something more formal? We will give a formal argument for it in a few paragraphs, after we have given a formal definition for limit.

What about convergence and bounded below?

What about the other way? If a sequence is bounded, does it necessarily follow that it must converge? Look at the examples in Activity 2 and see if you come to any conclusions.

Now let's change the question a little and ask if we can draw conclusions about a sequence if we know that it satisfies more than one property. Suppose that a sequence is monotone increasing and bounded above. Does it necessarily follow that the sequence must be bounded below? Must it converge?

Combinations to Make New Sequences — Arithmetic

It is possible to perform *termwise* arithmetical operations on sequences, as you did in Activity 6 and Activity 7. That means, for example, if you wish to add two sequences, then you form a new sequence by adding corresponding terms of two sequences. Here are formal definitions of arithmetic operations.

Addition. If $u = (u_n)_{n=1}^{\infty}$ and $v = (v_n)_{n=1}^{\infty}$ are two sequences, then their *sum* is the sequence $w = u + v$ where

$$w_n = u_n + v_n, \quad n = 1, 2 \ldots$$

Subtraction. If $u = (u_n)_{n=1}^{\infty}$ and $v = (v_n)_{n=1}^{\infty}$ are two sequences, then their *difference* is the sequence $w = u + v$ where

$$w_n = u_n - v_n, \quad n = 1, 2 \ldots$$

Multiplication. If $u = (u_n)_{n=1}^{\infty}$ and $v = (v_n)_{n=1}^{\infty}$ are two sequences, then their *product* is the sequence $w = u \cdot v$ where

$$w_n = u_n \cdot v_n, \quad n = 1, 2 \ldots$$

Division. If $u = (u_n)_{n=1}^{\infty}$ and $v = (v_n)_{n=1}^{\infty}$ are two sequences, then their *quotient* is the sequence $w = u/v$ where

$$w_n = \frac{u_n}{v_n}, \quad n = 1, \ 2\ldots$$

(of course, this does not work if any $v_n = 0$.)

Multiplication by a scalar. If $u = (u_n)_{n=1}^{\infty}$ is a sequence and t is a number, then the *scalar product* of u by t is the sequence $w = t \cdot u$

$$w_n = t \cdot u_n, \quad n = 1, \ 2\ldots$$

Combinations to Make New Sequences — Partial Sums

There is another way of forming new sequences from old sequence that will be very important in Section 2 when we move on to series. That is the sequence of partial sums that you worked with in Activity 8. If $u = (u_n)_{n=1}^{\infty}$ is a sequence then the corresponding *sequence of partial sums* is the sequence $s = (s_n)_{n=1}^{\infty}$ where,

$$s_n = u_1 + u_2 + \ldots + u_n = \sum_{i=1}^{n} u_i$$

Look at the sequences in Activity 2, parts (h), (i), and (j). Each of these is the sequence of partial sums of another certain sequence. Can you find the sequence in each of the three cases?

Recursively Defined Sequences

The simplest way of specifying a sequence is to give enough information about individual terms so that the general term is clear. This can be done as in Activity 2, by listing the first few terms so that the general pattern can be discerned. It can also be done as in Activity 5, by giving an expression that defines a function returning a value for any given index.

A less direct method of specifying a sequence that is very useful in certain applications is to express a general term in terms of the previous term, or the previous few terms. The following recursive sequence, given that f is a differentiable function,

$$x_{n+1} = x_n - \frac{f(x_n)}{f'(x_n)}$$

represents the Newton-Raphson Method for approximating zeros of a function (see page 291). It is also necessary to provide the value of the first term, or the first few terms. Thus, in Activity 9(a) and (c), the first term of the sequence is given, and there is a relation expressing a general term in terms of the previous term. Using this, you know the first term and so you can use the relation to obtain the second term. Then, knowing the second term, the relation gives you the third term. And so on.

For example, in Activity 9(a), the first term is 3. Using the relation, you can see that the next term is $\frac{3}{2}$. Repeating the procedure, the next term is $\frac{3}{4}$. You can continue this indefinitely or, as required in the activity, you can see an expression for the general term. The expression for the general term is called a *closed form* expression for the sequence.

In Activity 9(b), you are given the first two terms and the relation gives you the third term from the first and also the fourth term from the second. You can repeat the process to get the next two terms and continue it indefinitely getting two additional terms at every step.

Sequences defined in this way are called *recursively defined* or just *recursive sequences*. They are very useful in finding certain kinds of solutions to differential equations and we will explore this later in the chapter.

The recursive sequence considered in Activity 10, can be used to generate the coefficients of what is called a *Taylor series* for the function represented by **f**. The closed form you were asked to look for in that activity is quite simple and it may appear that it is simpler to work directly with it. In fact, however, using the recursive **func** can be more reasonable from a computation point of view.

The computational point of view can be extremely complicated as you may have gathered from looking at the effects of changing **h** in Activity 10. The trouble here comes from two features that tend to have opposite effects. The smaller is the value of **h**, the better is the approximation of the output of the **computer function** to the exact value given by the formula. That is, provided that the computer performs the arithmetic calculations accurately. Round-off error can lead to inaccurate calculations and this could make the approximation better or worse. Can you see why a smaller value of **h** could possibly lead to larger round-off errors? Could this explain what you saw in Activity 10?

Limit of a Sequence
The concept of the limit of a sequence is very close to the concept of the limit of a function at ∞ which you studied in Chapter 3. You might take a little time and review Chapter 3, Section 2. We considered there a function f, a number L, and the meaning of $\lim_{x \to \infty} f(x) = L$.

You may recall that this means that no matter how close we would like the values of $f(x)$ to be to L, we could do it by insisting on x being sufficiently large. The picture in Figure 6.4 may help you understand what is going on.

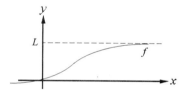

Figure 6.4. A geometrical interpretation of the limit of f at ∞.

Now with sequences, the situation is completely analogous. We have a function s, but it is a sequence so its domain is positive integers, not real numbers. We have a number L which is to be the limit (if the various criteria are satisfied). Just as weconsidered what happens as the domain values get arbitrarily large when we considered the limit at infinity of a function, we are now considering what happens as the domain values — the indices — get arbitrarily large, that is as they "go to ∞". Figure 6.5 suggests what is happening.

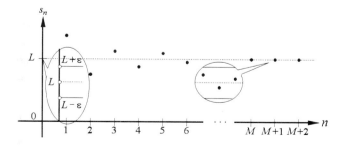

Figure 6.5. A sequence with a limit.

This is really very similar to what you studied in Chapter 3 under the paragraph, **Limits at ∞.**

So what we want to talk about is $\lim_{n\to\infty} s_n = L$, that is we want to give meaning to the idea of a sequence s converging to a limit L. Now is the time to think about Activity 11. In that activity, you were given a sequence, a candidate for the limit, and a criteria of closeness ε which was nothing more than a small positive number. Your task was to determine how far out in the sequence you had to go, that is, how large you had to require the index to be, in order to guarantee that the values of the terms of the sequence were within ε of L.

You may have found that this was sometimes possible and sometimes not. You may have found that it was possible with one value of ε, say $\varepsilon = 1$, but not with a smaller value, say $\varepsilon = 0.99$.

We are now very close to the definition of limit of a sequence. What we require in this definition is that the task of Activity 11, should always be possible, no matter what is the criteria of closeness, the value of ε. Here it is in formal language.

Definition 6.4

Let s be a sequence and L a number. We say that s converges to L, or that L is the limit of s if, for every $\varepsilon > 0$, there exists a positive number M such that if the index $n > M$, then $|s_n - L| < \varepsilon$.

If $s = (s_n)_{n=1}^{\infty}$ converges to L we write,

$$\lim_{n\to\infty} s_n = L$$

One way to get used to a complicated definition is to begin to use it. We can take sequences that we have already worked with and apply the definition to see whether or not they converge.

Let's consider the sequence given in Activity 11(f). The general term of this sequence is given by

$$s(n) = \begin{cases} 2 & \text{if } n = 1 \\ 1 & \text{if } n = 2 \\ \frac{1}{n} & \text{if } n > 2 \text{ is odd} \\ 0 & \text{if } n > 2 \text{ is even} \end{cases}$$

Make sure you understand how this formula gives the pattern of terms indicated in part (f) of Activity 11.

If you wrote a **computer function** to implement this sequence and evaluated terms for large values of n, you would find that the numbers were either equal to 0 or quite close to it. This suggests that the limit might be 0. Let's check it.

Here is our task. We are given a positive number ε. It is true that you might be more comfortable if we take an actual number, say 0.01, but our task is to show something is true no matter what value you took for ε. The best way to achieve this is to say as little as possible about ε, to keep it arbitrary. Of course, it might help you to try out a few specific values of ε before you tackle this task. That might show you how things go in preparation for the general analysis.

Now we must show that there exists a certain number M such that if $n > M$, then $|s_n - L| < \varepsilon$, or in this case, where $L = 0$, we must show that $|s_n| < \varepsilon$. We can ignore the first two terms (just make sure that $M > 2$). For the others, the value of the nth term is either 0 or $\frac{1}{n}$. In the former case, it is immediate that $|s_n| = 0 < \varepsilon$.

How about the case $s_n = \frac{1}{n}$? Can we guarantee that $|\frac{1}{n}| < \varepsilon$? What we know is that $n > M$ and we are free to choose M after looking at ε. Do you see that if we take $M \geq \frac{1}{\varepsilon}$, then everything is okay? We have shown that, for this sequence, $\lim_{n\to\infty} s_n = 0$.

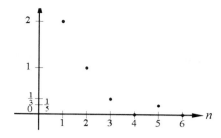

Figure 6.6. $s(1) = 2$, $s(2) = 1$; for $n > 2$, $s(n) = \frac{1}{n}$ if n is odd and $s(n) = 0$ if n is even.

Now let's take a look at the sequence in Activity 2 (d). That is, the sequence s where

$$s(n) = \begin{cases} \frac{n+1}{n} & \text{if } n \text{ is odd} \\ 0 & \text{if } n \text{ is even} \end{cases}$$

If you look at values of the terms in this case, you see that they oscillate between 0 and numbers that are close to 1. Do you see how this implies that no number can be the limit of this sequence? Surely 1 can't work because every other term is 0 which is distance 1 from 1. Also 0 can't work since every other term (after a while) will be within, say, 0.1 of 1 and hence at least 0.9 away from 0. (Compare this with what you saw in Chapter 3, Section 2, page 153, with

$$\lim_{t \to \infty} \frac{1+t}{t} \sin(t).)$$

Indeed, no number can work. Consider any number L. This number must be at least 0.5 distance from 0 or from 1. In applying the definition, take $\varepsilon = 0.4$. Then, no matter what choice you make for M, there will be both even and odd values of $n > M$. If L is 0.5 away from 0, then look at the even terms. These will not be within 0.4 of L. If L is 0.5 away from 1, then (once $n > 10$), the odd terms will not be within 0.4 of L.

We have shown that this sequence does not converge. See the graph of the sequence in Figure 6.7.

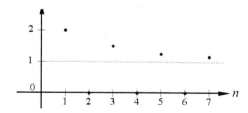

Figure 6.7. $s(n) = \frac{n+1}{n}$ if n is odd and 0 otherwise.

Limits of Combinations of Sequences
Limits of combinations of sequences are very simple. Everything that you would like to happen, happens in the way you expect. Specifically, the limit of the (pointwise) sum, difference, product, quotient (with the usual caveat) and scalar product of sequences is the sum, difference, product, quotient and scalar product of the limits. We will prove one of these statements in detail and leave the rest for the exercises.

Theorem 6.1

Let x, y be convergent sequences with $\lim_{n\to\infty} x_n = L$, $\lim_{n\to\infty} y_n = K$. Then the sequence $x + y$ is convergent and

$$\lim_{n\to\infty}(x_n + y_n) = L + K$$

Proof. We simply apply the definition three times. Given $\varepsilon > 0$ we apply the definition of the limit to x, its limit L, and the positive number $\frac{\varepsilon}{2}$. This gives M_1 such that if $n > M_1$ then $|x_n - L| < \frac{\varepsilon}{2}$. We do the same for y to obtain M_2 such that if $n > M_2$ then $|y_n - K| < \frac{\varepsilon}{2}$.

Now we take $M = \max\{M_1, M_2\}$. Then if $n > M$ it follows that both $n > M_1$ and $n > M_2$ and so we have,

$$|(x_n + y_n) - (L + K)| = |(x_n - L) + (y_n - K)| \le |(x_n - L)| + |(y_n - K)| < \frac{\varepsilon}{2} + \frac{\varepsilon}{2} = \varepsilon$$

$\square$

Examples of Limits of Sequences

There are a number of specific sequences whose limits you should commit to memory. For example, a few pages earlier on page 432, it was established that $\lim \frac{1}{n} = 0$ (all limits for sequences are taken as $n \to \infty$). Can you conclude from that anything about $\lim \frac{1}{2n}$ or $\lim(-\frac{3}{n})$? Does Theorem 6.1 help? What about $\lim \frac{1}{n+1}$?

The following is a general rule that is sometimes helpful in determining limits of sequences.

> If f is a function defined on the interval (a, ∞) for some real number a, then if $\lim_{x\to\infty} f(x) = L$, it follows that $\lim_{x\to\infty} f(n) = L$.

Do you see why the above rule is true? The proof is left as an exercise. What is the converse? Is the converse true?

In any case, this permits you to use anything you know about limits at infinity from Chapters 3 and 4 to derive something about limits of sequences. For example, how would you figure out $\lim \frac{n}{n+1}$? Can you use anything you learned in Chapter 3? Indeed, you should be able to figure out the limit of any sequence whose general term is a ratio of two polynomials.

Using what you know about finding limits of functions, together with another technique, you will be able to calculate many other important limits. Suppose you would like to find the limit L of the sequence whose general term is given by

$$L_n = \left(1 + \frac{1}{n}\right)^n.$$

Do you know this limit? Do you know how to calculate the limit? Consider the following,

$$\ln L = \lim_{n\to\infty} \ln\left(\left(1 + \frac{1}{n}\right)^n\right) = \lim_{n\to\infty} n \ln\left(1 + \frac{1}{n}\right) = \lim_{n\to\infty}\left(\frac{\ln(1 + \frac{1}{n})}{\frac{1}{n}}\right).$$

We have used the continuity of the natural logarithm function and a basic property of ln. By Theorem 3.7 of Chapter 3, Section 3, page 175, and the continuity of the natural logarithm function, ln, we can interchange the limit and ln. Hence, we can conclude that

$$\ln L = \lim_{n\to\infty}\left(\frac{\ln(1+\frac{1}{n})}{\frac{1}{n}}\right) = \lim_{h\to\infty}\frac{\ln(1+h)-\ln(1)}{h} = 1$$

where $h = \frac{1}{n}$. Do you see why this is true? Do you recognize the limit of the latter difference quotient as the derivative, $\frac{1}{x}$, of the natural logarithm function, ln, at $x = 1$? Yes, it is, and we have $\ln L = 1$. Now using the inverse relationship between ln and exp, or the definition of $\ln = \log_e$, it follows that $L = e$. Hence,

$$\lim_{n\to\infty}\left(1+\frac{1}{n}\right)^n = e$$

Relations Between Properties of Sequences — Proofs
Things get a little more challenging when we apply the definition to prove some of the facts about relations between properties of sequences. Again we will do one and leave the others for the exercises.

Theorem 6.2

A convergent sequence is bounded above.

Proof. Suppose that $\lim_{x\to\infty} x_n = K$. We apply the definition for a single ε, say $\varepsilon = 1$. The definition guarantees that we have a number M such that if $n > M$, then $|x_n - K| < 1$. In particular, this implies that $x_n < K + 1$. This takes care of the upper bound on x_n for $n > M$. What about $x_1, x_2, \dots x_{n_0}$ where the indices are less than or equal to M? Fortunately there are only finitely many of them, so there is a maximum. Let $A = \max\{x_1, x_2, \dots, x_{n_0}\}$. It then follows that $\max\{A, K+1\}$ is an upper bound for the sequence x.

□

In a similar manner it follows that a convergent sequence is bounded below. Hence, a convergent sequence is bounded.

There is one fact that is of fundamental importance so we state it here. Its proof is beyond the scope of the book, but you will have a chance to consider some aspects of it in the exercises.

Theorem 6.3

A sequence which is both monotone increasing and bounded above is necessarily convergent.

The Sequence r^n
Amongst the most important sequences are the geometric progressions, that is, sequences which are formed by taking successive powers of a single number. If r is the number, then we can write

$$a = (a_n)_{n=0}^{\infty} = (r_n)_{n=0}^{\infty} = 1, r, r^2, r^3, \dots$$

There is a notational convention here because, if $r = 0$ then r^0 has no meaning. Our convention is to take the first term to be 1 no matter what the value of r.

The first question for such a sequence is one which you investigated in Activity 13. Presumably you discovered there that the question of convergence depends entirely on the value of r. Here is a summary of the facts which you may already know.

- If $|r| < 1$ then $\lim_{n \to \infty} r^n = 0$

- If $r = 1$ then $\lim_{n \to \infty} r^n = 1$

- If $r = -1$ then the sequence does not converge.

- If $|r| > 1$ then the sequence does not converge.

Why do you think we gave the last two separately?

The second and third points in this list should be quite clear to you. To give careful proofs of the first and last would be just beyond the scope of this course.

6.1.4   Exercises

1. Give at least two examples of situations in mathematics, or science or any other walks of life that lead to sequences.

2. Consider the sequence

$$s = \left(\left(\frac{1 - (-1)^n}{2} \right) \frac{1}{n^2} \right)_{n=1}^{\infty}.$$

Write out the first 5 terms of this sequence.

3. Make your best guess at a possible n^{th} term for each of the following sequences.

   (a) $1, \dfrac{1}{8}, \dfrac{1}{27}, \dfrac{1}{64}, \ldots$

   (b) $\dfrac{1}{2}, \dfrac{2}{3}, \dfrac{3}{4}, \dfrac{4}{5}, \ldots$

   (c) $\dfrac{1}{2}, \dfrac{1}{2} + \dfrac{4}{3}, \dfrac{1}{2} + \dfrac{4}{3} + \dfrac{9}{4},$
   $\dfrac{1}{2} + \dfrac{4}{3} + \dfrac{9}{4} + \dfrac{16}{5}, \ldots$

   (d) $\dfrac{1}{3}, \dfrac{-8}{5}, \dfrac{27}{7}, \dfrac{-64}{9}, \ldots$

   (e) $\dfrac{3}{0.5}, \dfrac{4}{1.5}, \dfrac{5}{2.5}, \dfrac{6}{3.5}, \ldots$

   (f) $\dfrac{1}{0.5}, \dfrac{4}{1.5}, \dfrac{1}{27}, \dfrac{1}{64}, \ldots$

4. Each of the following expressions represents the nth term of a sequence. Determine whether each sequence converges or diverges. Give reasons for your answer. You may wish to check your answer (or see what is going on) by using your symbolic computer system.

   (a) -7 (b) π

   (c) $\dfrac{n-1}{5+3n}$ (d) $\dfrac{3-n^2}{n+1}$

   (e) $\dfrac{3n^2 - n + 1}{n(5-n)}$ (f) $\dfrac{n(1+n^3)}{2n+n^2}$

   (g) $(-1)^n \dfrac{n^2 - 1}{n + 3n^2}$

   (h) $(-1)^n \dfrac{n-1}{n(1+3n)}$

   (i) $(-1)^{n+1} \dfrac{3n}{\sqrt{2+n^4}}$

   (j) $\dfrac{5n^{5/2}}{\sqrt{1+n^2}}$

   (k) $(-1)^{n+1} \dfrac{1 + 3\sqrt{n}}{5 - n}$

   (l) $(-1)^n + 1$ (m) $-2 + (0.3)^n$

   (n) $\dfrac{1}{n} + \dfrac{3}{\pi}$

   (o) $\dfrac{3n^2}{1 + 2 + 3 + \ldots + n}$

   (p) $\dfrac{n}{3^n}$ (q) $n \sin\left(\dfrac{\pi}{2n} \right)$

   (r) $\sin(n)$ (s) $\cos^n(\pi n)$

(t) $\dfrac{3^{-n}}{\cos(\pi n)}$ (u) $\dfrac{n^2}{\sin(n^2)}$

(v) $\dfrac{1}{n}-\dfrac{1}{1+n}$ (w) $\dfrac{3^n}{n^3}$

(x) $\dfrac{\ln(n)}{n}$ (y) $\dfrac{\ln(n)}{\sqrt{1+n^3}}$

(z) $\dfrac{n+1}{e^n}$

5. A geometric sequence is a sequence of the form

$$a,\ ar,\ ar^2,\ ar^3,\ \dots,$$

where a and r are any real numbers and each term is obtained from the previous term by multiplying by the common *ratio* r. Notice that the ratio of any term of the sequence to the preceding term is always r. Let (x_n) be a geometric sequence with $x_3 = 8$ and $x_5 = 128$. What is the value of x_1?

6. Using the fact that $(-1)^n$ is 1 if n is even and -1 if n is odd you can write the sequence in Activity 2(a) as

$$\left(\dfrac{(-1)^n}{n}\right)_{n=1}^{\infty}$$

In a similar manner, rewrite the sequence in Activity 2(g), page 422, using this idea. Do the same for Activity 2(j), page 422.

7. Explain, in your own words, what it means for a sequence to be

(a) Not bounded above.

(b) Not bounded below.

(c) Not bounded.

(d) Not convergent.

8. A super elastic ball is dropped and rebounds one-half the distance it has fallen on each bounce. Find an expression for the total distance the ball has traveled after it has bounced n times where n is any integer with $n \geq 2$.

9. Show that each term of the sequence

$$1,\ 2,\ 6,\ 18,\ \dots,\ 2\cdot 3^n,\ \dots$$

is twice the sum of all the preceding terms of the sequence.

10. On page 432, the limit of the sequence $(s_n)_{n=1}^{\infty}$ such that for $n > 2$, $s_n = \frac{1}{n}$ if n is odd and $s_n = 0$ if n is even was considered using the definition of limit with $L = 0$. Explain why a choice of $M \geq \frac{1}{\varepsilon}$ shows that $\lim_{n\to\infty} s_n = 0$, that is, for $n > M$ if follows that $\left|\frac{1}{n}\right| < \varepsilon$.

11. Recall that for the sequence

$$(r_n)_{n=1}^{\infty} = 1,\ r,\ r^2,\ r^3,\ \dots$$

we have

- If $|r| < 1$ then $\lim_{n\to\infty} r^n = 0$
- If $r = 1$ then $\lim_{n\to\infty} r^n = 1$
- If $r = -1$ then the sequence does not converge.
- If $|r| > 1$ then the sequence does not converge.

Explain why you think we gave the last two statements separately.

12. If a sequence is bounded, does it necessarily follow that it must converge? Give reasons and examples for your answer.

13. Let x and y be convergent sequences with $\lim_{n\to\infty} x_n = L$ and $\lim_{n\to\infty} y_n = K$, and c a real number. Prove the following.

(a) The sequence $x - y$ is convergent and

$$\lim_{n\to\infty}(x_n - y_n) = L - K.$$

(b) The sequence cx is convergent and
$$\lim_{n\to\infty} cx_n = cL.$$

(c) The sequence xy is convergent and
$$\lim_{n\to\infty}(x_n y_n) = LK.$$

(d) The sequence $\frac{x}{y}$ is convergent and
$$\lim_{n\to\infty}\frac{x_n}{x_n} = \frac{L}{K}, \text{ provided } K \neq 0 \text{ and}$$
$$y_n \neq 0 \text{ for all } n.$$

14. Let (x_n) and (y_n) be sequences such that $\lim_{n\to\infty} x_n = 5$ and $1 \leq y_n - x_n \leq 2$ for all $n = 1, 2, \ldots$. Which of the following sequences has a finite limit? Give an explanation for your answer.

   (a) (y_n)

   (b) $\left(\frac{y_n}{\sqrt{n}}\right)$

   (c) $\left(\sqrt[n]{y_n}\right)$

   (d) $\left(\left(\frac{y_n}{6.5}\right)^n\right)$

15. Prove the following statement which allows you to use the limit theorems and theorems about properties of combinations of limits in Chapter 3.

   If f is a function defined on some interval (a, ∞) for some real number a, then if $\lim_{x\to\infty} f(x) = L$, it follows that $\lim_{x\to\infty} f(n) = L$.

   What is the converse? Is the converse true?

16. Prove each of the following if they are true. If not, give a counterexample.

   (a) If two sequences of positive numbers are monotone increasing, then the sum of the sequences is monotone increasing.

   (b) If two sequences are monotone increasing, then the product of the sequences is monotone increasing.

   (c) If two sequences of positive numbers are monotone decreasing, then the sum of the sequences is monotone decreasing.

   (d) If two sequences are monotone decreasing, then the product of the sequences is monotone decreasing.

   (e) If two sequences of negative numbers are monotone decreasing, then the product of the sequences is monotone increasing.

17. Recall that in Chapter 3 you learned that if f and g are functions such that $\lim_{x\to a} g(x) = L$ and f is continuous at L, then
$$\lim_{x\to a} f(g(x)) = f\left(\lim_{x\to a} g(x)\right) = f(L).$$

   Hence, the process of taking a limit commutes, in a sense, with a continuous function. This result was used to show that
$$\lim_{n\to\infty}\left(1+\frac{1}{n}\right)^n = e$$

   on page 164. Explain why the following statement is true.

   If f is a continuous function with positive values and $\lim_{x\to a} \ln(f(x)) = L$, then $\lim_{x\to a} f(x) = e^L$.

18. Is the meaning of the term, "bounded" according to our definition the same as the meaning of the phrase, "the sequence of absolute values is bounded above"? Explain why or give a counterexample.

19. For a sequence (a_n), show that if
$$\lim_{n\to\infty}|a_n| = 0, \text{ then } \lim_{n\to\infty} a_n = 0.$$

20. Is every unbounded sequence divergent? Give a proof or a counterexample.

21. Let $(a_n)_{n=1}^{\infty}$ be a sequence such that $a_n > 0$ for all n and
$$\frac{a_{n+1}}{a_n} \leq \frac{n+1}{2n+1} \text{ for all } n = 1, 2, \ldots.$$

   If possible, find $\lim_{n\to\infty} a_n$.

22. Look at the sequences in Activity 2 parts (h), (i), and (j). Each of them is the sequence of partial sums of a certain sequence. Find an expression for the nth term of each sequence in each of the three cases.

23. Use a hand-held calculator, to guess the following limit:

$$\lim_{n \to \infty} x^{\frac{1}{n}}$$

where $x > 0$.

24. Find the limits by hand, if possible, of the sequences with the following n^{th} terms. You may wish to use your symbolic computer system to see what is going on. Give reasons for your answer.

(a) $\dfrac{\sqrt{n} \cos(\pi e^n)}{2n+1}$

(b) $n \sin\left(\dfrac{1}{n}\right) + (-1)^n \dfrac{\cos(n)}{n}$

(c) $\sqrt{n+1} - \sqrt{n}$

(d) $\dfrac{n}{\sqrt{n}+1} - \dfrac{n}{\sqrt{n}-1}$

(e) $\ln(n) - \ln(n+1)$

(f) $\ln n - \ln\left(n + \sqrt{n^2+1}\right)$

(g) $\dfrac{\ln(n)}{n^p}$ where p is a positive real number

(h) $\dfrac{e^n}{n^p}$ where p is a positive real number

(i) $\dfrac{\cos(3n)}{n^p}$ where p is a positive real number

(j) $\dfrac{\cos(an) + \sin(bn)}{n!}$ where a and b are real numbers

(k) $\left(1 + \dfrac{2}{n}\right)^n$

(l) $\left(1 + \dfrac{2}{n}\right)^{n^2}$

(m) $\left(1 + \dfrac{a}{n}\right)^n$ where a is a real number

(n) $P\left(1 + \dfrac{r}{n}\right)^{nt}$ where P, r, and t are positive real numbers

(o) $\dfrac{2^n}{n^3}$

(p) $\dfrac{n^{100}}{n!}$

(q) $\dfrac{\pi^{2n}}{2^n 2n!}$

(r) $\dfrac{2^n}{n!}$

(s) $\ln(2n+1) - \ln(n+3)$

(t) $\dfrac{n^n}{n!}$

(u) $\dfrac{1 \cdot 4 \cdot 7 \cdots (3n+1)}{n! \, 2^n}$

(v) $a_n = 1 + 2\left(\dfrac{5}{6}\right) + 3\left(\dfrac{5}{6}\right)^2 + \ldots + n\left(\dfrac{5}{6}\right)^{n-1}$

(w) $a_n = \begin{cases} \frac{1+n}{n} & \text{if } n \text{ is odd} \\ \frac{3n-2}{3n} & \text{if } n \text{ is even} \end{cases}$

(x) $a_n = \begin{cases} 2n - \sqrt{4n^2 + n} & \text{if } n \text{ is odd} \\ \frac{1}{2} a_{n-1} & \text{if } n \text{ is even} \end{cases}$

25. For an n-sided regular polygon (i.e., one with n equal sides) inscribed in a circle of radius r, show that

(a) The perimeter is given by

$$2nr \sin\left(\dfrac{\pi}{n}\right).$$

(b) The limit of the sequence with n^{th} term given by the formula for the perimeter in part (a) is the circumference of the circle.

26. Show that if (a_n) is a sequence of real numbers and $\lim_{n \to \infty} a_n = L$, then $\lim_{n \to \infty} a_{n+m} = L$ where m is any integer.

27. Find expressions for the nth term of the following recursively defined sequences, if possible. You may wish to use the tool **tableint**, or a hand-held calculator to see what is going on.

    (a) $a_1 = 1$ and $a_{n+1} = \cos(a_n)$ for $n \geq 1$

    (b) $a_1 = 1$ and $a_{n+1} = a_n + \cos(a_n)$ for $n \geq 1$

    For the sequence in part (b), give a geometric interpretation of the nth term of the sequence for $n > 1$ and the resulting limit. (Hint: For a circle of radius one, let a_n be the angle made by a radius and the positive x-axis.)

    (c) $x_1 = 2$ and $x_{n+1} = \dfrac{1}{3}\left(2x_n + \dfrac{3}{(x_n)^2}\right)$ for $n > 0$

    (d) $x_1 = \dfrac{a}{2}$ and $x_{n+1} = \dfrac{1}{2}\left(x_n + \dfrac{a}{x_n}\right)$ for $n > 0$ and $a > 0$

    (e) $a_0 = 5$ and $a_n = \dfrac{1}{2}\left(a_{n-1} + \dfrac{8}{a_{n-1}}\right)$ for $n \geq 1$

    (f) $a_1 = 1$ and $a_n = 1 - \dfrac{a_{n-1}}{4}$ for $n \geq 2$

    (g) $a_1 = 1$ and $a_n = 2 + \dfrac{1}{2}a_{n-1}$ for $n \geq 2$

28. For each of the following recursive sequences, use the result of Exercise 26 to find the exact value of the limit of each sequence, assuming the limit of each sequence exists and equals some real number, L.

    (a) The sequence with n^{th} term defined in part (b) of Exercise 27.

    (b) The sequence with n^{th} term defined in part (c) of Exercise 27.

    (c) The sequence with n^{th} term defined in part (d) of Exercise 27.

    (d) The sequence with n^{th} term defined in part (e) of Exercise 27.

    (e) The sequence with n^{th} term defined in part (f) of Exercise 27.

    (f) The sequence with n^{th} term defined in part (g) of Exercise 27.

29. The number π can be approximated by the recursive sequence

    $$x_1 = 3 \text{ and } x_{n+1} = x_n - \tan x_n \text{ for } n \geq 1.$$

    Use the tool **tableint**, or a hand-held calculator to see that it is reasonable that the limit of the sequence is π. What would happen if $x_1 = 5$? Explain what happens if x_1 is not taken to be close to π.

30. The following statement is true (see page 435):

    A sequence which is both monotone increasing and bounded above is convergent.

    Although you cannot prove this statement with your knowledge from this text, you can show how to locate the limit of such a sequence using the *bisection method*. Suppose that L is the limit of the sequence (a_n) which is monotone increasing and bounded above, that is, $\lim_{n \to \infty} a_n = L$. You can locate L as follows:

    (a) Since $a_n \leq a_{n+1} \leq B$ for some number B for all n, show that L must be in the interval between a_1 and B.

    (b) Now consider the point

    $$B_1 = \frac{a_1 + B}{2}.$$

    Show that L must be between B_1 and B.

    (c) Continuing as in part (a) and (b), show that you obtain a sequence of "nested" closed intervals found (i.e., intervals containing the previous intervals). For example, show that the interval in part (a) is contained in the interval in part (b). It seems reasonable that the sequence of nested intervals "shrinks" down to a point and that point is a candidate for L. A complete proof requires knowledge about the *completeness property* of the real numbers and is usually done in a course in *Real Analysis*. Make a sketch showing a geometric interpretation of what is happening.

A similar statement is true for a monotone decreasing sequence which is bounded below and you can locate the limit of such a sequence in a similar manner.

31. For each of the following sequences, find a closed form (simple) expression for the nth term and prove that your expression is correct by using *mathematical induction*. Recall that to prove that a statement P_n about integers is true for all positive integers n, that is to do a proof by mathematical induction, you must do the following:

First, formulate the statement of the problem in terms of a function P whose domain is the positive integers and whose range is the set of two possible values, *true* or *false*. Next, show that the statement P_1 is true (or P_{n_0} is true for some positive integer value n_0 — the statement P_{n_0} is called the *base case*).

Then, show that the statement P_{n+1} follows from the statement P_n. The statement P_n is called the *induction hypothesis*.

(a) $\displaystyle\sum_{k=1}^{n} c = cn$

(b) $\displaystyle\sum_{k=1}^{n} k = \frac{n(n+1)}{2}$

(c) $\displaystyle\sum_{k=1}^{n} (2k-1)^2 = \frac{n(2n-1)(2n+1)}{3}$

(d) $\displaystyle\sum_{k=1}^{n} k^3 = \frac{n^2(n+1)^2}{4}$

32. An *arithmetic sequence* is a sequence with n^{th} term given by

$$a_n = a + (n-1)d$$

where a and d are real numbers. Find conditions on a and d so that an arithmetic sequence converges. What can you conclude about arithmetic sequences in general? Explain your answer.

33. The *Fibonacci sequence* is defined recursively as follows:

$$F_1 = 1, \ F_2 = 1, \text{ and } F_{n+1} = F_n + F_{n-1}$$
$$\text{for } n \geq 2.$$

(a) Write out the first 10 terms of the Fibonacci sequence by hand. Then use the tool **tableint** to find the first 50 terms of the sequence.

(b) Show that the Fibonacci sequence is divergent.

(c) Let (r_n) be the sequence with n^{th} term given by

$$r_n = \frac{F_{n+1}}{F_n}.$$

The limit of this sequence of ratios (r_n) is called the *Golden Ratio*. Use **tableint** to estimate the Golden Ratio to six decimal places.

(d) For the sequence of ratios (r_n) in part (c), show that

$$r_n = 1 + \frac{1}{r_{n-1}}.$$

(e) Use the result of part (d) to show that if $\lim_{n \to \infty} r_n = r$, then

$$r = \frac{1 + \sqrt{5}}{2}.$$

Use your hand-held calculator (or **tableint**) to estimate $(1+\sqrt{5})/2$ and compare the result with your answer in part (c).

34. (a) Use mathematical induction (see Exercise 31) to show that each of the following statements are true.

   i. For each positive integer $n > 1$

   $$\ln(n) < 1 + \frac{1}{2} + \frac{1}{3} + \frac{1}{4}$$
   $$+ \dots + \frac{1}{n} < 1 + \ln(n).$$

   ii. For each positive integer $n > 1$

$$1 + \frac{1}{2} + \frac{1}{3} + \frac{1}{4} + \dots + \frac{1}{n} - \ln(n) < 1.$$

   iii. The sequence

   $$1 + \frac{1}{2} + \frac{1}{3} + \frac{1}{4} + \dots + \frac{1}{n} - \ln(n)$$

   is monotone increasing.

(b) Explain why the sequence in part (iii) above is convergent using the results of parts (i)-(iii).

(c) The limit of the sequence in part (iii) above is *Euler's constant* which you may have seen in the exercises in Chapter 5. Approximate Euler's constant to six decimal places using the tool **tableint**, or a hand-held calculator.

▮ 6.2 SERIES OF NUMBERS

6.2.1 Overview

In this section, you will learn about the sources of infinite series, various types of infinite series, combinations and properties of combinations of infinite series. Sequences and series are useful in applications in such diverse fields as statics, medicine, economics and finance.

This material is not easy because there are two goals that require significant mental growth on your part. One goal is to understand about infinite series, convergence of such series, and tests for convergence and divergence of series. The other goal is to learn to apply various tests for convergence. Once you understand what is going on, a certain amount of drill and practice will be useful in order to learn to use the tests. The point is that not only do you have to learn each test, but you must learn to recognize, in a given situation, which test is most applicable.

Your knowledge of functions and limits from previous chapters will make this chapter easier to digest. In addition, your hands-on computer investigations will help you to achieve the goals mentioned above.

6.2.2 Activities

1. Consider the two series

$$\sum_{n=1}^{\infty} \left(\frac{1}{n} - \frac{1}{n+1} \right)$$

$$\sum_{n=0}^{\infty} \left(\frac{1}{2^n} \right)$$

Do the following for each of them,

(a) Write a **computer function** to represent the sequence of general terms.

(b) Apply your **computer function PS** from Section 1, Activity 8, page 423, to obtain the sequence of partial sums.

(c) Try to figure out a closed form expression for the nth partial sum.

(d) Using your closed form expression and/or values of the partial sums for large values of the index, try to guess the sum of this infinite series.

2. In each of the following two situations you are to add up the terms of a sequence to obtain an approximation to a number. Your MPL may have a convenient way of doing this. For example, in **ISETL** you can put the numbers in a tuple, say **t**, and then do **%+t**.

(a) Begin with the sequence in Chapter 2, Section 1, Activity 8, p. 72 and divide each term (after the first) by $J^3(J+1)$. Add up enough terms of the resulting sequence to get an approximation of the number to which the partial sums appear to converge. Referring back to the definition of this sequence if necessary, estimate the value of

$$\sum_{n=1}^{\infty} \frac{1}{n^2}.$$

(b) The value of the function represented by the expression $1 + 2x + 3x^2 + 4x^3 + 5x^4 + 6x^5$ from Activity 10, Section 6.1.2, page 424, at $x = 1$ can be approximated by adding up the coefficients of the series obtained from **taylco** for a sufficiently large value of n. Do it and compare your result with the evaluation of the original function at 1.

3. For each of the following series, write a **computer function** and apply **PS** as in Activity 1 to evaluate various partial sums as the index gets larger and larger and guess whether or not the series converges.

(a) $\sum_{n=0}^{\infty} 1.001^n$ (b) $\sum_{i=0}^{\infty} 3^i$ (c) $\sum_{x=0}^{\infty} \left(\frac{1}{2}\right)^x$ (d) $\sum_{m=0}^{\infty} 0.99^m$ (e) $\sum_{s=1}^{\infty} \frac{1}{s^{1.01}}$

(f) $\sum_{k=1}^{\infty} \frac{1}{k}$ (g) $\sum_{s=1}^{\infty} \frac{1}{s^2}$ (h) $\sum_{n=1}^{\infty} \frac{(-1)^n}{n^{0.9}}$ (i) $\sum_{s=1}^{\infty} \frac{1}{s^{0.9}}$ (j) $\sum_{n=1}^{\infty} \frac{(-1)^n}{n^3}$

4. Think about your experiences with Activity 3 and perhaps make up some examples of your own to try to find conditions on the real number r that guarantee convergence of the series

$$\sum_{k=0}^{\infty} r^k.$$

5. Consider the following pair of series.

$$\sum_{i=1}^{\infty} \frac{\sqrt{5i}}{i+1}, \quad \sum_{i=1}^{\infty} \frac{1}{\sqrt{i}}$$

Use the methods of Activity 3 to investigate their convergence and see if you can make any observations about the values of the terms of the two sequences and their convergence or divergence. Why do you think they were presented in a pair?

6. Do the same as the previous activity with the following pair.

$$\sum_{i=1}^{\infty} \frac{i}{(i+1)e^i}, \quad \sum_{i=1}^{\infty} \frac{1}{2^i}$$

7. Make a graph of the function f given by $f(x) = \frac{1}{x^2}$ considered as a function whose domain is a set of real numbers. Use your graphing utility to make a sketch of this function from 1 out to as far as you can. Print your graph and use rectangles to indicate the value of the partial sums of the series in Activity 3 (g).

   Can you use this situation to draw any conclusions about the convergence or divergence of the series?

8. Do the same as the previous activity for the series in Activity 3 (f). What will the function f be now? How about $f(x) = \frac{1}{x^2}$? Do you see why? The series in Activity 3 (f) is called the *harmonic series*.

9. Think about your experiences with Activities 3, 7, and 8. Make up some examples of your own to try to find conditions on the real number p that guarantee convergence of the series

$$\sum_{n=1}^{\infty} \frac{1}{n^p}.$$

10. Think about your experiences with Activity 3, and perhaps make up some examples of your own to try to find conditions on the real number p that guarantee convergence of the series

$$\sum_{n=1}^{\infty} (-1)^n \frac{1}{n^p}.$$

11. For each of the following two series, use the methods of Activity 3, to investigate their convergence. Also, for each series, form the sequence of ratios of successive terms, that is, the nth term of the new sequence is the ratio of the $(n+1)^{\text{st}}$ term to the nth term of the original sequence.

$$\sum_{n=1}^{\infty} \frac{2^n}{n!}, \quad \sum_{n=1}^{\infty} \frac{2^n}{n^3}$$

   Can you make any observations about connections between the convergence or divergence of a series and the sequence of ratios of its successive terms?

12. Consider the following two series.

$$\sum_{i=1}^{\infty} \frac{1}{\sqrt{i}}, \quad \sum_{i=1}^{\infty} \frac{(-1)^i}{\sqrt{i}}$$

   In fact, one of these series converges and the other diverges. The same is true of the following pair.

$$\sum_{n=2}^{\infty} \frac{1}{\ln n}, \quad \sum_{n=2}^{\infty} \frac{(-1)^n}{\ln n}$$

Now do the following.

(a) For each pair, given the information that one converges and the other doesn't, can you tell which is which without making a direct investigation?

(b) Investigate the convergence using the methods of Activity 3, to see if you were right.

(c) Make up a pair of your own that is like the pairs given here and perform the same analysis.

(d) Try to guess some general fact about convergence of series that is illustrated here.

13. Put together your combination **computer functions** from Activities 6 and 7, Section 1 (beginning on page 423), with your **PS** from Activity 8, Section 1 (page 423), and also include one other to obtain **computer functions** that will accept sequences of terms and produces sequences of partial sums for the following combinations of series.

(a) The sum of two series.

(b) The difference of two series.

(c) The termwise product of two series.

(d) The termwise quotient of two series (with the usual restriction).

(e) The termwise product of a series and a number (scalar).

(f) The *Cauchy Product* of two series: Given two series, $\sum_{n=0}^{\infty} a_n$ and $\sum_{n=0}^{\infty} b_n$, the Cauchy Product is the series $\sum_{n=0}^{\infty} c_n$ where

$$c_n = a_0 b_n + a_1 b_{n-1} + \ldots + a_{n-1} b_1 + a_n b_0.$$

(g) The series of absolute values of the terms of a series.

Apply your **computer functions** to various examples and come up with guesses as to whether the series that comes out is convergent, given that the series which go in are convergent. We warn you that the answer is not the same for all seven situations in (a)-(g).

14. With respect to the previous problem, make some computer investigations to help you get a feeling for what is going on, and try to guess answers to the following questions.

(a) What is the relation between the convergence of two series (or a series and a number) and the convergence of their termwise sum? difference? product? quotient? scalar product? Cauchy product?

(b) What is the relation between the sum of two series (or a series and a number) and the sum of their termwise sum? difference? product? quotient? scalar product? Cauchy product?

6.2.3 Discussion

Sources and Examples of Series   In a certain sense, series come from only one place — sequences. That is, the idea of a series is to begin with a sequence $x = (x_n)$ and, as was described in Section 1 (page 430), form the sequence $s = (s_n)$ of partial sums where

$$s_n = x_1 + x_2 + \cdots + x_n = \sum_{i=1}^{n} x_i$$

We say that s is the *series* whose terms are x_n, $n = 1, 2, \ldots$, and we write for the sum S of this series,

$$S = \sum_{n=1}^{\infty} x_n.$$

We will make considerable use of the fact that a series always has its source in a sequence. On the other hand, you will see that, in mathematics, series are very important in themselves. Perhaps their most important contribution, which you will study in the next two sections of this chapter, lies in the use of a special kind of series, *power series*, to represent functions.

Convergence of Series

We define the notion of convergence of a series in terms of convergence of its sequence of partial sums.

> **Definition 6.5**
>
> Let $\sum_{n=1}^{\infty} x_n$ be a series. We say that the series converges *and its sum is the number S if the sequence of partial sums*, $s = (s_n)$ *converges to S. In this case we say that S is the sum of the series and we write,*
>
> $$S = \sum_{n=1}^{\infty} x_n$$
>
> *If a series does not converge, then we say that it* diverges.

Although it may be a neat way to put it, this definition is perhaps not as useful for investigating convergence and sums of series as would be a more direct version that puts in all of the details. Before we work that out, try to think, on your own, how to express this definition without turning to partial sums.

All we need do to work it out is to replace the various phrases by what they mean. We begin with the statement that the sequence s converges to S. This means,

$$\lim_{n \to \infty} s_n = S$$

or, to put in the definition of limit,

for every $\varepsilon > 0$ there exists M such that if $n > M$ then $|s_n - S| < \varepsilon$

and, finally, substituting for s_n its definition as a partial sum,

for every $\varepsilon > 0$ there exists M such that if $n > M$ then

$$\left| \sum_{i=1}^{n} x_i - S \right| < \varepsilon$$

There is a way of thinking about this last expression that might be useful to you. It says that, if you take sufficiently many terms at the beginning of the sequence (x_n), then their sum is within ε of A.

Tests for Convergence — n^{th} Term Test

One reason that series are useful is that you can work with them. Think back about limits of functions. If a function f is continuous at a, then you can compute its limit at a simply by evaluating $f(a)$. Thus there is a mechanical procedure for finding out something about a limit.

There is nothing like this for sequences (although Theorem 6.3, p. 435 is a sort of procedure), but for series, there are a number of "tests for convergence". These are mechanical procedures that, when they work, will allow you to say something definite about whether or not a series converges.

The first test has to do with conclusions you can draw if you know already that a sequence converges. That may sound backwards for a convergence test, but it can be used to determine that a series does *not* converge.

Suppose you know that the series $\sum_{n=1}^{\infty} x_n$ converges. What can you say about the terms x_n? Take a look at Activity 3, separate the series that converge from those that do not and see if you can see anything about the terms x_n that is always true in one category and usually false in the other.

Now, think about what is going on. The partial sums, $s_n = \sum_{i=1}^{n} x_i$ are getting close to some number. So what if you compare two partial sums, say s_{n-1} and s_n. If they are both getting close to the same number, how far apart can they be? What can you say about the distance between them, the absolute value of their difference? On the other hand, can you see what the difference $s_n - s_{n-1}$ equals?

If you think about those questions, you should be able to understand the following theorem and even explain why it is true.

Theorem 6.4 (The nth term test)

If $\sum_{n=1}^{\infty} x_n$ *is a convergent series then* $\lim_{n \to \infty} x_n = 0$.

The proof of the above theorem is left as an exercise.

Check this theorem out with your categorization of the examples in Activity 3. Do you see that for every convergent series the nth term goes to zero?

The way you can use this theorem is to determine that a series does not converge, but diverges. You just check the nth term and see if it goes to zero. If it doesn't, you can be sure that the series diverges.

But suppose the nth term *does* go to zero. Can you then conclude that the series converges? Look at the example in Activity 3 (f). Does the nth term go to zero or not? Does the series converge or diverge?

> If the nth term goes to zero in a series, then you can draw no conclusion about whether the series converges or diverges.

However, we have the following useful test for divergence which is the "contrapositive" of Theorem 6.4.

Theorem 6.5

If $\lim_{n \to \infty} x_n \neq 0$, *then the series* $\sum_{n=1}^{\infty} x_n$ *is divergent.*

Series of Positive Terms

Some tests are only appropriate for series whose terms are positive. It will be helpful if we think a little about such series.

A very important observation that you can make about a series whose terms are positive is that the sequence of partial sums is monotone increasing. Do you see why?

This observation, together with Theorem 6.3 (p. 435) gives us a very simple criterion for convergence for series with positive terms. The point is that Theorem 6.3 tells us that if a sequence is both monotone increasing and bounded above, then it is convergent. Conversely, if a series is convergent, this means that its sequence of partial sums is convergent, so by Theorem 6.2, it is bounded above. Hence, with our observations, the only issue for series of positive terms is the boundedness (above) of the sequence of partial sums.

Theorem 6.6

Let $\sum_{n=1}^{\infty} x_n$ be a series of positive terms. Then, if the sequence of partial sums is bounded above, it follows that the series is convergent. Conversely, if the series converges, then the sequence of partial sums is bounded above.

This might seem a little vague and esoteric to you, but in a few paragraphs you will see how it can be used to establish an extremely powerful and practical test for convergence.

One way of thinking about this situation is the idea that, for a series of positive terms, the only possibilities are that the series converges or the sequence of partial sums goes to ∞. What other behaviors are possible for series whose terms are not positive?

Geometric Series

In a few, relatively rare but important, cases you can determine convergence or divergence of a series by computing an expression for the nth partial sum and determining its limit. One class of examples of such series are the *geometric series*. These are series of the form

$$\sum_{n=0}^{\infty} ar^n$$

where r and a are fixed numbers. In Activity 4, you had a chance to work out the facts about these series.

As we saw at the end of Section 1, (see page 435) the nth term ar^n fails to go to 0 in all cases except $|r| < 1$, so, by the nth term test, we know that a geometric series diverges if $|r| \geq 1$.

In the other case, $|r| < 1$, we can make use of an explicit expression for the nth partial sum. We have,

$$\sum_{i=0}^{n-1} ar^i = a + ar + ar^2 + \cdots + ar^{n-1} = \frac{a(1-r^n)}{1-r}$$

You can get this expression in several ways. One is to just multiply it out and check that it is correct. Can you figure out a direct way to derive this formula, that is, to find the closed form for the partial sum?

In any case, the formula is correct so we can take the limit of both sides as $n \to \infty$. Since $|r| < 1$ we have $\lim_{n \to \infty} r^n = 0$ and so,

$$\lim_{n \to \infty} \frac{a(1-r^n)}{1-r} = \frac{a}{1-r}$$

Therefore we may conclude that the series converges and, in fact, if $|r| < 1$,

$$\sum_{n=0}^{\infty} ar^n = \frac{a}{1-r}.$$

Hence, *the sum of an infinite geometric series is the first term divided by the quantity, one minus the ratio.*

Telescoping Series

Another example of a type of series for which we can not only determine convergence but also calculate the sum are the so-called *telescoping series*. An example of this is the first series in Activity 1. Here is why it is called telescoping. If you write out the first few terms of this series, you can see that almost all terms cancel. The only ones that don't are the first and last. As $n \to \infty$ the first term stays where it is and the last does whatever. Simple investigation of the last term will get the answer.

Let's work it out with general terms. A telescoping series is one of the form

$$\sum_{n=1}^{\infty}(a_{n-1}-a_n).$$

Writing out the nth partial sum, s_n, we have,

$$s_n = \sum_{i=1}^{n}(a_{i-1}-a_i) = (a_0-a_1)+(a_1-a_2)+\cdots+(a_{n-2}-a_{n-1})+(a_{n-1}-a_n) = a_0 - a_n.$$

Then taking the limit we have,

$$\lim_{n\to\infty} s_n = \lim_{n\to\infty}\sum_{i=1}^{n}(a_{i-1}-a_i) = \lim_{n\to\infty}(a_0-a_n) = a_0 - \lim_{n\to\infty} a_n.$$

Hence we may conclude that a telescoping series converges if the sequence $(a_n)_{n=1}^{\infty}$ converges. If it does and we call the limit a, then the sum of the telescoping series is $a_0 - a$.

Thus, the telescoping series in Activity 1 converges and its sum is 1.

If, in a telescoping series, the sequence $(a_n)_{n=1}^{\infty}$ does not converge, then the series diverges.

Tests for Convergence — Comparison

The comparison test is illustrated in Activities 5 and 6. The idea is really very simple. For a series of positive terms, the only possibilities are convergence, or getting large without bound. If you have two series with positive terms, then compare them, term by term. Suppose each term of one is less than or equal to the corresponding term of the other. If the larger one converges, then it must be bounded and so then must the smaller one be bounded and so it must converge also. Similarly, if the smaller one diverges, then so must the larger.

You can see this in Activities 5 and 6. The series $\sum_{i=1}^{\infty}\frac{1}{2^i}$ converges because it is a geometric series with $r = \frac{1}{2} < 1$ (page 448). You can also see that for each i, we have

$$\frac{i}{(i+1)e^i} < \frac{1}{e^i} < \frac{1}{2^i},$$

because $e = 2.718\ldots > 2$ and $\frac{i}{i+1} < 1$. Hence the series $\sum_{i=1}^{\infty}\frac{i}{(i+1)e^i}$ converges as well.

On the other hand, we will see below that the series $\sum_{i=1}^{\infty}\frac{1}{\sqrt{i}}$ diverges and you can see that, term-by-term, the series $\sum_{i=1}^{\infty}\frac{\sqrt{5i}}{i+1}$ is larger and hence it also diverges. The desired comparison can be seen as follows

$$\frac{\sqrt{5i}}{i+1} > \frac{2\sqrt{i}}{i+1} = \frac{2i}{\sqrt{i}(i+1)} = \frac{1}{\sqrt{i}}\frac{2i}{i+1} > \frac{1}{\sqrt{i}}.$$

We can now express all of this formally as a theorem and a proof.

Theorem 6.7 (The Comparison test.)

Let $\sum a_n$, $\sum b_n$ be two series of positive terms and suppose that for all n we have,

$$a_n \le b_n$$

Then if $\sum b_n$ converges, it follows that $\sum a_n$ converges. On the other hand, if $\sum a_n$ diverges, it follows that $\sum b_n$ diverges.

Proof. We will prove the first statement and leave the other for the exercises.

Suppose that $\sum b_n$ converges. Then by Theorem 6.6, page 448, the sequence of partial sums is bounded above. By comparison, since $a_n \leq b_n$, the same is true for the series $\sum a_n$. Hence, again by Theorem 6.6, the series $\sum a_n$ converges as well.

$\square$

Of course, if you compare two series and the smaller one converges or the larger one diverges, then this test does not help you at all.

The reason that the comparison test is so valuable is that we can build up a repertoire of specific series that we know converge or diverge. Then convergence or divergence can be determined for a much larger collection of series by making comparisons.

The following theorem gives us a slightly more powerful version of the comparison test in which you consider the limit of the ratio of corresponding terms of the two sequences. The proof is left as an exercise.

Theorem 6.8 (The Limit Comparison test.)

Let $\sum a_n$, $\sum b_n$ be two series of positive terms.

1. If $0 \leq \lim\limits_{n \to \infty} \dfrac{a_n}{b_n} < \infty$, then if the series $\sum b_n$ is convergent, it follows that the series $\sum a_n$ is convergent.

2. If $0 \leq \lim\limits_{n \to \infty} \dfrac{a_n}{b_n} < \infty$, then if the series $\sum b_n$ is divergent, it follows that the series $\sum a_n$ is divergent.

The limit comparison test is especially useful when it is not easy to establish an appropriate inequality in order to use the comparison test. Also, when the behavior of the nth term of a series is much like that of a series which is known to converge or diverge, then the limit comparison test can be used. For example, consider the series

$$\sum_{n=1}^{\infty} \frac{2n}{3n^2 + 1}.$$

For large values of n the nth term $\frac{2n}{3n^2+1}$ behaves like $\frac{2}{3n}$ and as you will see in the exercises, the series

$$\sum_{n=1}^{\infty} \frac{2}{3n}$$

diverges because its nth term is the product of a number times the nth term of the harmonic series (which is divergent as you saw in Activity 8. The fact that a series is divergent if its nth term is the product of a number times the nth term of a divergent series follows from what you know about sequences. Using the harmonic series as the comparison series in the limit comparison test, we see that the limit

$$\lim_{n \to \infty} \frac{\frac{2n}{3n^2+1}}{\frac{1}{n}} = \lim_{n \to \infty} \frac{2n^2}{3n^2+1} = \frac{2}{3} < 1$$

which shows that the series $\sum_{n=1}^{\infty} \frac{2n}{3n^2+1}$ diverges by part 2 of Theorem 6.8.

Tests for Convergence — Integral Test

If the expression for the general term of a series is considered to define a function whose domain consists, not just of integers, but also of real numbers in general, then there is an interesting and useful connection between the sum of the series and the integral of that function from 1 (or some other convenient starting point) to ∞. We will exploit this connection to derive another test for convergence. This is what you were looking for in Activities 7 and 8.

We will discuss the situation in general as a theorem and you can try to see how everything that is written here applies to the two specific examples in Activities 7 and 8.

Theorem 6.9

Let $\sum_{n=1}^{\infty} a_n$ be a series of positive terms and suppose that the sequence $(a_n)_{n=1}^{\infty}$ is monotone decreasing. Suppose further that f is a function whose domain includes the set of real numbers greater than or equal to 1, that f is also monotone decreasing and that $f(n) = a_n$, $n = 1, 2, \ldots$

Then the convergence or divergence of the series is the same as the convergence or divergence of the integral,

$$\int_1^{\infty} f(x)\,dx.$$

Proof. First, we remind you of the definition of an improper integral over an infinite interval on page 371. We have, continuing it one more step beyond the definition, and using the fact that a definite integral over a particular interval can be written as the sum of integrals over sub-intervals,

$$\int_1^{\infty} f(x)\,dx = \lim_{n\to\infty}\int_1^n f(x)\,dx = \lim_{n\to\infty}\sum_{i=1}^n \int_i^{i+1} f(x)\,dx = \sum_{n=1}^{\infty}\int_n^{n+1} f(x)\,dx.$$

The point is that these sums can be viewed as the partial sums of a series. That is, the infinite integral is exactly equal to the infinite series, $\sum_{n=1}^{\infty} b_n$ where $b_n = \sum_{n=1}^{\infty}\int_n^{n+1} f(x)\,dx$. To say that the improper integral converges is the same as saying that the series converges.

Now, the crux of the proof is that, using the hypotheses of the theorem, this series can be compared with the one whose convergence is being considered. The main fact that we will use is the estimation worked out in Chapter 5 (see page 327)

$$m \le \frac{1}{b-a}\int_a^b f(x)\,dx \le M$$

where m and M are, respectively, lower and upper bounds for the function f on the interval $[a, b]$.

Because of the assumption that f is monotonically decreasing, we can write this estimate, in our case for any n, by taking the value of f at the two endpoints. Since the length of the interval is 1, we have, for $i = 1, 2, \ldots$

$$a_{i+1} = f(i+1) \le \int_i^{i+1} f(x)\,dx \le f(i) = a_i$$

and so,

$$\sum_{i=1}^n a_{i+1} \le \sum_{i=1}^n \int_i^{i+1} f(x)\,dx \le \sum_{i=1}^n a_i$$

There are two inequalities in this expression and the middle term is equal to $\int_1^n f$ so we can make estimates.

Suppose, for example, we estimated the integral from 1 to n using the inequality on the left. The picture looks like that in Figure 6.8.

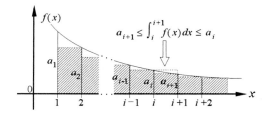

Figure 6.8. A geometric interpretation of the integral test.

If we know that the improper integral converges, we can then apply the comparison test to assert that the series $\sum_{i=1}^{\infty} a_{i+1}$ converges.

This proves half of the theorem. The proof of the other half is left as an exercise.

$\square$

How would this theorem apply to functions such as f and g given by

$$f(n) = \frac{1}{n \ln n}, \text{ and } g(n) = \frac{n}{e^{n^2}}?$$

Note that in the first function, you have to worry about the case $n = 1$. How do you deal with that?

p-Series   A series of the form

$$\sum_{n=1}^{\infty} \frac{1}{n^p}$$

where p is a fixed positive number is called a *p-series*. Why do we only consider a positive value for p?

The integral test can be used to answer the question of convergence for every p-series. We use the integral test with the function f given by

$$f(x) = \frac{1}{x^p}.$$

As required by Theorem 6.9, f is defined for $x \geq 1$, it is monotone decreasing and $f(n) = \frac{1}{n^p} = a_n$. Hence we need only evaluate the integral.

For $p \neq 1$ we have,

$$\int_1^{\infty} \frac{1}{x^p} = \lim_{n \to \infty} \int_1^n \frac{1}{x^p} = \lim_{n \to \infty} \left(\frac{1}{(1-p)n^{p-1}} - \frac{1}{1-p} \right)$$

and you can easily see that if $p > 1$ the improper integral converges, whereas it diverges if $p < 1$.

Finally, if $p = 1$ the integral is slightly different. We have,

$$\int_1^\infty \frac{1}{x} = \lim_{n\to\infty} \int_1^n \frac{1}{x} = \lim_{n\to\infty}(\ln n - \ln 1) = \lim_{n\to\infty} \ln n$$

which diverges.

Hence, we conclude that a p series converges if $p > 1$ and diverges if $p \le 1$.

Incidentally, the series $\sum_{n=i}^\infty \frac{1}{n}$ is very important and has its own name. It is called the *harmonic series*. It diverges, as you have seen in Activity 8, but in a few pages you will see a series that looks very much like it but converges.

Tests for Convergence — Ratio Test

Return to Activity 11 and recall what happened. Did you form the ratio of successive terms? Did you notice that in the first series these ratios were rather small but they were not so small in the second series? Indeed, the issue for what is called the *ratio test* is that the ratio of successive terms is, eventually, less than a number r which is strictly less than 1 or that, eventually, the ratios are greater than a number r which is strictly greater than 1. In the former case the series converges and in the latter case, it diverges.

The reason is very simple, although it can be obscured by the computations in a proof. The idea is to compare the series with a geometric series $\sum ar^n$. In the first case, the series is less than a convergent geometric series so it converges and in the second it is greater than a divergent geometric series so it diverges.

Theorem 6.10

Let $\sum_{n=1}^\infty a_n$ be a series of positive terms and suppose that

$$\lim_{n\to\infty} \frac{a_{n+1}}{a_n} = r.$$

Then if $r < 1$ it follows that the series converges; if $r > 1$ it follows that the series diverges.

Proof. If $r < 1$ then we can take any number R with $r < R < 1$ and for n sufficiently large it is the case that

$$\frac{a_{n+1}}{a_n} < R.$$

We will sweep some messy details under the rug and assume that this inequality holds for all n. You will get a chance to clean up the messy details in an exercise. Using the inequality for various values of n, we have, for any index n,

$$a_n = \frac{a_n}{a_{n-1}} \cdot \frac{a_{n-1}}{a_{n-2}} \cdots \frac{a_3}{a_2} \cdot \frac{a_2}{a_1} \cdot a_1 < R \cdot R \cdots R \cdot R \cdot a_1 = R^{n-1} \cdot a_1$$

which means that $\sum_{n=1}^\infty a_n$ is smaller than the series $\sum_{n=1}^\infty a_1 R^{n-1}$. Since $R < 1$ the latter series converges because it is a number times a convergent geometric series, and, as you will see later, a number times a convergent series gives a convergent series (this follows from Exercise 13, page 437).

The case of $r > 1$ is done in a similar manner and is left as an exercise.

We have not mentioned anything about what happens if $r = 1$, that is if we can only compare the ratio of successive terms with 1. In this case we can say nothing about convergence, no matter which direction the comparison goes.

In the exercises for this section, there is a slightly more complicated version of the ratio test in which you don't take any limits, but go directly to the inequality.

Tests for Convergence — Alternating Series

Take a look at what you observed in Activities 3, (especially parts (h) and (i)), 10 and 12. Why is it that in Activity 3 (h) the series converges, even though $p < 1$? In all of these examples, what is common about the series that converge?

A series whose terms alternate between positive and negative values is called an *alternating series*. Such a series is "more likely" to converge than one whose terms are all the same sign. Can you see why? Think about building a sum by adding now a positive, now a negative term. If the amount you add keeps getting smaller, then the sum you are building up is very likely going to oscillate around some point.

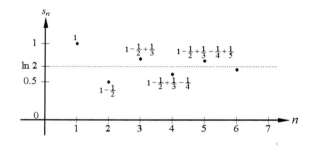

Figure 6.9. $\sum_{n=1}^{\infty} \frac{(-1)^{n+1}}{n} = ln(2)$

In Figure 6.9 you see how it looks with the series $\sum_{n=1}^{\infty} \frac{(-1)^{n+1}}{n}$ which is called the *alternating harmonic series*. What is going on here is that when you add a positive term, the cumulative sum increases. You then add a negative term, so it decreases. This negative term, however, is not as large as the positive term you just added, so it does not go back quite as far as it was. Then you add a positive term, so it increases again. But, once more, what you add is not as much as what you just subtracted so it doesn't go back as far as it came from. In this way you can see that the cumulative sum is closing in on a smaller and smaller region.

There are two issues. Does it actually get down to a single number or an interval containing many numbers? In fact, it can't be an interval because the amount you are adding goes to zero.

The second issue is whether there actually is a number that you are coming in on. This is a very deep question and the answer relies on how the real numbers are defined in the first place.

All of that may seem a little vague, but it can be made very precise and that leads to a very powerful test for alternating series. We leave the proof as an exercise.

> **Theorem 6.11 (Alternating series test)**
>
> *Let (x_n) be a sequence of positive terms. Assume that the values of the terms are monotonically decreasing with n and that $\lim_{n \to \infty} x_n = 0$. Then the series $\sum_{n=1}^{\infty}(-1)^n x_n$ converges. Moreover, when the nth partial sum s_n is used to approximate the sum of such a convergent alternating series, then the (absolute) error $|s - s_n|$ is less than or equal to the absolute value of the $(n+1)^{\text{st}}$ term of the series, that is, $|s - s_n| \leq x_{n+1}$.*

Absolute and Conditional Convergence

As we mentioned in the last paragraph, it is "more likely" that an alternating series converges than does a series with positive terms. Check that out with Activity 12. In both cases, the series with positive terms diverges, but the one with alternating signs converges.

Turning that around, suppose you have a series whose terms do not all have the same sign and replace it by a series whose terms are the absolute values of the terms of the series. Then, if the new series *does* converges, since that is "less likely", it stands to reason that the original series ought to converge.

This is completely true and we formalize it in a theorem. As in the following theorem, we sometimes write $\sum a_n$ as shorthand for the infinite series $\sum_{n=1}^{\infty} a_n$, or $\sum_{n=k}^{\infty} a_n$ where k is some integer.

> **Theorem 6.12**
>
> *Let $\sum a_n$ be a series and suppose that the series $\sum |a_n|$ converges. Then the original series $\sum a_n$ also converges.*

Proof. There is a neat way to prove this theorem. We'll talk it through and see if you understand.

First we replace the original series a with two series, b and c. The terms of b are the same as the terms of a when they are positive and otherwise the terms are 0. Similarly, the terms of c are the same as the *negatives* of the terms of a when they are negative (so these terms of c are positive!) and otherwise 0. Do you see that the series a is equal to the difference $b - c$ of the two new series?

Moreover, both series b and c are series of non-negative terms and they can be compared with the series whose terms are the absolute values of the terms of a. This latter series converges and the terms of b and c are smaller so they converge.

Finally, $a = b - c$ is the difference of two convergent series and as we will see in the next paragraph, this means that a converges as well.

$\square$

This theorem can be very useful for determining that a series converges. Consider for example,

$$\sum_{n=2}^{\infty} \frac{(-1)^n}{n^2 \ln n}$$

If we replace this series with the series whose terms are the absolute values, we get

$$\sum_{n=2}^{\infty} \frac{1}{n^2 \ln n}$$

The terms of this series are smaller than the terms of $\sum \frac{1}{n^2}$ and this is a p series with $p=2$ so it converges.   Hence the series with positive terms converges, so by our theorem, the original series converges.

There is some terminology in connection with the phenomena we have been discussing.

Definition 6.6

Let $\sum a_n$ be a series and suppose that the series $\sum |a_n|$ converges. Then we say that the series $\sum a_n$ is absolutely convergent. If a series converges, but is not absolutely convergent, then it is said to be conditionally convergent.

Can you find examples of series in the activities which are conditionally convergent? If you consider a series a and the series b formed by replacing each term by its absolute value, then, before thinking, there are four possible combinations of convergence and divergence of these two series. How many of these possibilities can actually happen?

Convergence of Combinations of Series
We hope that you discovered in Activity 13 that, even if you start with convergent series, some combinations will always converge, but for other combinations, they may not.

We will describe two cases, one which "works" and one which does not and leave the rest for the exercises.

Theorem 6.13

Let $\sum a_n$, $\sum b_n$, and $\sum (a_n + b_n)$ be series and suppose that the first two series converge with sums a and b, respectively. Then the third series converges and its sum is $a+b$.

Proof. Let (r_n), (s_n), (t_n) be the sequences of partial sums for the three series, respectively. Then we have,

$$r_n = \sum_{i=1}^{n} a_i$$

$$s_n = \sum_{i=1}^{n} b_i$$

so

$$t_n = \sum_{i=1}^{n} (a_i + b_i) = \sum_{i=1}^{n} a_i + \sum_{i=1}^{n} b_i = r_n + s_n$$

and it follows from Theorem 3.3 (see page 172) that

$$\lim_{n \to \infty} t_n = \lim_{n \to \infty} (s_n + r_n) = \lim_{n \to \infty} s_n + \lim_{n \to \infty} r_n = a + b$$

since the limit of a sum is the sum of the limits (provided the limits exist).

$\square$

We now consider a combination of convergent series that fails to converge. You should know by now that the harmonic series $\sum \frac{1}{n}$ diverges (see Activity 8, page 444). Can you write it as a termwise product of two series that converge? Here is a hint. The product of two positive numbers is a positive number and the product of two negative numbers is a positive number. You will get a chance to finish this as an exercise.

We close this discussion with a final remark about convergence of series. The following "tail end test" is useful in testing some series for convergence.

> A series $\sum_{n=1}^{\infty} a_n$ converges if and only if the series $\sum_{n=k}^{\infty} a_n$ converges, where k is a positive integer.

We leave the proof as an exercise. The above result says that in testing an infinite series for convergence, you can ignore any finite number of terms at the beginning of the series. In effect, you need only look at the "tail end" of the series to determine convergence (or divergence) of a series. However, we should note that although you can ignore any finite number of terms to determine convergence, you <u>cannot</u> ignore any of the terms of an infinite series when finding its sum.

Remark. At this point, you may find the various ways of trying to determine if a series converges somewhat confusing. It might help to make use of an explicit strategy. Here is one possibility. Learn it (by using it) and then modify it to make one of your own if you wish.

A Suggested Strategy for Testing a Series for Convergence

Ask yourself the following questions:

1. Does the nth term of the series converge to zero? If the nth term does not go to zero, then the series diverges. If the nth term goes to zero, then the series might converge and you should proceed to ask the following questions.

2. Is the series one of the following special types of series (or a combination of them)? Write out a few terms of the series to check the type of series.

   (a) Is the series a telescoping series? (If the series is a rational function whose denominator can be factored into linear factors, then you might try finding a partial fraction decomposition and then check to see if the series is a telescoping series.)

   (b) Is the series a geometric series? If it is geometric and the absolute value of the ratio is less than one, then the series converges and its sum is equal to the ratio of the first term to the quantity one minus the ratio.

   (c) Is the series a p-series? If so and $p > 1$, then the series converges. If not, i.e., $p \leq 1$, then the series diverges.

   (d) Is the series an alternating series? If so, check to see if the sequence of absolute values of the terms is monotonically decreasing (past some point in the sequence) to zero. If so, then the series converges.

   (e) Does the series have both positive and negative terms (but not alternating)? Test the series for absolute convergence using the tests in part (3) below.

3. Are all the terms of the series positive? If so, then ask yourself the following questions:

   (a) Can the terms of the series be easily compared to a known series which you know converges or diverges? Be sure to find the right comparison for convergence or divergence!

   (b) Do the terms of the series form a monotonically decreasing sequence (past some point in the sequence)? If so, and the nth term can be easily integrated, then apply the integral test.

   (c) Does the nth term of the series contain factorials, n^{th} powers, or powers of n and/or combinations of such expressions? If so apply the ratio test (or the root test; see Exercise 39, page 464).

6.2.4 Exercises

1. A very important observation that you can make about a series whose terms are positive is that the sequence of partial sums is monotone increasing. Explain why. What other behaviors are possible for series whose terms are not positive?

2. Show that
$$s_n = \sum_{i=0}^{n-1} ar^i = a + ar + ar^2 + \cdots + ar^{n-1} = \frac{a(1-r^n)}{1-r}.$$

3. If possible, find the sum of each of the following series. (You may wish to obtain a partial fraction decomposition for the nth term of some series. See Theorem 5.2 on page 367.)

   (a) $\displaystyle\sum_{n=1}^{\infty} \frac{1}{(n+1)(n+2)}$

   (b) $\displaystyle\sum_{n=1}^{\infty} \frac{1}{(2n+1)(2n-1)}$

   (c) $\displaystyle\sum_{n=1}^{\infty} \frac{3}{n^2+n}$

   (d) $\displaystyle\sum_{n=2}^{\infty} \frac{1}{n^2-1}$

   (e) $\displaystyle\sum_{n=1}^{\infty} \frac{2}{9-4n^2}$

   (f) $\displaystyle\sum_{n=3}^{\infty} \frac{1}{n^2-n-6}$

   (g) $\displaystyle\sum_{n=1}^{\infty} 3\left(\frac{1}{2}\right)^n$

   (h) $\displaystyle\sum_{n=1}^{\infty} 2\left(\frac{-2}{3}\right)^n$

   (i) $\displaystyle\sum_{n=1}^{\infty} \left(\frac{\pi}{3}\right)^n$

   (j) $\displaystyle\sum_{n=2}^{\infty} \frac{(-3)^n}{5^{n-1}}$

   (k) $\displaystyle\sum_{n=0}^{\infty} \frac{1}{3}\left(\frac{-4}{5}\right)^{n+1}$

   (l) $\displaystyle\sum_{n=1}^{\infty} 3e^{-n}$

   (m) $\displaystyle\sum_{n=0}^{\infty} \left(-\sqrt{0.3}\right)^n$

4. Suppose that in an certain country, approximately 95 percent of all income is spent and that the remaining income is saved. How much additional spending will result from a 30 billion dollar tax rebate if the spending habits of the country do not change in the future?

5. A patient is given an injection of 5 units of a certain drug every 24 hours. Suppose that the drug is eliminated from the body exponentially so that the fraction that remains in the patients body after t days is given by $e^{-0.2t}$. If the treatment is continued indefinitely, approximately how many units of the drug will there eventually be in the patient's body

   (a) just prior to an injection.

   (b) immediately after an injection.

6. Find a fraction which is equivalent to each of the following repeating decimals. Explain why your answer is correct by using an infinite series and finding its sum.

   (a) 0.99999...

   (b) 0.123123...

   (c) 0.262626...

   (d) 1.42854285...

(e) $-3.4747\ldots$

(z) $\displaystyle\sum_{n=1}^{\infty}(1+n)^{1/n}$

7. Use an appropriate test to determine whether or not each of the following series is convergent. You may wish to use the suggested strategy on page 457. Give reasons for your answer.

(a) $\displaystyle\sum_{n=1}^{\infty}\frac{-5}{n}$

(b) $\displaystyle\sum_{n=0}^{\infty}\frac{3}{1-(0.35)^{n+1}}$

(c) $\displaystyle\sum_{n=1}^{\infty}\frac{3^{n}}{n^{2}}$ (d) $\displaystyle\sum_{n=1}^{\infty}\frac{n}{n^{2}+1}$

((e) $\displaystyle\sum_{n=1}^{\infty}2ne^{-n^{2}}$ (f) $\displaystyle\sum_{n=1}^{\infty}\frac{e^{1/n}}{n^{2}}$

(g) $\displaystyle\sum_{n=1}^{\infty}\frac{1}{2+3^{n}}$ (h) $\displaystyle\sum_{n=1}^{\infty}\frac{\pi}{n(4n^{2}+n)}$

(i) $\displaystyle\sum_{n=1}^{\infty}\frac{3}{2n-5}$

(j) $\displaystyle\sum_{n=1}^{\infty}\frac{\ln(n)}{n^{3}}$ (Hint: Show that $\ln(n)<n$)

(k) $\displaystyle\sum_{n=1}^{\infty}\frac{\arctan(n)}{n^{2}+1}$ (l) $\displaystyle\sum_{n=1}^{\infty}\frac{3n}{\sqrt{1+n^{2}}}$

(m) $\displaystyle\sum_{n=1}^{\infty}\frac{5n^{2}-n}{3n^{4}-6n+1}$ (n) $\displaystyle\sum_{n=1}^{\infty}\sin\left(\frac{1}{n}\right)$

(o) $\displaystyle\sum_{n=1}^{\infty}ne^{-2n}$ (p) $\displaystyle\sum_{n=1}^{\infty}\frac{3}{\sqrt{n^{3}}}$

(q) $\displaystyle\sum_{n=1}^{\infty}\frac{1}{5\sqrt{n}-2}$ (r) $\displaystyle\sum_{n=1}^{\infty}\left(1-\frac{1}{n}\right)^{n}$

(s) $\displaystyle\sum_{n=1}^{\infty}(\ln(n)-\ln(n+1))$

(t) $\displaystyle\sum_{n=1}^{\infty}\frac{5}{1-4^{n}}$ (u) $\displaystyle\sum_{n=1}^{\infty}(-1)^{n+1}$

(v) $\displaystyle\sum_{n=2}^{\infty}\frac{1}{\sqrt{n-1}}$ (w) $\displaystyle\sum_{n=2}^{\infty}\frac{n^{2}}{\sqrt{n^{4}-1}}$

(x) $\displaystyle\sum_{n=1}^{\infty}\frac{1}{n+3^{n}}$ (y) $\displaystyle\sum_{n=1}^{\infty}\frac{1}{\sqrt[3]{n}}$

8. Use an appropriate test to determine whether or not each of the following series is convergent. Give reasons for your answer.

(a) $\displaystyle\sum_{n=2}^{\infty}\frac{3}{n\ln(n)}$ (b) $\displaystyle\sum_{n=1}^{\infty}\frac{|\cos(n)|}{n^{1.01}}$

(c) $\displaystyle\sum_{n=1}^{\infty}\frac{1\cdot4\cdot7\cdots(3n-2)}{5\cdot10\cdot15\cdots(5n)}$

(d) $\displaystyle\sum_{n=2}^{\infty}\frac{1}{\ln(n)}$ (Hint: Show that $\ln n<n$)

(e) $\displaystyle\sum_{n=1}^{\infty}\frac{1}{n3^{n}}$ (f) $\displaystyle\sum_{n=1}^{\infty}\frac{(-1)^{n}}{n+3}$

(g) $\displaystyle\sum_{n=1}^{\infty}\frac{1}{1+2+3+\cdots+n}$

(h) $\displaystyle\sum_{n=1}^{\infty}\frac{2}{n^{n}}$ (i) $\displaystyle\sum_{n=1}^{\infty}\frac{(-1)^{n}}{\ln(1+n)}$

(j) $\displaystyle\sum_{n=1}^{\infty}\frac{\sin(n^{3})}{3^{n}}$

(k) $\displaystyle\sum_{n=1}^{\infty}\frac{\ln(n)}{n^{2}}$ (Hint: Show that $\ln n<\sqrt{n}$, for $n>1$)

(l) $\displaystyle\sum_{n=1}^{\infty}\left(\frac{n!}{n^{n}}\right)^{2n}$ (m) $\displaystyle\sum_{n=1}^{\infty}\frac{n^{2}+n}{n!}$

(n) $\displaystyle\sum_{n=1}^{\infty}\frac{(-2.718)^{n}}{e^{n}}$ (o) $\displaystyle\sum_{n=1}^{\infty}\frac{2n}{4^{n}}$

(p) $\displaystyle\sum_{n=1}^{\infty}n\left(\frac{3}{4}\right)^{n}$ (q) $\displaystyle\sum_{n=1}^{\infty}\frac{n!5^{n}}{n^{n}}$

(r) $\displaystyle\sum_{n=1}^{\infty}\frac{\cos(\pi n)\ln(n)}{n+1}$

9. Use an appropriate test to determine whether or not each of the following combinations of series is convergent. Give reasons for your answer.

(a) $\sum_{n=1}^{\infty}\left(\left(\frac{1}{2}\right)^n+\left(\frac{3}{5}\right)^n\right)$

(b) $\sum_{n=1}^{\infty}\left(\left(\frac{-1}{3}\right)^n+\frac{3}{n}\right)$

(c) $\sum_{n=1}^{\infty}\frac{3^{n+1}+4^n}{5^{n+1}}$

(d) $\sum_{n=1}^{\infty}\left(2\left(\frac{3}{4}\right)^n-7\left(\frac{1}{2}\right)^{n-1}\right)$

(e) $\sum_{n=1}^{\infty}\left(\frac{3}{\sqrt{n}}+\frac{1}{n^5}\right)$

(f) $\sum_{n=0}^{\infty}\left(2^{-n-1}-3^{-n-1}\right)$

10. Can you find examples of series in the activities for this section which are conditionally convergent? If you consider a series *a* and the series *b* formed by replacing each term by its absolute value, then, before thinking, there are four possible combinations of convergence and divergence of these two series. How many of these possibilities can actually happen?

11. Write the harmonic series, which is divergent, as a termwise product of two series that converge. (Hint: The product of two positive numbers is a positive number and the product of two negative numbers is a positive number.)

12. Determine whether or not each of the following series is absolutely convergent, conditionally convergent, or divergent. Give reasons for your answer.

(a) $\sum_{n=0}^{\infty}\cos(\pi n)$

(b) $\sum_{n=1}^{\infty}(-1)^n\frac{n}{3n^2+1}$

(c) $\sum_{n=1}^{\infty}\frac{(-1)^{n+1}}{\sqrt{n}}$ (d) $\sum_{n=1}^{\infty}n\sin\left(\frac{2}{n}\right)$

(e) $\sum_{n=1}^{\infty}(-1)^n7e^{-3n}$ (f) $\sum_{n=1}^{\infty}\frac{\cos(\pi n)}{n}$

(g) $\sum_{n=1}^{\infty}\frac{(-1)^n}{n(1+n^2)}$ (h) $\sum_{n=1}^{\infty}\frac{(-1)^n}{(2n)!}$

(i) $\sum_{n=1}^{\infty}\frac{(-1)^{n+1}}{(1+n)^{4/3}}$ (j) $\sum_{n=1}^{\infty}(-1)^n\frac{n^n}{n!}$

(k) $\sum_{n=1}^{\infty}\frac{(-1)^n n}{\sqrt{n(n+1)}}$ (l) $\sum_{n=1}^{\infty}\frac{(-1)^{n+1}}{n^n}$

(m) $\sum_{n=1}^{\infty}\frac{(-3)^n}{(1+2n)!}$

(n) $\sum_{n=1}^{\infty}(-1)^n\left(\frac{2}{n^2}\right)^{1/3}$ (o) $\sum_{n=1}^{\infty}\frac{\sin(n^2)}{n!}$

(p) $\sum_{n=1}^{\infty}(-1)^{n+1}\left(\sqrt{n+1}-\sqrt{n}\right)$

(q) $\sum_{n=1}^{\infty}(-1)^n\left(\frac{n+1}{n}\right)^n$

(r) $\sum_{n=1}^{\infty}\cos(\pi n)(\ln(2n+1)-\ln(n+1))$

13. Show that the harmonic series $\sum_{n=1}^{\infty}\frac{1}{n}$ diverges as follows:

(a) Group the terms of the series as follows:

$$1+\frac{1}{2}+\left(\frac{1}{3}+\frac{1}{4}\right)+\left(\frac{1}{5}+\frac{1}{6}+\frac{1}{7}+\frac{1}{8}\right)+\cdots$$

where after the first two terms make groups of 2, 4, 8, …, 2^n, … terms.

(b) Replace all terms of each group by a number which is a smaller number than each term. Hence, you will end up with a series with a smaller sum.

(c) Show that the new series you formed in part (b) is divergent.

14. Using the method and result of the previous exercise, show that the comparison test can be used to show that a *p*-series diverges for $p<1$.

15. A square is inscribed in a square with side of length L by joining the midpoints of each side of the bigger square. This process is continued indefinitely forming a sequence of inscribed squares.

    (a) Find a formula for the area of the nth square.

    (b) Find the sum of the area of all the squares.

    (c) Find a formula for the perimeter of the nth square.

    (d) Find the sum of the perimeters of all the squares.

16. Suppose that

$$\sum_{n=1}^{\infty} a_n$$

is a divergent series of positive terms

    (a) Show that

$$\sum_{n=1}^{\infty} c a_n$$

is a divergent series, where $c \neq 0$ is a real number.

    (b) If the terms of the original series are not necessarily all positive is the series in part (a) divergent? If so, prove it, if not give a counterexample.

17. Show that if the series $\sum_{n=1}^{\infty} a_n$ converges and $a_n \neq 0$ for all n, then the following series is divergent

$$\sum_{n=1}^{\infty} \frac{1}{a_n}.$$

18. If the series

$$\sum_{n=1}^{\infty} (a_n + b_n)$$

is convergent, does it follow that the series

$$\sum_{n=1}^{\infty} a_n \text{ and } \sum_{n=1}^{\infty} b_n$$

both have to be convergent? If so prove it, if not give a counterexample.

19. Show that if $\sum_{n=1}^{\infty} a_n$ and $\sum_{n=1}^{\infty} b_n$ converges absolutely, then each of the following series converge absolutely.

    (a) $\sum_{n=1}^{\infty} (a_n + b_n)$

    (b) $\sum_{n=1}^{\infty} (a_n - b_n)$

    (c) $\sum_{n=1}^{\infty} k a_n$, where k is a real number.

    (d) $\sum_{n=1}^{\infty} (k a_n + m b_n)$, where k and m are real numbers.

20. Let (a_n) be the recursive sequence given by

$$a_0 = 1 \text{ and } (n^2 + 2) a_{n+1} - (n^2 - 1) k a_n = 0$$
$$\text{for } n \geq 0.$$

What are the values of k for which the series $\sum_{n=1}^{\infty} a_n$ is absolutely convergent?

21. Prove the Limit Comparison Test (page 450). (Hint: Choose numbers M and N such that $M < L < N$ for large n. Then show that $M < \frac{a_n}{b_n} < N$ for large n, and hence $M b_n < a_n < N b_n$ for large n. Apply the comparison test and finish the proof.)

22. Prove that if $x_n > 0$ for all n and $\lim_{n \to \infty} n x_n \neq 0$, then the series $\sum_{n=1}^{\infty} \frac{1}{x_n}$ is convergent.

23. Prove that if $x_n > 0$ and the series $\sum_{n=1}^{\infty} x_n$ converges, then the following series diverges:

$$\sum_{n=1}^{\infty} \frac{x_n}{n}.$$

24. Prove the other half of the integral test, page 451. That is, if $\int_1^\infty f(x)dx$ diverges, then $\sum_{n=1}^\infty a_n$ diverges.

25. Show that if the series $\sum_{n=1}^\infty a_n$ converges, then for $c \neq 0$ the series $\sum_{n=1}^\infty ca_n$ converges.

26. Finish the proof of the ratio test (page 453) for the case $\lim_{n\to\infty} \frac{a_{n+1}}{a_n} < 1$ as follows: Since $\lim_{n\to\infty} \frac{a_{n+1}}{a_n} < 1$ there is a number ρ such that $\lim_{n\to\infty} \frac{a_{n+1}}{a_n} = \rho < 1$. Explain why the definition of the limit of a sequence allows you to conclude that there is a number r and a positive integer N such that $\rho < r < 1$ and $\frac{a_{n+1}}{a_n} < r$ for $n \geq N$. Use this to "clean up" the "messy details" referred to in the text on page 453.

27. Prove the other half of the ratio test. That is, if $\lim_{n\to\infty} \frac{a_{n+1}}{a_n} > 1$, then $\sum_{n=1}^\infty a_n$ diverges. See the previous exercise.

28. (a) Prove the alternating series test (page 454) using the following:

    Since $x_{n+1} < x_n$ for all n, explain why $s_n < s_{n+k} < s_{n+1}$ for the partial sums s_n, s_{n+k} and s_{n+1} of $\sum_{n=1}^\infty (-1)^n x_n$ for $k > 1$.

    (b) Show that if the nth partial sum s_n is used to approximate the sum S of the alternating series $\sum_{n=1}^\infty (-1)^n x_n$ (where $x_n > 0$ for all n), then the absolute error $|R_n| = |S - s_n|$ in using s_n to approximate S is less than the absolute value of the first term omitted, i.e., $|R_n| = |S - s_n| \leq x_{n+1}$. (Hint: See part (a).)

    (c) For the situation in part (b), show that the sign of $R_n = S - s_n$ for an

alternating series is the same sign as the first term omitted. (Hint: Group the terms of the remainder in pairs.)

29. Use the result of part (b) in the previous problem to approximate the sum of each of the following series to the number of decimal places d indicated.

    (a) $\sum_{n=1}^\infty \frac{(-1)^{n+1}}{n^2}$, $d = 2$

    (b) $\sum_{n=1}^\infty \frac{(-1)^{n+1}}{n^3}$, $d = 3$

    (c) $\sum_{n=1}^\infty \frac{(-1)^{n+1}}{n^4}$, $d = 3$

30. Show that the following alternating series converges.

$$\frac{1}{3} - 1 + \frac{1}{9} - \frac{1}{4} + \frac{1}{27} - \frac{1}{8} + \cdots$$

31. Suppose that $\sum_{n=1}^\infty a_n$ and $\sum_{n=1}^\infty b_n$ converge, what can you say about the following. Give reasons and examples to illustrate your answer.

    (a) The termwise product of the two series.

    (b) The Cauchy Product of the two series (see part (f) of Activity 13 on p. 445).

32. (a) Prove the following "tail end" test for convergence which allows you to omit consideration of any finite number of terms when determining the convergence of an infinite series. Let k be any positive integer.

    A series $\sum_{n=1}^\infty a_n$ converges if and only if $\sum_{n=k}^\infty a_n$ converges.

    (b) This test is quite useful if a tail end series behaves rather nicely as compared to the original series. For example, suppose that the tail end of a series is alternating after a finite number of terms of the series are omitted. Then, if the tail end series satisfies the alternating series test, you can conclude that the original series converges.

Write a brief essay explaining at least three other situations in which the tail end test would be useful. Give examples illustrating the situation described here and illustrating your three situations.

33. Show that in the integral test,

$$\int_1^\infty f(x)dx$$

can be replaced by

$$\int_m^\infty f(x)dx,$$

where m is a positive integer. Hence, the integral test generalizes to series of the form

$$\sum_{n=m}^\infty a_n.$$

34. Apply the generalized integral test from the previous exercise to determine whether or not each of the following series is convergent.

(a) $\displaystyle\sum_{n=2}^\infty \frac{\ln n}{n}$ (b) $\displaystyle\sum_{n=4}^\infty \frac{3-n}{n+2}$

(c) $\displaystyle\sum_{n=2}^\infty \frac{2}{n(\ln n)^2}$ (d) $\displaystyle\sum_{n=2}^\infty \frac{1}{n\sqrt{\ln n}}.$

35. Let $\sum_{n=1}^\infty a_n$ be a convergent infinite series and suppose that f is a function defined by $f(n) = a_n$ for each index n which satisfies the hypothesis of the integral test (page 451).

(a) The remainder R_n when n terms of a convergent series $\sum_{n=1}^\infty a_n$ are used to approximate the sum A of the series, is given by

$$R_n = \sum_{n=1}^\infty a_n - \sum_{k=1}^\infty a_k = \sum_{k=n+1}^\infty a_k.$$

Moreover, if s_n is the nth partial sum of the series, then $R_n = S - s_n$. Show that the following is true:

$$0 < R_n < \int_n^\infty f(x)dx$$

where R_n is the remainder when n terms of the series have been used to approximate the sum of the series.

(b) Use the result of part (a) to estimate the remainder R_n of each of the following series and find an estimate of the sum of each series using the indicated value of n.

i. $\displaystyle\sum_{n=1}^\infty \frac{1}{n^2}$, $n = 8$

ii. $\displaystyle\sum_{n=1}^\infty \frac{1}{(1+n)^3}$, $n = 10$

iii. $\displaystyle\sum_{n=1}^\infty \frac{1}{e^n}$, $n = 8$

iv. $\displaystyle\sum_{n=1}^\infty \frac{1}{(n+1)(\ln(n+1))^2}$, $n = 6$

(c) Use the result of part (a) to determine the smallest value of n so that $0 \le R_n < \varepsilon$ for the specified series and indicated value of ε.

i. $\displaystyle\sum_{n=1}^\infty \frac{1}{n^2}$, $\varepsilon = 0.01$

ii. $\displaystyle\sum_{n=1}^\infty \frac{1}{(1+n)^3}$, $\varepsilon = 0.01$

iii. $\displaystyle\sum_{n=1}^\infty \frac{1}{e^n}$, $\varepsilon = 0.01$

iv. $\displaystyle\sum_{n=1}^\infty \frac{1}{(n+1)(\ln(n+1))^2}$, $\varepsilon = 0.05$

36. (a) Show that in a convergent p-series the remainder R_n (see the previous exercise) when the nth partial sum s_n

of the series is used to estimate the sum of the series, satisfies the following inequality:

$$R_n < \frac{1}{(p-1)n^{p-1}}.$$

(b) Use the result of part (a) to estimate the remainder R_n of each of the following series and find an estimate of the sum of each series using the first 10 terms of the series.

(c) Use the result of part (a) to determine how many terms should be used to estimate the sum of each series to within 0.01 of its actual sum.

i. $\sum_{n=1}^{\infty}\frac{1}{n^2}$ ii. $\sum_{n=1}^{\infty}\frac{1}{n^3}$

37. Find all values of p for which each of the following series is

(a) convergent.

(b) divergent.

i. $\sum_{n=1}^{\infty}\frac{(\ln(n))^p}{n}$

ii. $\sum_{n=2}^{\infty}\frac{1}{(n+1)(\ln(n+1))^p}$

38. Find all real numbers x so that the following series is convergent

$$\sum_{n=1}^{\infty}\frac{x^n}{n!}.$$

39. Another test for convergence is known as the *root test* and is as follows:

If $a_n \geq 0$ for all n, then the series $\sum_{n=1}^{\infty}a_n$ is

convergent if $\lim_{n\to\infty}(a_n)^{1/n} < 1$

and

divergent if $\lim_{n\to\infty}(a_n)^{1/n} > 1.$

If $\lim_{n\to\infty} a_n = 1$, then the series may either converge or diverge.

The root test is especially useful when the nth term of the series contains factors raised to the nth power.

(a) Prove the root test. (Hint: The proof is similar to that of the ratio test. Use the fact that if $0 \leq (a_n)^{\frac{1}{n}} < r$, it follows that $a_n < r^n$.)

(b) Use the root test above to determine whether or not each of the following series converges or diverges.

i. $\sum_{n=2}^{\infty}\frac{1}{e^n}$

ii. $\sum_{n=2}^{\infty}\left(\frac{1+2n}{3n-1}\right)^n$

iii. $\sum_{n=2}^{\infty}\frac{1}{n^{2n}}$ iv. $\sum_{n=2}^{\infty}\left(\frac{2}{3}\right)^n$

v. $\sum_{n=2}^{\infty}\frac{5^n}{2^{2n}}$

vi. $\sum_{n=2}^{\infty}\frac{(\ln(n))^{3n}}{n^n}$

vii. $\sum_{n=2}^{\infty}\frac{3e^n}{(n+1)^{2n}}$

viii. $\sum_{n=2}^{\infty}\frac{1}{(\ln(n))^n}$

ix. $\sum_{n=2}^{\infty}\frac{2^{3n}}{(1+\cos(\pi/n))^n}.$

40. For what values of k is the following series convergent.

$$\sum_{n=1}^{\infty}n^{kn}$$

41. Find a number $a > 0$, if possible, so that the following is true:

$$\sum_{n=2}^{\infty}\left(\frac{1}{1+a}\right)^n = 3.$$

42. Find the sum of each of the following series. (Hint: Write out several terms to see what is going on.) Give reasons for your answer.

    (a) $\displaystyle\sum_{n=1}^{\infty} \frac{n}{3^n}$

    (b) $\displaystyle\sum_{k=0}^{\infty} \frac{k+3}{2^k}$

43. The following is called Riemann's theorem:

    Part 1: Any rearrangement of an absolutely convergent series must converge to the sum of the original series.

    Part 2: A conditionally convergent series can be rearranged, by reordering the terms, so that the rearranged series converges to any given real number, diverges to ∞ or $-\infty$, or diverges by oscillating between any two real numbers.

    (a) In Section 4 (see page 490), it will be shown that

    $$\sum_{n=1}^{\infty} \frac{(-1)^{n-1}}{n} = \ln(2).$$

    Show that the series is conditionally convergent. Then rearrange the original series and insert parentheses as follows:

    $$\left(1 - \frac{1}{2}\right) - \frac{1}{4} + \left(\frac{1}{3} - \frac{1}{6}\right) - \frac{1}{8}$$
    $$+ \left(\frac{1}{5} - \frac{1}{10}\right) - \frac{1}{12} + \cdots$$

    where each odd term is followed by two even terms and parentheses are added as indicated. Show that the rearranged series converges to

    $$\frac{1}{2}\ln(2).$$

    (b) For the series

    $$\sum_{n=1}^{\infty} \frac{(-1)^{n-1}}{n}$$

    in part (a), show that the series can be rearranged so that the rearranged series has sum 3. You may wish to use your computer to help see what is going on. (Hint: First, take enough of the positive terms to get a sum ≥ 3; then take enough of the negative terms to get a total sum < 3; then take enough of the remaining positive terms to get a total sum ≥ 3; then take enough of the remaining negative terms to get a total sum < 3; continue in this manner and show that all the terms are used and hence the rearranged series has sum 3.)

(c) Find a rearrangement of the series in part (a) so that the rearranged series diverges to ∞ using an argument similar to that in part (b).

(d) Find a rearrangement of the series in part (a) so that the rearranged series diverges to $-\infty$ using an argument similar to that in part (b).

(e) Prove Part 1 of Riemann's theorem as follows: Suppose that the series $\sum_{n=1}^{\infty} a_n$ converges absolutely and has sum A. Let (b_n) be any rearrangement of the sequence of terms (a_n) of the original series. Show that $A = \sum_{n=1}^{\infty} b_n$.

    i. Show that for any $\varepsilon > 0$ there is an index N_1 so that $\sum_{n=N_1+1}^{\infty} |a_n| < \frac{\varepsilon}{2}$ and an index $N_2 \geq N_1$ so that $|s_{N_2} - A| < \frac{\varepsilon}{2}$.

    ii. Show that all the terms of the sequence $a_1, a_2, \ldots, a_{N_2}$ appear somewhere in the sequence (b_n), and there is an index $N_3 \geq N_2$ such that if $n \geq N_3$, then the difference $\sum_{m=1}^{n} b_m - s_{N_2}$ is at most a sum of terms of the form a_n with $n \geq N_1$.

iii. Finally, show that

$$\left| \sum_{m=1}^{\infty} b_m - A \right| \leq \left| \sum_{m=1}^{n} b_m - s_{N_2} \right| + \left| s_{N_2} - A \right| < \varepsilon.$$

(f) Prove Part 2 of Riemann's Theorem by generalizing the arguments you used in parts (b), (c), and (d) to any conditionally convergent series.

6.3 SERIES OF FUNCTIONS — POWER SERIES

6.3.1 Overview

In this section, you will study *power series* like

$$\sum_{n=0}^{\infty} a_n x^n \text{ and } \sum_{n=0}^{\infty} a_n (x-c)^n,$$

where $(a_n)_{n=0}^{\infty}$ is some sequence of real numbers. Such power series are "infinite analogues" of polynomial functions. A power series represents a function whose domain is the set of real numbers x for which the series converges. In fact, as you will see, a power series has domain consisting of a single point or an interval of real numbers. You will learn how to determine the domain, or *interval of convergence* of a power series, make combinations of power series and investigate the relationship between the properties of combinations, and differentiate and integrate power series. As you will see, power series can essentially be treated "like" polynomial functions. Also, the solutions of certain differential equations can be represented by power series.

6.3.2 Activities

1. Consider the following list of infinite series. Notice that in each series the general term is not a number, but an expression that represents a function. Thus, for each value of x, you have a different series.

   (a) $\sum_{n=0}^{\infty} x^n$ (b) $\sum_{n=0}^{\infty} n x^{n-1}$ (c) $\sum_{n=0}^{\infty} \frac{x^{n+1}}{n+1}$

   For each of these series, write a **computer function** that takes as the input parameter a value **x** and returns a **computer function** that represents the resulting sequence of partial sums.

2. Consider the series in Activity 1 (c). Check convergence for the values $x = -1.5, -1, -0.5, 0.5, 1, 1.5$.

3. For the series in Activity 1 (a), write a **computer function** that represents a function whose domain consists of real numbers. Your **computer function** should accept a value **x** as a parameter and produce a **computer function** that represents the sequence of general terms, for the given value of **x**. Apply your **PS** from the previous two sections to your **computer function** from Activity 1 (a) above to obtain a **computer function** that represents the sequence of partial sums. Finally, your **computer function** should return the value of the 50th partial sum.

   Use the tool **table** to generate a list of 20 values of your **computer function** including the six values of x chosen in Activity 2.

4. Repeat Activity 3 for the other two series in Activity 1. What is the same for all three? What is different?

   How would you explain the differences?

   What would you say about convergence of these three series?

5. Repeat Activity 3 for each of the following series, except that 50 is much too large. Replace it with 10. Also, run **table** over a wider range, say 10 values from -20 to 20.

   (a) $\displaystyle\sum_{n=0}^{\infty} \frac{x^n}{n!}$ (b) $\displaystyle\sum_{n=0}^{\infty} (-1)^n \frac{x^{2n}}{(2n)!}$ (c) $\displaystyle\sum_{n=0}^{\infty} n! \, x^n$

   What are the similarities between the results you got for all of the series you have been looking at. What are the differences?

   What would you say about convergence of these three series?

6. Use your graphing software to draw a graph of the function you represented in Activity 5 (b) on the domain $[-2, 2]$. Do you have any guess as to what this function is? Are you surprised?

7. So far in these activities you have dealt with six specific series and thought about convergence. Pick a number c (choose anything you like except 0) and replace x by $x = c$ in each of these series. What effect does this appear to have on the question of convergence?

8. Each of the series you have dealt with so far in this set of activities is called a power series. It represents a function, at least when it converges. Based on your experiences, can you make any general statement about the domain of a function represented by a power series?

9. Let f, g be the functions given by the power series

$$f(x) = \sum_{n=0}^{\infty} (-1)^n \frac{x^{2n}}{(2n)!}$$

$$g(x) = \sum_{n=0}^{\infty} (-1)^n \frac{x^{2n+1}}{(2n+1)!}$$

Looking at these expressions, make a guess of the value of

$$\lim_{x \to 0} \frac{g(x)}{f(x) - 1 + 2x}$$

(You might try writing out the first few terms.)

Use graphs or your **computer function lim** from Chapter 3 to check your guess.

10. Here are two more series so that you now have nine power series to investigate (see Activities 1, 5, and 9). Each of them represents a function.

$$\sum_{n=0}^{\infty}\left(\frac{x^n}{n!}+(-1)^n\frac{x^{2n}}{(2n)!}\right)$$

$$\sum_{n=0}^{\infty}\left(23.684(-1)^n\frac{x^{2n+1}}{(2n+1)!}+36.7432(-1)^n\frac{x^{2n}}{(2n)!}\right)$$

Of these nine functions, there are certain pairs that can be selected consisting of two functions f, g where g is the derivative of f, that is f is an anti-derivative of g. Find as many of these pairs as you can?

11. Suppose that you are given two power series representations of functions on some domain,

$$f(x)=\sum_{n=0}^{\infty}a_n(x-c)^n$$

$$g(x)=\sum_{n=0}^{\infty}b_n(x-c)^n$$

and λ, μ are real numbers. Based on your experiences with the previous activities, make a guess of what you think would be series representations for each of the following.

 - The function $\lambda f+\mu g$.

 - The derivative of f.

 - The anti-derivative of g.

 - The definite integral of f from a to b.

12. In this activity, you will write a **computer function** that represents the sequence of coefficients of a power series for a solution to an initial value differential equation problem.
   You are given an equation, and initial values of the function and of its derivative. The first thing you must do is derive a recursion relation for the coefficients. (See page 430 for a discussion of recursively defined sequences.) This is done by assuming that a function represented by a power series is a solution to the problem. Using the formula you derived for derivatives in Activity 11, you can substitute in the differential equation. Manipulating the resulting power series leads to the recursion relation.
   Then you must write a **computer function** that will accept a positive integer n and return a **computer function** which represents an approximation to the function represented by the sum of the first n terms of the power series.

Do this for the following problem.

$$y''-y=0,\ y(0)=1,\ y'(0)=1$$

The function y which solves this problem is given by

$$y(x) = e^x.$$

Why? Use graphs to compare the graph of your solution with the graph of the actual solution.

6.3.3 Discussion

Radius and Interval of Convergence

In this section, we turn from consideration of series of numbers to series in which the general term is a function. In other words, for each value of the independent variable you get a different series of numbers. This is the idea that you constructed in Activity 3.

The most important examples of series of functions are *power series*. These are series in which the nth term is a function given by an expression of the form,

$$a_n (x - c)^n$$

where c is a number and $(a_n)_{n=0}^{\infty}$ is a sequence of numbers. Note that here we are using the same convention that we used on page 435, that is, $(x - c)^0 = 1$ even if $x - c = 0$, or $x = c$. In other contexts in mathematics and its applications, 0^0 is undefined.

One thing should be very clear to you after working with the activities. The convergence of a series of functions depends on the value of the independent variable. For example, in Activity 2, we hope that you found that the series in Activity 1(c), converges for $x = -1, -0.5, 0.5$, but diverges for $x = -1.5, 1$ and 1.5.

If you tried some other points you would find that series converges when $|x| < 1$ and diverges when $|x| > 1$. In other words, there is an interval, $(-1, 1)$ in this case, such that the series converges for x in this interval and diverges for x outside this interval. The *base point* of this series is the number $c = 0$. The interval of convergence is centered at the base point and extends a distance of 1 on either side. Finally, we check the two endpoints and see that the series converges for $x = -1$ (do you see why?) and diverges for $x = 1$ (why?).

The analysis we just made can be done in pretty much the same way for any power series. Indeed, it is often possible to determine the region of convergence with a simple application of the ratio test. Consider, for example, the series

$$\sum_{n=0}^{\infty} n \left(3x + \frac{1}{2} \right)^n = \sum_{n=0}^{\infty} n 3^n \left(x + \frac{1}{6} \right)^n$$

(note that a coefficient of x does not destroy the general form). What is the base point c here? We can study the convergence of the series of absolute values of the terms,

$$\sum_{n=0}^{\infty} n \left| 3x + \frac{1}{2} \right|^n.$$

Applying the ratio test to this series of positive terms we have,

$$\lim_{n \to \infty} \frac{(n+1) \left| 3x + \frac{1}{2} \right|^{n+1}}{n \left| 3x + \frac{1}{2} \right|^n} = \lim_{n \to \infty} \frac{(n+1)}{n} \left| 3x + \frac{1}{2} \right| = \left| 3x + \frac{1}{2} \right|$$

and so we may conclude that this series is absolutely convergent for $\left| 3x + \frac{1}{2} \right| < 1$ and diverges for $\left| 3x + \frac{1}{2} \right| > 1$.

The endpoints can be checked separately and we see that the series diverges for $3x + \frac{1}{2} = -1, 1$ (why?).

Translating this to an interval, we can say that this series converges for $x \in (-\frac{1}{2}, \frac{1}{6})$ and diverges for any point that is not in this interval.

As we said, this is a completely general situation that works for any power series. We summarize the situation as follows. You will get a chance to (almost) prove the following characterization of the domain of a power series in the exercises.

Domain of a power series

For any power series $\sum_{n=0}^{\infty} a_n (x-c)^n$, there is a number $r \geq 0$ called the *radius of convergence*.

The series converges (absolutely) for $|x-c| < r$ and diverges for $|x-c| > r$.

The series may or may not converge at the endpoints, $x = c - r$ or $x = c + r$.

The interval from $c-r$ to $c+r$ (including those endpoints where the series converges) is called the *interval of convergence*.

Note that it may happen that $r = \infty$, hence the power series converges for all real numbers. Also, when $r = 0$, the interval of convergence is just the base point. We also note that the radius of convergence can usually be found using the ratio test or the root test (see Exercise 39, page 464).

The convergence at the two endpoints must be checked separately and all four possibilities can occur. We have just seen an example in which the series converges at the left endpoint of the interval of convergence and diverges at the right endpoint. We have also just seen an example in which it diverges at both endpoints. Can you find examples (from the activities or from your mind) of the other two possibilities?

Using Convergent Series to Define Functions   Reflect a moment on what is going on with a power series and what you did in Activities 3, 4 and 5. In those activities you produced six **computer functions**. Each **computer function** represents a function. You put in a value for x and, if the series converges for that value, then you get a value out. This is a function, but it only works when the series converges. In other words,

> *The domain of a function defined by a power series is the interval of convergence of the power series.*

Thus the domain of a function defined by a power series is always an interval. It may be open, closed, or half-open. Does it always have to be finite? What about the series in Activity 5? Can the interval of convergence be infinite? What about endpoints in that case? Can the interval of convergence be reduced to a single point? Can it be the empty set?

Does the domain of a function defined by a power series have to be centered at the origin? Can you think of one which is centered somewhere else? Which real numbers can be the center of such an interval?

Okay, so if a power series defines a function, can we discover what that function is? That is, will it always be a familiar function? What did you see in Activity 6? Does the graph look familiar? What about the first series in Activity 1?

You will see in the next section, that just about every function that is familiar to you can be represented as a power series. Could a power series ever give you a completely unfamiliar function? Do you think that any function can be represented as a power series? Suppose you draw, freehand, a curve on a rectangular two-dimensional coordinate system, taking care that your curve passed the vertical line test so it represents a function. Do you think that there is a power series representing a function whose graph is this curve?

In the next section we will look into the question of which functions can be represented as a power series. It is a vast topic and we will only make a beginning.

For the remainder of this section, we take a different point of view. Given a power series, we have its interval of convergence and a function whose domain is that interval. What can we say, regarding the topics we have been studying in this course about that function? Can we combine such functions? Can we take limits? What about derivatives and integrals?

Well, that's a lot of questions. Let's look for some answers.

Combinations of Power Series

Most combinations of power series are quite simple. You worked with this a little in Activity 11. Suppose that you have two power series representing functions,

$$f(x) = \sum_{n=0}^{\infty} a_n (x-c)^n$$

$$g(x) = \sum_{n=0}^{\infty} b_n (x-c)^n$$

and suppose that they have the same radius of convergence r. Then any linear combination of these power series (i.e., sums of constant multiples of power series) will represent the same linear combination of the functions (i.e., sums of constants multiples of the functions) and the new radius of convergence will be at least as large as r. In other words, given any real numbers λ, μ we have,

$$\lambda f(x) + \mu g(x) = \lambda \sum_{n=0}^{\infty} a_n (x-c)^n + \mu \sum_{n=0}^{\infty} b_n (x-c)^n = \sum_{n=0}^{\infty} (\lambda a_n + \mu b_n)(x-c)^n$$

and this last power series has radius of convergence $R \geq r$.

The reason for this is that, as you saw in the last section, (see Theorem 6.1 on page 434) such linear combinations of series will always converge if the original series converge. Moreover, they will converge to the corresponding linear combination of the sums of the original series.

The two power series that you worked with in Activity 10, are linear combinations of power series that you worked with in other activities. Can you see which ones?

Note that we have not said that the new radius of convergence of a linear combination is *equal* to the old radius of convergence; it could be larger so you would get more convergence. In other words, you don't destroy convergence by forming linear combinations, but you might increase it. Can you think of an example in which the radius of convergence of a linear combination could actually be larger?

The situation with products is the same except that the details are more complicated and we must use the Cauchy Product (see Activity 13, Section 6.2, page 445).

$$f(x)g(x) = \sum_{n=0}^{\infty} d_n (x-c)^n = \sum_{n=0}^{\infty} \left(\sum_{i=0}^{\infty} a_i b_{n-i} \right)(x-c)^n$$

In other words,

$$d_n = \sum_{i=0}^{\infty} a_i b_{n-i} = a_0 b_n + a_1 b_{n-1} + \cdots + a_{n-1} b_1 + a_n b_0$$

In case of the quotient, there is no simple formula.

Limits of Functions Defined by Power Series

We are going to study what are called the "analytic properties" of functions defined by power series. That means limits, derivatives and integrals. There are some really powerful and useful results here that have many applications throughout mathematics and science. The main point is that once you have functions represented as power series, then it is very easy to calculate derivatives and integrals — provided you are satisfied with a power series representation for your answer. We will even see at the end of this section that power series can be used to solve differential equations.

The reasons behind the facts about analytic properties of power series lie very deep, and their proofs are too difficult for the level of this course. But the calculations are well within your reach. In each case there is a theoretical fact from which straightforward calculations can be made. In general, the form of the theoretical fact is that calculations about a power series can be made by making the calculations term-by-term.

Theorem 6.14

A function defined by a power series is continuous at every point in its interval of convergence.

What this means is that in order to find the limit as, say, x approaches a, if a is in the interval of convergence, then it is enough to plug in the value of a for x in each term and evaluate the series (if possible). Notice that this includes the endpoints of the interval of convergence, although in this case the limit has to be taken from one side.

This may seem like a small thing, replacing a limit with an evaluation, but let's look at what it does for us in something like Activity 9. We are trying to compute

$$\lim_{x \to 0} \frac{g(x)}{f(x) - 1 + 2x}$$

where

$$f(x) = \sum_{n=0}^{\infty} (-1)^n \frac{x^{2n}}{(2n)!}$$

and

$$g(x) = \sum_{n=0}^{\infty} (-1)^n \frac{x^{2n+1}}{(2n+1)!}$$

Let's do some calculation. It can be hard to follow what is going on so you should take some paper and replace these big summations with the sums of the first few terms until you get familiar with the situation.

$$\lim_{x\to 0}\frac{g(x)}{f(x)-1+2x}=\lim_{x\to 0}\frac{\sum_{n=0}^{\infty}(-1)^{n}\frac{x^{2n+1}}{(2n+1)!}}{\sum_{n=0}^{\infty}(-1)^{n}\frac{x^{2n}}{(2n)!}-1+2x}$$

$$=\lim_{x\to 0}\frac{\sum_{n=0}^{\infty}(-1)^{n}\frac{x^{2n+1}}{(2n+1)!}}{\sum_{n=1}^{\infty}(-1)^{n}\frac{x^{2n}}{(2n)!}+2x}$$

$$=\lim_{x\to 0}\frac{\sum_{n=0}^{\infty}(-1)^{n}\frac{x^{2n+1}}{(2n+1)!}}{\sum_{n=0}^{\infty}(-1)^{n+1}\frac{x^{2n+2}}{(2n+2)!}+2x}$$

$$=\lim_{x\to 0}\frac{x\sum_{n=0}^{\infty}(-1)^{n}\frac{x^{2n}}{(2n+1)!}}{x\left(\sum_{n=0}^{\infty}(-1)^{n+1}\frac{x^{2n+1}}{(2n+2)!}+2\right)}$$

$$=\lim_{x\to 0}\frac{\sum_{n=0}^{\infty}(-1)^{n}\frac{x^{2n}}{(2n+1)!}}{\sum_{n=0}^{\infty}(-1)^{n+1}\frac{x^{2n+1}}{(2n+2)!}+2}$$

Now this last expression is the quotient of two power series. We know that the limit of a quotient is the quotient of the limits, provided the denominator is not 0. Also, Theorem 6.14 above lets us compute the limit of the numerator and of the denominator by just replacing x by 0. If we do this, every term of the series in the numerator is 0 except the first which is 1 and every term in the series in the denominator is 0, so the only thing that is left is 2. Hence,

$$\lim_{x\to 0}\frac{g(x)}{f(x)-1+2x}=\frac{1}{2}.$$

Is that what you got by experimenting?

You may find the calculations with series above somewhat technical and difficult to follow. One technique that can help with understanding such calculations is to write out all of the manipulations in two forms. One uses the formal summation notation as above and the other carries through the same calculations, but using only the first few terms of the series. Let us repeat the above calculations using this second method.

$$\lim_{x\to 0}\frac{g(x)}{f(x)-1+2x}=\lim_{x\to 0}\frac{x-\frac{x^3}{3!}+\frac{x^5}{5!}-\frac{x^7}{7!}\cdots}{1-\frac{x^2}{2!}+\frac{x^4}{4!}-\frac{x^6}{6!}\cdots-1+2x}$$

$$=\lim_{x\to 0}\frac{x-\frac{x^3}{3!}+\frac{x^5}{5!}-\frac{x^7}{7!}\cdots}{2x-\frac{x^2}{2!}+\frac{x^4}{4!}-\frac{x^6}{6!}\cdots}$$

$$=\lim_{x\to 0}\left(\frac{1-\frac{x^2}{3!}+\frac{x^4}{5!}-\frac{x^6}{7!}\cdots}{2-\frac{x}{2!}+\frac{x^3}{4!}-\frac{x^5}{6!}\cdots}\right)$$

$$=\frac{1-\lim_{x\to 0}\left(\frac{x^2}{3!}+\frac{x^4}{5!}-\frac{x^6}{7!}\cdots\right)}{2-\lim_{x\to 0}\left(\frac{x}{2!}+\frac{x^3}{4!}-\frac{x^5}{6!}\cdots\right)}$$

$$=\frac{1}{2}$$

Derivatives of Functions Defined by Power Series

The situation for derivatives is the same as for limits except that we must avoid the endpoints of the interval. In the interior of the interval of convergence, you can calculate the power series for the derivative by differentiating the terms

$a_n(x-c)^n$ of the series one by one. However, you must check for convergence at the endpoints for both the function and the derivative separately.

Theorem 6.15

Let $f(x) = \sum_{n=0}^{\infty} a_n(x-c)^n$ represent a function by a power series whose radius of convergence is $r > 0$. Then f is differentiable on the open interval $(c-r, c+r)$ and its derivative is represented by a power series with the same radius of convergence. Moreover, we have the formula for the derivative, f' of f,

$$f'(x) = \sum_{n=1}^{\infty} na_n(x-c)^{n-1}.$$

This is one of the formulas you were searching for in Activity 11. With this theorem, you should be able to answer part of the question in Activity 10, by calculating and comparing.

Examples of using this theorem to calculate derivatives will be given in the exercises.

Integrals of Functions Defined by Power Series

Once you have a formula for the derivative, then, applying the Fundamental Theorem of Calculus you can easily work out the anti-derivative and evaluate at the endpoints to obtain the formula in the following theorem.

Theorem 6.16

Let $f(x) = \sum_{n=0}^{\infty} a_n(x-c)^n$ represent a function by a power series whose radius of convergence is $r > 0$. Then the anti-derivatives of f all have the same radius of convergence and they are given by the formula

$$\int f(x)dx = \sum_{n=0}^{\infty} \frac{a_n}{n+1}(x-c)^{n+1} + K$$

where K is an arbitrary constant. Moreover, if α and β are in the interval of convergence, then

$$\int_{\alpha}^{\beta} f(x)dx = \sum_{n=0}^{\infty} \frac{a_n}{n+1}\left((\beta-c)^{n+1} - (\alpha-c)^{n+1}\right)$$

This is another one of the formulas you were searching for in Activity 11. With this theorem, you should be able to complete your answer to the question in Activity 10, by calculating and comparing. For indefinite integrals you must check for convergence at the endpoints for both the function and its anti-derivative separately.

Examples of using this theorem to calculate integrals will be given in the exercises.

Series Solutions of Differential Equations

We will run through the method that is outlined in Activity 12. Here are the conditions that the unknown function y must satisfy.

$$y'' - y = 0, \ y(0) = 1, \ y'(0) = 1$$

We begin by assuming that y has the form,

$$y(x) = \sum_{n=0}^{\infty} a_n(x-c)^n = \sum_{n=0}^{\infty} a_n x^n$$

where we have chosen the base $c = 0$ because the initial conditions are given at $x = 0$.

We can use the formula we obtained above in Theorem 6.15 for the derivative of a function given by a power series. Applying it twice we obtain,

$$y''(x) = \sum_{n=2}^{\infty} n(n-1)a_n x^{n-2} = \sum_{n=0}^{\infty} (n+2)(n+1)a_{n+2} x^n$$

and plugging this into the equation, we obtain,

$$\sum_{n=0}^{\infty} (n+2)(n+1)a_{n+2} x^n - \sum_{n=0}^{\infty} a_n x^n = 0$$

and, simplifying the left-hand side,

$$\sum_{n=0}^{\infty} \left((n+2)(n+1)a_{n+2} x^n - a_n x^n \right) = 0$$

or

$$\sum_{n=0}^{\infty} ((n+2)(n+1)a_{n+2} - a_n) x^n = 0$$

Now we come to a point that really deserves more attention than we will give it. This last equation says that a power series is equal to the zero function. In fact, this can only happen if every coefficient of the power series is zero. Can you imagine why this might be the case? Try to think about it and you will have an explanation in the next section. (See page 486.)

Accepting that fact for now, we can write, for $n = 0, 1, 2, \ldots$

$$(n+2)(n+1)a_{n+2} - a_n = 0$$

or

$$a_{n+2} = \frac{a_n}{(n+2)(n+1)}$$

which is the desired recursion relation.

Now you can go ahead and write the **computer function** requested in Activity 12. You may want to review the discussion of recursively defined sequences in Section 1, p. 429.

Actually, in this case, the situation is sufficiently simple that we can work it out. It is convenient to calculate the even and odd indexed terms separately. If n is even we have, using the given value for a_0,

$$a_n = \frac{a_{n-2}}{n(n-1)}$$

$$= \frac{a_{n-4}}{n(n-1)(n-2)(n-3)} = \cdots = \frac{a_2}{n(n-1)(n-2)(n-3)\cdots 3}$$

$$= \frac{a_0}{n!} = \frac{1}{n!}$$

and a similar argument yields the same value for a_n if n is odd. Hence we conclude that

$$y(x) = \sum_{n=0}^{\infty} \frac{x^n}{n!}$$

and you should compare function values, or the graph of this function, with the function given by the expression e^x. You will get a chance to do this in the exercises.

6.3.4 Exercises

1. Consider the series

$$\sum_{n=0}^{\infty} n\left(3x+\frac{1}{2}\right)^n = \sum_{n=0}^{\infty} n3^n\left(x+\frac{1}{6}\right)^n.$$

  (a) What is the base point c for this series?

  (b) Explain why the series diverges for $\left|3x+\frac{1}{2}\right|=1$.

2. For each of the following power series,

  (a) Find the radius of convergence.

  (b) Find the interval where the power series converges absolutely.

  (c) Find the interval of convergence.

    i. $\displaystyle\sum_{n=0}^{\infty} x^n$ ii. $\displaystyle\sum_{n=0}^{\infty} nx^n$

    iii. $\displaystyle\sum_{n=1}^{\infty} \frac{x^n}{n}$ iv. $\displaystyle\sum_{n=1}^{\infty} \frac{x^n}{n^2}$

    v. $\displaystyle\sum_{n=1}^{\infty} \frac{x^n}{\sqrt{n}}$ vi. $\displaystyle\sum_{n=0}^{\infty} n!\,x^n$

    vii. $\displaystyle\sum_{n=0}^{\infty} \frac{x^n}{(2n+1)!}$

    viii. $\displaystyle\sum_{n=0}^{\infty} \frac{(x-1)^n}{3^n}$

    ix. $\displaystyle\sum_{n=1}^{\infty} (x-1)^n$

    x. $\displaystyle\sum_{n=1}^{\infty} (2x)^n$

    xi. $\displaystyle\sum_{n=1}^{\infty} (3x+5)^n$

    xii. $\displaystyle\sum_{n=0}^{\infty} \frac{n!\,x^n}{1\cdot4\cdot7\cdots(3n+1)}$

    xiii. $\displaystyle\sum_{n=1}^{\infty} n^2(x+1)^n$

    xiv. $\displaystyle\sum_{n=0}^{\infty} \frac{(3x)^n}{4^{n+1}}$

    xv. $\displaystyle\sum_{n=0}^{\infty} \frac{1^2\cdot2^2\cdot3^2\cdots n^2 x^n}{1^2\cdot3^2\cdot5^2\cdots(2n+1)^2}$

    xvi. $\displaystyle\sum_{n=1}^{\infty} \left(\frac{x}{3}\right)^n$

    xvii. $\displaystyle\sum_{n=0}^{\infty} \frac{n!\,x^n}{(n+1)2^n}$

    xviii. $\displaystyle\sum_{n=2}^{\infty} \frac{x^n}{\ln(n)}$

    xix. $\displaystyle\sum_{n=1}^{\infty} \frac{(-1)^{n+1}x^n}{(3n)!}$

    xx. $\displaystyle\sum_{n=1}^{\infty} \frac{n+1}{2n}x^n$

    xxi. $\displaystyle\sum_{n=0}^{\infty} \frac{x^n}{1+3^n}$

    xxii. $\displaystyle\sum_{n=0}^{\infty} (-1)^n \frac{x^{2n}}{2^n}$

    xxiii. $\displaystyle\sum_{n=1}^{\infty} \frac{2\cdot4\cdot6\cdots(2n)x^n}{(n+1)!}$

    xxiv. $\displaystyle\sum_{n=0}^{\infty} (-1)^{n+1} \frac{(x-2)^n}{n\sqrt{n}}$

    xxv. $\displaystyle\sum_{n=1}^{\infty} \frac{(3-x)^n}{n2^n}$

    xxvi. $\displaystyle\sum_{n=1}^{\infty} \frac{(2x+1)^n}{n}$

3. Find the domain of the function f represented by each of the following power series.

  (a) $f(x) = \displaystyle\sum_{n=1}^{\infty} \frac{x^{2n}}{n^2\sqrt{n}}$

  (b) $f(x) = \displaystyle\sum_{n=0}^{\infty} \frac{2^n(2x-3)^n}{2^n+5}$

  (c) $f(x) = \displaystyle\sum_{n=1}^{\infty} \frac{(x+1)^n}{n^n}$

  (d) $f(x) = \displaystyle\sum_{n=1}^{\infty} \frac{n^n(2x+1)^n}{n!}$

  (e) $f(x) = \displaystyle\sum_{n=1}^{\infty} \frac{(-3)^n n!(x+1)^n}{3\cdot5\cdot7\cdots(2n+1)}$

  (f) $f(x) = \displaystyle\sum_{n=1}^{\infty} x^{n^2}$

(g) $f(x) = \sum_{n=0}^{\infty} x^{(2n)!}$

(h) $f(x) = \sum_{n=1}^{\infty} \frac{2^n(x-1)^n}{n}$

(i) $f(x) = \sum_{k=1}^{\infty} \frac{(x-1)^k}{k}$

4. Find the following limits. Write out the first few terms of the series to see what is happening.

(a) $\lim_{x \to 0} \dfrac{x(1-f(x))}{1-g(x)}$ where $f(x) = \sum_{n=0}^{\infty} \dfrac{x^n}{n!}$

and $g(x) = \sum_{n=0}^{\infty} (-1)^n \dfrac{x^{2n}}{(2n)!}$

(b) $\lim_{x \to 0} \dfrac{x^n}{h(x)}$ where $h(x) = \sum_{n=0}^{\infty} \dfrac{2^n x^n}{n!}$

(c) $\lim_{x \to 0} \left(\dfrac{1}{1-f(x)} - \dfrac{2}{x^2} \right)$ where

$f(x) = \sum_{n=0}^{\infty} (-1)^n \dfrac{x^{2n}}{(2n)!}$

(d) $\lim_{x \to 0} \dfrac{x - f(x)}{2x^3}$ where

$f(x) = \sum_{n=0}^{\infty} (-1)^n \dfrac{x^{2n+1}}{(2n+1)!}$

5. For the function f defined by

$$f(x) = \sum_{n=1}^{\infty} \frac{x^n}{n^2}$$

find

(a) The radius of convergence and the interval for the power series which defines f.

(b) The radius of convergence and the interval for the power series which defines f'. Compare with your answer in part (a).

(c) The radius of convergence and the interval for the power series which defines the indefinite integral of f.

Compare with your answers in parts (a) and (b).

6. For the function g defined by

$$g(x) = \sum_{n=0}^{\infty} \frac{(-1)^n x^n}{(n+1)2^n}$$

find

(a) The radius of convergence and the interval for the power series which defines g.

(b) The radius of convergence and the interval for the power series which defines g'. Compare to your answer in part (a).

(c) The radius of convergence and the interval for the power series which defines the indefinite integral of g. Compare to your answers in parts (a) and (b).

7. For each of the following power series, find

(a) the derivative

(b) the indefinite integral

(c) the definite integral on the indicated interval.

i. $f(x) = \sum_{n=0}^{\infty} \dfrac{x^n}{n!}$, $[0, 1]$

ii. $g(x) = \sum_{n=0}^{\infty} (-1)^n \dfrac{(x+1)^n}{n!}$, $[0, -1]$

iii. $f(x) = \sum_{n=0}^{\infty} \dfrac{(1+2x)^n}{n!}$, $[-0.5, 0]$

iv. $f(x) = \sum_{n=0}^{\infty} (-1)^n \dfrac{x^n}{n!}$, $[0, 2]$

v. $h(x) = \sum_{n=0}^{\infty} \dfrac{2^n x^n}{n!}$, $[0, 1]$

vi. $f(x) = \sum_{n=0}^{\infty} (-1)^n \dfrac{x^{2n}}{(2n)!}$, $[0, \pi]$

vii. $f(x) = \sum_{n=0}^{\infty} (-1)^n \frac{x^{2n+1}}{(2n+1)!}$,

[0, 1]

8. Find a power series solution for each of the following differential equation problems. What is the interval of convergence for the power series solution of each problem?

   (a) $y' + y = 0$, $y(0) = 1$

   (b) $y' - 2y = 0$, $y(0) = 1$

   (c) $y' - y = 0$, $y(1) = 0$

   (d) $y'' + y = 0$, $y'(0) = 1$, $y(0) = 0$

   (e) $y'' + y = 0$, $y'(0) = 0$, $y(0) = 1$

   (f) $y'' + 4y = 0$, $y'(0) = 0$, $y(0) = 1$

   (g) $y'' + xy = 0$, $y'(0) = 0$, $y(0) = 1$

   (h) $y'' + xy = 0$, $y'(0) = 2$, $y(0) = 0$

   (i) $y'' + 2y' + y = 0$, $y'(0) = 1$, $y(0) = 0$

9. The convergence at the two endpoints of the interval of convergence of a power series

$$\sum_{n=0}^{\infty} a_b (x-c)^n$$

   must be checked separately, and all four possibilities can occur for the interval of convergence: $[c-r, c+r]$, $(c-r, c+r)$, $(c-r, c+r]$, and $[c-r, c+r)$. Find examples (from the activities or from your mind) of power series for each of the four possibilities.

10. For a function defined by a power series, can we discover what that function is? That is, will it always be a familiar function? What did you see in Activity 6, page 467? Does the graph look familiar? What about the first series in Activity 1, page 466? Write a brief essay of your thoughts on these questions.

11. The two power series that you worked with in Activity 10, page 468, are linear

combinations of power series that you worked with in other activities. Which two power series are a linear combination of power series in other activities? Explain your answer.

12. The new radius of convergence of a linear combination of two power series could be larger than that of each power series, so you would get a larger interval of convergence for the linear combination. In other words, you don't destroy convergence by forming linear combinations, but you might increase it. Give an example in which the radius of convergence of a linear combination of two (or more) power series is larger than the radius of convergence of each series.

13. What can you say about the radius of convergence of the derivative of a power series in terms of the radius of convergence of the original series? What about the antiderivative?

14. Prove that if $\lim_{n \to \infty} \frac{a_{n+1}}{a_n} = \rho$ and r is the radius of converence of the power series $\sum_{n=0}^{\infty} a_n (x-c)^n$, then

   (a) $r = \dfrac{1}{\rho}$, if $\rho \neq 0$

   (b) $r = \infty$, if $\rho = 0$

   (c) $r = 0$, if $\rho = \infty$

15. Show that for all real numbers x and any positive integer k the following sequence converges and find its limit.

$$\left(\frac{x^n}{n!} \right)_{n=k}^{\infty}$$

16. Show that if the radius of convergence of the power series $\sum_{n=0}^{\infty} a_n x^n$ is $r = \infty$, then $\lim_{n \to \infty} a_n (x-c)^n = 0$, for all real numbers c.

17. Find the following limit, if possible.

$$\lim_{x \to 0} \sum_{n=0}^{\infty} \frac{x^2}{\left(1 + x^2\right)^n}$$

18. Find the interval of convergence of the following series.

$$\sum_{n=0}^{\infty} \frac{1}{2n+1} \left(\frac{x-5}{x-3} \right)^n$$

19. In this exercise you will (almost) prove the characterization of the domain of a power series on page 470.

    (a) Prove the following:

    > If a power series $\sum_{n=0}^{\infty} a_n (x - c)^n$ converges at a point x_1 then the power series is absolutely convergent for $|x - c| < |x_1 - c|$.

    Do so as follows.

    i. Explain why
       $\lim_{n \to \infty} a_n (x_1 - c)^n = 0$.

    ii. Using part i, explain why there is a number N so that $\left| a_n (x_1 - c)^n \right| < 1$ for $n \geq N$.

    iii. Using part ii, explain why it follows that

    $$\left| a_n (x - c)^n \right| = \left| a_n (x_1 - c)^n \right| \left(\frac{|x - c|}{|x_1 - c|} \right)^n$$

    $$< \left(\frac{|x - c|}{|x_1 - c|} \right)^n$$

    for $n \geq N$.

    iv. Explain why $\frac{|x-c|}{|x_1-c|} = r < 1$.

    v. Now use the comparison test to compare the original power series with a geometric series $\sum_{n=0}^{\infty} ar^n$ to finish the proof.

    (b) Explain why the following is an immediate consequence of part (a).

    > If a power series $\sum_{n=0}^{\infty} a_n (x - c)^n$ diverges at a point x_2 then the power series diverges for $|x - c| > |x_2 - c|$.

    (c) At $x = c$ the power series $\sum_{n=0}^{\infty} a_n (x - c)^n$ converges. Why?

    (d) If there is point $x_1 \neq c$ at which the power series $\sum_{n=0}^{\infty} a_n (x - c)^n$ converges and a point x_2 at which the power series diverges, then the power series does not converge for all real numbers x. Use parts (a) and (b) to show that the power series $\sum_{n=0}^{\infty} a_n (x - c)^n$ converges absolutely for all x with $|x - c| < |x_1 - c|$ and it is divergent for all x with $|x - c| > |x_2 - c|$. Show that $|x_1 - c| \leq |x_2 - c|$. Let x_3 be the number given by

    $$x_3 = \frac{|x_1 - c| + |x_2 - c|}{2}.$$

    Make a sketch illustrating the relationship between x_1, x_2, and x_3. If the series $\sum_{n=0}^{\infty} a_n (x - c)^n$ converges at x_3, then the series is absolutely convergent for $|x - c| < |x - x_3|$. Why? If the series is divergent for $x = x_3$, then it is divergent for $|x - c| > |x_3 - c|$. Why? Continue to bisect the distance between the interval of absolute convergence and the intervals of divergence in this manner to obtain a sequence (x_n) that converges to a number $r \geq 0$ which is the radius of convergence. Except for the justification of the last statement (which requires knowledge about the *completeness property* of the real numbers) the proof is done.

▌6.4 SERIES REPRESENTATIONS OF FUNCTIONS

6.4.1 Overview

In this section you will learn how to represent functions as power series, better known as *Taylor series*. Power series representations of functions can be used to approximate a function by a polynomial function, that is, a partial sum of a power series representation, called a *Taylor polynomial*. In effect, you will be able to use a "simple" polynomial function to replace a more complicated function. Power series representations are useful in making various approximations, including estimates of function values, limits of functions, and definite integrals (especially of functions which have no elementary functions as an anti-derivative). You will also investigate the error made when using a power series to make an approximation.

6.4.2 Activities

1. In this activity, you will work with two lists: a list of series and a list of functions. Take each of these series and produce a **computer function** that will represent a function which has a certain domain, so there should be a domain check. (You may use 50 terms in all cases.) Your task is to evaluate these functions at various points (for example, using **table**) and do the same for the functions in the second list and make match-ups where possible. When there is no match-up, try to guess what *would* match the series or the function.

   The series are the nine series that you worked with in Activities 1, 5, 9, and 10 beginning on page 466 in Section 3. For your convenience, we list the nine series below.

   (a) $\displaystyle\sum_{n=0}^{\infty} x^n$ (b) $\displaystyle\sum_{n=0}^{\infty} nx^{n-1}$ (c) $\displaystyle\sum_{n=0}^{\infty} \frac{x^{n+1}}{n+1}$ (d) $\displaystyle\sum_{n=0}^{\infty} \frac{x^n}{n!}$ (e) $\displaystyle\sum_{n=0}^{\infty} (-1)^n \frac{x^{2n}}{(2n)!}$

   (f) $\displaystyle\sum_{n=0}^{\infty} n! x^n$ (g) $\displaystyle\sum_{n=0}^{\infty} (-1)^n \frac{x^{2n+1}}{(2n+1)!}$ (h) $\displaystyle\sum_{n=0}^{\infty} \left(\frac{x^n}{n!} + (-1)^n \frac{x^{2n}}{(2n)!} \right)$

   (i) $\displaystyle\sum_{n=0}^{\infty} \left(23.684(-1)^n \frac{x^{2n+1}}{(2n+1)!} + 36.7432(-1)^n \frac{x^{2n}}{(2n)!} \right)$

   Here is the list of definitions of functions.

   i. $r(s) = \cos(s)$ ii. $f(a) = -\dfrac{1}{(1-a)^2}$ iii. $h(u) = \ln(1-u)$

   iv. $f(w) = -\cos(w)$ v. $g(t) = \dfrac{1}{t}$ vi. $f(x) = e^x$

   vii. $f(x) = 0$ viii. $h(x) = e^x + \cos(x)$ ix. $q(x) = \sin(x)$

2. In Section 1, Activity 10(a), page 424 you worked out a closed form expression for a sequence of numbers that could be computed, given a differentiable function. In this activity you will use the following modification of **taylco** from Activity 10. Here is the **ISETL** code. If you are using another MPL, then your instructor will provide the appropriate code.

```
taylco := func(f, c, n);
        if n = 0 then return f(c);
        else return (taylco(D(f), c, n-1))/n;
        end;
    end;
```

Compare the above **func** with the **func taylco** you used in Activity 10, page 424. Do you see what the differences are between the two versions of **taylco**? Do you see how the modification above will be useful?

Using the above modified **func taylco**, write a **computer function taylpol** that represents (approximately) a function which is the sum of the terms from 0 to n of the power series with base point c and coefficients coming from the expression you worked out in Activity 10(a), page 424.

Apply **taylpol** to the function f given by $f(x) = 1 + 2x + 3x^2 + 4x^3 + 5x^4 + 6x^5$ from Activity 10, page 424 with $c = 0$. Plot the resulting function and f on the same graph, using various values of n and of domain intervals centered at $x = 0$.

How would you describe what happened as both n and the domain interval get larger?

Now do the same thing for the function g, given by $g(x) = \frac{1}{x}$, $x \neq 0$, with base point $c = 1$. What restriction must there be on the domain interval?

3. Repeat the previous activity for the following functions using the base point 0. Plot, on a single graph, the original function and the function that you obtain. You only need use a single value of n. In each case, you decide, based on what happened in Activity 2, the smallest value of n that will give you the most interesting answer and use that one. Here is the list of definitions of functions.

   (a) $r(s) = \cos(s)$　　　(b) $f(a) = -\dfrac{1}{(1-a)^2}$　　　(c) $h(u) = \ln(1-u)$

   (d) $f(x) = e^x$　　　(e) $f(x) = 0$

4. Repeat Activity 3 above with the function f defined as follows.

$$f(x) = \begin{cases} e^{-1/x^2} & \text{if } x \neq 0 \\ 0 & \text{if } x = 0 \end{cases}$$

Can you figure out what happened and why?

5. Write a **computer function Rem** which will be used in the main portion of this activity. **Rem** accepts a **computer function** which represents a function f, numbers c, r determining an interval $[c-r, c+r]$ and a non-negative integer n. **Rem** returns an approximation, on the domain $[c-r, c+r]$, of the function R_n given by

$$R_n = \int_x^c \frac{f^{(n+1)}(t)}{n!} (x-t)^n \, dt$$

(You may use any of your **Riem functions** for the integral.)

Now, the main part of this activity is to write a **computer function** that will accept a **computer function** which represents a function f, numbers c, r determining an interval $[c-r, c+r]$ and a small positive number *eps*. Your **computer function** is to return a **computer function** which represents the function obtained from **taylpol**. The choice of n, the number of terms, is determined as follows.

Your **computer function** must include code that does something like the following (the sample syntax shown here should not be very different from what is used in your **MPL**)

$$
\begin{aligned}
&\textbf{n := 1;}\\
&\textbf{while |Rem(f, n, r, c)(x)| > eps do}\\
&\quad \textbf{n := n+1;}\\
&\textbf{end;}
\end{aligned}
$$

and after the **computer function** runs this code, n will have a good value.

Caution: If you allow your **computer function** to be given a value at which the original series does not converge, then you could get into an infinite loop.

Run your **computer function** on the functions and basepoint defined by the following. Use **eps=0.0001**. Decide on your own a reasonable value for **r** in each case.

(a) $\sin(2x),\ c = 0$ \qquad (b) $\frac{1-\cos(x)}{x^2},\ c = 0$ \qquad (c) $\ln(x),\ c = 1$

6. Write a **computer function powdiff** which accepts a representation of a function f, a base point c and a radius r. Your **computer function** should return a **computer function** which is a representation of the derivative of f. Do this by first applying what you wrote for **taylco** in Activity 2, (choose for yourself the number of terms to take, either by guessing or applying the **func** of Activity 5 above) and then using the formula in Theorem 6.15, page 474.

Put in a domain test to make sure that the **computer function** which your **computer function** produces is restricted to the appropriate interval of convergence.

In general, what is the difference between choosing the number of terms on your own and using Activity 5 above?

Apply your **computer function** to the function given by $\frac{1}{\sqrt{1-x^2}}$ with base point $c = 0$ and radius of convergence $r = 1$. Plot the derivative of this function and the approximation you obtained on a single graph.

7. Repeat the previous problem for the anti-derivative. Your **computer function** will take one more input which represents the value of the anti-derivative at the base point. Also, instead of Theorem 6.15, you will use Theorem 6.16, page 474. Apply your **computer function** to the function given by $\frac{1}{1-x^2}$ with base point $c = 0$ and radius of convergence $r = 1$. Plot the anti-derivative of this function and the approximation you obtained on a single graph.

8. The purpose of this activity is to test the method of finding derivatives and integrals of series by differentiating or integrating term-by-term.
Use this approach to calculate the derivative and anti-derivative of the function f given by the following series (using the value 0 for $x = 0$ in the anti-derivative.)

$$ f(x) = \sum_{n=1}^{\infty} \frac{\sin(n^2 x)}{n^2} $$

Compare what you get with the result obtained from **D** for the derivative and the modified form of **DefInt** for the integral.

Make your comparison either by using a table of values or graphs.

For what values of x does the original series converge? What happens with the derivative? The anti-derivative? Can you offer any explanation of what you are observing?

9. Obtain the first five terms of the powers series expansion about $c = 0$ for the function f given by

$$f(x) = \frac{x}{2 - 3x + x^2}.$$

You are to do this in two ways. First, calculate these terms by simply dividing the polynomial in the numerator by the polynomial in the denominator using long division. The other method is to apply **taylco**.

Compare the two approximations (obtained by using the power series with these coefficients) with each other and with f by using **table** and your graphics tools.

10. Write a **computer function binexp** which will take a real number α and return a **computer function** that will represent the sequence of coefficients of the Maclaurin series expansion (i.e., the expansion about 0) of the function given by $(1 + x)^\alpha$.

   (a) Use **binexp** to estimate

   $$1.02^{\sqrt{2}}$$

   correct to five decimal places.

   (b) Using your solution to Activity 10 (a), page 424, derive a formula for the Maclaurin series expansion of $(1 + x)^\alpha$.

   (c) Using your formula, work out the first three non-zero terms of the expansion for the following two functions.

   $$\sqrt{1 + 0.01 x^2}$$
   $$\sqrt{4 - x^2}$$

11. Calculate the following in two ways.

   $$\lim_{x \to 0} \frac{1 - \cos(2x)}{x^2}$$

   First, apply your **func lim** from Chapter 3, (See Chapter 3, Exercise 1, page 142.) Second, write out the series expansion (around 0) for the numerator, divide out the denominator and apply Theorem 6.14, p. 472 to get the limit.

6.4.3   Discussion

A Function and its Power Series   We begin our discussion with a review and then a somewhat subtle point. You have worked a great deal in this chapter with power series. You can think of a power series as a separate entity of its own. It looks like

$$\sum_{n=0}^{\infty} a_n (x-c)^n .$$

This series converges for x in the interval of convergence and so it defines a function say, f. If we give the name I to the interval of convergence we can write

$$f(x) = \sum_{n=0}^{\infty} a_n (x-c)^n, \quad x \in I$$

Thus any power series defines a function. Its domain is the interval of convergence.

It is nice when a function is given by a power series. In many ways, it is not very different from a function given by a polynomial. The domain is clear and you have a formula for at least approximating values of the function. Also, there are formulas for derivatives, anti-derivatives and definite integrals. Functions given by power series are relatively easy to work with.

So the question arises, if you have a function, can you find a power series that gives it, that is, can you *represent a given function as a power series*? This is where the subtlety is going to come in.

First of all, it is not hard to find a power series, given a function. All you need to know is that the function is differentiable to all orders at your chosen base point. In Section 1, Activity 10, page 424, you derived the following formula for a sequence of numbers, (the symbol $f^{(n)}$ stands for the nth derivative of f.)

$$a_n = \frac{f^{(n)}(0)}{n!}$$

and here in this section, in Activity 2, you used **taylco** to figure out the terms of a power series, given a function. This series is written

$$\sum_{n=0}^{\infty} \frac{f^{(n)}(0)}{n!} x^n .$$

It is called the *Maclaurin series* for f. If we change the base point to c, we write

$$\sum_{n=0}^{\infty} \frac{f^{(n)}(c)}{n!} (x-c)^n$$

and this is called the *Taylor series* for f at c.

Now we come to the subtlety. Notice that we never wrote an expression like

$$f(x) = \sum_{n=0}^{\infty} \frac{f^{(n)}(0)}{n!} (x-c)^n$$

that is, we have not said that the value of the function at a point is *equal* to the value of its Taylor series at that point. This is what you were working with in Activities 2, 3, and 4. Here is what you should have found.

In all of the examples in Activities 2 and 3, the values of each function were pretty close to the values of the partial sums of its Taylor series. The agreement was *very* good near the base point and began to deteriorate as you moved away. The situation improved if you increased the number of terms in the partial sums.

Was this the case in Activity 4? The power series that you got is a little strange and you might not have thought it is a real power series. But it is. What function does it define? Do you see how different that is from the function that gave rise to the series?

So here is the situation. Any function whose domain includes some interval centered at c and has all its derivatives at every point in this interval has a power series with base c. We will see in Theorem 6.17 that it is even unique (that is there is only one). The power series defines a function whose domain is the interval of convergence I. Thus you have two functions. In most cases they give exactly the same values on the interval of convergence. They are the same function and we can write,

$$f(x) = \sum_{n=0}^{\infty} \frac{f^{(n)}(c)}{n!}(x-c)^n, \ x \in I$$

This is true in most cases, but in some "pathological" examples such as the one in Activity 4 it may not be true.

In the next paragraphs we will formalize all this, prove some (but not all) of the facts, go over once again how you can work with derivatives and integrals of power series, and finally, you will see a few techniques and examples.

Finding the Power Series of a Function

Suppose a function, f, is equal to some power series based at some point. That is, we have

$$f(x) = \sum_{n=0}^{\infty} a_n (x-c)^n, \ x \in I$$

We can use Theorem 6.15, p. 474 to derive a formula for the a_n, $n = 0, 1, \dots$. First of all, it is immediate that

$$f(c) = a_0$$

Now use the formula in Theorem 6.15 to calculate f'.

$$f'(x) = \sum_{n=1}^{\infty} n a_n (x-c)^{n-1}, \ x \in I$$

and evaluating at c gives

$$f'(c) = a_1$$

because every term except the first is 0. Do it again.

$$f''(x) = \sum_{n=2}^{\infty} n(n-1) a_n (x-c)^{n-2}, \ x \in I$$

so

$$f''(c) = 2a_2$$

Can you see that next time we get

$$f'''(c) = 6a_3$$

Indeed, if you keep on going, you get

$$f^{(n)}(c) = n! a_n$$

or

$$a_n = \frac{f^{(n)}(c)}{n!}, \ n = 0, 1, 2, \dots$$

This is the formula you should have come up with in Section 1, Activity 10 (a), p. 424. The above formula for a_n is known as *Taylor's formula* for the nth coefficient of the power series expansion of the function f about c.

What we have done is to prove the following "uniqueness theorem".

Theorem 6.17 (Uniqueness of power series)

If f is a function which is equal to a power series based at c, then this power series must be the Taylor series of f at c.

Thus we see that the Taylor series is "the only game in town." If you are trying to represent a function as a power series, then your only chance is its Taylor series. You can compute the Taylor series by taking derivatives (or using **taylco**) — indeed you have done quite a few such computations in this chapter.

Do you see what happens if the original function f is the zero function? What do you know about the coefficients in this case? Do you see how this relates to the point that was discussed on p. 475 in connection with series solutions of differential equations?

An Example

Because of Theorem 6.17, it is not always necessary to go back to the definition and work very hard to calculate the Taylor series of a function. For example, consider finding the Maclaurin series (Taylor series with base $c = 0$) for the function f given by $f(x) = \sin(2x)$.

We begin by finding the Maclaurin series for the function represented by $\sin x$. For this one we do have to go to the definition, but it is standard and you don't have to do this very often. Also, it is not too difficult. We have to evaluate, $f^{(n)}(0)$ when f is given by, $f(x) = \sin x$. Run through a few of these and see if what you get is not described by the following.

> For every even integer n, $f^{(n)}(0)$, and for the odd integers, the values of $f^{(n)}(0)$ oscillate between 1 and -1, starting with 1.

Now put this into a mathematical formula for the nth Taylor coefficient for sin. You have to divide by $n!$ so,

$$a_{2k+1} = \frac{(-1)^k}{(2k+1)!}$$

and so the Maclaurin series for the function represented by $\sin x$ is

$$\sum_{k=0}^{\infty} \frac{(-1)^k}{(2k+1)!} x^{2k+1}.$$

Using the ratio test it is easy to show that the radius of convergence of this power series is infinite. That does not guarantee that the function is equal to it. You will see in the next two paragraphs how to determine that in this case, sin *is* equal to its power series.

But our point here is to see how to use this formula and the uniqueness theorem to get the power series for the function given by $\sin(2x)$. Do you see what to do?

Because (as we will see below), the sine function equals its power series, we have, for all x,

$$\sin(x) = \sum_{k=0}^{\infty} \frac{(-1)^k}{(2k+1)!} x^{2k+1}$$

and so we can substitute $2x$ for x to obtain,

$$\sin(2x) = \sum_{k=0}^{\infty} \frac{(-1)^k}{(2k+1)!} (2x)^{2k+1} = \sum_{k=0}^{\infty} \frac{(-1)^k 2^{2k+1}}{(2k+1)!} x^{2k+1}$$

This is a power series representation of the function and so it follows from the uniqueness theorem that it *must* be theMaclaurin series — provided we can show that the sin function is equal to its power series.

Taylor Polynomials and Remainder Terms

Now, the only thing left to deal with is whether the Taylor series of a function really represents it. That is, is a function equal to its power series?

Actually, there are two questions. You can find the power series, but does it converge at a particular point? Then, if you know it converges, is the sum equal to the value of the original function at that point? These are really different questions. You saw in Activity 4, page 481, that even if the Taylor series converges, it may not equal the function at any point except the base point. Do you see why a function is always equal to its Taylor series at the base point?

There is no general way to deal with this issue, but there is a method and you were actually introduced to it in Activity 5. If you take the $(n+1)^{st}$ partial sum of the Taylor series at c of a function f with domain $(c-r, c+r)$, then you can write, for any $x \in (c-r, c+r)$ at which $f^{(k)}$ exists for $k = 1, 2, \ldots, (n+1)$,

$$f(x) = \sum_{k=0}^{n} \frac{f^{(k)}(c)}{k!} (x-c)^k + R_n(f, n, r, c)(x).$$

This equation does not say very much. It only says that a function is equal to something plus what is left over. R_n is a function that, given a function f, an index n, a radius r and a base point c, returns a function whose value at x is the difference of the value of the original function and the value of the $(n+1)^{st}$ partial sum of its Taylor series at c.

For simplicity we give a name to the $(n+1)^{st}$ partial sum,

$$P_n(f, n, r, c) = \sum_{k=0}^{n} \frac{f^{(k)}(c)}{k!} (x-c)^k$$

and we call $P_n(f, n, r, c)$ the nth *Taylor polynomial of f at c*.

Thus, we can write, for short,

$$f(x) = P_n(x) + R_n(x).$$

The issue of a Taylor series converging and being equal to the original function can now be all wrapped up in the question of whether the remainder term goes to 0. That is, we can write,

$$f(x) = \sum_{n=0}^{\infty} \frac{f^{(n)}(c)}{n!} (x-c)^n$$

for a particular x, if and only if it is the case for that x that

$$\lim_{n \to \infty} R_n(x) = 0.$$

Working with R_n

You might think that everything in the previous paragraph is just a bunch of words, without any useful consequences. You would be right — so far. All of that discussion is to put things in an appropriate framework so that the mathematical facts that come next can be more easily understood.

The reason it is helpful to shift attention from the function and its Taylor series to the remainder term R_n is that there are several formulas for R_n that make it convenient to work with. You saw one such formula in Activity 5 and we repeat it here.

$$R_n(x) = \int_c^x \frac{f^{(n+1)}(t)}{n!}(x-t)^n\, dt$$

Well this is pretty complicated, but there are some functions for which you can work with it.

Look at the sin function for example. If we want to show that the remainder term goes to 0, then it is enough to take the absolute value, replace the formula by a quantity that is larger and show that the larger one goes to zero. We do this by making a series of estimates. All derivatives of the sin function are sin and cos, and these functions never get bigger in absolute value than 1. That and the estimate for the absolute value of a definite integral in Chapter 5 (see page 323) is all we need to make the following computation.

$$|R_n(x)| \le \int_c^x \left| \frac{(x-t)^n}{n!} \right| dt \le \frac{|x-c|^n}{n!}|x-c| = \frac{|x-c|^{n+1}}{n!}$$

where the last sequence goes to zero as n goes to ∞. That is a fact that you learned in Section 1. It is a special case of the following very useful general fact.

Suppose that $k > 1$ is some positive number and the sequence (a_n) is given by

$$a_n = \frac{k^n}{n!}$$

Then $\lim_{n\to\infty} a_n = 0$.

We now give an argument that suggests why the above statement is true. The argument goes like this. Think about the terms of the sequence. Each one is a fraction. The numerator is computed by multiplying k by itself n times. The denominator is also obtained by multiplying n factors but they differ. At first, they are smaller than k, but after a few factors they are larger and then much larger. We can make a more precise estimate. After some fixed number of terms, say, n_0, the factors are more than $2k$. Thus you can write, for $n > n_0$,

$$a_n = \frac{k^{n_0} k^{n-n_0}}{n_0!(n-n_0)!}$$

and writing the constant $M = k^{n_0}/n_0$, we can estimate,

$$a_n = M\frac{k^{n-n_0}}{(n-n_0)!} \le M\left(\frac{k}{n}\right)^{(n-n_0)} \le \frac{M 2^{n_0}}{2^n}$$

and this last goes to 0. Hence the remainder term goes to 0 and we can write, using the formula for the coefficients that was worked out earlier,

$$\sin x = \sum_{n=0}^{\infty} a_n x^n = \sum_{k=0}^{\infty} \frac{(-1)^k}{(2k+1)!} x^{2k+1}$$

This establishes the fact that we used to obtain the Maclaurin series for the function given by $\sin 2x$ on p. 487.

Derivatives and Integrals of Functions Given by Series

In the previous section, in Theorems 6.15 (page 474) and 6.16 (page 474) you worked with formulas for the derivative and integral of a function given by power series. Now you have some of the theoretical basis for using those formulas. If you begin with a function, figure out its Taylor series and determine that the remainder term goes to zero, then you can use the Taylor series formulas to obtain Taylor series for the derivative and integral of the original function. This is exactly what you did in Activity 11, page 483.

A theoretical analysis is really necessary because, quite often, things will not work so nicely. Look at the function f you considered in Activity 8, where

$$f(x) = \sum_{n=1}^{\infty} \frac{\sin(n^2 x)}{n^2}$$

Does this series converge for every value of x? It does. Why?

It is also quite easy to calculate the derivatives of the individual terms of the series. You get,

$$\sum_{n=1}^{\infty} \cos(n^2 x).$$

Based on your work in Activity 8, do you think this series is equal to f'? Does it even converge?

Calculating Power Series by Hand

The theoretical foundation we have been developing permits us to get power series representations of many functions *and be sure that the function is equal to its power series.* For example, we obtained the power series representation for sin,

$$\sin x = \sum_{k=0}^{\infty} \frac{(-1)^k}{(2k+1)!} x^{2k+1}$$

so we can differentiate to obtain,

$$\cos x = \sum_{k=0}^{\infty} \frac{(-1)^k}{(2k+1)!} (2k+1) x^{2k} = \sum_{k=0}^{\infty} \frac{(-1)^k}{(2k)!} x^{2k}$$

which is valid for all real numbers x by the ratio test.

These formulas are valid for all values of x. We may also consider power series with finite radii of convergence. Consider the following simple algebraic formula,

$$\frac{1-x^n}{1-x} = 1 + x + x^2 + \cdots + x^{n-1}.$$

We can rewrite this as

$$\frac{1}{1-x} = 1 + x + x^2 + \cdots + x^{n-1} + \frac{x^n}{1-x}.$$

Now, if $|x| < 1$ it is easy to check that

$$\lim_{n \to \infty} \frac{x^n}{1-x} = 0$$

and so we have

$$\frac{1}{1-x} = \sum_{n=0}^{\infty} x^n, \, |x| < 1$$

which is, by the uniqueness theorem, the Maclaurin series.

Now, you can differentiate again, as many times as you like, to get Taylor series for the functions given by

$$\frac{1}{(1-x)^n}, \, n = 1, 2, \ldots$$

In each case, the radius of convergence is 1.

It is even more interesting if you integrate. First, let's replace x by $-x$ to obtain,

$$\frac{1}{1+x} = \sum_{n=0}^{\infty} (-1)^n x^n, \quad |x| < 1$$

and so, with a constant K to be determined later, we can take anti-derivatives of both sides to obtain,

$$\ln(1+x) = \sum_{n=0}^{\infty} \frac{(-1)^n x^{n+1}}{n+1} + K, \quad |x| < 1.$$

The constant K is determined by checking the values at $x = 0$. The conclusion is that $K = 0$ (Why?), so (rewriting so that the index starts at $n = 1$) we have

$$\ln(1+x) = \sum_{n=1}^{\infty} \frac{(-1)^{n-1} x^n}{n}, \, |x| < 1.$$

Checking the endpoints of the interval $|x| < 1$, i.e., $-1 < x < 1$, we see that the series converges at $x = 1$ (why?) and diverges at $x = -1$ (why?). Hence, the power series representation of $\ln(1+x)$ is valid on the interval $-1 < x \le 1$. It follows that for $x = 1$ (as promised in part (a) of Exercise 43, Section 6.2, page 465), we have

$$\sum_{n=1}^{\infty} \frac{(-1)^{n-1}}{n} = \ln(2).$$

An Application to Statics

A problem in Statics (the subject which is the study of force diagrams where the forces are in equilibrium, i.e., balanced) gives rise to a simple application of power series. In the design of a suspension bridge it is important to determine how the cable hanging in the shape of a parabola sags under a particular loading distribution.

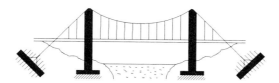

Figure 6.10. A suspension bridge.

In situations such as that depicted in Figure 6.11 it is necessary to know the length of the cable, $s_A + s_B$.

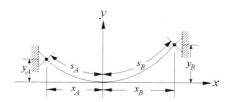

Figure 6.11. A cable with length $s_A + s_B$.

Calculating the length s_B, for example, leads to an integral of the form,

$$s_B = \int_0^{x_B} \left(1 + (y_B')^2 \, x^2\right)^{\frac{1}{2}} dx.$$

It is possible to calculate a closed form for this integral, or to use Riemann sums to estimate it (see the discussion of Riemann sums and arc length beginning on page 381). A method that is more convenient in some situations is to expand the integrand in a power series and integrate. This requires the Maclaurin series for the function g given by,

$$g(x) = (1+x)^{\frac{1}{2}}.$$

It is useful to write out the first few derivatives.

$$g'(x) = \frac{1}{2}(1+x)^{-\frac{1}{2}}$$

$$g''(x) = -\frac{1}{4}(1+x)^{-\frac{3}{2}}$$

$$g'''(x) = \frac{3}{8}(1+x)^{-\frac{5}{2}}$$

$$g^{(iv)}(x) = -\frac{1 \cdot 3 \cdot 5}{2^4}(1+x)^{-\frac{7}{2}}$$

$$g^{(v)}(x) = \frac{1 \cdot 3 \cdot 5 \cdot 7}{2^5}(1+x)^{-\frac{9}{2}}$$

It is easy to see how this looks in general and evaluating at $x = 0$, we get the general term,

$$g^{(n)}(0) = (-1)^n \frac{1 \cdot 3 \cdots (2n-3)}{2^n}.$$

It can be shown that for $|x| < 1$ the remainder term goes to zero and so,

$$(1+x)^{\frac{1}{2}} = 1 + \frac{x}{2} + \sum_{n=2}^{\infty} (-1)^n \frac{1 \cdot 3 \cdots (2n-3)}{2^n n!} x^n, \ |x| < 1.$$

For our cable, we must replace x by $(y_B x)^2$, integrate, and evaluate at the endpoints to obtain,

$$s_B = x_B \left(1 + (x_B)^2 (y_B)^3 + \sum_{n=2}^{\infty} (-1)^n \frac{1 \cdot 3 \cdots (2n-3)}{2^n n!} (x_B)^{2n} (y_B)^{2n+1} \right)$$

for $|x_B y_B| < 1$.

You will get a chance to use this result in the exercises.

Binomial Series

The example in the last paragraph is a special case of the *Binomial series* which you worked with in Activity 10. The **computer function binexp** will produce a **computer function** that will give you the binary expansion of any function given by an expression of the form $(1+x)^\alpha$. As you saw in Activity 10, this can be used to approximate the value of a number close to 1 raised to a power. You also saw in the previous paragraph how this can be used to approximate certain integrals.

It is not very difficult to generalize the computations of the previous paragraph to write out a formula for the Binomial series with any value of α. The formula is then an alternative to the **computer function binexp**.

As in the special case, the best way to work on the general formula is to begin by writing out the first few terms. If we have a function g given by $g(x) = (1+x)^\alpha$, then we can write,

$$g(x) = (1+x)^\alpha$$
$$g'(x) = \alpha(1+x)^{\alpha-1}$$
$$g''(x) = \alpha(\alpha-1)(1+x)^{\alpha-2}$$
$$g'''(x) = \alpha(\alpha-1)(\alpha-2)(1+x)^{\alpha-3}$$
$$g^{(iv)}(x) = \alpha(\alpha-1)(\alpha-2)(\alpha-3)(1+x)^{\alpha-4}$$

and in general, this can be written,

$$g^{(n)}(x) = \alpha(\alpha-1)\cdots(\alpha-(n-1))(1+x)^{\alpha-n}.$$

Therefore, evaluating at $x = 0$ we get the Maclaurin series for $(1+x)^\alpha$,

$$1 + \sum_{n=1}^{\infty} \frac{\alpha(\alpha-1)\cdots(\alpha-(n-1))}{n!} x^n.$$

The analysis of the remainder term can be made to determine that this series converges and is equal to the function for $|x| < 1$ and so we have,

$$(1+x)^\alpha = 1 + \sum_{n=1}^{\infty} \frac{\alpha(\alpha-1)\cdots(\alpha-(n-1))}{n!} x^n, \ |x| < 1, \alpha \neq -1.$$

What happens to the binomial series when α is zero or a positive integer?

We can make one more application of the Binomial series using the fact that

$$\arcsin(x) = \int_0^x \frac{1}{\sqrt{1-t^2}} \, dt, \, |x| < 1.$$

If we use $\alpha = -\frac{1}{2}$, replace x by $-t^2$ in the Binomial series and integrate, we obtain,

$$\arcsin(x) = x + \sum_{n=1}^{\infty} (-1)^n \frac{\frac{1}{2}\left(\frac{1}{2}-1\right)\cdots\left(\frac{1}{2}-(n-1)\right)}{n!(2n+1)2^n} x^{2n+1} + k, \, |x| < 1.$$

Checking at $x = 0$ establishes that $k = 0$, since $\arctan(0) = 0$. We can now simplify to obtain,

$$\arcsin(x) = x + \frac{x^3}{6} + \sum_{n=2}^{\infty} (-1)^n \frac{1 \cdot 3 \cdot 5 \cdots (2n-3)}{n!(2n+1)2^n} x^{2n+1}, \, |x| < 1.$$

Indeterminate Forms

You may have noticed that the problem you worked with in Activity 11 was a simple example in which you could use L'Hôpital's rule (see Chapter 4, Section 3, page 252). In fact, with the work you have done, we can give another way of proving that this technique really works. Not only can you prove the rule with series, but in some cases, the method of proof can actually be used to get the limit in a way that might be more convenient than using L'Hôpital's rule.

We begin by recalling the general situation in which L'Hôpital's rule is applied: you have two functions f and g with $f(c) = g(c) = 0$ and you wish to determine $\lim_{x \to c} \frac{f(x)}{g(x)}$. Now, for our new method, suppose that both of these functions are equal to their Taylor series expansion about c in some interval of convergence. Because the value of these functions at c is 0, then the first term of their Taylor series at c is also 0 and so we have,

$$f(x) = \sum_{n=1}^{\infty} a_n (x-c)^n$$

$$g(x) = \sum_{n=1}^{\infty} b_n (x-c)^n$$

Let us suppose further, that $b_1 \neq 0$. Then we can write, for $x \neq c$,

$$\frac{f(x)}{g(x)} = \frac{(x-c)\sum_{n=1}^{\infty} a_n (x-c)^{n-1}}{(x-c)\sum_{n=1}^{\infty} b_n (x-c)^{n-1}}$$

$$= \frac{\sum_{n=1}^{\infty} a_n (x-c)^{n-1}}{\sum_{n=1}^{\infty} b_n (x-c)^{n-1}}$$

$$= \frac{a_1 + \sum_{n=2}^{\infty} a_n (x-c)^{n-1}}{b_1 + \sum_{n=2}^{\infty} b_n (x-c)^{n-1}}$$

and, applying Theorem 6.14, page 472, we have,

$$\lim_{x \to c} \frac{f(x)}{g(x)} = \frac{a_1}{b_1}$$

What makes this calculation particularly useful is that it is not necessary to work out the power series in order to do it. We can make use of the fact that $a_1 = f'(c)$ and $b_1 = g'(c)$. In other words, we have the following very important principle.

If f and g are two functions which are equal to their Taylor series with base c in some interval of convergence and $\lim_{x \to c} f(x) = f(c) = 0$ *and* $\lim_{x \to c} g(x) = g(c) = 0$, *but* $\lim_{x \to c} g'(x) = g'(c) \neq 0$, *then*

$$\lim_{x \to c} \frac{f(x)}{g(x)} = \lim_{x \to c} \frac{f'(x)}{g'(x)} = \frac{f'(c)}{g'(c)}.$$

The above result is another form of *L'Hôpital's rule* for the indeterminate form $\frac{0}{0}$ which you studied in Chapter 4.

Using Known Maclaurin Series   A list of some known Maclaurin series of common elementary functions is given in the Chapter Summary, Section 6.5. Such a list of known power series can be quite useful and could save you a lot of work in finding other Maclaurin series. For example, do you see how you could find the Maclaurin series for the function represented by $\arctan(3x)$? What combination of functions is used to form this function? Did you say composition? Yes, the composition of arctan and the function represented by $3x$. Now do you see what to do? Yes, just simply substitute $3x$ for x in the known Maclaurin series for $\arctan(x)$ and the condition for convergence given by

$$\arctan(x) = \sum_{n=0}^{\infty} \frac{(-1)^n x^{2n+1}}{2n+1}, \quad \text{for } |x| \leq 1.$$

Doing so, we get

$$\arctan(3x) = \sum_{n=0}^{\infty} \frac{(-1)^n (3x)^{2n+1}}{2n+1}, \quad \text{for } |3x| \leq 1.$$

or

$$\arctan(3x) = \sum_{n=0}^{\infty} \frac{(-1)^n 3^{2n+1}}{2n+1} x^{2n+1}, \quad \text{for } |x| \leq \frac{1}{3}.$$

What about the power series for the function represented by $x \arctan(3x)$? Yes, just do the same as you did for $\arctan(3x)$ and then multiply your answer by x and simplify to get,

$$x \arctan(3x) = x \left(\sum_{n=0}^{\infty} \frac{(-1)^n 3^{2n+1}}{2n+1} x^{2n+1} \right) = \sum_{n=0}^{\infty} \frac{(-1)^n 3^{2n+1}}{2n+1} x^{2n+2}, \text{ for } |x| \leq \frac{1}{3}.$$

You can also make more complicated calculations. Suppose, for example, you wanted the Maclaurin series for the function given by $\frac{1}{(1-x)^2}$. You know that this function is the derivative of the function given by $\frac{1}{1-x}$ and that is on the list. So you take the term-by-term derivative of the known series, and simplify a little to get,

$$\frac{1}{(1-x)^2} = \sum_{n=0}^{\infty} (n+1)x^n, \text{ for } |x| < 1.$$

What about the condition for convergence, or the interval of convergence? Is it the same? Why or why not?

What happens to the interval of convergence if you find a power series by integrating a known series term-by-term?

6.4.4 Exercises

1. What function does the power series you found in Activity 4, page 481, define? Do you see how different that is from the function that gave rise to the series?

2. What power series do you get if the original function f is the zero function? What are the coefficients in this case? How does this relate to the point that was discussed on p. 475 in connection with series solutions of differential equations? Explain your answer.

3. Explain why a function is always equal to its Taylor series at the base point?

4. Use your computer to make a sketch of the function represented by $\exp(x) = e^x$ and the Taylor (Maclaurin) polynomial $P_{10}(x)$ represented by the first eleven terms of the Maclaurin series for e^x. Also make a table using **table2** for $\exp(x)$ and $P_{10}(x)$ for x in the interval $[-0.5, 0.5]$ with $n = 50$. Explain what is going on here.

5. Consider the function f given by $f(x) = \frac{1}{1-x}$. The value of f at $x = 2$ is given by $f(2) = -1$. However, the known power series expansion for f about 0 is $\sum_{n=0}^{\infty} x^n$ and at $x = 2$ the series is $\sum_{n=0}^{\infty} 2^n$, which is divergent (why?) and hence not equal to $f(2) = -1$. Explain what is going on here.

6. Find the Maclaurin series expansion for the function represented by each of the following expressions

   (a) using Taylor's formula for the nth coefficient;

   (b) using the known power series page 487.

   Here are the expressions.

   i. $\cos(2)$

   ii. $x^4 - x^3 + 3x^2 + 2x - 1$

   iii. $\exp(-x) = e^{-x}$

   iv. $\ln(1-x)$ v. $\frac{1}{(x-1)^2}$

   vi. $\frac{1}{1-2x}$

   vii. $\sinh(x) = \frac{e^x - e^{-x}}{2}$

   viii. $\cosh(x) = \frac{e^x + e^{-x}}{2}$

7. Find the Taylor series expansion for each of the following functions about the indicated point c using Taylor's formula for the nth coefficient.

   (a) $x^3 - 3x^2 + x - 5$, $c = -1$

   (b) $\ln(x)$, $c = 1$

   (c) $\exp(x) = e^x$, $c = 2$

   (d) $\sin(x)$, $c = \frac{\pi}{2}$

   (e) $\cos(x)$, $c = -\frac{\pi}{4}$

   (f) $\sin(x)$, $c = -\frac{\pi}{6}$

   (g) $\sin(2x)$, $c = \frac{\pi}{12}$

8. For the function represented by each of the following expressions,

   (a) Find the nth Maclaurin polynomial for the indicated value of n.

   (b) Approximate $f(x_0)$ using your answer in part (a). Check your approximation using your computer or hand-held calculator using the appropriate functions.

   Here are the expressions.

   i. $\sin(x)$, $n = 5$, $x_0 = 0.3$

   ii. $\tan(x)$, $n = 3$, $x_0 = 0.2$

    iii. e^{3x}, $n = 4$, $x_0 = 0.01$

    iv. $\ln(1 + x)$, $n = 5$, $x_0 = 0.3$

    v. $\arctan(x)$, $n = 4$,
       $x_0 = 0.5$

    vi. $\sec(x)$, $n = 4$, $x_0 = 1.5$

    vii. $\arcsin(x)$, $n = 5$,
       $x_0 = 0.15$

9. For the functions represented by the following expressions,

   (a) Find the nth Taylor polynomial about the specified point c.

   (b) Approximate $f(x_0)$ using your answer in part (a). Check your approximation using your computer or hand-held calculator using the appropriate functions.

   Here are the expressions.

    i. e^x, $n = 4$, $c = 1$, $x_0 = 0.9$

    ii. $\ln(x)$, $n = 5$, $c = 2$, $x_0 = 2.3$

10. Show that the Maclaurin series for the function represented by $\cos(x)$ can be obtained from the Maclaurin series for $\sin(x)$ by

   (a) Differentiation.

   (b) Indefinite integration.

11. Use the known power series expansions on page 494 and composition of functions to do each of the following.

   (a) Find a power series expansion for the function represented by each of the following expressions.

   (b) Find the interval of convergence of the power series found in part (a).

   Here are the expressions.

    i. $\sin(2x)$ ii. e^{3x}

    iii. $\cos(x^2)$

    iv. $\ln(1 + 2x)$

    v. $\sqrt{1 + x}$

    vi. $\dfrac{1}{\sqrt{1 + 2x}}$

    vii. $\dfrac{1}{\sqrt{4 + x^2}}$ viii. $\dfrac{1}{1 - x^2}$

    ix. e^{-x^2}

    x. $\sin(x) + \cos(x)$

    xi. $2e^x - \sin(x)$

    xii. $\sin(x) + e^x - e^{-x}$

    xiii. $\dfrac{\sin(x)}{x} - \cos(x)$

    xiv. $\dfrac{1 - \cos(x)}{x^2}$

    xv. $\dfrac{x^2 - 2\sin(x)}{x^3}$

12. Using the known power series on page 494 find the first four nonzero terms of the Maclaurin series for the functions represented by the following products.

   (a) $x^2 e^x$ (b) $(1 + x)\cos(x)$

   (c) $e^x \cos(2x)$ (d) $e^{-x} \sin(x)$

   (e) $\sin^2(x)$ (f) $1 - \cos^2(x)$

13. Using the known power series on page 494 and long division find the first four nonzero terms of the Maclaurin series for the functions represented by the following quotients.

   (a) $\dfrac{e^x}{x + 1}$ (b) $\dfrac{\sec(x)}{\cos(x)}$

   (c) $\dfrac{\cos(x)}{1 - x}$ (d) $\dfrac{e^x}{\sin(x)}$

   (e) $\dfrac{\sin(x)}{\sqrt{1 - x}}$

14. Find the following limits using power series. You may wish to check your answers using L'Hôpital's Rule, your SCS or any other reasoning.

(a) $\lim\limits_{x\to 0}\dfrac{\sin(5x)}{3x}$

(b) $\lim\limits_{\theta\to 0}\dfrac{1-\cos(3\theta)}{2\theta}$

(c) $\lim\limits_{x\to 0}\dfrac{1-e^x}{\sin(2x)}$

(d) $\lim\limits_{u\to 0}\dfrac{1-u-e^u}{5u^2}$

(e) $\lim\limits_{x\to -1}\dfrac{\ln(x^2)}{x+1}$

(f) $\lim\limits_{x\to 0}\dfrac{1-x-e^{-x}}{x^2}$

(g) $\lim\limits_{x\to 0}\left(\dfrac{1}{x}-\dfrac{1}{\sin(x)}\right)$

(h) $\lim\limits_{y\to 0}\dfrac{y(1-e^y)}{1-\cos(y)}$

(i) $\lim\limits_{x\to \pi}\dfrac{1-\cos(x)}{(x-\pi)^2}$

(j) $\lim\limits_{t\to -\infty}\dfrac{1+e^{-2t}}{t}$

(k) $\lim\limits_{x\to 0}\dfrac{3-3\cos(x)}{x^2}$

(l) $\lim\limits_{x\to 0}\dfrac{1-e^x}{1-\cos(x)}$

(m) $\lim\limits_{x\to \infty} x^2\sin\left(\dfrac{1}{x}\right)$

15. Using the known series on page 494, find the first three nonzero terms of the Maclaurin series expansion of the functions represented by the following expressions.

(a) $\sqrt{1+\cos(x)}$

(b) $(1-\sin(x))^{1/3}$

(c) $\sqrt{2-e^x}$

16. What is the interval of convergence of the power series

$$\sum_{n=1}^{\infty}\dfrac{(\sin(x))^n}{n}?$$

17. Find $f^{(10)}(0)$ for the function f represented by the following power series

$$f(x)=\sum_{n=0}^{\infty}\dfrac{(-1)^n 2^n x^{2n+3}}{(2n+1)!}.$$

18. Find $f^{(15)}(0)$ for the function f given by

$$f(x)=\dfrac{x-\sin(x)}{2x^3}.$$

19. Find $f^{(5)}\left(\tfrac{1}{10}\right)$ for the function f represented by the following power series

$$f(x)=\sum_{n=0}^{\infty}\dfrac{x^{n+1}}{2(n+1)}.$$

20. Find $h^{(10)}(1)$ for the function h represented by the following power series

$$h(x)=\sum_{n=4}^{\infty}\dfrac{(x-1)^n}{n(n-1)(n-2)(n-3)}.$$

21. Find $g^{(100)}(1)$ for the function g given by

$$g(x)=e^{-x^2}.$$

22. (a) Differentiate the Maclaurin series for the natural exponential function represented by $\exp(x)=e^x$. What function does the resulting series represent?

(b) Find the indefinite integral of the Maclaurin series for the natural exponential function represented by $\exp(x)=e^x$ and use the fact that $\exp(0)=1$ to find the constant of integration. What function does the resulting series represent?

(c) Are you surprised by what you saw in parts (a) and (b)? Explain your answer.

23. Using the Maclaurin series for the function represented by

$$\dfrac{x^2}{x^3+1},$$

find the Maclaurin series for

$$\ln(1+x^3).$$

24. Using the Maclaurin series for the function represented by

$$\arctan\left(\sqrt{x}\right),$$

find the Maclaurin series for

$$\frac{1}{\sqrt{x}+\sqrt{x^3}}.$$

25. Using the Maclaurin series for the function represented by

$$\ln\left(1+x^2\right),$$

find the Maclaurin series for

$$\frac{x}{x^2+1}.$$

What is the interval of convergence for the Maclaurin series you found?

26. Find the Maclaurin series for the function represented by

$$\frac{1}{(1-x)^4}$$

by differentiating an appropriate Maclaurin series.

27. Using an appropriate power series expansion, approximate each definite integral to three decimal places, that is, an error less than 0.0005. (Hint: Use the error estimate in Theorem 6.11, page 455, to determine the number of terms n to use in your series expansion.)

(a) $\int_0^{0.01} \sin\left(x^2\right)dx$ (b) $\int_0^1 e^{-x^2}dx$

(c) $\int_0^{0.3} \sqrt{1+x^3}\,dx$ (d) $\int_0^{0.1} \frac{1}{\sqrt{1+x^4}}dx$

(e) $\int_0^{0.1} \cos\left(x^3\right)dx$

(f) $\int_0^{0.1} \arctan\left(x^2\right)dx$

(g) $\int_0^{0.1} h(x)dx$ where

$$h(x) = \begin{cases} \frac{\sin(x)}{x} & \text{if } x \neq 0 \\ 1 & \text{if } x = 0 \end{cases}$$

(h) $\int_0^{0.5} g(x)dx$ where

$$g(x) = \begin{cases} \frac{1-\cos(x)}{x} & \text{if } x \neq 0 \\ 0 & \text{if } x = 0 \end{cases}$$

28. Estimate π to two decimal places by approximating the definite integral

$$\int_0^1 \frac{4}{1+x^2}\,dx$$

using a known Maclaurin series. Explain now this leads to an approximation to π.

29. Estimate $\ln(2)$ to four decimal places by approximating the sum of a known Maclaurin series. (See Figure 6.9, page 454.)

30. Show that the Maclaurin series for the function represented by each expression is equal to the function on the interval of convergence by showing that $\lim_{n\to\infty} R_n(x) = 0$.

(a) $\cos(x)$ (b) e^x (c) $\ln(1+x)$

31. Prove that if two power series $\sum_{n=0}^{\infty} a_n(x-c)^n$ and $\sum_{n=0}^{\infty} b_n(x-c)^n$ are equal for all x in some open interval I, then $a_n = b_n$ for all n.

32. Suppose that the function f represented by the power series $\sum_{n=0}^{\infty} a_n(x-c)^n$ has a radius of convergence $0 < r < \infty$. Show that if the power series for $f'(x)$ converges absolutely on the interval $[c-r, c+r]$, then the power series for f converges absolutely on the interval $[c-r, c+r]$.

33. Find a Maclaurin series for the function represented by $\arctan(x)$ using a known Maclaurin series and the fact that

$$\arctan(x) = \int_0^x \frac{1}{1+t^2}\,dt.$$

34. Find the Maclaurin series expansion for the function represented by

$$\frac{\ln(1-x)}{\ln(x+1)}.$$

What is the interval of convergence of the series you found?

35. For this exercise refer to the discussion of the situation in Statics involving a suspension bridge beginning on page 491. A suspension cable hangs in the shape of a parabola and supports a bridge with length 2 miles and having suspension towers 600 feet high. Approximate the total length of the cable using three terms of an appropriate power series expansion.

36. Consider the following power series which defines a function f.

$$f(x) = \sum_{n=1}^{\infty} \frac{\sin(n^2 x)}{n^2}$$

(a) Does this series converge for every value of x? Explain your answer.

(b) It is also quite easy to calculate the derivatives of the individual terms of the series. You get,

$$\sum_{n=1}^{\infty} \cos(n^2 x).$$

Based on your work in Activity 8, page 482, do you think this series is equal to f'? Does it even converge? Explain your answer.

37. Find the following limit, if possible.

$$\lim_{x \to \infty} \sum_{n=1}^{\infty} \frac{1}{n}\left(\frac{1-x}{x}\right)^{n-1}$$

38. Consider the following function f from Activity 4, page 481.

$$f(x) = \begin{cases} e^{-1/x^2} & \text{if } x \neq 0 \\ 0 & \text{if } x = 0 \end{cases}$$

Show that the power series for f is zero, that is, $f''(0) = 0$ for $n = 1, 2, \dots$. (Hint: Use the limit definition of the derivative to show that $f'(0) = 0$. Then show that $f''(0) = 0$ in a similar manner. Use mathematical induction (see Exercise 32, Section 6.1.4, page 441) to show that $f''(0) = 0$ for all positive integers n).

39. Explain what happens to the Binomial series (page 492) when α is zero or a positive integer. What is the resulting formula for any nonnegative integer? It is called the Binomial formula. Explain what happens when $\alpha = -1$. Is the resulting series familiar to you? Explain your answer.

40. In this exercise you will find power series solutions of differential equations as was described in Section 6.3

For each of the following differential equations problems, find

(a) the first five nonzero terms of a power series solution near $c = 0$.

(b) the general term of the power series solution near $c = 0$.

    i. $y' + y = 0$, $y(0) = 1$

    ii. $y' + xy = 0$, $y(0) = 2$, $y'(0) = -1$

    iii. $y' - x^2 y = 0$, $y(0) = 1$, $y'(0) = 0$

    iv. $y' + y = 0$, $y'(0) = -1$

41. In this exercise you will prove and use the following formula known as *Taylor's formula with remainder*.

Let c be a point in the domain of the function f and suppose that the $(n+1)^{\text{st}}$ derivative of f, $f^{(n+1)}$, exists and is continuous in an open interval I about c. Then for each x in I, it follows that

$$f(x) = f(c) + f'(c)(x - c)$$
$$+ \frac{f''(c)}{2!}(x - c)^2 + \cdots$$
$$+ \frac{f^{(n)}(c)}{n!}(x - c)^n$$
$$+ \frac{f^{(n+1)}(z)}{(n + 1)!}(x - z)^{n+1}$$

where z is between c and x.

(a) Prove Taylor's formula using the following steps.

    i. There is nothing to prove if $x = c$. Why? Assuming that $x \neq c$ and x is fixed, let $R_n(x) = f(x) - P_n(x)$ where P_n is the nth Taylor (or Maclaurin) polynomial for f. It follows that you must show that for some z between c and x

$$R_n(x) = \frac{f^{(n+1)}(z)}{(n + 1)!}(x - z)^{n+1}.$$
$$\text{Why}$$

    ii. Consider the function F defined as follows:

$$F(t) = (f(t) + f'(t)(x - t)$$
$$+ \frac{f''(t)}{2!}(x - t)^2 + \cdots$$
$$+ \frac{f^{(n)}(t)}{n!}(x - t)^n)$$
$$+ R_n(x)\frac{(x - c)^{n+1}}{(x - t)^{n+1}} - f(x).$$

    Show that $F(c) = F(x) = 0$ and

$$F'(t) = \frac{f^{(n+1)}}{n!}(x - t)^n$$
$$- \frac{(n + 1)R_n(x)}{(x - c)^{n+1}}(x - t)^n.$$

(Hint: The derivative of the expression in the parentheses in the expression for $F(t)$ is a telescoping series.)

    iii. Apply Rolle's Theorem, page 277, to F on the interval with endpoints c and x to complete the proof.

(b) For the function F defined in part i of part (a) and its derivative F', show that $\int_c^x F'(t)dt = 0$. Use this result to obtain the integral formula for the remainder $R_n(x)$ which you used in Activity 5, page 481 (also see page 488).

(c) The following term in Taylor's formula with remainder

$$\frac{f^{(n+1)}(z)}{(n + 1)!}(x - z)^{n+1},$$

where z is between c and x, is known as *Lagrange's form* of the remainder $R_n(x)$ (see page 488). Use Lagrange's form of the remainder to show that the MacLaurin Series for the function represented by each expression converges for the indicated real numbers x.

    i. $\cos(x)$ for all real numbers x

    ii. $\sin(x)$ for all real numbers x

    iii. e^x for all real numbers x

    iv. $\ln(1 + x)$ for $-1 < x \leq 1$

    v. $\arctan(x)$ for $|x| \leq 1$

    vi. $\arcsin(x)$ for $|x| < 1$

(d) Use Lagrange's form of the remainder in part (c) to estimate the error in using the nth Maclaurin polynomial P_n to estimate the value of the function represented by each expression at the point x_0 for the indicated value of n.

i. $\sin(x)$, $x_0 = 0.5$, $n = 5$

ii. $\cos x$, $x_0 = 0.3$, $n = 4$

iii. $\ln(1+x)$, $x_0 = 1$, $n = 4$

iv. e^x, $x_0 = 0.1$, $n = 5$

v. $\arctan(x)$, $x_0 = 0.3$, $n = 5$

42. In this exercise you will prove that the number $e = \exp(1)$ is irrational. Do so as follows.

    (a) Assume that e is a positive rational number, that is, there are positive integers p and q with $q > 1$ such that $e = p/q$, and then show that this leads to a contradiction. Assuming that $e = p/q$ as above, use the Maclaurin series for e^x to show that

    $$\frac{p}{q} = \sum_{n=0}^{\infty} \frac{1}{n!}.$$

    Write the infinite series without summation notation showing the first $q+2$ terms of the series and then show that

    $$(q-1)!\, p = q!\left(1 + 1 + \frac{1}{2!} + \ldots + \frac{1}{q!}\right)$$
    $$+ q!\left(\frac{1}{(q+1)!} + \ldots\right).$$

    (b) Using the result of part (a) show that

    $$(q-1)!\, p - q!\left(1 + 1 + \frac{1}{2!} + \ldots + \frac{1}{q!}\right)$$

    is an integer.

    (c) Show that series

    $$q!\left(\frac{1}{(q+1)!} + \ldots\right)$$

    satisfies the inequality

    $$q!\left(\frac{1}{(q+1)!} + \ldots\right) < \frac{1}{q+1} + \frac{1}{(q+1)^2}$$
    $$+ \frac{1}{(q+1)^3} + \ldots = \sum_{n=1}^{\infty} \frac{1}{(q+1)^n}.$$

    (d) Show that

    $$\sum_{n=1}^{\infty} \frac{1}{(q+1)^n} = \frac{1}{q}.$$

    (e) Explain why parts (a)-(d) show that there is a positive integer less than one.

    (f) Explain why you can assume that $q > 1$ in part (a) and also explain why the proof is now complete.

43. In this exercise you will prove *Picard's iteration procedure*. It is based on the recursive process $x_{n+1} = g(x_n)$ where g is a suitable function. An example of this type of procedure is the *Newton-Raphson method* (see page 291). Under the right conditions on the function g, Picard's iteration procedure converges to a solution of the equation $x = g(x)$. That is, the sequence (x_n) converges to a "fixed point" of the function g. (See Exercise 27 on page 296).

Picard's iteration procedure is as follows:

Let g be a continuous function on a closed interval $[a, b]$, with g' defined on the open interval (a, b). Let (x_n) be a sequence which is in (a, b) for each $n = 0, 1, \ldots$, and satisfying $x_{n+1} = g(x_n)$. If $|g'(x)| \le M < 1$ for all $x \in (a, b)$, then the sequence (x_n) converges to a solution of the equation $x = g(x)$.

Do the following to prove Picard's iteration procedure.

    (a) Let (a_n) be the sequence defined by $a_0 = x_0$ and $a_n = x_n - x_{n-1}$ for $n \ge 1$. Show that $x_n = \sum_{k=0}^{n} a_k$.

    (b) Show that

    $$a_2 = g(x_1) - g(x_0)$$
    $$= g'(c_1)(x_1 - x_0)$$

    for some number c_1 in between x_0 and x_1.

(c) Show that $|a_2| \le M|x_1 - x_0| = M|a_1|$, $|a_3| \le M|a_2| \le M^2|a_1|$, and in a similar manner $|a_n| \le M^{n-1}|a_1|$ for $n \ge 1$.

(d) Using the result of part (c), show that the series

$$\sum_{n=1}^{\infty} |a_n|$$

converges and then explain why the series $\sum_{n=0}^{\infty} a_n$ converges.

(e) Use the fact that $x_n = \sum_{k=0}^{n} a_k$ to show that the sequence (x_n) converges to a number x in $[a, b]$.

(f) Finish the proof by using the continuity of the function g and the fact that $x_{n+1} = g(x_n)$ to show that $x = \lim_{n \to \infty} g(x_n) = g(x)$.

44. An acid is heated with a catalyst to form a polymer containing some number of units. An important parameter of this reaction is the probability p that a particular unit reacts to become what is called *esterified*. One can show that (using appropriate units), the "number average molecular mass" is given by

$$m = M(p) = \sum_{r=1}^{\infty} r(1-p)p^{(r-1)}$$

What this means is that the function M is obtained by adding up infinitely many functions given by the expressions,

$$1-p, \, 2(1-p)p, \, 3(1-p)p^2,$$
$$4(1-p)p^3, \, 5(1-p)p^4, \ldots$$

In **ISETL**, for example, we can represent this with a **tuple** of functions as follows.

Mseq := [func(p); value r; return r*(1-p)*p**(r-1); end :

r in [1..N]];

There are a few comments to make about this code. First of all, the value **N** should be set to some large positive integer before running the code. The larger the value, the more of the functions will be included. The second concerns a couple of syntax quirks. Note the **value r** declaration and the fact that there is no semicolon after **end**. These are technical matters you will not have to worry about.

Enter this code (with a value for **N**) and compute, the following,

Mseq(1)(0.4), Mseq(2)(0.4),
Mseq(3)(0.4), Mseq(4)(0.4)

Give an explanation of the meaning of these quantities.

Now you can define a **func** which represents M as follows.

M := func(p);
return %+[f(p) : f in Mseq];
end;

Graph the function represented by this **func** (think about what should be its domain) and see if you can guess an expression for some simple function that might represent this situation.

6.5 CHAPTER SUMMARY

In this chapter you studied sequences and series of numbers, and series of functions. The first two are the foundation for the third and series of functions are important because of their value in representing functions. Using series, many complicated functions can be represented by special series of functions called power series. A power series is very convenient to work with because its behavior is very similar to the behavior of polynomials. We now summarize the main ideas, definition, theorems and some other useful information you studied in this chapter.

Sequences of Numbers   So, in the beginning, there is the sequence of numbers. A sequence of numbers is a function whose domain is, for example, the set of positive integers. You can represent a sequence of numbers with a computer function, just like you represented functions with continuous domains. The only difference is that the input must be a positive integer. You can also represent a sequence as a tuple, except that it has infinite length, if not on the computer then in your mind.

The main issue regarding sequences is their convergence and here again is the definition. As you read it, try to put it into your own words; think of some examples of sequences which converge and those which do not.

Let s be a sequence and L a number. We say that s converges to L, or that L is the limit *of s if, for every $\varepsilon > 0$, there exists a positive number M such that if the index $n > M$, then $|s_n - L| < \varepsilon$.*

If $s = (s_n)_{n=1}^{\infty}$ converges to L we write,

$$\lim_{n \to \infty} s_n = L$$

There were also a number of properties which a sequence might or might not have. Here is a list. Again, put the definitions in your own words and think of examples.

1. *A sequence $s = (s_n)_{n=1}^{\infty}$ is* monotone increasing *if $s_n \leq s_{n+1}$ for all $n = 1, 2, \ldots$. It is* monotone decreasing *if $s_n \geq s_{n+1}$ for all $n = 1, 2, \ldots$. The adjective* strictly *is used if the inequalities $\leq, \geq$ are replaced by $<, >$ respectively.*

2. *A sequence $s = (s_n)_{n=1}^{\infty}$ is* oscillating *if $s_n \leq s_{n+1}$ and $s_{n+1} \geq s_{n+2}$ for all $n = 1, 2, \ldots$, or if $s_n \geq s_{n+1}$ and $s_{n+1} \leq s_{n+2}$ for all $n = 1, 2, \ldots$.*

3. *A sequence $s = (s_n)_{n=1}^{\infty}$ is said to be* bounded above *if there is a number B such that $s_n \leq B$ for all $n = 1, 2, \ldots$. It is said to be* bounded below *if there is a number b such that $s_n \geq b$ for all $n = 1, 2, \ldots$. A sequence is said to be* bounded *if it is bounded above and bounded below.*

You also learned how to combine sequences to make new ones.

An important part of your study of sequences concerned the relationships among these properties, combinations of sequences, and convergence of sequences. There are several facts that you considered in the discussion and in the exercises. Here are three important theorems from this section.

1. *Let x, y be convergent sequences with $\lim_{n \to \infty} x_n = L$, $\lim_{n \to \infty} y_n = K$.*
   Then the sequence $x + y$ is convergent and

$$\lim_{n \to \infty} (x_n + y_n) = L + K$$

   (There are similar results for other combinations of sequences.)

2. *A convergent sequence is bounded above.*

3. *A sequence which is both monotone increasing and bounded above is necessarily convergent.*

Special kinds of sequences can be useful. The most important example of the sequence is the geometric progression. This is the sequence obtained by taking a positive number r and forming,

$$\left(r^n\right)_{n=1}^{\infty}$$

For such sequences the complete story about convergence is known.

- If $|r| < 1$ then $\lim_{n\to\infty} r^n = 0$
- If $r = 1$ then $\lim_{n\to\infty} r^n = 1$
- If $r = -1$ then the sequence does not converge.
- If $|r| > 1$ then the sequence does not converge.

Series of Numbers

A series is constructed by beginning with a sequence and constructing the corresponding sequence of partial sums. The limit of the sequence of partial sums is then, by definition, the sum of the series. Thus, once you understand convergence of sequences and can make the conversion from a sequence to a series, there is nothing conceptually new in the idea of the sum of an infinite series of numbers.

That is as far as theory goes — which is pretty far, but not even close to all the way. There is also practice. In practice, convergence of series introduces a whole new list of issues connected with tests for convergence. These are sufficiently complicated that we have suggested the use of an explicit strategy.

Here is the strategy we gave earlier on p. 457.

A Suggested Strategy for Testing a Series for Convergence

Ask yourself the following questions:

1. Does the nth term of the series converges to zero? If the nth term does not go to zero, then the series diverges. If the nth term goes to zero, then the series might converge and you should proceed to ask the following questions.

2. Is the series one of the following special types of series (or a combination of them)? Write out a few terms of the series to check the type of series.

   (a) Is the series a telescoping series? (If the series is a rational function whose denominator can be factored into linear factors, then you might try finding a partial fraction decomposition and then check to see if the series is a telescoping series.)

   (b) Is the series a geometric series? If it is geometric and the absolute value of the ratio is less than one, then the series converges and its sum is equal to the ratio of the first term to the quantity one minus the ratio.

   (c) Is the series a p-series? If so and $p > 1$, then the series converges. If not, i.e., $p \le 1$, then the series diverges.

   (d) Is the series an alternating series? If so, check to see if the sequence of absolute values of the terms is monotonically decreasing (past some point in the sequence) to zero. If so, then the series converges.

   (e) Does the series have both positive and negative terms (but not alternating)? Test the series for absolute convergence using the tests in part (3) below.

3. Are all the terms of the series positive? If so, then ask yourself the following questions:

   (a) Can the terms of the series be easily compared to a known series which you know converges or diverges? Be sure to find the right comparison for convergence or divergence!

   (b) Do the terms of the series form a monotonically decreasing sequence (past some point in the sequence)? If so, and the nth term can be easily integrated, then apply the integral test.

   (c) Does the nth term of the series contain factorials, n^{th} powers, or powers of n and/or combinations of such expressions? If so apply the ratio test (or the root test; see Exercise 39, page 464).

Series of Functions

When you move from series of numbers to series of functions, a new horizon appears. You can use a series of functions to define a function. This has some theoretical value and reversing the process, representing a function by a series of (simple) functions, has practical value.

Defining a function by a series raises issues such as how this is actually done. You start with a number in the domain of every function which is a term of the series, apply each of these functions to it to get a number and take the (infinite) sum of those numbers. This gives you a number. That is the process of a function defined by a series. When this is done with a power series, then a lot of information is available as indicated in the chart below.

One of the values of representing a function by a power series is that the main operations of calculus: limits, derivatives, and anti-derivatives can be computed easily, at least if you are satisfied with a power series representation for an answer. Here are the relevant theorems.

1. *A function defined by a power series is continuous at every point in its interval of convergence.*

2. *Let $f(x) = \sum_{n=0}^{\infty} a_n (x-c)^n$ represent a function by a power series whose radius of convergence is $r > 0$. Then f is differentiable on the open interval $(c-r, c+r)$ and its derivative is represented by a power series with the same radius of convergence. Moreover, we have the formula for the power series,*

$$f'(x) = \sum_{n=1}^{\infty} n a_n (x-c)^{n-1}.$$

3. *Let $f(x) = \sum_{n=0}^{\infty} a_n (x-c)^n$ represent a function by a power series whose radius of convergence is $r > 0$. Then the anti-derivatives of f all have the same radius of convergence and they are given by the formula*

$$\int f(x)dx = \sum_{n=0}^{\infty} \frac{a_n}{n+1}(x-c)^{n+1} + K$$

where K is an arbitrary constant. Moreover, if α and β are in the interval of convergence, then

$$\int_{\alpha}^{\beta} f(x)dx = \sum_{n=0}^{\infty} \frac{a_n}{n+1}\left((\beta-c)^{n+1} - (\alpha-c)^{n+1}\right)$$

Domain of a power series

For any power series $\sum_{n=0}^{\infty} a_n (x - c)^n$, there is a number $r \geq 0$ called the *radius of convergence*.

The series converges (absolutely) for $|x - c| < r$ and diverges for $|x - c| > r$.

The series may or may not converge at the endpoints, $x = c - r$ or $x = c + r$.

The interval from $c - r$ to $c + r$ (including those endpoints where the series converges) is called the *interval of convergence*.

When you think about trying to represent a given function as a power series, based at a certain point, then two questions arise: existence, or can you actually find a power series; and uniqueness, or how many are there?

The problem of existence is complicated. First, you need to be sure that the function has derivatives of any order, then you must make some computations from a formula, and finally you must check if the remainder term goes to zero. This whole procedure must be done for each specific function you wish to represent by a series.

The problem of uniqueness is much simpler. A function can have only one power series based at a given point. Thus if, somehow or other, you find a power series for a function, then that is the only one you will ever find.

All of this leads to a formula for the power series based at c for a function f. It is,

$$\sum_{n=0}^{\infty} \frac{f^{(n)}(c)}{n!} (x - c)^n.$$

This is called a *Taylor Series* for f at c. In the case of $c = 0$ then we call it a *Maclaurin Series* and the formula simplifies to

$$\sum_{n=0}^{\infty} \frac{f^{(n)}(0)}{n!} x^n$$

It is not always necessary to go through the procedure of finding all of the items in this formula to find a power series. Another method that works very well in practice is to take some known series and perform manipulations on the series to get the function you want. This requires knowing a lot of series. Below is a list for the Maclaurin case: some we have derived in this section, others you may have derived in the exercises.

Some Known Maclaurin Series

$$\frac{1}{1-x} = \sum_{n=0}^{\infty} x^n, \text{ for } |x| < 1$$

$$\sin(x) = \sum_{n=0}^{\infty} \frac{(-1)^n x^{2n+1}}{(2n+1)!}, \text{ for all } x$$

$$\cos(x) = \sum_{n=0}^{\infty} \frac{(-1)^n x^{2n}}{(2n)!}, \text{ for all } x$$

$$\ln(1+x) = \sum_{n=1}^{\infty} \frac{(-1)^{n-1} x^n}{n}, \text{ for } -1 < x \le 1$$

$$\exp(x) = e^x = \sum_{n=0}^{\infty} \frac{x^n}{n!}, \text{ for all } x$$

$$\arctan(x) = \sum_{n=0}^{\infty} \frac{(-1)^n x^{2n+1}}{2n+1}, \text{ for } |x| \le 1$$

$$\arcsin(x) = x + \frac{x^3}{6} + \sum_{n=2}^{\infty} \frac{(1 \cdot 3 \cdot 5 \cdots (2n-3)) x^{2n+1}}{n!(2n+1)2^n}, \text{ for } |x| < 1$$

$$(1+x)^{\alpha} = \sum_{n=1}^{\infty} \frac{(\alpha(\alpha-1)(\alpha-2)\cdots(\alpha-(n-1)))x^n}{n!}, \text{ for } |x| < 1$$

Conclusion   In Chapter 2 you made a serious study of functions. Mainly you worked with functions having continuous domains, that is, intervals of real numbers. In the succeeding three chapters you learned to use three really powerful tools for analyzing and applying functions with continuous domains: limits, derivatives and integrals. Now, in this chapter, the emphasis changed. The object of study became functions with discrete domains, that is, a subset of the positive integers. A function with discrete domain is a sequence and sequences, together with addition, can be used to build series. Up to this point the co-domain of the functions (actually of almost all of the functions you have studied so far) were always sets of real numbers.

The level of mathematics takes a big jump when you begin to think about a function whose co-domain is a set of functions. For this, it is necessary to think about functions themselves as objects, which you worked on back in Chapter 2. Thus, a sequence of functions is a function which accepts a positive integer and gives back, as the "answer", a whole function. A sequence of functions can be used to form partial sums and construct a series of functions. The sum of a series of functions then gives back a function with continuous domain and the circle is complete.

In all of this, you saw how the idea of the limit of a sequence is a critical tool that forms the theoretical basis for all of the powerful calculations you can make.

CURVES, VECTORS, SURFACES AND CALCULUS

■ 7.1 CURVES IN POLAR COORDINATES

7.1.1 Overview

In this section you will study *polar coordinates.* You will investigate the graphs of some curves, including the conic sections (lines, circles, parabolas, ellipses and hyperbolas) and other special curves of interest in polar coordinates. You will also find areas bounded by polar curves and arc lengths of such curves. In the exercises you will have the opportunity to develop other calculus ideas as they relate to polar coordinates, for example, slope and areas of surfaces of revolution.

In the remaining three sections of this chapter you will study vectors, curves, and surfaces in two and three dimensions. You will learn to represent curves and surfaces in terms of a variable called a *parameter* by specifying several parametric equations. You will also study calculus as it relates to parametric equations and the study of motion along curves or what is called *curvilinear motion.*

7.1.2 Activities

1. Figure 7.1 shows the geometric relation between rectangular xy-coordinates and polar $r\theta$-coordinates. Write a **computer function** that will accept a tuple representing the rectangular coordinates of a point and return a tuple with its polar coordinates. Test your **computer function** with the four points $P = [1, 1]$, $Q = \left[-1, \sqrt{3}\right]$, $R = [-1, -1]$, and $S = \left[-1, \sqrt{3}\right]$.

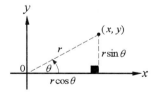

Figure 7.1. The relationship between rectangular and polar coordinates.

2. Write a **computer function** that will accept a tuple representing the polar coordinates of a point and return a tuple with its rectangular coordinates. Test your **computer function** with the four points $P = \left[\sqrt{2}, \frac{\pi}{4}\right]$, $Q = \left[2, \frac{2\pi}{3}\right]$, $R = \left[2, \frac{5\pi}{4}\right]$, and $S = \left[2, \frac{5\pi}{3}\right]$.

3. Many families of curves have names suggestive of their shapes. Following is a list of such names along with two or three examples of a polar equation for each category.

   (a) For each equation in polar form, select an appropriate domain for the angle that will produce the entire curve if it has finite length, or all essential features if not. You may find, in some cases, that because of scaling, it is necessary to experiment a little with your choice of domain for the angle.

   (b) Use the polar coordinate graphing facility in your MPL or SCS to draw a graph of each equation. For example, in the MPL **ISETL** you can use **graphpol** and in **MapleV** you can use **polarplot**. Notice how the curve is traced out by the radius vector as the angle θ increases.

   (c) For each family of curves, write a few sentences describing the characteristics of all curves in that family.

   (d) For each family, describe the characteristics of all equations in that family.

   (e) In as many cases as you can, analyze the situation by explaining how the features of the curve and of the equation are related.

   Here are the families with three or four examples of each. If you are using the MPL **ISETL**, then the **computer functions** you use with **graphpol** will have to deal with the situation in which the corresponding function is not defined.

   i. Straight line
      A. $r\cos(\theta) - 1 = 0$
      B. $r\sin(\theta) + 1 = 0$
      C. $r(5\cos(\theta) + 7\sin(\theta)) = 40$

   ii. Circle
      A. $r = 4\sin(\theta)$
      B. $r = 4\cos(\theta)$
      C. $r = 4$

   iii. Ellipse
      A. $r(5\cos(\theta) + 7) = 40$
      B. $r(7 - 5\sin(\theta)) = 40$

   iv. Parabola
      A. $r(7\cos(\theta) + 7) = 40$
      B. $r(7\sin(\theta) + 7) = 40$
      C. $r(7 - 7\sin(\theta)) = 40$

   v. Hyperbola
      A. $r(7\cos(\theta) + 5) = 40$
      B. $r(7\sin(\theta) + 5) = 40$
      C. $r(5 - 7\sin(\theta)) = 40$

   vi. Spiral
      A. $r = \theta$
      B. $r = -\theta$
      C. $r^2 = \theta$

   vii. Rose
      A. $r = \sin(2\theta)$
      B. $r = \cos(3\theta)$
      C. $r = \cos\left(\frac{2\theta}{3}\right)$

   viii. Limaçon
      A. $r = 2 + 4\cos(\theta)$
      B. $r = 4 - 2\sin(\theta)$

    ix. Cardioid

        A. $r = 2 - 2\cos(\theta)$

        B. $r = 2 + 2\sin(\theta)$

    x. Lemniscate

        A. $r^2 = 4\cos(2\theta)$

        B. $r = -4\cos(2\theta)$

        C. $r^2 = 4\sin(2\theta)$

4. In this activity, you are to write a version of the **Riem computer functions** you worked with in previous chapters which can be used with a function given in polar coordinates. To help you decide what to include, here is a summary of the situation for functions which are graphed in rectangular coordinates.

> Let f be a continuous function and $[a, b]$ an interval contained in its domain. The left Riemann sum is obtained as follows: First, the interval $[a, b]$ is partitioned into n subintervals of equal length according to the following formula.

$$x = \left(a + \frac{(b-a)}{n}(i-1) \right)_{i=1}^{n+1}$$

> Then, the following quantity is computed.

$$\sum_{i=1}^{n} f(x_i)(x_{i+1} - x_i).$$

> For example, these two operations can be performed in the MPL **ISETL** using the following code.

$$\text{x := [a+((b-a)/n)*(i-1) : i in [1..n+1]];}$$

$$\text{\%+[f(x(i))*(x(i+1)-x(i)) : i in [1..n]];}$$

> The sum in the above line of code is computed for larger and larger values of n, and finally the limit is taken as n goes to ∞.

> The interpretation of all this is that the original Riemann sum, for a fixed value of n and a nonnegative function f, is the sum of the areas of rectangles that make up a region which closely approximates the region between the graph of f and the x-axis between $x = a$ and $x = b$. The sum is an approximation to the area of the region under the curve. The limit gives the exact value of the area.

Do all of this in full detail for a function in polar coordinates, that is, the length of the radius r is given as a function of the angle θ which the radius makes with the positive x-axis (i.e., where θ is positive in the counterclockwise direction). Include in your discussion a graph of some curve in polar coordinates and an indication on the picture of what the sum is calculating. (Look at the examples in Activity 3 and note that it is not really appropriate to talk about the "area under the curve". For polar coordinates, find a suitable replacement for the phrase "area under the curve".)

Also include the analogue for polar coordinates of the expression

$$\int_a^b f(x)dx.$$

Note, the polar version of this expression is not obtained merely by substituting r and θ for x and y. The expression will be a somewhat different. Explain why the expression for polar coordinates is different from that in rectangular coordinates.

5. For each of the following two curves (parts vii B and ix B of Activity 3):

$$r = \cos(3\theta) \text{ and } r = 2 + 2\sin(\theta)$$

   do the following.

   Choose an appropriate interval and run the **computer function** you constructed in Activity 4, using some large value of n. Also, write down the integral in polar form for your example and compute it. Compare the two answers you get.

   In making your choice of the interval, let one of the curves be "the whole curve" and the other, a "part of the curve".

6. Use your **computer function** from Activity 4 to estimate the area of the region which is the intersection of the region bounded by the circle $r = \cos(\theta)$ and the region bounded by the cardioid $r = 1 - \cos(\theta)$.

   Calculate the area again by integrating and compare your answers.

7. Repeat the analysis and construction of Activity 4, not for the area swept out by the radius vector, but this time for the length of the arc swept out by the radius vector over an appropriate interval $\alpha \le \theta \le \beta$.

   Use the circle $r = 4\sin(\theta)$ (part ii A of Activity 3) and calculate its entire arc length. Again, use your **computer function** to get an approximate value for the arc length. Then, set up an integral and compute it. Compare your two answers. Can you think of a easy way to find the exact arc length of the circle by hand?

8. Sketch a graph of the following ellipse, given in polar coordinates by

$$r = \frac{10}{3 + 2\cos(\theta)}.$$

   Now transform the equation by replacing θ with $\theta - 2$. Try to figure out what the graph will look like before you use your computer to sketch it.

7.1.3   Discussion

The Polar Coordinate System   Instead of using rectangular coordinates to specify a point in the plane, an alternate way of specifying points in two dimensions is the *polar coordinate system*. Here, the system consists of a single point, the origin or *pole*, and a single axis, or ray, emanating from the origin in a single direction as shown in Figures 7.2a and 7.2b. Again there are two components used in representing a point. One is the (directed) distance from the pole to the point and the other is an angle which the ray containing the point and the origin makes with the horizontal axis. In Figure 7.2b, the angle is taken in the "counter-clockwise" direction and is measured in radians. In general, the angle can be positive (counterclockwise) or negative (clockwise). The component r can also be positive or negative. What does this mean geometrically? See Figure 7.2a. Do you see how to represent the point $\left[1, \frac{\pi}{6}\right]$ another way? How about as $\left[-1, \frac{7\pi}{6}\right]$ or $\left[1, -\frac{11\pi}{6}\right]$? Can you think of other ways to represent this point? The (directed) distance from the point to the origin is usually called r and the angle is usually called θ (read "theta"). The entire system is sometimes referred to as an $r\theta$-polar coordinate system. The polar coordinates of the pole, are the values in the pair $[0, \theta]$, where θ is any real number.

One way to think about representing a point in this way is by imagining a circle centered at the pole whose circumference passes through the point. Then the coordinate r can be thought of as the radius of the circle and the angle θ gives the position of the point on the circumference. See Figure 7.2b. Can you think of another way to locate a point in polar coordinates? Look again at Figure 7.2a. We can take the positive direction along the ray representing the polar axis to be from the pole toward the tip of the ray and we choose a unit of distance along the polar axis. Then we can take the negative direction as the opposite direction from the origin away from the tip of the polar axis (along the extension of the polar axis) as shown in Figure 7.2a. For a point $[r, \theta]$ in polar coordinates, we can first rotate the polar axis an angle θ and then locate the point using the (directed) distance r from the pole.

What about the fact that the pole has more than one specification or set of coordinates (in fact, infinitely many)? What is the meaning of this? What about other points? Are there any points that don't have coordinates at all?

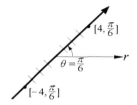

Figure 7.2a. A polar coordinate system.

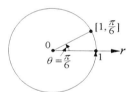

Figure 7.2b. A polar coordinate system.

Coordinate Transformations

Why do we use different coordinate systems? In many cases, the rectangular system is the simplest and most direct. But if you are looking at a radar screen, you might want to use polar coordinates. Can you think of other situations in which different systems are useful?

Sometimes you don't have a choice. You might have coordinates available in one system, but prefer to work with them in another system. For this we have formulas that transform the coordinates. In Activities 1 and 2 you had a chance to think about what these formulas might be for polar coordinates. Hopefully you can derive them yourself from the descriptions given above and the picture in Figure 7.1, page 509. We list them below for reference. But we don't suggest that you memorize them. Instead, see if you can close the book and work them out. Trying to do this and working with the formulas in problems will lead to your coming to remember them naturally, and with understanding.

Transformation equations for polar and rectangular coordinates.

<u>polar to rectangular</u>

$$x = r\cos(\theta) \quad \text{and} \quad y = r\sin(\theta)$$

<u>rectangular to polar</u>

$$r^2 = x^2 + y^2 \quad \text{and} \quad \tan(\theta) = \frac{y}{x}$$

What can you say about the ambiguity of the angle?

Conic Sections in Polar Coordinates

In Activity 3, the first five categories of curves are all familiar to you as *conic sections*, so-called because they come from passing a plane through a cone. You probably had difficulty with some of the examples in choosing appropriate domains for the angle. For example, the straight line may not show very many points and the first two parabolas require choices that avoid extremely small values of r.

For the conic sections, it is probably best to think simultaneously in terms of the values of the radius, r, for given values of the angle, θ, and a transformation to rectangular coordinates where the expressions may be more familiar.

For example, consider first the hyperbola in Activity 3 v, part B, given by

$$r(7\cos(\theta) + 5) = 40 \quad \text{or} \quad 7r\cos(\theta) + 5r = 40.$$

Using the standard formulas for transformation of coordinates, $x = r\cos(\theta)$, $y = r\sin(\theta)$, and $r^2 = x^2 + y^2$, the equation transforms into

$$7x + 5\sqrt{x^2 + y^2} = 40$$

or

$$5\sqrt{x^2 + y^2} = 40 - 7x$$

and squaring both sides gives

$$25(x^2 + y^2) = (40 - 7x)^2.$$

After completing the square and simplifying, we have

$$24\left(x - \frac{35}{3}\right)^2 - 25y^2 = \frac{5000}{3}$$

which we can recognize as a hyperbola.

There is another way we can look at the five conic section categories. Each formula (except for two of the circles) is a special case of the general form,

$$r = \frac{a}{b\cos(\theta) + c\sin(\theta) + d}.$$

Can you see how to put conditions on a, b, c and d so as to specify which of the five categories the formula represents?

We can use a computational manipulation to obtain an even simpler form. Did you know that for non-zero constants b, c and d any expression of the form

$$b\cos(\theta) + c\sin(\theta) + d$$

can be put into the form

$$d(1 + e\cos(\alpha))$$

for appropriate choices of e and α? We will give you a few hints and let you work it out in the exercises.

- Use one of the following trigonometric identities,

$$\cos(\beta + \gamma) = \cos(\beta)\cos(\gamma) - \sin(\beta)\sin(\gamma)$$

or

$$\cos(\beta - \gamma) = \cos(\beta)\cos(\gamma) + \sin(\beta)\sin(\gamma).$$

- The numbers $\frac{b}{d}$ and $\frac{c}{d}$ may not be such that their squares add up to 1, but they do add up to *something*.

- If s and t are two numbers whose squares add up to 1, then there is an angle whose cosine is s and sine is t.

We leave the messy details to you and turn our attention to something a little more interesting. Using these facts, we can write the general form of the conic sections as

$$r = \frac{ke}{1 - e\cos(\theta)} \quad \text{or} \quad r = \frac{ke}{1 + e\cos(\theta)}$$

and the only constants that must be chosen to specify a particular curve are k and e. What choices must be made to each of the five categories in Activity 3?

This form is particularly convenient for reading off information about the conic. All of the conics in this form have a *focus* at the pole (origin) as shown in Figure 7.3. The directrix of the conic is a line with equation $x = r\cos(\theta) = k$ (i.e., a vertical line $|k|$ units from the focus), and the number e, called the *eccentricity*, determines which category the conic is. (The word "eccentric" means "off center.")

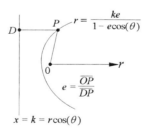

Figure 7.3. A portion of a conic with a focus at the pole and a vertical directrix.

The eccentricity e of a conic is the ratio of the distance from any point P on the conic to a focus to the distance from P to the directrix. If $0 < e < 1$ then the curve is an ellipse. If $e = 1$, it is a parabola and if $e > 1$, it is a hyperbola.

How do you get a straight line out of all this? How do you get a circle? What happened to the two circles of the form $r = a\cos(\theta)$ and $r = a\sin(\theta)$ that did not get included in the above discussion?

Special Curves in Polar Coordinates

Mathematics has existed for at least 3000 years. For a large part of that time, mathematicians have been concerned with varieties of curves. A major step forward occurred when people began to realize that the interesting shapes and twists and turns that one saw in these curves could often be described by properties of the expressions in the equations that define the curves. Even today, the connection between graphical and analytic representations is a major concern to mathematicians, and others.

In the categories of Activity 3 we have tried to introduce you to some of the curves that mathematicians have studied. Conic sections are of major importance, as are the other five categories in Activity 3.

The spiral is pretty easy. As the angle goes round and round, the radius gets larger and larger. This is a verbal description of what is happening and you can see that it applies to either the picture or the expression.

Look at the roses. What determines how many petals there are? Why is the third one so different from the other two? What would happen if you multiplied θ by an irrational number — say $\sqrt{2}$ or π?

Notice how similar are the equations of limaçons and cardiods. In fact, a cardiod is a special kind of limaçon. How does a cardioid differ from other limaçons? How is this reflected in the equation of the curve? Can you express precisely why this one change in the equation makes the corresponding change in the curve?

Lemniscates are curves whose polar equations are of the form $r^2 = a \sin(2\theta)$ or $r^2 = a \sin(2\theta)$ where a is a real number. What effect does the r^2 in the equation have on the graph of such curves? What happens when $\sin(2\theta)$ or $\cos(2\theta)$ is negative?

In all of these curves, what is the effect of replacing cosine by sine?

Area in Polar Coordinates   In polar coordinates, the issue of area is not necessarily about the area between two curves, but rather about the area swept out by the radius vector as it runs along the curve from one angle to another as shown in Figure 7.4.

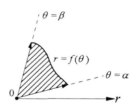

Figure 7.4. An area swept out by the radius vector.

How did you develop a formula for such an area in Activity 4? Were you able to apply the general method of Chapter 5 to obtain, first a method of approximation, and then an "exact" value in the form of a definite integral?

Here is what was always done in Chapter 5 to use the integral to compute something. There are three steps.

1. Break the desired quantity (like area) into little pieces.

2. Approximate the quantity you are after on each little piece, compute the value, and add up over the pieces.

3. Take the limit as the size of the pieces goes to zero.

Figure 7.5 shows what the partitioned area will look like for the area swept out by the radius vector running along the curve from angle α to β. Look now at a single piece of the partition in Figure 7.6. If you replace the varying value of the function with a single value, what do you get? It is not a rectangle, but some other figure. Do you see what the area of that figure is? Do you see why the area of the ith "circular-like" segment can be approximated by the following quantity?

$$\pi(f(\theta_i))^2\left(\frac{\theta_{i+1}-\theta_i}{2\pi}\right).$$

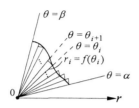

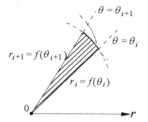

Figure 7.5. The area of Figure 7.4 partitioned into "circular-like" segments.

Figure 7.6. A single piece of the partition of area.

It's because we can approximate the area of each piece of the partition by the area of a circular segment. Since the area of the circle with radius $f(\theta_i)$ is $\pi(f(\theta_i))^2$, we can multiply this area by $\frac{\theta_{i+1}-\theta_i}{2\pi}$, which is just the fraction of the circular region traced out as θ increases from θ_i to θ_{i+1}. When you add up the individual areas and take the limit you get

$$\frac{1}{2}\int_\alpha^\beta (f(\theta))^2\, d\theta$$

where the interval $[\alpha, \beta]$ is such that the curve (or piece of the curve) is traced out exactly once. Armed with this formula, you should have no difficulty setting up an integral for an area swept out by the radius vector along a curve given by a particular polar expression from one given angle to another. Once you have this integral, you can evaluate the actual area either by hand or by using your Symbolic Computer System.

There can be a small difficulty if you are not actually given the angles to use in the limits of integration and must figure them out yourself. Suppose, for example, your problem is to find the area swept out by the larger loop in the graph of the function given in polar coordinates by

$$r = 1 + 2\sin(\theta).$$

Figure 7.7 shows what the graph looks like.

It is necessary to try to analyze this curve. What is the value of r when $\theta = 0$ and what happens to r as θ increases from 0? Which loop is traced out first? Can you figure out the value of θ when the large loop first returns to the intersection point? What was the value when the curve was first at the intersection point? These values of θ give the limits of integration. Do you see that they are $-\frac{\pi}{6}$ and $\frac{7\pi}{6}$?

Can you work out how these can be found? Try setting $r = 1 + 2\sin(\theta)$ equal to 0 and solving for θ. Why does this work? You'll get a chance to answer this question in the exercises. What are the intervals for θ for which the curve traces out the outer loop? the inner loop?

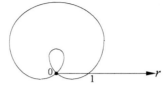

Figure 7.7. A graph of $r = 1 + 2\sin(\theta)$.

A problem like this requires just a little more thought than merely finding the values of θ where the curve intersects itself. You must also imagine the radius vector tracing out the curve and understand just when these intersection points occur.

Intersection of Two Regions   Finding the area between two curves traced out by a radius vector following along the curves amounts to computing two areas and subtracting them. This can be done by plugging in a formula. The only analysis that is required comes with figuring out the limits of integration. For this you must find the intersection points and see how the radius vector reaches them.

Suppose, for example, you wanted to find the area between the curves defined by $r = 1 + \cos(\theta)$ and $r = 1$. The picture looks like that in Figure 7.8.

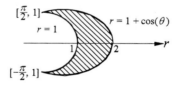

Figure 7.8. The area between the curves $r = 1 + \cos(\theta)$ and $r = 1$.

From the equations you can conclude that the curves intersect when θ is an angle for which $\cos(\theta) = 0$. From the picture you can see that this happens for an angle of $-\frac{\pi}{2}$ and $\frac{\pi}{2}$, and the area is given by

$$\frac{1}{2}\int_{-\frac{\pi}{2}}^{\frac{\pi}{2}} \left((1 + \cos(\theta))^2 - 1\right)d\theta = 2 + \frac{\pi}{4} \approx 2.785398164.$$

Are you at all surprised by the integrand? Why don't we have to compute two integrals? Can you go back to the Riemann sum situation and see that this is the right integrand? Or can you see it another way?

Arc Length in Polar Coordinates

There is a very natural way that you might try to modify the above discussion to compute arc length. Everything is exactly the same as in the above analysis, except that instead of looking at the *area* of a circular segment, you are computing the *length of an arc* of a circle. Thus, the expression

$$\pi(f(\theta_i))^2 \left(\frac{\theta_{i+1} - \theta_i}{2\pi} \right).$$

is replaced in the Riemann sums by

$$2\pi(f(\theta_i)) \left(\frac{\theta_{i+1} - \theta_i}{2\pi} \right) = f(\theta_i)(\theta_{i+1} - \theta_i).$$

Hence after adding up and then taking the limit, the integral for arc length becomes

$$\int_\alpha^\beta f(\theta)d\theta.$$

It would not be surprising if you made such an analysis in doing Activity 7. But did you notice that the example you were applying your calculations to is a circle of radius 2? You know the length of such a curve — it is 4π. Is that what you got? Is anything wrong? You will have the opportunity to answer these questions in the exercises.

The upshot of all this is that applying the same method we have been using from the very beginning of this book with partitions, Riemann sums and integrals, breaks down when we come to arc length in polar coordinates. This is something that happens throughout mathematics and science. Perfectly good methods work in a multitude of situations but can break down if you carry them too far. This is actually a good thing because it gives you an opportunity to look more deeply into the situation and learn some more.

We won't be able to do that here. Instead, we will describe what does work for arc length and leave it at that as, perhaps, a motivation for you in studying more advanced mathematics.

The point of our explanation is that the arc of a circle is not the only possibility for estimating the arc length. You could also use the chord connecting the two points. You saw this ambiguity before, in Chapter 5. There, in rectangular coordinates, there were several possibilities and we also chose the chord connecting the two points. Figure 7.9 shows two ways of estimating an arc length.

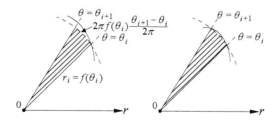

Figure 7.9. Some estimates of arc length.

Reading off the figure, you see that the second approximation is the length of the line from the point with polar coordinates $(f(\theta_i), \theta_i)$ to the point $(f(\theta_{i+1}), \theta_{i+1})$. As we did in Chapter 5, we will omit the calculations. The result is that the formula for the arc length of a curve given by $r = f(\theta)$ from the angle $\theta = \alpha$ to the angle $\theta = \beta$ is

$$\int_{\alpha}^{\beta} \sqrt{(f(\theta))^2 + (f'(\theta))^2}\, d\theta.$$

You will get a chance to derive the above formula in the exercises in Section 7.4.

Rotation of Polar Coordinates

In Activity 8, what happened to the graph of the ellipse given by

$$r = \frac{10}{3 + 2\cos(\theta)}$$

when you replaced θ by $\theta - 2$? Yes, the ellipse was rotated by an angle of 2 radians in the counterclockwise direction. You can see that it is much easier to rotate a conic section using the general form in polar coordinates. If the equation of a conic section is

$$r = \frac{ke}{1 + e\cos(\theta)}.$$

then the same figure rotated through an angle α in the counterclockwise direction has the equation

$$r = \frac{ke}{1 + e\cos(\theta - \alpha)}.$$

7.1.4 Exercises

1. For each of the following points in polar coordinates, convert to rectangular coordinates. Then plot and label each point on a polar coordinate system.

   (a) $[3, 0°]$ (b) $[2, 90°]$

   (c) $[5, \pi]$ (d) $[2.5, 270°]$

   (e) $[7, -3\pi/4]$ (f) $[3, 30°]$

   (g) $[2, 120°]$ (h) $[1.5, 4\pi/3]$

   (i) $[4, 300°]$ (j) $[4, -\pi/3]$

2. For each of the following points in rectangular coordinates, convert to polar coordinates. Give two answers for each point, one answer using degree measure for the angle and the other using radian measure.

   (a) $[3, 0]$ (b) $[-2, 0]$

   (c) $[3, 3]$ (d) $[0, -3]$

   (e) $[-2, 2]$ (f) $[\sqrt{3}, 1]$

   (g) $[2, -1]$ (h) $[-3, 2]$

   (i) $[-4, -7]$

3. For each of the following situations, give two different polar coordinate pairs for each point.

   (a)

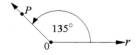

(b)

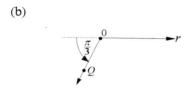

4. For each of the following rectangular equations in $\mathcal{R}^2$, convert to a polar coordinate equation.

   (a) $y = 2$ (b) $4x + y + 3 = 0$

   (c) $x^2 + y^2 = 5$ (d) $x^2 + 4y^2 = 4$

   (e) $x^2 + y^2 - 2x = 0$

   (f) $(x - 1)^2 + y^2 = 1$

   (g) $x^2 + y^2 = \left(x^2 + y^2 - 2x\right)^2$

5. For each of the following polar coordinate equations in $\mathcal{R}^2$, convert to a rectangular coordinate equation.

   (a) $\theta = \pi/4$ (b) $r = 2$

   (c) $r\cos(\theta) = 3$ (d) $r = 3\sin(\theta)$

   (e) $r\sin(\theta) = 2$ (f) $r = 3\sec(\theta)$

   (g) $r\sin(\theta) + 2r\cos(\theta) = 4$

   (h) $r = -\sin(\theta)$

   (i) $r = \dfrac{3}{\sin(\theta) - 2\cos(\theta)}$

   (j) $r = 2\cos(\theta) + \sin(\theta)$

6. Answer the following questions about polar coordinates.

   (a) What are the units for the two coordinates in polar coordinates?

   (b) Polar coordinates for a point in two dimensions are not unique, i.e., there is more than one way to specify the coordinates of a point in polar coordinates. Give an explanation of this ambiguity for representing a point in a plane using polar coordinates.

7. Sketch and shade the "region between" the indicated curves in polar coordinates, clearly labeling each boundary curve.

   (a) $0 \le r \le 1$ and $0 \le \theta \le \pi/3$

   (b) $1 \le r \le 3$ and $\pi/4 \le \theta \le \pi/2$

   (c) $0 \le r \le \cos(\theta)$ and $0 \le \theta \le \pi/2$

8. Sketch and shade the "region between" the indicated curves, clearly labeling each boundary curve. Find all points of intersection of the curves (if any) and describe the region between them using inequalities with r and θ.

   (a) $r = 3$, $\theta = \pi/4$ and $\theta = 2\pi/3$.

   (b) $r = 4\sin(\theta)$, $\theta = \pi/6$ and $\theta = \pi/2$.

   (c) $r = \sin(\theta)$ and $r = 3\sin(\theta)$.

   (d) $r = \cos(\theta)$ and $r = \sin(\theta)$.

9. A rose curve has a polar equation of the form $r = a\cos(b\theta)$, or $r = a\sin(b\theta)$, where a and b are real constants. You may wish to consider what you did in Activity 3, page 510 and look at other sketches to help answer the following questions. Give an explanation for your answers analytically, i.e., based on the properties of the equations.

   (a) What determines the length of each petal? What is the length of each petal?

   (b) What determines how many petals there are when b is an integer?

   (c) What happens when b is a rational number which is not an integer?

   (d) Show that if b is an irrational number — say $\sqrt{2}$ or π, then the graph has an infinite number of "overlapping" leaves. Explain why using your computer will not be helpful to illustrate this case.

10. Limaçons and cardioids are curves whose polar equations are of the form $r = a + b\sin(\theta)$, or $r = a + b\cos(\theta)$, where a and b are real constants. How does a cardioid differ from other limaçons? How is this reflected in the equation of the curve? Can you express precisely why this one change in the equation makes the corresponding change in the curve?

11. Lemniscates are curves whose polar equations are of the form $r^2 = a\cos(2\theta)$, or $r^2 = a\sin(2\theta)$ where a is a real number. What effect does the r^2 in the equation have on the graph of such curves? What happens when $\sin(2\theta)$ or $\cos(2\theta)$ is negative?

12. For a polar curve which is defined in terms of only a sine or cosine function, like those in Exercises 9, 10, and 11, what is the effect of replacing cosine by sine, or vise-versa?

13. Consider the curve given by $r = 1 + 2\sin(\theta)$.

    (a) What is the value of r when $\theta = 0$? What happens to r as θ increases from 0? Which loop is traced out first?

    (b) Can you figure out the value of θ when the large loop first returns to the intersection point? What was the value when the curve was first at the intersection point? Show that these values of θ are $-\frac{\pi}{6}$ and $\frac{7\pi}{6}$ by solving the equation $r = 1 + 2\sin(\theta) = 0$ for θ. Why does this work? What are the intervals for θ for which the curve traces out the outer loop? inner loop?

14. On page 519, we wanted to find the area between the curves defined by $r = 1 + \cos(\theta)$

and $r = 1$ see Figure 7.8 on page 518). Explain why the area is given by

$$\frac{1}{2}\int_{-\frac{\pi}{2}}^{\frac{\pi}{2}} \left((1 + \cos(\theta))^2 - 1\right)d\theta.$$

Are you at all surprised by the integrand? Why don't we have to compute two integrals? Explain why the integral gives the area using Riemann sums or some other way.

15. On page 519 we suggested that to find a formula for the arc length of a polar curve, you might try replacing the expression

$$\pi(f(\theta_i))^2\left(\frac{\theta_{i+1} - \theta_i}{2\pi}\right)$$

which was used in finding areas by the expression

$$2\pi(f(\theta_i))\left(\frac{\theta_{i+1} - \theta_i}{2\pi}\right) = f(\theta_i)(\theta_{i+1} - \theta_i).$$

Doing so leads to the integral

$$\int_{\alpha}^{\beta} f(\theta)d\theta.$$

Try finding the length of the circle given by $r = 4\sin(\theta)$ using this formula. Be careful with your choice of the interval of integration — be sure to pick one for which the curve is traced out exactly once. Notice that the circle in question has radius 2, hence its length is 4π. Explain what's wrong. Now try using the formula for finding the arc length of the polar curve $r = f(\theta)$ on page 520.

16. Sketch a graph of each of the following equations by hand. You may wish to use your computer to help you see what is going on. For each equation, give an interval for the angle θ for which the complete (or entire) curve is traced out exactly once.

    (a) $r = 2 - \sin(\theta)$ (b) $r = 1 - 2\sin(\theta)$

    (c) $r = \cos(3\theta)$ (d) $r = 2$

(e) $r\theta = 1$

(f) $r = \tan(\theta)$

(g) $r = 4\sin\left(\frac{3\theta}{2}\right)$

17. Show that on <u>any</u> interval of length 4π the curve given by $r = a + b\cos(\theta)$ is traced out exactly twice.

18. Show that an ellipse with one focus at the origin, to the right of the center of the ellipse, and a directrix at $x = k$ with $k > 0$ has an equation of the form

$$r = \frac{ke}{1 + e\cos(\theta)}$$

where e is the eccentricity of the ellipse with $0 < e < 1$.

19. Show that a conic section with one focus at the origin, eccentricity e, and whose directrix parallel to the line $\theta = \frac{\pi}{2}$ and $|k|$ units above the focus has an equation of the form

$$r = \frac{ke}{1 \pm e\sin(\theta)}.$$

20. Consider conic sections represented by an equation of one of the following forms:

$$r = \frac{ke}{1 \pm e\cos(\theta)} \quad or \quad r = \frac{ke}{1 \pm e\sin(\theta)}.$$

By evaluating the four r values for $\theta = 0, \frac{\pi}{2}, \pi,$ and $\frac{3\pi}{2}$, show that such a conic can be classified according to the following three cases. Give an equation and make a sketch of your equation for each case.

(a) If one of the four r values is not defined, then the conic is a parabola.

(b) If all four of the r values have the same sign, then the conic is an ellipse.

(c) If three of the four r values have the same sign and the other is of the opposite sign, then the conic is a hyperbola.

21. Classify each of the following conics as a parabola, an ellipse, or a hyperbola. Give a reason for your answer. Do so in two ways: using the results of the previous exercise and also using what you know about the eccentricity e. Make a sketch of each conic.

(a) $r = \dfrac{4}{1 + \cos(\theta)}$

(b) $r = \dfrac{4}{2 + \cos(\theta)}$

(c) $r = \dfrac{-3}{1 + \sin(\theta)}$

(d) $r = \dfrac{10}{4 + 5\cos(\theta)}$

(e) $r = \dfrac{12}{-2 + 4\cos(\theta)}$

(f) $r = \dfrac{1}{-3 + 3\cos(\theta)}$

(g) $r = \dfrac{1}{1 + 3\cos\left(\theta - \frac{\pi}{4}\right)}$

(h) $r = \dfrac{4}{2 - \sin\left(\theta + \frac{\pi}{3}\right)}$

(i) $r = \dfrac{-1}{3 + 3\cos\left(\theta - \frac{\pi}{6}\right)}$

22. Convert each of the following rectangular equations to a polar equation using the transformation equations:

$$x = r\cos(\theta), \quad y = r\sin(\theta), \quad \text{and} \quad r^2 = x^2 + y^2.$$

(a) $4x + y - 1 = 0$

(b) $9x^2 + 25y^2 - 72x - 81 = 0$

(c) $x^2 + 4 = 4y^2$

(d) $x^2 + y^2 - 4x = 0$

(e) $4x^2 - 9y^2 = 36$

(f) $16x^2 + 9y^2 = 144$

23. Convert each of the following polar equations to an xy-rectangular equation using the transformation equations:

$$x = r\cos(\theta), \quad y = r\sin(\theta), \quad r^2 = x^2 + y^2$$
$$\text{and} \quad \tan(\theta) = \tfrac{y}{x}.$$

(a) $r(3\sin(\theta) - 2\cos(\theta)) = 3$

(b) $\theta = \frac{\pi}{3}$ (c) $r\sec(\theta) = 3$

(d) $r(3 + \sin(\theta)) = 4$

(e) $r\sin(\theta) + 1 = 0$

(f) $r(5\cos(\theta) + 7\sin(\theta)) = 40$

(g) $r = 4\cos(\theta)$ (h) $r = 4$

(i) $r(5\cos(\theta) + 7) = 40$

(j) $r(7 - 5\sin(\theta)) = 40$

(k) $r(7\sin(\theta) + 7) = 40$

(l) $r(5 - 4\cos(\theta)) = 9$

(m) $r(7\sin(\theta) + 5) = 40$

(n) $r(5 - 7\sin(\theta)) = 40$

(o) $r = \theta$ (p) $r^2 = \theta$

(q) $r = e^{\theta}$ where $e = \exp(1)$ is the natural base.

(r) $r = \sin(2\theta)$ (Use the identity $\sin(2\theta) = 2\sin(\theta)\cos(\theta)$.)

(s) $r = 2 + 4\cos(\theta)$

(t) $r = 4 + 2\sin(\theta)$

(u) $r = 2 - 2\cos(\theta)$

(v) $r^2 = 4\cos 2(\theta)$ (Use the identity $\cos(2\theta) = \cos^2(\theta) - \sin^2(\theta)$.)

(w) $r^2 = -4\cos 2(\theta)$

(x) $r^2 = 4\sin 2(\theta)$ (Use the identity $\sin(2\theta) = 2\sin(\theta)\cos(\theta)$.)

24. Show that the graph of the equation $r = a\sin(\theta)$ is a circle with radius $\frac{|a|}{2}$ and center at the point $\left(0, \frac{a}{2}\right)$.

25. The strength s, in millivolts, of a signal generated by a radio antenna is given by

$$s = S(\theta) = 120 + 60\sin(\theta)$$

Find the rate of change of s with respect to θ for any angle θ. What is the rate of change of s with respect to θ when $\theta = 0$? $\theta = \frac{\pi}{3}$?

26. The *extreme points* of a polar equation $r = f(\theta)$ are the points on the graph of the equation at which the function $f(\theta)$ has extrema. For each of the following curves, find the extreme points of the graph of the equation using the derivative

$$\frac{dr}{d\theta} = f'(\theta).$$

Also find all points, if any, on the graph where the tangent line is vertical. Make a sketch of the graph of each equation, clearly labeling the polar coordinates of the extreme points and points at which the tangent line is vertical.

(a) $r = 3 + \sin(\theta)$ (b) $r = 2 + 2\sin(\theta)$

(c) $r = 2 - 4\sin(\theta)$ (d) $r = 4\sin(\theta)$

(e) $r = -2\cos(3\theta)$

27. A satellite in orbit around the earth travels in the path of the ellipse given by

$$r = \frac{80000}{10 - 9\cos(\theta)}.$$

Use calculus to find the polar coordinates of the highest and lowest points the satellite is above the surface of the earth, assuming that the radius of the earth is 4000 miles and r is measured in miles.

28.   (a) Show that the slope $\frac{dy}{dx}$ of a polar curve $r = f(\theta)$ is given by

$$\frac{dy}{dx} = \frac{f'(\theta)\sin(\theta) + f(\theta)\cos(\theta)}{f'(\theta)\cos(\theta) - f(\theta)\sin(\theta)}.$$

Use the chain rule in the form

$$\frac{dy}{d\theta} = \frac{dy}{dx}\frac{dx}{d\theta}$$

provided $\frac{dx}{d\theta} \neq 0$, and the polar coordinate transformation equations

$$x = r\cos(\theta) = f(\theta)\cos(\theta)$$

and

$$y = r\sin(\theta) = f(\theta)\sin(\theta).$$

(b) Write a brief essay discussing the difference between the slope $\frac{dy}{dx}$ of a polar curve $r = f(\theta)$ and the derivative $\frac{dr}{d\theta} = f'(\theta)$. Give examples to illustrate any differences between these two derivatives.

(c) Use the result of part (a) to find the slope of the tangent line to each polar equation at the indicated value of θ. Make a sketch of the curve and the tangent line to the curve at the point determined by θ.

     i. $r = 1 + 2\cos(\theta)$ at $\theta = \frac{\pi}{6}$.

     ii. $r = 6\sin(\theta)$ at $\theta = \frac{\pi}{3}$.

     iii. $r = \frac{1}{1-\sin(\theta)}$ at $\theta = \frac{\pi}{6}$

     iv. $r^2 = 4\sin(2\theta)$ at $\theta = \frac{\pi}{4}$

     v. $r^2 = -3\cos(2\theta)$ at $\theta = \frac{\pi}{3}$

29. In this exercise you investigate and find tangents to polar curves which pass through the pole. Such tangents are referred to as "tangents at the pole."

(a) Consider the following statement.

Let $r = f(\theta)$ be the polar equation of a curve, where f is a function for which $f'(\theta)$ exists, $f'(\alpha) \neq 0$, and $f(\alpha) = 0$.

Use the result of the previous exercise to show that the line $\theta = \alpha$ is tangent to the graph of $r = f(\theta)$ at the pole. Hence, to find tangents at the pole to the graph of a polar

equation $r = f(\theta)$ it suffices to find all angles θ for which $r = f(\theta) = 0$. Such angles are useful in making careful sketches of polar curves and in finding limits of integration for polar integrals which give areas, arc lengths and other such quantities.

(b) Use the result of part (a) to find all tangents at the pole, if any, for each of the following polar curves. Make a sketch of each curve and clearly label all tangents at the pole on your sketch.

     i. $r = 4\sin(\theta)$

     ii. $r = -3\cos(\theta)$

     iii. $r = 4\sin(2\theta)$

     iv. $r = -5\cos(3\theta)$

     v. $r = 1 - 2\sin(\theta)$

     vi. $r = 2 + 2\cos(\theta)$

     vii. $r = 3 - 2\sin(\theta)$

     viii. $r = 3 + 5\cos(\theta)$

     ix. $r = 3 - 3\sin(\theta)$

     x. $r^2 = 4\sin(2\theta)$

     xi. $r^2 = -9\cos(2\theta)$

30. Find the area of each region. You may wish to use the result of part (a) of the previous exercise to help you find some limits of integration.

(a) The region bounded by the graph of $r = 4\cos(\theta)$ from $\theta = -\frac{\pi}{6}$ to $\theta = \frac{\pi}{6}$.

(b) The region bounded by the curve $r = 6\sin(\theta)$ from $\theta = 0$ to $\theta = \frac{\pi}{4}$.

(c) The region bounded by the graph of $r = 4 + 2\sin(\theta)$.

(d) The region bounded by the graph of the curve $r = 3 - \cos(\theta)$.

(e) The region bounded by one leaf of the curve $r = 4\sin(3\theta)$.

(f) The region bounded by one leaf of the curve $r = -5\cos(2\theta)$.

(g) The region bounded by the curve $r^2 = 9\cos(2\theta)$.

(h) The region bounded by the curve $r^2 = -3\sin(2\theta)$.

(i) The region inside both the curves $r = \sin(\theta)$ and $r = 1 - \sin(\theta)$.

(j) The region inside $r = 4\cos(\theta)$ and outside $r = 1 + 2\cos(\theta)$.

(k) The region bounded by the inner loop of the curve $r = 1 - 2\sin(\theta)$.

(l) The region inside of $r = 4 + 4\sin(\theta)$ and outside $r = 8\sin(\theta)$.

(m) The region inside the curve $r = \sqrt{2} - \cos(\theta)$.

(n) The region inside the curve $r = 3$ and outside the curve $r = 1$.

(o) The region inside the curve $r = -2 + 2\sin(\theta)$ and outside the curve $r = 2 + \sin(\theta)$.

(p) The region inside both $r = 3\cos(\theta)$ and $r = 2$.

(q) The region inside both $r = 2 - \cos(\theta)$ and $r = 1 + \cos(\theta)$.

(r) The region inside $r = a$ and outside $r = a\sin(\theta)$ where $a > 0$.

(s) The region bounded by the curve $r = e^{\theta}$ where $0 \leq \theta \leq 2\pi$.

(t) The region inside both $r = -2\cos(\theta)$ and $r = 2\sin(2\theta)$.

(u) The region outside the curve $r = \cos(\theta)$ and inside the inner loop of $r = 1 + 2\sin(\theta)$.

(v) The region inside $r = 3\cos(\theta)$ and outside $r = 3\sin(2\theta)$.

(w) The region outside the curve $r = \cos^2(\theta)$ and inside the curve $r = \sin^2(\theta)$.

31. The pattern of radiation of a radio antenna is the graph of the region bounded by the cardioid $r = 30 + 30\cos(\theta)$. Find the area of the region covered by the antenna where the units of distance are measured in miles (i.e., r is measured in miles.)

32. In this exercise you will investigate the symmetry of polar equations with respect to some lines and points.

(a) Show that a polar curve $r = f(\theta)$ is symmetric with respect to the line $\theta = \frac{\pi}{2}$ if the graph of the equation $r = f(\theta)$ contains the point $(r, \pi - \theta)$ whenever it contains the point (r, θ). Show that the graph of the equation $r = a + b\sin(\theta)$ is symmetric with respect to the line $\theta = \frac{\pi}{2}$.

(b) Show that a polar curve $r = f(\theta)$ is symmetric with respect to the line $\theta = 0$ if the graph of the equation $r = f(\theta)$ contains the point $(r, -\theta)$ whenever it contains the point (r, θ). Show that the equation

$$r = \frac{ke}{1 + e\cos(\theta)}$$

is symmetric with respect to the line $\theta = 0$.

(c) Show that a polar curve $r = f(\theta)$ is symmetric with respect to the pole if the graph of the equation $r = f(\theta)$ contains the point $(r, \pi + \theta)$ whenever it contains the point (r, θ). Show that the curve represented by the equation

$$r = a\cos(2\theta)$$

is symmetric with respect to the pole.

(d) What can you say, if anything, about the symmetry of the polar equation $r = a\cos(b\theta)$ where a is a real constant and b is a positive integer constant? Give reasons for your answer.

(e) What can you say, if anything, about the symmetry of the polar equation $r = a\sin(b\theta)$ where a is a real constant and b is a positive integer constant? Give reasons for your answer.

33. Find all points of intersection of the pair of polar curves given by $r = 2 + \sin(\theta)$ and $r = -2 + 2\sin(\theta)$. You may wish to make a sketch by hand (or use a computer) to see what is going on and help find the points of intersection.

34. Show that the tangent of the angle ψ formed by the line tangent to the graph of $r = f(\theta)$ at a point $(f(\theta), \theta)$ and the radial line from the origin to the point $(f(\theta), \theta)$ is given by

$$\tan(\psi) = \frac{f(\theta)}{(\theta)}.$$

Make a sketch illustrating this situation.

35. Consider the general form of a conic

$$r = \frac{a}{b\cos(\theta) + c\sin(\theta) + d}$$

where a, b, c, and d are real numbers.

(a) How do you get a straight line out of the general form?

(b) How do you get a circle out of the general form?

(c) What are the conditions on a, b, c and d which specify each of the following five conic categories: lines, circles, parabola, ellipse and hyperbola?

(d) Show that for $a \neq 0$, $b \neq 0$, $c \neq 0$, and $d \neq 0$ the general form above can be put in the form

$$r = \frac{ke}{1 + e\cos(\theta)}$$

by showing that

$$b\cos(\theta) + c\sin(\theta) + d$$
$$= d(1 + e\cos(\alpha))$$

for the following choices of e and α:

$$e^2 = \left(\tfrac{b}{d}\right)^2 + \left(\tfrac{c}{d}\right)^2 \text{ and } \alpha = \psi - \theta,$$

where $\cos(\alpha) = \frac{b}{de}$ and $\sin(\alpha) = \frac{c}{de}$.

Use the following fact:

$$b\cos(\theta) + c\sin(\theta) + d$$
$$= d\left(1 + \frac{b}{d}\cos(\theta) + \frac{c}{d}\sin(\theta)\right)$$

and the three hints given on page 515.

36. Show that a conic with an equation of the general form

$$r = \frac{ke}{1 + e\cos(\theta)}$$

is

(a) an ellipse when $0 < e < 1$.

(b) a parabola if $e = 1$.

(c) a hyperbola if $e > 1$.

37. Find the length of each of the following polar curves using the formula for arc length in polar coordinates given on page 520.

(a) $r = 5$ for $\frac{\pi}{6} \leq \theta \leq \pi$

(b) $r = 4\sin(\theta)$ for $0 \leq \theta \leq \pi$

(c) $r = 1 + \sin(\theta)$ for $0 \leq \theta \leq 2\pi$

(d) $r = e^\theta$ for $0 \leq \theta \leq 2\pi$

(e) $r = 2\theta$ for $0 \leq \theta \leq \pi$.

38. Show that the shape of a limaçon given by $r = a + b\cos(\theta)$ is determined by the absolute value of the ratio of a to b for each of the following cases. Make sketches by hand or using a computer illustrating each case. A similar classification occurs for limaçons given by $r = a + b\sin(\theta)$.

    (a) When $\left|\frac{a}{b}\right| = 1$ the curve is a cardioid.

    (b) When $0 < \left|\frac{a}{b}\right| < 1$ the curve has an inner loop.

    (c) When $\left|\frac{a}{b}\right| \geq 2$ the curve has no inner loop and its graph is "flat" at the point closest to the pole.

    (d) When $1 < \left|\frac{a}{b}\right| < 2$ the curve has no inner loop and its graph has a "dimple" at the point closest to the pole.

39. A portion of a polar curve $r = f(\theta)$ is revolved about the line $\theta = 0$.

    (a) Use the equation for arc length in polar coordinates given on page 520, the formula for the area of a surface of revolution obtained by revolving a curve about the x-axis given on page 383, and the transformation equations $x = r\cos(\theta)$, $y = r\sin(\theta)$ to find a formula for the area of the surface of revolution. What restrictions might you make, if any, on the graph of the curve? Explain your answer.

    (b) Find the area of the following surfaces of revolution obtained by revolving the curve about the polar axis.

        i. $r = a\sin(\theta)$, where $a > 0$

        ii. $r = a + a\cos(\theta)$, where $a > 0$ and $0 \leq \theta \leq \pi$.

40. A portion of a polar curve $r = f(\theta)$ is revolved about the line $\theta = \frac{\pi}{2}$.

    (a) Find a formula for the area of the surface of revolution obtained by revolving a portion of a polar curve $r = f(\theta)$ about the line $\theta = \frac{\pi}{2}$. (Hint: See part (a) of the previous exercise.) What restrictions must you make, if any, on the graph of the curve?

    (b) Find the area of the following surfaces of revolution obtained by revolving the curve about the line $\theta = \frac{\pi}{2}$.

        i. $r = a\cos(\theta)$, where $a > 0$

        ii. $r = a + a\sin(\theta)$, where $a > 0$ and $0 \leq \theta \leq \frac{\pi}{2}$.

41. Describe geometrically the following polar equations where a, b and c are real constants.

    (a) $r = \dfrac{c}{a\sin(\theta) + b\cos(\theta)}$

    (b) $r = a\cos(\theta) + b\sin(\theta)$

    (c) $r^2 \sin(2\theta) = 2a$ (d) $r = a\sec^2\left(\frac{\theta}{2}\right)$

42. (a) Show that the polar coordinate formula for the distance D between two points (r_1, θ_1) and (r_2, θ_2) is given by

    $$D = \sqrt{r_1^2 + r_2^2 - 2r_1 r_2 \cos(\theta_1 - \theta_2)}.$$

    (b) Using the formula for the distance in part (a), find the distance between the following pairs of points.

        i. $[2, \pi]$ and $[3, \pi/2]$

        ii. $[1, \pi/6]$ and $[4, \pi/3]$

        iii. $[3, -\pi/2]$ and $[2, 5\pi/6]$

    (c) Convert the pairs of points above to rectangular coordinates and check your answer in part (b) using the distance formula in rectangular coordinates.

7.2 VECTORS IN $\mathcal{R}^2$ AND $\mathcal{R}^3$

7.2.1 Overview

In this section you will study the concept of *vector* in two dimensional space, $\mathcal{R}^2$, and in three dimensional space, $\mathcal{R}^3$, respectively. As you will see, a *vector* is a quantity that has both a *magnitude*, or *length*, together with an associated *direction*. Thus, we say that a *vector quantity* has magnitude and direction, whereas a *scalar quantity*, or *scalar* (e.g., a number) has only a magnitude. Vectors are among the most important mathematical quantities, or objects, in physics, engineering and other many other disciplines.

Once you have a feel for vectors, you might ask the following question: What can be done with them? It has been said that to know something is to act on it, to transform it. In this section you get to know these mathematical objects better and better by looking at some of their properties and learning to perform certain operations with them.

Vectors are not the same as numbers, but they are similar in that you can add them and (in at least two useful senses) take the product of two of them. Since vectors express both length and direction you can use them to calculate angles as well as length, area and volume. One important tool for doing this is the *dot product* and another tool is the *cross-product*. The latter product is also useful in obtaining various representations of lines and planes in three dimensions.

Once again we advise you to look ahead through the text as you work on the activities. If you see something you don't know anything about or something you don't understand, look it up in the index, or thumb through the discussion following the activities to get some ideas and any necessary formulas that might be of help.

7.2.2 Activities

1. Write a **computer function** that will accept two tuples representing two points on a Cartesian coordinate system and will return the following data (listed in a tuple, for example or just printed on the screen):

    (a) The distance between the two points in $\mathcal{R}^2$, that is, the length of the line segment connecting them.

    (b) Coordinates of the midpoint of the line segment connecting the two points in $\mathcal{R}^2$.

    Apply your **computer function** to pairs of points taken from the points $P = [2, 3]$, $Q = [-1, 2]$, $R = [-2, -2]$, and $S = [4, 3]$.

2. To do this activity and others below, use an appropriate vector graphing facility, such as the **ISETL** tool **vectors** in **ISETL**'s graphing facility, to plot a set of vectors with initial point at the origin and terminal point any different point in the plane. If you use the **ISETL** tool **vectors**, you can do so by typing the command

    $$\text{vectors(S);}$$

    where **S** must be a set of ordered pairs — that is, **tuples**, each consisting of two numerical components. Thus for example, you could use **ISETL** vector graphing facility and enter the following in **ISETL**,

$$S := \{[3,2], [-1,2.7], [-0.1, 5.83]\};$$
$$\text{vectors}(S);$$

Use this vector graph facility to produce screens with the vectors from the origin to the point for each of the points $P = [2, 3]$, $Q = [-1, 2]$, $R = [-2, -2]$, and $S = [4, 3]$.

3. Use a vector graphing facility (or do so by hand) to produce sketches with the vectors connecting the points at the heads of the vectors in the previous activity. (How many are there?). Produce various screens showing, for example, 1, 2, 3, 6 and all of these connecting vectors. Assign these vectors (represented as pairs of pairs) to variables such as **PQ, QP, PR, QS,...**.

If you are using the **ISETL tool vectors**, then you can plot a vector connecting two points, neither of which is the origin, using **vectors**, by including pairs of pairs in the set S on which **vectors** is to act. For example, if you say

$$S := \{[[3,2], [-1,2.7], [-0.1,5.83]]\};$$

then **vectors** will plot one vector from **[3,2]** to **[-1,2.7]** and another vector from the origin to **[-0.1,5.83]**.

4. Write **computer function**s which do the following.

   (a) Accepts a vector in $\mathcal{R}^2$ with tail at the origin and returns the length of the vector.

   (b) Accepts a vector in $\mathcal{R}^3$ with tail at the origin and returns the length of the vector.

   (c) Accepts a vector in $\mathcal{R}^3$ with tail at the origin and a scalar and returns the vector which is the product of the original scalar and vector.

   (d) Accepts two vectors in $\mathcal{R}^2$ with the tails at the origin and returns the vector sum of these two vectors.

   (e) Accepts two vectors in $\mathcal{R}^2$ with the tails at the origin and returns the dot product of these two vectors.

   (f) Accepts two vectors in $\mathcal{R}^3$ with the tails at the origin and returns the dot product of these two vectors.

   Test your **computer function**s on the following vectors and scalars:

   $$[3, 4], [-7, 24], [1, 2, 2], [-3, 0, 4], -1, 2, 0.5, \text{ and } -0.5$$

5. Use a vector graphing facility, like **vectors** in **ISETL**, to produce the parallelogram determined by two vectors in $\mathcal{R}^2$ originating at the origin. Have your picture include the altitude of the parallelogram. Do this for the vectors from the origin to $P = [2, 3]$ and $Q = [-1, 2]$. Refer to this picture, if necessary and write an expression for the area of the parallelogram determined by two vectors. Your expression should be in terms of the dot product of the two vectors.

6. Show that any vector **u** in $\mathcal{R}^2$ can be written in the following form

$$\mathbf{u} = |\mathbf{u}|\cos(\theta)\mathbf{i} + |\mathbf{v}|\sin(\theta)\mathbf{j}$$

where θ is the angle between **u** and the positive *x*-axis (or the unit vector **i**). See Figure 7.1 on page 509. When a vector **u** is written in this manner, we say that **u** is *resolved* into its **component vectors**,

$$|\mathbf{u}|\cos(\theta)\mathbf{i} \text{ and } |\mathbf{v}|\sin(\theta)\mathbf{j}.$$

The **i** and **j** *components*, respectively, for **u** are $|\mathbf{u}|\cos(\theta)$ and $|\mathbf{u}|\sin(\theta)$.

7. Write a **computer function** that accepts two non-zero and non-parallel vectors in $\mathcal{R}^3$ with tails at the origin and returns the area of the parallelogram determined by the vectors.

8. Use a vector graphing facility to do this activity. If you use the vector graphing facility in **ISETL**, then the following line of **ISETL** code generates a set of pairs of pairs [[*a*, *b*], [*c*, *d*]].

**S := {[[p-sin(p),1-cos(p)],[1+p-cos(p)-sin(p),1+sin(p)-cos(p)]]:
p in [0,1..3*Pi]};**

Each pair of pairs [[*a*, *b*], [*c*, *d*]] represents the vectors from the point [*a*, *b*] to [*c*, *d*]. Each of these vectors is tangent to a certain curve. The goal of this activity is to use a vector graphing facility, like **vectors** in **ISETL**, to make a picture containing all these vectors. You can get a better idea of what the curve looks like if you increase the number of vectors. This can be accomplished very easily by changing the definition of **S** slightly. Print out your picture and try to draw a hand sketch of the curve determined by these tangent lines.

9. The angles α, β and γ between a vector **u** in $\mathcal{R}^3$ and the vectors **i**, **j** and **k**, respectively, are called the *direction angles* of **u**. The cosines of the angles are called the *direction cosines* of **u**. Do the following for this situation.

   (a) Make a sketch illustrating the direction angles of a vector **u** which lies in the first octant and not along any of the coordinate axes.

   (b) For any non-zero vector **u**, show that

$$\frac{1}{|\mathbf{u}|}\mathbf{u} = \frac{\mathbf{u}}{|\mathbf{u}|} = \cos(\alpha)\mathbf{i} + \cos(\beta)\mathbf{j} + \cos(\gamma)\mathbf{k}.$$

   (c) Write a **computer function** which accepts any non-zero vector and returns the direction angles of the vector.

10. For two non-zero vectors **u** and **v**, in $\mathcal{R}^2$ (or $\mathcal{R}^3$), (i.e., two vectors such that $|\mathbf{u}| \neq 0$ and $|\mathbf{v}| \neq 0$) the *projection of* **u** *along* **v**, denoted **proj$_\mathbf{v}$u**, is given by

$$\mathbf{proj_v u} = |\mathbf{u}|\cos(\theta)\frac{\mathbf{v}}{|\mathbf{v}|} = \left(\frac{\mathbf{u}\cdot\mathbf{v}}{\mathbf{v}\cdot\mathbf{v}}\right)\mathbf{v}$$

where θ is the (smallest) angle between **u** and **v**. The number given by $|\mathbf{u}|\cos(\theta)$ is called the *component of* **u** *along* **v**. Do the following for this situation.

   (a) Make a sketch illustrating this situation in two dimensions for the vectors $\mathbf{u} = \mathbf{i} + 2\mathbf{j}$ and $\mathbf{v} = 3\mathbf{i} - 4\mathbf{j}$. Be sure to indicate the (smallest) angle θ between the two vectors and the meaning of $|\mathbf{u}|\cos(\theta)$ on your sketch.

(b) Write a **computer function** which accepts two non-zero vectors **u** and **v** in $\mathcal{R}^2$ and returns the projection of **u** along **v**. Test your **computer function** on the vectors $\mathbf{u} = \mathbf{i} + 2\mathbf{j}$ and $\mathbf{v} = 3\mathbf{i} - 4\mathbf{j}$.

(c) For the vectors $\mathbf{u} = \mathbf{i} + 2\mathbf{j}$ and $\mathbf{v} = 3\mathbf{i} - 4\mathbf{j}$, show that the dot product of $\mathbf{u} - \mathbf{proj_v u}$ and **v** is zero. What does this result say about the vectors $\mathbf{u} - \mathbf{proj_v u}$ and **v**? Make a sketch of the vectors $\mathbf{u} - \mathbf{proj_v u}$ and **v** for this situation.

(d) For the vectors $\mathbf{u} = \mathbf{i} + 2\mathbf{j}$ and $\mathbf{v} = 3\mathbf{i} - 4\mathbf{j}$, use the result of part (e) to find two vectors $\mathbf{u}_1$ and $\mathbf{u}_2$ such that $\mathbf{u} = \mathbf{u}_1 + \mathbf{u}_2$ where $\mathbf{u}_1$ is parallel to **v** and $\mathbf{u}_2$ is perpendicular to **v**.

11. In addition to the dot product, there is another way of taking the product of two vectors in $\mathcal{R}^3$ which produces a vector rather than a scalar. In this activity, you will investigate the *cross-product* of two vectors in $\mathcal{R}^3$. For two non-zero vectors $\mathbf{u} = [u_1, u_2, u_3]$ and $\mathbf{v} = [v_1, v_2, v_3]$ in three dimensions, the cross-product is denoted $\mathbf{u} \times \mathbf{v}$ and it is a vector given by

$$\mathbf{u} \times \mathbf{v} = \begin{vmatrix} u_2 & u_3 \\ v_2 & v_3 \end{vmatrix} \mathbf{i} - \begin{vmatrix} u_1 & u_3 \\ v_1 & v_3 \end{vmatrix} \mathbf{j} + \begin{vmatrix} u_1 & u_2 \\ v_1 & v_2 \end{vmatrix} \mathbf{k}$$

or

$$\mathbf{u} \times \mathbf{v} = (u_2 v_3 - v_2 u_3)\mathbf{i} - (u_1 v_3 - v_1 u_3)\mathbf{j} + (u_1 v_2 - v_1 u_2)\mathbf{k}$$

where we have used the notation for two-by-two *determinants* in this definition. Note that any two-by-two determinant

$$\begin{vmatrix} a & b \\ c & d \end{vmatrix}$$

where a, b, c and d are any real numbers is evaluated follows:

$$\begin{vmatrix} a & b \\ c & d \end{vmatrix} = ad - bc.$$

The cross-product of two vectors is often referred to as the *vector product* of the vectors to distinguish it from the dot (or scalar) product which is a scalar (or number) quantity.

(a) Write a **computer function** which computes the cross-product of any two vectors in $\mathcal{R}^3$. Test your **computer function** on the following vectors

$$\mathbf{u} = 3\mathbf{i} + \mathbf{j} + 2\mathbf{k} \quad \text{and} \quad \mathbf{v} = 2\mathbf{i} - 5\mathbf{j} + \mathbf{k}.$$

(b) For the vectors **u** and **v** in part (a), find $\mathbf{v} \times \mathbf{u}$. Do you notice anything interesting? What does this tell you about the order of taking the cross-product?

(c) For the vectors **u** and **v** in part (a), find $\mathbf{u} \cdot (\mathbf{u} \times \mathbf{v})$ and $\mathbf{v} \cdot (\mathbf{u} \times \mathbf{v})$. Do you notice anything interesting?

7.2.3 Discussion

Three Dimensions — Rectangular Coordinates The *rectangular coordinate system* in three dimensions is completely analogous to the two-dimensional situation and you should have no difficulty with it. It is sometimes called an *xyz*-coordinate system. See Figure 7.10.

We denote the set of all triples of real numbers by $\mathcal{R}^3$.

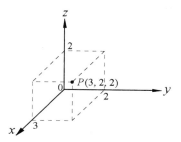

Figure 7.10. *A three dimensional rectangular coordinate system.*

Distance and Midpoint in $\mathcal{R}^3$

As for two dimensions, the Pythagorean Theorem helps us to derive a formula for the distance between two points in three dimensions. If S and T are the points with coordinates as shown in Figure 7.11, then the length of the line segment connecting them is

$$\sqrt{(x_1 - x_2)^2 + (y_1 - y_2)^2 + (z_1 - z_2)^2}.$$

Do you see the similarity between the distance formula for two and three dimensions? How are they related?

Once again in three dimensions, the midpoint of a line segment between its two endpoints is, essentially, the average of the endpoints. Thus its coordinates are obtained by averaging the coordinates of the endpoints. As you can see in the picture in Figure 7.11, the midpoint U of the line connecting S and T has coordinates,

$$\left[\frac{x_1 + x_2}{2}, \frac{y_1 + y_2}{2}, \frac{z_1 + z_2}{2} \right]$$

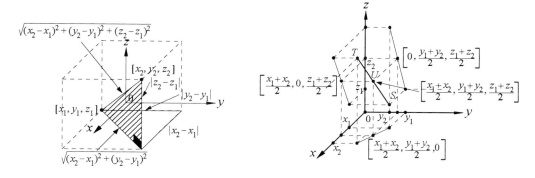

Figure 7.11. *Distance and midpoint formulas in three dimensions.*

You will see how to represent lines and planes using rectangular *xyz*-coordinate equations at the end of this section.

Vectors

In Activities 2 through 11, you worked with the idea of a vector — a line segment with an arrow on one end. Each *directed line segment* from one point, its *tail* or *initial point*, to another point, its *head* or *terminal point*, determines a vector; different line segments determine the same vector if and only if they have the same length and direction. It is important to keep both the tail and head in mind when thinking about a vector geometrically. When representing vectors analytically, however, we often move them to the origin and just specify the head. Thus a vector in two (respectively three) dimensions can be represented by a list (or tuple) of two (in the plane) or three (in space) coordinates.

Representations of Vectors — Components

We use the same notation for a vector and a point in $\mathcal{R}^2$ and also in $\mathcal{R}^3$. That is, in $\mathcal{R}^2$ we use an ordered pair or tuple of numbers. For example, the ordered pair $[2, -1]$ can represent a point, or the vector $\mathbf{v} = [2, -1]$ with initial point the origin, $[0, 0]$, and terminal point at $[2, -1]$. Similarly, in $\mathcal{R}^3$, for example, the ordered triple $[3, -2, 1]$ can represent a point or a vector $\mathbf{u} = [3, -2, 1]$ with initial point the origin, $[0, 0, 0]$, and terminal point at $[3, -2, 1]$. The meaning will be clear from the context in which the notation is used. However, in some texts a different notation may be used to distinguish a vector from a point. For example, the vector from the point P with rectangular coordinates $[-2.5, -1]$ to Q with coordinates $[1, 1]$ can be referred to as the vector $\mathbf{v} = [3.5, 2]$ emanating from P to Q, where $[3.5, 2] = [1 - (-2.5), 1 - (-1)]$. That is, the coordinates of $\mathbf{v}$ are just the coordinates of Q minus the corresponding coordinates of P. Do you see why this is so? What about the coordinates of the vector $\mathbf{w} = -\mathbf{v}$ from Q to P? See Figure 7.12.

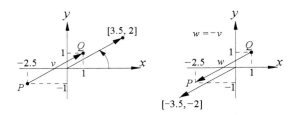

Figure 7.12. The vectors v from P to Q and –v from Q to P.

Another way of specifying vectors that is often used in the physical sciences is in terms of two "unit vectors" **i** and **j**. These are the vectors whose rectangular coordinates are $[1, 0]$ and $[0, 1]$, respectively. Hence, $\mathbf{i} = [1, 0]$ and $\mathbf{j} = [0, 1]$. As you saw in Activity 6, every vector in two dimensions can be written as a "linear combination" (a sum of multiples) of these two vectors. In Activity 6, you showed that any vector $\mathbf{u}$ in $\mathcal{R}^2$ can be written as a linear combination of the unit vectors **i** and **j** as follows:

$$\mathbf{u} = |\mathbf{u}|\cos(\theta)\mathbf{i} + |\mathbf{v}|\sin(\theta)\mathbf{j}$$

where θ is the angle between $\mathbf{u}$ and the positive x-axis (or the unit vector **i**). When a vector $\mathbf{u}$ is written in this manner, we say the $\mathbf{u}$ is *resolved* into its **component vectors**,

$$|\mathbf{u}|\cos(\theta)\mathbf{i} \quad \text{and} \quad |\mathbf{v}|\sin(\theta)\mathbf{j}.$$

The **i** component and **j** component for $\mathbf{u}$ are, respectively, $|\mathbf{u}|\cos(\theta)$ and $|\mathbf{u}|\sin(\theta)$. See Figure 7.13.

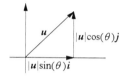

Figure 7.13. *A vector* $\mathbf{u} = |\mathbf{u}|\cos(\theta)\mathbf{i} + |\mathbf{v}|\sin(\theta)\mathbf{j}$.

It is similar in three dimensions. The two most common ways of expressing a vector in three dimensions is as a triple of numbers or as a linear combination of the unit vectors **i**, **j**, and **k**. For a vector $\mathbf{u} = a\mathbf{i} + b\mathbf{j} + c\mathbf{k}$, the coefficients a, b and c of **i**, **j** and **k** are referred to as the **i**, **j** and **k** components of **u**, respectively. Similarly, any vector **v** in $\mathcal{R}^2$ can be written as $a\mathbf{i} + b\mathbf{j}$ for some numbers a and b called the **i** and **j** components, respectively, of **v**. See the vector in Figure 7.14. We will return to this point later in this section when we discuss the concepts of the direction angles and the direction cosines of a vector in $\mathcal{R}^3$.

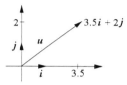

Figure 7.14. *The vector* $\mathbf{v} = 3.5\mathbf{i} + 2\mathbf{j}$.

Can you think of anything interesting about vectors in one dimension?

Vector Operations — Scalar Multiplication   There are several ways in which you can combine vectors and numbers to get new vectors. You can multiply a vector by a number. You can add two vectors.

As we said earlier, in mathematics, when talking about vectors and numbers it is common practice to use the term *scalar* to refer to the number. Multiplying a vector **v** by a number (or scalar) t has a specific effect. To see this, let $\mathbf{v} = a\mathbf{i} + b\mathbf{j}$. Then $t\mathbf{v} = ta\mathbf{i} + tb\mathbf{j}$. Thus, if $t > 0$ then $t\mathbf{v}$ is a vector in the same direction with its length stretched (or contracted) by an amount t. If $t < 0$ then $t\mathbf{v}$ is a vector in the opposite direction with its length stretched (or contracted) by an amount $|t|$. See Figure 7.15. Can you specify a value of t for which there is a stretching? a contracting?

It is exactly the same in three dimensions.

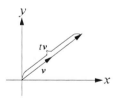

Figure 7.15. *Scalar multiplication of vectors.*

Vector Operations — Addition and Subtraction

We can analyze the sum or difference of two vectors in a similar manner. The sum of two vectors is also called the *resultant* vector. (See the exercises for some applications of this operation.) This time let's use a tuple of coordinates and work in three dimensions.

Let $\mathbf{v} = [a_1, a_2]$ and $\mathbf{w} = [b_1, b_2]$. Then we take

$$\mathbf{v} + \mathbf{w} = [a_1 + b_1, a_2 + b_2].$$

You can see by plotting points that $\mathbf{v} + \mathbf{w}$ is the diagonal of the parallelogram determined by $\mathbf{v}$ and $\mathbf{w}$. Can you prove this? See Figure 7.16.

You can also see from the picture that another way of visualizing $\mathbf{v} + \mathbf{w}$ is to imagine sliding one of the vectors, say $\mathbf{w}$, parallel to itself so that its tail coincides with the head of the other, $\mathbf{v}$. Then $\mathbf{v} + \mathbf{w}$ is the vector with tail at the tail of $\mathbf{v}$ and head at the head of the vector resulting from sliding $\mathbf{w}$. See Figure 7.17.

Vector addition is exactly the same in three dimensions, that is, you add component-wise.

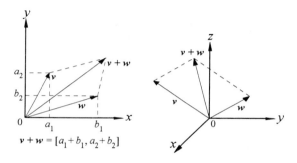

Figure 7.16. Vector addition in two and three dimensions.

How would you subtract two vectors? Recall that subtraction of two real numbers a and b is given by $a - b = a + (-b)$. In a similar manner, we define vector subtraction of the vectors $\mathbf{v} = [a_1, a_2]$ and $\mathbf{w} = [b_1, b_2]$ by

$$\mathbf{v} - \mathbf{w} = \mathbf{v} + (-\mathbf{w}) = [a_1, a_2] + [-b_1, -b_2] = [a_1 - b_1, a_2 - b_2].$$

See Figure 7.18.

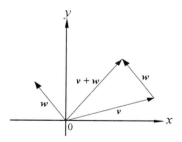

Figure 7.17. Vector addition geometrically.

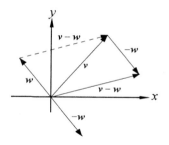

Figure 7.18. Vector subtraction.

Vector subtraction in three dimensions is also done by subtracting component-wise.

Vector Representations of Lines in $\mathcal{R}^2$ and $\mathcal{R}^3$

Any line L in two dimensions can be represented in the vector form, $\mathbf{v} = \mathbf{a} + t\mathbf{b}$ for some vectors $\mathbf{a}$ and $\mathbf{b}$ where the terminal point or tip of $\mathbf{a}$ is any point on the line L and b is any vector which lies along the line L. Are the vectors $\mathbf{a}$ and $\mathbf{b}$ unique?

There are two main points here. First, that the multiples of a vector comprise the points on a line through the origin and in the direction of the vector. Thus we consider all real numbers t as giving multiples $t\mathbf{b}$ of the vector $\mathbf{b}$. And, second, we use the fact that adding the vector $\mathbf{a}$ has the effect of sliding along the vector $\mathbf{a}$. See Figure 7.19.

We can summarize this discussion as follows. Now that you have an understanding of the idea of scalar multiplication and addition of vectors, you can see how the expression $\mathbf{v} = \mathbf{a} + t\mathbf{b}$ represents a straight line through the head of $\mathbf{a}$ and parallel to $\mathbf{b}$. The interpretation of this expression is that the vectors $\mathbf{a}$ and $\mathbf{b}$ are fixed, while the variable t runs through all real numbers. The values of $\mathbf{v}$ that are obtained consist of all multiples of $\mathbf{b}$ (which amounts to the line through the origin in the direction of $\mathbf{b}$) translated by $\mathbf{a}$. Look once again at Figure 7.19.

The vector form in three dimensions is identical to the two dimensional situation, so it also has the form, $\mathbf{v} = \mathbf{a} + t\mathbf{b}$ for some vectors $\mathbf{a}$ and $\mathbf{b}$ in $\mathcal{R}^3$. See Figure 7.20.

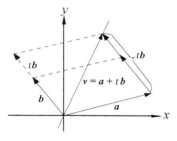

Figure 7.19. A portion of a line in $\mathcal{R}^2$ represented by the vector form $\mathbf{v} = \mathbf{a} + t\mathbf{b}$.

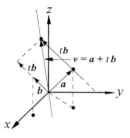

Figure 7.20. A portion of a line $\mathcal{R}^3$ represented by the vector form $\mathbf{v} = \mathbf{a} + t\mathbf{b}$.

We can summarize this discussion as follows. Now that you have an understanding of the idea of scalar multiplication and addition of vectors, you can see how the expression $\mathbf{v} = \mathbf{a} + t\mathbf{b}$ represents a straight line through the head of $\mathbf{a}$ and parallel to $\mathbf{b}$. The interpretation of this expression is that the vectors $\mathbf{a}$ and $\mathbf{b}$ are fixed, while the variable t runs through all real numbers. The values of $\mathbf{v}$ that are obtained consist of all multiples of $\mathbf{b}$ (which amounts to the line through the origin in the direction of $\mathbf{b}$) translated by $\mathbf{a}$. Look once again at Figure 7.19.

The vector form in three dimensions is identical to the two dimensional situation, so it also has the form, $\mathbf{v} = \mathbf{a} + t\mathbf{b}$ for some vectors $\mathbf{a}$ and $\mathbf{b}$ in $\mathcal{R}^3$. See Figure 7.20.

Length of a Vector

The length of a vector $\mathbf{v}$ is the distance between its initial point (or tail) to its terminal point (or head). It is denoted $|\mathbf{v}|$. Thus if $\mathbf{v} = [v_1, v_2]$ and $\mathbf{w} = [w_1, w_2, w_3]$ are vectors in two and three dimensions respectively with their tails at the origin, then we have the following formulas by the *Pythagorean Theorem*. See Figure 7.21.

$$|\mathbf{v}| = \sqrt{v_1^2 + v_2^2} \,.$$

$$|\mathbf{w}| = \sqrt{w_1^2 + w_2^2 + w_3^2} \,.$$

Angles — the Dot Product of Two Vectors   A standard trigonometric identity can be used to derive a formula for the cosine of the smallest angle, or what we will commonly refer to as "the angle between two non-zero vectors". Let $\mathbf{x} = [x_1, x_2]$ and $\mathbf{y} = [y_1, y_2]$ as shown in Figure 7.22.

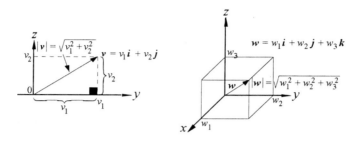

Figure 7.21. The length of a vector.

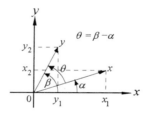

Figure 7.22. The (smallest) angle between two vectors.

Referring to the diagram in Figure 7.22, we have,

$$\cos(\theta) = \cos(\beta - \alpha)$$

$$= \cos(\beta)\cos(\alpha) + \sin(\beta)\sin(\alpha)$$

$$= \frac{x_1}{|\mathbf{x}|}\frac{y_1}{|\mathbf{y}|} + \frac{x_2}{|\mathbf{x}|}\frac{y_2}{|\mathbf{y}|}$$

$$= \frac{x_1 y_1 + x_2 y_2}{|\mathbf{x}||\mathbf{y}|}$$

If we define the *dot product* of the vectors $\mathbf{x}$, $\mathbf{y}$ by

$$\mathbf{x} \cdot \mathbf{y} = x_1 y_1 + x_2 y_2$$

then we have the simple formula

$$\cos(\theta) = \frac{\mathbf{x} \cdot \mathbf{y}}{|\mathbf{x}||\mathbf{y}|}$$

or

$$\mathbf{x} \cdot \mathbf{y} = |\mathbf{x}||\mathbf{y}|\cos(\theta)$$

where θ is the angle between the vectors $\mathbf{x}$, $\mathbf{y}$.

This formula has many applications. One is to the question of slopes of orthogonal lines that we discussed before. Suppose that you have two lines that intersect at the point S. To study the (smallest) angle between them and their slopes, we may translate everything to the origin. See Figure 7.23.

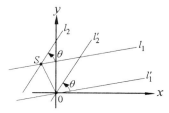

Figure 7.23. *An angle between two lines.*

Now take any points $v = [a, b]$ and $w = [c, d]$ on the lines l_1, l_2, respectively. (You can't use the origin. Why?) The slope of l_1 is then $m_1 = b/a$ and the slope of l_2 is $m_2 = c/d$. To say that the two lines are orthogonal means that the cosine of the angle between them is zero. By the above formula, this is exactly the same as saying that the dot product of the two vectors is 0. That is, $ac + bd = 0$ or,

$$1 + \frac{bd}{ac} = 0.$$

(What happens if a or c is 0?)

By substitution you can see that this amounts to the relation,

$$1 + m_1 m_2 = 0$$

or

$$m_1 = -\frac{1}{m_2}$$

In other words, *two lines in the plane are perpendicular if and only if their slopes are negative reciprocals of each other.*

Dot Products in Three Dimensions

The analysis is more complicated in three dimensions. We have

$$\mathbf{v} \cdot \mathbf{w} = v_1 w_1 + v_2 w_2 + v_3 w_3$$

and we still get the formula,

$$\mathbf{v} \cdot \mathbf{w} = |\mathbf{v}||\mathbf{w}|\cos(\theta)$$

where $\mathbf{v}$ and $\mathbf{w}$ are vectors and θ is the angle between them.

    Although a line does not have a unique perpendicular in three dimensions, this formula still tells us that two vectors are orthogonal if and only if their dot product is 0 and they are parallel (angle between them is 0) if and only if their dot product is equal to the product of their lengths.

    There is also a geometric interpretation of dot product which works the same in two or three dimensions. Looking at the picture you can see that the dot product of **v** and **w** is the length of the "projection" of **v** onto **w**, $|\mathbf{v}|\cos(\theta)$, times the length of **w**, $|\mathbf{w}|$. See Figure 7.24.

Area of the Region Determined by Two Vectors

In Activity 5 you investigated the area resulting from two different vectors in $\mathcal{R}^2$ (or $\mathcal{R}^3$). The length and dot product of vectors can be used to give a very simple formula for the area of the parallelogram determined by two non-zero and non-parallel vectors **v** and **w**. This parallelogram is obtained by sliding each of the two vectors parallel to itself until its tail coincides with the head of the other as shown in Figure 7.25. It can be shown that the resulting area is given by

$$\sqrt{\left(|\mathbf{v}|^2|\mathbf{w}|^2 - (\mathbf{v}\cdot\mathbf{w})^2\right)}.$$

You will get a chance to derive the latter formula in the exercises.

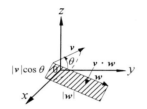

Figure 7.24. $\mathbf{v}\cdot\mathbf{w} = |\mathbf{v}||\mathbf{w}|\cos(\theta)$

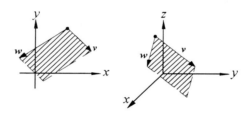

Figure 7.25. The area of a parallelogram.

    Notice that this formula works just as well in three dimensions as in two.

Direction Cosines and Angles

In Activity 9 you investigated the direction cosines and direction angles of a vector in $\mathcal{R}^3$. You saw in Activity 9 that any vector **u** in $\mathcal{R}^3$ can be written in the following form

$$\mathbf{u} = |\mathbf{u}|\cos(\alpha)\mathbf{i} + |\mathbf{u}|\cos(\beta)\mathbf{j} + |\mathbf{u}|\cos(\gamma)\mathbf{k}$$

where $|\mathbf{u}|\cos(\alpha)$, $|\mathbf{u}|\cos(\beta)$, and $|\mathbf{u}|\cos(\gamma)$ are called the **i**, **j** and **k** components, respectively, of the vector **u**.

    The *direction cosines* of any non-zero vector in $\mathcal{R}^3$ can be obtained by forming the *unit vector* (a vector of length one) given by

$$\frac{1}{|\mathbf{u}|}\mathbf{u} = \frac{\mathbf{u}}{|\mathbf{u}|} = \cos(\alpha)\mathbf{i} + \cos(\beta)\mathbf{j} + \cos(\gamma)\mathbf{k}.$$

The direction cosines of the vector **u** in $\mathcal{R}^3$ are given by $\cos(\alpha)$, $\cos(\beta)$ and $\cos(\gamma)$. The angles α, β, and γ are the angles made by the vector **u** and the vectors **i**, **j**, and **k**, respectively. See Figure 7.26.

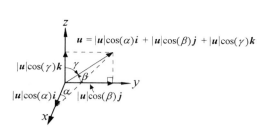

Figure 7.26. Direction angles and direction cosines.

Note that any vector **u** in $\mathcal{R}^2$ can be written in the form $\mathbf{u} = |\mathbf{u}|\cos(\theta)\mathbf{i} + |\mathbf{u}|\sin(\theta)\mathbf{j}$ where θ is the (smallest) angle between **u** and the positive x-axis (or the vector **i**). The direction cosines for such a vector **u** in $\mathcal{R}^2$ are given by $\cos(\theta)$ and $\cos(\frac{\pi}{2} - \theta)$. Do you see why? Do you recall the following trigonometric (co-function) identity?

$$\sin(\theta) = \cos\left(\frac{\pi}{2} - \theta\right)$$

Projections   In Activity 10, you investigated the notion of a projection of one vector onto another. For two non-zero vectors **u** and **v**, in $\mathcal{R}^2$ (or $\mathcal{R}^3$), (i.e., two vectors such that $|\mathbf{u}| \neq 0$ and $|\mathbf{v}| \neq 0$) the *projection of* **u** *in the direction of* **v**, denoted $\mathbf{proj_v u}$, is given by

$$\mathbf{proj_v u} = |\mathbf{u}|\cos(\theta)\frac{\mathbf{v}}{|\mathbf{v}|}$$

where θ is the angle between **u** and **v**. The number given by $|\mathbf{u}|\cos(\theta)$ is called the *component of* **u** *along* **v**.

For any two non-zero vectors **u** and **v** in $\mathcal{R}^2$ or $\mathcal{R}^3$, the projection of **u** in the direction of **v** is also given by

$$\mathbf{proj_v u} = \left(\frac{\mathbf{u} \cdot \mathbf{v}}{\mathbf{v} \cdot \mathbf{v}}\right)\mathbf{v} = \left(\frac{\mathbf{u} \cdot \mathbf{v}}{|\mathbf{v}|^2}\right)\mathbf{v}.$$

Do you see why this is true? This can be shown using the fact that

$$\mathbf{u} \cdot \mathbf{v} = |\mathbf{u}||\mathbf{v}|\cos(\theta)$$

where θ is the (smallest) angle between **u** and **v**. It follows from this formula for the dot product that

$$\frac{\mathbf{u} \cdot \mathbf{v}}{|\mathbf{v}|} = |\mathbf{u}|\cos(\theta).$$

So, substituting from the latter equality for $\mathbf{u} = \cos(\theta)$ and using the definition of projection, we have

$$\mathbf{proj_v u} = |\mathbf{u}|\cos(\theta)\frac{\mathbf{v}}{|\mathbf{v}|} = \frac{\mathbf{u} \cdot \mathbf{v}}{|\mathbf{v}|}\frac{\mathbf{v}}{|\mathbf{v}|} = \left(\frac{\mathbf{u} \cdot \mathbf{v}}{|\mathbf{v}|^2}\right)\mathbf{v}$$

or

$$\mathbf{proj_v u} = \left(\frac{\mathbf{u} \cdot \mathbf{v}}{\mathbf{v} \cdot \mathbf{v}}\right)\mathbf{v}.$$

Do you see why the latter equation is true? Do you see that $|\mathbf{v}|^2 = \mathbf{v} \cdot \mathbf{v}$? Suppose that $\mathbf{v} = v_1\mathbf{i} + v_2\mathbf{j}$. Then, $|\mathbf{v}|^2 = v_1^2 + v_2^2 = \mathbf{v} \cdot \mathbf{v}$. The same is true for vectors in $\mathcal{R}^3$.

As you saw in Activity 10, the dot product of $\mathbf{u} - \text{proj}_v\mathbf{u}$ and $\mathbf{v}$ is zero. What does this result say about the vectors $\mathbf{u} - \text{proj}_v\mathbf{u}$ and $\mathbf{v}$? Yes, the vectors $\mathbf{u} - \text{proj}_v\mathbf{u}$ and $\mathbf{v}$ are perpendicular.

As you also saw in Activity 10, the two vectors $\mathbf{u}_1 = \text{proj}_v\mathbf{u}$ and $\mathbf{u}_2 = \mathbf{u} - \text{proj}_v\mathbf{u}$ are such that $\mathbf{u} = \mathbf{u}_1 + \mathbf{u}_2$ with $\mathbf{u}_1 = \text{proj}_v\mathbf{u}$ parallel to $\mathbf{v}$ and $\mathbf{u}_2 = \mathbf{u} - \text{proj}_v\mathbf{u}$ perpendicular to $\mathbf{v}$. This provides another useful way of resolving a vector into component vectors.

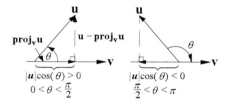

Figure 7.27. Projections of vectors.

Cross-Products

In Activity 11, you investigated the notion of the cross-product of vectors in $\mathcal{R}^3$. For any two non-zero vectors $\mathbf{u} = [u_1, u_2, u_3]$ and $\mathbf{v} = [v_1, v_2, v_3]$ in three dimensions, the cross-product is denoted $\mathbf{u} \times \mathbf{v}$ and it is a vector given by

$$\mathbf{u} \times \mathbf{v} = \begin{vmatrix} u_2 & u_3 \\ v_2 & v_3 \end{vmatrix}\mathbf{i} - \begin{vmatrix} u_1 & u_3 \\ v_1 & v_3 \end{vmatrix}\mathbf{j} + \begin{vmatrix} u_1 & u_2 \\ v_1 & v_2 \end{vmatrix}\mathbf{k}$$

or symbolically by

$$\mathbf{u} \times \mathbf{v} = \begin{vmatrix} \mathbf{i} & \mathbf{j} & \mathbf{k} \\ u_1 & u_2 & u_3 \\ v_1 & v_2 & v_3 \end{vmatrix} = (u_2v_3 - v_2u_3)\mathbf{i} - (u_1v_3 - v_1u_3)\mathbf{j} + (u_1v_2 - v_1u_2)\mathbf{k}$$

where we have used the notation for two-by-two *determinants* in this definition and you should also remember that the cross-product of two vectors is actually defined by the latter of the three expressions above. The notation

$$\begin{vmatrix} \mathbf{i} & \mathbf{j} & \mathbf{k} \\ u_1 & u_2 & u_3 \\ v_1 & v_2 & v_3 \end{vmatrix}$$

actually has no meaning. The symbolism of three-by-three determinants and how they are evaluated using two-by-two determinants is used to help remember the definition of the cross-product.

The cross-product of two vectors is often referred to as the *vector product* of the vectors to distinguish it from the dot (or scalar) product which is a scalar (or number) quantity.

The cross-product of two non-zero vectors in $\mathcal{R}^3$ yields a vector $\mathbf{u} \times \mathbf{v}$ which is perpendicular to the plane determined by $\mathbf{u}$ and $\mathbf{v}$. Moreover, the vector $\mathbf{u} \times \mathbf{v}$ points upward in the direction of the thumb on your right hand if you curl your fingers from $\mathbf{u}$ to $\mathbf{v}$. This is called the *right-hand rule* for determining the direction of $\mathbf{u} \times \mathbf{v}$. What about $\mathbf{v} \times \mathbf{u}$? It is not hard to show that $\mathbf{u} \times \mathbf{v} = -(\mathbf{v} \times \mathbf{u})$. See Figure 7.28.

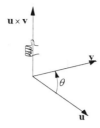

Figure 7.28. Cross-products and right-hand rule.

For two different non-zero vectors **u** and **v** in $\mathcal{R}^3$, we have the following formula which is reminiscent of the formula for the dot product of two vectors.

$$|\mathbf{u} \times \mathbf{v}| = |\mathbf{u}||\mathbf{v}|\sin(\theta)$$

where θ is the (smallest) angle between two non-zero vectors **u** and **v**. We leave the derivation of the formula as an exercise.

You will get a chance to investigate and derive other important properties of the cross-product in the exercises.

Volumes of Regions Determined by Three Vectors

Suppose you have three different vectors **u**, **v** and **w** in space. Then the three vectors can be thought of as sides of a parallelopiped as shown in Figure 7.29. Do you see how to find the volume of the parallelepiped? Look at Figure 7.29 again.

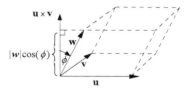

Figure 7.29. The volume of a parallelepiped.

It turns out that the volume V of the parallelepiped with base formed by the vectors **u** and **v**, like that shown in Figure 7.29, is given by

$$V = |\mathbf{u} \times \mathbf{v}||\mathbf{w}||\cos(\phi)| = |(\mathbf{u} \times \mathbf{v}) \cdot \mathbf{w}|$$

where ϕ is the (smallest) angle made by the vector $\mathbf{u} \times \mathbf{v}$ and **w**. The derivation of the above formula is left as an exercise.

Rectangular Representation of Planes — Dot Product

The dot product can be used to derive the rectangular representation of a plane Π when a (normal) vector $\mathbf{n} = [a, b, c]$ perpendicular to the plane Π is known. Do you see how to derive the equation of a plane through the point $P_0 = [x_0, y_0, z_0]$ when $P = [x, y, z]$ is any other point on the plane different than P_0 in this situation? See Figure 7.30.

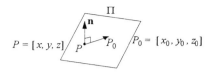

Figure 7.30. Rectangular representation of planes.

Yes, you just take the dot product of the (normal) vector $\mathbf{n} = a\mathbf{i}+b\mathbf{j}+c\mathbf{k}$ and the vector $\mathbf{v} = (x-x_0)\mathbf{i}+(y-y_0)\mathbf{j}+(z-z_0)\mathbf{k}$ to obtain the vector equation

$$\mathbf{n}\cdot\mathbf{v} = (a\mathbf{i}+b\mathbf{j}+c\mathbf{k})\cdot((x-x_0)\mathbf{i}+(y-y_0)\mathbf{j}(z-z_0)\mathbf{k}) = 0.$$

Taking the dot product and simplifying, we obtain

$$a(x-x_0)+b(y-y_0)+c(z-z_0) = 0$$

as the equation of a plane for the situation shown in Figure 7.30. Simplifying the latter equation, we obtain

$$ax+by+cz+(-ax_0-by_0-cz_0) = 0.$$

Now letting $d = -ax_0-by_0-cz_0$, the latter equation gives us the general equation of a plane in rectangular coordinates

$$ax+by+cz+d = 0.$$

In the spirit of Chapter 1, we refer to the latter form of an equation of a plane in rectangular coordinates as the *relational form* of a plane, since a point $[x, y, z]$ is on the plane if and only if the equation $ax+by+cz+d = 0$ is satisfied.

Rectangular Representation of Planes — Cross-Product
Suppose you know three different points which lie on a plane Π. Can you think of a way to use the cross-product of two vectors to find an equation of the plane? For example, consider any three different points $P_0 = [x_0, y_0, z_0]$, $P_1 = [x_1, y_1, z_1]$, and $P_2 = [x_1, y_1, z_1]$. Then, you can find two vectors which lie in Π, say the vector $\mathbf{u}$ from P_0 to P_1 and the vector $\mathbf{v}$ from P_0 to P_2. Now, how can you find a vector $\mathbf{n}$ perpendicular to the plane Π? Yes, that's right, you can use the cross-product of the vectors $\mathbf{u} = [x_1-x_0, y_1-y_0, z_1-z_0]$ and $\mathbf{v} = [x_1-x_0, y_1-y_0, z_1-z_0]$ which produces the (normal) vector $\mathbf{n} = \mathbf{u}\times\mathbf{v}$ that is perpendicular to the plane Π containing $\mathbf{u}$ and $\mathbf{v}$. Now, you can use the method in the previous discussion on the dot product and rectangular representations of planes to find the equation of the plane Π. You will get a chance to work with this method in the exercises.

7.2.4 Exercises

1. Sketch the indicated vectors on the same coordinate system.

   (a) $3\mathbf{i}$, $5\mathbf{j}$ and $3\mathbf{i}+5\mathbf{j}$

   (b) $\mathbf{i}+2\mathbf{j}$ and $2\mathbf{i}+4\mathbf{j}$

   (c) $\mathbf{i}-3\mathbf{j}$ and $-2\mathbf{i}+6\mathbf{j}$

   (d) $2\mathbf{i}$, $-3\mathbf{j}$ and $2\mathbf{i}-3\mathbf{j}$

2. In each of the following situations, represent the point or vector in the indicated form(s).

    (a)

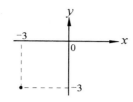

      rectangular and polar coordinates

    (b)

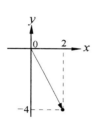

        i, **j** notation

    (c)

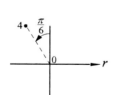

      polar and rectangular coordinates

3. In each of the following situations, find an analytic description of the line L in vector form.

    (a)

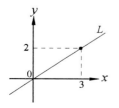

(b)

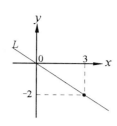

(c)

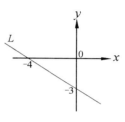

(d)

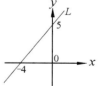

(e)

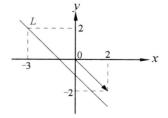

(f)

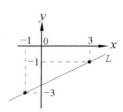

4. What are the rectangular coordinates of the head of each of the vectors **i**, **j** and **k** in three dimensions?

5. Explain why it is the case that every point in the plane has a unique representation as a linear combination of **i** and **j**.

6. Explain why it is the case that every point in space has a unique representation as a linear combination of **i**, **j** and **k**.

7. Compute the indicated quantity in each of the following situations where $\mathbf{a} = [2, -1]$, $\mathbf{b} = [3, 4]$, $\mathbf{u} = [1, 0, 2]$ and $\mathbf{v} = [-2, 1, 3]$

 (a) $\mathbf{a} + \mathbf{b}$ (b) $\mathbf{b} - \mathbf{a}$ (c) $3\mathbf{a}$

 (d) $\mathbf{a} - 2\mathbf{b}$ (e) $|\mathbf{a}|$ (f) $|\mathbf{a} - \mathbf{b}|$

 (g) $|\mathbf{u}|$ (h) $|-2\mathbf{v}|$ (i) $\mathbf{a} \cdot \mathbf{b}$

 (j) $\mathbf{u} \cdot \mathbf{v}$ (k) $2\mathbf{u} - 3\mathbf{v}$

8. Compute and sketch the vectors $\mathbf{a} + \mathbf{b}$, $\mathbf{a} - \mathbf{b}$, $-\mathbf{a}$ and $-\mathbf{b}$, for the indicated vectors.

 (a) $\mathbf{a} = \mathbf{i} - \mathbf{j}$ and $\mathbf{b} = 2\mathbf{i}$

 (b) $\mathbf{a} = 3\mathbf{i} + \mathbf{j}$ and $\mathbf{b} = \mathbf{i} - 2\mathbf{j}$

 (c) $\mathbf{a} = \mathbf{i} + \mathbf{j}$ and $\mathbf{b} = 2\mathbf{i} + \mathbf{j}$

 (d) $\mathbf{a} = \mathbf{i} - 2\mathbf{j}$ and $\mathbf{b} = -2\mathbf{i} + \mathbf{j}$

 (e) $\mathbf{a} = \mathbf{i} - \mathbf{j} + \mathbf{k}$ and $\mathbf{b} = 2\mathbf{i} + \mathbf{j} - \mathbf{k}$

 (f) $\mathbf{a} = \mathbf{i} + \mathbf{j} + \mathbf{k}$ and $\mathbf{b} = 2\mathbf{i} - \mathbf{j} + 3\mathbf{k}$

9. For each pair of points P, Q,

 (a) Find the vector with tail at the origin which is parallel to the vector with

 tail at the point P and head at the point Q.

 (b) Find the vector with tail at the origin which is parallel to the vector with tail at the point Q and head at the point P. How does this vector differ from the one in part (a)?

 (c) Find the length of the vectors PQ and QP. What can you conclude in each case?

  i. $P = [0, 0]$ and $Q = [4, 1]$

  ii. $P = [0, 1]$ and $Q = [2, -1]$

  iii. $P = [-1, 3]$ and $Q = [5, 4]$

  iv. $P = [0, 0, 0]$ and $Q = [3, -1, 4]$

  v. $P = [1, 0, -2]$ and $Q = [3, -1, 4]$

  vi. $P = [2, 3, -1]$ and $Q = [4, 1, 5]$

10. Find the angle between the indicated vectors.

 (a) $\mathbf{a} = [2, -1]$ and $\mathbf{b} = [3, 4]$

 (b) $\mathbf{u} = 3\mathbf{i} - 4\mathbf{j}$ and $\mathbf{v} = -8\mathbf{i} + 6\mathbf{j}$

 (c) $\mathbf{a} = [1, 0, 2]$ and $\mathbf{b} = [-2, 1, 3]$

 (d) $\mathbf{u} = \mathbf{i} - \mathbf{j} + 2\mathbf{k}$ and $\mathbf{v} = -2\mathbf{i} + \mathbf{i} + \mathbf{j}$

11. Find two unit vectors (vectors of length one) orthogonal to each vector.

 (a) $-3\mathbf{j}$ (b) $5\mathbf{i}$

 (c) $[-2, 1]$ (d) $3\mathbf{i} - 4\mathbf{j}$

 (e) $12\mathbf{i} + 5\mathbf{j}$ (f) $[-2, 1, 3]$

12. For the forces $\mathbf{F}_1$ and $\mathbf{F}_2$ in each of the following situations, find the *resultant force*.

 (a) $\mathbf{F}_1 = 3\mathbf{j}$ and $\mathbf{F}_2 = 2\mathbf{i} - 3\mathbf{j}$

 (b) $\mathbf{F}_1 = \mathbf{i} - 2\mathbf{j}$ and $\mathbf{F}_2 = 3\mathbf{i} + 5\mathbf{j}$

(c) $\mathbf{F}_1 = \mathbf{i} - 2\mathbf{j} + \mathbf{k}$ and $\mathbf{F}_2 = 3\mathbf{i} + 5\mathbf{k}$

(d)

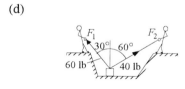

(e)

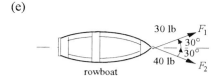

rowboat

13. Find the force which balances the resultant of the forces in each of the following situations.

(a) The forces in part (a) of Exercise 12.

(b) The forces in part (b) of Exercise 12.

(c) The forces in part (c) of Exercise 12.

(d) The forces in part (d) of Exercise 12.

(e) The forces in part (e) of Exercise 12.

14. The *work w* done by a constant force $\mathbf{F}$ as its point of application moves along a *displacement vector* $\mathbf{D}$ is given by

$$w = \mathbf{F} \cdot \mathbf{D} = |\mathbf{F}||\mathbf{D}|\cos(\theta)$$

where θ is the (smallest) angle between the vectors $\mathbf{F}$ and $\mathbf{D}$. Do the following for this situation.

(a) Find the work done by the force $\mathbf{F} = 30\mathbf{i} + 40\mathbf{j}$, where the magnitude of $\mathbf{F}$ is measured in newtons, as the point of application of $\mathbf{F}$ moves along the line segment from the point $[0, 2]$ to the point $[3, 10]$, where the distance is measured in meters.

(b) A constant force $\mathbf{F}$, with a magnitude of 20 newtons, acts in the direction of the positive z-axis. Do the following for these situations. Assume that distance is measured in meters.

  i. Find the work done by $\mathbf{F}$ as the point of its application moves along the line segment from the origin to the point $[2, 1, 3]$.

  ii. Find the work done by $\mathbf{F}$ as the point of its application moves along the line segment from the point $[2, 1, 3]$ to the origin.

  iii. Compare your answers in parts (a) and (b). Do you notice anything interesting? What general statement do you think you can make about the work done by a force as its point of application moves along a line segment and the work done by the force as its point of application moves in the opposite direction the same distance?

15. Show that the area A of a parallelogram determined by two vectors $\mathbf{v}$ and $\mathbf{w}$ is given by the formula

$$A = \sqrt{|\mathbf{v}|^2 |\mathbf{w}|^2 - (\mathbf{v} \cdot \mathbf{w})^2}.$$

(Hint: First derive a formula for the area A in terms of $|\mathbf{v}|$ and $|\mathbf{w}|$ and the $\sin(\theta)$ where θ is the (smallest) angle between $\mathbf{v}$ and $\mathbf{w}$.)

16. For any two non-zero vectors $\mathbf{u}$ and $\mathbf{v}$, show that

$$\mathbf{u} \times \mathbf{v} = -(\mathbf{v} \times \mathbf{u}).$$

What does this tell you about the order of taking the cross-product?

17. Find the dot product of $\mathbf{u} - \text{proj}_v\mathbf{u}$ and $\mathbf{v}$. What can you say about the vectors $\mathbf{u} - \text{proj}_v\mathbf{u}$ and $\mathbf{v}$?

18. Use the result of the previous exercise to show that for any non-zero vectors $\mathbf{u}$ and $\mathbf{v}$, there are two vectors $\mathbf{u}_1$ and $\mathbf{u}_2$ such that $\mathbf{u} = \mathbf{u}_1 + \mathbf{u}_2$ where $\mathbf{u}_1$ is parallel to $\mathbf{v}$ and $\mathbf{u}_2$ is perpendicular to $\mathbf{u}$.

19. Any three non-zero vectors $\mathbf{u}$, $\mathbf{v}$ and $\mathbf{w}$ which do not lie in the same plane form a parallelepiped with three edges lying along the vectors. Do the following for this situation.

    (a) Show that the volume V of the parallelepiped with base formed by the vectors $\mathbf{u}$ and $\mathbf{v}$ is given by

    $$V = |\mathbf{u} \times \mathbf{v}||\mathbf{w}||\cos\phi| = |(\mathbf{u} \times \mathbf{v}) \cdot \mathbf{w}|$$

    where ϕ is the (smallest) angle made by the vector $\mathbf{u} \times \mathbf{v}$ and $\mathbf{w}$.

    (b) Use the result of part (a) to find the volume of the parallelepiped which is formed by the vectors $\mathbf{u} = 3\mathbf{i} + \mathbf{j} + 2\mathbf{k}$, $\mathbf{v} = 2\mathbf{i} - 5\mathbf{j} + \mathbf{k}$ and $\mathbf{w} = \mathbf{i} + 2\mathbf{j} + 3\mathbf{k}$.

    (c) Show that the volume V of the parallelepiped does not depend on which pair of vectors is taken to form the base.

    (d) Show that $(\mathbf{u} \times \mathbf{v}) \cdot \mathbf{w} = \mathbf{u} \cdot (\mathbf{v} \times \mathbf{w})$. The scalar $(\mathbf{u} \times \mathbf{v}) \cdot \mathbf{w}$ is called the *triple product* of the vectors $\mathbf{u}$, $\mathbf{v}$ and $\mathbf{w}$.

    (e) Illustrate the result of part (d) for the vectors $\mathbf{u} = 3\mathbf{i} + \mathbf{j} + 2\mathbf{k}$, $\mathbf{v} = 2\mathbf{i} - 5\mathbf{j} + \mathbf{k}$ and $\mathbf{w} = \mathbf{i} + 2\mathbf{j} + 3\mathbf{k}$.

20. Find a rectangular equation of the plane with given vector $\mathbf{n}$ perpendicular to the plane and passing through the specified point P_0.

    (a) $\mathbf{n} = [2, -1, 5]$ and $P_0 = [1, 0, 2]$

    (b) $\mathbf{n} = [2, -1, 5]$ and $P_0 = [0, 0, 0]$

    (c) $\mathbf{n} = 3\mathbf{i} + 2\mathbf{j} - \mathbf{k}$ and $P_0 = [3, -1, 2]$

21. Find a rectangular equation of the plane that contains the indicated three points.

    (a) $[0, 0, 0]$, $[-1, 3, 2]$, and $[1, 2, 0]$

    (b) $[1, 0, 1]$, $[3, -1, 2]$ and $[1, 2, -1]$

22. Show that the following are true: $\mathbf{i} \times \mathbf{j} = \mathbf{k}$, $\mathbf{j} \times \mathbf{k} = \mathbf{i}$, and $\mathbf{k} \times \mathbf{i} = \mathbf{j}$.

23. Show that the cross-product satisfies the following properties:

    (a) $\mathbf{0} \times \mathbf{u} = \mathbf{u} \times \mathbf{0} = \mathbf{0}$.

    (b) $\mathbf{u} \times \mathbf{u} = \mathbf{0} = 0\mathbf{i} + 0\mathbf{j} + 0\mathbf{k}$.

    (c) $\mathbf{u} \times (\mathbf{v} + \mathbf{w}) = \mathbf{u} \times \mathbf{v} + \mathbf{u} \times \mathbf{w}$.

    (d) $(\mathbf{u} + \mathbf{v}) \times \mathbf{w} = \mathbf{u} \times \mathbf{w} + \mathbf{v} \times \mathbf{w}$.

    (e) $(c\mathbf{u}) \times \mathbf{v} = c(\mathbf{u} \times \mathbf{v}) = \mathbf{u} \times (c\mathbf{v})$ where c is any scalar.

    (f) $(\mathbf{u} \times \mathbf{v}) \cdot \mathbf{w} = \mathbf{u} \cdot (\mathbf{v} \times \mathbf{w})$.

24. For two non-zero and non-parallel vectors $\mathbf{u}$ and $\mathbf{v}$, where θ is the angle between $\mathbf{u}$ and $\mathbf{v}$, do the following.

    (a) Show that the area A of a parallelogram with two sides lying along the vectors $\mathbf{u}$ and $\mathbf{v}$ is given by $A = |\mathbf{u} \times \mathbf{v}| = |\mathbf{u}||\mathbf{v}||\sin(\theta)|$. (Hint: First show that $|\mathbf{u} \times \mathbf{v}|^2 = |\mathbf{u}|^2|\mathbf{v}|^2 - (\mathbf{u} \cdot \mathbf{v})^2$.)

    (b) Use the result of part (a) to find the area of the parallelogram with two sides lying along the vectors $\mathbf{u} = 3\mathbf{i} + \mathbf{j} + 2\mathbf{k}$ and $\mathbf{v} = 2\mathbf{i} - 5\mathbf{j} + \mathbf{k}$.

    (c) Use the result of part (a) to show that the area of a triangle with two sides lying along $\mathbf{u}$ and $\mathbf{v}$ is given by $A = \frac{1}{2}|\mathbf{u} \times \mathbf{v}| = \frac{1}{2}|\mathbf{u}||\mathbf{v}||\sin(\theta)|$.

    (d) Find the area of the triangle with two sides lying along the vectors $\mathbf{u} = 3\mathbf{i} + \mathbf{j} + 2\mathbf{k}$ and $\mathbf{v} = 2\mathbf{i} - 5\mathbf{j} + \mathbf{k}$.

(e) Use the result of part (c) to find the area of the triangle formed by the three points $[0, 0, 0]$, $[-1, 3, 2]$ and $[1, 2, 0]$.

(f) Find the area of the triangle formed by the three points $[1, 0, 1]$, $[3, -1, 2]$ and $[1, -2, 1]$.

■ 7.3 CURVES AND SURFACES REPRESENTED BY PARAMETRIC EQUATIONS

7.3.1 Overview

In this section you will investigate curves and surfaces represented by *parametric equations*. The concept of a *parameter* refers to an intermediate variable which relates two or more other variables. Equations which relate two or more variables are known as parametric equations. You will investigate the notions of reparametrizing a curve (changing the parameter), parametric representations of lines, planes, and various other curves and surfaces.

We note that although the work of several mathematicians involved the notion of parametric representations of curves, it was the famous mathematician Leonhard Euler who first gave a development of parametric equations in his book "Introduction in analysis infinitorum" in 1748. This book was one of the most influential books on the development of mathematics in the last two centuries.

7.3.2 Activities

1. In addition to using a graphing facility to plot (explicit) functions, implicit functions, and polar functions, you can also use your computer to graph *parametric equations* using a parametric graphing facility, like **graphpar** in **ISETL** or the **plot[parametric]** command in **MapleV**. Your instructor will give you documentation explaining how to use your parametric graphing facility. Each of the expressions below is a way of specifying a curve. Use the appropriate graphing facility to obtain the graph. Before you can use the graphing facilities, you have to figure out the numerical values of the constants (everything but x, y, r, θ, and t). Use the notational definitions and your symbolic computing system (or a hand-held calculator) to figure them out.

Notational Definitions for Activity 1

$$A = 4\cos^2(\alpha) + 25\sin^2(\alpha)$$
$$B = 4\sin^2(\alpha) + 25\cos^2(\alpha)$$
$$C = 42\sin(\alpha)\cos(\alpha)$$
$$D = 100$$
$$\alpha = 0.62832$$

List of Expressions

(a) $Ax^2 + By^2 + Cxy = D$

(b) $y = \dfrac{-Cx + \sqrt{4BD + (C^2 - 4AB)x^2}}{2B}$, $y = \dfrac{-Cx - \sqrt{4BD + (C^2 - 4AB)x^2}}{2B}$

(c) $r = \dfrac{10}{\sqrt{4\cos^2(\theta-\alpha)+25\sin^2(\theta+\alpha)}}$

(d) $x(t) = 5\cos\alpha\cos t + 2\sin\alpha\sin t,\; y(t) = -5\sin\alpha\cos t + 2\cos\alpha\sin t$

What happens if you use other values of α? For example $\frac{\pi}{4}$, $\frac{\pi}{2}$, π, and 0?

2. For each of the following pairs of parametric equations and their domain intervals, make a hand sketch of what you think the graph will look like. You may consider that the first equation refers to the x-coordinate and the second to the y-coordinate.

   Next, use the parametric graphing facility on your computer to make a sketch of each pair of parametric equations and indicated interval(s). For example, you can use the **graphpar** in **ISETL** or **plot[parametric]** in **MapleV**. Compare your hand sketch with the one you obtained using your computer.

   Be sure that you make your guess *before* you use a parametric graphing facility on your computer.

   (a) $x = f(t) = \cos(2t),\; y = g(t) = 2\sin^2(t),\; -\frac{\pi}{2} \le t \le \frac{\pi}{2}.$

   (b) $x = f(t) = \cos(2t),\; y = g(t) = 2\sin^2(t),\; 0 \le t \le \frac{\pi}{2}.$

   (c) $x(t) = 3.6\sin(t),\; y(t) = 4.8\cos(t),\; 0 \le t \le \pi.$

   (d) $x(t) = 3.6\sin(t),\; y(t) = 4.8\cos(t),\; 0 \le t \le 2\pi.$

   (e) $p(r) = \dfrac{\sin(r)}{\sin(r)+\cos(r)},\; q(r) = \dfrac{\cos(r)}{\sin(r)+\cos(r)},\; 0 \le r \le \dfrac{\pi}{3}$

   (f) $x(u) = u^3,\; y(u) = u^6,\; -1.5 \le u \le 1.5.$

   (g) $x(t) = t - \sin(t),\; y(t) = 1 - \cos(t),\; 0 \le t \le 10$

3. In watching the sketch that is made on your computer using a parametric graphing facility, like **graphpar** in **ISETL** or **plot[parametric]** in **MapleV**, you will notice that the curve is traced out in a particular direction. For the graphs in parts (a), (c) and (g) of Activity 2, figure out a way to change the situation so that the same curve will be traced out, but in the opposite direction.

4. In principle, it is also possible to change the speed at which the point moves along the curve. The trouble is that, often, the computer is so fast that you can't observe the change. Try to imagine what you would do to change the speed and see if you can get something noticeable in parts (c) and (f) of Activity 2.

5. The purpose of this activity is to review a point that you worked with in Chapter 2 concerning composition of functions that will be used in this chapter.

   Write a **computer function comp** that will accept two **computer functions** representing functions and will return a **computer function** that represents the composition of the two original functions. Be sure to test your **computer function comp**.

6. For each of the following curves (taken from Activity 2), use your **computer function comp** to construct a *reparametrization* by composing each of the two original functions with the new function α. Then apply a parametric graphing facility to sketch the reparametrized curve. In each

case, what is it about the original curve and the reparametrized curve that is the same? What is different about the two curves, if anything? What is it about α that causes trouble or doesn't cause trouble?

(a) The original curve is given by

$$f(t) = \cos(2t), \quad g(t) = 2\sin^2(t), \quad -\frac{\pi}{2} \le t \le \frac{\pi}{2}$$

The reparametrization function is $\alpha : [-1, 1] \to \left[-\frac{\pi}{2}, \frac{\pi}{2}\right]$ given by $\alpha(t) = \frac{\pi t^3}{2}$.

(b) The original curve is given by

$$f(t) = \cos(2t), \quad g(t) = 2\sin^2(t), \quad -\frac{\pi}{2} \le t \le \frac{\pi}{2}$$

The reparametrization function is $\alpha : [-1, 1] \to \left[-\frac{\pi}{2}, \frac{\pi}{2}\right]$ given by $\alpha(t) = t^2$.

(c) The original curve is given by

$$p(r) = \frac{\sin(r)}{\sin(r) + \cos(r)}, \quad q(r) = \frac{\cos(r)}{\sin(r) + \cos(r)}, \quad 0 \le r \le \frac{\pi}{3}.$$

The reparametrization function is $\alpha : \left[0, \frac{1}{2}\right] \to \left[0, \frac{\pi}{3}\right]$ given by $\alpha(t) = 2t$.

(d) The original curve is given by

$$p(r) = \frac{\sin(r)}{\sin(r) + \cos(r)}, \quad q(r) = \frac{\cos(r)}{\sin(r) + \cos(r)}, \quad 0 \le r \le \frac{\pi}{3}.$$

The reparametrization function is $\alpha : \left[0, \frac{1}{3}\right] \to \left[0, \frac{\pi}{3}\right]$ given by $\alpha(t) = \pi t$.

7. What would a parametric form of a curve in three dimensions look like? Begin with the curve in Activity 2 part (c) and write a parametric form of a curve in three dimensions that would represent a vertical helix that, if squashed down, would look exactly like the ellipse in Activity 2 part (c). Draw a hand sketch of your curve in three dimensions.

8. Draw a hand sketch, in three dimensions, of the figure given in parametric form by the parametric equations given below. Then check your hand sketch using a computer.

$$x = u\cos(v), \, y = u\sin(v), \, z = u^2, \, 0 \le u \le 1, \, 0 \le v \le \pi.$$

7.3.3 Discussion

Representations of Curves in $\mathcal{R}^2$ and $\mathcal{R}^3$ Conceptually, curves are represented in the same way as lines. The details are much more complicated and a lot of mathematics is devoted to studying them, but the basic idea of representation is the same. We can represent a curve in the same three forms: generative, relational, and vector. These were all illustrated in Activity 1. Which is which? Did you find that the curve was the same in all five cases? It is an ellipse rotated through an angle α (read "alpha").

In Activity 1, the generative form is that given in part (b) (two equations are needed for the two halves of the ellipse) and (c) (this is using polar coordinates). The relational form is that given in part (a). The vector form is that given in part (d) except that we have written it in the more common *parametric* form. The strictly vector notation would look like this.

$$v(t) = [5\cos\alpha\cos t + 2\sin\alpha\sin t, \; 5\cos\alpha\cos t + 2\sin\alpha\sin t]$$

or,

$$v(t) = (5\cos\alpha\cos t + 2\sin\alpha\sin t)\mathbf{i} + (5\cos\alpha\cos t + 2\sin\alpha\sin t)\mathbf{j}$$

In three dimensions, the same form is used to represent a curve in vector form or in parametric form, but a third term is added.

The Parametric Form of a Curve in $\mathcal{R}^2$ You may recall that in Chapter 1, we referred to the equation of a line in the form $y = mx + b$ as the generative form of an equation of a line. Well, in a similar manner, we say that the parametric form of a curve is generative in nature. We write it as

$$x = x(t) = \phi(t), \; y = y(t) = \psi(t), \; a \le t \le b$$

where ϕ and ψ are functions of t. The variable t is called a parameter. As t runs through its domain, the interval $[a, b]$, the two functions ϕ and ψ give the x- and y-coordinates of a particle, respectively. In turn, the point $(\phi(t), \psi(t))$ traces out a curve in $\mathcal{R}^2$.

The parametric form of a curve is a very dynamic representation and can be interpreted in terms of the motion of a particle. Indeed, you might think about the parameter t as representing time and then the pair $(\phi(t), \psi(t))$ is the position of the particle at time t. However, the parameter need not represent time. Indeed, the parameter may represent an angle or some other quantity. In this interpretation, how would you get the velocity of the particle? The speed of the particle?

One way to develop an understanding of how the parametric form of a curve works is to look at examples and try to make the connection, in your mind, between the action of the two functions and the position of the point. What you have to think about is very similar to what was going on with polar coordinates. For example, consider

$$x = \phi(t) = 3t - 2, \; y = \psi(t) = 4t + 1, \; -1 \le t \le 2.$$

Do you see why this gives a line segment? Think about the rise and run. If t changes by a certain amount, how do x and y change? What is the slope of this line? Can you read it off the equations? What could be done to get the whole line and not just a segment?

Here is another example.

$$x = \phi(t) = \cos(\pi t), \; y = \psi(t) = \sin(\pi t), \; 0 \le t \le 2.$$

As t runs through the values from 0 to 2, the x-coordinate is the cosine of some angle and the y-coordinate is the sine of the same angle. This means that the point is at distance 1 from the origin and makes an angle of πt with the horizontal axis. In other words, the curve is a circle.

How would the previous example change if the domain, $[0, 2]$, of t was replaced by $[0, 1]$? By $[-1, 1]$? By $[0, 4]$?

How would you change the previous example to get a circle of radius a? How would you change the previous example to get an ellipse instead of a circle?

Conversion to Standard Form

If you have a parametric representation of the form,

$$x = x(t) = \phi(t),\ y = y(t) = \psi(t),\ a \le t \le b$$

can you replace it with a single function f which represents the curve in the usual way — that is, with the x- and y-coordinates related by the equation $y = f(x)$? Or put another way, given ϕ and ψ, how could you find f?

One way to think of this question is to reflect on the fact that the parametric form is a situation in which values of t are put in and values of x and y come out. How could this be changed to a situation in which values of x go in and values of y come out?

You could do this by starting with a value of x and asking what is the t that would give it. This is the same as finding the inverse function ϕ^{-1} of the function ϕ. If you did this, you could then get the y value by applying ψ.

In other words, the function f that we are looking for is exactly the composition of ψ with the inverse of ϕ. That is,

$$f = \psi \circ \phi^{-1}.$$

Can you explain what is going on in terms of a picture of the curve? See the picture in Figure 7.31.

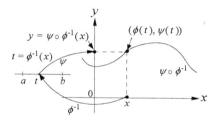

Figure 7.31. *A geometric illustration of the conversion from parametric to rectangular form.*

Now there is a difficulty here. What if the function ϕ does not have an inverse? How could you fix things up in such a case?

Consider, for example, the following curve given parametrically.

$$x = \phi(t) = 3.6\cos(t),\ y = \psi(t) = 4.8\sin(t),\ 0 \le t \le 2\pi.$$

If you begin with x and apply ϕ^{-1}, you must divide by 3.6 and then use the arccosine function. This requires the choice of a branch of that function. Look at the graph in Figure 7.32.

You might decide to look at the branch below the horizontal axis. This means that t (which represents not only the time, but the angle) must be in the interval $[\pi, 2\pi]$. Then you can apply ψ to obtain

$$\psi \circ \phi^{-1}(x) = 4.8\sin\left(\arccos\left(\frac{x}{3.6}\right)\right) = -4.8\sqrt{1 - \left(\frac{x}{3.6}\right)^2}$$

(why the minus sign?) and, simplifying, the equation becomes,

$$y = f(x) = \left(\psi \circ \phi^{-1}\right)(x) = -\frac{4}{3}\sqrt{12.96 - x^2}$$

which is the lower half of the ellipse whose equation in the relational form is

$$16x^2 + 9y^2 - 207.36 = 0.$$

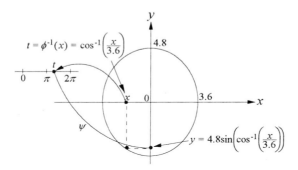

Figure 7.32. A conversion from parametric to rectangular form.

Thus, you can see how this conversion can be used to look at a curve in the more familiar form of $y = f(x)$. The point, however, is that using the parametric form will allow us to make investigations that are not so simple in the old form.

Notice that if you solve for $\cos(t)$ and $\sin(t)$, then you get

$$\cos(t) = \frac{x}{3.6} \quad \text{and} \quad \sin(t) = \frac{y}{4.8}.$$

Now, square $\cos(t)$ and $\sin(t)$, and then add the results to get

$$\frac{x^2}{(3.6)^2} + \frac{y^2}{(4.8)^2} = \cos^2(t) + \sin^2(t) = 1.$$

After doing a little more algebra you will see that the latter equation is equivalent to what we obtained above, $16x^2 + 9y^2 - 207.36 = 0$. Hence, we can use algebra and trigonometric identities to help make the conversion from parametric form to rectangular form.

The relation $f = \psi \circ \phi^{-1}$ is not so important for working out details about particular curves. Its main role is as a tool in determining things like tangents and normals to these curves. This will be discussed in the next section.

Reparametrization of a Curve in $\mathcal{R}^2$ Focus now on the interpretation of a parametric curve as a time record of a particle moving around in $\mathcal{R}^2$. For each t there is a point in the plane. Thus you have two functions (which we are calling ϕ and ψ) which produce, for a given t, the x-coordinate and the y-coordinate of the point as illustrated in Figure 7.33.

You might think of this in terms of readings. The interval $[a, b]$ could be a clock. You read the time on the clock and then at that time you check the position of the point. The coordinates give the two functions.

This particle is moving at a certain rate (that may not be constant, but might vary with time according to a certain function). Suppose that the rate changes. Then the path of the particle will not change, but all the readings will be different.

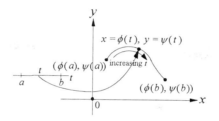

Figure 7.33. A parametric curve $x = \phi(t)$ and $y = \psi(t)$ for $a \leq t \leq b$.

How can this change be captured mathematically? Do you see how to use a function that transforms t to capture a change in rate?

Look at Activity 2 (f). Suppose that the particle were tracing the same path over a time interval, not from -1.5 to 1.5, but from -3 to 3. Suppose further that the point were only moving "half as fast". You can implement this in two ways. You can try to figure out how to go directly from the interval $[-3, 3]$ to the path. Alternatively, you could go first from the interval $[-3, 3]$ to the interval $[-1.5, 1.5]$ and then use the same pair of functions you already had to get to the curve.

Using the second alternative, what would be the function from $[-3, 3]$ to $[-1.5, 1.5]$? There are lots of ways to get it, but a very simple one would be division by 2. Thus, we can define the function α by

$$\alpha(r) = \frac{r}{2}, \quad -3 \leq r \leq 3$$

and then if the original functions ϕ, ψ were given as in Activity 2 (f) by

$$\phi(u) = u^3, \quad \psi(u) = u^6, \quad -1.5 \leq u \leq 1.5$$

then the new pair of functions, which we call a *reparametrization* of the curve, is given by $\phi \circ \alpha$ and $\psi \circ \alpha$. Thus we have,

$$\phi \circ \alpha(r) = \frac{r^3}{8}, \quad \psi \circ \alpha(r) = \frac{r^6}{64}, \quad -3 \leq r \leq 3$$

Now you can go back and look at Activities 3 and 4, and see how to change the direction of a point, or change its speed.

The Parametric Form of a Curve in $\mathcal{R}^3$

The idea for parametric curves in $\mathcal{R}^3$ is the same as for $\mathcal{R}^2$. You just need to add a third function for the third coordinate. Thus, a general formulation is

$$x = \phi(t), \, y = \psi(t), \, z = \lambda(t), \, a \leq t \leq b$$

It is perhaps a little harder to see what the graphs will look like. Consider the following example.

$$x = \phi(t) = t^2, \, y = \psi(t) = t, \, z = \lambda(t) = t^3, \, -\infty < t < \infty.$$

If you "smashed" everything down in the y direction to the xz-plane, you would get a curve, given parametrically by the equations

$$x = t^2, \, z = t^3, \, -\infty < t < \infty.$$

You can use your computer to graph this, or eliminate t to get the two equations

$$z = x^{3/2} \text{ and } z = -x^{3/2}.$$

The curve looks like the one shown in Figure 7.34.

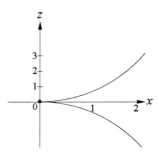

Figure 7.34. The curve $x = \phi(t) = t^2$ and $z = \lambda(t) = t^3$ for $-\infty < t < \infty$.

But it is necessary to also take into account the variation in y. This has the effect of spreading this curve out in the y direction and the resulting curve looks like the one in Figure 7.35.

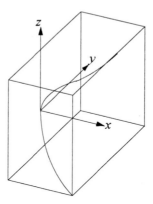

Figure 7.35. The curve $x = t^2$, $y = t$, and $z = t^3$ for $-\infty < t < \infty$.

You can do the same thing with reparametrization as was done in two dimensions. What about the conversion to a single equation? Does that make any sense here?

The Parametric Form of a Line in $\mathcal{R}^3$　Consider the following special case of a curve in $\mathcal{R}^3$,

$$x = \phi(t) = a + bt, \quad y = \psi(t) = c + dt, \quad z = \lambda(t) = e + ft, \quad -\infty < t < \infty$$

where a, b, c, d, e, and f are constants. This is the parametric form of the equation of a line. Can you see why this is a line? What properties does a straight line have that will be satisfied by a curve whose parametric equations have this "linear" form?

Suppose you were given two points $P_1 = [x_1, y_1, z_1]$ and $P_2 = [x_2, y_2, z_2]$ in $\mathcal{R}^3$ and you wanted to find the parametric form of the equation of the line passing through these two points. You can work with the idea that the line you are after consists of precisely those points P whose coordinates differ from the corresponding coordinates of P_1 by amounts which are a fixed multiple of the differences between corresponding coordinates of P_2 and P_1. See the picture in Figure 7.36.

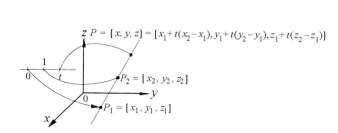

Figure 7.36. *A line through the points $P_1 = [x_1, y_1, z_1]$ and $P_2 = [x_2, y_2, z_2]$ in $\mathcal{R}^3$.*

What we are saying is that a typical point $P = [x, y, z]$ in $\mathcal{R}^3$ is on this line precisely when the three ratios

$$\frac{x - x_1}{x_2 - x_1} = \frac{y - y_1}{y_2 - y_1} = \frac{z - z_1}{z_2 - z_1}$$

are the same. They can be any value, but they must be the same.

In other words, these three ratios have a value t which can be any real number. This can be expressed analytically as,

$$x = x_1 + (x_2 - x_1)t, \ y = y_1 + (y_2 - y_1)t, \ z = z_1 + (z_2 - z_1)t, \ -\infty < t < \infty.$$

Do you see that this is precisely the general form we gave at the beginning of this paragraph? Solving the parametric equations for t we obtain what is commonly called a *symmetric form* of the line through the points P_1 and P_2:

$$\frac{x - x_1}{x_2 - x_1} = \frac{y - y_1}{y_2 - y_1} = \frac{z - z_1}{z_2 - z_1}.$$

Notice that the symmetric form of a line really tells us three things:

$$\frac{x - x_1}{x_2 - x_1} = \frac{y - y_1}{y_2 - y_1}, \quad \frac{x - x_1}{x_2 - x_1} = \frac{z - z_1}{z_2 - z_1}, \quad \text{and} \quad \frac{y - y_1}{y_2 - y_1} = \frac{z - z_1}{z_2 - z_1}.$$

What do each of these three equations represent? Did you say that each equation represents a plane? Yes, that's right! So what do the three equations in a symmetric form of a line L tell you? Did you say that the line L is the intersection of the planes represented by the three equations? If so, you're right!

Do you see how you could obtain the parametric form of a line in $\mathcal{R}^3$ (or $\mathcal{R}^2$) from its vector form? We leave this as an exercise.

The Parametric Form of a Surface in $\mathcal{R}^3$

By now you should be noticing a pattern. If the parametric form of a curve in $\mathcal{R}^2$ is a pair of functions whose domain and range are in $\mathcal{R}^1$ and the parametric form of a curve in $\mathcal{R}^3$ is a triple of functions whose domain and range are in $\mathcal{R}^1$, then what do you think is the parametric form of a surface in $\mathcal{R}^3$?

Take a look back in Section 2 of this chapter. Do you recall the vector form of the equation of a line? (See page 536.) This is not really very different from what we are doing now. Instead of two or three functions from $\mathcal{R}^1$ (in the case of a curve) or $\mathcal{R}^2$ (in the case of a surface) to $\mathcal{R}^1$ there is just one "vector-valued" function from $\mathcal{R}^1$ or $\mathcal{R}^2$ to the set of vectors of two or three dimensions.

So, a surface in $\mathcal{R}^3$ can be given in parametric form by specifying three functions from $\mathcal{R}^2$ to $\mathcal{R}^1$. For example, can you work out what the surface given by the following equations would look like?

$$x = \phi(s, t) = -1 - s + t, \; y = \psi(s, t) = 2 + s + 2t, \; z = \lambda(s, t) = s, \; -\infty < s, t < \infty.$$

You might take as a hint the fact that all of the expressions are linear. No, the graph is not a line.

What about the surface given by the following equations?

$$x = s\cos(2\pi t), \; y = \sqrt{s}, \; z = s\sin(2\pi t), \text{ for } 0 < s < \infty, \; 0 \leq t \leq 1.$$

What does it look like? You can try to analyze this surface by looking at values of s and t and trying to see what happens as they vary in certain ways.

For example, if you hold s constant, then the y-coordinate is constant. This means that you are looking at the xz-plane. Here you have the parametric equations,

$$x = s\cos(2\pi t), \; z = s\sin(2\pi t), \text{ for } 0 \leq t \leq 1.$$

This is a parametric form of an equation of a circle centered at the "origin" (that is, the point $\left[0, \sqrt{s}, 0\right]$) with radius s.

To see this surface geometrically, you can imagine a plane parallel to the xz-plane at $y = \sqrt{s}$ as shown in Figure 7.37.

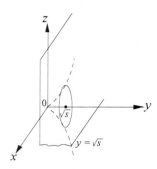

Figure 7.37. The circle $x = s\cos(2\pi t), \; y = \sqrt{s}, \; z = s\sin(2\pi t), \;$ ***for*** $0 \leq t \leq 1$***.***

Now, as you let s vary, you still get a circle centered at the point $\left[0, \sqrt{s}, 0\right]$, but two things change. The radius r of the circle changes according to the formula $r = s$ and the plane moves (always parallel to the xz-plane) according to the value of s. Thus the full picture looks like that in Figure 7.38.

The Parametric Form of a Plane in $\mathcal{R}^3$ Let's return to the example on page 558,

$$x = \phi(s, t) = -1 - s + t, \; y = \psi(s, t) = 2 + s + 2t, \; z = \lambda(s, t) = s, \; -\infty < s, t < \infty$$

or the more general form,

$$x = \phi(s, t) = a + bs + ct, \; y = \psi(s, t) = d + es + ft, \; z = \lambda(s, t) = g + hs + it, \; -\infty < s, t < \infty$$

where a, b, c, d, e, f, g, h, and i are constants. This is the parametric form of the equation of a plane in $\mathcal{R}^3$.

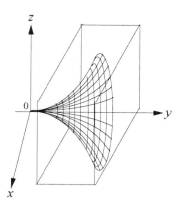

Figure 7.38. The helix $x = s\cos(2\pi t)$, $y = \sqrt{s}$, $z = s\sin(2\pi t)$, **for** $0 < s < \infty$, $0 \le t \le 1$

How do we decide that it is a plane? It depends to some extent on how we define a plane in the first place. Would it be enough to say that a surface is a plane if whenever it contains two points, it also contains the line segment connecting them? Does every plane satisfy this property? Can you think of a surface which satisfies this property but is not a plane?

Go back and read over the discussion of a line in $\mathcal{R}^3$ (see page 556). Can you make an analogous analysis for a plane? What would you look at instead of two points?

7.3.4   Exercises

1. Sketch a graph of each curve. Indicate on the curve the direction of motion along the curve (i.e., indicate the direction along the curve as the parameter increases).

    (a) $x = 2t$, $y = \sqrt{t}$

    (b) $x = t^2$, $y = t^3$

    (c) $x = t + 1$, $y = t^2 - t$

    (d) $x = 1 - 2t$, $y = 3 - t^2$

    (e) $x = 3\sin^2(t)$, $y = -4\cos^2(t)$

    (f) $x = 2\sin(t)$, $y = \cos(t)$

    (g) $x = -3\cos(t)$, $y = -3\sin(t)$

    (h) $x = 5\sin(0.5t)$, $y = -5\cos(0.5t)$

    (i) $x = 3\sin(2t)$, $y = \cos^2(2t)$

    (j) $x = 2\tan(t)$, $y = 2\sec^2(t)$

    (k) $x = e^{3t}$, $y = 2e^t$

    (l) $x = 4t - t^3$, $y = t^2$

    (m) $x = \sqrt[3]{t}$, $y = 2t - 1$

(n) $x = 2 - \sec^2(t),\ y = -1 + \tan(t)$

(o) $x = 1 + \sin(t),\ y = 1 + \cos(t)$

(p) $x = 2 - \sin(t),\ y = 2 - \cos(t)$

(q) $x = 1 - 2\sin(t),\ y = 1 - 2\cos(t)$

(r) $x = 3\cos(\theta) + \cos(3\theta),$
     $y = 3\sin(\theta) + \sin(3\theta)$

(s) $x = \cos^3(t),\ y = \sin^3(t)$

2. Convert each pair of the parametric equations to an equivalent rectangular equation. Sketch a graph of the resulting rectangular equation and also the curve represented by the parametric equations. You may wish to check your sketch on a computer. What are the restrictions, if any, on the parameter in each case? What are the restrictions, if any, on the rectangular coordinates in each case?

(a) $x = 2t,\ y = \sqrt{t}$

(b) $x = t^2,\ y = t^3$

(c) $x = t + 1,\ y = t^2 - t$

(d) $x = 3\sin^2(t),\ y = -4\cos^2(t)$

(e) $x = 2\sin(t),\ y = \cos(t)$

(f) $x = -3\cos(\theta),\ y = -3\sin(\theta)$

(g) $x = 5\sin(0.5t),\ y = -5\cos(0.5t)$

(h) $x = 3\sin(2t),\ y = \cos^2(2t)$

(i) $x = 2\tan(u),\ y = 2\sec^2(u)$

(j) $x = e^{3t},\ y = 2e^t$

(k) $x = \sin^2(t),\ y = 3\cos(t)$

(l) $x = 3\cos(\theta),\ y = 3\sin(\theta)$

(m) $x = 2\sin(5t),\ y = 3\cos(5t)$

(n) $x = \dfrac{2t}{1+t},\ y = \dfrac{1}{t}$

(o) $x = \dfrac{u}{1+u},\ y = \dfrac{1}{1+u}$

(p) $x = \dfrac{t^2}{t^2+1},\ y = \dfrac{t}{t^2+1}$

3. Find parametric equations which represent the curve described in each of the following situations. Make a sketch of each curve.

(a) The line containing the points $[-1, 2]$ and $[3, -4]$.

(b) The line segment with endpoints $[-1, 2]$ and $[3, -4]$.

(c) The line containing the points $[0, 2]$ and $[-1, 2]$.

(d) The (open) line segment between the points $[0, 2]$ and $[-1, 2]$.

(e) The line containing the points $[3, 4]$ and $[3, -5]$.

(f) The line segment with endpoints $[3, 4]$ and $[3, -5]$.

(g) The line parallel to the line $3x - 2y = 6$ and passing through the point $[-2, 3]$.

(h) The line perpendicular to the line $3x - 2y = 6$ and passing through the point $[-2, 3]$.

(i) The circle with radius 3 and center at the origin.

(j) The circle with radius $\sqrt{5}$ and center at the point $[-2, 3]$.

(k) The ellipse with center at the origin with major axis horizontal with length 5 and minor axis vertical with length 4.

(l) The parabola with vertex at the origin, axis along the y-axis and containing the point $[-1, 3]$.

4. (a) Find parametric equations which represent the line containing the two points $[x_0, y_0]$ and $[x_1, y_1]$ in each of the following cases. What are the

restrictions on x_0, y_0, x_1, and y_1 in each case?

    i. Horizontal.

    ii. Vertical.

    iii. Neither horizontal nor vertical.

(b) Show that parametric equations of a line in $\mathcal{R}^2$ can be obtained from the *vector form*, $\mathbf{v} = \mathbf{a} + t\mathbf{b}$, of a line (see page 537 in Section 2 of this chapter). What would you do differently for a line in $\mathcal{R}^3$?

5. Find parametric equations which represent the curve described in each of the following situations.

(a) A circle with center at $[h, k]$ with radius a.

(b) An ellipse with center at $[h, k]$, axes parallel to the x-and y-coordinate axes, with major axis with length a and minor axis of length b. How many cases are there in this situation? Explain your answer.

6. Sketch a graph of each curve in $\mathcal{R}^3$. Indicate on the curve the direction of motion along the curve (i.e., indicate the direction along the curve as the parameter increases).

(a) $x = t - 1$, $y = 2t$, $z = t + 3$, $0 \le t \le 4$.

(b) $x = t^3$, $y = t$, $z = t^2$, $0 \le t \le 1$.

(c) $x = \sin(\pi t)$, $y = \cos(\pi t)$, $z = 2$

(d) $x = -3$, $y = 2\sin(t)$, $z = 3\cos(2t)$

(e) $x = 2\sin(t)$, $y = 2\cos(t)$, $z = t$ for $0 \le t \le 2\pi$.

(f) $x = 3\cos(2t)$, $y = 3t$, $z = 4\sin(2t)$ for $0 \le t \le \pi$.

(g) $x = t^2$, $y = t$, $z = \sin(t)$ for $0 \le t \le 2\pi$.

(h) $x = e^t$, $y = t$, $z = e^{-t}$ for $0 \le t \le \pi$.

7. Do the lines l_1 and l_2 intersect? If so, find the point of intersection. Explain how you arrived at your answer.

(a) $l_1 : x = 13 - 9t$, $y = 3 - 15t$, $z = 6t - 4$ and $l_2 : x = 4 - 3s$, $y = 12 + 5t$, $z = 2t$

(b) $l_1 : x = 3 + t$, $y = 1 - t$, $z = -3 - 3t$ and $l_2 : x = -1 + 4u$, $y = 2 + u$, $z = 3 - 2u$

8. Do the following for each of the indicated surfaces below.

(a) Sketch the surface.

(b) Convert the parametric equations to an xyz-rectangular equation.

Here are the surfaces.

    i. $x = t - s$, $y = s$, $z = t + s$

    ii. $x = t^2$, $y = 3$, $z = 2t$

    iii. $x = s + t$, $y = s - t$, $z = s$

    iv. $x = \sin(t)$, $y = \cos(t)$, $z = 3t$

    v. $x = s\cos(t)$, $y = 2s\sin(t)$, $z = s^2$

    vi. $x = u\sin^2(t)$, $y = u\cos^2(t)$, $z = 3u$

    vii. $x = \sqrt{s}\sin(t)$, $y = \sqrt{s}\cos(t)$, $z = s$

    viii. $x = \sin u \sin(v)$, $y = \sin(u)\cos(v)$, $z = \cos(u)$

    ix. $x = u\sin(t)$, $y = u\cos(t)$, $z = u$

    x. $x = u\tan(t)$, $y = u\sec(t)$, $z = u$

    xi. $x = 2u\sec^2(3v)$, $y = 2u\tan^2(3v)$, $z = u$

    xii. $x = 3\cos(2u)$, $y = 3t$, $z = 4\sin(2u)$

    xiii. $x = s$, $y = st$, $z = s^2\cos(t)$

xiv. $x = e^{-t}$, $y = 2s$, $z = se^{t}$

9. Consider the parametric equations

$$x = \phi(t) = a\cos(\pi t), \ y = \psi(t) = a\sin(\pi t),$$
$$0 \le t \le 2.$$

   (a) Explain why the curve is a circle. In what way(s) can you change the equations and still have a circle? Give reasons for your answers.

   (b) How does the curve change if the domain is $0 \le t \le 1$? What if the domain is the interval $[0, 0.5]$? $[-1, 1]$? $[0, 4]$? Give reasons for your answers.

   (c) Explain how would you change the equation to get an ellipse instead of a circle.

10. Find a symmetric form for the line passing through the point $[2, -3, 4]$ and parallel to the line given by

$$x = 2t - 3, \ y = 1 + t, \ z = \pi t.$$

11. Show that the curve represented by the parametric equations

$$x = \frac{a}{\sqrt{1+t^2}}, \ y = \frac{at}{\sqrt{1+t^2}}$$

   is a circle. Use your parametric graphing facility to make a sketch of the curve for at least two choices of a.

12. Are parametric equations of a line in $\mathcal{R}^2$ unique; is there only one set of parametric equations to describe each line in $\mathcal{R}^2$? What about the xy-rectangular equation of a line; is it unique? Explain your answer and give examples to illustrate your answers to these questions.

13. Is a symmetric form of a line in $\mathcal{R}^3$ unique, that is, is there only one such equation? Are parametric equation of a line unique? Explain your answers to these

questions, discuss the relationship between these two questions, and give examples to illustrate your answer.

14. For a line through two points $P_1 = [x_1, y_1, z_1]$ and $P_2 = [x_2, y_2, z_2]$ are the direction numbers $x_2 - x_1$, $y_2 - y_1$, $z_2 - z_1$ unique; that is, is there only one triple of numbers which gives the direction of the line? Explain your answer and give examples to illustrate your answers to these questions.

15. For each of the following curves represented by the parametric equations and the reparametrization function α, make a sketch of the original curve and sketch the curve obtained by reparametrization with α. What is the difference, if any, between the original curve and the reparametrized curve? What is it about α that causes trouble, or does not cause trouble? What is it about α that changes the curve, or changes how it is traced out? Explain your answers.

   (a) $x = \phi(t) = \sin(\pi t)$,
       $y = \psi(t) = \cos(\pi t)$, $0 \le t \le 1$
       $\alpha : [0, 0.5] \to [0, 1]$ is given by
       $\alpha(t) = 2t$.

   (b) $x = \phi(t) = \sin(\pi t)$,
       $y = \psi(t) = \cos(\pi t)$, $0 \le t \le 1$
       $\alpha : [-1, 1] \to [0, 1]$ is given by
       $\alpha(t) = t^2$.

   (c) $x = \phi(t) = \sin(\pi t)$,
       $y = \psi(t) = \cos(\pi t)$, $0 \le t \le 1$
       $\alpha : [-1, 0] \to [0, 1]$ is given by
       $\alpha(t) = -t$.

   (d) $x = \phi(t) = \sin(\pi t)$,
       $y = \psi(t) = \cos(\pi t)$, $0 \le t \le 1$
       $\alpha : [0, 1] \to [0, 1]$ is given by
       $\alpha(t) = 1 - t$.

16. In this exercise you will investigate curves from the family known as *cycloids*. The curves you will investigate are represented by parametric equations of the form

$$x = \phi(\theta) = a(\theta) + b\sin(\theta),$$
$$y = \psi(\theta) = a + b\cos(\theta).$$

Use a computer to investigate the graphs of the following three types of cycloids.

   (a) The case $a > b$ known as the *prolate cycloid*.

   (b) The case $a < b$ known as the *curtate cycloid*, or *trochoid*.

   (c) The case $a = b$ known simply as the *cycloid*.

Write a brief essay discussing the differences between these three types of cycloids. Illustrate with appropriate examples and graphs, clearly labeled.

We remark that a cycloid is a curve with practical applications. For example, the *brachistochrone problem*, first posed by the mathematician Johann Bernoulli in 1696, is as follows: You are to connect two points P_1 and P_2 in a plane with wire so that a bead sliding down on a wire from P_1 to P_2 travels in the least time, assuming that the wire is well greased enough so that only the force of gravity and the shape of the wire effect the speed of the bead. What is the shape of the curve requiring the least amount of time for the bead to travel? The solution of this problem is a curve in the shape of a cycloid. (Note that the Greek words *brachistos* and *chronos* mean "shortest" and "time", respectively. Hence, the name brachistochrone problem.) Leonard Euler was the first to develop parametric equations for cycloids in his textbook "Introductio in analysis infinitorum" in 1748.

17. A reflector attached to the spoke of a wheel with radius a on a motorcycle is b units from the center of the wheel. Find parametric equations for the path of the center of the reflector as the motorcycle moves over a level roadway. Make a sketch of the path of the center of the reflector. The path traveled by the center of the reflector is called a *cycloid* (see Exercise 16).

18. A curve traced out by a point on a circle of radius b as it rolls on the outside of a larger circle of radius a is called an *epicycloid*. Let θ be the angle, measured counterclockwise, made by the positive x-axis and the line segment between the centers of the two circles. Assume that the center of the larger circle is at the origin. Make a sketch illustrating this situation.

   (a) Show that parametric equations for an epicycloid are given by

   $$x = (a+b)\cos(\theta) - b\cos\left(\frac{(a+b)}{b}\theta\right),$$
   $$y = (a+b)\sin(\theta) - b\sin\left(\frac{(a+b)}{b}\theta\right).$$

   (b) Use your parametric graphing facility to make a sketch of an epicycloid for three different choices of two values of a and b. What interval for the parameter θ is needed to trace out the curve exactly once?

19. A curve traced by a point on a circle of radius b that rolls on the inside of a larger circle of radius a is called a *hypocycloid*. Let θ be the angle, measured counterclockwise, made by the positive x-axis and the line segment between the centers of the two circles. Assume that the center of the larger circle is at the origin. Make a sketch illustrating this situation.

   (a) Show that parametric equations for a hypocycloid are given by

$$x = (a-b)\cos(\theta) + b\cos\left(\frac{(a-b)}{b}\theta\right),$$

$$y = (a-b)\sin(\theta) - b\sin\left(\frac{(a-b)}{b}\theta\right).$$

(b) Use your parametric graphing facility to make a sketch of a hypocycloid with $a = 4b$ for two choices of b. What can you say about such a curve? Do the following for such a curve.

   i. Show that parametric equations for the curve are $x = a\cos^3(\theta)$ and $y = a\sin^3(\theta)$. What interval for the parameter θ is needed to trace out the curve exactly once?

   ii. Convert the parametric equations in part (i) to an equivalent xy-rectangular equation.

20. A curve C is given by parametric equations $x = \phi(t)$, $y = \psi(t)$, $a \le t \le b$. Based on your experiences in the activities for this section, or further investigations of your own, answer the following. Give reasons for your answers and give examples which illustrate your answers.

   (a) How could you reparametrize the curve so that the direction of the curve (i.e., the direction along the curve corresponding to increasing parameter) is reversed?

   (b) How could you reparametrize the curves so that the rate at which a point traces out the curve increases?

   (c) How could you reparametrize the curve so that the rate at which a point traces out the curve decreases?

   (d) How could you reparametrize the curve so that the rate at which a point traces out the curve increases and the

curve is traced out in the opposite direction?

21. Find parametric equations of the plane which contains the three points $P_1 = [-1, 0, 2]$, $P_2 = [1, -1, 0]$, and $P_3 = [2, 1, 2]$.

22. Consider the parametric equations

$$x = \phi(t) = a + bt, \; y = \psi(t) = c + dt$$

where a, b, c, and d are real constants and $-\infty < t < \infty$. Do you see why this gives a line as long as one of the numbers a, b, c or d is not 0? Think about the rise and run. If t changes by a certain amount, how do x and y change? What is the slope of this line? Can you read the slope off from the equations? What could be done to get just a segment? Give reasons for your answers.

23. The idea for parametric curves in $\mathcal{R}^3$ is the same as for $\mathcal{R}^2$. You just need to add a third function for the third coordinate. Thus, a general formulation is

$$x = \phi(t), \; y = \psi(t), \; z = \lambda(t), \; a \le t \le b.$$

Does it make sense to talk about the conversion to a single equation? Explain your answer.

24. Consider the following parametric equations which represent a line L in $\mathcal{R}^3$: $x = \phi(t) = a + bt$, $y = \psi(t) = c + dt$, $z = \lambda(t) = e + ft$ where $-\infty < t < \infty$, and a, b, c, d, e and f are real constants. If $[x_1, y_1, z_1]$ and $[x_2, y_2, z_2]$ are distinct points on L, then the above equations are equivalent to the set of equations:

$$x = x_1 + (x_2 - x_1)t, \; y = y_1 + (y_2 - y_1)t,$$
$$z = z_1 + (z_2 - z_1)t, \; -\infty < t < \infty$$

25. Go back and read over the discussion of a line in $\mathcal{R}^3$ (beginning on page 537). Make an analogous analysis for a plane. What

would you look at instead of two points? Explain your answer.

26. Consider the parametric equations,

$$x = a + bs + ct, \quad y = d + es + ft,$$
$$z = g + hs + it, \quad -\infty < s, \, t < \infty$$

where a, b, c, d, e, f, g, h and i are real constants. Explain why these equations represent a plane in $\mathcal{R}^3$. Is it enough to say that a surface is a plane if whenever it contains two points, it also contains the line segment connecting them? Explain your answer. Does every plane satisfy this property? If possible, give an example of surface which satisfies this property but is not a plane.

■ 7.4 CALCULUS AND PARAMETRIC EQUATIONS

7.4.1 Overview

In this section you will investigate calculus as it relates to curves and surfaces represented by parametric equations. You will study derivatives of functions called *vector-valued functions*, or simply *vector functions*, from $\mathcal{R}^1$ to $\mathcal{R}^2$ or $\mathcal{R}^3$. In addition, you will learn about slope and arc length for curves represented by parametric equations. Finally, you will study some basic notions of curvilinear motion, that is, motion along curves, and you will have the opportunity to further extend the ideas of calculus as they relate to vector-valued functions and their applications in the exercises.

7.4.2 Activities

1. If you are using **ISETL** you can use the **computer functions graphpt, graphpn,** and **graphptn** to draw graphs for the three curves below which are given parametrically. If you are not using **ISETL** as your MPL, then your instructor will give you instructions on how to complete this activity.

   (a) $x(t) = 2\cos(t)$, $y(t)\sin(t)$, $-\pi \le t \le \pi$

   (b) $x(t) = 1 + t^3$, $y(t) = 1 + t^2$, $-1 \le t \le 1$

   (c) $x(t) = t - \sin(t)$, $y(t) = 1 - \cos(t)$, $0 \le t \le 10$

   If you are using **ISETL**, explain what you think the **computer functions graphpt, graphpn,** and **graphptn** are doing?

2. Consider the following reparametrizations of the curves in Activity 1.

   (a) The reparametrization α of the curve in Activity 1 (a) where

   $$\alpha : [0, 50] \to [-\pi, \pi] \text{ with } \alpha(t) = 0.04\pi t$$

   (b) The reparametrization α of the curve in Activity 1 (b) where

   $$\alpha : [-1, 1] \to [-1, 1] \text{ with } \alpha(t) = -t$$

   (c) The reparametrization α of the curve in Activity 1 (c) where

   $$\alpha : [0, 2] \to [0, 10] \text{ with } \alpha(t) = 10 - 5t.$$

Once again, if you are using **ISETL** as your MPL, use **graphpt**, **graphpn**, and **graphptn** to each curve. If you are not using **ISETL** your instructor will give you directions on how to complete this activity. For each graph, state what the reparametrization has changed and what it has not changed. Give an explanation in each case.

3. Write a **computer function arclength** which approximates the length of a curve given in parametric form,

$$x = x(t) = \phi(t), \ y = y(t) = \psi(t), \ a \le t \le b.$$

This can be done using the following steps.

- Pick a value for n and partition the domain $[a, b]$ into n equal subintervals.
- If $[t_i, t_{i+1}]$ is a typical subinterval, add up the lengths of the chords from the point $[\phi(t_i), \psi(t_i)]$ to the point $[\phi(t_{i+1}), \psi(t_{i+1})]$.
- Take large values of n and estimate a limit.

Apply your **computer function arclength** to estimate the lengths of the following curves,

(a) $\phi(t) = 4t$, $\psi(t) = 3t$, $-1.5 \le t \le 1.5$

(b) $\phi(s) = 3\cos 2s$, $\psi(s) = 3\sin 2s$, $0 \le s \le \frac{\pi}{2}$

(c) $\phi(u) = u^3$, $\psi(u) = u^6$, $-1.5 \le u \le 1.5$

4. Estimate the arc lengths of the curves in Activity 3, except this time use the following formula,

$$L = \int_a^b \sqrt{(\phi'(t))^2 + (\psi'(t))^2}\, dt$$

which gives the arc length L of a parametric curve given by $x = \phi(t)$ and $y = \psi(t)$. You may perform the calculations by hand, use your symbolic computer system, or use your approximate derivative **computer function D** together with your modified **computer function DefInt** to approximate the above expression. Compare the results you get with what you got in Activity 3.

5. Use a method analogous to that in Activity 3 to estimate the arc length of a curve in $\mathcal{R}^3$. Apply your method to the arc given by

$$x = \phi(t) = t^2, \ y = \psi(t) = t, \ z = \lambda(t) = t^3, \ 0 \le t \le 2.$$

6. Use a method analogous to that in Activity 4 to obtain the arc length of a curve in $\mathcal{R}^3$. Apply your method to the arc given by

$$x = \phi(t) = t^2, \ y = \psi(t) = t, \ z = \lambda(t) = t^3, \ 0 \le t \le 2.$$

Compare the answer you get here with your answer in the previous activity.

7. Write a **computer function arclength** that will accept two **computer functions** and numbers **a** and **b** that represent a curve given parametrically. Your **computer function** should return a **computer function** which represents the function that computes, for a given **t**, the length of the given curve with the "time" variable restricted to the interval $[a, t]$.

What is the domain of the function represented by **arclength(f,g,a,b)**? What is its range?

8. Consider the curve given parametrically by $x = t^2$, $y = t^3$, for $1 \le t \le 3$. Apply your **computer function arclength** from Activity 7 to this curve on the indicated interval for the parameter and then graph the resulting arc length function. What can you say about the monotonicity of this function? Is there anything that will be true in general? What about the inverse?

9. Consider the *vector function* **R** given by $\mathbf{r} = \mathbf{R}(t) = 2t\,\mathbf{i} + 3\sqrt{t}\,\mathbf{j}$. Use a vector graphing facility, like **vectors** in **ISETL**, to produce a sketch of the set of vectors represented by

$$r := \{[2*t,\, 3*t**(1/2)] : t \text{ in } [0..10]\};$$

Next, by hand connect the tips of the vectors on your sketch. Can you describe what $\mathbf{R}(t)$ represents relative to the graph of the curve C given by $x = 2t$, $y = 3\sqrt{t}$?

Now consider the vector function **V** given by

$$\mathbf{v} = \mathbf{V}(t) = 2\mathbf{i} + \frac{3}{2\sqrt{t}}\,\mathbf{j}.$$

Use a vector graphing facility, like **vectors** in **ISETL**, to produce a sketch containing the vectors $\mathbf{V}(t)$ for $t = 1, 2, \ldots, 10$. Why do we omit the vector at $t = 0$?

Now produce a sketch containing the following set of vectors

$$rv := \{[[2*t,\, 3*t**(1/2)],\, [2*t+2,\, 3*t**(1/2)+3/(2*t**(1/2))]] : t \text{ in } [1..10]\};$$

Can you explain what the set of vector rv represents and where is comes from? Can you explain how the vector $\mathbf{V}(t)$ is related to the vector $\mathbf{R}(t)$? What do you think $\mathbf{V}(t)$ represents?

7.4.3 Discussion

Derivatives of Functions from $\mathcal{R}^1$ to $\mathcal{R}^2$ A curve in $\mathcal{R}^2$ is a function whose domain is a set in $\mathcal{R}^1$ and whose range is $\mathcal{R}^2$. What would be the meaning of the derivative of such a function? You can ask the same question for a curve in $\mathcal{R}^3$.

Sometimes a problem in several dimensions can be reduced to several problems in one dimension. For example, we have written curves in $\mathcal{R}^2$ as a pair of functions, (ϕ, ψ), each from $\mathcal{R}^1$ to $\mathcal{R}^1$. There is no trouble in taking the derivative of each of these functions individually, provided that the derivatives exist. Thus we can have the new pair of functions, (ϕ', ψ'). One major concern of this section is to consider interpretations of this pair of derivatives.

The first interpretation we can try to make is to relate the situation to the single function representation of a curve in the form of a function f and an equation $y = f(x)$. Recall (page 553) that this function is given by

$$f = \psi \circ \phi^{-1}.$$

Do you see how to get the derivative of f from the derivatives of the functions ϕ and ψ?

We can use the chain rule and the rule for the derivative of the inverse of a function. Here are the relevant formulas for these two rules.

$$(g \circ h)'(x) = g'(h(x))h'(x)$$

$$(g^{-1})'(x) = \frac{1}{g'(g^{-1}(x))}$$

We apply these two rules to compute the derivative of $f = \psi \circ \phi^{-1}$ as follows.

$$f'(x) = (\psi \circ \phi^{-1})'(x) = \left(\psi'\left(\phi^{-1}(x)\right)\right) \cdot (\phi^{-1})'(x) = \frac{\psi'\left(\phi^{-1}(x)\right)}{\phi'\left(\phi^{-1}(x)\right)}$$

Thus, we can write,

$$x = x(t) = \phi(t), \; y = y(t) = \psi(t), \; a \leq t \leq b$$

$$f(x) = \psi\left(\phi^{-1}(x)\right)$$

$$\frac{dy}{dx} = f'(x) = \frac{\psi'\left(\phi^{-1}(x)\right)}{\phi'\left(\phi^{-1}(x)\right)} = \frac{\psi'(t)}{\phi'(t)} = \frac{\frac{dy}{dt}}{\frac{dx}{dt}}$$

and our first interpretation is that the derivative of the function f is the ratio of the derivative of the function y to the derivative of the function x.

In particular, this means that if a particle is moving in time along a curve, then the rate of change of its y position with respect to its x position is the ratio of the rate of change of its y position with respect to time to the rate of change of its x position with respect to time.

Tangents and Normals to Curves in Parametric Form   Continuing the discussion of the previous paragraph, we can obtain the equation of the tangent and normal lines to a curve given parametrically.   We consider such a curve given by the pair of functions (ϕ, ψ) and the function $f = \psi \circ \phi^{-1}$.   The only thing that is needed to find equations of tangent and normal lines at a point $P = [x_0, y_0] = [\phi(t_0), \psi(t_0)]$ is the slope of the tangent.   This is given, as usual, by $f'(x_0)$.   Hence, as we saw above, in terms of the parametric form, the slope is given by

$$f'(x_0) = \frac{\psi'(t_0)}{\phi'(t_0)}$$

and the equation of the tangent line through P is

$$\phi'(t_0)(y - y_0) - \psi'(t_0)(x - x_0) = 0.$$

Immediately then you can write the equation of the normal line by using the negative reciprocal of the slope of the tangent.   The result is,

$$\psi'(t_0)(y - y_0) - \phi'(t_0)(x - x_0) = 0.$$

There is one sticky point here that we sort of sloughed over.   In order to get the slope of the tangent, we passed to the function f and took its derivative to get the slope.   This requires that whole business of cutting down to a piece of the curve, taking inverse functions and using rules for computing derivatives. All of that is perfectly correct, but it does not come into the calculations.   The slope of the tangent can be computed directly from the parametric form.

Go back to basics and ask yourself, what is the slope of the tangent at a point?   It is the limit of slopes of chords from the point to nearby points, as the nearby points approach the point in question. Figure 7.39 shows how it looks in a picture.

The chord from the point $P_0 = [\phi(t_0), \psi(t_0)]$ to the point $P = [\phi(t_0 + h), \psi(t_0 + h)]$ has slope,

$$\frac{\psi(t_0 + h) - \psi(t_0)}{\phi(t_0 + h) - \phi(t_0)}$$

Do you see how to get from here to a ratio of derivatives?

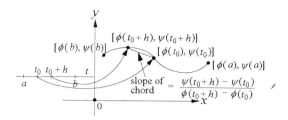

Figure 7.39. The slope of a chord on a curve.

We divide numerator and denominator by h and take the limit as h goes to zero. Hence, we find that the slope of the tangent to the curve at P_0 is

$$\lim_{h \to 0} \frac{\psi(t_0 + h) - \psi(t_0)}{\phi(t_0 + h) - \phi(t_0)} = \lim_{h \to 0} \left(\frac{\frac{\psi(t_0 + h) - \psi(t_0)}{h}}{\frac{\phi(t_0 + h) - \phi(t_0)}{h}} \right) = \frac{\lim_{h \to 0} \left(\frac{\psi(t_0 + h) - \psi(t_0)}{h} \right)}{\lim_{h \to 0} \left(\frac{\phi(t_0 + h) - \phi(t_0)}{h} \right)} = \frac{\psi'(t_0)}{\phi'(t_0)}$$

which is the same as the expression for the slope we found above.

Lengths of Curves Given in Parametric Form

On two occasions in this book we avoided rigorously developing the formula for arc length. In both cases we discussed the main ideas and then jumped to the formula without actually deriving it. There was a reason for that and now it can be told. The reason is that the computations were too messy. Moreover, there is a simpler way to go about it using the parametric form of a curve. We will go through that now and then show you how the formulas that we obtained earlier come out of the parametric form.

Actually, you already made the calculations in Activity 3. Compare your calculations in that activity with the picture in Figure 7.40.

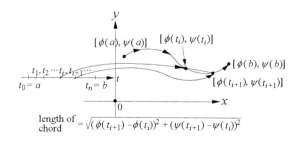

Figure 7.40. An approximation to the arc length of a curve.

Hopefully you obtained something like the following as your estimate for the length of the arc,

$$\sum_{i=1}^{n} \sqrt{(\phi(t_{i+1}) - \phi(t_i))^2 + (\psi(t_{i+1}) - \psi(t_i))^2}.$$

Now, before passing to the limit, we can multiply and divide the ith term by $t_{i+1} - t_i$ to obtain,

$$\sum_{i=1}^{n} \sqrt{\left(\frac{\phi(t_{i+1}) - \phi(t_i)}{t_{i+1} - t_i}\right)^2 + \left(\frac{\psi(t_{i+1}) - \psi(t_i)}{t_{i+1} - t_i}\right)^2} \, (t_{i+1} - t_i).$$

Now if we pass to the limit as $n \to \infty$, expressions such as

$$\frac{\phi(t_{i+1}) - \phi(t_i)}{t_{i+1} - t_i}$$

and

$$\frac{\psi(t_{i+1}) - \psi(t_i)}{t_{i+1} - t_i}$$

can be approximated by $\phi'(t_i)$, $\psi'(t_i)$, respectively. Hence, the expression is approximated by,

$$\sum_{i=1}^{n} \sqrt{(\phi'(t_i))^2 + (\psi'(t_i))^2} \, (t_{i+1} - t_i)$$

which is a left Riemann sum and so, taking the limit, the formula for arc length is obtained as,

$$\int_a^b \sqrt{(\phi'(t_i))^2 + (\psi'(t_i))^2} \, dt.$$

Now we can do some simple calculations to obtain a formula for finding the arc length of a curve given by $y = f(x)$ where f is a function whose derivative f' is continuous. First, factor out $(\phi'(t))^2$ in the integrand,

$$\int_a^b \sqrt{1 + \left(\frac{\psi'(t)}{\phi'(t)}\right)^2} \, \phi'(t) dt.$$

Now, make the substitution $x = \phi(t)$, $dx = \phi'(t)dt$ to obtain

$$\int_{\phi(a)}^{\phi(b)} \sqrt{1 + \left(\frac{\psi'(\phi^{-1}(x))}{\phi'(\phi^{-1}(x))}\right)^2} \, dx = \int_{\phi(a)}^{\phi(b)} \sqrt{1 + (f'(x))^2} \, dx.$$

The continuity of f' guarantees the existence of the definite integral.

Finally, we can consider what happens for curves in $\mathcal{R}^3$. What we did for curves in $\mathcal{R}^2$ has two parts. First we used the parametric form to obtain approximations to the arc length by adding up lengths of chords. We took the limit and this gave us the integral formula. That is the first part. The second part is to transform the formula by replacing the parametric form for the curve by the standard form. You can apply an analogous derivation for curves in $\mathcal{R}^3$. What formulas do you get?

Reparametrizations and Tangents

In Activity 2 you looked at the effect on the tangent lines to a curve when you made a reparametrization. Did you see what happened? The length of the line segments you got on the graph changed, but the direction did not. This means that the tangent line will not change at all. We can see analytically why this is the case.

Suppose we have a curve in $\mathcal{R}^2$ given parametrically by the pair of functions (ϕ, ψ) and we use the function α to make a reparametrization. The reparametrized curve is given parametrically by the pair of functions $(\phi \circ \alpha, \psi \circ \alpha)$.

The slope of the tangent line at some point $(\phi(t),\ \psi(t))$ in the original parametrization is given by the ratio,

$$\frac{\psi'(t)}{\phi'(t)}.$$

Applying the chain rule, we obtain the slope of the tangent at the same point $(\phi(\alpha(s)),\ \psi(\alpha(s)))$ (where $t = \alpha(s)$),

$$\frac{\psi'(\alpha(s))\alpha'(s)}{\phi'(\alpha(s))\alpha'(s)} = \frac{\psi'(\alpha(s))}{\phi'(\alpha(s))} = \frac{\psi'(t)}{\phi'(t)}.$$

In other words, the factor $\alpha'(s)$ cancels and the slope is unchanged. Our conclusion is that reparametrizing a curve does not change the tangent line. Consequently, the normal line will also be unchanged.

What about three dimensions? Do you see what happens there? You will get an opportunity to investigate what happens in $\mathscr{R}^3$ in the exercises.

Parametrization with Respect to Arc Length

There is one particular reparametrization that is of special interest. We begin by considering the **computer function arc length** which you worked with in Activities 7 and 8. Consider a generic curve given in parametric form.

$$x = \phi(t),\ y = \psi(t),\ a \le t \le b.$$

You can use the formula we derived to compute its arc length.

$$l = \int_a^b \sqrt{(\phi'(t))^2 + (\psi'(t))^2}\, dt.$$

This is the length of the entire curve. But we can also make a function by computing the length of the arc obtained as the "time" variable runs from a to a variable point t. Thus we can write

$$L(t) = \int_a^t \sqrt{(\phi'(t))^2 + (\psi'(t))^2}\, dt$$

What is the domain of the function L? Can you see why this function is strictly increasing? What about its maximum and minimum values? How much of the curve is $L(a)$? $L(b)$?

You can use the Fundamental Theorem of Calculus to compute the derivative of L as you did in Chapter 5 (see page 343). It is,

$$L'(t) = \sqrt{(\phi'(t))^2 + (\psi'(t))^2}.$$

Notice that $L'(t) \ge 0$. If we assume that it never happens with this curve that both $\phi'(t) = 0$ and $\psi'(t) = 0$, then we can even say that $L'(t) > 0$ provided $t > a$.

A curve with this property has special interest so we single it out and give it a name.

Definition 7.1

A curve given in parametric form by

$$x = \phi(t),\ y = \psi(t),\ a \le t \le b$$

is said to be **non-singular** *if it is never the case, for any* $t \in [a, b]$, *that both* $\phi'(t) = 0$ *and* $\psi'(t) = 0$.

Moreover, a non-singular curve for which $\phi'(t)$ and $\psi'(t)$ are continuous on the interval $[a, b]$ is said to be a *smooth* curve. Such curves have no sharp corners.

If you think about a curve as the path of a point moving through $\mathcal{R}^2$ in time, then to say that the curve is non-singular means that it never completely stops moving. What can you say about the tangent line to a curve at a point where the condition of non-singularity is violated?

Now suppose we have a non-singular curve with length l. Then we have a function $L : [a, b] \to [0, l]$. This function is strictly increasing and has a strictly positive derivative given by the above formula.

Under these conditions, the function L has an inverse $\alpha : [0, l] \to [a, b]$. It is also strictly increasing and you can compute its derivative using the rule for derivatives of inverses. We use the function $\alpha = L^{-1}$ to reparametrize the original curve. Such a reparametrization is called the *parametrization with respect to arc length*. Certain formulas expressing properties of curves become considerably simpler when this parametrization is used. Let's work out the details for the following example which you worked with in Activity 8.

$$x = \phi(t) = t^2, \ y = \psi(t) = t^3 \text{ for } 1 \le t \le 3.$$

First we can compute the arc length function L. Its domain is the interval $[1, 3]$ and,

$$L(t) = \int_1^t \sqrt{(\phi'(\tau))^2 + (\psi'(\tau))^2} \, d\tau$$
$$= \int_1^t \sqrt{(2\tau)^2 + (3\tau)^2} \, d\tau$$
$$= \int_1^t \tau\sqrt{4 + 9\tau^2} \, d\tau$$
$$= \frac{1}{27}(4 + 9t^2)^{3/2} - \frac{13^{3/2}}{27}$$

If you write $s = L(t)$ and solve for t in terms of s you get,

$$t = L^{-1}(s) = \frac{\sqrt{(27s + 13^{3/2})^{2/3} - 4}}{3}$$

and this is the function used for the reparametrization with respect to arc length. Substituting the expression for $L^{-1}(s)$ in for t in the original parametric equations, we obtain parametric equations for the curve reparametrized with the arc length parameter,

$$x = \frac{(27s + 13^{3/2})^{2/3} - 4}{9}, \ y = \frac{\left((27s + 13^{3/2})^{2/3} - 4\right)^{3/2}}{27}$$

where

$$0 \le s \le \frac{(85)^{3/2} - (13)^{3/2}}{27}.$$

You usually will be pleased to know that in practice, one does not really compute this parametrization very often. It is used for general investigations, to understand and simplify various properties of curves.

Vector Functions — Unit Tangent and Normal Vectors   We can express some of
the ideas about curves by using vector notation. Consider, once again, a smooth curve C given in
parametric form by

$$x = \phi(t), \; y = \psi(t), \; a \le t \le b.$$

Assume that this curve C is non-singular. We can define a *vector-valued function*, or simply *vector
function*), **R** by taking, for each $t \in [a, b]$, the vector $\mathbf{r} = \mathbf{R}(t) = \phi(t)\mathbf{i} + \psi(t)\mathbf{j}$. As you saw in Activity 9,
the vector $\mathbf{R}(t_0)$ gives the position of a particle (with respect to the origin) along the curve C at $t = t_0$.
We can think of a vector function **R** as describing the motion of a moving particle as it travels along the
curve C given by $x = \phi(t)$, $y = \psi(t)$, for $a \le t \le b$.

Of course, a similar notion can be considered in $\mathcal{R}^3$, as well as the notion of the velocity vector.

Let us define another vector function **V** by taking, for each $t \in [a, b]$, the vector $\mathbf{v} = \mathbf{V}(t)$ in $\mathcal{R}^2$
given by

$$\mathbf{v} = \mathbf{V}(t) = \phi'(t)\mathbf{i} + \psi'(t)\mathbf{j}.$$

This vector is the velocity vector associated with $\mathbf{r} = \mathbf{R}(t)$ and its magnitude, $|\mathbf{V}(t)|$, is the particle's
(linear) *speed*, $\frac{ds}{dt}$. Can you explain why? Do you see that this vector is in the same direction as the
tangent to the curve for each t and it points in the direction of motion? We are particularly interested in
determining, for each $t \in [a, b]$, the unit vector $\mathbf{T}(t)$ in the direction of the velocity vector $\mathbf{V}(t)$. We can
get this by dividing by the magnitude of the vector $\mathbf{V}(t)$ to obtain the vector function **T** given by

$$\mathbf{T}(t) = \frac{\mathbf{V}(t)}{|\mathbf{V}(t)|},$$

provided $\mathbf{V}(t) \ne 0$. How can you be sure that this is always possible, that is, that the denominator can not
be 0?

To study *curvilinear motion*, that is, motion along curves, we are also interested in the unit vector
perpendicular to the tangent. There are two such unit vectors. One unit normal vector always points to
the concave side of the curve (i.e., in the direction that the curve represented by **R** is bending). Someone
once made a neat discovery. Define

$$\mathbf{N}(t) = \frac{\mathbf{T}'(t)}{|\mathbf{T}'(t)|},$$

provided $|\mathbf{T}'(t)| \ne 0$. Here, the expression $\mathbf{T}'(t)$ is the vector obtained by differentiating each component
of $\mathbf{T}(t)$.

Now, obviously, $\mathbf{N}(t)$ is a unit vector. Why? How can you be sure it is perpendicular to the tangent?
You can compute the dot product $\mathbf{T}(t) \cdot \mathbf{N}(t)$ directly and see that it is 0. The computations are a bit
messy and we don't want to go through them here. You will get a chance to do it in the exercises. It can
be shown that $\mathbf{N}(t)$ always points in the direction the curve is bending (i.e., it always points toward the
concave side of the curve). Can you explain why?

We call **T** the *unit tangent vector* and **N** the *(principal) unit normal vector* of the curve. Figure 7.41
shows the relationship between **R**, **T**, and **N**. Notice that **N** always points toward the concave side of the
curve C associated with $\mathbf{R}(t)$. You will see the reason for this later. In $\mathcal{R}^3$ the vectors $\mathbf{T}(t)$ and $\mathbf{N}(t)$
are defined in an entirely similar manner.

If you actually want to compute these quantities in an example, the computations can be formidable.
The main complication comes from the necessity of dividing by the magnitude. This is bad enough for
the tangent vector, but it can become totally uncivilized for the normal. One application of
reparametrization with respect to arc length is to simplify these formulas.

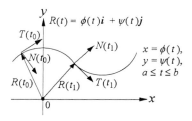

Figure 7.41. The relationship between R, T, and N.

Suppose that you start with a generic smooth curve

$$x = \phi(t), \; y = \psi(t), \; a \le t \le b$$

and let $L : [a, b] \to [0, l]$ be its arc length function so that the reparametrization function α is given by $L^{-1} : [0, l] \to [a, b]$. As on page 572, we write $s = L^{-1}(t)$ for the arc length parameter s.

Now let's compute $|\mathbf{V}(t)|$. We have,

$$|\mathbf{V}(t)| = |\phi'(t)\mathbf{i} + \psi'(t)\mathbf{j}| = \sqrt{(\phi'(t))^2 + (\psi'(t))^2} = L'(t).$$

Hence we have,

$$\begin{aligned}
\mathbf{T}(t) &= \frac{1}{L'(t)}(\phi'(t)\mathbf{i} + \psi'(t)\mathbf{j}) \\
&= \frac{\phi'(t)}{L'(t)}\mathbf{i} + \frac{\psi'(t)}{L'(t)}\mathbf{j} \\
&= \frac{\phi'(L^{-1}(s))}{L'(L^{-1}(s))}\mathbf{i} + \frac{\psi'(L^{-1}(s))}{L'(L^{-1}(s))}\mathbf{j} \\
&= (\phi \circ L^{-1})'(s)\mathbf{i} + (\psi \circ L^{-1})'(s)\mathbf{j}
\end{aligned}$$

since

$$(L^{-1})'(s) = \frac{1}{L'(L^{-1}(s))}.$$

The last expression for $\mathbf{T}(t)$ is exactly what would be obtained by starting with the curve reparametrized with respect to arc length and just computing the derivatives, without having to divide by the absolute value.

Actually, you can see that the unit tangent is a property of the curve so it will not be changed if we make the reparametrization and compute it replacing ϕ and ψ by $\phi \circ L^{-1}$ and $\psi \circ L^{-1}$, respectively. What this means is that the reason it is not necessary to divide by the magnitude is that the magnitude of $\mathbf{T}(s)$, that is, the unit tangent computed from the parametrization with respect to arc length, is already one.

There is a convention in general use by which the expression $\mathbf{T}(s)$ refers to the unit tangent computed when the curve is parametrized with respect to arc length.

You don't get this simplification for the normal. The magnitude of $\mathbf{T}'(s)$ is not necessarily one. In fact, the magnitude, $|\mathbf{T}'(s)|$, is a measure of how much the curve is turning. It is called the curvature and is often denoted by the symbol κ.

For theoretical derivations it is convenient to use the parametrization with respect to arc length. For calculations in examples it is usually more trouble than it is worth to compute the reparametrization. Hence it is of value to have a formula for the curvature in terms of any given parametrization. A straightfoward derivation yields the following formula in $\mathcal{R}^2$.

$$\kappa = \kappa(t) = \frac{|\phi'(t)\,\psi''(t) - \phi''(t)\,\psi'(t)|}{\left((\phi'(t))^2 + (\psi'(t))^2\right)^{\frac{3}{2}}}$$

In the exercises you will be given a list of hints in order to help you derive this formula for the curvature of a curve.

The *acceleration vector* $\mathbf{a} = \mathbf{A}(t)$ associated with a vector function $\mathbf{R}$ given by $\mathbf{r} = \mathbf{R}(t) = \phi(t)\mathbf{i} + \psi(t)\mathbf{j}$ is defined as follows:

$$\mathbf{a} = \mathbf{A}(t) = \phi''(t)\mathbf{i} + \psi''(t)\mathbf{j}.$$

It turns out that the following important relationship between the acceleration vector $\mathbf{A}$, the unit tangent, $\mathbf{T}$, and the (principal) unit normal vector, $\mathbf{N}$, is true

$$\mathbf{a} = \mathbf{A}(t) = \frac{d^2 s}{dt^2}\mathbf{T}(t) + \kappa\left(\frac{ds}{dt}\right)^2 \mathbf{N}(t)$$

where

$$\frac{ds}{dt} = |\mathbf{V}(t)|$$

is the speed of a particle moving along the curve represented by $\mathbf{R}(t)$ (provided the curve represented by $\mathbf{T}(t)$ is nonsingular, i.e., $\mathbf{T}'(t) \neq 0 = 0\mathbf{i} + 0\mathbf{j}$). See the picture in Figure 7.42. We omit the proof.

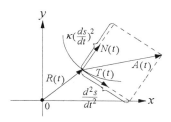

Figure 7.42. The relationship between A, T, and N.

The (scalar) components of $\mathbf{a} = \mathbf{A}(t)$ relative to this decomposition with respect to $\mathbf{T}(t)$ and $\mathbf{N}(t)$ are called the *tangential component of acceleration*, $\mathbf{a}_T$, and the *normal component of acceleration*, $\mathbf{a}_N$, respectively. That is, the acceleration vector $\mathbf{a}$ can be written as the linear combination

$$\mathbf{a} = \mathbf{a}_T\mathbf{T} + \mathbf{a}_N\mathbf{N}$$

and it is easy to show the following:

$$\mathbf{a}_T = \mathbf{A}\cdot\mathbf{T}, \quad \mathbf{a}_N = \mathbf{A}\cdot\mathbf{N}, \quad \text{and} \quad |\mathbf{a}| = \sqrt{(\mathbf{a}_T)^2 + (\mathbf{a}_N)^2}$$

where $|\mathbf{a}| = |\mathbf{A}(t)|$ is the magnitude of the acceleration vector $\mathbf{a}$. The latter relationship tells us that

$$\mathbf{a_N} = \sqrt{|\mathbf{a}|^2 - (\mathbf{a_T})^2}.$$

Hence, in order to compute the normal component of acceleration, $\mathbf{a_N}$, it suffices to compute the tangential component, $\mathbf{a_T}$, and the magnitude of the acceleration vector, $|\mathbf{a}|$. This fact is quite useful in practice, since the computation of $\mathbf{N}(t)$ by its definition can be quite messy.

    You will get a chance to derive the above relationships, study their consequences and applications, and further investigate vector functions and related calculus notions in the exercises. As you will see in the exercises, most of what we have done with vector functions in $\mathcal{R}^2$ is completely analogous in $\mathcal{R}^3$. However, the notion of curvature and the associated formulas to compute it are a bit more complicated and we leave that for you to study in a more advanced course.

7.4.4 Exercises

1. For each curve represented by each pair of parametric equations or vector-valued functions below, do the following.

    (a) Find an equation of the tangent line to each curve indicated below at the indicated value of the parameter.

    (b) Find an equation of the normal line to the curve at the indicated value of the parameter.

    (c) Find an expression for the curvature κ at any value of the parameter.

    (d) Find the curvature at the indicated value of the parameter.

    (e) Find all points, if any, where the curve has a horizontal tangent.

    (f) Find all points, if any, where the curve has a vertical tangent.

    Here are the pairs of parametric equations or vector-valued functions.

    i. $x = 2t$, $y = \sqrt{t}$, $t = 4$

    ii. $\mathbf{R}(t) = t^2\mathbf{i} + t^3\mathbf{j}$, $t = -1$

    iii. $x = \dfrac{2t}{1+t}$, $y = \dfrac{1}{t}$, $t = 1$

    iv. $x = 1 - 2v$, $y = 3 - v^2$, $v = 2$

    v. $x = 2\sin(t)$, $y = \cos(t)$, $t = \frac{\pi}{6}$

    vi. $x = 3\sin^2(\pi\theta)$,
    $y = -4\cos^2(\pi\theta)$, $\theta = 0.5$

    vii. $\mathbf{R}(t) = 3\sin(2t)\mathbf{i} + 3\cos(2t)\mathbf{j}$,
    $t = \frac{\pi}{6}$

    viii. $x = 3\sin(u)$, $y = \cos^2(u)$,
    $u = \frac{\pi}{4}$

    ix. $x = 2\tan(t)$, $y = \sec^2(t)$,
    $t = \frac{\pi}{4}$

    x. $\mathbf{R}(t) = e^{3t}\mathbf{i} - 2e^t\mathbf{j}$, $t = 0$

    xi. $x = 3t - 1$, $y = \sqrt[3]{1 - 2t}$, $t = 1$

    xii. $\mathbf{R}(t) = (1 - 2\sin(\theta))\mathbf{i} +$
    $(1 - 2\cos(\theta))\mathbf{j}$, $\theta = \pi$

    xiii. $x = \cos^3(t)$, $y = \sin^3(t)$, $t = \frac{\pi}{4}$

    xiv. $\mathbf{R}(t) = 3\ln(2t)\mathbf{i} + t^2\mathbf{j}$, $t = 1$

2. For the parametric equations $x = x(t)$ and $y = y(t)$, you saw on page 568 that the slope to the curve represented by the parametric equations, $y' = \frac{dy}{dx}$, is given by

$$y' = \frac{\frac{dy}{dt}}{\frac{dx}{dt}}.$$

(a) Find a formula for

$$y'' = \frac{d^2y}{dx^2} = \frac{dy'}{dx}.$$

(b) Find a formula for

$$y^{(n)} = \frac{d^ny}{dx^n}$$

by generalizing the result of part (a).

3. Use the result of part (a) of the previous exercise to determine whether or not the curve represented by the parametric equations is concave up or concave down at the indicated value of the parameter. Use a computer to make a sketch of the curve to verify your answer.

 (a) $x = 2t$, $y = \sqrt{t}$, $t = 4$

 (b) $x = t^2$, $y = t^3$, $t = -1$

 (c) $x = t + 2$, $y = 3t + \dfrac{1}{t}$, $t = -1$

 (d) $x = \dfrac{2t}{1+t}$, $y = \dfrac{1}{t}$, $t = 1$

 (e) $x = 1 - 2t$, $y = 3 - t^2$, $t = 2$

 (f) $x = 2\sin(t)$, $y = \cos(t)$, $t = \frac{\pi}{6}$

 (g) $x = 3\sin(u)$, $y = \cos^2(u)$, $u = \frac{\pi}{4}$

 (h) $x = 2\tan(t)$, $y = \sec^2(t)$, $t = \frac{\pi}{4}$

 (i) $x = e^{3t}$, $y = -2e^t$, $t = 0$

 (j) $x = 3t - 1$, $y = \sqrt[3]{1 - 2t}$, $t = 1$

 (k) $x = \sin(2t)$, $y = e^{-t}$, $t = 0$

 (l) $x = 1 - 2\sin(\theta)$, $y = -2\cos(\theta)$, $\theta = \pi$

4. Does the slope of the curve represented by $x = 1 - 3t^2$, $y = 2t + 2t^3$ have a minimum? a maximum? Give reasons for your answers.

5. Does the slope of the curve represented by $x = 4t + t^3$, $y = 1 - t^3$ have a minimum? a maximum? Give reasons for your answers.

6. What do you think is the curvature of a straight line in $\mathcal{R}^2$? Prove your conjecture.

7. Consider the circle represented by $x = a\sin(t)$, $y = a\cos(t)$ where $a > 0$.

 (a) Find the curvature of the circle.

 (b) Are you surprised by the result of part (a)? Explain your answer.

8. Find a formula for the curvature of the ellipse represented by $x = a\sin(t)$, $y = b\cos(t)$.

9. Find the arc length of each curve represented by the parametric equations or vector function.

 (a) $x = 3\cos(2t)$, $y = 3\sin(2t)$

 (b) $\mathbf{R}(t) = t^3\mathbf{i} + t^2\mathbf{j}$, $0 \le t \le 2$

 (c) $x = 2\cos^2(t)$, $y = 2\sin^2(t)$, $0 \le t \le 3$

 (d) $\mathbf{R}(t) = (\cos(t) + t\sin(t))\mathbf{i} + (\sin(t) - t\cos(t))\mathbf{j}$, $0 \le t \le 1$

10. For the ellipse represented by $x = 3\cos(t)$, $y = 4\sin(t)$

 (a) Write a definite integral which represents the length of the ellipse.

 (b) Try using your Symbolic Computer System to evaluate the integral in part (a). Explain what happens. Can you find an antiderivative of the integrand function in your integral?

 (c) The integral in part (b) is called an *elliptical integral* and it can be shown that the integrand function has no elementary function as an antiderivative. Use a computer to estimate the length of the ellipse.

11. For a smooth curve represented by the parametric equations

$$x = \phi(t),\ y = \psi(t),\ a \le t \le b,$$

that is, a curve for which the functions ϕ', ψ', and γ' are continuous

(a) Show that a formula for the length of the curve is given by

$$\int_a^b \sqrt{(\phi'(t))^2 + (\psi'(t))^2 + (\gamma'(t))^2}\,dt.$$

(b) Using the formula in part (a), find the length of each curve represented by the parametric equations or vector function

     i. $x = 3 + t,\ y = 2t + 1,\ z = 5t,$ $-1 \le t \le 3$

     ii. $\mathbf{R}(t) = 3\sin(t)\mathbf{i} + 3t\mathbf{j}$ $+ 3\cos(t)\mathbf{k},\ 0 \le t \le \pi$

     iii. $x = t^2,\ y = 2\sqrt{t},\ z = \ln(t^2),$ $1 \le t \le 4$

12. Let C be a smooth curve in $\mathcal{R}^2$ represented by the parametric equations

$$x = \phi(t),\ y = \psi(t),\ a \le t \le b.$$

Consider the arc length function L given by

$$L(t) = \int_a^t \sqrt{(\phi'(\tau))^2 + (\psi'(\tau))^2}\,d\tau.$$

What is the domain of the function L? Can you see why this function is strictly increasing? What about its maximum and minimum values? How much (i.e., what fraction) of C is $L(a)$? How much of C is $L(b)$?

13. If we think about a curve as the path of a particle moving through the plane $\mathcal{R}^2$ (or in $\mathcal{R}^3$) in time, then to say that the curve is non-singular means that it never completely stops moving. What can you say about the tangent line to a curve at a point where the condition of non-singularity is violated? Explain your answer and illustrate it with appropriate examples.

14. For each curve represented by the parametric equations or vector function, find the reparametrization with respect to the arc length parameter $t = L^{-1}(s)$ where L is the arc length function.

(a) $x = 2t + 1,\ y = 2 - 3t,\ -1 \le t \le 3$

(b) $\mathbf{R}(t) = 2\sin(t)\mathbf{i} + 2\cos(t)\mathbf{j},\ 0 \le t \le 2\pi$

(c) $x = 2t\sqrt{t},\ y = 3t,\ 0 \le t \le 8$

(d) $x = a\sin(bt),\ y = a\cos(bt),$ $0 \le t \le \frac{2\pi}{b}$

15. Let f be a function such that f' is continuous on the interval $[a, b]$. Show that the curvature κ of the curve represented by $y = f(x)$ is given by

$$\kappa = \frac{|y''|}{\left(1 + (y')^2\right)^{3/2}} = \frac{|f''(x)|}{\left(1 + (f'(x))^2\right)^{3/2}}$$

using the parameterization $x = t,\ y = f(t)$.

16. Use the result of Exercise 15 to do the following for each curve listed below.

(a) Find a formula for the curvature of each curve.

(b) Find the curvature for the curve at the indicated value of the independent variable or point on the curve.

     i. $y = x^2 - 3x,\ x = -1$

     ii. $y = 3\sin(2t),\ t = \frac{\pi}{3}$

     iii. $y = \ln(2x),\ x = 0.5$

     iv. $y = 3e^x,\ x = 0$

     v. $y - xe^y = 1,\ [-1, 0]$

     vi. $x^2 + y^3 = 2x + 1,\ [0, 1]$

17. Does the curvature of the curve $y = e^x$ have a maximum value? a minimum value? If so, find the maximum and minimum values

of the curvature. If not, give an explanation.

18. Use the formula

$$\int_a^b \sqrt{(\phi'(t))^2 + (\psi'(t))^2}\, dt$$

for arc length of a smooth curve given by the parametric equations $x = \phi(t)$, $y = \psi(t)$, $a \le t \le b$, to derive the polar coordinate formula

$$\int_\alpha^\beta \sqrt{(f(\theta))^2 + (f'(\theta))^2}\, d\theta = \int_\alpha^\beta \sqrt{r^2 + (r')^2}\, d\theta$$

for the arc length of a polar curve $r = f(\theta)$ for $\alpha \le \theta \le \beta$. Use the polar coordinate transformation equations $x = r\cos(\theta)$ and $y = r\sin(\theta)$.

19. Use the formula in Exercise 18 to find the length of the polar curve $r = e^{2\theta}$, $0 \le \theta \le \pi$

20. Use the result of Exercise 18 to find the reparameterization in terms of the arc length parameter of the polar curve

$$r = ae^{b\theta}, 0 \le \theta \le \frac{2\pi}{b}.$$

21. Use the result of Exercise 18 to find a formula for the area of the surface of revolution obtained when a polar curve $r = f(\theta)$, $\alpha \le \theta \le \beta$, is revolved about the polar axis. Assume that the curve does not cross over the polar axis.

22. Show that the formula for the curvature κ of a polar curve $r = f(\theta)$ is given by

$$\kappa = \frac{\left|r^2 + 2(r')^2 - rr''\right|}{\left(r^2 + (r')^2\right)^{3/2}}$$

where the derivatives of r are with respect to θ.

23. Use the formula in Exercise 22 to find the curvature of the polar curve $r = a + b\cos(\theta)$.

24. Use the formula in Exercise 22 to find the curvature at the pole of the polar curve $r = ae^{b\theta}$.

25. Use the formula in Exercise 22 to show that the curvature at the pole of a polar curve $r = f(\theta)$ which passes through the pole is given by

$$\kappa = \frac{2}{|r'|} = \frac{2}{|f'(\theta)|}.$$

26. For the curve represented by each vector function below do the following.

   (a) Find $\mathbf{R}(t_0)$, $\mathbf{V}(t_0)$, and $\mathbf{A}(t_0)$.

   (b) Find the speed associated with a particle moving along the curve at t_0.

   (c) Find an expression for the unit tangent vector $\mathbf{T}(t)$. Then find the vector $\mathbf{T}(t_0)$.

   (d) Find an expression for the (principal) unit normal vector $\mathbf{N}(t)$. Then find the vector $\mathbf{N}(t_0)$.

   (e) Use a vector graphing facility, like **vectors** in **ISETL**, to make a screen containing the vectors $\mathbf{R}(t_0)$, $\mathbf{T}(t_0)$, $\mathbf{N}(t_0)$, and $\mathbf{A}(t_0)$ so that the latter three vectors have their initial point located at the tip of $\mathbf{R}(t_0)$. See Activity 9, page 567.

   (f) Find the tangential component of acceleration at any t. Then find it at t_0.

   (g) Find the normal component of acceleration at any t. Then find it at t_0.

       i. $\mathbf{R}(t) = (t^2 + 1)\mathbf{i} - 2t\mathbf{j}$, $t_0 = -1$

       ii. $\mathbf{R}(t) = t^3\mathbf{i} - t^2\mathbf{j}$, $t_0 = 2$

       iii. $\mathbf{R}(t) = (1 - t^2)\mathbf{i} - 4\sqrt{t}\,\mathbf{j}$, $t_0 = -1$

iv. $\mathbf{R}(t) = 2\cos(t)\mathbf{i} + 2\sin(t)\mathbf{j}$,
$t_0 = \frac{\pi}{6}$

v. $\mathbf{R}(t) = e^{-t}\mathbf{i} - 2t\mathbf{j}$, $t_0 = -1$

vi. $\mathbf{R}(t) = 2\ln(t)\mathbf{i} + t^2\mathbf{j}$, $t_0 = 2$

vii. $\mathbf{R}(t) = 4t\mathbf{i} + (t^2 - \ln(t^2))\mathbf{j}$,
$t_0 = 1$

viii. $\mathbf{R}(t) = e^t\sin(t)\mathbf{i} + e^t\cos(t)\mathbf{j}$,
$t_0 = 0$

27. For the vector function given by $\mathbf{R}(t) = \phi(t)\mathbf{i} + \psi(t)\mathbf{j} + \gamma(t)\mathbf{k}$, how would you define the velocity and acceleration vectors $\mathbf{V}(t)$ and $\mathbf{A}(t)$? Give an explanation for your answer.

28. For each curve represented by the parametric equations or vector function below do the following.

(a) Find an expression for the unit tangent vector $\mathbf{T}(t)$. Then find the vector $\mathbf{T}(t_0)$.

(b) Find an expression for the (principal) unit normal vector $\mathbf{N}(t)$. Then find the vector $\mathbf{N}(t_0)$.

i. $x = t-1$, $y = 2t$, $z = t+3$,
$t = 2$

ii. $\mathbf{R}(t) = t^3\mathbf{i} + t\mathbf{j} + t^2\mathbf{k}$, $t = 1$

iii. $x = \sin(\pi t)$, $y = \cos(\pi)t$,
$z = 2$, $t = 0$

iv. $\mathbf{R}(t) = e^t\mathbf{i} + 2t\mathbf{j} + e^{-t}\mathbf{k}$, $t = 0$

29. For a vector function $\mathbf{R}(t) = \phi(t)\mathbf{i} + \psi(t)\mathbf{j}$ in $\mathcal{R}^2$, explain why it is reasonable to define the limit as $t \to a$ of $\mathbf{R}(t)$ as follows:

$$\lim_{t \to a}\mathbf{R}(t) = \left(\lim_{t \to a}\phi(t)\right)\mathbf{i} + \left(\lim_{t \to a}\psi(t)\right)\mathbf{j}.$$

A similar definition can be made for one-sided limits, limits at infinity and minus infinity. An entirely similar definition can be made in $\mathcal{R}^3$. Hence, limits of vector functions are taken "component-wise".

Use the definition of limit above to find the following limits.

(a) $\lim_{t \to 0}\left(3\cos(t)\mathbf{i} - \frac{\sin(2t)}{t}\mathbf{j}\right)$

(b) $\lim_{t \to -1}\left(\sqrt{8-t^3}\,\mathbf{i} - \frac{1-t^2}{t-1}\mathbf{j}\right)$

(c) $\lim_{t \to \infty}\left(\frac{3t^2-t}{1-2t^2}\mathbf{i} - \frac{t^2+1}{1+e^{3t}}\mathbf{j} + \frac{3\cos(\pi t)}{t^2}\mathbf{k}\right)$

30. In this exercise you are to use the definition of limit in Exercise 29. For a vector function $\mathbf{R}(t)$ in $\mathcal{R}^2$ or $\mathcal{R}^3$, the derivative $\mathbf{R}'$ is the vector function defined by

$$\mathbf{R}'(t) = \lim_{h \to 0}\frac{\mathbf{R}(t+h) - \mathbf{R}(t)}{h}$$

where

$$\frac{\mathbf{R}(t+h) - \mathbf{R}(t)}{h} = \frac{1}{h}(\mathbf{R}(t+h) - \mathbf{R}(t)).$$

Do the following.

(a) Suppose that $\mathbf{R}(t) = \phi(t)\mathbf{i} + \psi(t)\mathbf{j}$ where ϕ and ψ are differentiable functions. Use Exercise 29, and the fact that vector subtraction and scalar multiplication are done component-wise to show the following.

$$\mathbf{R}'(t) = \phi'(t)\mathbf{i} + \psi'(t)\mathbf{j}.$$

(b) Make a sketch of a curve described by a vector function $\mathbf{R}(t)$ and on it show the difference quotient

$$\frac{\mathbf{R}(t+h) - \mathbf{R}(t)}{h}.$$

Use your sketch to help explain why you think it is reasonable that the velocity vector $\mathbf{V}$ associated with $\mathbf{R}$ is given by $\mathbf{V}(t) = \mathbf{R}'(t)$.

(c) For a vector function $\mathbf{R}$ in $\mathscr{R}^3$ given by $\mathbf{R}(t) = \phi(t)\mathbf{i} + \psi(t)\mathbf{j} + \gamma(t)\mathbf{k}$ where ϕ, ψ and γ are differentiable functions, generalize the result of part (a).

(d) Make a sketch of a curve described by a vector function $\mathbf{R}(t)$ and on it show the difference quotient

$$\frac{\mathbf{T}(t+h) - \mathbf{T}(t)}{h}.$$

Use your sketch to help explain why you think it is reasonable that the (principal) unit normal vector $\mathbf{N}$ associated with $\mathbf{R}$ always points in the direction of motion (i.e., toward the concave side of the curve).

(e) Use the results of parts (a), (b), and (c) to find the derivatives of the following vector functions. Then find the derivative at t_0.

  i. $\mathbf{R}(t) = \sin(2t)\mathbf{i} - e^{2t}\mathbf{j}$, $t_0 = 0$

  ii. $\mathbf{R}(t) = 2\ln(3t)\mathbf{i} + (t^2 - \sqrt{t})\mathbf{j}$, $t_0 = 1$

  iii. $\mathbf{R}(t) = (2t - \frac{3}{t})\mathbf{i} + \cos^2(\pi t)\mathbf{j}$, $t_0 = -1$

  iv. $\mathbf{R}(t) = t\cos(t)\mathbf{i} - \frac{t}{1+t}\mathbf{j} + e^{-t}\mathbf{k}$, $t_0 = 0$

31. For a vector function $\mathbf{R}(t) = \phi(t)\mathbf{i} + \psi(t)\mathbf{j}$ in $\mathscr{R}^2$, the indefinite integral $\int \mathbf{R}(t)dt$ is the vector function defined by

$$\int \mathbf{R}(t)dt = \left(\int \phi(t)dt\right)\mathbf{i} + \left(\int \psi(t)dt\right)\mathbf{j}.$$

A similar definition can be made in $\mathscr{R}^3$. Do the following.

(a) Show that

$$\int (\phi'(t)\mathbf{i} + \psi'(t)\mathbf{j})dt$$
$$= \phi(t)\mathbf{i} + \psi(t)\mathbf{j} + \mathbf{C}$$

where $\mathbf{C} = C_1\mathbf{i} + C_2\mathbf{j}$ with C_1 and C_2 arbitrary real constants.

(b) Generalize the result of part (a) to $\mathscr{R}^3$.

(c) Use the results of parts (a) and (b) to find the indefinite integral of each of the following vector functions.

  i. $\mathbf{R}(t) = (t - \sin(t))\mathbf{i}$
  $+(1 - e^{2t})\mathbf{j}$

  ii. $\mathbf{R}(t) = te^t\mathbf{i} + \frac{t}{\sqrt{1-t^2}}\mathbf{j}$

  iii. $\mathbf{R}(t) = (1 + 2t)\mathbf{i}$
  $+t^2\sqrt{1+t^3}\mathbf{j} - 3\mathbf{k}$

  iv. $\mathbf{R}(t) = \sqrt{1+2t}\mathbf{i} + \frac{3}{t^2}\mathbf{k}$

(d) Use the results of parts (a) and (b) to find a vector function $\mathbf{R}$ which satisfies the following conditions.

  i. $\mathbf{R}'(t) = \frac{1}{1+t^2}\mathbf{i} + \sin(\pi t)\mathbf{j}$ and $\mathbf{R}(1) = \mathbf{i} - 2\mathbf{j}$.

  ii. $\mathbf{R}'(t) = \sqrt{t+1}\mathbf{i} + (1 + \cos(t))\mathbf{j}$
  $+\sec^2(t)\mathbf{k}$
  and $\mathbf{R}(0) = \mathbf{i} + 3\mathbf{j} - 2\mathbf{k}$.

32. Use the results of parts (a) and (b) of Exercise 31 to find the position vector $\mathbf{R}(t)$ when $\mathbf{A}(t) = 2\sin(t)\mathbf{i} - \cos(t)\mathbf{j}$, $\mathbf{V}(0) = -\mathbf{i} + \mathbf{j}$ and $\mathbf{R}(0) = 3\mathbf{j}$.

33. A projectile is fired upward at an angle θ made with the horizontal and with speed $v_0 = |\mathbf{V}(0)|$. Assume that the only force acting on the projectile is that of gravity pulling it back toward the earth's surface, that is, the acceleration vector $\mathbf{A}$ at any time t is given by $\mathbf{A}(t) = -g\mathbf{j}$ where g is a gravitational constant. Use the result of part (a) of Exercise 31 to do the following.

(a) Show that the velocity vector $\mathbf{V}$ is given by

$$\mathbf{V}(t) = v_0\cos(\theta)\mathbf{i} + (v_0\sin(\theta) - gt)\mathbf{j}.$$

(b) Show that the position vector $\mathbf{R}$ is given by

$$\mathbf{R}(t) = ((v_0 \cos(\theta))t + C_1)\mathbf{i}$$

$$+\left((v_0 \sin(\theta))t - \frac{1}{2}gt^2 + C_2 \right)\mathbf{j}$$

where C_1 and C_2 are constants.

(c) Find a formula similar to the one in part (b) when the projectile is fired with an initial position of $\mathbf{R}(0) = a\mathbf{i} + b\mathbf{j}$.

34. Let $\mathbf{R}(t) = \phi(t)\mathbf{i} + \psi(t)\mathbf{j} + \gamma(t)\mathbf{k}$ where ϕ, ψ, and γ are integrable functions. Use Riemann sums, Exercise 31, and the fact that vector addition and scalar multiplication are done component-wise to explain why the following definition of the definite integral of the vector function $\mathbf{R}$ is reasonable.

$$\int_a^b \mathbf{R}(t)dt = \left(\int_a^b \phi(t)dt \right)\mathbf{i} + \left(\int_a^b \psi(t)dt \right)\mathbf{j}$$

$$+\left(\int_a^b \gamma(t)dt \right)\mathbf{k}$$

35. In this exercise you are to use the results of Exercise 30 to prove the following rules for finding derivatives of vector functions. Use the fact that the vector operations (addition, subtraction, scalar multiplication) are done component-wise. Let $\mathbf{R}(t) = f_1(t)\mathbf{i} + g_1(t)\mathbf{j}$ and $\mathbf{S}(t) = f_2(t)\mathbf{i} + g_2(t)\mathbf{j}$ be two vector functions where f_1, f_2, g_1, and g_2 are differentiable functions.

(a) Constant rule: $(\mathbf{C})' = \mathbf{0}$ where $\mathbf{C}$ is a constant vector and $\mathbf{0} = 0\mathbf{i} + 0\mathbf{j}$ is the zero vector.

(b) Scalar multiplication rule: $(k\mathbf{R}(t))' = k\mathbf{R}'(t)$ where k is any real constant.

(c) Sums and differences rules:

$$(\mathbf{R}(t) \pm \mathbf{S}(t))' = \mathbf{R}'(t) \pm \mathbf{S}'(t)$$

(d) Multiplication by a scalar function rule:

$$(h(t)\mathbf{R}(t))' = h(t)\mathbf{R}'(t) + h'(t)\mathbf{R}(t)$$

where $h(t)$ is a scalar valued function.

(e) Dot product rule:

$$(\mathbf{R}(t) \cdot \mathbf{S}(t))' = \mathbf{R}(t) \cdot \mathbf{S}'(t) + \mathbf{R}'(t) \cdot \mathbf{S}(t)$$

(f) Chain rule:

$$(\mathbf{R}(h(t)))' = h'(t)\mathbf{R}'(h(t))$$

where $h(t)$ is a scalar valued function. In other notation, we have

$$\frac{d\mathbf{R}}{dt} = \frac{d\mathbf{R}}{dh}\frac{dh}{dt}.$$

36. In this exercise you will prove the formula for curvature given on page 575. The curvature function κ is defined as follows:

$$\kappa = \kappa(s) = |\mathbf{T}'(s)| = \left| \frac{d\mathbf{T}}{ds} \right|.$$

You will use this definition of curvature to derive the curvature formula for a curve represented by parametric equations $x = \phi(t)$, $y = \psi(t)$ in $\mathcal{R}^2$ (or equivalently by the vector function $\mathbf{R}(t) = \phi(t)\mathbf{i} + \psi(t)\mathbf{j}$). Do the following to derive the formula for κ

(a) Let θ be the angle that the unit tangent vector $\mathbf{T}$ makes with the horizontal. Make a sketch illustrating θ. Show that

$$\mathbf{T} = \cos(\theta)\mathbf{i} + \sin(\theta)\mathbf{j} \text{ and } \left| \frac{d\mathbf{T}}{d\theta} \right| = 1.$$

(b) Think of θ as a (scalar) function of s, i.e., $\theta = \theta(s)$. Then by the chain rule for vector functions (part (f) of Exercise 35) it follows that

$$\frac{d\mathbf{T}}{ds} = \frac{d\mathbf{T}}{d\theta}\frac{d\theta}{ds}.$$

Use the above formula and the result of part (a) to show that

$$\kappa = \left| \frac{d\theta}{ds} \right|.$$

This result says that κ is a measure of the rate at which a curve is turning, or equivalently, the rate at which the unit tangent vector is turning.

(c) Explain why

$$\tan(\theta) = \frac{\psi'(t)}{\phi'(t)}.$$

Hence, it follows that

$$\theta = \arctan\left(\frac{\psi'(t)}{\phi'(t)}\right)$$

provided $\phi'(t) \neq 0$. Use this fact to show that

$$\left|\frac{d\theta}{dt}\right| = \frac{|\phi'(t)\psi''(t) - \psi'(t)\phi''(t)|}{(\phi'(t))^2 + (\psi'(t))^2}.$$

(d) By the chain rule

$$\frac{d\theta}{ds} = \frac{\frac{d\theta}{dt}}{\frac{ds}{dt}}.$$

Now use the above formula, part (c) and the fact that

$$\frac{ds}{dt} = |\mathbf{V}(t)| = \sqrt{(\phi'(t))^2 + (\psi'(t))^2}$$

to show that

$$\kappa = \kappa(t) = \frac{|\phi'(t)\psi''(t) - \psi'(t)\phi''(t)|}{\left((\phi'(t))^2 + (\psi'(t))^2\right)^{\frac{3}{2}}}.$$

37. Use the dot product rule for vector functions in part (e) of Exercise 35 to prove that the (principal) unit normal vector $\mathbf{N}(t)$ is perpendicular to the unit tangent vector $\mathbf{T}(t)$ at any point where $|\mathbf{T}'(t)| \neq 0$. That is, show that $\mathbf{T}(t) \cdot \mathbf{N}(t) = 0$. Do so in the following two steps.

(a) Explain why the definition of $\mathbf{N}(t)$ implies that this can be done by showing that $\mathbf{T}(t)$ is perpendicular to $\mathbf{T}'(t)$, i.e., $\mathbf{T}(t) \cdot \mathbf{T}'(t) = 0$.

(b) Now finish the proof by showing that $\mathbf{T}(t) \cdot \mathbf{T}'(t) = 0$.

38. Consider a curve C represented by the parametric equations

$$x = \phi(t), \; y = \psi(t), \; a \leq t \leq b.$$

Assume that this curve is non-singular.

(a) Explain why the vector function $\mathbf{V}$ in $\mathcal{R}^2$ given by $\mathbf{v} = \mathbf{V}(t) = \phi'(t)\mathbf{i} + \psi'(t)\mathbf{j}$ represents the velocity vector of a particle moving along the curve C. Explain why $|\mathbf{V}(t)|$ represents the speed of the particle.

(b) Explain why the vector $\mathbf{V}(t)$ in part (a) is in the same direction as the tangent to the curve for each t and it points in the direction of motion.

(c) Consider the unit tangent vector function $\mathbf{T}$ given by $\mathbf{T}(t) = \frac{\mathbf{V}(t)}{|\mathbf{V}(t)|}$. Under what circumstances can you be sure that a curve has a unit tangent vector for each value of t? Explain your answer and illustrate it with appropriate examples.

39. Consider the unit normal vector function $\mathbf{N}$ given by

$$\mathbf{N}(t) = \frac{\mathbf{T}'(t)}{|\mathbf{T}'(t)|}.$$

(a) For a position vector $\mathbf{R}$ which represents a curve C given by $\mathbf{r} = \mathbf{R}(t) = \phi(t)\mathbf{i} + \psi(t)\mathbf{j}$, consider the acceleration vector $\mathbf{a} = \mathbf{A}(t)$ and the relationship

$$\mathbf{a} = \mathbf{A}(t) = \frac{d^2 s}{dt^2}\mathbf{T}(t) + \kappa\left(\frac{ds}{dt}\right)^2 \mathbf{N}(t)$$

where $\frac{ds}{dt} = |\mathbf{V}(t)|$ is the (linear) speed of a particle, moving along the curve represented by $\mathbf{R}(t)$.

i. Using the above decomposition of $\mathbf{A}(t)$, explain why the vector $\mathbf{N}(t)$ always points in the direction the curve is bending (i.e., toward the concave side of the curve).

ii. Using the above decomposition of $\mathbf{A}(t)$, explain why a car traveling on a curve described by the position function $\mathbf{R}$ should slow down in order to make a sharp turn where the curvature κ is "relatively large"?

iii. The (scalar) components of the acceleration vector $\mathbf{A}(t)$ relative to the decomposition with respect to $\mathbf{T}(t)$ and $\mathbf{N}(t)$ are called the *tangential component of acceleration,* $\mathbf{a_T}$, and the *normal component of acceleration,* $\mathbf{a_N}$, respectively, where $|\mathbf{a}| = |\mathbf{A}(t)|$. Show that

A. $\mathbf{a_T} = \mathbf{A} \cdot \mathbf{T}$

B. $\mathbf{a_N} = \mathbf{A} \cdot \mathbf{N}$

C. $|\mathbf{a}| = \sqrt{(\mathbf{a_T})^2 + (\mathbf{a_N})^2}$ and

$$\mathbf{a_N} = \sqrt{|\mathbf{a}|^2 - (\mathbf{a_T})^2}$$

40. Show that the only curves of constant curvature in $\mathcal{R}^2$ are lines and circles.

41. For a curve in $\mathcal{R}^2$, what can you say about the principal unit normal at a point of inflection on the curve? Explain your answer. Give an example(s) which illustrates your answer.

42. For a curve in $\mathcal{R}^2$, what can you say about the curvature at a point of inflection on the curve? Explain your answer. Give an example(s) which illustrates your answer.

7.5 CHAPTER SUMMARY

In this chapter you studied polar coordinates in the plane, vectors in two and three dimensions, parametric representation of curves and surfaces, and some calculus of parametric equations. You learned about some basic curves in polar coordinates and how to find areas bounded by polar curves and arc lengths of polar curves. Your investigations included graphs of curves in the plane, some surfaces in space, and an introduction to vector-valued functions. You learned about calculus as it relates to parametric representations, vector-valued functions, and the study of curvilinear motion. Below you will find the major results of this chapter.

Polar Coordinates   Transformation equations for polar and rectangular coordinates.

polar to rectangular

$$x = r\cos(\theta) \text{ and } y = r\sin(\theta)$$

rectangular to polar

$$r^2 = x^2 + y^2 \text{ and } \tan(\theta) = \frac{y}{x}$$

Some Polar Curves

Lines

$\theta = c$ where c is any real number — a line through the origin, or pole

$r\cos(\theta) = a$ where a is any real number — a line parallel to the y-axis or perpendicular to the x-axis

$r\sin(\theta) = a$ where a is any real number — a line parallel to the x-axis or perpendicular to the y-axis

$r(a\cos(\theta) + b\sin(\theta)) = c$ where a, b, and c are real numbers with at least two of a, b, and c non-zero.

Circles

$r = a\sin(\theta)$ where a is any real number — a circle with center at $(\frac{a}{2}, \frac{\pi}{2})$ and radius $\frac{|a|}{2}$.

$r = a\cos(\theta)$ where a is any real number — a circle with center at $(0, \frac{a}{2})$ and radius $\frac{|a|}{2}$.

$r = a$ where a is any real number — a circle with center at the origin and radius $|a|$.

Roses

$r = a\sin(b\theta)$ where a and b are any non-zero real numbers.

$r = a\cos(b\theta)$ where a and b are any non-zero real numbers.

$r = \cos(\frac{a\theta}{b})$ where a and b are any non-zero real numbers.

Spirals

$r = a\theta$ where a is any real number

$r^2 = a\theta$ where a is any real number.

Limaçons

$r = a + b\cos(\theta)$ where a and b are real numbers with $a \neq b$.

$r = a + b\sin(\theta)$ where a and b are real numbers with $a \neq b$.

Cardioids

$r = a + b\cos(\theta)$ where a and b are any real numbers with $|a| = |b|$ or $a = \pm b$.

$r = a + b\sin(\theta)$ where a and b are any real numbers with $|a| = |b|$.

Lemniscates

$r^2 = a\cos(2\theta)$ where a is any non-zero real number.

$r^2 = a\sin(2\theta)$ where a is any non-zero real number.

Conic sections in polar coordintes

The graphs of the polar curves of the following forms are conic sections, circles when $e = 0$, ellipses when $0 < e < 1$, and hyperbolas when $e > 1$.

$$r = \frac{ke}{1 - e\cos(\theta)} \text{ or } r = \frac{ke}{1 + e\cos(\theta)}$$

$$r = \frac{ke}{1 - e\sin(\theta)} \text{ or } r = \frac{ke}{1 + e\sin(\theta)}$$

Areas in Polar Coordinates.
The area bounded by a polar curve $r = f(\theta)$ over the interval $\alpha \le \theta \le \beta$ is given by

$$\frac{1}{2}\int_\alpha^\beta (f(\theta))^2 \, d\theta$$

where the interval $[\alpha, \beta]$ is such that the curve (or piece of the curve) is traced out exactly once.

Arc Lengths in Polar Coordinates.
The arc length of a polar curve given by $r = f(\theta)$ from the angle $\theta = \alpha$ to the angle $\theta = \beta$ is given by

$$\int_\alpha^\beta \sqrt{(f(\theta))^2 + (f'(\theta))^2} \, d\theta.$$

Distance and Midpoint in $\mathcal{R}^3$.
The *distance* between two points $S = [x_1, x_2, x_3]$ and $T = [y_1, y_2, y_3]$ is the length of the line segment connecting them and is given by

$$\sqrt{(x_1 - x_2)^2 + (y_1 - y_2)^2 + (z_1 - z_2)^2}.$$

The *midpoint* of a line segment between with endpoints S and T is given by

$$\left[\frac{x_1 + x_2}{2}, \frac{y_1 + y_2}{2}, \frac{z_1 + z_2}{2}\right].$$

Vectors in $\mathcal{R}^2$.
Any vector $\mathbf{u}$ can be written as a linear combination of the unit vectors $\mathbf{i}$ and $\mathbf{j}$ as follows:

$$\mathbf{u} = |\mathbf{u}|\cos(\theta)\mathbf{i} + |\mathbf{v}|\sin(\theta)\mathbf{j}$$

where θ is the angle between $\mathbf{u}$ and the positive x-axis (or the unit vector $\mathbf{i}$). When a vector $\mathbf{u}$ is written in this manner, we say that $\mathbf{u}$ is resolved into its $\mathbf{i}$-$\mathbf{j}$ *component vectors*,

$$|\mathbf{u}|\cos(\theta)\mathbf{i} \text{ and } |\mathbf{v}|\sin(\theta)\mathbf{j}.$$

The $\mathbf{i}$ component and $\mathbf{j}$ component for $\mathbf{u}$ are, respectively, $|\mathbf{u}|\cos(\theta)$ and $|\mathbf{u}|\sin(\theta)$.

Scalar Multiplication of Vectors.
For any vector $\mathbf{u} = [a_1, a_2]$ in $\mathcal{R}^2$ the scalar t times the vector $\mathbf{u}$ is given by $t\mathbf{u} = [ta_1, ta_2]$ and a completely analogous formula holds for scalar multiplication in $\mathcal{R}^3$. Scalar multiplication is done component-wise.

We can illustrate scalar multiplication geometrically as follows.

Addition and Subtraction of Vectors.
For two vectors $\mathbf{v} = [a_1, a_2]$ and $\mathbf{w} = [b_1, b_2]$ in $\mathcal{R}^2$, we have

$$\mathbf{v} + \mathbf{w} = [a_1 + b_1, a_2 + b_2] \text{ and } \mathbf{v} - \mathbf{w} = [a_1 - b_1, a_2 - b_2]$$

Completely analogous formulas hold for vector addition and subtraction in $\mathcal{R}^3$. Addition and subtraction of vectors is done component-wise.

Two vectors are added and subtracted geometrically as shown in the following diagram.

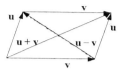

Vector addition can be done geometrically by sliding the vector **u** parallel to itself along **v** until the tail (initial point) of **u** coincides with the head (terminal point) of **v**. Then the *sum*, or *resultant vector*, **u** + **v** has its initial point at the initial point of **v** and its terminal point at the terminal point of **u**. Since **v** − **u** = **v** + (−**u**), you can subtract **u** from **v** by just adding **v** and −**u**. As shown in the diagram above, you can simply draw the diagonals of the parallelogram formed by **u** and **v** and label the tails and heads appropriately to obtain **u** + **v**, **u** − **v** and **v** − **u**.

Vector Form of a Line in $\mathcal{R}^2$ and $\mathcal{R}^3$.

The *vector form* of a line L in $\mathcal{R}^2$ or $\mathcal{R}^3$ is an expression of the form **v** = **a** + *t***b** for some vectors (not necessarily unique) **a** and **b** where the terminal point or tip of **a** is any point on the line L and **b** is any vector which lies along the line L.

The Length of a Vector.

For any vector **v** = $[v_1, v_2]$ in $\mathcal{R}^2$, the *length*, |**v**|, of **v** is given by

$$|\mathbf{v}| = \sqrt{v_1^2 + v_2^2}.$$

A completely analogous formula holds for the length of a vector in $\mathcal{R}^3$.

The Dot Product of Two Vectors — Angles.

For any vectors **u** and **v** in $\mathcal{R}^2$, the *dot product* of **u** and **v** is given by

$$\mathbf{u} \cdot \mathbf{v} = |\mathbf{u}||\mathbf{v}|\cos(\theta)$$

where θ is the (smallest) angle between **u** and **v**. Moreover, if **u** = $[a_1, a_2]$ and **v** = $[b_1, b_2]$, then

$$\mathbf{u} \cdot \mathbf{v} = a_1 b_1 + a_2 b_2.$$

Completely analogous formulas for the dot product hold in $\mathcal{R}^3$.

For any two non-zero vectors in $\mathcal{R}^2$ or $\mathcal{R}^3$, the cosine of the (smallest) angle θ between two vectors **u** and **v** is given by

$$\cos(\theta) = \frac{\mathbf{u} \cdot \mathbf{v}}{|\mathbf{u}||\mathbf{v}|}.$$

The angle θ can easily be found using the inverse cosine function, that is,

$$\theta = \arccos\left(\frac{\mathbf{u} \cdot \mathbf{v}}{|\mathbf{u}||\mathbf{v}|}\right).$$

Areas and the Dot Product.

The area A of a parallelogram determined by two vectors $\mathbf{v}$ and $\mathbf{w}$ is given by

$$A = \sqrt{\left(|\mathbf{v}|^2|\mathbf{w}|^2 - (\mathbf{v} \cdot \mathbf{w})^2\right)}\,.$$

Direction Cosines and Angles.

Any vector $\mathbf{u}$ in $\mathcal{R}^3$ can be written in the following form

$$\mathbf{u} = |\mathbf{u}|\cos(\alpha)\mathbf{i} + |\mathbf{u}|\cos(\beta)\mathbf{j} + |\mathbf{u}|\cos(\gamma)\mathbf{k}$$

where $|\mathbf{u}|\cos(\alpha)$, $|\mathbf{u}|\cos(\beta)$, and $|\mathbf{u}|\cos(\gamma)$ are called the $\mathbf{i}$, $\mathbf{j}$ and $\mathbf{k}$ components, respectively, of $\mathbf{u}$.

The direction cosines $\cos(\alpha)$, $\cos(\beta)$, and $\cos(\gamma)$ of a vector in $\mathcal{R}^3$ are obtained by forming the *unit vector* given by

$$\frac{\mathbf{u}}{|\mathbf{u}|} = \cos(\alpha)\mathbf{i} + \cos(\beta)\mathbf{j} + \cos(\gamma)\mathbf{k}.$$

The direction angles α, β, and γ are the angles made by the vector $\mathbf{u}$ and the vectors $\mathbf{i}$, $\mathbf{j}$, and $\mathbf{k}$, respectively.

Projections.

For two non-zero vectors $\mathbf{u}$ and $\mathbf{v}$, in $\mathcal{R}^2$ or $\mathcal{R}^3$, (i.e., two vectors such that $|\mathbf{u}| \neq 0$ and $|\mathbf{v}| \neq 0$) the *projection of* $\mathbf{u}$ *along* $\mathbf{v}$, denoted $\mathbf{proj}_v\mathbf{u}$, is given by

$$\mathbf{proj}_v\mathbf{u} = |\mathbf{u}|\cos(\theta)\frac{\mathbf{v}}{|\mathbf{v}|} = \left(\frac{\mathbf{u} \cdot \mathbf{v}}{\mathbf{v} \cdot \mathbf{v}}\right)\mathbf{v} = \left(\frac{\mathbf{u} \cdot \mathbf{v}}{|\mathbf{v}|^2}\right)\mathbf{v}.$$

where θ is the angle between $\mathbf{u}$ and $\mathbf{v}$. The number $|\mathbf{u}|\cos(\theta)$ is called the *component of* $\mathbf{u}$ *along* $\mathbf{v}$.

For any two non-zero vectors in $\mathcal{R}^2$ or $\mathcal{R}^3$, the two vectors $\mathbf{u}_1 = \mathbf{proj}_v\mathbf{u}$ and $\mathbf{u}_2 = \mathbf{u} - \mathbf{proj}_v\mathbf{u}$ are such that $\mathbf{u} = \mathbf{u}_1 + \mathbf{u}_2$ with $\mathbf{u}_1 = \mathbf{proj}_v\mathbf{u}$ parallel to $\mathbf{v}$ and $\mathbf{u}_2 = \mathbf{u} - \mathbf{proj}_v\mathbf{u}$ perpendicular to $\mathbf{v}$. This provides another way of resolving a vector into component vectors.

The Cross-Product of Two Vectors.

For two non-zero vectors $\mathbf{u} = [u_1, u_2, u_3]$ and $\mathbf{v} = [v_1, v_2, v_3]$ in three dimensions, the *cross-product* of $\mathbf{u}$ and $\mathbf{v}$ is denoted $\mathbf{u} \times \mathbf{v}$ and it is a vector given symbolically by

$$\mathbf{u} \times \mathbf{v} = \begin{vmatrix} \mathbf{i} & \mathbf{j} & \mathbf{k} \\ u_1 & u_2 & u_3 \\ v_1 & v_2 & v_3 \end{vmatrix} = \begin{vmatrix} u_2 & u_3 \\ v_2 & v_3 \end{vmatrix}\mathbf{i} - \begin{vmatrix} u_1 & u_3 \\ v_1 & v_3 \end{vmatrix}\mathbf{j} + \begin{vmatrix} u_1 & u_2 \\ v_1 & v_2 \end{vmatrix}\mathbf{k}$$

or

$$\mathbf{u} \times \mathbf{v} = (u_2v_3 - v_2u_3)\mathbf{i} - (u_1v_3 - v_1u_3)\mathbf{j} + (u_1v_2 - v_1u_2)\mathbf{k}\,.$$

The cross-product of two vectors is often referred to as the *vector product* of the vectors to distinguish it from the dot product which is a scalar (or number) quantity.

The cross-product of two non-zero vectors in $\mathcal{R}^3$ yields a vector $\mathbf{u} \times \mathbf{v}$ which is perpendicular to the plane of $\mathbf{u}$ and $\mathbf{v}$. Moreover, $\mathbf{u} \times \mathbf{v} = -(\mathbf{v} \times \mathbf{u})$.

For two different non-zero vectors $\mathbf{u}$ and $\mathbf{v}$ in $\mathcal{R}^3$, we have

$$|\mathbf{u} \times \mathbf{v}| = |\mathbf{u}||\mathbf{v}|\sin(\theta)$$

where θ is the (smallest) angle between the vectors $\mathbf{u}$ and $\mathbf{v}$.

Volumes and Cross-Products.

For three different vectors **u**, **v** and **w** in space, the volume V of the parallelepiped with base formed by the vectors **u** and **v**, like that shown in Figure 7.29, is given by

$$V = (|\mathbf{u} \times \mathbf{v}|)|\mathbf{w}||\cos(\phi)| = |(\mathbf{u} \times \mathbf{v}) \cdot \mathbf{w}|$$

where ϕ is the angle made by the vector $\mathbf{u} \times \mathbf{v}$ and **w**.

Representations of Planes in Rectangular Coordinates.

The general equation of a plane in rectangular coordinates is given by

$$ax + by + cz + d = 0$$

where any multiple of the vector $\mathbf{n} = a\mathbf{i} + b\mathbf{j} + c\mathbf{k}$ is normal (perpendicular) to the plane.

The Parametric Form of a Curve in $\mathcal{R}^2$ and $\mathcal{R}^3$.

The parametric form of a curve in $\mathcal{R}^2$ is given by

$$x = x(t) = \phi(t), \; y = y(t) = \psi(t)$$

for an appropriate interval of a parameter t, where ϕ and ψ are functions of t. As t runs through its domain, the interval $[a, b]$, the two functions ϕ and ψ give the x- and y-coordinates of a point, respectively. In turn, the point $[\phi(t), \psi(t)]$ traces out a curve in $\mathcal{R}^2$.

The idea for parametric curves in $\mathcal{R}^3$ is the same as for $\mathcal{R}^2$. You just need to add a third function for the third coordinate. Thus, a general formulation is

$$x = \phi(t), \; y = \psi(t), \; z = \lambda(t)$$

for an appropriate interval of the parameter t.

Conversion from Parametric Form to Standard Form in $\mathcal{R}^2$.

If you have a parametric representation of a curve in the form,

$$x = x(t) = \phi(t), \;\; y = y(t) = \psi(t)$$

for an appropriate interval of the parameter t, then you can replace it with a single function f which represents the curve in the usual way — that is, with the x- and y-coordinates related by the equation $y = f(x)$ by finding the inverse function ϕ^{-1} of the function ϕ to get the y value by applying ψ. In other words, the function f that we are looking for is exactly the composition of ψ with the inverse of ϕ. That is,

$$f = \psi \circ \phi^{-1}.$$

You may also be able to use algebra and trigonometry to help convert from parametric to rectangular form (and visa-versa).

Reparametrization of a Curve in $\mathcal{R}^2$ or $\mathcal{R}^3$.

For a curve with parametric form given by

$$x = \phi(t), \;\; y = \psi(t), \;\; z = \lambda(t)$$

for an appropriate interval on the parameter t, a reparametrization of the curve is made by using an appropriate function α of a new parameter, say r, with $t = \alpha(r)$, to obtain the parametric form

$$x = \phi(\alpha(r)), \;\; y = \psi(\alpha(r)), \;\; z = \lambda(\alpha(r))$$

for an appropriate interval of the parameter r.

The Parametric Form of a Line in $\mathcal{R}^2$ and $\mathcal{R}^3$.

The parametric form of the equation of a line in $\mathcal{R}^3$ is a special case of a curve in $\mathcal{R}^3$ given by

$$x = \phi(t) = a + bt, \ y = \psi(t) = c + dt, \ z = \lambda(t) = e + ft, \ -\infty < t < \infty$$

where a, b, c, d, e and f are constants.

If you know two points $P_1 = [x_1, y_1, z_1]$ and $P_2 = [x_2, y_2, z_2]$ in $\mathcal{R}^3$, then parametric equations for the line through P_1 and P_2 are given by

$$x = x_1 + (x_2 - x_1)t, \ \ y = y_1 + (y_2 - y_1)t, \ \ z = z_1 + (z_2 - z_1)t, \ \ -\infty < t < \infty.$$

Solving the parametric equations for t we obtain a *symmetric* (or *relational*) *form* of the line through the points P_1 and P_2 given by

$$\frac{x - x_1}{x_2 - x_1} = \frac{y - y_1}{y_2 - y_1} = \frac{z - z_1}{z_2 - z_1}.$$

The triple of numbers $x_2 - x_1$, $y_2 - y_1$, $z_2 - z_1$ are called the direction numbers of the line. These three numbers are not unique.

In $\mathcal{R}^2$, parametric equations of a line are given by

$$x = \phi(t) = a + bt, \ \ y = \psi(t) = c + dt, \ \ -\infty < t < \infty$$

where a, b, c, d and e are (non-unique) constants.

Parametric equations of line in $\mathcal{R}^2$ (or $\mathcal{R}^3$) can be obtained from the vector form of the line, $\mathbf{v} = \mathbf{a} + t\mathbf{b}$ where $\mathbf{a}$ is any vector whose tip is on the line and $\mathbf{b}$ is any vector parallel to the line. (The vectors $\mathbf{a}$ and $\mathbf{b}$ are not unique.) In $\mathcal{R}^2$, we would have $\mathbf{v} = [x, y]$, $\mathbf{a} = [a_1, a_2]$ and $\mathbf{b} = [b_1, b_2]$, and then $\mathbf{v} = [x, y] = [a_1, a_2] + t[b_1, b_2] = [a_1 + tb_1, a_2 + tb_2]$, or $x = a_1 + tb_1$ and $y = a_2 + tb_2$ where $-\infty < t < \infty$. The situation in $\mathcal{R}^3$ is entirely similar.

The Parametric Form of a Surface in $\mathcal{R}^3$.

A surface in $\mathcal{R}^3$ can be given in parametric form by specifying three functions from $\mathcal{R}^2$ to $\mathcal{R}^1$, that is,

$$x = \phi(s, t), \ \ y = \psi(s, t), \ \ z = \lambda(s, t)$$

for appropriate intervals of the parameters s and t.

The Parametric Form of a Plane in $\mathcal{R}^3$.

The parametric form of the equation of a plane in $\mathcal{R}^3$ is given by

$$x = \phi(s, t) = a + bs + ct, \ \ y = \psi(s, t) = d + es + ft, \ \ z = \lambda(s, t) = g + hs + it, \ \ -\infty < s, t < \infty$$

where a, b, c, d, e, f, g, h, and i are constants.

Derivatives of Functions from $\mathcal{R}^1$ to $\mathcal{R}^2$.

For a curve given parametrically in $\mathcal{R}^2$

$$x = x(t) = \phi(t), \ y = y(t) = \psi(t), \ a \le t \le b,$$

we have

$$f(x) = \psi\big(\phi^{-1}(x)\big)$$

and

$$\frac{dy}{dx} = \frac{\frac{dy}{dt}}{\frac{dx}{dt}}.$$

Hence, the derivative of the function f is the ratio of the derivative of the function y to the derivative of the function x.

This means that if a particle is moving in time along a curve, then the rate of change of its y position with respect to its x position is the ratio of the rate of change of its y position with respect to time to the rate of change of its x position with respect to time.

Tangents and Normals to Curves in Parametric Form in $\mathscr{R}^2$.

For a curve given by the pair of functions (ϕ, ψ) and the function $f = \psi \circ \phi^{-1}$ and a point $P_0 = [x_0, y_0] = [\phi(t_0), \psi(t_0)]$ the slope of the curve is given by

$$f'(x_0) = \frac{\psi'(t_0)}{\phi'(t_0)}$$

and the equation of the tangent line through P is

$$\phi'(t_0)(y - y_0) - \psi'(t_0)(x - x_0) = 0.$$

The equation of the normal line is given by

$$\psi'(t_0)(y - y_0) + \phi'(t_0)(x - x_0) = 0.$$

Non-Singular and Smooth Curves.

A curve given in parametric form by

$$x = \phi(t), \; y = \psi(t), \; a \le t \le b$$

is said to be *non-singular* if it is never the case, for any $t \in [a, b]$, that both $\phi'(t) = 0$ and $\psi'(t) = 0$. A non-singular curve for which $\phi'(t)$ and $\psi'(t)$ are continuous on the interval $[a, b]$ is said to be a *smooth curve*. Such curves have no sharp corners. In a similar way we can speak of non-singular and smooth curves in $\mathscr{R}^3$.

Lengths of Curves Given in Parametric Form.

The *arc length* L of a smooth curve in parametric form $\mathscr{R}^2$ where

$$x = \phi(t), \; y = \psi(t), \; a \le t \le b$$

where $\phi'(t)$ and $\psi'(t)$ are both continuous on the interval $[a, b]$ is given by

$$L = \int_a^b \sqrt{(\phi'(t))^2 + (\psi'(t))^2} \, dt.$$

An entirely similar integral gives the arc length of as smooth curve in parametric form in $\mathscr{R}^3$.

Parametrization with Respect to Arc Length.

For a smooth curve given in parametric form.

$$x = \phi(t), \; y = \psi(t), \; a \le t \le b$$

we define the arc length function L by

$$L = \int_a^t \sqrt{(\phi'(\tau))^2 + (\psi'(\tau))^2}\, d\tau.$$

The derivative of L is given by

$$L'(t) = \sqrt{(\phi'(t))^2 + (\psi'(t))^2}.$$

The function $\alpha(r) = L^{-1}(r)$ can be used to reparametrize the original curve for an appropriate interval of the arc length parameter r. Such a reparametrization is called the *parametrization with respect to arc length*. Certain formulas expressing properties of curves become considerably simpler when the parametrization with respect to arc length is used.

Vector-Valued Functions and Curvilinear Motion.   A *vector-valued function*, or *vector function*, $\mathbf{R}$ is given by

$$\mathbf{r} = \mathbf{R}(t) = \phi'(t)\mathbf{i} + \psi'(t)\mathbf{j}$$

where $x = \phi(t)$, $y = \psi(t)$, for $a \le t \le b$ is a curve in parametric form. The vector function $\mathbf{V}$ which gives the *velocity vector* $\mathbf{v}$ at any point t is given by

$$\mathbf{v} = \mathbf{V}(t) = \mathbf{R}'(t) = \phi'(t)\mathbf{i} + \psi'(t)\mathbf{j}$$

and the magnitude of V given by

$$|\mathbf{V}(t)| = |\mathbf{R}'(t)| = \frac{ds}{dt}$$

represents the speed of a particle moving along the curve represented by the graph of $\mathbf{R}$.

The *acceleration vector* a associated with the vector function $\mathbf{R}$ is the vector function given by

$$\mathbf{a} = \mathbf{A}(t) = \mathbf{V}'(t) = \mathbf{R}''(t) = \phi''(t)\mathbf{i} + \psi''(t)\mathbf{j}.$$

The *unit tangent vector*, $\mathbf{T}(t)$, in the direction of the velocity vector $\mathbf{V}(t)$ is obtained by dividing $\mathbf{V}(t)$ by the magnitude of the vector $\mathbf{V}(t)$ to obtain the vector function $\mathbf{T}$ given by

$$\mathbf{T}(t) = \frac{\mathbf{V}(t)}{|\mathbf{V}(t)|},$$

provided $\mathbf{V}(t) \ne 0$.

To study *curvilinear motion*, or motion along curves, we are also interested in the unit vector perpendicular to the tangent. There are two such unit vectors. One (*principal*) *unit normal vector* always points to the concave side of a curve (i.e., in the direction that the curve represented by $\mathbf{R}$ is bending). We define the (principal) unit vector, $\mathbf{N}(t)$, by

$$\mathbf{N}(t) = \frac{\mathbf{T}'(t)}{|\mathbf{T}'(t)|},$$

provided $|\mathbf{T}'(t)| \ne 0$. This will be the case when the curve represented by $\mathbf{R}(t) = \phi(t)\mathbf{i} + \psi(t)\mathbf{j}$ is a smooth curve. Here, the expression $\mathbf{T}'(t)$ is the vector obtained by differentiating each component of $\mathbf{T}(t)$. The vector $\mathbf{N}(t)$ unit normal vector always is a unit normal vector which points to the concave side of the curve.

In $\mathcal{R}^3$ the vectors $\mathbf{T}(t)$ and $\mathbf{N}(t)$ are defined in an entirely similar manner.

Curvature. The magnitude, $|\mathbf{T}'(s)|$, is a measure of how much the curve is turning and is called the *curvature* which is often denoted by the symbol κ.

For a smooth curve given by

$$\mathbf{r} = \mathbf{R}(t) = \phi'(t)\mathbf{i} + \psi'(t)\mathbf{j}$$

where $x = \phi(t)$, $y = \psi(t)$, for $a \leq t \leq b$ is a curve in parametric form, the curvature $\kappa(t)$ is given by

$$\kappa(t) = \frac{|\phi'(t)\,\psi''(t) - \phi''(t)\,\psi'(t)|}{\left(\phi'(t)^2 + (\psi'(t))^2\right)^{\frac{3}{2}}}.$$

The Relationship Between A, T, and N.

The following important decomposition of the acceleration vector $\mathbf{A}$ in terms of the unit tangent, $\mathbf{T}$ and the (principal) unit normal vector, $\mathbf{N}$, is useful in the study of curvilinear motion

$$\mathbf{a} = \mathbf{A}(t) = \frac{d^2s}{dt^2}\mathbf{T}(t) + \kappa\left(\frac{ds}{dt}\right)^2\mathbf{N}(t)$$

where $\frac{ds}{dt} = |\mathbf{V}(t)|$ is the *(linear) speed* and $\frac{d^2s}{dt^2}$ is the *(linear) acceleration* of a particle moving along the curve represented by $\mathbf{R}(t)$ provided the curve represented by $\mathbf{T}'(t)$ is nonsingular, i.e., $\mathbf{T}'(t) \neq \mathbf{0} = 0\mathbf{i} + 0\mathbf{j}$.

The (scalar) components of the acceleration vector $\mathbf{a} = \mathbf{A}(t)$ relative to the above decomposition with respect to $\mathbf{T}(t)$ and $\mathbf{N}(t)$ are called the *tangential component of acceleration*, $\mathbf{a_T}$, and the *normal component of acceleration*, $\mathbf{a_N}$, respectively.

The acceleration vector $\mathbf{a}$ can be written as the linear combination

$$\mathbf{a} = \mathbf{a_T}\mathbf{T} + \mathbf{a_N}\mathbf{N}$$

and it is easy to show the following:

$$\mathbf{a_T} = \mathbf{A}\cdot\mathbf{T},\ \mathbf{a_N} = \mathbf{A}\cdot\mathbf{N},\ \text{and}\ |\mathbf{a}| = \sqrt{(\mathbf{a_T})^2 + (\mathbf{a_N})^2}$$

where $|\mathbf{a}| = |\mathbf{A}(t)|$ is the magnitude of the acceleration vector $\mathbf{a}$. The latter relationship tells us that

$$\mathbf{a_N} = \sqrt{|\mathbf{a}|^2 - (\mathbf{a_T})^2}.$$

Conclusion.

Congratulations! You have completed your introduction to differential and integral calculus. At times, it must have seemed as if this point would never come. As you progressed through the text, you saw the standard problems and techniques of calculus previewed, visited, and very often, re-visited. Each time you re-visited a problem, it was a little more familiar, a little more your own. Through the Activities, you developed an experiential base for your subsequent learning. It was through the use of a mathematical programming language that you were able to have concrete experiences with the fundamental objects and processes of Calculus. We hope that you will agree that the struggle and frustration which inevitably must accompany any meaningful learning of such deep mathematical concepts was well worth it. You have learned by doing, by thinking about what you did, and by discussing this with other students. And in doing all of this, you have had the opportunity to learn successfully and meaningfully one of the most beautiful, albeit difficult, pieces of mathematics.

APPENDIX A

A BRIEF REVIEW OF TRIGONOMETRY

A.1 Overview

This appendix deals with that branch of mathematics which involves the measurement of the angles and sides of triangles known as trigonometry (which means "triangle measurement"). In this appendix you will use trigonometry in applications involving triangles and you will also see how trigonometry can be used in many applications related to cyclic (or periodic) phenomena. You will review the basic relationships among the trigonometric and inverse trigonometric functions. Then, you will be able to solve problems involving trigonometry with your knowledge of calculus concepts. You are probably familiar with angles and the degree measure of angles from geometry — if not this appendix will provide all the information you will need for use in this text. As you will see, calculus of the trigonometric functions is simpler when radian measures of angles are used. As always, we advise you to read through all of the activities and also read ahead in the discussions before you begin to work the activities.

A.2 Activities

1. Suppose that you make a (central) angle θ with vertex at the origin of an xy-rectangular coordinate system by rotating the positive x-axis in the counterclockwise (or positive) direction as shown in Figure A.1a below. (Negative angles are those made by rotating the positive x-axis in the clockwise direction.) One *degree* is defined to be 1/360 of a revolution, so the angle made by one complete revolution of the positive x-axis has a measure of 360°. There is also another commonly used measure of angles, especially in calculus, and that measure is called the *radian* measure of an angle. The radian measure of a central angle θ, as shown in Figure A.1b below, is defined to be the ratio of the arc length s associated with θ to the radius r of the circle shown.

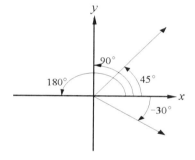

Figure A.1a. Degree measure.

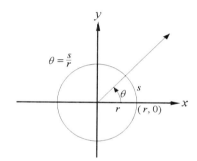

Figure A.1b. Radian measure.

Can you explain why the relationship between degree and radian measures is as follows?

$$2\pi \text{ radians} = 360 \text{ degrees} \quad \text{or} \quad \pi \text{ radians} = 180 \text{ degrees}$$

Now do the following.

(a) Construct a **computer function degtorad** that converts the degree measure of an angle to the radian measure of the angle. Test your **computer function** on the angles 1°, 30°, 45°, 90°, and 180°.

(b) Construct a **computer function radtodeg** that converts the radian measure of an angle to the degree measure of the angle. Check your **computer function** on the angles with radian measures 1, $\frac{\pi}{6}$, $\frac{\pi}{4}$, $\frac{\pi}{2}$ and π.

2. Construct a **computer function ratios** whose inputs are the lengths of the two legs of a right triangle and whose output is a **tuple** that contains the six possible ratios of the three sides of the right triangle associated with an angle θ as shown in Figure A.2 below.

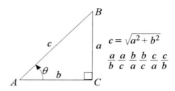

Figure A.2. The ratios of lengths of the sides of a right triangle.

Run your **computer function ratios** with the following sets of inputs: 3, 4; 6, 8; 9, 12. Is there anything interesting about the output in these three cases? What's going on here?

3. In the previous activity you investigated the six possible ratios of the three sides of the right triangle. These six ratios associated with the angle θ shown in Figure A.2 are used to define the following six trigonometric functions as follows:

$$\sin(\theta) = \frac{a}{c} \quad \cos(\theta) = \frac{b}{c} \quad \tan(\theta) = \frac{a}{b}$$
$$\csc(\theta) = \frac{c}{a} \quad \sec(\theta) = \frac{c}{b} \quad \cot(\theta) = \frac{b}{a}$$

(a) Do you see why these are the only six possible ratios associated with θ? Do you see why each ratio defines a function of θ? Why is there only one output for each input? (Think about your work in the previous activity.)

(b) The trigonometric functions can be used in problem solving. Consider the following situation.

You measure the distance from a point A to the base of a building to be 100 feet and the angle made with the horizontal (ground level) from A to the top of the building to be 56°.

Use one of the trigonometric functions to help answer the following questions.

Give your answers to the nearest foot. You may use a calculator if you wish (be sure to have your calculator set to degree mode). If you use a computer, check with your instructor to see if you need to convert angles to radian measure for the software you are using.

    i. What is the height of the building?

    ii. What is the distance from point A to the top of the building?

    iii. Suppose you did not know the distance from the point A to the baseof the building, but instead you knew the height of the building to be 210 feet. In this case, what is the distance from point A to the base of the building?

4.   In this activity you will investigate two special right triangles and the values of the six trigonometric functions.

    (a) Draw an equilateral triangle with sides of length two. Now draw a line segment from the vertex of one angle which is perpendicular to the side opposite the angle. Label the degree measure of the acute angle and label the lengths of the sides of each triangle. Write out a table showing the values of the six trigonometric functions of each acute angle.

    (b) Draw a right triangle with two legs of length one. Label the degree measure of each acute angle and label the lengths of the sides of the triangle. Write out a table showing the values of the six trigonometric functions of each acute angle.

5. For the triangle in Figure A.2, write out all six ratios which define the trigonometric functions of the angle $(90° - \theta)$. Do you see any relationships between the trigonometric functionsof θ on page A-2 and the trigonometric functions of the angle $(90° - \theta)$? What are all the relationships?

6. For the six trigonometric ratios of the angle θ in Figure A.2, answer the following questions.

    (a) Do you see any relationships between the reciprocal of a trigonometric function of θ and any of the other trigonometric functions of θ? What are all the relationships?

    (b) Do you see any relationships between the two possible ratios of $\sin(\theta)$ and $\cos(\theta)$ and any other trignometric functions of θ?

7. For the sine and cosine functions defined by the ratios associated with the acute angle θ in Figure A.2, do the following.

    (a) Use the trigonometric ratios for $\sin(\theta)$ and $\cos(\theta)$ to compute and simplify an expression for the following sum: $(\sin(\theta))^2 + (\cos(\theta))^2$. This sum is usually written as $\sin^2(\theta) + \cos^2(\theta)$. What relationship do you see?

    (b) Use your computer or calculator to make a sketch of the expression $(\sin(x))^2 + (\cos(x))^2$. How does this relate to what you observed in part (a)?

8. You may recall that the domain of the trigonometric functions can be extended to angles other than acute angles. Use your computer to make a sketch of each of the six trigonometric functions on the interval $[-2\pi, 2\pi]$. Also make a sketch of the following pairs of functions on the interval $[-2\pi, 2\pi]$ on the same coordinate system: $\sin, \csc$ and $\cos, \sec$. (In most mathematical

programming languages and symbolic computer systems, π is represented by **Pi**.) Do you see anything interesting about the graphs? Do you see why these functions are said to be cyclic or periodic? Give an explanation for your answer.

9. Use your computer to make a sketch of the functions represented by the following expressions on an interval which shows a complete graph. Print a copy of each sketch and then describe as many of the differences as you can between the graph of each expressionand the graph of $\sin(x)$ you made in Activity 8. Explain any differences you see.

$$0.5\sin(x), \quad \sin\left(\frac{x}{2}\right), \quad 2\sin(x), \quad \sin(2x), \quad 2\sin(2x), \quad 1+\sin(x)$$

$$1+2\sin(x), \quad 1+2\sin(2x), \quad \sin\left(x-\frac{\pi}{2}\right), \quad \sin(2x-\pi), \quad 1+2\sin(2x-\pi)$$

10. Many phenomena in the world we live in follow a *cyclic*, or *periodic*, pattern. Use your computer to make the indicated sketches for the functions in each of the following situations.

    (a) A ferris wheel has a radius of 20 feet and the height h of a seat above the ground at the loading point is located three feet above ground level. During a ride on the ferris wheel your height above the level ground is given by

    $$h = H(t) = 23 - 20\cos(2\pi t) \text{ feet}$$

    where t is the time, in minutes, after you leave (or pass) the loading point. Make a sketch of the function H on the interval $0\,\text{min} \leq t \leq 2\,\text{min}$. (In most mathematical programming languages and symbolic computer systems, π is represented by **Pi**.) What height are you above the ground for each of the following times after your ride begins? 15 sec, 30 sec, 45 sec, 60 sec, 75 sec, 90 sec, 105 sec, 120 sec.

    (b) The current i in an electric circuit is given by

    $$i = I(t) = \sin^2(\pi t) \text{ amperes}$$

    where t is in seconds. Make a sketch of the function I on the time interval $0 \leq t \leq 2$. What is the current in the circuit after 0.5 sec, 0.75 sec, 1 sec, 1.25 sec, 1.5 sec, 1.75 sec, and 2 sec?

11. In this activity you will consider an "inverse" problem: Given the lengths of two sides of a right triangle, find the measures of the unknown acute angles. You can solve an equation like

$$\sin(\theta) = \frac{a}{c}$$

if you know a and c by using the "inverse sine" function. That is,

$$\theta = \arcsin\left(\frac{a}{c}\right) = \sin^{-1}\left(\frac{a}{c}\right)$$

where arcsin or $\sin^{-1}$ are notations commonly used for the inverse sine function. Similar situations and notations are used for the other five inverse trigonometric functions. Consider the following situation.

You know the height of a building to be 100 feet and you measure the distance from a point A to the base of the building to be 80 feet.

Use your calculator or computer, and one of the inverse trigonometric functions to answer the following questions. If you use a computer, check with your instructor to see if you need to convert your answers to degrees for the software you are using. (In some mathematical programming languages and symbolic computer systems, the arcsin function is represented by asin. Similar notations are used for the other five inverse trigonometric functions.) Give your answers to the nearest degree.

(a) What is the degree measure of the angle made with the horizontal and the line of sight from point A to the top of the building?

(b) What is the degree measure of the angle made by the building and the line of sight from the top of the building to the point A?

12. Find all solutions, by hand, to the following equations on the interval $[0, 2\pi]$. You may wish to check your answers using your computer or calculator.

$$2\sin^2(x) - \sin(x) = 0, \quad \sin^2(x) = 0.25, \quad \cos^2(x) - \cos(x) = 2$$

A.3 Discussion

Angles and Measures of Angles An angle θ is made by two rays (or sides) emanating from a common point called the *vertex* of the angle. See Figure A.3.

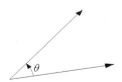

Figure A.3. An angle θ.

We say that an angle is in *standard position* when it is represented in an xy-rectangular coordinate system with its vertex at the origin and one side along the positive x-axis as shown in Figure A.4.

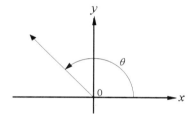

Figure A.4. An angle θ in standard position.

We can think of an angle θ as being formed by rotating the positive *x*-axis about the origin. In this situation, we say that the positive *x*-axis is the *initial side* of θ and the other side of θ is called the *terminal side*. An angle formed in this manner is said to be *positive* if the rotation is in the counterclockwise direction and *negative* if the rotation is in the clockwise direction. Two angles in standard position are said to be *coterminal* when they have the same terminal side. See Figure A.5.

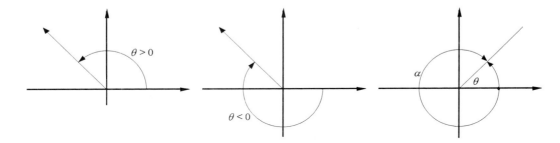

Figure A.5. Positive, negative and coterminal angles in standard position.

Measures of Angles — Degrees and Radians

In Activity 1, you worked the relationship between degree and radian measures of angles. As stated in Activity 1, one *degree* is defined to be 1/360 of a revolution, so the angle made by one complete (counterclockwise) revolution of the positive *x*-axis has a measure of 360°. The commonly used measure of angles, especially in calculus, is the *radian* measure of an angle. The use of radian measure leads to simpler formulas for derivatives and integrals of trigonometric functions. The radian measure of a central angle θ is the ratio of the arc length *s* associated with θ to the radius *r* of the circle shown in Figure A.6. When $r = 1$, we have $\theta = s/1 = s$. Hence, the numerical value of θ is the same as the numerical value of the arc length *s* when $r = 1$. Note that the ratio of arc length *s* to the length of the radius *r* provides a numerical value that has no unit of measure — a numerical value we call the radian measure of θ. This will be useful in problem solving.

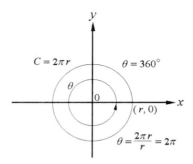

Figure A.6. Radian measure.

Since one complete (counterclockwise) revolution gives an angle of 360° and the circumference (arc length) of a circle of radius *r* is $2\pi r$, do you see way the following is true? $360° = 2\pi$ radians See Figure A.7. Hence, we have π radians = 180 degrees

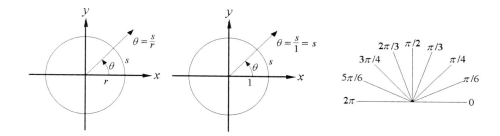

Figure A.7. The relationship between degree and radian measure.

Figure A.7 shows how you can relate degree measure of an angle to the corresponding radian measure. Do you see an easy way to think of the radian measure of an angle and the measure of a 180° or π radians? Look again at Figure A.7.

In Activity 1 you wrote **computer functions** to convert from degree measure to radian measure and for radian measure to degree measure. These conversions can be made using the relationship 2π radians $= 180°$. For example, how would you convert 60° to radians? We can do so as follows.

$$60° = 60° \cdot \frac{\pi \text{ radians}}{180°} = \frac{\pi}{3} \text{ radians}$$

How would you convert 2 radians to degrees? How about as follows?

$$2 \text{ radians} = 2 \text{ radians} \cdot \frac{180°}{\pi \text{ radians}} = \frac{360°}{\pi} = 114.59°$$

where the answer is rounded to two decimal places. From this, it is not hard to see that you must be careful when labeling the measure of an angle, since we have

$$1 \text{ radian} \approx 57.3° \quad \text{and} \quad 1° = 0.01745 \text{ radians}.$$

The Trigonometric Functions

In Activities 2 and 3, you investigated the six possible ratios of the three sides of the right triangle. These six ratios associated with the acute angle θ shown in Figure A.8a below are used to define the six basic trigonometric functions as follows:

$$\sin(\theta) = \frac{a}{c} \quad \cos(\theta) = \frac{b}{c} \quad \tan(\theta) = \frac{a}{b}$$
$$\csc(\theta) = \frac{c}{a} \quad \sec(\theta) = \frac{c}{b} \quad \cot(\theta) = \frac{b}{a}$$

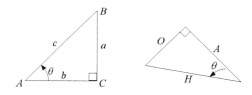

Figure A.8a. Some right triangles.

Another way to express the trigonometric functions of an acute angle is as follows. See Figure A.8a.

$$\sin(\theta) = \frac{opposite}{hypotenuse} = \frac{O}{H} \qquad \cot(\theta) = \frac{adjacent}{opposite} = \frac{A}{O}$$

$$\cos(\theta) = \frac{adjacent}{hypotenuse} = \frac{A}{H} \qquad \sec(\theta) = \frac{hypotenuse}{adjacent} = \frac{H}{A}$$

$$\tan(\theta) = \frac{opposite}{adjacent} = \frac{O}{A} \qquad \csc(\theta) = \frac{hypotenuse}{opposite} = \frac{H}{O}$$

We will be using these trigonometric ratios in many problem situations.

In Activity 2, you constructed a **computer function** whose inputs were the lengths of two legs of a right triangle, and whose outputs were the six possible ratios of the three sides of that triangle. Your **computer function ratios** might have looked something like the following:

```
ratios := func (a, b);
   c := sqrt(a**2 + b**2);
   return [a/b, a/c, b/a, b/c, c/a, c/b];
end;
```

When you ran your **computer function ratios** for Activity 3, did you notice that for the three triangles whose legs were 3 and 4, 6 and 8, and 9 and 12, the outputs were all the same? What do you think is going on? Would you get the same results if the inputs were 5 and 12?

Triangles whose sides are proportional are called *similar* triangles. One property concerning pairs of similar triangles is that their *corresponding angles* are *equal*. Notice that similar triangles can be "nested" inside of each other as in Figure A.8b. When you formed the six ratios for the triangles in Activity 2, these ratios turned out to be equivalent fractions. In other words, $\frac{a}{b} = \frac{3}{4}$ (in the first triangle), $\frac{a}{b} = \frac{6}{8}$ (in the second triangle), and $\frac{a}{b} = \frac{9}{12}$ (in the third triangle). Now do you see why the six ratios of the sides of a right triangle can be used to define six corresponding functions? That is, do you see why there is only one "output" for each "input" θ for each of the six ratios?

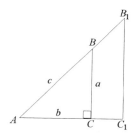

Figure A.8b. Similar triangles.

In Activity 7 you saw an important fact concerning right triangles. For right triangles, in addition to the ratios associated with each of the acute angles, there is a special relationship among the sides of the triangle: *The sum of the squares of the lengths of the two legs of the triangle is equal to the square of the length of the hypotenuse.* This is know as the *Pythagorean Theorem*. See Figure A.9. Can you see why this shows (or proves) that for any acute angle θ, $\sin^2(\theta) + \cos^2(\theta) = 1$?

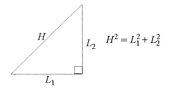

Figure A.9. The Pythagorean Theorem.

Indirect Measurement   One important application of trigonometry is to help us measure things that we cannot actually reach. For example, we can use triangles to help measure tall buildings, redwood trees, or the distance between two satellites. The basic idea is that if we know a few pieces of data about a right triangle, say the measure of one of the acute angles and the length of the hypotenuse, then we can easily figure out the measure of the remaining angle and the remaining two sides. This is what you were doing in part (b) of Activity 3. For example, do you see that the following is true for the situation in Figure A.10?

$$\tan(56°) = \frac{h}{100}$$

This is so by the definition of the tangent function.   Solving for h gives, to the nearest foot, $h = 100\tan(56°) = 148$ feet .

What is the minimum amount of data that you would need to find the measures of all the sides and all the angles in a right triangle?

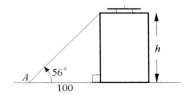

Figure A.10. The situation in Activity 3.

Two Special Right Triangles 30°–60°–90° and 45°–45°–90°   There are two special right triangles which show up in many problem situations. These are the 30°–60°–90° and 45°–45°–90° right triangles.   They show up frequently enough that it is worth remembering the values of the trigonometric ratios of these two triangles as special examples. Knowing (and recognizing) the values of trigonometric functions involving these special angles can be very helpful in some problem situations.

In Activity 4(a), you were asked to sketch an equilateral triangle, and then bisect it by drawing one of its altitudes. Your sketch might have looked something like the one in Figure A.11a.

Since the equilateral triangle has sides of length two, the hypotenuses of both 30°– 60°– 90° triangles shown in Figure A.11a are also of length two. The shorter leg of each 30°– 60°– 90° triangle is half as long as the side of the equilateral triangle,or one unit long. Using the Pythagorean Theorem, the length of the longer leg is $\sqrt{3}$ units. Since all 30°– 60°– 90° triangles are similar, their sides will always be in the ratio of a: $2a$: $a\sqrt{3}$ where a is the length of one of the legs. That is, if you draw a 30°– 60°– 90° right triangle with one leg, say, five units long, then the length of the hypotenuse will be 10 units and the length of the other leg will be $\sqrt{3}$ units — a ratio of 5: 10: $5\sqrt{3}$.

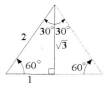

Figure A.11a. A 30°– 60°– 90° triangle.

In part (b) Activity 3, you were asked to sketch an isosceles right triangle, or a 45°– 45°– 90° triangle. If the legs are both one unit long, then (again using the Pythagorean Theorem) you should have found the hypotenuse to be $\sqrt{2}$ units long. Since all isosceles right triangles are similar to each other, their sides will always be in the ratio of a: a: $a\sqrt{2}$ where a is the length of each leg. That is, if you draw an isosceles right triangle whose legs are each, say, five units long, then the hypotenuse will have length $5\sqrt{2}$ — a ratio of 5: 5: $5\sqrt{2}$.

Figure A.11b. A 45°– 45°– 90° triangle.

These two special triangles give us three special angles: 30°, 45°, 60°. It may be useful to be able to give the "exact" values (that is, in fractional — rather than decimal — form) of the six basic trigonometric functions of these special angles. Can you complete Table A.1?

ratio	30°	45°	60°
sin	$\frac{1}{2}$	$\frac{1}{\sqrt{2}}$	
cos	$\frac{\sqrt{3}}{2}$		$\frac{1}{2}$
tan		1	
cot	$\sqrt{3}$		
sec		$\sqrt{2}$	
csc			$\frac{2}{\sqrt{3}}$

Table A.1. Values of the trigonometric functions for 30°, 45° and 60°.

Trigonometric functions of any angle or real number

The trigonometric functions sine and cosine can easily be extended to all angles using a circle of radius one with center at the origin. Such a circle is usually called the *unit circle*. See Figure A.12.

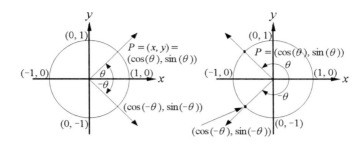

Figure A.12. Trigonometric functions of any angle.

Do you see why the point of intersection P of the terminal side of the angle θ shown in Figure A.12 with the unit circle has coordinates $(\cos(\theta), \sin(\theta))$? Hence, for the angle θ shown in Figure A.12, $\cos(\theta)$ is the x-coordinate of the point P and $\sin(\theta)$ is the y-coordinate of P. That is,

$$\cos(\theta) = x \ \text{ and } \ \sin(\theta) = y$$

where x and y are the coordinates of the point P. What are the values of the other four trigonometric functions of θ in terms of x and y? What if $x = 0$ or $y = 0$? Do you see how to find the values of the other four trigonometric functions in terms of the function values $\cos(\theta)$ and $\sin(\theta)$? Do you see that these values are given as follows?

$$\tan(\theta) = \frac{y}{x} = \frac{\sin(\theta)}{\cos(\theta)} \qquad \cot(\theta) = \frac{x}{y} = \frac{\cos(\theta)}{\sin(\theta)}$$

$$\csc(\theta) = \frac{1}{y} = \frac{1}{\sin(\theta)} \qquad \sin(\theta) = y = \frac{1}{\csc(\theta)}$$

$$\sec(\theta) = \frac{1}{x} = \frac{1}{\cos(\theta)} \qquad \cos(\theta) = x = \frac{1}{\sec(\theta)}$$

provided the denominators are not 0. From the last two identities we also have

$$\sin(\theta) = y = \frac{1}{\csc(\theta)} \qquad \cos(\theta) = x = \frac{1}{\sec(\theta)}.$$

Do you see that these are the same six basic trigonometric relationships or "trigonometric identities" you found for right triangles in Activity 6?

From the discussion above it follows that the following *Pythagorean identity* is true not just for acute angles θ as we saw in Figure A.9, but for all real numbers θ?

$$\sin^2(\theta) + \cos^2(\theta) = 1$$

Do you see why this is true? Do you see why $x^2 + y^2 = 1$? Why do you think this identity is referred to as a *Pythagorean identity*? Do you see why the following Pythagorean identities are also true?

$$1 + \tan^2(\theta) = \sec^2(\theta) \quad \text{ and } \quad 1 + \cot^2(\theta) = \csc^2(\theta)$$

The first identity follows from the identity $\sin^2(\theta) + \cos^2(\theta) = 1$ by dividing each side by $\cos^2(\theta)$, provided $\cos(\theta) \neq 0$, and the second identity follows by dividing each side by $\sin^2(\theta)$, provided $\sin(\theta) \neq 0$.

Now let's consider the case when $x = 0$, that is, when the coordinates of the corresponding points on the unit circle are either $(0, 1)$ or $(0, -1)$. We will consider the point $(0, 1)$ and leave the point $(0, -1)$ for you to think about. For the point $(0, 1)$, it follows that for an angle whose terminal side lies along the positive y-axis, we have $\cos(\theta) = 0$ and $\sin(\theta) = 1$. For example, when $\theta = 90° = \pi/2$ radians, we have

$$\cos(\pi/2) = 0, \ \sin(\pi/2) = 1, \ \csc(\pi/2) = 1, \ \cot(\pi/2) = 0$$

and

$$\tan(\pi/2) \quad \text{and} \quad \csc(\pi/2) \quad \text{are both undefined.}$$

Similarly, we have $\cos(-3\pi/2) = 0$ and $\cos(-3\pi/2) = 1$. Do you see why?

For an angle θ whose terminal side lies along the negative y-axis, positive x-axis, or negative x-axis, the values of the trigonometric functions of θ can be found in a similar manner using the point of intersection on the unit circle and the terminal side of θ. Can you complete Table A.2 of function values?

	0	$\frac{\pi}{6}$	$\frac{\pi}{4}$	$\frac{\pi}{3}$	$\frac{\pi}{2}$	$\frac{5\pi}{6}$	$\frac{3\pi}{4}$	$\frac{2\pi}{3}$	π	$\frac{7\pi}{6}$	$\frac{5\pi}{4}$	$\frac{4\pi}{3}$	$\frac{3\pi}{2}$	$\frac{5\pi}{3}$	$\frac{7\pi}{4}$	$\frac{11\pi}{6}$	2π
sin					1												
cos					0												
tan					—												
cot					0												
sec					—												
csc					1												

Table A.2. *Trigonometric function values of 0, $\pi/6$, $\pi/4$, $\pi/3$, ..., $5\pi/3$, $7\pi/4$, $11\pi/6$, 2π*

The Signs of the Trigonometric Functions

Sometimes it is only necessary to know the sign of a trigonometric function value and not the function value itself when solving a problem. From the previous discussion, can you determine the sign (i.e., plus or minus sign) of a trigonometric function value of an angle θ with its terminal side in a given quadrant? Figure A.13 shows a diagram which indicates when a trigonometric function has a positive value. Can you guess what the labels A, S, T, and C represent? Think about the signs of the coordinates of a point in one of the four quadrants.

What about the sign of a trigonometric function value for an angle with terminal side along one of the coordinate axes?

Do you see how the trigonometric functions can be said to have domains which are appropriate sets of real numbers? Think about how the radian measure of an angle is defined. Radian measure was defined to be the ratio of two lengths, hence a number without units. The units associated with the corresponding central angle are called radians. However, we can think of the number associated with a central angle as a real number. Hence, we can use the name or label radians when appropriate, or we can just think of a real number if we wish. Hence, sin(2) can be thought of as the sine of two radians, or the sine of the real number 2. This is what you were doing when you made graphs of the trigonometric expression $\sin^2(x) + \cos^2(x)$ in Activity 7, and when you made graphs of the trigonometric functions in Activity 9.

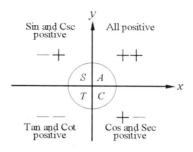

Figure A.13. "Signs" of the trigonometric functions.

Some Trigonometric Identities

In Activities 5, 6, and 7 you investigated various relationships between the trigonometric functions. The relationships you investigated, among others, are referred to as trigonometric identities. Such relationships are true for all angles, or numerical values, where they are defined. Figure A.14 gives a list of some useful identities. Consult a handbook of trigonometry or a trigonometry text for other useful identities. Can you think of a way to derive some of the identities in Figure A.14 from a diagram of a unit circle?

$$\sin(\theta) = \frac{1}{\csc(\theta)} \quad \cos(\theta) = \frac{1}{\sec(\theta)} \quad \tan(\theta) = \frac{\sin(\theta)}{\cos(\theta)}$$

$$\csc(\theta) = \frac{1}{\sin(\theta)} \quad \sec(\theta) = \frac{1}{\cos(\theta)} \quad \cot(\theta) = \frac{1}{\tan(\theta)}$$

$$\sin^2(\theta) + \cos^2(\theta) = 1$$

$$1 + \tan^2(\theta) = \sec^2(\theta)$$

$$1 + \cot^2(\theta) = \csc^2(\theta)$$

$$\sin(\alpha + \beta) = \sin(\alpha)\cos(\beta) + \cos(\alpha)\sin(\beta)$$

$$\sin(\alpha - \beta) = \sin(\alpha)\cos(\beta) - \cos(\alpha)\sin(\beta)$$

$$\cos(\alpha + \beta) = \cos(\alpha)\cos(\beta) - \sin(\alpha)\sin(\beta)$$

$$\cos(\alpha - \beta) = \cos(\alpha)\cos(\beta) + \sin(\alpha)\sin(\beta)$$

$$\sin(2\theta) = 2\sin(\theta)\cos(\theta), \quad \cos(2\theta) = \cos^2(\theta) - \sin^2(\theta)$$

$$\sin(\theta + 2\pi) = \sin(\theta), \quad \cos(\theta + 2\pi) = \cos(\theta), \quad \tan(\theta + \pi) = \tan(\theta)$$

$$\sin\left(\frac{\pi}{2} - \theta\right) = \cos(\theta), \quad \cos\left(\frac{\pi}{2} - \theta\right) = \sin(\theta), \quad \tan\left(\frac{\pi}{2} - \theta\right) = \cot(\theta)$$

$$\sin(-\theta) = -\sin(\theta), \quad \cos(-\theta) = \cos(\theta), \quad \tan(-\theta) = -\tan(\theta)$$

Figure A.14. Some trigonometric identities.

Note that the use of the Greek letter θ is not really necessary. For example, the first identity in Figure A.14 can also be written as follows:

$$\sin(x) = \frac{1}{\csc(x)} \quad \text{or} \quad \sin(t) = \frac{1}{\csc(t)}.$$

Graphs of Trigonometric Functions

Figure A.15 shows a portion of the graph of each of the six basic trigonometric functions. Notice that the graphs of the sine, cosine, cosecant, and secant functions repeat themselves on every interval in the domain of length 2π. Do you see that 2π is the length of the shortest such interval for which these four graphs repeat themselves? We call 2π the *period* of these functions (i.e., the shortest interval for which the graphs repeat themselves). Similarly, the period of the tangent and cotangent functions is π. Do you see why?

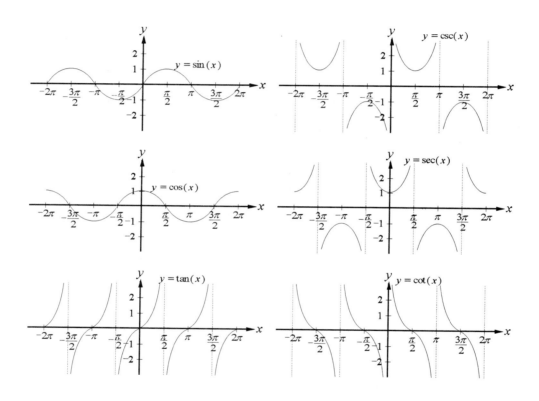

Figure A.15. Portions of the graphs of the six basic trigonometric functions.

Graphs of $A\sin(Bx+C)+D$ and $A\cos(Bx+C)+D$.

Graphs of many common trigonometric functions can be obtained from the graphs of the six basic trigonometric functions using various transformations of the basic graphs. For example, in Activities 9 and 10, you used your computer to make sketches of several trigonometric functions. In particular, graphs of the functions represented by the following expressions

$$A\sin(Bx+C)+D \quad \text{and} \quad A\cos(Bx+C)+D$$

where A, B, C and D are constants, are useful in many applications, especially in science, business, and technology. Figure A.16 shows a graph of a portion of the function f given by $f(x) = 1 + 2\sin(2x - \pi)$.

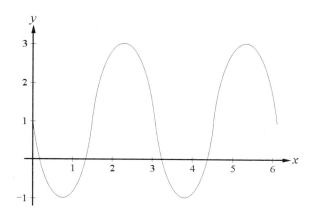

Figure A.16. A portion of the graph of $f(x) = 1 + 2\sin(2x - p)$.

Notice that one-half of the distance between the high and low points, or the *amplitude* of the graph of f, is 2. The interval of the shortest length for which the graph repeats itself, or the *period* of f, is π. The graph looks like the graph of the sine function represented by $\sin(x)$, but it is shifted up one unit upward in the vertical direction and then it is shifted in the horizontal direction to the right $\pi/2$ units. The graph also "wiggles" twice as fast as the sine function, i.e., it has period π. The horizontal shift of the sine function is commonly referred to as the *phase shift* of the graph. Entirely analagous statements can be made of the graph of the function represented by the expression $1 + 2\cos(2x - \pi)$.

Notice that we need only consider the following expressions $A\sin(Bx + C) + D$ and $A\cos(Bx + C) + D$ where $B > 0$, since if $B < 0$ we can rewrite the expression using one of the identities $\sin(-\theta) = -\sin(\theta)$ or $\cos(-\theta) = \cos(\theta)$ and if $B < 0$, we get something entirely different. (What?) For example, the expression $\sin(-2x + \pi)$ is equivalent to $-\sin(2x - \pi)$. Do you see why?

In general, for a function represented by one of the expressions

$$A\sin(Bx + C) + D \text{ or } A\cos(Bx + C) + D$$

where A, B, C and D are constants with $B > 0$, we can make the following summary.

The amplitude of the graph is given by $|A|$.

The graph is shifted in the vertical direction upward by D units when $D > 0$ and downward by $|D|$ units when $D < 0$.

The period of the graph is given by $2\pi/B$.

The phase shift of the graph is $-C/B$ where the graph is shifted C/B units to the right when $C > 0$ and the graph is shifted $-C/B$ units to the left when $C < 0$. Notice that the phase shift can be found by setting the expression $Bx + C$ equal to zero and solving for x to get $-C/B$. Do you see why this is so?

Inverse Trigonometric Functions

Recall that for the exponential function exp given by $\exp(x) = e^x$, the natural logarithm function ln with base $e \approx 2.71828$ is defined to be the inverse of exp. That is, $\ln(x) = \log_e(x)$ where

$$\exp(\ln(x)) = x \text{ for } x > 0 \text{ and } \ln(\exp(x)) = x \text{ for all } x.$$

An inverse for the function exp can be defined since for each y, there is only one x such that $x = e^y$. The latter relationship is written as $y = \ln(x)$. Hence, we have the following inverse relationship.

$$y = \ln(x) \text{ is equivalent to } x = e^y$$

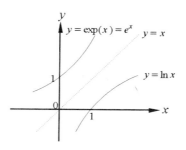

Figure A.17. Graphs of $y = \exp(x) = e^x$ and $y = \ln(x)$.

A quick glance at the graphs of the $y = \sin(x)$ and $x = \sin(y)$ shown in Figures A.18 and A.19, for example, shows that there are many values of x in the interval $[-1, 1]$ such that $x = \sin(y)$. Does this cause any problems if we try to define an inverse function for $y = \sin(x)$? Well, not a problem that can't be overcome, as you will see below.

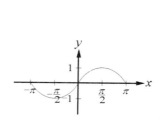

Figure A.18. A portion of $y = \sin(x)$.

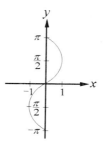

Figure A.19. A portion of $x = \sin(y)$.

Just as we dealt with multiple values in implicit functions by cutting off a branch of the function, we can restrict our attention to just a portion of the sin function to define an inverse of the sine function. Consider the function given by $y = \sin(x)$, for example, on the interval $-\frac{\pi}{2} \leq x \leq \frac{\pi}{2}$ as shown in Figure A.20.

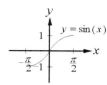

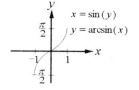

Figure A.20. $y = \sin(x)$, $-\frac{\pi}{2} \leq x \leq \frac{\pi}{2}$.

Figure A.21. $x = \sin(y)$, $-1 \leq x \leq 1$,

$$-\frac{\pi}{2} \leq x \leq \frac{\pi}{2}.$$

The graphs in Figure A.20 and A.21 certainly suggest that the sine function when restricted to the domain interval $-\frac{\pi}{2} \leq x \leq \frac{\pi}{2}$ has an inverse with domain $-1 \leq x \leq 1$ and co-domain, or range, restricted to the interval $-\frac{\pi}{2} \leq y \leq \frac{\pi}{2}$. We denote this inverse function arcsin, or $\sin^{-1}$, and refer to the choice of the interval on which it exists as the *principal branch* of the arcsin function. Hence, we can define an inverse sine function by

$$y = \arcsin(x) \text{ is equivalent to } x = \sin(y) \text{ with } -1 \leq x \leq 1 \text{ and } -\frac{\pi}{2} \leq y \leq \frac{\pi}{2}.$$

The latter definition is not unique. Why? However, it is the usual definition made in trigonometry. We should also note that it is often helpful know that the equation

$$x = \arcsin(y)$$

says that "x is the angle, with $-\frac{\pi}{2} \leq x \leq \frac{\pi}{2}$, whose sine is y" since $x = \arcsin(y)$ is equivalent to $\sin(x) = y$.

A similar equivalence can be written for each of the other five inverse trigonometric functions with a proper choice of a co-domain (or range) for each function. Table A.3 gives the usual co-domains which have been chosen as convention in mathematics for the inverse trigonometric functions (where $\leftrightarrow$ means "is equivalent to").

Inverse Function Relationship	Co-domain (or range)	Domain
$y = \arcsin(x) \leftrightarrow \sin(y) = x$	$-\frac{\pi}{2} \leq y \leq \frac{\pi}{2}$	$-1 \leq x \leq 1$
$y = \arccos(x) \leftrightarrow \cos(y) = x$	$0 \leq x \leq \pi$	$-1 \leq x \leq 1$
$y = \arctan(x) \leftrightarrow \tan(y) = x$	$-\frac{\pi}{2} < y < \frac{\pi}{2}$	$-\infty < x < \infty$
$y = \operatorname{arcsec}(x) \leftrightarrow \sec(y) = x$	$0 \leq y < \frac{\pi}{2}$ or $\frac{\pi}{2} < y \leq \pi$	$-\infty < x \leq -1$ or $1 \leq x < \infty$
$y = \operatorname{arccsc}(x) \leftrightarrow \csc(y) = x$	$-\frac{\pi}{2} \leq y < 0$ or $0 < y \leq \frac{\pi}{2}$	$-\infty < x \leq -1$ or $1 \leq x < \infty$
$y = \operatorname{arccot}(x) \leftrightarrow \cot(y) = x$	$0 < y < \pi$	$-\infty < x < \infty$

Table A.3. The inverse trigonometric functions.

You can use your computer, or a graphics calculator, to make sketches of the inverse trigonometric functions in Table A.3. These functions are the same as those pre-defined in symbolic computer systems (like **MapleV**), in mathematical programming languages (like **ISETL**), and in scientific graphics calculators.

Do you think there are other possible choices of an interval for a co-domain that will give an inverse of the sine function? Can you make other choices of a co-domain that will give an inverse for each of the other five trigonometric functions?

Note that the notation $\sin^{-1}(x)$ for the inverse sine function should <u>not</u> be confused with the fraction $\frac{1}{\sin(x)}$. A similar remark holds for the other five inverse trigonometric functions.

As with the trigonometric functions, the inverse trigonometric functions can be used in problem solving. You did just that in Activity 11. Recall that in part (a) of Activity 11 you were given the following situation (see Figure A.22): You know the height of a building to be 100 feet and you measure the distance from a point A on the (level) ground to the base of a building to be 80 feet.

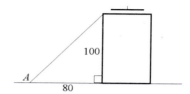

Figure A.22. A sketch of the situation in Activity 11.

What is the degree measure of the angle made with the horizontal and the lineof sight from a point A to the top of a building? For this situation, do you see why the following is equation is true?

$$\theta = \arctan\left(\frac{100}{80}\right)$$

Using a calculator or computer, we can use the latter equation to find that $\theta = 51°$ to the nearest degree.

Periodic Phenomena   Many phenomena in our world follow a *cyclic* or *periodic* pattern. Have you ever ridden on a ferris wheel? Suppose that you are riding a ferris wheel with a radius of 20 feet and the wheel turns at a rate of one revolution per minute. Then your position with respect to the ground is rising and falling. If the loading platform is three feet above ground level, then your lowest position as you ride the ferris wheel should be about three feet. And at the highest point, you would be about three feet plus the diameter of the wheel above the ground. Once you reach the top, you would come down, and then continue on back up again; "up and down and up and down" until the ride comes to an end. Do you see why your height h above the ground is given by the values of a function H represented by the expression $23 - 20\cos(2\pi t)$ feet? That is, why is it true that

$$h = H(t) = 23 - 20\cos(2\pi t) \text{ feet}?$$

Solving Trigonometric Equations   In Activity 12, you were to solve trigonometric equations. For example, let's find the solutions to the following equationin the interval $[0, 2\pi]$, or $0 \le x \le 2\pi$.

$$\cos^2(x) = \sin(x)\cos(x)$$

Do you see how we might proceed? Did you say to divide by $\cos(x)$? If you do you will be dividing by zero when $\cos(x) = 0$. So we better not divide by $\cos(x)$, or we won't find any solutions which correspond to $\cos(x) = 0$. Certainly any x for which $\cos(x) = 0$ is a solution of the equation. Do you see why?

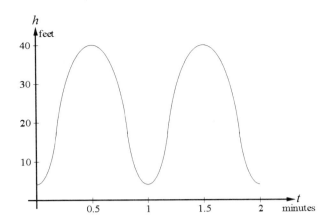

Figure A.23. A portion of the graph of h = H(t) = 23 − 20cos(2πt) feet.

Well, we can rewrite the equation to obtain $\cos^2(x) - \sin(x)\cos(x) = 0$. Now do you see what to do? That's right factor! Factoring, we obtain

$$\cos(x)(\cos(x) - \sin(x)) = 0.$$

Now, we can set each term equal to zero and then solve for x. Hence, we have

$$\cos(x) = 0 \text{ and } \cos(x) = \sin(x).$$

It is not too hard to see that the solutions to these first equation are $x = \pi/2$ and $x = 3\pi/2$ and the solutions to the second equation are $x = \pi/4$ and $x = 5\pi/4$.

Now, let's consider the following equation.

$$\sin(2x) + \cos(x) = 0$$

What are all the solutions to this equation on the interval $[0,\ 2\pi]$? How does this situation differ from the one above? Yes, we have a $\sin(2x)$ in the equation. We can take care of this difficulty by using the identity $\sin(2x) = 2\sin(x)\cos(x)$. That is, we can substitute $2\sin(x)\cos(x)$ for $\sin(2x)$ in the equation $\sin(2x) + \cos(x) = 0$ to obtain the following equivalent equation (i.e., an equation with the same solution set)

$$2\sin(x)\cos(x) + \cos(x) = 0.$$

Now as we did above, we factor and set each term equalto zero to obtain,

$$\cos(x) = 0 \text{ and } 2\sin(x) + 1 = 0.$$

Hence, we need to find all values of x for which $\cos(x) = 0$ and $\sin(x) = -1/2$. Do you see that the solutions to these equations, and hence the original equation, are given by $x = \pi/2$, $x = 3\pi/2$, $x = 7\pi/6$, and $x = 11\pi/6$? We leave it for you to verify that these are the solutions.

A.4   Exercises

1. For each angle with the indicated degree measure, do the following.

(a) Draw each angle as a central angle in an xy-coordinate system with vertex at the origin and one leg along the positive x-axis.

(b) Convert to radian measure. Leave your answer in terms of π.

Here are the angles.

i. 30°   ii. −60°

iii. 72°   iv. 105°

v. 45°   vi. 395°

vii. 135° viii. −240°

ix. 210°

2. For each angle with the indicated radian measure, do the following.

(a) Draw each angle as a central angle in an xy-coordinate system with vertex at the origin and one leg along the positive x-axis.

(b) Convert to degree measure to the nearest degree.

Here are the angles.

i. $\frac{\pi}{3}$ ii. $\frac{7\pi}{6}$

iii. $\frac{-3\pi}{4}$ iv. -2

v. $\frac{5\pi}{2}$ vi. $\frac{-2\pi}{3}$

vii. 3   viii. $\frac{11\pi}{4}$

ix. $\frac{-5\pi}{6}$

3. Compute the following function values by hand.

(a) $\sin\left(\frac{5\pi}{6}\right)$ (b) $\cos\left(\frac{2\pi}{3}\right)$

(c) $\tan\left(\frac{3\pi}{4}\right)$ (d) $\cos(3\pi)$

(e) $\tan(-\pi)$ (f) $\sec(0)$

(g) $\sin\left(\frac{-7\pi}{3}\right)$ (h) $\cos\left(-\frac{\pi}{3}\right)$

(i) $\tan\left(\frac{\pi}{3}\right)$ (j) $\sec(\pi)$

(k) $\cot\left(\frac{3\pi}{2}\right)$ (l) $\cos(3\pi)$

(m) $\tan\left(\frac{-\pi}{4}\right)$ (n) $\sin\left(\frac{-3\pi}{2}\right)$

(o) $\arcsin(0)$ (p) $\arccos(0.5)$

(q) $\arcsin(-1)$ (r) $\arctan(1)$

(s) $\arccos(1)$ (t) $\arcsin(0.5)$

(u) $\arctan(\sqrt{3})$ (v) $\arctan(-1)$

(w) $\cos(\arcsin(0.5))$

(x) $\tan(\arccos(1))$ (y) $\sin\left(\arctan\left(\frac{3}{4}\right)\right)$

4. Complete Table A.1 on page A-10.

5. Complete Table A.2 on page A-12.

6. Use the identity $\sin^2(\theta) + \cos^2(\theta) = 1$ to derive the following identities.

(a) $1 + \tan^2(\theta) = \sec^2(\theta)$

(b) $1 + \cot^2(\theta) = \csc^2(\theta)$

7. Use a diagram like the one in Figure A.12, page A-12, to derive each of the following trigonometric identities.

(a) $\sin(\theta + 2\pi) = \sin(\theta)$

(b) $\cos(\theta + 2\pi) = \cos(\theta)$

(c) $\tan(\theta + \pi) = \tan(\theta)$

(d) $\sin(\pi/2 - \theta) = \cos(\theta)$

(e) $\cos(\pi/2 - \theta) = \sin(\theta)$

(f) $\tan(\pi/2 - \theta) = \cot(\theta)$

(g) $\cos(-\theta) = \cos(\theta)$

(h) $\tan(-\theta) = -\tan(\theta)$

8. Through how many radians does the minute hand of a clock turn from 2 PM to 5:45 PM? The hour hand during the same period of time?

9. A right triangle has a hypotenuse of 20.8 centimeters and an angle of 30°. Find the lengths of the two legs of the triangle.

10. A right triangle has one leg of length 8 inches and a hypotenuse of 15 inches.

    (a) Find the exact values of the six trigonometric functions for the angle opposite the leg 8 inches long.

    (b) Find the degree measures of the acute angles of the triangle to the nearest tenth of a degree.

11. A guy wire is attached to the top of a 200 foot vertical metal pole and to a stake at a point A on the level ground so that the angle of elevation from the point A to the top of the pole is 62.3°

    (a) What is the length of the guy wire?

    (b) What is the distance from the stake to the base of the metal pole?

12. A section at the top of a red barn is painted white. From a point on the (level) ground which is 95 feet from the base of the barn, you measure the angle of elevation to the top of the barn to be 50° and the angle of elevation to the bottom of the painted section to be 42°. What is the height of the painted section?

13. The current i in an alternating circuit is given by

    $$i = I(t) = 30\sin(120\pi t) \text{ amperes}$$

    where t is measured in seconds.

    (a) Determine the period of the alternating current.

    (b) Determine the number of cycles (periods) per minute of the current.

    (c) What is the maximum strength of the current?

    (d) Make a sketch of two cycles (periods) of the current, label your axes.

14. A ferris wheel has a radius of 20 feet and a seat at the loading point is located three feet

above ground level. During a ride on the ferris wheel your height above the level ground is given by

$$h = H(t) = 23 - 20\cos(2\pi t) \text{ feet}$$

where t is the time, in minutes, after you leave (or pass) the loading point. What is the amplitude of the function H? The period of H? How do the amplitude and period relate to this situation? Give correct units with your answers.

15. Simplify the following expressions using basic trigonometric identities. Then find all values of x, if any, where the original expression and simplified expression are (i) zero and (ii) undefined. Explain any differences between your answers in (i) and (ii).

    (a) $\cos(x)\cot(x) + \sin(x)$

    (b) $\cos(x) + \sin(x)\tan(x)$

    (c) $\cos^2(x)\tan(x) + \cot(x)\sin^2(x)$

    (d) $\dfrac{1}{\tan(x)\sin(x) + \cos(x)}$

    (e) $\dfrac{\sin(x) - \cos(x)}{\cos(x)} + 1$

16. Use the identity
    $\sin(\alpha+\beta) = \sin(\alpha)\cos(\beta) + \cos(\alpha)\sin(\beta)$
    to derive the following identity:
    $\sin(\alpha-\beta) = \sin(\alpha)\cos(\beta) - \cos(\alpha)\sin(\beta)$.

17. Use the identity
    $\sin(\alpha+\beta) = \sin(\alpha)\cos(\beta) + \cos(\alpha)\sin(\beta)$
    to derive the following identity:
    $\sin(2x) = 2\sin(x)\cos(x)$.

18. Use the identity
    $\cos(\alpha+\beta) = \cos(\alpha)\cos(\beta) - \sin(\alpha)\sin(\beta)$
    to derive the following identity:
    $\cos(2x) = \cos^2(x) - \sin^2(x)$.

19. Find the indicated function value from the specified information.

(a) $\cos(x)$, $\sin(x) = -\frac{3}{5}$ and $\cos(x) > 0$.

(b) $\sin(x)$, $\tan(x) = \frac{1}{2}$ and $\sin(x) > 0$.

(c) $\tan(x)$, $\cos(x) = \frac{1}{3}$ and $\tan(x) > 0$.

(d) $\sec(x)$, $\sin(x) = 0.4$ and $\sec(x) > 0$.

(e) $\tan(x)$, $\sec(x) = -4$ and $\tan(x) < 0$.

(f) $\sin(x)$, $\cos(x) = \frac{2}{3}$ and $\sin(x) < 0$.

(g) $\sec(x)$, $\csc(x) = -2$ and $\tan(x) > 0$.

20. Find the indicated function values, assuming in each case that x is to be an angle in a triangle.

   (a) Find $\tan(x)$ given that $\cos(x) = -\frac{1}{2}$.

   (b) Find $\cos(x)$ given that $\sin(x) = 0.6$.

   (c) Find $\sin(x)$ given that $\tan(x) = -1$.

21. A triangle has sides of length 10 and 7 inches. Do the following for this situation.

   (a) Express the area of the triangle as a function of the angle between the two sides whose lengths are given.

   (b) Express the perimeter of the triangle as a function of the angle between the two sides whose lengths are given.

22. A radio transmitter sends out signals at certain voltage strengths. The amount of voltage, measured in volts, will usually change as time goes on. Consider a particular signal which varies as follows. It is considered that nothing happens before time begins. In the first second, the voltage is equal to twice the time. In the next second it is 3 volts minus the time. From then until the fourth second, the voltage is sinusoidal, that is, it is given by the expression $\frac{\pi}{2}(t-2)$, after which it disappears completely.

   (a) For the above situation, write (i) a mathematical expression (ii) a

computer function which represents the function which gives voltage as a function of time.

   (b) Where is the voltage increasing? Decreasing?

23. A slender bar of mass m and length L is at an angle of elevation θ in a static equilibrium position. The spring is unstretched in the position where $\theta = 60°$. Suppose that a force is applied vertically downward at the point F as indicated in Figure A.24 below.

   (a) For the above situation, write (i) a mathematical expression (ii) a computer function which represents the amount by which the spring is shortened in terms of the angle of the bar in its new position.

   (b) What is the amount by which the spring is shortened at angle positions of 60°, 30°, and 0°. Use a mass of 0.25 kg and length of 10 cm.

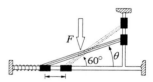

Figure A.24. A slender bar in static equilibrium.

24. In the context of the previous problem, the stiffness k of the compressed spring is a physical quantity defined in terms of work and is given by

$$k = \frac{mg\cos(\theta)}{L\sin(\theta)(2\cos(\theta)-1)}$$

where m is the mass of the bar, g is the gravitational constant and L is the length of the bar.

(a) Write an expression (ii) a **computer function** which represents the stiffness of the spring in terms of the angle of the bar. Use the values $m = 5$ kilograms, $L = 0.36$ meters and $g = 9.8$ meters$/$sec^2. What are the units of the stiffness of the spring?

(b) What is the stiffness at angle positions of $60°$, $30°$, and $0°$?

25. For the expression representing the stiffness k in the previous exercise, the expression will not "work" if $\theta = 60°$. The numbers for which the calculation can be made form the domain of the function. How does this relate to the situation which gave rise to the function? How do the other values of θ where the expression will not "work" relate to the situation?

26. Make a graph of two cycles (periods) of the graph of each function represented by the following expressions. For each function give the amplitude, and period. Also, in parts (g)-(j) give the phase shift.

   (a) $3\cos(x)$ (b) $2\sin(3x)$

   (c) $-\sin(0.5x)$ (d) $1 + \sin(2x)$

   (e) $-2\sin(\pi x)$ (f) $-\sin\left(\frac{\pi}{3}\right)$

   (g) $\cos(2x + \pi)$ (h) $1 - \cos(3x - \pi)$

   (i) $3\sin(2x + 1)$ (j) $1 + 2\sin\left(x - \frac{\pi}{6}\right)$

   (k) $\sin^2(x)$ (l) $\cos^2(2x)$

   (m) $|2\cos(x)|$

27. Find all solutions to each of the following equations in the interval $[0, 2\pi]$.

   (a) $2\sin(x)\cos(x) = \sqrt{3}\cos(x)$

   (b) $\sin(x) + \cos(x) = 0$

   (c) $2\sin(x) = \sin(2x)$

   (d) $3\sin^2(x) + \cos(2x) = 0$

   (e) $2\cos^2(x) + \sin(x) = 1$

   (f) $\sin^2(x) - \cos^2(x) = 0$

28. The line $2x - 3y = 0$ lies along the terminal side of an angle θ in the first quadrant. Find the values of the six trigonometric functions of θ.

29. The line $3x + 4y = 0$ lies along the terminal side of an angle θ in the second quadrant. Find the values of the six trigonometric functions of θ.

30. Make a sketch of the function represented by the expression $\sin(x) + \cos(x)$ on the interval $0 \le x \le 2\pi$ from the graphs of $\sin(x)$ and $\cos(x)$ by adding graphically.

31. Make a sketch of the function represented by the expression $\sin(2x) - \cos(x)$ on the interval $0 \le x \le 2\pi$ from the graphs of $\sin(x)$ and $\cos(x)$ by adding graphically.

32. Rewrite each of the following expressions as an algebraic expression in x using the inverse relationship of an inverse trigonometric function.

   (a) $\sin(\arccos(x))$ (b) $\sec(\arctan(x))$

   (c) $\cos(\arcsin(x))$

   (d) $\sin(\arctan(2x))$ (e) $\tan\left(\cos\left(\frac{\pi}{2}\right)\right)$

33. Give values for A, B, C, and D so the function represented by the indicated expression has a portion of its graph as shown in Figure A.25 below. Explain how you arrived at your answers.

   (a) $f(x) = A\sin(Bx + C) + D$

   (b) $g(x) = A\cos(Bx + C) + D$

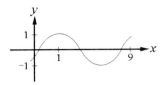

Figure A.25. The graph for Exercise 33.

34. A triangle has sides with lengths a, b, and c where the corresponding angles opposite the sides are, respectively, A, B, and C as shown in Figure A.26 below.

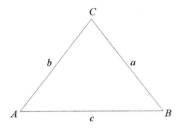

Figure A.26. The triangle for Exercises 34 and 35.

For any triangle like the one shown in Figure A.26, the *law of sines* is given by

$$\frac{\sin(A)}{a} = \frac{\sin(B)}{b} = \frac{\sin(C)}{c}.$$

That is, *the ratio of the sine of a given angle of a triangle to the length of the side opposite the angle is the same as the ratio of the sine of any other angle to the length of the corresponding side opposite the other angle.* Solve the following problem situations.

(a) Given $A = 30°$, $B = 50°$ and $a = 10$, if possible, find b, c and C.

(b) Given $B = 45°$, $b = 5$ and $a = 10$, if possible, find c, A and C.

(c) Given $B = 45°$, $b = 10$ and $a = 8$, if possible, find c, A and C.

(d) Given $A = 30°$, $a = 3$ and $b = 6$, if possible, find c, B and C.

(e) Two satellites are above the earth orbiting at 1600 km and 1760 km, respectively. At noon, the angle between tracking beams from a tracking station to the two satellites is 113.4°. How far apart are the two satellites?

35. For any triangle like the one shown in Figure A.26, the *law of cosines* is given by

$$c^2 = a^2 + b^2 - 2ab\cos(C), \text{ or}$$
$$b^2 = a^2 + c^2 - 2ac\cos(B), \text{ or}$$
$$a^2 = b^2 + c^2 - 2bc\cos(A).$$

That is, *the square of the length of a given side of a triangle is equal to the sum of the squares of the other two sides minus twice the product of the other two sides and the cosine of the given angle.* Solve the following problem situations.

(a) Given $B = 120°$, $a = 5$ and $c = 10$, find b, A and C.

(b) Given $A = 30°$, $b = 10$ and $c = 5$, find a, B and C.

(c) Given $c = 8$, $a = 3$ and $b = 5$, find C, A and B.

(d) A surveyor takes measurements for a tunnel under construction from a point 456 meters from one end of the tunnel and 605 meters from the other end of the tunnel. The surveyor measures the angle between the lines of sight to the two ends of the tunnel to be 57.3°. How long will the tunnel be when it is completed?

INDEX